# PRECAMBRIAN SEDIMENTARY ENVIRONMENTS:
# A MODERN APPROACH TO ANCIENT DEPOSITIONAL SYSTEMS

SPECIAL PUBLICATION NUMBER 33 OF
THE INTERNATIONAL ASSOCIATION OF SEDIMENTOLOGISTS

# Precambrian Sedimentary Environments: A Modern Approach to Ancient Depositional Systems

EDITED BY

WLADYSLAW ALTERMANN &
PATRICIA L. CORCORAN

Blackwell Science

Editorial Offices:
Osney Mead, Oxford OX2 0EL, UK
Tel: +44 (0)1865 206206
Blackwell Science, Inc., 350 Main Street, Malden,
MA 02148-5018, USA
Tel: +1 781 388 8250
Blackwell Science Asia Pty, 54 University Street, Carlton,
Victoria 3053, Australia
Tel: +61 (0)3 9347 0300
Blackwell Wissenschafts-Verlag, Kurfürstendamm 57,
10707 Berlin, Germany
Tel: +49 (0)30 32 79 060



First published 2002 by Blackwell Science Ltd

*Library of Congress Cataloging-in-Publication Data*

Precambrian sedimentary environments:
a modern approach to ancient depositional systems/
edited by Wladyslaw Altermann & Patricia L. Corcoran.
p. cm. —(Special publication number 33 of
the International Association of Sedimentologists)
Includes bibliographical references and index.
ISBN 0-632-06415-3 (pb.: alk. paper)
1. Geology, Stratigraphic—Precambrian.
2. Sedimentation and deposition.
I. Altermann, Wladyslaw, 1954– II. Corcoran,
Patricia L. III. Special publication . . . of the International
Association of Sedimentologists; no. 33.

QE653 .P742 2002
551.7′1—dc21

2001043209

ISBN 0-632-06415-3

A catalogue record for this title is available from the British Library.

Set in 9/11pt Times
by Graphicraft Limited, Hong Kong
Printed and bound in Great Britain
by MPG Books Ltd, Bodmin, Cornwall

For further information on Blackwell Science, visit our website: www.blackwell-science.com

## Contents

*Spec. Publs int. Ass. Sediment.* (2002) **33**, vii–xi

# Introduction

W. ALTERMANN* *and* P. L. CORCORAN†

**Institut für Allgemeine und Angewandte Geologie, Ludwig-Maximilians-Universität München, Luisenstrasse 37, D-80333 Munich, Germany; and*

*†Department of Earth Sciences, Dalhousie University, Halifax, Nova Scotia, B3H 3J5, Canada*

Why are Precambrian sedimentary basins important?

At sedimentological conferences, the message conveyed from the number of presentations concerning Precambrian rocks is deceiving: Precambrian geology is poorly represented and, thus, can only be of minor significance to the scientific community. Similarly, the number of Precambrian sedimentology publications is insubstantial compared with Phanerozoic counterparts, and discussions of Precambrian sedimentary deposits in Earth science textbooks are less voluminous than those of other chapters. The plethora of Phanerozoic sedimentary studies is misleading for two main reasons: (i) to most geologists living in the northern hemisphere, Phanerozoic rocks are more widespread, closer to the Earth's surface and more accessible, making them easier to investigate and understand; and (ii) the economic power of the hydrocarbon industry leads to a concentration of geoscientists on rocks potentially suitable for oil and gas reservoirs. As a consequence, basic sedimentological models were developed by, or in cooperation with, the hydrocarbon industry and were only later applied to other rocks. In addition, the increasing significance of environmental geology and the enormous job market offered by this discipline attract many geologists to young sedimentary deposits. The wealth of the world appears to be grounded on Phanerozoic formations and certainly our knowledge of the Phanerozoic is several orders of magnitude greater than that of the Precambrian.

Despite its underrepresentation in the scientific literature, the Precambrian constitutes about 85% of the Earth's history. Precambrian sedimentary deposits can be traced back for more than 4.3 billion years, to the time when the Earth was approximately 250 million years young (Mojzsis *et al.*, 2001; Wilde *et al.*, 2001). Such traces of the Earth's early history contain copious, but concealed, information concerning ancient atmospheric and environmental conditions. About 3.75 Gyr of the Precambrian are generally accessible for scientific investigation, almost seven times longer than the entire Phanerozoic. Some Precambrian sedimentary basins are composed of strata, tens of kilometres thick, recording continuous sedimentation over hundreds of millions of years of Earth's history. A prime example is the 2.6 to *c.*1.9 Ga Transvaal basin of the Kaapvaal craton, where approximately 600 Myr of sedimentation and erosion are recorded and preserved with a greater continuity than in Phanerozoic counterparts (Eriksson *et al.*, 2001).

From the economic point of view, the Precambrian is at least as important as the Phanerozoic. Mining exploration in Precambrian sedimentary and sediment-hosted ore deposits offers world-wide job opportunities for sedimentologists. Most of the world's mineral deposits are derived from Precambrian shields, and some important hydrocarbon, coal and graphite deposits are also Precambrian in age. Precambrian sequences supply the world with almost 75% of important resources, such as Au, Pt and Cr, and many of these elements are associated with sedimentary rocks and their processes. Almost the entire supply of iron and manganese is found in Precambrian sedimentary rocks, whose deposits are large enough to supply these basic resources for another 200 years. This makes attempts to explore deep sea manganese nodules an interesting scientific experiment, but with only minor economic value (Altermann & Eriksson, 1995). Understanding Precambrian geological processes, and the Earth's early evolution and its hydro-, atmo- and biosphere, is the key to comprehending present-day geological and ecological systems. Therefore, although 'the present is the key to the past', as stated by Sir Charles Lyell, more significantly, the Precambrian past is the key to the present and to the true understanding of the Phanerozoic. This idea was the motivation for this special publication, in addition to studies in countries prospering from the mining industry, such as South Africa, Australia, Canada, Russia and Brazil.

This special volume was first considered during the Eighteenth Regional European Meeting of

Sedimentology in Heidelberg, Germany, in 1997. This international meeting of sedimentologists hosted only a few presentations concerning Precambrian rocks, whereas hundreds of geologists concentrated on the remaining 15% of the Earth's history. The fact that most sedimentologists focus on problems of the 'geological yesterday', while the big questions of the Earth's history remain virtually untouched, is both disappointing and challenging. A Precambrian session for one of the following IAS European Meetings of Sedimentology was organized to resolve this issue. The Nineteenth Regional European Meeting of Sedimentology, in August 1999 in Copenhagen, Denmark, presented an excellent opportunity to organize a Precambrian special session because Denmark, Greenland, Finland, Sweden and Norway are well represented by Precambrian sedimentologists. The Precambrian session was strongly supported by IAS officials such as Andre Strasser, and also by the Scientific Committee of the Copenhagen Conference, especially its chairman Lars Clemmensen. Alec F. Trendall, one of the specialists on banded iron-formations and an expert on Precambrian sedimentary basins, was invited as a keynote speaker. An excursion to the Neoproterozoic Sparagmite basin outcrops in southern Norway was superbly led by Johan Peter Nystuen. Although the session was a success, Precambrian sedimentology still needs to be promoted. We are certain that the Sixteenth International Sedimentological Congress in Johannesburg, South Africa, in 2002 will further shift Precambrian sedimentology towards the centre of attention of the scientific community. The initiation of this special volume occurred during the Copenhagen conference, where all presenters were invited to contribute. Some additional workers who could not attend the conference, but had original material related to the theme, were also invited to publish in the volume. There are two main types of papers: (i) introductory articles discussing special Precambrian conditions; and (ii) studies from different Precambrian shields and successions. The overall aim of this volume is to demonstrate that the Precambrian rock record holds important keys to understanding present-day processes and that modern techniques used in Phanerozoic sequences can also be applied with accuracy to ancient sedimentary deposits.

This volume contains five introductory papers and fourteen studies. J.A. Donaldson, P.G. Eriksson and W. Altermann discuss actualistic versus non-actualistic conditions in the Precambrian sedimentary record. In this 'reappraisal of an enduring discussion' the authors advocate the actualistic point of view of Sir Charles Lyell and his followers and suggest that although some atmospheric conditions were different, the same processes and natural laws were active in the past and today. The introductory paper by W. Altermann addresses the problems of reconstructing life's early evolution, and the possible influence the evolving and advancing biosphere had on sedimentation and sediment preservation processes. A.F. Trendall presents 'The significance of iron-formation in the Precambrian stratigraphic record'. This contribution provides an extensive overview of Precambrian iron-formation and discusses the genetic models considered for its formation. In addition, new aspects, such as possible microbial involvement in iron-formation genesis, are included, together with influences of the evolution of Archaean continental crust and of oceanic circulation. It is suggested that iron-formations were deposited when oceanic circulation was facilitated by deepening of the basins, after pronounced periods of volcanism. Banded iron-formation deposition ceased when the oceans were flushed of their iron content and the supply of iron by volcanic activity was negligible. H. Strauss presents the evolution of sedimentary sulphides from magmatic to bacterial sulphate reduction origin as reflected in the variability of $\delta^{34}S$ values. The paper also includes the history of the sulphur cycle in Precambrian basins, abundant new data from sedimentary pyrite and a comparison of these data to the isotopic composition of Precambrian seawater sulphate, as determined from marine evaporites. The data indicate that the modern style of bacterial sulphate reduction emerged in the Neoarchaean. Proterozoic bacterial sulphate reduction is characterized by additional sulphate limitation, as seen in frequently positive sulphur isotope values. This is attributed to effective sulphate turnover as a consequence of readily metabolizable organic matter and possibly a lower sulphate concentration in the ocean. The Neoproterozoic sulphur cycle appears to respond to perturbations of the global ocean/atmosphere system as a result of repeated glaciations and subsequent biogeochemical changes. The introductory papers close with a contribution from D.Y. Sumner on 'Decimetre-thick encrustations of calcite and aragonite on the sea floor and implications for Neoarchaean and Neoproterozoic ocean chemistry'. Neoarchaean subtidal aragonite fans are attributed to high aragonite saturation states in sea water and to the presence of chemical inhibitors slowing down crystal nucleation and micrite sedimentation rates. In contrast, the Neoproterozoic aragonite pseudomorphs are present only in association with glacial deposits and are accompanied by abundant micrite deposition.

The studies in this volume are divided on the basis of three main themes: (i) depositional settings and associated processes in Archaean terranes; (ii) depositional settings and processes in Palaeoproterozoic sequences, with some emphasis on lack of vegetation and comparisons with modern successions; and (iii) depositional settings and processes of late Proterozoic sequences, with some emphasis on glacial deposits.

Three studies were conducted in the Archaean Slave Province. S.J. Pehrsson describes the formations comprising the Archaean Indin Lake supracrustal belt and interprets them as the results of marginal basin–arc development on Mesoarchaean crust, followed by marine foredeep construction. Abundant mass and sediment gravity flows are attributed to elevated erosion rates in non-vegetated environments. W.U. Mueller, P.L. Corcoran and J.A. Donaldson provide a detailed facies analysis of a tectonically influenced sedimentary basin and provide a model for its formation. The model of the Jackson Lake Formation (cover page) illustrates a high-relief basin developed along a rugged coastline in which the sedimentary deposits were strongly influenced by waves and macrotidal conditions. Tide- and wave-influenced sedimentation associated with tectonism may be attributed to unique Archaean Earth–Moon dynamics. The third contribution from the Slave Province is by P.L. Corcoran and W.U. Mueller. The authors present geochemical and clast count data from unconformity-bound, late orogenic, tectonically controlled sedimentary sequences to determine the main factors influencing the composition of sedimentary deposits in three basins. The results indicate that the sedimentary sequences were controlled by source rock composition, chemical weathering and sorting during transport and at the depositional site to varying degrees.

Two Archaean studies were conducted in Australia. G. Pike and R. Cas discuss the stratigraphy and lithology of the 3.0–2.9 Ga Whim Creek volcano-sedimentary succession and present event stratigraphy, basin evolution and possible modern and Phanerozoic analogues. They conclude that the Whim Creek Belt was deposited in two stratigraphic successions separated by a low-angle disconformity. The stratigraphic architecture of the Whim Creek Belt was controlled by voluminous volcanism that overwhelmed the tectonic and eustatic influences on the basin. A second contribution concerning the Precambrian of Western Australia is presented by J.L. Hand, R.A.F. Cas, L. Ong, S.J.A. Brown, B. Krapez and M.E. Barley, in which detailed facies descriptions and a discussion of submarine fan models and gravity flow deposits are included. The study, based mainly on the results from 10 diamond drill cores, reveals that five distinct facies deposited below wave base and from mass flows constitute the succession of the Late Archaean Black Flag Group. In addition, three depositional systems of a progradational submarine fan were identified in the Eight Mile Dam succession.

Two interesting contributions concerning poorly known Precambrian basins in the former Soviet Union are presented in this volume. These papers are excellent examples of how the mineral composition of highly metamorphosed sedimentary rocks, calculated by the MINLITH computer program, can be used to reconstruct sedimentary protoliths and their depositional facies. The paper by O.M. Rosen, V.T. Safronov and A.A. Abbyasov on the Early Proterozoic Keiva and Kukas basins (NE Baltic Shield) calculates primary mineral compositions of schists that were originally composed mainly of kaolinite and quartz fragments with an iron oxide admixture, whereas dolomite and feldspar fragments predominated in meta-carbonates. The protoliths are interpreted as shallow open-shelf clastic sediments, derived from humid weathering of continental crust and as dolomite that formed in a closed basin, under arid conditions. The contribution by V.L. Zlobin, O.M. Rosen and A.A. Abbyasov, although concerning Early Proterozoic deposits, also employs the MINLITH program to study sedimentary rocks from the Anabar Shield (polar Siberia), which were metamorphosed to granulite facies. The 2.4 Ga Vyurbyur Group was deposited under shelf conditions on the active continental margin of the Archaean Magan micro-continent, whereas the 2.1 Ga Hapschan Group formed under shelf conditions along a passive margin of the Birekte micro-continent.

An example from the Palaeoproterozoic of North America is offered by L.B. Aspler and J.R. Chiarenzelli, who meticulously describe the siliciclastic and carbonate deposits comprising a basin in the western Churchill Province of northern Canada. The stratigraphy and detailed sedimentology are consistent with development of a siliciclastic–carbonate ramp characterized by alluvial processes, marine flooding, suspension sedimentation and mass flows. The deposits compare favourably with Phanerozoic siliciclastic–carbonate systems. The South American Precambrian is represented by a contribution concerning the sedimentology of Late Palaeoproterozoic deposits from the Roraima Supergroup, south-eastern Venezuela, by D.G.F. Long. The author demonstrates that the fluvial-dominated sequence was deposited in a vegetation-free foreland basin and warns the reader that similar sedimentary structures in younger vegetated settings could be easily misinterpreted as beach facies. K. Strand presents the pyroclastic and volcaniclastic deposits from the central part of the Fennoscandian Shield in western Finland. These Palaeoproterozoic volcanic-related deposits are overlain by fan-delta to fluvial conglomerates in which the clast compositions reflect erosion of the underlying volcanic pile. The volcano-sedimentary succession is interpreted as an intra-arc basin fill sequence related to early development of a juvenile continent. From the southern African Kaapvaal craton, P.G. Eriksson, W. Altermann, L. Eberhardt, S. Ahrend-Heidbrinck and A.J. Bumby describe the deposits of Palaeoproterozoic epeiric seas in the Transvaal Supergroup. The authors contend that epeiric seas were common depositional environments during the Precambrian, in contrast to modern sedimentary basins. The deposits are consistent with development under transgressive and highstand conditions, followed by regressive, tidally reworked braid-delta sandstones marking the retreat of the epeiric sea embayment off the Kaapvaal craton.

An additional contribution concerning epeiric seas, but from the late Proterozoic, is provided by S. Sarkar, S. Chakraborty, S. Banerjee and P.K. Bose. The authors present a facies and sequence analysis of epeiric basin sediments deposited in an overall prograding highstand systems tract comprising a number of parasequences in the Sirbu Shale, Central India. Six facies of cyclical interbedded shale, siltstone, fine sandstone and coarse sandstone are recognized. The coarse interbeds contain storm signatures in all facies. Fischer curves are correlated with regional tectonic features and the synsedimentary deformation history of the basin is discussed in a sedimentological context. The deposition took place in a westward-opening intracratonic sag basin. Additional studies from the late Proterozoic deal with glacial deposits. M.A. Martins-Neto and C.M. Hercos describe Early Neoproterozoic sedimentary deposits and their tectonic setting in the São Francisco Craton, south-eastern Brazil. Stratigraphic, sedimentological and structural data indicate ice-proximal and proglacial outwash deposition of the Macaúbas Megasequence in sub-basins which were partitioned by rifting processes. It is suggested by the authors that rift-related uplift contributed to the *c.*900 Ma glaciation of this South American Precambrian craton. K. Laajoki provides new data supporting a glacial origin for a regional angular unconformity which separates late Proterozoic and late Vendian rocks in northern Norway. Late Proterozoic glacial striations and small grooves indicate glacial abrasion of the unconformity. A glacial origin is further corroborated by the fact that the classical unconformity surface is actually a striated trough and not a plane as was previously thought.

A review of the contributions in this volume demonstrates that Precambrian sedimentary deposits are often similar to their Phanerozoic counterparts in terms of composition, sedimentary processes and depositional setting, but may differ significantly as a result of lack of vegetation, climatic and biological constraints, composition and circulation of sea water and the secular involvement of continental crust. Essential clues to the development of Phanerozoic and Modern sedimentary successions are hidden in the Precambrian rock record, waiting to be uncovered.

## ACKNOWLEDGEMENTS

Many people have contributed to this special volume and we would like to thank all of them for their suggestions, discussions, editorial work and most of all for the constructive peer reviews of more than 20 manuscripts originally submitted to this volume. In alphabetical order they are: L.B. Aspler, Ottawa; J. Bragin, CSEOL, UCLA; C. Busby, University of California at Santa Barbara; J.R. Chiarenzelli, State University of New York, Oswego; E.H. Chown, Kingston, Ontario; K. Condie, New Mexico Institute of Mining and Technology; M. Coniglio, University of Waterloo; P. Cousineau, Université du Québec à Chicoutimi; C. deWet, Franklin and Marshall College; J.A. Donaldson, Carleton University; J. Dostal, St Mary's University; R.M. Easton, Ontario Geological Survey; P.G. Eriksson, University of Pretoria; M. Gibling, Dalhousie University; Li Guo, Cambridge University, UK; C. Klein, University of New Mexico; D.D. Klemm, Ludwig-Maximilians-University of Munich; H. Hofmann, McGill University; P.F. Hoffman, Harvard University; L. Kah, University of Tennessee; K. Laajoki, University of Oulu, Finland and Geological Museum, Oslo, Norway; T.W. Lyons, University of Missouri; B. Mayer, University of Calgary; J. McPhie, CODES, University of Tasmania; H. Miller, Ludwig-Maximilians-University of Munich; W.U. Mueller, Université du Québec à Chicoutimi; D.R. Nelson, Geological Survey of Western Australia and Curtin University, Perth; R.W. Ojakangas, University of Minnesota, Duluth;

R. Pflug, University of Freiburg; B.R. Pratt, University of Saskatchewan; J. Schieber, University of Texas at Arlington; B. Simonson, Oberlin College Ohio; E. Simpson, Kutztown University; S.J. Sowerby, University of Stockholm; H. Stollhofen, University of Würzburg; V. Testa, IGEO-Universidade Federal da Bahia; H. Tirsgaard, Maersk Oil and Gas AS; A.M. Thorne, Geological Survey of Western Australia; P. Thurston, Laurentian University; J.C. Tipper, University of Freiburg; M. Tucker, Durham University; J. White, University of Otago; B. Wilkinson, University of Michigan; G. Young, University of Western Ontario. Heidi Felske, L-M-U, Munich, helped in reshaping and drawing many figures and tables. Last, but certainly not least, we are indebted to the series editor, Guy Plint, for his support, guidance and rigorous editing thoughout this endeavour.

## REFERENCES

Altermann W. & Eriksson P.G. (Eds) (1995) South African Early Proterozoic ore deposits. *Mineral. Deposita*, **30**(2), 85–198.

Eriksson, P.G., Altermann, W. & Hartzer, F.J. (2001) The Transvaal Supergroup and precursors. In: *The Geology of Africa* (Eds Thomas, R.J. & Hunter, D.R.). Geological Society of South Africa, Johannesburg.

Mojzsis, S.J., Harrison, M.T. & Pidgeon, R.T. (2001) Oxygen-isotope evidence from ancient zircons for liquid water at earth's surface 4300 Myr ago. *Nature*, **409**, 178–181.

Wilde, S.A., Valley, J.W., Peck, W.H. & Graham, C.M. (2001) Evidence from detrital zircons for the existence of continental crust and oceans on the Earth 4.4 Gyr ago. *Nature*, **409**, 175–178.

# Introductory Papers

*Spec. Publs int. Ass. Sediment.* (2002) **33**, 3–13

# Actualistic versus non-actualistic conditions in the Precambrian sedimentary record: reappraisal of an enduring discussion

J. A. DONALDSON*, P. G. ERIKSSON† *and* W. ALTERMANN‡

**Ottawa-Carleton Geoscience Centre, Department of Earth Sciences, Carleton University, Ottawa, ON, K1S 5B6, Canada;*
*†Department of Earth Sciences, University of Pretoria, Pretoria 0002, South Africa; and*
*‡Institut für Allgemeine und Angewandte Geologie, Ludwig-Maximilians-Universität München, Luisenstrasse 37, D-80333 Munich, Germany*

## ABSTRACT

Actualistic models can be successfully applied to virtually all sedimentary successions preserved in the Precambrian rock record, allowing for differences in the relative rates and intensities of those processes that control weathering, erosion, transportation, deposition and lithification. As the basis for this statement, we understand actualism to be the principle that the same processes and natural laws applied in the past as those active today. This definition readily accommodates events of even catastrophic character, such as bolide impacts and tsunamis. Non-actualistic models introduce semantic problems and are of little help for understanding geological processes. A review of the English and German literature on actualism reveals that many misconceptions can arise because of the two principal meanings of the term 'actual': (i) real, factual; (ii) present-day. However, if the full range of possible outcomes subject to natural laws is considered, rather than just the expression of these laws as observed in present environments, all Precambrian settings can be investigated and understood by applying the actualistic approach, which is fundamental but not unique to the geological sciences. Thus, all studies involving Precambrian sedimentary rocks can be approached by means of comparisons to present-day environments. Non-actualistic models should be reserved for speculations about early Hadean environments, cosmology, religion and philosophical conjectures about 'parallel' worlds (such as those inhabited by unicorns).

## INTRODUCTION

Throughout the history of geology, new terms have been introduced, old terms redefined or abandoned and familiar terms assigned new meaning in the scientific literature. Sedimentological publications contain recent examples of evolving terminology, including: paradigm (example, pattern or model), architecture (arrangement, structure), model or system (complex entity). This paper deals specifically with two terms that have appeared in a number of recent papers on Precambrian sedimentology: actualistic and non-actualistic. Although the former is clearly understood, the latter is commonly used with meanings that seem to be at variance with the meaning initially assigned to it. This may stem from the various meanings assigned to the root word, actual: existing in fact and not merely potential or possible; real (as distinct from ideal); current, acting at the present moment.

Some geologists, primarily sedimentologists, palaeontologists and geochemists, in their attempt to emphasize differences between Precambrian and present-day conditions, landscapes, biotic communities and environments, have begun to refer to non-actualistic Precambrian settings, biota and sediments (e.g. Pflüger, 1995; Pflüger & Gresse, 1996). Although there is a need to draw attention to the numerous differences between the Precambrian and the present day, many are simply differences of degree, and we argue here that application of the adjective non-actualistic is confusing, at times misleading, and therefore inappropriate for the purpose of stressing different and/or unique events that occurred, and conditions that prevailed, during segments of Precambrian time. Where past palaeoenvironmental conditions have resulted in the formation of rocks or structures for which modern analogues are difficult or impossible to find (e.g. Trendall, this volume, pp. 33–66; Sumner, this volume, pp. 107–122; Strauss, this volume, pp. 67–105), terms such as unique or distinctive can serve to draw attention to inferences about conditions unlike those observed today. In the light of the duration of the Precambrian Era for almost 90% of Earth's history, it is surprising how rarely such unique rock records have been preserved. Of course, the present surface and interior of the Earth do not necessarily reflect all past environments, because with time, new environments continuously evolved which were not present in early Earth history. Conditions prevailing today are thus the result of the evolution of our planet constrained by invariant physical, chemical and biological laws. Many evolutionary events (Altermann, this volume, pp. 15–32), such as the appearance of land plants, the development of coral reefs and the ascendancy of silica-secreting organisms, all took place during the Phanerozoic, thus significantly diminishing the value of using the term 'non-actualistic' to stress differences between Precambrian and modern environments.

## DEFINITIONS AND RELATED TERMINOLOGY

Actualism is not listed in all standard language dictionaries, but where it is, the meaning provided generally relates to its usage in philosophy. Because of entrenchment in the geological literature, however, it appears in most geoscience dictionaries. For example, the *Oxford Dictionary of Earth Sciences* (Allaby & Allaby, 1999) defines actualism as: 'The theory that present-day processes provide a sufficient explanation for past geomorphological phenomena, although the rate of activity of these processes may have varied.' Another example is provided by *Challinor's Dictionary of Geology* (Wyatt, 1986): 'Actualism was the geologic parallel to the 18th Century postulate of universal physical determinacy: invariance of the laws of nature through time and space; as stated by Holmes (1965) it is the principle that the same processes and natural laws prevailed in the past as those we can now observe or infer from observations.' Thus, these two well known sources provide definitions that correspond, in terms of natural laws and processes, to accepted definitions of modern uniformitarianism. Unfortunately, the widely used *Glossary of Geology* (Jackson, 1997) defines actualism as: 'the same forces presently in evidence, acting over time with energies and frequencies now observable, are sufficient causes of all geological results, relationships and configurations'. The qualification 'with energies and frequencies now observable', which is at variance with the views of many authorities, including those presented in the papers reviewed below, may have unduly influenced the use of this term. In this regard it is interesting to note that Preston Cloud (1988), in his book *Oasis in Space*, nowhere uses the term actualism, but instead describes as fundamental to geology the 'principle of natural causes'. He argues that, averaged over sufficient time, all rates are essentially uniform, thus resolving the apparent conflict between uniformitarianism and catastrophism. In other words, given enough time, anything can happen that is consistent with natural law.

Paradoxically, although the definitions discussed above would appear to characterize actualism as the basis of all science, the term seems not to be used in other scientific disciplines. For instance, it is not listed in the *McGraw-Hill Encyclopedia of the Geological Sciences* (Parker, 1991) or in *Van Nostrand's Scientific Encyclopedia* (Considine, 1995). Perhaps the principal doctrine of geology that 'the present is the key to the past' is so self-evident in other natural sciences that it does not require any definition. It seems an irony in science that Sir Charles Lyell's (1830) actualism-based theory of uniformitarianism, from which actualism was formed, and its application by Darwin (1859), were systematically fought by esteemed natural scientists like Lord Kelvin. Only the discovery of radioactivity, as a new source of energy, provided novel arguments in the calculation of the age of the Earth by Charles Darwin's son George in 1903, and eventually forced the physicists to acknowledge Lyell's achievements (Stinner, 2000). Physicists, biologists and chemists can continuously observe the phenomena of acting forces and reactions. The demonstrable reproducibility of scientific experiments is an indispensable prerequisite to acceptance of the validity of hypotheses arrived at through deduction. In geology, natural processes that tend to be extremely slow are commonly difficult to evaluate, and can rarely be directly observed. Their products, however, can be explained in accordance with natural laws by inductive reasoning, related as closely as possible to presently observed processes.

In more than 30 recent (post-1970) English texts on physical geology consulted for this review, all discuss uniformitarianism and catastrophism, but actualism is mentioned in only two (Dott & Batten, 1980; Judson

& Richardson, 1995). Dott & Batten (1980, reprinted from first edition, 1971), state that the 'necessity for clear distinction of modern from Lyellian uniformitarianism is evidenced by the impact of neo-catastrophists, such as Immanuel Velikovsky, author of *Worlds in Collision* (1950), and *Earth in Upheaval* (1955). Such writers repeatedly misinterpret modern uniformitarianism, or more correctly, actualism. They take as authority Lyell's (1830) principles, assuming that the strict uniformity (gradualism) expressed therein is still the guiding doctrine of geology. The problem is compounded by confusion of what is meant by catastrophic processes and by a lack of appreciation of geologic time. Geologists today routinely accept sudden, violent, and even certain unique events as perfectly consistent with contemporary Earth history. Only by substituting the term actualism, in the sense outlined herein, can misconceptions be minimized.'

A similar definition of actualism appears in the earlier, widely used text by Arthur Holmes (1965), who stated that 'actualism . . . conveys much more appropriately [than does uniformitarianism] the real meaning of Hutton's (1788) inspired appeal to actual causes: the principle that the same processes and natural laws prevailed in the past as those we can now observe or infer from observations'.

One author of a physical geology textbook out of the 30 examined (Hamblin, 1991) suggested that the word naturalism should take the place of uniformitarianism. However, the term naturalism does carry other meanings, as listed below, from the *Canadian Oxford Dictionary* (Barber, 1998):

**1** The theory or practice in art and literature of representing nature, character, etc., realistically and in great detail.
**2** A theory of the world that excludes the supernatural or spiritual.
**3** Any moral or religious system based on this theory.
**4** Action based on natural instincts.
**5** Indifference to conventions.

The first meaning is in accord with actualism, but the other four hardly provide a successor term to uniformitarianism that clarifies the essence of geological methodology.

In contrast to the scarcity of references to actualism in English physical geology texts, 18 of 30 recent historical geology texts consulted contain discussions of the term, and provide definitions that closely follow the Holmes (1965) and Dott & Batten (1980) definitions. For instance, in *The Earth through Time*, Levin (1999) defines actualism as the principle that natural laws governing both past and present processes on Earth have been the same, and in *Origin and Evolution of the Earth: Principles of Historical Geology*, Condie & Sloan (1997) define actualism as 'the principle by which modern geologic processes are used to understand ancient ones'. The relationship of actualism to uniformitarianism is emphasized by Stanley (1989) in his *Earth and Life through Time*: 'The principle of uniformitarianism, sometimes called actualism, governs geologists' interpretations of the most ancient rocks on Earth.' In addition, the *Encyclopedia of Evolution* (Milner, 1990) defines actualism as the concept that 'geological processes are governed by natural physical and chemical laws', attributing this definition to the scientist L.C. Prevost (1825). Thus, Prevost appears to have discussed actualism eight years before Lyell's expansion and promulgation of the Hutton-Playford enunciations on uniformitarianism, and 80 years before Geike's (1903) application of the phrase 'the present is the key to the past' to uniformitarianism. As pointed out by Gordeev (1961) and Muller *et al.* (1991), the methodology of actualism was introduced in this sense much earlier by M.V. Lomonosov (1763), and even earlier by de Buffon (1749), who thus both preceded Hutton's application of actualistic concepts, which were subsequently incorporated in the principle of uniformitarianism. Similarly, von Hoff (1822) accumulated substantial data to document actualism; in a review of publications and correspondence, Hamm (1993) argued that von Hoff's writings influenced Lyell's (1830) pronouncements on uniformitarianism.

In consulted German geology books (published from 1970 to 2000), only one out of eight does not include an explanation of actualism. Even in the German translation of Press and Siever's (1994) *Understanding the Earth*, the term actualism (*Aktualismus*) was added to the glossary, although it is missing in the English original. This contrast underlines the long-lasting debate on the principle of actualism in Germany. In German literature the rise of the idea of actualism is credited mainly to von Hoff (1822), who concluded that all presently active forces can adequately explain all changes on the Earth's surface, especially the migration of land and sea, if unlimited time for Earth history is provided. Some German authors distinguish between von Hoff's 'actualistic method' and Lyell's 'principle of actualism' (e.g. Schwartz, 1978). According to this dichotomy, the principle of actualism proclaims that, throughout Earth history, all processes and forces have remained unchanged, and therefore no causes and effects have been active in the past other than those observed in the present.

From the 1920s on into the 1940s, a wide and prolonged discussion of actualism transpired in German geological publications. Strict followers of the principle, as defined above (e.g. Salomon, 1926, who demanded that only processes observable today can be accepted for the past), were soon succeeded by subscribers to the actualistic method who allowed for new ideas and new observations. The most powerful and influential among them were Erich Kaiser (1931) and von Bubnoff (1937). Kaiser discussed pre-vegetational Earth environments, concluding that primitive Earth deserts must have included humid climatic realms, with different sedimentation and erosion patterns than are known from recent examples. Von Bubnoff questioned whether the sediment accumulation rates observed in modern environments are applicable to old sedimentary rocks, which he inferred to have accumulated at slower rates in early Earth history.

Discussion of actualism as a principle was also triggered by observations that glacial and interglacial periods are not the 'usual' state in Earth history, but are simply exceptions from long-lasting warmer time intervals. Accordingly, Kummerow (1932) concluded that the present is not the key to the past, and that actualism is not a valid principle applicable to periodic events. Strangely enough, the same phenomena (e.g. the climate of the Tertiary) served Lyell as argument for his *Principles of Geology* and his discussion of 'how far the former changes of the Earth surface are referable to causes now in operation'. The discussion went as far as to construct a contradiction between actualism and the historical character of the geological sciences. For example, Beurlen (1935, and several subsequent papers) proclaimed that actualism was non-historic and dogmatic. In his criticisms of actualism, Beurlen (1935) applied the political ideas of the national socialistic regime, extending his arguments and those of his followers *ad absurdum*.

## ACTUALISM AND CATASTROPHISM

Because of the initially established polarity between uniformitarianism and catastrophism, some authors have suggested a restricted meaning for actualism. For instance, Stanley (1989) offered bolide impact as 'a case [for which] actualism cannot be applied'. This stance presumably was taken because of the widely held view that gradualism is implicit in Lyellian uniformitarianism. In a recent discussion, however, Stanley (personal communication, 1999) would now accept a broad definition of actualism that includes events of geological catastrophism such as bolide impacts, because such impacts are governed by the natural laws of physics and chemistry. In fact, we have recently been able to observe planetary bolide impacts directly, with world-wide television coverage of the Shoemaker–Levy event on the surface of Jupiter (Shoemaker, 1995). The surfaces of the Moon and of several planets and their moons provide analogues for Hadean and early Archaean history. In the Proterozoic record, few would dispute origin by bolide impact for the Acraman structure of South Australia (Gostin *et al.*, 1986), or for the *c*.2.0 Ga Vredefort structure of the Kaapvaal craton (Therriault *et al.*, 1997).

Catastrophism initially designated the doctrine that Earth history can be explained in terms of a succession of sudden and rapid large-scale but short-lived violent events due to unknown causes, or, as proposed by Georges Cuvier (1817), Divine Intervention. As now used by earth scientists, catastrophism is applied without the Divine Intervention rider. Examples of geologically recognized catastrophic events include: bolide impacts; hurricanes, cyclones and tornadoes; tsunamis; flash floods; landslides and avalanches; earthquakes, volcanic eruptions and lahars. To this list might be added long-term catastrophic events such as global continental glaciations (snowball-earth states). Hsu (1983) revived the phrase actualistic catastrophism (introduced by Hooykaas, 1956, 1970), applying it to his recognition of the creation of vast saline basins in the Mediterranean, as a result of the Messinian event. Other examples suggested by Hsu are extreme landslides and large impact craters. To these examples can be added extreme floods, such as the superfloods that produced the Channeled Scablands of Washington state (Baker & Bunker, 1985), and superfaults, such as the one which created glassy fusion sheets during instantaneous slip of hundreds of metres during the Langtang Slide in Nepal (Spray, 1997).

Just as the term uniformitarianism carries baggage, so does catastrophism. For instance, an internet check turns up an interesting collection of cross-listings for the latter term: Velikovsky, Bible studies, rebuttals to uniformitarianism, creationism. Because of these linkages, some have introduced alternative terms, referring to rare, uncommon, sporadic, episodic, cataclysmic, profound or convulsive geologic events. However, some of these terms pose problems: cataclysmic is derived from the Greek word for flood; profound has philosophical and theological connotations; and convulsive geologic event (Clifton, 1988) is suggestive of organic movement; most of the other terms lack a sense of energy and magnitude.

From the discussion above it appears that the terms 'event' and 'catastrophic event' demand some clarification as well because definitions by various authors range from events of local to global impact, and from events of short to long duration. In our understanding, event describes a state or process deviating from the normal, average condition; catastrophic or extreme geological events include incidents of violent, disastrous changes to the environment that occur rapidly, not allowing for adaptation of the milieu to new conditions, or for escape of coexisting organic populations to new habitats.

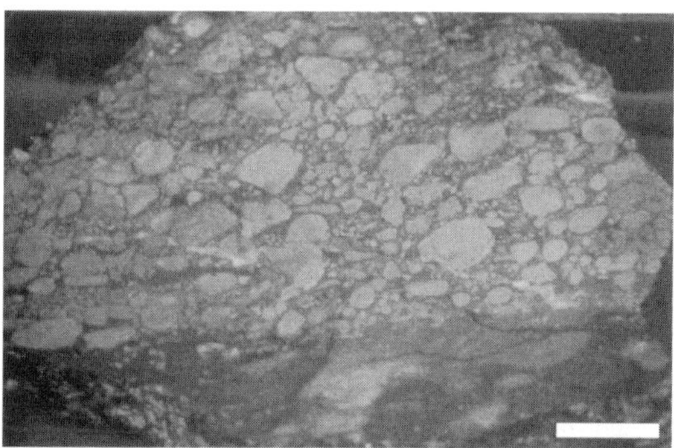

**Fig. 1.** Conglomerate composed of pyrite clasts in a black shale matrix; from *c.*2.9 Ga Witwatersrand Supergroup, South Africa. In this rare sample, imbrication of the subangular to rounded pyrite clasts provides evidence of aquatic transport. The conglomeratic layer has an erosional contact with the underlying lenticular-textured black shale bed. Although the composition of the clasts indicates unusual chemical conditions of weathering and transportation, the sedimentary structures and textures can be matched with younger conglomerates and modern gravel beds (e.g. imbrication, roundness, matrix to clast support and erosional base), thus providing evidence of actualistic processes in the Precambrian. Scale bar is 2 cm.

## EXTREME AND UNIQUE EVENTS AND ENVIRONMENTS IN EARTH HISTORY

The Precambrian rock record contains numerous examples of extreme geologic events (e.g. astroblemes, such as the Sudbury Structure in Canada). In addition, geoscientists have been able to infer some unusual and/or unique conditions and events in the Precambrian, based on observations of the rock record, e.g.

**1** Early reducing atmosphere (Fig. 1).
**2** Unknown differences in the early hydrosphere.
**3** Origin of life.
**4** Simple biotic assemblages dominated by cyanobacteria in the oceans and lack of metazoan fauna.
**5** Lack of terrestrial plant cover.
**6** Extrusion of large volumes of ultramafic lava and tephra.
**7** Accumulation of vast deposits of banded iron-formation.

Some have suggested that conditions not directly comparable to those of the present day should be designated as non-actualistic. As we have argued above, however, this is contrary to the widely held definition of geological actualism. The term non-actualistic seems to have taken hold in the sedimentological literature in 1995, in a paper by Friedrich Pflüger, in which he designated five different classes of sedimentary structures:

**1** Strictly historic, of no preservation potential.
**2** Non-actualistic *sensu strictu*, i.e. structures that under no circumstances can be produced today.
**3** Crypto-actualistic, i.e. structures that are probably formed today in inaccessible environments and therefore cannot be observed.
**4** Actualistic *sensu strictu*, i.e. structures that are formed today and can be found in the rock record.
**5** Strictly modern, i.e. structures of no preservation potential.

It is obvious that this classification is not very helpful for understanding the formation of sedimentary structures, and that all categories can be explained by valid physical and chemical laws. Examples of strictly historic structures are no longer observable results of past planetary collisions, and the trace structures produced by extinct organisms in unpreserved facies. Dinosaur traces and pyritic conglomerates (Fig. 1) are listed as examples of non-actualistic structures. The origin of both examples, however, can easily be explained by comparison with modern analogues. Seilacher (1957) has already discussed preservational and outcrop problems in recent and various Phanerozoic ichnocoenoses, and concluded that many apparently different trace fossil assemblages are the product of preferential exposure and preservation, and not of different grazing habits of the various sediment-dwelling organisms. Further, Pflüger & Gresse's (1996) prime case of ancient sedimentary structures attributable to the former presence of biofilms does not stand the test of non-actualistic proof. Microbial mats are relatively widespread today (Fig. 2; e.g. Donaldson, 1967; Witkowski, 1990; Schieber, 1998). Pflüger's use of non-actualism is also at odds with the

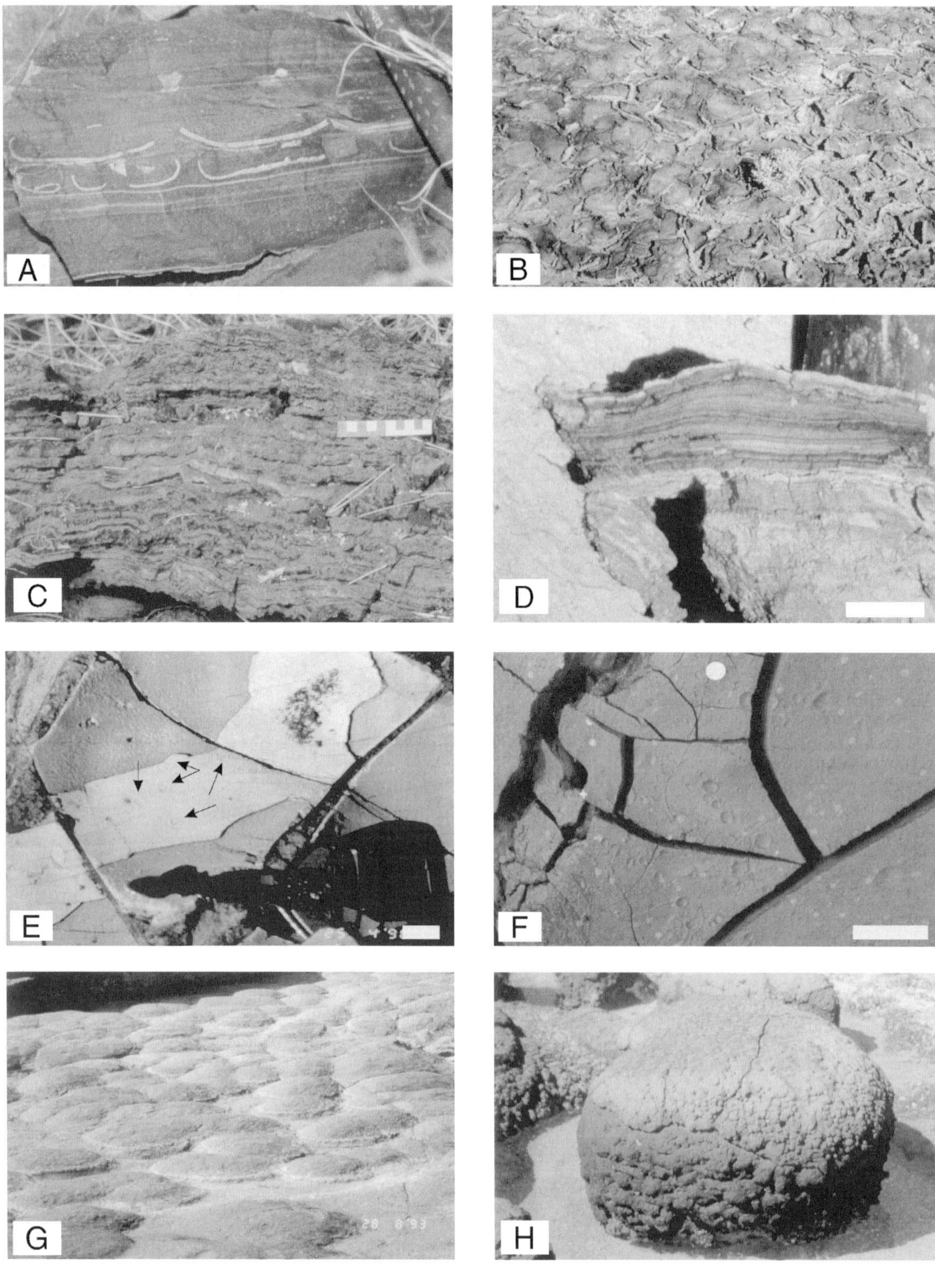
A
B
C
D
E
F
G
H

earlier literature. Richter (1929) introduced the term actuo-palaeontology, and Schafer (1959) subsequently defined actuo-geology as the study of physical and chemical processes and products. This earlier terminology seems to be more in accord with the actualism of Holmes (1965) and Dott & Batten (1980) than with Pflüger's (1995) definition.

In an excellent challenge to the meanings of both uniformitarianism and actualism, Shea (1982) suggested abandonment of both terms. Although we disagree with this proposal, we do support most of his arguments, including the importance of following the rule of simplicity in evaluating hypotheses about the Earth and its evolution. We especially endorse his defence of a rigorous approach to examining processes, conditions and results by means of empirical observation, which remains fundamental to the natural sciences. A lucid defence of empiricism has been presented by McLaren (1994) in a discussion of extinctions in relation to the geological record of bolide impacts.

**Fig. 2.** (*opposite*) (A) Microbial chips in the Ventersdorp Supergroup, South Africa, *c*.2.7 Ga old. The chips are composed of fine-grained volcanic ash particles bound by microbial mat, now disintegrated. These roll up structures are embedded in fine lapilli tuff. Field of view about 25 cm wide. (B) Desiccated, leathery microbial mat, in Laguna Mormona, Baja California, Mexico, which serves as an example for actualistic processes leading to the formation of ancient microbial chips. The microbial mat flakes are rolled up and bind fine sand sediment that can be preserved in this form, as a result of burial before disintegration of the binding organic matter. Field of view about 1 m wide. (C) Example of an ancient sabkha environment from the *c*.3.5 Ga Warrawoona Group, Western Australia. The crinkled lamination is inferred to represent *in situ* desiccation of microbial mats. Barite pseudomorphs after gypsum, barite lenses and mini stromatolites are abundant in this facies. Scale is 10 cm. (D) Vertical section cut into the Jerba marsh sediments of Tunisia, showing intercalated gypsum layers and crinkled mudstone layers. The dark laminae are rich in organic matter (remnants of sulphur bacteria). Scale bar is 5 cm. (E) Rain drop imprints (arrows) on desiccation polygons, preserved on a fine ash surface, from the *c*.2.7 Ga Ventersdorp Supergroup, South Africa. Scale bar is 1 cm. (F) Raindrop imprints on the surface of an unconsolidated, desiccated mud pit in Germany. Scale bar is 1 cm. (G) Stromatolites formed in an intertidal environment, as seen on a bedding surface in the *c*.2.6 Ga Campbellrand Subgroup, South Africa. The average diameter of these domical stromatolites is about 35 cm. (H) Modern intertidal stromatolite, Shark Bay, Western Australia, showing size and shape similar to the ancient stromatolites seen in (G).

## THE PRECAMBRIAN ROCK RECORD: THE CASE FOR ACTUALISM

In a recent review of the Precambrian clastic sedimentary record (Eriksson *et al.*, 1998), an attempt was made to define differences between Precambrian sedimentary processes and systems and their Phanerozoic–modern counterparts. Some of the differences noted in that study included: the paucity of aeolian deposits prior to *c*.1.8 Ga (Eriksson & Simpson, 1998); the abundance of braided alluvial and deltaic channel systems; possibly more uniform shallow marine circulation patterns related to wide and gently sloping marine shelves; an apparent scarcity of Precambrian foreshore deposits, particularly of barrier island deposits; and poorly confined tidal flat drainage channels which commonly formed sheet sandstones. Nevertheless, similarities with Phanerozoic–modern sedimentary processes and products (Fig. 2) were inferred to outweigh greatly the apparent differences (Eriksson *et al.*, 1998).

Actualism thus appears to hold good for the Precambrian clastic sedimentary record. Deserts today typically occur within the interiors of large cratons and at subtropical latitudes; the period after *c*.2.0 Ga was typified by the formation of supercontinents, such as Laurentia, and the Eburnean supercontinent. The controls on the earliest known Precambrian ergs were thus most probably no different from those prevailing today. Similarly, braided alluvial and delta channels, widespread in the Precambrian, are also abundant today. No special non-actualistic case needs to be made for the apparent character of Precambrian coastal to shelf environments either. It should be borne in mind that most Precambrian cratons which exist today represent the interior portions of stable continental crustal plates; their margins have been repeatedly reworked due to plate tectonic processes. Given that continental crustal growth, plate tectonics and continental freeboard are the most important controls on eustasy (e.g. Eriksson, 1999), the shallow marine deposits preserved on most Precambrian cratons are dominated by epeiric sea sediments. Epeiric seas are characterized by tidal rather than wave action (Eriksson *et al.*, this volume, pp. 351–368), especially along their shorelines, thus promoting an absence of common foreshore and barrier island deposits. High sedimentation rates in a pre-vegetative world would have promoted unconfined tidal channels and resultant sheet-like sandstone deposits.

Among the sedimentary environments, those of glaciogenic affinity show the least significant differences;

however, their genesis has been ascribed to a more uniquely Precambrian mix of craton emergence above sea level and formation of the first supercontinents, the development of widespread stromatolitic shelf carbonates, early weathering processes and atmospheric compositions, and changes to the Earth's obliquity and rotation rate (Williams, 1998). Although the latter combination of factors may have generated more extreme global glaciations in the Precambrian, possibly resulting in a snowball Earth (e.g. Hoffman, 1999), supercontinents and their dispersion, amalgamation and concomitant eustatic changes have been very much a part of Phanerozoic geology as well.

Obviously, oceans that were globally frozen for periods of millions of years would have a tremendous influence on the Earth's climate and on worldwide weathering, erosion and sedimentation patterns. In our present state of knowledge, such conditions diverge so strongly from any situation known to us from the Phanerozoic to modern, that any comparison with younger glaciations seems equivocal. Nevertheless, sediments that accumulated during the snowball Earth periods were recognized as glaciogenic by comparisons to modern deposits, perhaps with the exception of cap carbonates and post-glacial banded iron-formations that are explained on a theoretical basis (Hoffman *et al.*, 1998). Physical and chemical laws thus allow us to understand the formation of such sediments, although their genesis is still a matter of debate, as is the entire scenario of the snowball Earth.

An example of problematic Precambrian sediments, which some workers would categorize as non-actualistic, is provided by certain carbonate rocks. Large, platformal accumulations of carbonates are unknown from the sedimentary record prior to the Wit Mfolozi Formation of the 3.0 Ga old Pongola Supergroup, and become volumetrically important only in the Neoarchaean (Altermann, this volume, pp. 15–32). Nevertheless, carbonate rocks produced in similar, shallow marine environments are known from the oldest sedimentary record, and microbial mediation in their production has been demonstrated based on carbon and sulphur isotopic content (Strauss, this volume, pp. 67–105). Microbial mediation in early diagenetic dolomitization of Archaean carbonates also has been discussed as a process known from younger sediments (Wright & Altermann, 2000). Accumulation rates of Archaean carbonates match those of modern carbonate platforms, when compared over equally long periods of time, as discussed by Altermann & Nelson (1998). Stromatolitic, biotically induced carbonate precipitation led to the formation of extensive reefs in the Archaean. These reefs, although lacking the biotic diversity of their younger counterparts, evolved to an enormous richness of forms and occupied various marine and lacustrine settings, just as they do today. Shelf carbonates, although no longer predominantly stromatolitic in character, nevertheless still remove large quantities of carbon dioxide from the atmosphere, and thus serve to reduce the atmospheric greenhouse potential. Extremes in modern weathering regimes define a continuum of conditions which overlaps with the extreme denudation rates of the Archaean and Palaeoproterozoic (Corcoran & Mueller, this volume, pp. 183–212). Changes to the Earth's obliquity have occurred in the past, and the rate of the Earth's axial rotation appears to have decreased continuously throughout the time represented by the known geological record.

With regard to the chemical composition of the atmosphere and hydrosphere in the early Precambrian, possible $O_2$ enrichment of the Archaean and early Proterozoic atmosphere as high as ±50% of the present atmospheric level (PAL) has been recently considered (e.g. Holland, 1999; Ohmoto, 1999; Rasmussen *et al.*, 1999). The question remains unresolved, although a general acceptance of low $O_2$ atmospheric conditions still prevails. For the hydrosphere, models of Ca-saturated ocean water (Sumner, this volume, pp. 107–120) and of a soda ocean (Kempe & Degens, 1985) have been proposed. Such varying chemistries must have led to different modes of precipitation, and sediments fitting such models have been found in the Precambrian rock record. Among these sediments are banded iron-formations, magadiite pseudomorphs (Martini, 1990) and zebraic or herringbone calcites (Sumner, this volume, pp. 107–120). However, all these sediments, including BIF (Trendall, this volume, pp. 33–66), can also occur in the Phanerozoic rock record. There is no evidence to suggest that the Precambrian atmosphere and hydrosphere did not obey the same chemical equilibria laws that apply to present-day systems, and thus any possible compositions of the atmosphere and hydrosphere can be evaluated in terms of actualistic models.

The much debated (e.g. Goodwin, 1996; Kusky, 1997; Hamilton, 1998) issue of Precambrian (especially Archaean) plate tectonics is also pertinent to the concept of actualism and non-actualism. These arguments are hindered by the small volumes of early Precambrian crust which have survived and the fact that, ophiolites apart, Precambrian oceanic crust has not been preserved. A currently prevailing school of thought argues that plate tectonics was indeed operative in the Middle Archaean, and that intra-oceanic generation of continental crust through island arc obduction was an important process (e.g. de Wit *et al.*, 1992). De Wit & Ashwal (1997) also provide strong evidence for greenstones still forming in the Phanerozoic–modern period; greenstone successions may thus not constitute a unique Precambrian occurrence. Nelson (1998) suggested that global catastrophic magmatic events may have been superimposed on plate tectonic processes, and that these influences also played a prominent role in Archaean greenstone generation. To what extent such magmatic events differed from events triggered by mantle plumes, for which there is a convincing Phanerozoic record, remains to be evaluated.

## CONCLUSIONS

We have argued that all geological processes and products are embraced by actualism (see Figs 1 & 2 for a few examples), whereas non-actualism applies to unique pre-geological events that were not governed by the extant physical and chemical laws applicable to actualistic processes. We further contend that actualism should be accepted as the fundamental premise of the geosciences (as well as the other natural sciences) which serves to amplify the principle of uniformitarianism. For earth scientists, non-actualism would thus apply to events unlikely to have contributed to the rock record, and organic-mat-rich Precambrian environments that have been recently described as non-actualistic can be simply designated as pre-metazoan environments or biofilm-dominant environments. As a final observation, it is interesting to note that in modern philosophy (e.g. Hazen, 1996; Jubien, 1996; Tomberlin, 1998) there is a substantial literature regarding actualism and non-actualism. In most of the philosophical literature, actualism is equated with reality, whereas non-actualism deals with 'parallel worlds' that are inhabited, for instance, by unicorns. To advance clarity in the geosciences, we suggest that the term non-actualism should be graciously relinquished to the mystical realm of such 'parallel worlds'. If this term is retained, we suggest that non-actualistic geological models should be reserved for speculations about early Hadean environments (the first few microseconds?), and for cosmological problems concerned with topics such as black holes, wormholes and the 'Big Bang' origin of the universe, involving as they do time–space regions for which the laws of nature, as we presently understand them, may not be applicable.

As our knowledge of natural laws continues to expand, and new processes and environments are discovered, we can vastly widen our perspectives, and appropriately modify our paradigms and theories. With our increasing understanding of the present world, the application of actualism (both method and principle) to geological problems will gain greater utility, allowing us to provide new and better explanations for our geological observations.

## ACKNOWLEDGEMENTS

The idea for this paper was generated by the participation of one of us (J.A.D.) in a Special Session on Actualistic and Non-actualistic Precambrian Sedimentary Styles that took place at the Annual Meeting of the Geological Association of Canada (May 1999, in Sudbury, Ontario). The wide range of views expressed by participants in this session suggested the need for a more extensive international discussion of actualism/non-actualism, resulting in the collaboration of the present authors. In preparing this paper we have drawn from numerous discussions with colleagues about their use of the terms reviewed herein, but we especially thank Darrel Long, Rob Rainbird and Bryan Krapez for their input. Comments by Larry Aspler and Keith Bell provided many useful ideas for revision of our initial draft of this paper. Critical reveiws by Mario Coniglio and an anonymous reviewer provided guidance for additional improvements.

## REFERENCES

ALLABY, A. & ALLABY, M. (Eds) (1999) *Oxford Dictionary of Earth Sciences*. Oxford University Press, Oxford, 619 pp.

ALTERMANN, W. & NELSON, D.R. (1998) Sedimentation rates, basin analysis, and regional correlations of three Neoarchean and Paleoproterozoic sub-basins of the Kaapvaal Craton as inferred from precise U–Pb zircon ages from volcaniclastic sediments. *Sediment. Geol.*, **120**, 225–256.

BAKER, V.R. & BUNKER, R.C. (1985) Cataclysmic late Pleistocene flooding from glacial Lake Missoula: a review. *Quat. Sci. Rev.*, **4**, 1–41.

BARBER, K. (Ed.) (1998) *The Canadian Oxford Dictionary*. Oxford University Press, Toronto, 1707 pp.

BEURLEN, K. (1935) Der Aktualismus in der Geologie, eine Klarstellung. *Zeit. Mineral. Abteil.*, **B12**, 520–525.

BUFFON, G.L.L., COMTE DE (1749) *Theorie de la Terre: Epoques de la Nature*. Histoire naturelle, generale at particulaire. Tome 1. L'Imprimerie Royale, Paris.

CLIFTON, H.E. (Ed.) (1988) Sedimentologic consequences of convulsive geologic events. Spec. Paper geol. Soc. Am., Boulder, **229**, 157 pp.

Cloud, P.E. (1988) *Oasis in Space: Earth History from the Beginning.* W.W. Norton & Company, New York, 508 pp.

Condie, K.C. & Sloan, R.E. (1997) *Origin and Evolution of Earth: Principles of Historical Geology.* Prentice Hall, Upper Saddle River, NJ, 498 pp.

Considine, D.M. (Ed.) (1995) *Van Nostrand's Scientific Encyclopedia*, 8th edn. Van Nostrand Reinholt, New York.

Cuvier, G. (1817) *Essay on the Theory of the Earth* (English translation). Wm Blackwood, Edinburgh.

Darwin, C. (1859) *On the Origin of Species by Means of Natural Selection.* John Murray, London, 490 pp.

de Wit, M.J. & Ashwal, L.D. (1997) *Greenstone Belts.* Clarendon Press, Oxford, 809 pp.

de Wit, M.J., Roering, C., Hart, R.J. *et al.* (1992) Formation of an Archaean continent. *Nature*, **357**, 553–562.

Donaldson J.A. (1967) Precambrian vermiform structures: a new interpretation. *Can. J. Earth Sci.*, **4**, 1273–1276.

Dott, R.H. Jr & Batten, R.L. (1980) *Evolution of the Earth*, 3rd edn. McGraw-Hill, New York, 649 pp.

Eriksson, K.A. & Simpson, E.L. (1998) Controls on spatial and temporal distribution of Precambrian eolianites. *Sediment. Geol.*, **120**, 275–294.

Eriksson, P.G. (1999) Sea level changes and the continental freeboard concept: general principles and application to the Precambrian. *Precam. Res.*, **97**, 143–154.

Eriksson, P.G., Condie, K.C., Tirsgaard, H. *et al.* (1998) Precambrian clastic sedimentation systems. *Sediment. Geol.*, **120**, 5–53.

Geikie, A. (1903) *Text-book of Geology*, Vol. 1, 4th edn. Macmillan & Co., London, 702 pp.

Goodwin, A.M. (1996) *Principles of Precambrian Geology.* Academic Press, London, 327 pp.

Gordeev, D.I. (1961) Stikhiinaya materialisticheskaya dialektika sochineniyakh M.V. Lomonosova (Dialectical materialism in the works of M.V. Lomonosov). *Moscow Univ. Vest. Geol.*, **5**, 7–26.

Gostin, V.A., Haines, P.W., Jenkins, R.J.F., Compston, W. & Williams, I.S. (1986) Impact ejecta horizon within Late Precambrian Shales, Adelaide Geosyncline, South Australia. *Science*, **233**, 198–200.

Hamblin, W.K. (1991) *Introduction to Physical Geology.* Macmillan, New York, 378 pp.

Hamilton, W.B. (1998) Archean tectonics and magmatism. *Int. geol. Rev.*, **40**, 1–39.

Hamm, E.P. (1993) Bureaucratic statistic or actualism? *Hist. Sci.*, **31**, 151–176.

Hazen, A.P. (1996) Actualism again. *Phil. Stud.*, **84**, 155–181.

Hoffman, P.F. (1999) The break-up of Rodinia, birth of Gondwana, true polar wander and the snowball Earth. *J. Afr. Earth Sci.*, **28**, 17–33.

Hoffman, P.F., Kaufman, A.J., Halverson, G.P. & Schrag, D.P. (1998) A Neoproterozoic snowball Earth. *Science*, **281**, 1342–1346.

Holland, H.D. (1999) When did the earth's atmosphere become oxic? A reply. *Geochem. News*, **100**, 20–22.

Holmes, A. (1965) *Principles of Physical Geology.* Ronald Press, New York, 1288 pp.

Hooykaas, R. (1956) The principal of uniformity in geology, biology and theology. *Vict. Inst. J. Trans.*, **88**, 101–116.

Hooykaas, R. (1970) Catastrophism in geology; its scientific character in relation to actualism and uniformitarianism. *Konink. Nederl. Akad. Wetenschap. Lett. Med.* (n.r.), **33**, 271–316.

Hsu, K.J. (1983) Actualistic catastrophism. *Sedimentology*, **30**, 3–9.

Hutton, J. (1788) Abstract of a dissertation concerning the system of the Earth, its duration and stability. *Trans. R. Soc. Edinburgh*, **1**, 209–304.

Jackson, J.A. (Ed.) (1997) *Glossary of Geology*, 4th edn. American Geological Institute, Alexandria, VA, 769 pp.

Jubien, M. (1996) Actualism and iterated modalities. *Phil. Stud.*, **84**(2/3), 109–125.

Judson, S. & Richardson, S.M. (1995) *Earth: an Introduction to Geologic Change.* Prentice Hall, Engelwood Cliffs, NJ, 551 pp.

Kaiser, E. (1931) Der Grundsatz des Actualismus in der Geologie. *Zeit. Deutsch. Geol. Gesell.*, **83**, 389–407.

Kempe, S. & Degens, E.T. (1985) An early soda ocean? *Chem. Geol.*, **53**, 95–108.

Kummerow, E. (1932) Die aktualistische Methode in der Geologie. *Deutsch. Geol. Gesell. Zs.*, **84**, 563–565.

Kusky, T.M. (1997) Principles of Precambrian Geology (book review). *GSA Today*, **7**(5), 29–34.

Levin, H.L. (1999) *The Earth through Time*, 3rd edn. Saunders College, Philadelphia, 593 pp.

Lomonosov, M.V. (1763) See: Muller, D.W., McKenzie, J.A. & Weissert, H. (1991) *Controversies in Modern Geology.* Academic Press, New York; and Gordeev, D.I. (1961) Stikhiinaya materialisticheskaya dialektika sochineniyakh M.V. Lomonosova (Dialectical materialism in the works of M.V. Lomonosov). *Moscow Univ. Vest. Geol.*, **5**, 7–26.

Lyell, C. (1830) *Principles of Geology*, Vol. 1, 1st edn. John Murray, London, 511 pp.

McLaren, D.J. (1994) Impacts and extinctions: science or dogma? In: *Mass Exctintion Debates: How Science Works* (Ed. Glen, W.), pp. 121–131. Stanford University Press, Stanford, CA.

Martini, J.E.J. (1990) An early Proterozoic playa in the Pretoria Group, Transvaal, South Africa. *Precam. Res.*, **46**, 341–351.

Milner, R. (1990) *The Encyclopedia of Evolution: Humanity's Search for Its Origins.* Facts on File, New York, 481 pp.

Muller, D.W., McKenzie, J.A. & Weissert, H. (Eds) (1991) *Controversies in Modern Geology.* Academic Press, London, 490 pp.

Nelson, D.R. (1998) Granite–greenstone crust formation on the Archaean Earth—a consequence of two superimposed processes. *Earth planet. Sci. Lett.*, **158**, 109–119.

Ohmoto, H. (1999) Redox state of the Archean Atmosphere: Evidence from detrital heavy minerals in *c.*3250–2750 Ma sandstones from the Pilbara Craton, Australia: comment. *Geology*, **27**, 1151–1152.

Parker, S.P. (Ed.) (1991) *The McGraw-Hill Encyclopedia of the Geological Sciences*, 2nd edn. McGraw-Hill, New York.

Pflüger, F. (1995) Morphodynamik, Aktualismus und Sedimentstrukturen. *Neues Jahrb. Geol. Paläontol. Abh.*, **195**, 75–83.

Pflüger, F. & Gresse, P.G. (1996) Microbial sand chips—a non actualistic sedimentary structure. *Sediment. Geol.*, **102**, 263–274.

Press, F. & Siever, R. (1994) *Allgemeine Geologie (Understanding Earth).* Spektrum Akademischer Verlag, Heidelberg, 602 pp.

Prevost, L.C. (1825) De la formation des terrains des environs de Paris. *Soc. Philomath. Bull.*, 74–77.

RASMUSSEN, B., BUICK, R. & HOLLAND, H.D. (1999) Redox state of the Archean atmosphere: evidence from detrital heavy minerals in *c.*3250–2750 Ma sandstones from the Pilbara Craton, Australia: reply. *Geology*, **27**, 1152.

RICHTER, R. (1929) Gründung und Aufgaben der Forschungsstelle für Meeresgeologie 'Senkenberg' in Wilhelmshaven. *Natur u. Museum*, **59**, 1–30.

SALOMON, W. (1926) Gibt es Gesteine die für bestimmte Erdperioden charakteristisch sind? *Sitzber. Heidelberger Akad. Wissenschaft. math-naturw.* Kl., 9 Abh.

SCHAFER, W. (1959) Elements of actuo-paleontology. *Salt Marsh Conf. 1958 Proc., Univ. Georgia, Marine Inst.*, pp. 122–125.

SCHIEBER, J. (1998) Possible indicators of microbial mat deposits in shales and sandstones: examples from the Mid-Proterozoic Belt Supergroup, Montana, USA. *Sediment. Geol.*, **120**, 105–124.

SCHWARTZ, H.-U. (1978) Die Grenzen der aktualistischen Arbeitsweise in der modernen Geologie. *Mitteil. Geol.-Paläontol. Inst. Univ. Hamburg*, **48**, 1–15.

SEILACHER, A. (1957) An-aktualistisches Wattenmeer? *Paläontol. Zeit.*, **31**, 198–206.

SHEA, J.H. (1982) Twelve fallacies of uniformitarianism. *Geology*, **10**, 455–460.

SHOEMAKER, E.M. (1995) Comet Shoemaker-Levy 9 at Jupiter. *Geophys. Res. Lett.*, **22**, 1555.

SPRAY, J.G. (1997) Superfaults. *Geology*, **25**, 579–582.

STANLEY, S.M. (1989) *Earth and Life through Time*, 2nd edn. W.H. Freeman & Company, New York, 689 pp.

STINNER, A. (2000) Lord Kelvin and the age-of-the-Earth debate: the rise and fall of Kelvin's calculations of the age of the Earth and the Sun. *Phys. Can.*, Nov./Dec., 312–332.

THERRIAULT, A.M., REIMOLD, W.U. & REID, A.M. (1997) Geochemistry and impact origin of the Vredefort Granophyre. *S. Afr. J. Geol.*, **100**, 115–122.

TOMBERLIN, J.E. (1998) Naturalism, actualism and ontogeny. *NOUS*, Supp. S, 489–498.

VELIKOVSKY, I. (1950) *Worlds in Collision*. Doubleday, Garden City, NJ, 401 pp.

VELIKOVSKY, I. (1955) *Earth in Upheaval*. Doubleday, Garden City, NJ, 301 pp.

VON BUBNOFF, S. (1937) Die historische Betrachtungsweise in der Geologie. *Geistige Arbeit*, **4**(13), 1–3.

VON HOFF, K.E.A. (1822) *Geschichte der durch Überlieferung nachgewiesenen natürlichen Veränderungen der Erdoberfläche*, Vol. 1, Gotha, 489 pp.

WILLIAMS, G.E. (1998) Precambrian tidal and glacial clastic deposits: implications for Precambrian Earth–Moon dynamics and palaeoclimate. *Sediment. Geol.*, **120**, 55–74.

WITKOWSKI, A. (1990) Fossilization processes of the microbial mat developing in clastic sediments of the Puck Bay (southern Baltic Sea, Poland). *Acta Geol. Pol.*, **40**, 3–27.

WRIGHT, D.T. & ALTERMANN, W. (2000) Microfacies development in Late Archean stromatolites and ooids. In: *Carbonate Platform Systems. Components and Interactions* (Eds Insalco, E., Skelton, P.W. & Palmer, T.J.). Spec. Publs geol. Soc. London, No. 178, pp. 51–70. Geol. Soc. London, Bath.

WYATT, A. (Ed.) (1986) *Challinor's Dictionary of Geology*, 6th edn. University of Wales Press, Cardiff, 374 pp.

*Spec. Publs int. Ass. Sediment.* (2002) **33**, 15–32

# The evolution of life and its impact on sedimentation

W. ALTERMANN

*Institut für Allgemeine und Angewandte Geologie, Ludwig-Maximilians-Universität München, Luisenstrasse 37, D-80333 Munich, Germany*

## ABSTRACT

In this overview, Precambrian Earth environments are discussed together with the pathways of the abiotic synthesis of life and its early evolution. Consideration of the possible steps in the assembly of amino acids, protein polymers, RNA strings or DNA helixes provides contradictory inferences about environmental conditions at the time of their formation, varying between cold and hot aqueous settings. Problems in identification of microbial life in Precambrian sedimentary rocks are grounded in insufficient preservation, metamorphic alteration and lack of criteria other than isotopic and morphological similarity to extant microbes. The earliest microfossiliferous rocks hint at extreme hypersaline and thermophilic conditions of microbial life, although at 3.5 Ga, life was already widespread in aquatic environments, and was sufficiently advanced, being capable of photosynthesis. As life has arisen on Earth, it has gradually gained an increasing influence on sedimentary environments. This influence was initially mediated through weathering and depositional processes, governed by the chemistry of the atmosphere and oceans and by the direct biochemical precipitation of sediments. Stabilization of siliciclastic sediments by microbial mats appears to have been an important process in the Proterozoic, but is difficult to prove because of organic degradation. Only at the terminal Neoproterozoic did the direct influence on sediments by grazing and burrowing organisms become apparent in the rock record. Stromatolitic carbonates of the Archaean and early Proterozoic are recognized as *bona fide* biogenic sediments, although microbial mediation in their formation can rarely be proven. Stromatolites and bacterial microfossils are considered unsuitable for stratigraphic subdivison of the Precambrian because of the primarily environmental significance of stromatolite morphology, and the evolutionary conservatism in bacterial morphology. Calcification of stromatolites might have been induced directly by the bacterial communities or by chemical precipitation from Ca-oversaturated water. The chemistry of ocean water in the Precambrian is equally controversial, as the oxygen levels in the Precambrian atmosphere, and their influence on chemical sediments, are difficult to ascertain. Examples of the influence of the atmosphere and hydrosphere on the evolution of life demonstrate the interrelated dependence of the Precambrian biosphere, hydrosphere, atmosphere and lithosphere.

## INTRODUCTION

Examination of Precambrian sedimentary environments, especially those of the Archaean Eon, is intriguing because next to physical and chemical processes visible in the rock record and comparable to those acting in the Phanerozoic, a striking difference is implied by the apparent lack of fossilized organisms. In all Phanerozoic environments, life had a strong and direct impact on erosion, sediment transport and deposition. This is evident not only in sediments, such as carbonate rocks, that are the direct results of biologic activity. In the Phanerozoic, living organisms were able to form skeletal and precipitational reefs that could control the sedimentology, morphology and bathymetry of carbonate platforms. Pelitic and coarser siliciclastic sedimentation is also strongly, although less noticeably, influenced by biological processes. For example, erosion, and thus the availability of clastic debris, is directly dependent on the nature of the vegetation in the source area. Destruction of the vegetation cover can induce catastrophic sedimentation events. Rapid delivery of large amounts of clastic debris into basin margins can disturb the sensitive balance of microbial, faunal and floral symbioses and eventually erase complex ecological communities, resulting in dramatic changes in sediment distribution. The absence of communities of sediment feeders and grazers can eventually lead to preservation of sedimentary structures usually not preservable under modern normal marine or lacustrine conditions.

The above examples are easily recognized in modern environments. Today we are concerned with the destruction of ecosystems, and the impact this has on other biologic and/or physical realms. Were these processes different in early geological history, with an Earth barren of life and slowly evolving to increasingly complicated ecological organization?

Donaldson *et al.* (this volume, pp. 3–14) emphasize that an actualistic approach is required to understand Precambrian sedimentary environments. Clearly, some processes must have nevertheless acted at different scales and different rates when life was only in its infancy. Early life altered the atmospheric composition, from one dominated by $CO_2$ to an oxygen-rich atmosphere (Holland, 1984), and thus affected oceanic and atmospheric currents through temperature gradients, but also weathering patterns in terrestrial realms (cf. Corcoran & Mueller, this volume, pp. 183–212; Eriksson & Cheney, 1992). As another consequence of atmospheric change, life influenced the ability of water to transport chemical complexes at different oxidation stages and, thus, contributed to the genesis of some important mineral deposits. The presence or absence of biota in the sedimentary system and the delivery of varying amounts of organic matter into the basin also control the geochemistry of the resulting sediment to some degree, and thus may be important for mineral deposit-forming processes (cf. Strauss, this volume, pp. 67–105; Trendall, this volume, pp. 33–66). For example, diagenetic and hydrothermal fluids tend to precipitate their metal ion wealth when entering different oxygenation and pH states, and metal ions can, vice versa, be mobilized from their chemical complexes by microbial mediation.

Because sedimentology is the main subject of the present special publication on Precambrian sedimentary environments, this review is limited to the most important facts and theories of the evolution of life. A comprehensive review of the biological aspects would be beyond the scope of this review, which aims at bridging between Precambrian sedimentary environments and the advancement of living organisms. This paper presents an unbiased account of data and interpretations given by other workers. Because of the vast amount of original literature, however, parts of the important discussion of biochemistry and theories on the very early history of the Earth are not fully covered in this review. Such theories and discussions take into account, for example, questions regarding the nature of early plate tectonics, the origin of the oceans and atmosphere, the origin of the first living cell and the last common ancestor to all living organisms. These subjects are regarded as not of primary importance to geologists working on Precambrian rocks because no rocks older than 4.0 Ga are known. This appraisal serves as an introduction for readers usually not concerned with Precambrian sedimentary rocks and Precambrian palaeontology. It is based on many recent papers in sedimentological and palaeontological journals, but mainly on the publications of the IGPP Center for the Study of the Evolution of Life (CSEOL) at the University of California, Los Angeles (Schopf, 1983, 1992, 1999; Schopf & Klein, 1992), de Duve's (1991) book *Blueprint for a Cell*, the Proceedings of Nobel Symposium No. 84 (Bengtson, 1994) and Knoll's (1996) overview of Archaean and Proterozoic palaeontology.

## THE SYNTHESIS AND EARLY EVOLUTION OF LIFE

Some modern forms of life are adapted to extreme surroundings, such as those that prevail around deep submarine vents and continental thermal springs, or in hypersaline lagoons. Many of these macrocosms, generally regarded as hostile to life, resemble in some respects conditions widespread on the Archaean Earth and therefore it is likely that some of these environments may have been the sites of the origin of life. This is supported by investigations showing that organisms inhabiting such extreme environments (i.e. bacteria, archaea or cyanobacteria) are some of the most primitive in the evolutionary diversification path. Sequential analyses of DNA imply that thermophily, acidophily and sulphur-based metabolism are original primitive attributes of the oldest eubacteria and archaea, and were probably present in the earliest prokaryotes (Woese, 1987; Stetter, 1994). In fact, eubacteria are found fossilized in similar conditions, e.g. hypersaline lagoons or close to hydrothermal springs, in rocks as old as 3.5 Ga (Schopf, 1993; Lowe, 1994).

The major unresolved question of studies concerning the evolution of life is how and where life began. It is generally accepted that life arose early in Earth history, as soon as conditions became favourable for preservation of simple organic compounds. The basic precondition for the synthesis, preservation and accumulation of organic molecules, such as amino acids, is the formation of a stable crust on Earth wrapped in an atmosphere and oceans. Conditions in which water was stable and gases were not overheated probably already prevailed 100–300 Myr after the accumulation of cosmic dust to form the Earth, before 4.55 Ga (Mojzsis *et al.*, 2001). The heavy bombardment of large (>50 km) meteorites and asteroids on the young Earth, however, might have led to the complete vapourization of the earliest atmosphere and hydrosphere. No form of life would survive a catastrophe of this extent. During the early period of Earth formation an impact of a Mars-size body probably formed the Moon and had a catastrophic impact on the surface shells of the Earth, down to its core (Newsom & Jones, 1990). As the heavy bombardment came to an end (*c*.4.0 Ga), conditions stabilized to allow for the formation of oceans, where organic compounds would have been synthesized and accumulated. The water necessary for the formation of oceans might have been derived from the degassing of incoming planetesimals and, later, from the degassing of the Earth's interior. A major contribution to the water reservoir of the Earth was probably from comets (Oro & Lazcano, 1996). As soon as the Earth cooled and slabs of crust were formed, plate tectonic processes must have started to accommodate the convective heat transfer from the interior to the surface. Plate tectonic processes together with tidal forces created environments where organic compounds could concentrate sufficiently to trigger reactions necessary for the formation of complex structures. The energy necessary for the synthesis of such complex organic molecules was provided by the sun's luminosity, by atmospheric electric discharge and by geothermal activity. Light is the most abundant energy source, and therefore photochemical reactions probably played a major role in the pre-biotic synthesis of organic compounds. The organic carbon participating in these reactions could have been a remnant from the primary accretion phase of the Earth, or could have originated from extra-terrestrial reactions and carbonaceous chondrites (McClendon, 1999).

Four stages of development were essential in the origin of life (Schopf, 1992, 1999). First, specific environmental conditions must have made self-organization of the major elements of life (C, H, O, N) to complex molecules possible and compulsory. As a result, simple organic compounds originated. Second, such compounds must have stabilized by using energy from the environment and by building more complex compounds (nutrition and growth), and thus became able to protect themselves from hydrolysis and other forms of energy loss. Third, the invention of a self-replication process, a basic requirement in the definition of life, was probably the most enigmatic event in the history of life. In the fourth stage, identical replication was abolished in favour of mutation, to allow for Darwinian evolution. This evolutionary process had to be strong enough to generate new species, but the reproduction also had to be sufficiently stable to prevent total molecular disorder and extinction. None of these steps, nor the processes linking them along the evolutionary path, can yet be fully explained, and although significant progress has been made in their study, we are still a long way from acquiring unequivocal answers to the numerous questions. Even the chronological order of the four stages listed above is uncertain. Self replication and mutation, for example, might have been simultaneous and could have preceded the second stage of compound stabilization.

The synthesis of organic compounds in a primitive, reducing atmosphere of methane ($CH_4$), ammonia ($NH_3$), $H_2O$ vapour and molecular oxygen, required for stage 1 above, was first demonstrated in an experiment by Miller (1953), following theories on the origin of life by Oparin (1938), Urey (1952) and others. As a result of subsequent experiments with electric discharge (spark) or pyrolysis as an energy source in an atmosphere containing only traces of $NH_3$ together with $CH_4$, $N_2$ and $H_2O$, or an admixture of $CO_2$, CO and $H_2S$, many important biological compounds, including 17 of the 20 amino acids found in modern proteins, have been synthesized *in vitro*.

Oparin (1938) was the first to propose that the earliest organisms were heterotrophic. They thus were independent of complicated, evolutionarily advanced photosynthesis and were able to use organic compounds. More recent atmospheric models support a $CO_2$-rich early Archaean atmosphere rather than a strongly reducing one, composed mainly of $CH_4$ and $NH_3$, as simulated in the original experiment by Miller (1953). High concentrations of $CO_2$ in the Archaean atmosphere, of over 100 PAL (present atmospheric level; Kasting, 1987), but also of water vapour and methane, probably compensated for the predicted lower luminosity of the Sun (15–18% below the present luminosity, at 2.5 Ga; Gough, 1981). Therefore, the Earth's surface could have been subjected to higher temperatures and humidity than today, considering the higher geothermal gradient and the early greenhouse atmosphere. Early Archaean surface temperatures could have been as high as 80–100°C (Kasting & Ackerman, 1986), dropping to 30–50°C at 3.5 Ga (Lowe, 1992). At such temperatures, high concentrations of organic macro-molecules in the pre-biotic broth would not have been probable because of fast hydrolysis. For this and other reasons, some workers (e.g. de Duve, 1991; Wächtershäuser, 1994) believe that the first organisms were not heterotrophic but chemoautotrophic. In any case, the first living cells were probably composed of self-assembled complex molecular systems of polymers, with a self-replicating mechanism, encapsulated by a membrane of lipids and proteins (Deamer, 1993). Numerous models for their formation exist, but the chemical evolution to form the first living cell and the beginning of biological evolution remains a mystery.

The conditions for the pre-biotic formation of protein polymers, RNA strings or DNA helixes are not well understood. Nevertheless, such structures shed some light on the possible environments of their origin. For the removal of $H_2O$ from monomers and the linkage of monomers, a thermodynamic barrier must be overcome. Experiments on synthesis of peptides (OC–NH bonds that link amino acids) and of nucleotides (carriers of genetic code in DNA) yielded good results at temperatures near 100°C. Such results hint at thermal springs as possible sites of early polymerisation (i.e. stage 2 above). On the other hand, according to Miller (1992), the necessity of gathering a high concentration of organic compounds for the synthesis of the earliest organisms requires low temperatures. The rates of hydrolysis of peptide and polynucleotide polymers and of the decomposition of sugars are so high that significant accumulation of such compounds in aqueous solutions was unlikely unless temperatures were close to 0°C. Another reason for the requirement of a low-temperature early Earth is the melting temperature of double helix structures, which ranges from below 0 to a maximum of 35°C. However, low temperatures slow down not only decomposition, but also the reactions necessary for the formation of organic compounds. Therefore, at some temperature the ratio of synthesis and decomposition must have reached an equilibrium, supporting a self-maintaining reactive process. An early atmosphere low in $O_2$ is essential, because $O_2$ rapidly reacts with organic compounds, especially under high ultraviolet radiation, as must have been active without the protecting shield of the atmospheric ozone layer. Because of fast decomposition, life must have originated and became stable within a relatively short time span of perhaps less than a few hundred thousand years.

An alternative way of polymerizing monomers and of replication (synthesis stages 3 and 4), is by using templates. It is envisaged that some minerals may have served as templates on which nucleotides formed chains. Cairns-Smith (1982) proposed the origin and replication of nucleic acids and other organic molecules on the surfaces of clay minerals. Mutations necessary for evolutionary variation of such mineral-bound organic molecules could conceivably have occurred due to lattice defects. Wächtershäuser (1994) suggested an alternative way in which the first organisms used the formation of pyrite ($Fe_2S$) from hydrogen sulphide ($H_2S$) and FeS as a source of energy. Pyrite crystals have positively charged surfaces and can bond negatively charged products of $CO_2$ fixation. Such surface bonding is selective, favouring polymers over monomers, and therefore RNA, DNA and other organic molecules are seen by Wächtershäuser as the result of chemical selection without any participation of a pre-biotic broth. Although replication of molecules in this way is plausible, the molecules bonded to the crystal surface remain 'two-dimensional' (single layer) and their evolution into a three-dimensional cell, independent of the electrostatic force of the pyrite, is difficult to conceive. Interesting experimental work pointing in the template theory direction was performed by Sowerby & Heckl (1998). These authors have demonstrated that a mineral template surface can spontaneously self-assemble DNA bases and form a primitive coding mechanism for the subsequent polymerization of amino acids.

Gilbert (1986) proposed that primordial organisms were composed of RNA, capable of catalytic activity and of carrying genetic information, and that biosynthesis of proteins and DNA was only developed later. In living organisms, dehydration–condensation reactions are supported by energy delivered from adenosine triphosphate (ATP) molecules. Cells with primitive ATP molecules and DNA genomes would have had the capacity for biotic protein synthesis, and therefore must have evolved from RNA cells of the last common ancestor of the three main branches of the evolutionary tree, namely eubacteria, archaea and eukaryotes (Woese *et al.*, 1990). Nevertheless, proteins must have been abundant in the RNA world because amino acids are polymerized non-biotically more readily than RNA.

## FIRST EVIDENCE OF MICROBIAL LIFE AND PROBLEMS IN ITS IDENTIFICATION

Reports of biological activity in the Earth's earliest sediments are scarce and equivocal because of the metamorphic and diagenetic alteration such rocks experienced and the fact that sedimentary deposits of early Archaean age are extremely rare. Schidlowski (1988) reported a whole rock carbon isotopic signature typical of biogenic $^{12}C/^{13}C$ ratios from the *c*.3.8 Ga metasediments of Isua, Greenland. However, because these rocks have been altered to amphibolite facies, this evidence for the earliest known life on Earth remained strongly disputed until Mojzsis *et al.* (1996) reported similar results, obtained by an ion microprobe analysis on apatite mineral grains which are thought to have survived the metamorphism, from Isua metasediments. In a subsequent study by Rosing (1999), graphite globules in the pelagic Isua metasediments were interpreted as plankton remnants, and have an isotopic composition of $\delta^{13}C$ of about −19‰ (versus PDB), verifying the results of Mojzsis *et al.* (1996). Findings like those of Schidlowski, Mojzsis, Rosing or Schopf suggest that microbial life was already widespread in the Archaean oceans between 3.8 and 3.5 Ga. However, new investigations by Myers (2001) again cast severe doubts on the authenticity of the geochemical 'fossils' from the Isua rocks.

The appearance of microfossils in the rock record is sudden and bewildering. The earliest optically recognizable microfossils were reported from the 3.46 Ga Apex Chert of the Warrawoona Group, Australia. From these most ancient sedimentary rocks, 11 taxa of cellularly preserved filamentous microbes are described (Schopf, 1993). Rubisco-type (ribulose biphosphate carboxylase/oxygenase enzyme, which fixes $^{12}C$ in higher proportions than $^{13}C$ in all photosynthesizing organisms) carbon isotopic composition and morphological affinity to recent cyanobacteria and bacteria indicate that at least some of the earliest taxa from the Apex Chert were capable of photosynthesis. The carbon-isotopic signature from the apatite in the Isua metasediments in Greenland indicates that life around 3850 Ma was already capable of chemoautotrophic or photoautotrophic metabolism (Mojzsis *et al.*, 1996). Therefore, life in a more primitive form must have existed in the oceans before that time, and the key to proving it lies in finding older rocks or new methods, allowing for identification of cryptic organic remnants.

## THE PRECAMBRIAN FOSSIL RECORD

The microfossil record of the Proterozoic is incomparably greater than that of the Archaean. Approximately 2800 authentic Proterozoic microfossil occurrences contrast with 26 findings for the Archaean. The appearance of eukaryotic cells and, most significantly, of sexual reproduction in the Proterozoic led to a rapid diversification of life. The considerably higher number of known Proterozoic microfossils, when compared to the Archaean, although undoubtedly real, reflects not only the evolutionary progress, but also non-biological factors, of which the preservational effect of rocks is the most important. About 30 Archaean geologic units are known world-wide to contain stromatolites (Hofmann, 2000). From six of these units, microfossils have been described by 10 different workers. In the other units, the search for microfossils was either unsuccessful or not conducted. From the Proterozoic, over 600 stromatolitic units world-wide are known. Schopf and Klein (1992, Table 22.3) presented a list of authentic microfossils from the Proterozoic, described from a total number of 328 units, by approximately 200 authors, in hundreds of publications. The preservation state of rocks and the intensity of investigations performed on these formations have contributed greatly to the overwhelmingly better record of Proterozoic microfossils.

Eukaryotes first appeared in the geologic record as suddenly as the emergence of prokaryotes at 3.5 Ga. Megascopic spiral-shaped fossils (*Grypania*) were reported from the 2.1 Ga Neguanee BIF, in Michigan, by Han & Runnegar (1992) and have been classified as probable eukaryotic algae because of their size of *c*.1 mm width and up to 90 mm filament length. Walter *et al.* (1990) classified similar, but younger, *Grypania* fossils as probable multicellular algae. Yet another possibility is that the *Grypania* from the Neguanee BIF represents a large cyanobacterium (J.W. Schopf, personal communication, 1996) or even belongs to an extinct biological group (Knoll, 1996). Coccoid microfossils larger than 60 μm in diameter are also usually regarded as eukaryotes because of their size. They only became abundant in the Middle Proterozoic.

Phylogenetic trees (dendrograms) calculated from the analysis of nucleotides of ribosomal RNA molecules show that eukaryotes are more closely related to archaea (archaebacteria) than to eubacteria. If Han & Runnegar's (1992) interpretation of *Grypania* is correct, then the division of the phylogenetic tree separating prokaryotic eubacteria from archaea and eukaryotes must have occurred long before 2.1 Ga. This is surprisingly early, as all eukaryotes are strict aerobes and require a relatively high oxygen concentration in the atmosphere to maintain respiration, and thus could not have emerged before the oxygen levels reached 1–2% PAL (Chapman & Schopf, 1983). Towe (1990) and Ohmoto (1999), however, argued that such high levels could have already existed during the Archaean. New and startling findings by Brocks *et al.* (1999) hint at an even more unexpected scenario. In 2.7 Ga prehnite-pumpellyite-metamorphosed shales of the Fortescue Group from the Pilbara Craton, Western Australia, molecular evidence was found for oxygenic photosynthesis and for the existence of eukaryotes. In these ancient rocks molecular fossils of biological lipids like methylhopanes or steranes are preserved. Such findings in rocks 600 Myr older than the spiral-shaped *Grypania* findings by Han & Runnegar (1992) require an explanation. Although it might be possible to explain the presence of certain steranes, as reported by Brocks *et al.* (1999), other than by their derivation from Archaean eukaryotic cells, no such explanation has yet been presented.

Acritarchs, eukaryotic algae of unknown biological affinity, are the most widespread fossils in Meso- and Neoproterozoic rocks. They first appeared in 1.75 Ga rocks and reached their maximum diversity around 600 Ma, after the Varanger ice age, allowing for a biostratigraphic subdivision of the upper Proterozoic. Eukaryotic organisms must have arisen from prokaryotes which may have lived in symbiosis with other eubacteria (Margulis, 1981). According to the endosymbiosis theory, mitochondria originated by incorporating purple eubacteria and chloroplasts of cyanobacteria into a host cell (summarized in Lipps, 1993).

Cyst-like structures, interpreted as reproductive bodies and evidence of meiotic cell division, first appeared in microfossils at around 1.1 Ga. At that time, a rapid diversification of eukaryotic phytoplankton in the fossil record occurred, reaching a maximum at about 900 Ma, followed by a major decline from 800 to 700 Ma. This period appears to reflect a major decrease in atmospheric $CO_2$ and an increase of $O_2$ (Holland, 1984), conditions that were unfavourable for photosynthetic activity of algae and which were followed by a major decline in stromatolite abundance and diversity.

Similar to the prokaryotes and eukaryotes, metazoan fossils appeared in the geologic record, unannounced by any possible links to their ancestors. Seilacher *et al.* (1998) reported trace fossils from a 1.1 Ga sandstone formation from India, thus setting back the metazoan appearance into the Mesoproterozoic. Trace fossils interpreted as burrows produced by bilateralian animals are known from several occurrences in the terminal Neoproterozoic and Lower Cambrian, but the producers of these traces are not known. The oldest of such trace fossils (*Treptichnus pedum*) occur in the 548–545 Ma shallow-marine siliciclastic rocks of the Nama Group, Namibia (Jensen *et al.*, 2000), and are age equivalent to the Ediacara fauna that range from about 565 Ma into the Cambrian. The relationships of the Vendian Ediacara fauna to living organisms are highly disputed and the classification is uncertain, hence even their incorporation in the animal kingdom is equivocal (Seilacher, 1989). The reader is referred to Runnegar & Fedonkin (1992) for a comprehensive overview of the Vendozoa problem. Multicellular fossils appear preserved in phosphorites of the Neoproterozoic (570 Ma) Doushantuo Formation of southern China (Xiao *et al.*, 1998). According to these authors, fossil animal embryos preserved in early cleavage stages indicate that the divergence of lineages leading to bilateralians may have taken place long before the macroscopic remains of body fossils appear in the rock record.

## CLASSIFICATION OF PRECAMBRIAN MICROFOSSILS

The systematic classification of Precambrian microfossils relies on morphometrics of often badly preserved species, and on morphological comparisons to extant species. The main characteristics used in classification include, according to Schopf (1992):

**1** Shape and size of cells.
**2** Form of filament and thallus.
**3** Patterns of cell growth and division.
**4** Presence or absence of extracellular sheets or envelopes.
**5** Form of extracellular structures and of wall ornamentation.
**6** Presence or absence of colonial organization.

Schopf (1992) suggested that fossil septate filaments <1.5 μm wide can be regarded as probable bacteria and those >3.5 μm wide as probable cyanobacteria. The range between these two classes is referred to as 'undifferentiated prokaryotes'. Coccoid microfossils larger than 60 μm in diameter were assigned to 'assured eukaryotes'. However, because the size ranges of extant and fossil bacteria and cyanobacteria and alga overlap, this subdivision is not inevitably correct, nor do morphological similarities necessarily imply similar metabolism. It has been shown that the early evolution of life is characterized by morphological conservatism, evidenced in the extremely slow evolutionary progress of the prokaryotes. Schopf (1995) named this 'arrested' or 'hypobradytelic evolution', and suggested that generalists with the ability to survive in almost any environment had no need for further development. For this reason, prokaryotic organisms like oscillatoriacean cyanobacteria have remained virtually unchanged during the Earth's history. Another problem in classifying Archaean microfossils is the lack of a continuous fossil record in the Archaean, which does not allow possible morphological changes to be traced. For sedimentary rocks formed during the first billion years after initial microfossil appearance, reports of authentic microfossils are extremely sparse, due to the scarcity of unmetamorphosed stromatolitic formations. A slightly better record exists in the 2.6 Ga Neoarchaean and younger sedimentary rocks (Altermann & Schopf, 1995). Consequently, for the reasons of hypobradytely and deficient fossil record, prokaryotes, like cyanobacteria, are of no stratigraphic significance.

## THE FIRST EVIDENCE OF BIOSEDIMENTATION

The first directly visible impact of life on sedimentation was the appearance of stromatolites. It is surely not coincidental that the earliest stromatolites (and the earliest microfossils) in the rock record are found exactly in the environments, as envisaged above, for the possible sites of the origin of life. The 3.5 Ga stromatolites from the Warrawoona Group, Pilbara Craton, Australia (Fig. 1), bear evidence of shallow lagoonal and evaporitic conditions (Dunlop, 1978; Barley *et al.*, 1979). Lowe's (1994) alternative suggestion, that these earliest stromatolites might be partly hydrothermal precipitates, points towards another possible environment for the origin of life. New reports of complex stromatolitic structures from the Warrawoona Group demonstrate an astonishing variety of early Archaean biosediments, in part comparable to those typical of the Proterozoic (Hofmann *et al.*, 1999) and Neoarchaean (Höferle *et al.*, 2002).

The Warrawoona stromatolites form small structures of limited lateral extent. Large continuous carbonate accumulations appear first in the Wit Mfolozi Formation of Pongola Supergroup (3.0 Ga). The formation comprises laterally extensive, stromatolitic reefs up to 30 m thick that interfinger with peritidal dolarenites, marking for the first time in the Earth's history the ability of microbial communities to form large bioherms (Walter, 1983). However, this ability might well have been acquired much earlier, and the occurrence of the Wit Mfolozi platform may indicate stable tectonic conditions within the rigid craton margin rather than biological evolution of microbial communities, as suggested by Grotzinger (1994). It also may be a preservational artefact that resulted from older platforms being either covered by sediments or destroyed by plate tectonic processes.

The gross morphology of Precambrian carbonate depositional systems was reviewed by Grotzinger (1989), who found wide analogies to Phanerozoic ramps and rimmed shelf environments. Obviously the absence of eukaryotes and metazoans and the prevalence of microbially induced carbonate precipitation are less important factors than eustacy and tectonics in the development of shelf morphology. At 2.6 Ga, giant carbonate platforms emerged in intracratonic basins. The Transvaal–Griqualand West stromatolitic carbonate platform extended over more than 200 000

**Fig. 1.** Stromatolites in the evaporitic environment of the 3460 Ga Warrawoona Group, Western Australia, North Pole locality. The stromatolites (A) are preserved in chert (diagenetic replacement of carbonate) with abundant blades of baryte (B), probably replacing former gypsum crystals. Scale bars in centimetres.

**Fig. 2.** Giant stromatolitic reefs of the Ghaap Plateau Platform of the 2.6 Ga Campbellrand Subgroup, Griqualand West, South Africa. The individual stromatolitic domes attain up to 30 m height, over 50 m length and 20 m width. They occupied the photic zone of the shallow intracratonic Transvaal basin, extending for over 200 000 km$^2$ on the Kaapvaal Craton. The domes are organized in bed-packages of alternating cycles of stromatolites that can be laterally traced for tens of kilometres. Such deepening or shallowing upward cycles can be over hundreds of metres thick, where stromatolite morphology varies with depth and hydrodynamic conditions.

km$^2$, building a mammoth reef complex, where biogenic activity governed sedimentation (Fig. 2). At approximately the same time, an equally large carbonate platform existed on the Pilbara Craton of Western Australia (Nelson *et al.*, 1999). On these carbonate platforms sediment accumulation and rates of organic production were comparable to those on modern carbonate platforms and in microbial mats (Lanier, 1986;

Altermann & Nelson, 1998). These huge intracratonic basins contained a wide variety of facies, ranging from supratidal to deep subtidal, below storm wave base. The variety of depositional realms is reflected by different assemblages of stromatolites and by different morphologies of bioherms and biostromes.

## THE NATURE OF PRECAMBRIAN BIOSEDIMENTATION

By definition, stromatolites are lithified organosedimentary structures produced by sediment trapping, binding and/or precipitation, as a result of metabolic activity of cyanophyta and other single-celled micro-organisms. This broad definition, adapted from Walter (1976), includes the microbolites, microbialites and clotted fabrics of thrombolites. Because stromatolites very rarely contain preserved microfossils, a long dispute on their biological origin was resolved only with the discovery of living stromatolites and cyanobacterial mats. For example, the superbly preserved and morphologically diverse stromatolites of the Neoarchaean Chuniespoort–Ghaap Groups, Transvaal Supergroup, South Africa, were initially interpreted as pressure dissolution phenomena (Young, 1928). Only after the description of recent microbially formed limestones did Young (1932) recognize the biosedimentary origin of these structures. Discussion on the biogenicity of many ancient stromatolites still persists, and although for the vast majority of Precambrian stromatolites biogenicity is not directly demonstrable, it is well accepted among stromatolite specialists (Walter, 1976; Buick *et al.*, 1981). Appealing attempts to simulate mathematically the growth of stromatolite-like structures generated by fallout of sediment from suspension, by downslope gliding of settled sediment and by chemical precipitation on sediment surfaces or surface tension effects in chemical sediments resulted in structures that were astonishingly similar to stromatolitic lamination (Grotzinger & Rothmans, 1996). However, these authors also agree on the microbial mediation in the construction of most known Precambrian stromatolitic build-ups.

Most living microbial mats are organized in three layers of unequal thickness. The uppermost layer with the growth surface is colonized by an interwoven mesh of filamentous and/or coccoidal, oxygen-producing, photosynthesizing cyanobacteria and aerobic microbes. The middle layer is colonized by non-oxygen producing, photosynthesizing bacteria and facultative aerobic microbes. They use oxygen where available, grow by fermentation or, like some purple bacteria, move upward and downward following the nocti-diurnal changes in $O_2$ content of the layer. The thickest layer is the oxygen-depleted zone at the base of the mat, where anaerobic microbiota decompose the organic remnants of the upper layers buried by sediment. Following the phototactic growth of the uppermost layer, the other two layers prograde upward. It can thus be concluded that ancient microbial mats had a similar threefold subdivision. From the Precambrian rock record, however, microbial mats thriving in deep, aphotic conditions were reported (Simonson *et al.*, 1993) and simple anaerobic mats can thus be expected to have covered the sediment surface in the deep oceanic realms of the Precambrian.

Calcification of stromatolites presents another unsolved problem. Extant self-calcifying cyanobacteria are known from hypersaline or freshwater environments only, but because of their wide distribution in exceptionally large basins, most of the Precambrian stromatolites presumably thrived under normal marine conditions. Many early Precambrian stromatolites are extremely finely laminated and do not show much evidence of sediment trapping and binding. Because of diagenetic recrystallization and poor preservation, direct calcification is, however, impossible to demonstrate in Precambrian cyanobacteria. Good evidence for biogenic calcification is found in the superbly preserved filamentous microbial sheaths of *Siphonophycus transvaalensis*, described by Klein *et al.* (1987), which display minute needles of mineral crystals (±0.1 μm across), probably originally composed of aragonite (Fig. 3). Where the sheaths have disintegrated, the silicified needles float freely in the diagenetic chert matrix.

To explain the finely laminated texture of Precambrian stromatolites, Grotzinger (1990) proposed a model of Ca-saturated or nearly saturated sea water, from which $CaCO_3$ is precipitated along or within stromatolitic laminae when $CO_2$ is extracted from the water and the pH is lowered by photosynthesis and by degradation of organic matter. Direct inorganic precipitation of (high-Mg) calcite or aragonite from Ca-saturated sea water within stromatolitic frameworks, but also directly on the sea-floor, was interpreted by Sumner & Grotzinger (1996) from the occurrence of fan-shaped calcite crystals (herringbone calcite) in the Neoarchaean Transvaal Supergroup of South Africa and in other sedimentary rocks (compare Sumner, this volume, pp. 107–120). Such cements of botryoidal aragonite fans are, however, present in many carbonate environments (e.g. Davaud *et al.*, 1994) for which no special conditions such as those extant in the

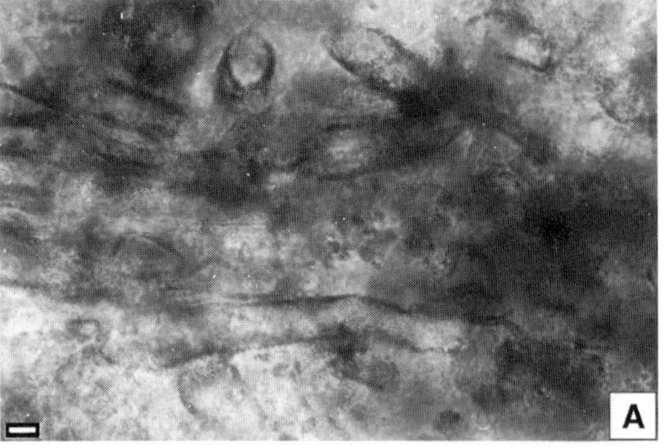


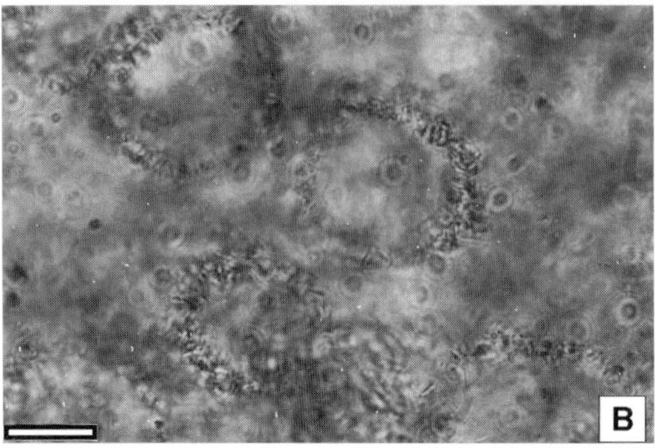


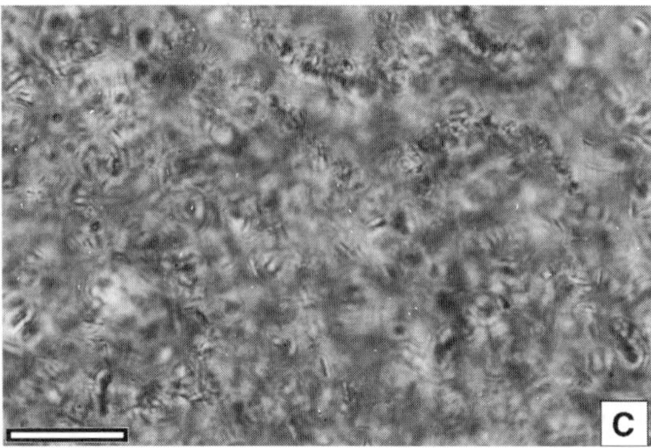


**Fig. 3.** Mineralic needles, probably originally aragonite, in *Syphonophycus transvaalensis* filament sheaths, from the Tsineng Member of the Gamohaan Formation (2516 Ga), Campbellrand Subgroup, South Africa. In (A) longitudinal, oblique and cross sections of the interwoven filaments are recognizable in a cryptocrystalline quartz matrix. Higher magnification in (B) exhibits the encrustation of the sheaths by mineralic needles (micrite). The needles are of higher refraction than the matrix, but their mineralogy is not clear owing to the minute size. In (C) the needles float freely in the chert matrix, where the filaments have disintegrated. Remnants of encrusted sheath walls are, however, recognizable in the upper part of the micrograph. Scale bars equal 25 μm. Sample CK#2, courtesy Bill Schopf, UCLA, CSEOL.

Precambrian can be claimed (compare also Grotzinger & Knoll, 1995).

Most of the Precambrian stromatolitic platforms are thoroughly dolomitized, and dolomitization of Precambrian carbonates is equally problematic as in modern dolomite formation (Wright, 2000). From the 2.6 Ga Campbellrand dolomites of South Africa, microbial mediation of dolomitization was reported by Wright & Altermann (2000), who observed nucleation of dolmicrite along the outer margins of cyanobacterial sheaths and progressive dolomitization with increasing anoxic degradation of the sheaths. They concluded that the degree and type of organic degradation was a major controlling factor on carbonate mineralogy. Active bacterial sulphate reduction, already active in the Neoarchaean (cf. Strauss, this volume, pp. 67–105), probably removed iron as pyrite from the environment, with calcite being precipitated, whereas dolomite formed below the sulphate reduction zone.

## STROMATOLITES AND PRECAMBRIAN STRATIGRAPHY

The absence of Precambrian body fossils has prompted numerous schemes that attempt to use stromatolites as biostratigraphic markers for the Archaean and Proterozoic. Based on the assumption that different stromatolite morphologies reflect different biology, a classification of stromatolites using binominal biological nomenclature was established (Hall, 1883). Such an assumption and the use of binominal nomenclature imply that the morphological evolution of the organo-sedimentary structures is the direct result of the evolution of the involved microbiota. Hundreds of stromatolite taxa were described and the usefulness of (primarily columnar) stromatolites in stratigraphy and basin correlation was demonstrated mainly by Russian workers for many Proterozoic formations of the former Soviet Union (e.g. Serebryakov, 1976) and also by Walter (1972) for Cambrian and Precambrian stromatolites of Australia, and Bertrand-Sarfati & Trompette (1976) for Proterozoic stromatolites in northern Africa. Such correlations and subdivisions are based on the relative diversity and the occurrence of entire taxonomic groups rather than on the appearance or disappearance of single taxa. The reasons for such an apparent stratigraphic usefulness of stromatolites are probably grounded in the evolution of sedimentary facies belts and in cyclicity of sediments (i.e. regressive and transgressive cycles) rather than in biological evolution of microbial communities. Grotzinger & Knoll (1999) noted that Archaean and Palaeoproterozoic stromatolites are largely formed through *in situ* precipitation of laminae, whereas younger stromatolites tend to form laminae by a physical, microbially mediated trapping and binding of sediment. This fundamental difference in stromatolite growth modi was also attributed to evolutive differences of Earth environments rather than to the evolution of microbial communities. According to Fedonkin (1996), it is possible to divide the Proterozoic into stromatolitic zones, which correlate with a precision of 100–300 Myr. However, at this precision, stromatolites rarely contribute to the geochronology of the formation in question, and it seems ambiguous whether the morphologic diversity reflects a taxonomic diversity or merely different environmental factors.

Frequency distribution curves of Precambrian stromatolite taxa suggest an explosion of stromatolite forms at 2200 Ma (Awramik, 1992). The same curve shows a decline of stromatolite taxa at the end of the Proterozoic, which is attributed to the appearance of grazing and browsing metazoa, to climatic changes (glaciation), or to changes in the Ca saturation state of sea water. Pratt (1982) interpreted the late Proterozoic decline simply as an artefact of the incomplete knowledge of Phanerozoic stromatolites. Equally, the poor preservation of stromatolitic formations in the Archaean may be responsible for the apparent explosion in the Proterozoic, where the rock record is an order of magnitude better. Consequently, some Proterozoic stromatolitic formations have been meticulously investigated and described by numerous authors, who tend to introduce new taxa (compare discussion of microfossil preservation above). Some of the formations listed as Proterozoic (Awramik, 1992) were later shown to be Neoarchaean in age by application of new dating techniques (Simonson *et al.*, 1993; Barton *et al.*, 1994; Altermann & Nelson, 1998). Because of rare fossil preservation in stromatolites, the relationships of bioherm or biostrome structure and biology are difficult to demonstrate, and satisfactory biological explanations for the putative time-restricted stromatolite morphology are not yet available. On the contrary, for Proterozoic stromatolites, identical microfossil taxa are reported from different stromatolites of various ages (Schopf & Klein, 1992, Table 22.3).

Black (1933) and Logan (1961) observed a strong environmental impact on the structure and morphology of recent organo-sedimentary build-ups, and clear relationships of different stromatolite forms to distinctive environments have been observed in many Precambrian successions (e.g. Donaldson, 1976a). Logan *et al.* (1964) demonstrated the dependence of the binominal stromatolite nomenclature on the size and specific portion of the observed bioherm, and on the scale of the observation. Such findings illustrate the equivocal character of stromatolite stratigraphy and put the results of many stratigraphic correlations in question. Some stromatolite forms, such as conical columns of the *Conophyton* type, usually interpreted as reflecting a subtidal to shelf facies assemblage (Donaldson, 1976b), are thought to be restricted to the Precambrian (Walter, 1989). Other forms, like the *Cryptozoon* form of intertidal facies, are described from the Precambrian (Walcott, 1906) and from recent deposits in Shark Bay, Australia (Logan, 1961). Although such club-shaped columns are formed mainly by the activity of cyanobacteria (*Entophysalis*, *Schizotrix* and others), club-shaped stromatolites, constructed mainly by the encrusting activity of red algae (rhodophyta, *Neogoniolithon*), were described by Thornton *et al.* (1978) from Bahiret el Bibane, Tunisia, and cyanobacterial and serpulid *Cryptozoon*-shaped bioherms have been described by Davaud *et al.* (1994) from the Holocene Sabkha el Melah, Tunisia. Such morphological similarities, although superficial, strongly imply environmental rather than biological control over the shape of stromatolites (for an alternative view, see discussion by Monty, 1977).

## ENVIRONMENTAL DISTRIBUTION OF PRECAMBRIAN STROMATOLITES

The usefulness of stromatolites as universal indicators of tidal facies is also disputed. Although recent stromatolites are almost exclusively restricted to the peritidal environment, sedimentary facies associations imply that some Precambrian stromatolites grew even in deep subtidal regimes, between fair weather and storm wave base (*c.*100 m deep). This is surprising considering the lower luminosity of the early Precambrian sun, and the higher suspension load of the water in the absence of filtering organisms and biomineralization. Photosynthesis, as evidenced by Rubisco-type carbon isotopic composition, must have been greatly hindered under such depth conditions, where only a small part of the remaining blue light radiation prevailed. Some investigators propose that certain microbial mats may have been non-photosynthesizing and could grow at depths below the photic zone (Simonson *et al.*, 1993). It is possible that in the absence of oxygen, which is toxic to anaerobic microbes, stromatolites without the upper photosynthetic layer could have developed below the photic zone. However, no unequivocal examples of such non-photosynthesizing mats have yet been demonstrated from the Precambrian. Nevertheless, the wide depth range of stromatolites and particularly the large variety of hydrodynamic and environmental conditions undoubtedly contributed to stromatolite diversity in the Precambrian. Whatever factors forced the stromatolites into protected niches in the late Precambrian, the resultant restricted variety of sedimentary environments occupied by stromatolites must have limited the diversity of forms, leading to the total disappearance ('extinction') of taxa typical of the environments that they formerly occupied.

From the foregoing discussion we can conclude that stromatolites were already diverse in the Archaean and occupied a wide range of environments of chemical sedimentation. Precambrian stromatolites formed at sites of carbonate precipitation and were preserved as dolomites, limestones, diagenetic cherts or even BIF deposits. However, stromatolites and microbial communities probably also had an effect on siliciclastic sedimentation. High siliciclastic influx usually prevents microbial growth on a sediment surface, but microbial communities might also have existed below the sediment–water interface. Probable stromatolitic structures are known from the Proterozoic Great Slave Supergroup, where conical columns are embedded in turbidites (Hoffman, 1974). Probably the best indicators for microbial colonization of terrigenous clastic sedimentary environments were described by Schieber (1998), from the Middle Proterozoic Belt Supergroup, Montana. Here domal build-ups resembling hemispheroidal stromatolites were found in shales and sandstones, next to mat-like crinkly lamination, cohesive deformed laminae, inverted hemispheroidal structures (such as found in polygonal stromatolites) and many other criteria for the recognition of microbial mats. The potential for preservation of such microbial mats is poor because, after disintegration of the organic layer, the cohesiveness of the stratum disappears and the clastic grains are subject to reworking and compaction. Therefore, microbial mats colonizing terrigenous clastic sediments have only rarely been observed in the rock record, requiring fast burial, low compaction and early diagenetic cementation in order to be preserved. In the absence of grazers, sediment feeders and burrowers, however, the Precambrian microbial mats thriving on terrigenous clastic sediments may have had a higher preservation potential relative to Phanerozoic counterparts.

It is difficult to assess when the microbial mats entered non-marine realms, because it is often impossible to distinguish between Archaean marine and non-marine environments. Stromatolites survived strong siliciclastic and volcaniclastic sedimentation within lacustrine, fluvial and deltaic settings within the late Archaean Ventersdorp Supergroup and Wolkberg Group of South Africa (Buck, 1980; Bosch *et al.*, 1993) and in the Tumbiana Formation of Western Australia (Buick, 1992). Microbial communities colonized supratidal flats (Altermann & Herbig, 1991) and from there might also have inhabited moist soils on land. The spread of microbial life into isolated basins or lakes was probably induced by winds eroding desiccated stromatolitic mats. Microbial binding of sand deposits has been recognized in the form of intraclasts comprising sand grains that appear to have been sufficiently cohesive to undergo desiccation and ductile soft sediment deformation (the 'microbial sand chips' of Pflüger & Gresse, 1996). Desiccated sand chips with raindrop imprints are known, for example, from non-marine sedimentary rocks in the 2.7 Ga Ventersdorp Supergroup (van der Westhuizen *et al.*, 1989; Donaldson *et al.*, this volume, pp. 3–13). A carbon-rich palaeosol below the Black Reef Quartzite Formation of the Transvaal Supergroup, and thus older than 2.64 Ga, was described by Martini (1994), who concluded that the carbon was derived from microbial mats thriving in the terrestrial environment during formation of the palaeosol. Martini (1994) suggested the possible protective role of microbial mats against soil erosion in the Precambrian.

Unequivocal evidence for the Archaean colonization of continental environments by microbes has still to be found. 'Fossilized plant remains' in Archaean rocks (e.g. Hallbauer *et al.*, 1977) can certainly be regarded as contaminants or misinterpretations (Schopf & Klein, 1992). Some geochemical evidence for widespread microbial communities in Neoarchaean and Palaeoproterozoic palaeosols has been presented by Ohmoto (1996), who advocated the importance of organic acids as agents leaching ferric oxides in Precambrian weathering profiles. It is generally accepted that the oxygen level in the atmosphere only rose to values of 0.03 atm in the time period from 2.25 to 2.0 Ga, when the earliest oxygenated palaeosols appeared in the geological record and detrital uraninite and pyrite placers disappeared from sedimentary rocks (for a contrary point of view see Ohmoto, 1999).

Microbial mats on soils would give only weak protection from the erosion of terrestrial realms. The siliciclastic influx into the Precambrian sedimentary basins must therefore have been strong and the sedimentation rates high (e.g. Schumm, 1968; Corcoran *et al.*, 1998; Corcoran & Mueller, this volume, pp. 183–211). Under such conditions, stromatolites could not have spread significantly until tectonism stabilized enough to allow for the development of extensive cratonic areas and wide shelves at the end of the Neoarchaean (Altermann & Nelson, 1998).

## PRECAMBRIAN LIFE IN RELATION TO SEDIMENTARY IRON, SILICON, OXYGEN AND OTHER MAJOR ELEMENTS

The most conspicuous type of Precambrian sedimentary deposits, genetically often related to the evolution of the atmosphere (and thus the biosphere), are banded iron formations (BIF). Trendall (this volume, pp. 33–66) presents an extensive review and reappraisal of BIF. Thus, only a short paragraph in this paper discusses the involvement of the evolution of life in BIF sedimentation. Cloud (1973) noted that the global maximum in BIF sediments at approximately 2.5–2.0 Ga coincides with the rise of oxygen in the atmosphere, as a consequence of evolution of photosynthesizing organisms. BIF sedimentation was explained by oxidation of ferrous iron dissolved in the ocean and precipitation as ferric iron.

The 'Cloud hypothesis' (see Trendall, this volume, pp. 00–00), however, is a much more complex and far reaching concept in which the evolution from anoxygenic to oxygenic photosynthesis plays a major role. Cloud (1988) assumed that the development of oxygenic photosynthesis and the ability of cyanobacteria to thrive in the oxidizing environment, toxic to anaerobes, led to an explosion of cyanobacteria in the oceans and resulted in a rapid decrease of $CO_2$ and increase of $O_2$ in the atmosphere, following the precipitation of the large BIF deposits. Today, it is clear that the apparently simultaneous precipitation of large BIF deposits is untrue and an artefact of imprecise dating techniques. Nevertheless, microbial mediation in precipitation of BIF is still a valid concept.

LaBerge *et al.* (1987) suggested that iron formations originated almost entirely as a biological precipitate, altered during diagenesis to its present banded appearance and mineralogy. The primary deposit consisted of ferrihydrite sheaths produced by iron-stripping bacteria and of silica microspheres precipitated as frustules by the coccoid cyanobacterium *Eosphera*. Unfortunately, the microfossils presented by Laberge *et al.* (1987), Robbins *et al.* (1987) and Robbins (1987) are most probably mineral artefacts, which can only be considered as dubiofossils *sensu* Schopf & Klein (1992). Konhauser & Ferris (1996) suggested that oversaturation of the ocean water with silica could have been responsible for almost continuous precipitation of chert (Beukes & Klein, 1990), while iron was deposited episodically, driven by planktonic $O_2$ productivity, leading to the banding in BIF. Anoxygenic bacterial photosynthesis, however, probably played a more important role in this process than oxygenic photosynthesis (Widdel *et al.*, 1993). Biogenic silica precipitation in BIF was proposed by several authors, among them LaBerge (1973) and Klemm (1979), although silica-precipitating organisms are unknown in the Precambrian.

It is interesting to consider that, in the same way that the evolution of life had an impact on sedimentary environments and sedimentation processes, the evolution of the atmosphere and oceans influenced the evolution of life, forming a magic circle of self-driving processes. Kempe & Degens (1985) suggested that the Precambrian oceans, until about 1.0 Ga, may have been bicarbonate and soda dominated, with low chloride concentrations, in analogy to modern soda lakes. This soda ocean would have vanished gradually by removal of sodium carbonates to the crust and slow accumulation of NaCl from hydrothermal leaching of the ocean floor. High Na relative to Cl concentrations would not only favour Na carbonate instead of Na chloride precipitation, but also allow for high concentrations of phosphate owing to low $Ca^{2+}$ content. The high pH would allow for high concentrations of organic complexes, making them available as nutrients for primitive life. The calcification capability of modern cyanobacteria is also dependent on Ca saturation and on the phosphate concentration of sea water. It is thus possible that ancient cyanobacteria, like their modern counterparts, were only capable of calcium precipitation in phosphate-poor environments (Merz-Preiß, 2000). Kempe & Degens (1985) also suggested that the gradually increasing calcium concentration eventually led to the appearance of bio-calcification by the increasing Ca stress on organisms. Kazmierczak & Degens (1986) argued that the increasing $Ca^{2+}$ pressure in the oceans made the aggregation of cells possible, finally resulting in the development of multicellular life.

The Ca content of Precambrian oceans is, however, strongly disputed. Direct inorganic precipitation of calcite from Ca-saturated sea water on the sea-floor was suggested by Sumner & Grotzinger (1996). In his review of Precambrian carbonate sediments, Grotzinger (1994) argued that Archaean carbonates differ from Proterozoic carbonate deposits based on the occurrences of carbonate cements precipitated directly, as thick crusts, on the sediment surfaces in many Archaean environments. In the middle Proterozoic, carbonate rocks show a decreasing tendency to precipitate cements directly on to the sea-floor through time. This development occurs parallel to the decrease in taxonomic diversity and abundance of stromatolites, and with increasing evidence for the formation of sulphate evaporites from about 1.8 Ga on. Grotzinger (1994) suggested that the switch from a soda ocean to a halite ocean occurred when the calcitic sea-floor precipitates gave way to sulphate evaporites, at about 1.8 Ga. He concluded further that the middle Proterozoic sea water probably suffered from a substantial decrease in the calcium carbonate content, but also a decrease in the total $HCO_3^-$ to $Ca_2^+$ ratio, and that the decreased saturation may have led to the subsequent stromatolite decline. Grotzinger & Kasting (1993) related the absence of sulphate evaporites in rocks older than 1.8 Ga to low levels of oceanic sulphate, in turn explained by a low concentration of oxygen in the atmosphere, thus indirectly pointing towards another possible way in which life influenced the Precambrian sedimentary record.

In the context of the present synopsis, it should also be noted that Runnegar (1992) emphasized the nearly contemporaneous emergence of the earliest trace fossils, metazoan fossils and eukaryotic phytoplankton with mineralized skeletons, with the first appearance of sedimentary phosphate deposits. The increased $O_2$ and reduced $CO_2$ content of the atmosphere at the end of the Precambrian, the possible global glaciation, the decline in stromatolite diversity and the evolutionary changes documented in the fossil record appear to be related to Neoproterozoic processes, most probably amplifying each other. The late Precambrian biological crisis might have been self-produced, being triggered by photosynthesis and by contemporary fixing of carbon via burial of organic matter in sediments. The rise in $O_2$ and fall in $CO_2$ partial pressure could have lead to an 'inverse greenhouse' and eventually to global glaciation and unfavourable conditions for microbial growth (Hoffman *et al.*, 1998). Under such conditions, microbial mats would have been forced into niches of hot springs, subterranean aquifers or subglacial water bodies, from where they again had to conquer the world until more advanced forms of life emerged at the terminal Proterozoic, finally to occupy virtually all sedimentary environments.

The stabilization of sediment by microbial mats and the burial of carbon in carbonate rocks are the most apparent examples of the direct influence of life on sediments and sedimentary basins. These processes are recorded in the earliest sedimentary rocks. The successive precipitation of large iron deposits (BIF), the disappearance of uraninite and pyrite conglomerates from the sedimentary record and the emergence of the first red beds and oxygenated palaeosols in the lower Proterozoic (Eriksson *et al.*, 1998) are further evidence of the influence that life had on the atmosphere and hydrosphere. The subsequent development of metazoans not only altered the biotopes of microbial mats but also introduced pelletal and skeletal sediments as new types to the sedimentary record. The emergence of filterers and suspension feeders probably cleaned ocean water and thereby increased the depth of the photic zone, giving way to an extended colonization of the sediment surface. The development of sediment burrowers and grazers must have led to widespread oxygenation of the sediment subsurface and to changes in diagenetic conditions (Fedonkin, 1996). The strong diversification of eukaryotic life in the Neoproterozoic had far reaching consequences for biostratigraphy. Combined with better preserved facies belts, it made it possible to distinguish between facies-typical palaeoecological communities and, for the first time, the Neoproterozoic fossil record offers a biostratigraphic resolution that permits inter-basinal correlations.

## ACKNOWLEDGEMENTS

I would like to express my sincere thanks to Pat Eriksson for encouraging me to write this synthesis. John Bragain and Bill Schopf constructively discussed earlier versions of the manuscript. Al Donaldson and Larry Aspler helped with many ideas and references. Bill Schopf kindly provided sample CK#2, containing Neoarchaean microbial filaments of *Siphonophycus transvaalensis* for further investigation.

## REFERENCES

Altermann, W. & Herbig, H.-G. (1991) Tidal flat deposits of the Lower Proterozoic Campbell Group along the southwestern margin of the Kaapvaal Craton, Northern Cape Province, South Africa. *J. Afr. Earth Sci.*, **13**(3/4), 415–435.

Altermann, W. & Nelson, D.R. (1998) Sedimentation rates, basin analysis and regional correlations of three

Neoarchean and Palaeoproterozoic sub-basins of the Kaapvaal craton as inferred from precise U–Pb zircon ages from volcaniclastic sediments. *Sediment. Geol.*, **120**, 225–256.

Altermann, W. & Schopf, J.W. (1995) Microfossils from the Neoarchean Campbell Group, Griqualand West Sequence of the Transvaal Supergroup, and their paleoenvironmental and evolutionary implications. *Precam. Res.*, **75**, 65–90.

Awramik, S.M. (1992) The history and significance of stromatolites. In: *Early Organic Evolution: Implications for Mineral and Energy Resources* (Eds Schidlowski, M., Golubic, S., Kimberley, M.M., McKirdy, D.M. & Trudinger, P.A.), pp. 435–449. Springer-Verlag, Berlin.

Barley, M.E., Dunlop, J.S.R., Glover, J.E. & Groves, D.I. (1979) Sedimentary evidence for an Archaean shallow-water volcanic-sedimentary facies, eastern Pilbara Block, Western Australia. *Earth planet. Sci. Lett.*, **43**, 74–84.

Barton, E.S., Altermann, W., Williams, I.S. & Smith, C.B. (1994) U–Pb zircon age for a tuff in the Campbell Group, Griqualand West Sequence, South Africa: implications for Early Proterozoic rock accumulation rates. *Geology*, **22**(4), 343–346.

Bengtson, S. (1994) *Early Life on Earth*. Nobel Symposium No. 84. Columbia University Press, New York, 630 pp.

Bertrand-Sarfati, J. & Trompette, R. (1976) Use of stromatolites for intrabasinal correlation: example from the Late Proterozoic of the northwestern margin of the Taoudenni Basin. In: *Stromatolites* (Ed. Walter, M.R.), pp. 517–522. Elsevier, Amsterdam.

Beukes, N.J. & Klein, C. (1990) Geochemistry and sedimentology of a facies transition—from microbanded to granular iron-formation—in the early Proterozoic Transvaal Supergroup, South Africa. *Precam. Res.*, **47**, 99–139.

Black, M. (1933) The algal sediments of Andros Island, Bahamas. *R. Soc. phil. Trans.*, *B*, **122**, 169–192.

Bosch, P.J.A., Eriksson, P.G. & Snyman, C.P. (1993) The Wolkberg Group in the northern Transvaal: palaeoenvironment derived from sedimentology and geochemistry. *S. Afr. J. Geol.*, **96**(4), 190–204.

Brocks, J.J., Logan, G.A., Buick, R. & Summons, R.E. (1999) Archean molecular fossils and the early rise of eukaryotes. *Science*, **285**, 1033–1036.

Buck, S.G. (1980) Stromatolite and ooid deposits within the fluvial and lacustrine sediments of the Precambrian Ventersdorp Supergroup of South Africa. *Precam. Res.*, **12**, 311–330.

Buick, R. (1992) The antiquity of oxygenic photosynthesis: evidence from stromatolites in sulfate-deficient Archean Lakes. *Science*, **255**, 74–77.

Buick, R., Dunlop, J.S.R. & Groves, D.I. (1981) Stromatolite recognition in ancient rocks: an appraisal of irregularly laminated structures in an early Archaean chert-barite unit from North Pole, Western Australia. *Alcheringa*, **5**, 161–181.

Cairns-Smith, A.G. (1982) *Genetic Takeover and the Mineral Origins of Life*. Cambridge University Press, Cambridge, 477 pp.

Chapman, D.J. & Schopf, J.W. (1983) Biological and biochemical effects of the development of an aerobic environment. In: *Earth's Earliest Biosphere: Its Origin and Evolution* (Ed. Schopf, J.W.), pp. 302–320. Princeton University Press, Princeton, NJ.

Cloud, P.E. (1973) Paleoecological significance of banded iron-formation. *Econ. Geol.*, **68**, 1135–1143.

Cloud, P.E. (1988) *Oasis in Space*. Norton & Co., New York.

Corcoran, P.L., Mueller, W.U. & Chown, E.H. (1998) Climatic and tectonic influences on fan deltas and wave- to tide-controlled shorface deposits: evidence from the Archean Keskarrah Formation, Slave Province, Canada. *Sediment. Geol.*, **120**(1–4), 125–152.

Davaud, E., Strasser, A. & Jedoui, Y. (1994) Stromatolite and serpulid bioherms in a Holocene restricted lagoon (Sabkha el Melah, southeastern Tunisia). In: *Phanerozoic Stromatolites II* (Eds Bertrand-Sarfati, J. & Monty, C.), pp. 131–151. Kluwer Academic, Dordrecht.

Deamer, D.W. (1993) Prebiotic conditions and the first living cells. In: *Fossil Prokaryotes and Protists* (Ed. Lipps, J.H.), pp. 11–18. Blackwell Scientific Publications, Oxford.

Donaldson, J.A. (1976a) Aphebian stromatolites in Canada: implications for stromatolite zonation. In: *Stromatolites* (Ed. Walter, M.R.), pp. 371–380. Elsevier, Amsterdam.

Donaldson, J.A. (1976b) Paleoecology of conophyton and associated stromatolites in the Precambrian Dismal Lakes and Rae Groups, Canada. In: *Stromatolites* (Ed. Walter, M.R.), pp. 523–534. Elsevier, Amsterdam.

Dunlop, J.S.R. (1978) Shallow water sedimentation at North Pole, Pilbara, Western Australia. In: *Archean Cherty Metasediments: Their Sedimentology, Micropalaeontology, Biochemistry and Significance to Mineralization* (Eds Gloverand, J.E. & Groves, D.I.), pp. 30–38. Geology Department & Extension Service, University of Western Australia, Perth.

Duve, C. de (1991) *Blueprint for a Cell: the Nature and Origin of Life*. Neil Patterson Publications, Burlington, NC, 296 pp.

Eriksson, P.G., Condie, K.C., Tirsgaard, H. *et al.* (1998) Precambrian (pre-vegetational) clastic sedimentation systems. *Sediment. Geol.*, **120**(1–4), 5–53.

Eriksson, P.G. & Cheney, E.S. (1992) Evidence for the transition to an oxygen-rich atmosphere during the evolution of red beds in the Lower Proterozoic sequences of southern Africa. *Precam. Res.*, **54**, 257–269.

Fedonkin, M.A. (1996) Geobiological trends and events in the Precambrian Biosphere. In: *Global Events and Event Stratigraphy* (Ed. Walliser, O.H.), pp. 89–112. Springer, Berlin.

Gilbert, W. (1986) The RNA world. *Nature*, **319**, 618.

Gough, D.O. (1981) Solar interior structure and luminosity variations. *Solar Phys.*, **74**, 21–34.

Grotzinger, J.P. (1989) Facies and evolution of Precambrian carbonate depositional systems: emergence of modern platform archetype. In: *Controls on Carbonate Platform and Basin Development* (Eds Crevello, P.D, Wilson, J.L., Sarg J.F. & Read, J.F.), Spec. Publ. Soc. econ. Paleont. Miner., Tulsa, **44**, 79–106.

Grotzinger, J.P. (1990) Geochemical model for Proterozoic stromatolite decline. *Am. J. Sci.*, **290A**, 80–103.

Grotzinger, J.P. (1994) Trends in Precambrian carbonate sediments and their implication for understanding evolution. In: *Early Life on Earth* (Ed. Bengtson, S.), pp. 245–258. Nobel Symposium **84**. Columbia University Press, New York.

Grotzinger, J.P. & Kasting, J.F. (1993) New constraints on Precambrian ocean composition. *J. Geol.*, **101**, 235–243.

Grotzinger, J.P. & Knoll, A.H. (1995) Anomalous carbonate precipitates: Is the Precambrian the key to the Permian? *Palaios*, **10**, 578–596.

Grotzinger, J.P. & Knoll, A.H. (1999) Stromatolites in Precambrian carbonates: evolutionary mileposts or environmental dipsticks? *Ann. Rev. Earth planet. Sci.*, **27**, 313–358.

Grotzinger, J.P. & Rothmans, D.H. (1996) An abiotic model for stromatolite morphogenesis. *Nature*, **383**, 423–425.

Hall, J.D. (1883) Cryptozön (poliferum) n.g. and s.p. *Rep. NY State Museum*, **36**, pl. 6.

Hallbauer, D.K., Jahns, H.M. & Beltmann, H.A. (1977) Morphological and anatomical observations on some Precambrian plants from the Witwatersrand, South Africa. *Geol. Rundsch.*, **66**, 478–491.

Han, T.M. & Runnegar, B. (1992) Megascopic Eukariotic Algae from the 2.1-billion-year-old Neguanee Iron-Formation, Michigan. *Science*, **275**, 232–235.

Höferle, R., Haller, D., Tetzlaff, A. & Altermann, W. (2000) The facies assemblage of elongated, coniform stromatolites in the NeoArchaean Campbellrand Subgroup, southwestern Kaapvaal Craton, South Africa. *J. Afr. Earth Sci.*, **33**(1).

Hofmann, H.J. (2000) Archean stromatolites as Archean archives. In: *Microbial Sediments* (Eds Riding R.E. & Awramik S.M.), pp. 315–327. Springer-Verlag, Heidelberg.

Hofmann, H.J., Grey, K., Hickmann, A.H. & Thorpe, R.I. (1999) Origin of 3.45 Ga coniform stromatolites in Warrawoona Group, Western Australia. *GSA Bull.*, **111**, 1256–1262.

Hoffman, P.F. (1974) Shallow and deepwater stromatolites in lower Proterozoic platform-to-basin facies change, Great Slave Lake, Canada. *AAPG Bull.*, **58**, 856–867.

Hoffman, P.F., Kaufman, A.J., Halverson, G.P. & Schrag, D.P. (1998) A Neoproterozoic Snowball Earth. *Science*, **281**, 1342–1346.

Holland, H.D. (1984) *The Chemical Evolution of the Atmosphere and the Oceans*. Princeton University Press, Princeton, NJ, 582 pp.

Jensen, S., Saylor, B.Z., Gehling, J.G. & Germs, G.J.B. (2000) Complex trace fossils from the terminal Proterozoic of Namibia. *Geology*, **28**, 143–146.

Kasting, J.F. (1987) Theoretical constraints on oxygen and carbon dioxide concentrations in the Precambrian atmosphere. *Precam. Res.*, **34**, 205–229.

Kasting, J.F. & Ackerman, T.P. (1986) Climatic consequences of very high $CO_2$ levels in Earth's early atmosphere. *Science*, **234**, 1383–1385.

Kazmierczak, J. & Degens, E.T. (1986) Calcium and the early eukaryotes. *Mitt. Geol. Paläont. Inst. Univ. Hamburg*, **61**, 1–20.

Kempe, S. & Degens, E.T. (1985) An early soda ocean? *Chem. Geol.*, **53**, 95–108.

Klein, C., Beukes, N.J. & Schopf, J.W. (1987) Filamentous microfossils in the Early Proterozoic Transvaal Supergroup: their morphology, significance, and paleoenvironmental setting. *Precam. Res.*, **36**, 81–94.

Klemm, D.D. (1979) A biogenic model for the formation of banded iron formation in the Transvaal Supergroup, South Africa. *Miner. Deposita*, **14**, 381–385.

Knoll, A.H. (1996) Archean and Proterozoic paleontology. In: *Palynology: Principles and Applications*, Vol. 1 (Eds Jansonius, J. & McGregor, D.C.), pp. 51–80. AASP Foundation, Salt Lake City.

Konhauser, K.O. & Ferris, F.G. (1996) Diversity of iron and silica precipitation by microbial mats in hydrothermal waters, Iceland: implications for Precambrian iron formations. *Geology*, **24**, 323–326.

LaBerge, G.L. (1973) Possible biological origin of Precambrian iron formations. *Econ. Geol.*, **68**, 1098–1019.

LaBerge, G.L., Robbins, E.I. & Han, T.-M. (1987) A model for the biological precipitation of Precambrian iron-formations. A: geological evidence. In: *Precambrian Iron-formations* (Eds Appel, P.W.U. & LaBerge, G.L.), pp. 69–96. Theoprastus Publications, Athens.

Lanier, W.P. (1986) Approximate growth rates of Early Proterozoic microstromatolites as deduced by biomass productivity. *Palaios*, **1**, 525–542.

Lipps, J.H. (1993) Introduction to fossil prokaryotes and protists. In: *Fossil Prokaryotes and Protists* (Ed. Lipps, J.H.), pp. 1–10. Blackwell Scientific Publications, Oxford.

Logan, B.W. (1961) Cryptozoon and associated stromatolites from the Recent, Shark Bay, Western Australia. *J. Geol.*, **69**, 517–533.

Logan, B.W., Rezak, R. & Ginsburg, R.N. (1964) Classification and environmental significance of algal stromatolites. *J. Geol.*, **72**, 68–83.

Lowe, D.R. (1992) The Proterozoic sedimentary record. In: *The Proterozoic Biosphere* (Eds Schopf, J.W. & Klein, C.), pp. 53–58. Cambridge University Press, Cambridge.

Lowe, D.R. (1994) Abiological origin of described stromatolites older than 3.2 Ga. *Geology*, **22**, 387–390.

McClendon, J.H. (1999) The origin of life. *Earth Sci. Rev.*, **47**, 71–93.

Margulis, L. (1981) *Symbiosis in Cell Evolution*. Freeman and Co., San Francisco, 419 pp.

Martini, J.E.J. (1994) A late Archaean—Palaeoproterozoic (2.6 Ga) palaeosol on ultramafics in the Eastern Transvaal, South Africa. *Precam. Res.*, **67**, 159–180.

Merz-Preiß, M. (2000) Calcification in cyanobacteria. In: *Microbial Sediments* (Eds Riding, R.E. & Awramik, S.M.), pp. 51–56. Springer-Verlag, Berlin.

Miller, S.L. (1953) A production of amino acids under possible primitive Earth conditions. *Science*, **117**, 528–529.

Miller, S.L. (1992) The prebiotic synthesis of organic compounds as a step toward the origin of life. In: *Major Events in the History of Life* (Ed. Schopf, J.W.), pp. 1–28. Jones & Bartlett, Boston.

Mojzsis, S.J., Arrhenius, G., McKeegan, K.D., Harrison, T.M., Nutman, A.P. & Friend, C.R.L. (1996) Evidence for life on Earth before 3800 million years ago. *Nature*, **384**, 55–59.

Mojzsis, S.J., Harrison, M.T. & Pidgeon, R.T. (2001) Oxygen-isotope evidence from ancient zircons for liquid water at earth's surface 4300 Myr ago. *Nature*, **409**, 178–181.

Monty, C. (1977) Evolving concepts on the nature of ecological significance of stromatolites, a review. In: *Fossil Algae* (Ed. Flügel, E.), pp. 15–35. Springer-Verlag, Berlin.

Myers, J. (2001) Protoliths of the 3.8–3.7 Ga Isua greenstone belt, West Greenland. *Precam. Res.*, **105**(2–4), 129–141.

Nelson D.R., Trendall, A.F. & Altermann, W. (1999) Chronological correlations between the Pilbara and Kaapvaal cratons. *Precam. Res.*, **97**, 165–169.

Newsom, H.E. & Jones, J.H. (1990) *Origin of the Earth.* Oxford University Press, Oxford.

Ohmoto, H. (1996) Evidence in pre 2.2 Ga paleosols for the early evolution of atmospheric oxygen and terrestrial biota. *Geology*, **24**, 1135–1138.

Ohmoto, H. (1999) Redox state of the Archean atmosphere: evidence from detrital heavy minerals in ca. 3250–2750 Ma sandstones from the Pilbara Craton, Australia: comment. *Geology*, **27**, 1151–1152.

Oparin, A.I. (1938) *Origin of Life.* Macmillan, New York, 270 pp.

Oro, J. & Lazcano, A. (1996) Comets and the origin and evolution of life. In: *Comets and the Origin of Life* (Eds Thomas, P.J., Chyba, C.F. & McKay, C.P.), pp. 3–27. Springer-Verlag, Berlin.

Pflüger, F. & Gresse, P.G. (1996) Microbial sand chips—a non-actualistic sedimentary structure. *Sediment. Geol.*, **102**, 263–274.

Pratt, B.R. (1982) Stromatolite decline—a reconsideration. *Geology*, **10**, 512–515.

Robbins, E.I. (1987) Appelella ferrifera, a possible new iron-coated microfossil in the Issua iron-formation, southwestern Greenland. In: *Precambrian Iron-formations* (Eds Appel, P.W.U. & LaBerge, G.L.), pp. 141–154. Theoprastus Publications, Athens.

Robbins, E.I., LaBerge, G.L. & Schmidt, R.G. (1987) A model for the biological precipitation of Precambrian iron-formations. B: Morphological evidence and modern analogs. In: *Precambrian Iron-formations* (Eds Appel, P.W.U. & LaBerge, G.L.), pp. 97–139. Theoprastus Publications, Athens.

Rosing, M.T. (1999) $^{13}$C-depleated carbon microparticles in >3700 Ma sea-floor sedimentary rocks from West Greenland. *Science*, **283**, 674–676.

Runnegar, B. (1992) Evolution of the earliest animals. In: *Major Events in the History of Life* (Ed. Schopf, J.W.), pp. 65–94. Jones & Bartlett, Boston.

Runnegar, B. & Fedonkin, M. (1992) Proterozoic metazoan body fossils. In: *The Proterozoic Biosphere* (Eds Schopf, J.W. & Klein, C.), pp. 369–388. Cambridge University Press, Cambridge.

Schidlowski, M. (1988) A 3800-million-year isotopic record of life from carbon in sedimentary rocks. *Nature*, **333**, 313–318.

Schieber, J. (1998) Possible indicators of microbial mat deposits in shales and sandstones: examples from the Mid-Proterozoic Belt Supergroup, Montana, USA. *Sediment. Geol.*, **120**, 105–124.

Schopf, J.W. (1983) *Earth's Earliest Biosphere: Its Origin and Evolution.* Princeton University Press, Princeton, NJ, 543 pp.

Schopf, J.W. (1992) The oldest fossils and what they mean. In: *Major Events in the History of Life* (Ed. Schopf, J.W.), pp. 29–64. Jones & Bartlett, Boston.

Schopf, J.W. (1993) Microfossils of the Early Archean Apex Chert: new evidence of the antiquity of life. *Science*, **260**, 640–646.

Schopf, J.W. (1995) Metabolic memories of Earth's earliest biosphere. In: *Evolution and Molecular Revolution* (Ed. Marshall, C.), pp. 73–107. Jones & Bartlett, Boston.

Schopf, J.W. (1999) *Cradle of Life.* Princeton University Press, Princeton, NJ, 367 pp.

Schopf, J.W. & Klein, C. (1992) *The Proterozoic Biosphere: a Multidisciplinary Study.* Cambridge University Press, Cambridge, 1348 pp.

Schumm, S.A. (1968) Speculations concerning paleohydrologic controls of terrestrial sedimentation. *Geol. Soc. Am. Bull.*, **79**, 1573–1588.

Seilacher, A. (1989) Vendozoa: organism construction in the Proterozoic biosphere. *Lethaia*, **22**, 229–239.

Seilacher, A., Bose, P.K. & Pfluger, F. (1998) Triploblastic animals more than 1 billion years ago: trace fossil evidence from India. *Science*, **282**, 80–84.

Serebryakov, S.N. (1976) Biotic and abiotic factors controlling the morphology of Riphean stromatolites. In: *Stromatolites* (Ed. Walter, M.R.), pp. 321–336. Elsevier, Amsterdam.

Simonson, B.M., Schubel, K.A. & Hassler, S.W. (1993) Carbonate sedimentology of the early Precambrian Hamersley Group of Western Australia. *Precam. Res.*, **60**, 287–335.

Stetter, K.O. (1994) The lesson of Archaebacteria. In: *Early Life on Earth* (Ed. Bengtson, S.), pp. 143–151. Nobel Symposium No. 84. Columbia University Press, New York.

Sowerby, S.J. & Heckl, W.M. (1998) The role of self-assembled monoleyers of the purine and pyrimidine bases in the emergence of life. *Orig. Life Evol. Biosphere*, **28**, 283–310.

Sumner, D.Y. & Grotzinger, J.P. (1996) Herringbone calcite: petrography and environmental significance. *J. sediment. Res. A*, **66**(3), 419–429.

Thornton, S.E., Pilkey, O.H. & Lynts, G.W. (1978) A lagoonal crustose coralline algal micro-ridge: Bahiret el Bibane, Tunisia. *J. sediment. Petrol.*, **48**, 743–750.

Towe, K.M. (1990) Aerobic respiration in the Archean? *Nature*, **348**, 54–56.

Urey, H.C. (1952) *The Planets: Their Origin and Development.* Yale University Press, New Haven, CT, 245 pp.

Wächtershäuser, G. (1994) Vitalysts and virulysts: a theory of self-expanding reproduction. In: *Early Life on Earth* (Ed. Bengtson, S.), pp. 124–132. Nobel Symposium No. 84. Columbia University Press, New York.

Walcott, C.D. (1906) Algonkian Formations of Northwestern Montana. *Geol. Soc. Am. Bull.*, **17**, 1–28.

Walter, M.R. (1972) Stromatolites and the biostratigraphy of the Australian Precambrian and Cambrian. Spec. Paper Paleont. Ass. London, **11**, 190 pp.

Walter, M.R. (1976) *Stromatolites.* Developments in Sedimentology, 20. Elsevier, Amsterdam, 790 pp.

Walter, M.R. (1983) Archean stromatolites: evidence of the Earth's earliest benthos. In: *Earth's Earliest Biosphere, Its Origin and Evolution* (Ed. Schopf, J.W.), pp. 187–213. Princeton University Press, Princeton, NJ.

Walter, M.R. (1989) Major features in record of Proterozoic stromatolites. *IGC Abstr.*, **3**, 318.

Walter, M.R., Rulin, D. & Horodyski, R.J. (1990) Coiled carbonaceous megafossils from the Middle Proterozoic of Jixian (Tianjin) and Montana. *Am. J. Sci.*, **290**-A, 133–148.

Westhuizen, W.A. van de, Grobler, N.J., Loock, J.C. & Tordiffe, E.A.W. (1989) Raindrop imprints in the late Archean–early Proterozoic Ventersdorp Supergroup, South Africa. *Sediment. Geol.*, **61**, 303–309.

Widdel, F., Schnell, S., Heising, S., Ehrenreich, A., Assmus, B. & Schink, B. (1993) Ferrous iron oxidation by anoxygenic phototrophic bacteria. *Nature*, **362**, 834–836.

Woese, C.R. (1987) Bacterial evolution. *Microbiol. Rev.*, **51**, 221–271.

Woese, C.R., Kandler, O. & Wheelis, M.L. (1990) Towards a natural system of organisms: proposal for the domains Archaea, Bacteria, and Eukarya. *Proc. natl Acad. Sci. USA*, **87**, 4576–4579.

Wright, D.T. (2000) Benthic microbial communities and dolomite formation in marine and lacustrine environments—a new dolomite model. In: *Marine Authigenesis: from Global to Microbial* (Eds Glenn, C.R., Lucas, J. & Prevot-Lucas, L.). Spec. Publ. Soc. econ. Petrol. Miner., Tulsa.

Wright, D.T. & Altermann, W. (2000) Microfacies development in Late Archean stromatolites and ooids. In: *Carbonate Platform Systems. Components and Interactions* (Eds Insalco, E., Skelton, P.W. & Palmer, T.J.). Spec. Publs geol. Soc. London, No. 178, pp. 51–70. Geol. Soc. London, Bath.

Xiao, S., Zhang, Y. & Knoll, A.H. (1998) Three-dimensional preservation of algae and animal embryos in a Neoproterozoic phosphorite. *Nature*, **395**, 553–558.

Young, R.B. (1928) Pressure phenomena in the dolomitic limestones of the Campbell Rand Series, Griqualand West. *Trans. geol. Soc. S. Afr.*, **31**, 157–165.

Young, R.B. (1932) The occurrence of stromatolitic or algal limestones in the Campbell Rand Series, Griqualand West. *Trans. geol. Soc. S. Afr.*, **35**, 29–36.

*Spec. Publs int. Ass. Sediment.* (2002) **33**, 33–66

# The significance of iron-formation in the Precambrian stratigraphic record

A. F. TRENDALL

*School of Physical Sciences, Curtin University of Technology, GPO Box 1987, Perth 6000, Western Australia*

## ABSTRACT

Iron-formation (IF) is an iron-rich (±30% Fe) and siliceous (±50% $SiO_2$) sedimentary rock which results from extreme compaction and diagenesis of a chemical precipitate in which those components were major constituents. It is widely, but irregularly, distributed within Precambrian volcano-sedimentary successions. During most of the period from *c.*3.8 Ga until *c.*2.4 Ga IF was formed in greenstone successions, in relatively thin, tectonically deformed and metamorphosed units whose poor preservation makes the nature of their deposition uncertain. A number of major IF units of the Gondwana continents deposited in the later part of that period are better preserved, thicker and more areally extensive, and occur within successions whose depositional environments can be interpreted with more confidence. The IF of those units, and of all older IFs, has a centimetre-scale alternation of iron-rich and silica-rich bands (mesobands) whose presence justifies its usual designation as banded iron-formation (BIF); characteristically, the mesobanding of BIF has a high degree of lateral stratigraphic continuity. By contrast, the IF typically represented in the circum-Ungava belt of North America is not only significantly different in sedimentological characteristics but also younger (*c.*1.8 Ga). Although it is also banded, the banding is coarser and less regular that that of BIF, and the material of the bands is often finely granular. The name granular iron-formation (GIF) distinguishes it from the typical BIF of earlier sequences. Following deposition of these there is a hiatus in IF deposition until the later Neoproterozoic. The BIFs of the *c.*2.6–2.45 Ga Hamersley Group of Western Australia are by far the largest (in terms of contained iron), most extensive and thickest known, only the roughly coeval and lithologically near-identical BIFs of the Transvaal Supergroup of South Africa rivalling them in these respects. The Hamersley BIFs are therefore taken as archetypes for comparison with others. The mainly older (*c.*3.8–2.5 Ga) BIFs of greenstone belts are sufficiently similar to suggest that they have the same origin. The formation of such BIFs by precipitation of dissolved deep ocean ferrous iron by photosynthesizing organisms is consistent with all available evidence; and geochemical evidence is consistent with the mantle as an iron source for the oceans. The necessary and sufficient conditions for the deposition of BIF are the formation of depositories which: (i) remained tectonically stable for periods approaching $10^6$ years; (ii) were deep enough both to avoid contamination with epiclastic material and to be free of bottom disturbance; and (iii) had dispositions such that deep ocean water was able to circulate freely into and out of them. Control of the time-distribution of Precambrian IFs solely by abrupt steps in biochemical evolution, closely linked to the increasing oxygen content of the atmosphere, cannot now be totally accepted; although aspects of Preston Cloud's model proposing this are believed to remain valid, particularly his enthusiasm for a biogenic involvement in IF deposition. An additional factor controlling the early distribution of BIFs is thought to be the evolving architecture of volcano-sedimentary depositories related to the evolution of continental crust, a new model for which is outlined. The late Archaean peak of BIF deposition can be interpreted, to the extent that its existence is real, by gradually increasing tectonic suitability of depositories for BIF deposition as the structure of continents evolved. IFs are unusually sensitive indicators of depositional environment. Their distribution in the Precambrian stratigraphic record indicates that, from a sedimentological viewpoint, conditions on the Precambrian Earth were sufficiently different from those now obtaining to make the simplistic application of uniformitarian principles misleading.

## INTRODUCTION

The significance of iron-formation (IF) in the Precambrian stratigraphic record cannot be fully understood without taking into account many different aspects of Earth's evolution, each normally the preserve of separate specialist groups. The purpose of this paper is to describe a hypothesis that relates the deposition of IF not only to the evolution of life, the oceans and the atmosphere, but also to the origin and growth of continents. This holistic approach follows a philosophy established earlier in the words 'It is useless in any present considerations of the origin of iron formation to confine attention to the iron formation itself' (Trendall, 1965). Some of the components of the hypothesis are long-established or 'mainstream' beliefs; others are new. The views put forward here have been developed following a 37-year involvement with IF research, and their presentation here in an integrated synthesis has resulted in an emphasis, in parts of the text, on personal preferences for particular interpretations; I have tried to flag these as they arise, and to indicate where alternative views may be found.

This paper flows from a keynote address at the Nineteenth Regional European Meeting of the International Association of Sedimentology held in Copenhagen in August 1999, and the eight headings below largely follow those of the oral presentation. Their logic is as follows. Under the first heading the definition, nomenclature and classification of IFs are discussed, and their occurrence in space and time summarized. Then the banded IFs (BIFs) of the Hamersley Basin are described in greater detail, before I examine to what extent they can be accepted as type examples for BIFs in general. The following three headings deal with when, how and why IFs were deposited.

The paper offers neither a comprehensive survey of published work on IF nor a catalogue of described occurrences. Extensive references to work prior to their publication dates can be found in a number of publications dedicated to IF (James & Sims, 1973; UNESCO, 1973; Mel'nik, 1982; Trendall & Morris, 1983; Radhakrishna, 1986; Appel & La Berge, 1987); and two papers later than these provide excellent brief accounts of most aspects of IF occurrence and deposition (Beukes & Klein, 1992; Klein & Beukes, 1992).

## WHAT IS IRON-FORMATION, AND WHERE DOES IT OCCUR?

### Definition and typical characteristics

IF is an iron-rich sedimentary rock largely confined to the Precambrian stratigraphic record. The term 'iron-formation' originated in the Lake Superior area as a contraction of the 'iron-bearing formation' of Van Hise & Leith (1911). James (1954), whose work had focused on the same area, first formally defined IF as 'a chemical sediment, typically thin-bedded or laminated, containing 15 percent or more iron of sedimentary origin, commonly but not necessarily containing layers of chert'.

For present purposes it will be useful to summarize the typical characteristics of IF as the term is now applied globally, rather than focus on its formal definition. In terms of chemical composition most IFs contain about 30% Fe (Gole & Klein, 1981); see also Davy (1983) for a review and summary of analytical data to that date. Well over 90% of rocks that would be called IF in the field have between 25 and 35% Fe. Thus about half the rock by weight is iron oxides; the other half is mostly silica. Carbon dioxide is present as a significant minor constituent in many BIFs, and is a major constituent in some, but all other oxides (e.g. $Al_2O_3$, MgO, alkalies) are quite minor, and 'trace' elements are just that—these are chemically very 'clean' rocks. Haematite ($Fe_2O_3$) and magnetite ($Fe_3O_4$) are the principal iron minerals. Others that may or may not be present are carbonates (ankerite, siderite) and silicates (stilpnomelane, greenalite, riebeckite). The silica normally occurs as microcrystalline quartz, usually called chert.

All the mineral components of IF (and particularly of BIF; see below) are typically fine-grained, and it is characteristically a hard, heavy, tough material exceptionally resistant to both hammering and weathering. In landscapes cut into Precambrian rocks throughout the world, IFs characteristically form conspicuous resistant ridges. The two major mineral constituents (quartz and iron oxides) are normally concentrated in alternating iron-rich and silica-rich bands on the mesoscopic scale; often these are brightly coloured —red, black or white.

### Nomenclature and classification

Historically a number of local names have been applied to IF in different continents (Trendall, 1983a). Examples include the 'itabirite' of Brazil, the 'BHQ' (banded haematite quartzite) of India, the 'taconite' of the Lake Superior ranges, the 'ironstone' of South Africa and the 'jaspilite' of Australia. All these are now subsumed under the generic name IF.

Attempts have been made to classify and name different types of IF, but none of them has been universally adopted. The earliest suggested classification was that of James (1954), who described and named four 'facies' of IF in the Lake Superior area. The terms he used (oxide facies, carbonate facies, silicate facies and sulphide facies) are useful as names for chemical variants but tend to have fallen out of use, in part because James's (1954) suggestion that the four facies are lateral depth-related stratigraphic equivalents has been neither demonstrated in the Lake Superior area nor shown to hold elsewhere. Gross (1980) proposed a first-order twofold subdivision of IFs into Superior type and Algoma type; note that while Gross's (1980) subdivision differentiated two types of IF as stratigraphic units, and was based primarily on depositional environments, the twofold subdivision emphasized in this paper (banded IF, or BIF, and granular IF, or GIF) is a purely descriptive subdivision into lithological types. While Gross's (1980) division seemed to have some validity for North America at the time it tends to break down if a global view is taken. Most seriously, as Ewers (1983) has pointed out, major IFs such as the Dales Gorge Member of the Brockman Iron Formation, in the Hamersley Basin, have been classed as both Lake Superior type (Gross, 1980) and Algoma type (Dimroth, 1976); and if the BIFs of the Hamersley Basin are carefully evaluated against Gross's (1980) criteria they still do not fall clearly into either of his categories. Other classifications have been put forward (e.g. Kimberley, 1978; Beukes, 1980) but despite persuasive logical arguments for their adoption none has found widespread favour.

I believe that the most significant division of IF is that between BIF, which includes most occurrences older than *c*.2.0 Ga, and the type of IF characteristically present in the circum-Ungava belt of North America, which is distinguished as GIF. The criteria on which this twofold distinction is based are discussed further below; however, the nomenclatural point is emphasized here that, although in this paper the terms BIF and GIF are applied to separate lithological subtypes of IF, in much current and historical literature on IF the term BIF is also used indiscriminately as a synonym for IF, and when so used may include GIF. And a final nomenclatural point is that, although BIF and GIF are descriptive lithological terms, both names can also be used stratigraphically to denote an IF unit which consists solely or predominantly of either lithology; most IFs consist mainly of one or other type, but many contain both.

## Global distribution

IFs occur in all continents, in all major areas of early Precambrian rocks, as well as in many younger Precambrian sequences. IFs closely similar in composition and lithology to those typical of the early Precambrian even extend into the Palaeozoic (Kalugin, 1969, 1973). The locations of selected widely known Precambrian IFs are shown in Fig. 1. This map is included to emphasize their wide global distribution, and does not attempt to show all described occurrences.

IFs of the older cratons include, notably, the oldest known BIF, at Isua in Greenland, whose age is about 3.8 Ga. BIFs also occur in the greenstone belt sequences of all the main old cratons. Examples include the Abitibi belt of the Superior Province, the greenstone belts of the Yilgarn and Pilbara Cratons of Australia, the greenstone belts of the Baltic Shield (Finland and Karelia), An Shan in China, the Amazon Craton of Brazil and the Kaapvaal and West African Cratons. The ages of most that I have mentioned lie between about 2.8 and 2.5 Ga. The BIFs of Krivoi Rog and Kursk in Ukraine probably also belong here. All these early Precambrian BIFs are tectonically deformed, and to a variable extent metamorphosed, so that the nature of their depositional basins cannot confidently be reconstructed.

In four of the Gondwana continents (South America, southern Africa, India and Australia) BIFs occur in well preserved, little metamorphosed, supracrustal sequences, rather than in greenstone belts. In each case the sequences are quite mildly deformed, so that the flat to gently dipping BIFs form extensive and conspicuous topographic plateaus; for present convenience I will call these the 'Great Gondwana BIFs'. They include, in Brazil, the Carajás Formation of the Grão Pará Group of the Amazon Craton and the Cauê Itabirite of the Itabira Group of the São Francisco Craton. In South Africa the Kuruman Iron Formation and some overlying units of the Transvaal Supergroup in the Griqualand West Basin and the Penge Iron Formation of the Transvaal Basin belong in this category. In India the Mulaingiri Formation of the Bababudan Basin, in the Karnataka Craton, also belongs here. And the BIFs of the Hamersley Basin of Western Australia, which are described in greater detail below, are also included among the Great Gondwana BIFs. Their ages are discussed below, but

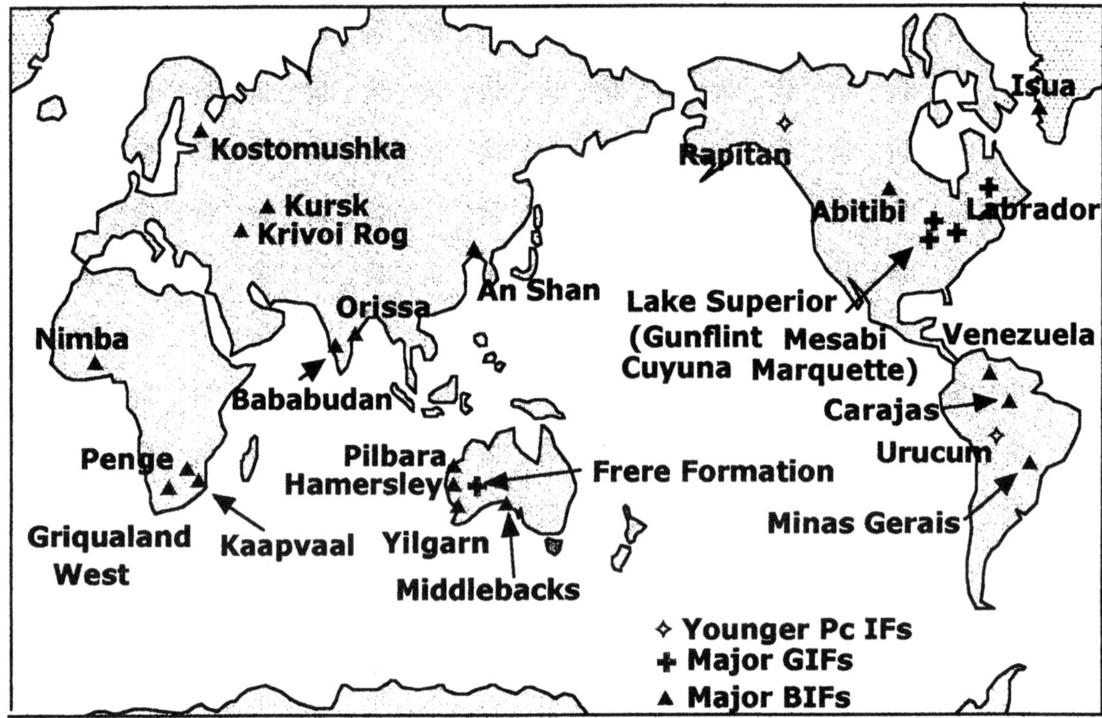


**Fig. 1.** Global occurrences of Precambrian BIF. The map emphasizes the wide distribution of BIF, and shows only a selection of well known occurrences of different ages and kinds.

some have the same age as the greeenstone belt BIFs, while others are slightly younger, perhaps extending to 2.2 Ga.

IFs also occur in the basins associated with younger Precambrian terranes. Of these some of the best known are the Lake Superior, or circum-Ungava, IFs of Canada and the United States. They include the I Fs of Labrador and the Gunflint Formation, famous for its well preserved biota, as well as those of the Lake Superior Ranges: the Biwabik Formation of the Mesabi Range and the IFs of the Cuyuna, Menominee, Gogebic and Marquette Ranges, long considered stratigraphic equivalents but recently reinterpreted (Morey & Southwick, 1995). All these consist mainly of GIF, although some also contain BIF. There is a precisely similar GIF (Frere Formation) in the Naberru Basin of Western Australia.

Finally, Fig. 1 includes a number of later Precambrian occurrences (700–600 Ma). These include Urucum, in Brazil, Rapitan, in the Yukon, and those of the Damara Belt, in southern Africa. These are a special case, as explained under the later heading 'Neoproterozoic IFs: a different problem'.

## THE HAMERSLEY GROUP BIFS AS EXAMPLES

### The Hamersley Basin

The BIFs of the Hamersley Basin of Western Australia are the largest (in terms of contained iron), most extensive and thickest known in the Precambrian stratigraphic record; only the roughly coeval and lithologically near-identical BIFs of the Transvaal Supergroup of South Africa rival them in these respects. Partly for this reason they are used here as a yardstick against which to compare other occurrences; other reasons to select them for this purpose include the fact that they have been studied in exceptional detail, and are probably the best exposed BIFs known. The descriptive account of their stratigraphy and petrography given by Trendall & Blockley (1970) provides the basis of the following summary, with additional references to subsequent work which augments it, notably that of Morris (Ewers & Morris, 1981; Morris, 1993).

The Hamersley Basin (Trendall, 1983b, 1990a) has a floor, or 'basement', of older granite-greenstone terrane, and its outcrop edges are hidden by younger rocks. Apart from very low-grade metamorphism, minor folding and uplift, the rocks of the basin are essentially preserved as they were when deposited between *c*.2.8 and 2.4 Ga. The stratified volcano-sedimentary succession of the basin—the Mount Bruce Supergroup—is formally subdivided into three successive groups. From the base upwards these are the Fortescue Group, the Hamersley Group and the Turee Creek Group. The basal Fortescue Group, which locally reaches a thickness of 7 km, consists mainly of mafic volcanic rocks. The median 2.5 km thick Hamersley Group has abundant thick BIF units, and is the focus of interest here. The topmost unit, the Turee Creek Group, consists mainly of epiclastic sediments.

Figure 2 is a geological sketch map of the basin showing the present outcrop areas of the Fortescue and Hamersley Groups. For the Hamersley Group a minimum original depositional area of about $10^5$ $km^2$ is a reasonable first-order estimate from an envelope enclosing areas of present outcrop.

The sedimentary rocks of the Hamersley Group are broadly divisible into three lithologies: BIF, shale and carbonate. However, a major part of its total thickness consists of intrusive igneous rocks, both mafic and silicic. The changes of lithology which define the stratigraphy are sharply defined. Figure 3 shows the lithostratigraphy, but for simplicity most of the formal stratigraphic names have been omitted. BIF is concentrated in five major thick units, as distinct from being thinly intercalated with other rocks. The BIF of each of these units is subtly different from that of the other four (Trendall & Blockley, 1970), but these subtleties do not decrease the significance of their common characteristics, which are now emphasized by selecting one of them—the Dales Gorge Member of the Brockman Iron Formation—for more detailed description; the justification for this is that it is the best exposed and most closely studied of the five.

## Scales of banding in the Dales Gorge Member

There are three scales of banding in the 140 m thick Dales Gorge Member, to which Trendall & Blockley (1970) applied the names macrobanding, mesobanding (the eponymous banding of BIF) and microbanding (Fig. 3). There are 33 numbered macrobands, 17 BIF macrobands numbered upwards from 0 to 16, and 16 intercalated S (or shale) macrobands similarly numbered from 1 to 16.

The S macrobands are generally thinner than the BIF macrobands, and most consist principally of microcrystalline stilpnomelane, with variable subordinate amounts of biotite, chlorite, carbonates (siderite, ankerite, dolomite, calcite), quartz, feldspar, pyrite, magnetite, tourmaline, haematite, apatite, zircon and chalcopyrite. Some contain carbon. Some of the S macrobands have volcanoclastic textures, including bubble-wall shards, texturally preserved by microcrystalline stilpnomelane. Most of them, but not all, are fine-grained ash-fall tuffs (La Berge, 1966).

Each of the BIF macrobands consists of BIF with the typical characteristics already noted: it consists of silica-rich (iron-poor) mesobands of chert alternating with iron-rich (silica-poor) mesobands, to which Trendall & Blockley (1970) applied the name chert-matrix. Within many, but not all, chert mesobands there is a regular small-scale lamination defined by some Fe mineral; this may be haematite, magnetite, carbonate (siderite or ankerite), stilpnomelane or some combination of these. These laminae (and *only* these laminae) are called microbands. In Fig. 3 the three microbanded chert mesobands represented have between six and 16 microbands; in fact such mesobands commonly contain up to 40 microbands, with a maximum recorded of 236.

The representation of both mesobands and microbands in Fig. 3 is idealized, and omits many of the textural and compositional complexities described and illustrated by, for example, Trendall & Blockley (1970), Ewers & Morris (1981) and Morris (1993). This omission is consistent with the view that the simplified picture is the one on which depositional interpretation must focus; and that this focus is blurred by excessive attention to detail. It is characteristic of microbanding that the thickness of successive microbands varies little within any one microbanded chert mesoband. By contrast, the microband thickness may vary substantially from one chert mesoband to another: microband intervals are mainly in the range 1.6–0.2 mm, with a mean about 0.5 mm.

Trendall & Blockley (1970) showed that the total Fe content of microbanded chert mesobands is inversely related to the microband thickness, varying from about 30% Fe when the microband thickness is 0.3 mm, through about 20% Fe when the microband thickness is 0.2 mm, and finally to about 5% Fe when the microband thickness is 1.5 mm. Put another way, the iron content of microbands is constant, and variations in the silica content control both thickness and relative iron content.

The dark iron-rich mesobands between the chert mesobands consist of fine-grained quartz and iron

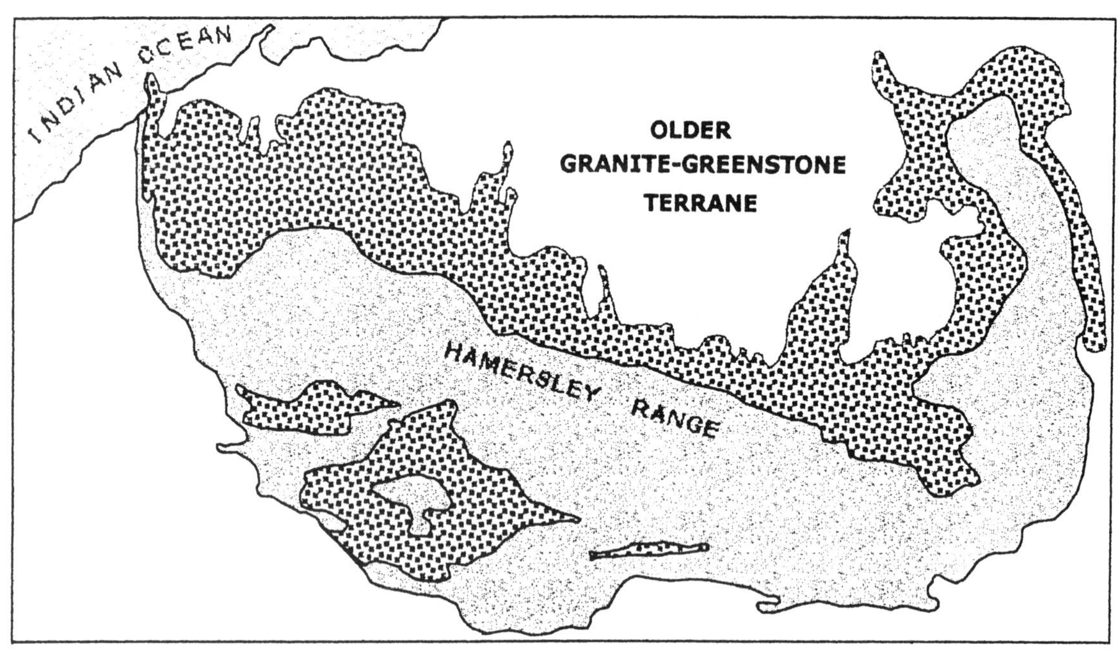


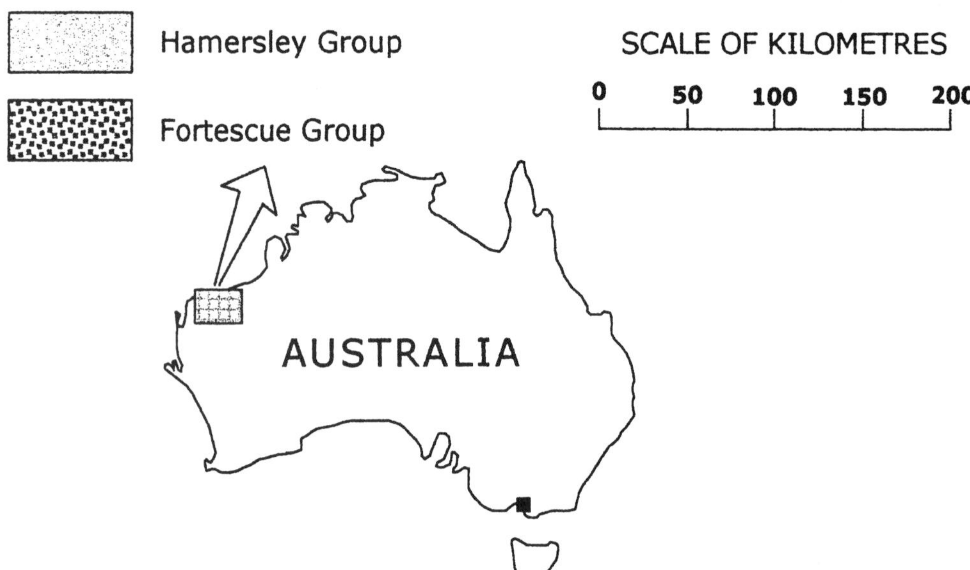


**Fig. 2.** Simplified geological sketch map of the Hamersley Basin, showing the outcrop extent of the Fortescue and Hamersley Groups.

oxides, for which reason we first referred to them by the acronym QIO, but later changed it to chert-matrix, because they form a matrix within which the chert mesobands lie. Chert-matrix has upwards of 40% Fe, and has no microbanding, although there is normally a faint internal lamination parallel to the band edges.

Before I discuss the depositional significance of these different scales of banding it is relevant to emphasize a significant characteristic of them: the lateral stratigraphic continuity at all scales in the Hamersley Basin.

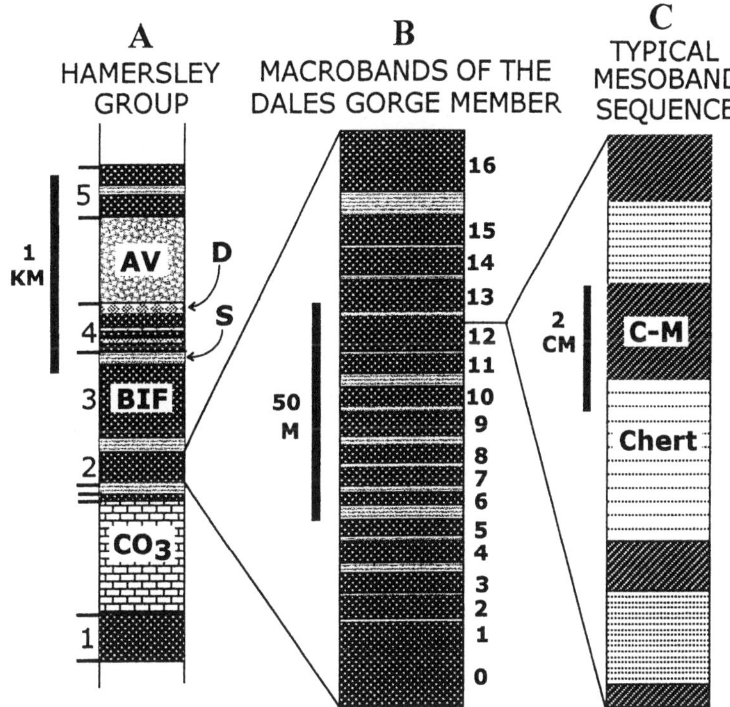


**Fig. 3.** Stratigraphy of the Hamersley Group (A) and scales of banding in the BIF of the Dales Gorge Member (B and C). Intrusive acid igneous rock, carbonate and BIF are labelled AV, $CO_3$ and BIF in A; intrusive mafic rocks (dolerite) and shale are marked D and S respectively. Shale and BIF in B are similarly ornamented. Column C has only two mesoband components, chert and chert-matrix (C–M), which are labelled. The numbers 1–5 on the left-hand side of column A show the five main lithostratigraphic units of BIF within the Hamersley Group; the Dales Gorge Member is no. 2. Short projecting lines on the same side show formally defined lithostratigraphic divisions (Formations). The numbers to the right of column B show BIF macroband numbers. The thin dotted lines within chert mesobands in column C represent the iron-rich component of microbands.

### Lateral stratigraphic continuity

First, all the formal stratigraphic subdivisions—the Formations and Members—within the Hamersley Group are easily recognizable, and of course mappable, throughout the area of the basin. Second, within the Dales Gorge Member, all of the 33 numbered macrobands can also be identified over the whole outcrop area. Third, within individual BIF macrobands, some individual mesobands can be identified and correlated; provided that the quality of exposure is there. Drillcore is of course best. Individual chert mesobands are demonstrably continuous over the whole area of outcrop.

Fourth and finally, when the microbands within individually correlated chert mesobands are examined closely, we find that individual microbands—that is, iron-rich laminae at the millimetre to sub-millimetre scale—can also be correlated. The achievement of persuasive microband correlation is a challenge in view of their repetitive regularity and (within a single chert mesoband) constant thickness. It is necessary to find slight thickness or textural variations that form a recognizable pattern. Trendall & Blockley (1970, p. 107) illustrated microband correlation between boreholes about 80 km apart, and reported correlation of microbanding in surface samples over a distance of about 300 km. Ewers & Morris (1981) have also documented microband correlation in samples 130 km apart.

### Microbands as varves: the key to a depositional model

From the features of the BIFs of the Dales Gorge Member described above, Trendall & Blockley (1970) developed a hypothesis for the depositional parameters of the Dales Gorge Member BIFs; it has since been summarized in more widely available publications (Trendall, 1972, 1983b). Before I briefly restate their hypothesis, a major difficulty in interpreting BIF needs to be emphasized. This is that it is not obvious, unlike with most other sedimentary rocks, what was actually deposited. Clearly, the tough silica/iron cherty material we see today (of any BIF, not just the Dales Gorge Member) has no petrographic elements that are self-evidently the components which were laid down in the depositional basin. For that reason the nature of the deposited material also has to be one of the elements of a complete depositional hypothesis.

The steps of Trendall & Blockley's (1970) argument for a depositional model followed the numbered sequence below:

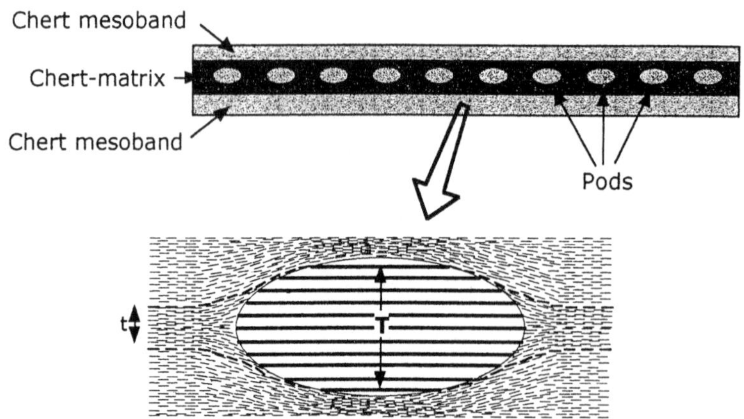


**Fig. 4.** Diagrammatic relationships between a microbanded chert pod and the surrounding chert-matrix in Dales Gorge Member BIF. The upper diagram shows a podded mesoband and the lower one shows textural detail at the lateral termination of a single pod. The microbands of the pod, indicated by solid lines, fade out into the slight lamination normally present in chert-matrix, so that a thickness 't' of chert-matrix at the lateral pod terminations appears to be derived from the compaction of a thickness 'T' of microbanded chert. The pecked lines in the chert-matrix emphasize this apparent continuity diagrammatically. Photographs of actual pod terminations are given by Trendall & Blockley (1990).

**1** It was accepted initially that the parent material of the BIF was a chemical precipitate.

**2** As a second step it was accepted that the present chemical constituents of the BIF reflected the composition of the primary precipitate; it was also suggested that the only other important constituent was water, so that the precipitated material was in a hydrous (?90% $H_2O$) colloidal state. It may now seem pedantic of Trendall & Blockley (1970) to have made this point, but there was in the 1970s a school of thought which regarded all the iron minerals of cherty iron-formation as of secondary (diagenetic or metamorphic) rather than primary origin (e.g. Kimberley, 1974; Dimroth, 1975).

**3** Noting the extraordinarily regular spacing of microbands within microbanded cherts, they postulated that each iron-rich/iron-poor microband couplet may result from one year of chemical deposition: each microband, they suggested, was an annual layer—a chemical varve. Trendall & Blockley (1970) were not totally committed to a biogenic mechanism for precipitation, but Trendall (1983b) later expressed a preference for oxidation of dissolved ferrous iron by photosynthesizing algae as a depositional process. An alternative suggestion that evaporation may have been an important trigger for deposition (Trendall, 1973b) was by implication abandoned.

**4** They then pointed to the variation of microband thickness in different chert mesobands, and its inverse correlation with Fe content; and they postulated first that finely microbanded chert mesobands were simply the more compacted equivalents of coarsely microbanded chert, and second that while the iron had remained fixed during the compaction process some of the more soluble silica had been removed in the course of diagenetic dehydration.

**5** After step 4 a working hypothesis existed that accounted for microbanding, and for different microband thickness in different chert mesobands. But the hypothesis stopped at the margins of mesobands, and did not explain the origin of chert-matrix.

**6** To include chert-matrix in the depositional hypothesis Trendall & Blockley (1970) drew attention to a feature of the banding of Dales Gorge Member BIFs that they called podding. From time to time, where a chert mesoband might be expected to occur in well exposed vertical faces, there is instead a line of discontinuous, lenticular, concretion-like, cherty lumps or pods. Pods also have internal microbanding. At the lateral termination of pods the microbands are not sharply terminated, but seem to fade out into the slight lamination normally present in chert-matrix, as shown in Fig. 4, where it appears that a thickness 't' of chert-matrix at the lateral pod terminations is derived from the compaction of a thickness 'T' of microbanded chert. Since the chert-matrix to the side of a pod is indistinguishable from the chert-matrix above and below it, Trendall & Blockley (1970) extended their depositional model to postulate that chert-matrix was the simple outcome of ongoing compression of what was formerly microbanded chert, again with the iron remaining stable. They thus supposed that the whole mesoband sequence resulted from gross compaction and diagenesis of an originally continuous succession of relatively thick annual layers of precipitate.

Applying this six-step argument to analyses of microbanded chert mesobands of measured microband thicknesses it was possible for Trendall & Blockley (1970) to calculate the amount of iron in each microband, and to estimate that 225 g of Fe were precipitated per square metre of basin area each year. Using this figure, and also the known total Fe content and density of BIF, they were able to estimate the gross rate of deposition of the (compacted) BIF. It is an implicit corollary of Trendall & Blockley's (1970) depositional model, perhaps insufficiently emphasized by them at the time of publication, that the rate of deposition of microbanded BIFs is controlled by the biological productivity in the basin, itself dependent on mean annual insolation.

Not all subsequent workers accept the Trendall & Blockley (1970) model in its entirety, and a number of alternative interpretations have been made of its various components. Nealson & Myers (1999), for example, suggested that the low iron content of iron-poor bands in BIF is due to post-depositional reduction and dissolution of iron; but they did not relate their hypothesis clearly to the real banding of a particular BIF. Carey (1976) proposed that the iron of microbands is derived from windborne dust, but the regular basin-wide deposition of equal annual increments of iron seems inconsistent with this possibility, and it is hard to explain why iron, any form of which would be expected to be in a heavy component of dust, should be preferentially winnowed from a land surface. In a later reaffirmation of his earlier proposal, Carey (1996) has linked the formation of all the major Precambrian BIFs to a single brief wind-dominated climatic episode; this proposal is not consistent with the known wide range of BIF ages in the early Precambrian. The proposal of Cisne (1984) that microbands are diurnal is also unpersuasive, because it implies unrealistically rapid depositional rates for BIF: a minimal rate of 7 km $Myr^{-1}$ would be required for Hamersley Basin BIF (see Fig. 5). Castro (1994) has recently exhumed the idea of diurnal control of iron precipitation for BIF deposition, but it is unclear whether he envisages it as the control of microbands (which have not been described from the Cauê Itabirite) or mesobands.

Walker & Zahnle (1986), while accepting an annual origin for microbanding, interpreted a 23.3-year cyclicity in BIF microbanding of the Weeli Wolli Formation, earlier suggested by Trendall (1973a) to reflect the Hale sunspot cycle, as an expression of the lunar nodal cycle. Williams (2000) has made further counts in Weeli Wolli Formation cherts with microband cyclicity, and has found between 28 and 30 microband couplets per cycle. He suggests that the microband couplets are either diurnal increments arranged in monthly cycles or fortnightly increments arranged in annual cycles.

More notably, Ewers & Morris (1981) have put forward a persuasively argued case against several aspects of the Trendall and Blockley model, which has been reinforced by Morris (1993). Although accepting the concept that microbands represent varves, Ewers & Morris (1981) are uncomfortable with Trendall & Blockley's (1970) interpretation of mesobanding, which they see as primary depositional banding related to variations in magmatic activity along mid-ocean ridges (MORs). At times of low hot-spot or MOR hydrothermal activity water in the BIF depository had a relatively low iron content, and the precipitated material was mainly silica; the microbands within consequent chert mesoband were caused mainly by direct photo-oxidation. By contrast, 'during periods of more violent MOR or hot-spot activity, higher levels of Fe(II) reached the depository by convection-driven upwelling, with increased nutrients . . . possibly triggering a parallel growth of organisms' (Morris, 1993, p. 368); the iron-rich chert-matrix mesobands were deposited during these periods.

In presenting their preferred model, Ewers & Morris (1981) and Morris (1993) also discussed the possible depositional significance of textural and compositional variation within microbands; their detailed descriptions of these built on those of Trendall & Blockley (1970), and represent an outstanding contribution to the study of BIF. However, in this paper, the identity of the microband as a chemical varve is regarded as the key factor in depositional modelling. Morris & Trendall's (1988) contribution to the discussion emphasized their unanimity on that interpretation; and apart from the carefully reasoned arguments of Williams (2000), there has been no challenge to it since its publication.

## Rate of deposition

Whether microbands reflect diurnal, fortnightly or annual depositional layers must remain an open question until definitive evidence is found. An important contribution to this interpretational problem can clearly be made by independent determination of the rate of BIF deposition. Similarly, although the disparity between Trendall & Blockley's (1970) and Ewers & Morris's (1981) views on the origin of mesobanding is largely independent of many other arguments affecting interpretation of the depositional environment of BIF, a critical difference between the two

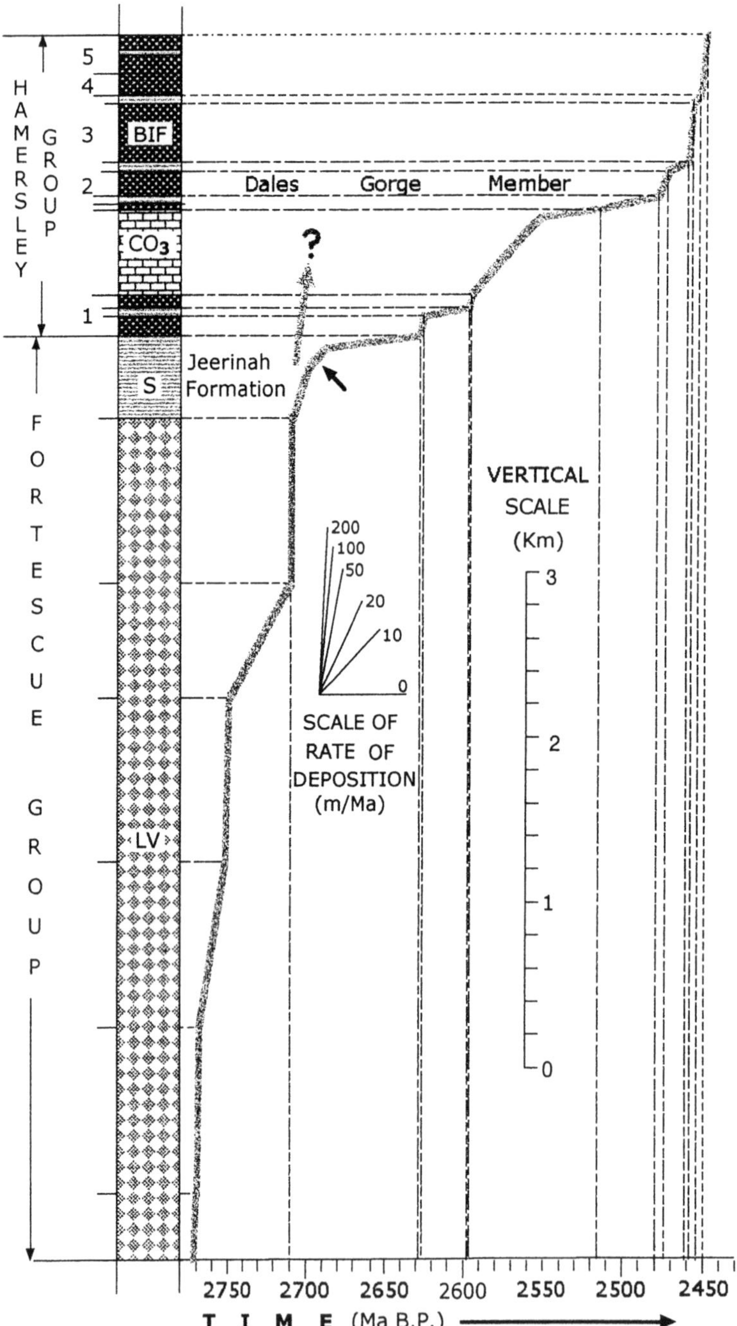


**Fig. 5.** Time calibration of deposition in the Hamersley Basin. Lines on the left-hand side of the stratigraphic column show formal lithostratigraphic boundaries (Formations and Members) within the Hamersley and Fortescue Groups; intrusive igneous rocks have been taken out of the Hamerlsey Group (cf. Fig. 3) and Fortescue Group thicknesses have been similarly modified according to estimated thicknesses of contained sills. Most names are omitted, as they are not relevant in the context of this paper, but the position of the Dales Gorge Member and the Jeerinah Formation, which are referred to in the text, are shown. As in Fig. 3 the ascending numbers 1–5 show the main BIF units of the Hamersley Group. Horizontal lines on the right-hand side of the columns extend the lithostratigraphic boundaries out to their intersection with the broad grey line, which shows the progress of basin infilling. The corresponding vertical lines show the best estimate of the ages of stratigraphic boundaries. The grey arrow with a terminal query mark shows the possible continuation of basin sinking, as infill rate slowed during deposition of the Jeerinah Formation, at the point indicated by the short black arrow; these features are discussed in the text. The lithologies of the stratigraphic column have been generalized to four components: BIF, carbonate ($CO_3$), shale (S) and mixed lavas and volcaniclastic rocks (LV).

models is the depositional rate they predict. Morris's (1993, p. 276) analysis of this point concluded that these outcomes differed by a factor of approximately four, specifically between about 1120 years to form a metre of BIF in the Ewers & Morris model and 4400 years for the Trendall & Blockley model (these figures correspond to depositional rates of 893 and 227 m $Myr^{-1}$ in the format preferred in this paper). Morris

(1993, p. 276) correctly concluded that 'The major conflict between these differently based estimates will need to be resolved.'

As Morris (1993) pointed out, Trendall & Blockley (1970) did not commit themselves to a single depositional rate for Hamersley Group BIF. Their implied values ranged from a minimum of about 23 m $Myr^{-1}$ (Trendall & Blockley, 1970, p. 262, assuming the higher estimate of 3000 years for a Knox cyclothem 7 cm thick) to a maximum of about 230 m $Myr^{-1}$ (Trendall & Blockley, 1970, p. 264, assigning 44 years to 1 cm of BIF, based on the mean Fe content of microbands).

Immediately following publication of Trendall & Blockley's (1970) estimates of depositional rates, attempts were made to check them using isotopic methods of age determination. Pilot use of the Rb–Sr method (Trendall & de Laeter, 1972) on tuffs (porcelanites) within BIF showed that this system was insufficiently robust to retain depositional ages, even at the low metamorphic level of the Hamersley Group. Compston *et al.* (1981) reported a $^{207}Pb/^{208}Pb$ age of 2490 ± 20 Ma from multi-grain fractions of zircon extracted from one of the S macrobands of the Dales Gorge Member, which clearly indicated that the zircon of these ash-fall tuffs had the potential to provide reliable depositional ages for the enclosing BIF. However, accurate determination of the depositional rate of BIF units of the thicknesses of those of the Hamersley Group clearly requires lower errors than ±20 Ma.

Another decade was to go by before the necessary resolution and precision were reached for effective single-zircon analysis; a vital step was the development of the Sensitive High-Resolution Ion MicroProbe (SHRIMP) by Professor Bill Compston at the Australian National University. There are now about 20 precise SHRIMP zircon U–Pb ages from the Hamersley Basin, and these have been used in the construction of Fig. 5. That figure shows a simple plot of time against the accumulated thickness of material (volcanic and sedimentary rocks) in the central part of the basin. The sedimentary thickness is compacted thickness, not deposited thickness, since this is hypothetical for BIF.

In general terms, for the basin as a whole, we see an immense thickness of mainly mafic volcanic rocks, emplaced rapidly, and roughly in step with the sinking of the basin, as the lavas and tuffs are either shallow-water or terrestrial. Towards the top of the Fortescue group this regime changes, and the *integrated* accumulation rate of the Hamersley Group became smaller. However, the available SHRIMP data give estimates for the five major thick units of BIF in the Hamersley Basin ranging from 19 to 225 m $Myr^{-1}$ (Trendall, 1998). This is in rather good agreement with the 1970 estimates of between 23 and 230 m $Myr^{-1}$. While it has still not been proven that microbands are varves, it does indicate that the best resources of current isotope geochronology give results that are completely consistent both with the 'microbands-are-varves' hypothesis and with Trendall & Blockley's (1970) proposals for the origin of mesobands.

**Some basic depositional parameters**

If the microbands-are-varves hypothesis is true it must be envisaged that during Hamersley Group BIF deposition thin, alternating and continuous layers of iron- and silica-rich precipitate were deposited annually, and evenly, across the entire $10^5$ $km^2$ area of the basin. The thicknesses of individual BIF units means that the environment in which this happened remained stable for periods of at least $10^6$ years.

These very basic parameters of deposition have been derived solely from the detailed structure of the banding. Other parameters, such as the mechanism of precipitation, are pursued below, but it is relevant to consider here what conclusions may be drawn from two other features of Dales Gorge Member BIFs: the absence of current-generated structures and the lack of epiclastic material.

The absence of current-generated structures affecting the banding sets a minimum water depth for the basin during BIF deposition. Johnson & Baldwin (1996) suggest a depth of 200 m for modern storm wave base, although on the present Earth strong currents can be generated by exceptional events at much greater depths. Although 200 m can be regarded as a minimum depth of deposition for the presumably rather incoherent primary precipitate of BIF, other arguments discussed further below ('Quantitative aspects of BIF deposition') imply a depositional depth substantially greater than this.

The significance of the absence of epiclastic material from the basin waters during BIF deposition was discussed by Trendall & Blockley (1970), and later by Ewers & Morris (1981) and Morris (1993). Although it is easy to speculate on this topic, it is more difficult to find evidence which can differentiate between the possibilities. These include: a climate sufficiently arid to preclude any erosional clastic influx; deposition beyond the range of effective offshore epiclastic deposition, such as the offshore platform of Morris &

Horwitz (1983); and such extreme possibilities as a totally ice-covered Earth.

There is a presumption from Fig. 5 that the absence of epiclastic material in Hamersley Group BIFs may be associated with deep-water deposition, providing a direct link between both features. The plot shows that the integrated accumulation rate of the volcanoclastic Fortescue Group succession was relatively fast, averaging about 90 m $Myr^{-1}$ (cf. the 40–120 m $Myr^{-1}$ of Arndt *et al.*, 1991); and the generally shallow-water or subaerial facies of those rocks (Thorne & Trendall, 2001) indicates clearly that basin-floor depression was closely synchronized with accumulation. The uppermost volcano-sedimentary unit of the Fortescue Group (Jeerinah Formation) shows a steady increase from base to top of the apparent water depth: a basal sandstone gives way to black shale, with a decreasing epiclastic component, and there is locally intercalated pillow basalt. Clearly, if the Fortescue Group rate of basin-floor depression continued during deposition of the Hamersley Group, as the rate of deposition slowed to the approximate integrated rate of <20 m $Myr^{-1}$, then the onset of BIF deposition (unit 1 of Fig. 5—Marra Mamba Iron Formation) may be attributable to the achievement of some minimal water depth.

## HOW REPRESENTATIVE ARE THE HAMERSLEY BIFs?

### Early Precambrian BIF

It is vitally important to examine to what extent the BIFs of the Hamersley Basin are representative of other IFs, since if they are not, then any depositional hypothesis for the Hamersley BIFs cannot be generally applied; and conversely, if they are, then our understanding of Hamersley BIFs can be applied to BIFs in general. Occurrences of IF are so numerous that it is not practicable here to evaluate in detail the similarities and differences of each one from those of the Hamersley area. Nor has any systematic comparison of all known IF occurrences been published by others. The most serious attempt to do so is that of Gole & Klein (1981); and Gole (1981), in a descriptive review of the BIFs of the Yilgarn Craton of Western Australia, specifically commented on their similarities with those of the Hamersley Group. My own comparative evaluation of the characteristics of Precambrian IFs in general, using those of the Hamersley Group as a reference standard, follows. Some of the criteria used for this comparison were not used by Gole & Klein (1981), but wherever similar criteria are used, my comparative judgement concurs with theirs.

The vast majority of early (>2000 Ma) Precambrian IFs, including (to the extent that can be judged through tectonism and metamorphism) the oldest known at Isua (Allaart, 1976; Appel, 1980; Dymek & Klein, 1988; Ohtake *et al.*, 1992), consist of BIF, and have the following similarities with Hamersley BIFs.

First, the *broad stratigraphic characters* of BIF as described for the Hamersley Basin are characteristic of most early Precambrian IFs. That is, they generally form discrete, sharply bounded units mostly less than 100 m thick, as distinct from being thinly interstratified, or interdigitating, with other lithologies. No BIF of substantial thickness has ever been shown to be laterally gradational into another sedimentary lithology. They are commonly associated with a variety of volcanic rocks, as well as shales and carbonates, but also occur in association with clastic sequences.

There are exceptions to the generalization that BIF is never thinly interstratified with other rocks. Thus Eriksson's (1983) idealized cross-sections of early Archaean sequences of the Pilbara and Barberton greenstone belts show IF intercalations less than a metre thick capping the graded turbidite beds of clastic units, and in other special situations. Similarly, in the IF ('banded ironstone') of the Mount Belches area of the Western Australian Yilgarn Craton, Dunbar & McCall (1971) used a discrete BIF unit ('Santa Claus Ironstone Member') as a structural marker, although noting that minor intercalations of IF less than a metre thick are intercalated within the adjacent, mainly clastic, sequence. Similar thin intercalations of IF have also been noted within the clastic sediments of the Beardmore–Geraldton greenstone belt of Ontario by Barrett & Fralick (1985, 1989). Despite these examples, it remains the case that major IF units of both areas are discrete, well defined and relatively thick and extensive.

Second, as far as *banding* is concerned, the mesobanding of most BIFs follows the pattern described for the Hamersley BIFs rather closely in respect to thickness, sharpness of mesoband boundaries and simple division into Fe-poor (chert) mesobands alternating with Fe-rich chert-matrix. Mesobanding of this type may be taken as the defining characteristic of BIF. Microbanding exactly similar to Hamersley microbanding is also present in equal abundance in the Transvaal Supergroup BIFs of South Africa (e.g. Beukes, 1973). Among other Great Gondwana BIFs it is present in chert mesobands of the Carajas Formation of Brazil and the Mulaingiri Formation of India, although in neither is it as conspicuous as in the Hamersley BIFs. I have seen it in many other BIFs (Trendall, 1973b, Fig. 5), but it tends not to be described in published papers, as it is not specifically sought. Gole (1981) has provided figures of excellent examples from BIFs of the Yilgarn Craton of Western Australia, and Matin & Mukhopadhyay (1992) describe microbands from an IF in the Sandur Schist Belt of the Indian Karnataka Craton.

Third, *current-generated structures*, such as cross-bedding, ripple-marks and current scours, are so rare in BIFs as to be virtually absent. The unequivocally current-generated bedding features described as primary by Gross (1972) in the IFs of Canada come from GIFs. Other irregularities of banding illustrated by Gross, like those described by Majumder & Chakraborty (1977) and Chakraborty & Majumder (1992) from Orissa, as well as those illustrated by Matin & Mukhopadhyay (1992) from the Sandur Schist Belt, are probably attributable (as the latter authors state) to post-depositional modification of the banding when the deposited material was still hydroplastic; Trendall & Blockley (1970) described a range of comparable structures from the Dales Gorge Member. Nevertheless, some S macrobands of the Dales Gorge Member have current-related ripple-scale cross-laminations, indicating that gentle bottom currents were sufficient to disturb the fine-grained volcanic ash from which they formed, and it would not be surprising if some current-generated structures were shown to exist in the adjacent BIF, as they have been in the Transvaal (Beukes, 1973).

Fourth, demonstrably *epiclastic (as distinct from volcaniclastic) components are virtually absent*. Examples are so exceptional as to prove the rule. Moreover, there are no textural features in Hamersley or similar BIFs that can be convincingly interpreted as primarily deposited components of the rock: the whole banded material is so modified and recrystallized during diagenesis and lithification that (like many carbonates) nothing in the texture of the rock can convincingly be interpreted as a primary depositional component.

Fifth, I have already noted that *chemical composition is very uniform*. However, there is substantial *mineralogical variation*: some BIFs have no magnetite, but haematite as the only Fe oxide, and carbonate content tends to be variable.

Sixth, the *lateral continuity* of Hamersley detail has not been recorded in other BIFs, but I believe largely because insufficient attention has been devoted to seeking it. On outcrop scale, none of the many BIFs that I have seen shows significant lateral impersistence of mesobanding.

Seventh, the *depositional rates* of other BIFs have not been measured with the same reliability as those of the Hamersley BIFs. However, if the main mechanism of BIF deposition involved the oxidation of dissolved oceanic iron by photosynthesizing phytoplankton, as discussed further below, the simple assumption is that depositional rates would not have varied greatly.

Eighth, much the same problem applies to comparison of *depositional basin sizes*. The Great Gondwana BIFs demonstrably cover large areas, but many older Precambrian BIFs of granite-greenstone terranes have much smaller outcrops. To what extent this may be an effect of tectonic dismemberment is not certain, but the balance of evidence suggests that most BIFs of greenstone belts were deposited in much smaller basins.

## Granular iron-formation

The one major group of IFs fundamentally dissimilar to the Hamersley BIFs and their like are those typified by the circum-Ungava examples of Canada and the USA, including those of the Lake Superior ranges and Labrador. Good accounts of these include the classic descriptions of Van Hise & Leith (1911) and such later acccounts as those of James (1954), Dimroth & Chauvel (1973) and Zajac (1974), and many papers cited in Morey (1983) and Simonson (1987). These IFs consist predominantly of GIF, which has the following characteristics:

**1** In *broad stratigraphic parameters* GIF also tends to form discrete sharply bounded units. However, in terms of lithological association GIF is commonly interstratified with coarse or medium-grained epiclastic sediments, and a volcanic association, although usually present, is smaller relative to the volume of IF present.

**2** In terms of *banding* the GIF lacks the regular mesobanding typical of BIF. There is a comparable alternation of iron-rich and iron-poor bands, but this tends to be coarser and less regular. The coarsely crystalline cherts tend to be wavy or lenticular, probably reflecting the rippled nature of the primary sandy material. Both iron-rich and silica-rich bands may be granular (see point 4), more particularly the latter. Neither the cherts nor the iron-rich bands have a distinctive microbanding or lamination.

**3** The presence of *current-generated structures* such as cross-bedding, ripple-marks and current scours is widespread in GIF (e.g. Gross, 1972; Ojakangas, 1983; Simonson, 1985).

**4** The iron-rich bands of GIF, as the name implies, often consist of a close-packed and lithified mass of granules or ooliths, about 1 mm across. These are made up of iron oxides with or without quartz, and their interstices are filled by the same minerals, but usually with a lower iron content. The granules have the appearance of primary depositional components, and the material has been referred to as 'a special type of sandstone' (Mengel, 1965). The textural distinction between GIF and BIF is critical, and it should be emphasized that BIF and GIF may occur within the same stratigraphically defined IF. Thus Simonson & Goode (1989) have described ferruginous granules from the Hamersley Group (though not from a BIF unit), and BIFs with banding exactly like that characteristic of Hamersley Group BIFs certainly occur locally in the Lake Superior ranges; a spectacular example in the Negaunee Iron Formation of the Marquette Range appears as Fig. 1 in James & Sims (1973). Beukes & Klein (1990) have described a transition from microbanded BIF to GIF in a borehole section from the Transvaal Supergroup; and Bunting (1986) records the presence of both BIF and GIF within the Frere Formation of Western Australia.

**5** Despite their major textural differences, the *chemical composition* of GIF is rather uniform, and closely comparable with that of BIF. However, as in BIF there are substantial *mineralogical variations*.

**6** *Lateral continuity* of banding comparable to that of the Hamersley BIFs has not been demonstrated in GIFs, and in view of their deposition in relatively shallow water it is unlikely to be found.

**7** The *depositional rates* of GIFs have not been measured.

**8** *Depositional basin sizes*: the major GIFs appear in well preserved basins of substantial size, comparable to the depositories of the Great Gondwana BIFs.

Thus the principal features of GIF that differentiate it from BIF are the granular, sandstone-like texture, the presence of current-generated structures and coarser banding. All these suggest a shallow-water, high-energy environment.

In summary, the vast majority of *early* (older than 2000 Ma) Precambrian BIFs, but *not* the granular iron formations, so closely resemble those of the Hamersley Basin that any compelling depositional model for the Hamersley BIFs is likely to be equally applicable to early Precambrian BIFs as a whole. It does not of course follow that every detail of Hamersley Basin geology needs to have existed to produce any other BIF, but we do need to consider what the necessary and sufficient conditions for Precambrian BIF deposition might be. This topic is taken up below, and it will be useful to discuss beforehand both the distribution of IF in time and a possible explanation of that distribution.

## WHEN WAS IF DEPOSITED?

### Distribution of IF in time

The ages of some IFs have already been noted in passing; their distribution in time now needs to be examined more closely. During the 1960s and 1970s, as the results of isotope geochronology were increasingly accepted and applied by geologists, it came to be widely believed that many major IFs were deposited around 2000 Ma. Goldich (1973) noted, for example, that 'Most geologists will agree that a remarkable development of banded iron-formations occurred during the Middle Precambrian, or Lower Proterozoic, 2600 to 1600 m.y. ago.' Most of the data underpinning this belief came from the whole-rock Rb–Sr isochron method. As this isotopic system was shown to be sensitive to metamorphic updating, geochronologists turned successively to the whole-rock Pb–Pb, the Sm–Nd and finally to the zircon U–Pb isotopic system, first in multi-grain samples and then in single crystals; we have already seen how this has been applied to the Hamersley Group.

As data from more robust isotopic systems accumulated, Goldich's views were refined, and a number of authors published diagrams showing the estimated abundance of IF in time (e.g. James, 1983; Klein & Beukes, 1992). In almost all such diagrams the *Y*-axis is unquantified, so that they represent the respective authors' subjective estimates of relative abundance. James (in James & Trendall, 1982) made an attempt at quantification, which served to illustrate the difficulties of accurately estimating the total deposited iron content of most IFs. A further difficulty with earlier appraisals of relative abundance of IF deposition in time is that very few IFs were (and still are) precisely dated, estimates of their age being largely based on the ages of datable rocks above and below them in the sequences in which they lie. Most recently, Isley & Abbott (1999) have adopted an innovative approach to the temporal distribution of IF deposition. Their statistical analyses are less concerned with the distribution of IF in time than with studying the correlation of IF deposition with magmatic activity.

With the proviso already noted, I accept the diagram of Klein & Beukes (1992, Fig. 4.2.1) as a reasonable representation of the distribution of Precambrian IF distribution in time. Figure 6 is essentially identical, but incorporates a differentiation into different IF types believed to be insufficiently emphasized in the past.

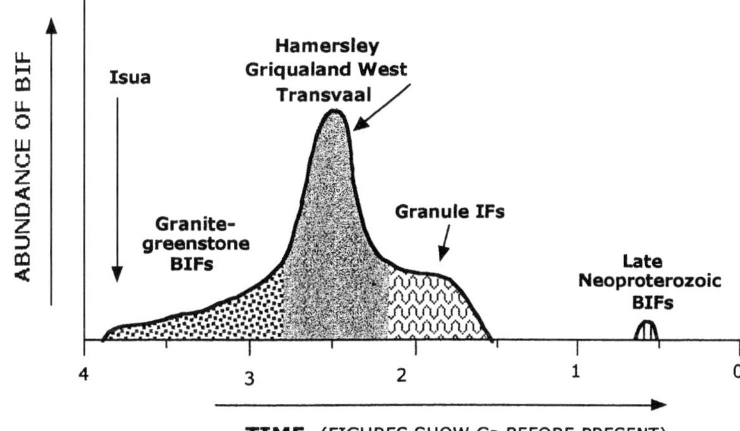


**Fig. 6.** Time distribution of BIF in the stratigraphic record. Like almost all other published diagrams of this type the curve is highly generalized and subjective.

### Correlation of type with time

The key difference of Fig. 6 is that it shows, by differing ornaments, characteristic differences in the IF deposited at different times during the Precambrian. Throughout early Precambrian time individual IFs tend to be thinner, laterally less extensive and often closely associated with volcanism in greenstone belts. There then seems to be a peak at *c*.2.5 Ga, to which a very significant contribution is made by two of the Great Gondwana BIFs: those of the Hamersley and Transvaal–Griqualand West basins. As noted above, these BIFs have lithologies closely similar to those of the greenstone belts, but individual IF units are thicker and more extensive. There also appears to be a later period of abundant IF deposition, possibly around 1.8 Ga, to which the main contributors are the GIFs. And finally there is a latest Precambrian scattering of mainly small IFs of various types. These last are mentioned briefly below under the heading 'Neoproterozoic IFs: a different problem'.

Before the question of what this time-distribution of IF means is addressed it will be useful to consider another question: how was IF deposited?

## HOW WAS IF DEPOSITED?

### The Cloud hypothesis

By 'how was IF deposited' is meant: what chemical mechanism caused the local annual precipitation of Fe-rich bands on so widespread a scale during such long periods?

The first compelling explanation for the 2.5–2.0 Ga IF peak was that of Preston Cloud. His hypothesis for an interrelationship between IF, the biochemical evolution of terrestrial life and the chemical evolution of the atmosphere and oceans was an intellectual construct of outstanding elegance. The outline form in which it is summarized below is a composite derived from published accounts (Cloud, 1968, 1972, 1973, 1976, 1988) and personal discussions; see Fig. 7.

Cloud took as one of his starting points the idea (a commonly accepted and reasonable one: e.g. Holland, 1984) that the earliest atmosphere was likely to be $CO_2$-rich, and reducing, and that the resultant acid weathering of the early continents would have, *inter alia*, carried immense quantities of dissolved ferrous iron into the oceans, where they would have remained in solution. The next step was to accept that the depositional mechanism for IF was precipitation of this oceanic iron due to the oxygen produced by photosynthesizing organisms (phytoplankton) living in the basin water.

Cloud then linked the observed time-distribution of IF with stages in the biochemical evolution of life. He proposed that the earliest organisms were chemoautotrophs, reliant on a variety of chemical reactions to obtain their energy. Some chemoautotrophs later evolved to use sunlight as an energy source: they became photosynthesizers, or photoautotrophs. The next step in the hypothesis was to point out that the earliest photoautotrophs, because they had evolved in a reducing environment, would not have possessed the oxygen-mediating mechanisms that most modern organisms rely on for survival. They would therefore have been reliant on the ability of their environment

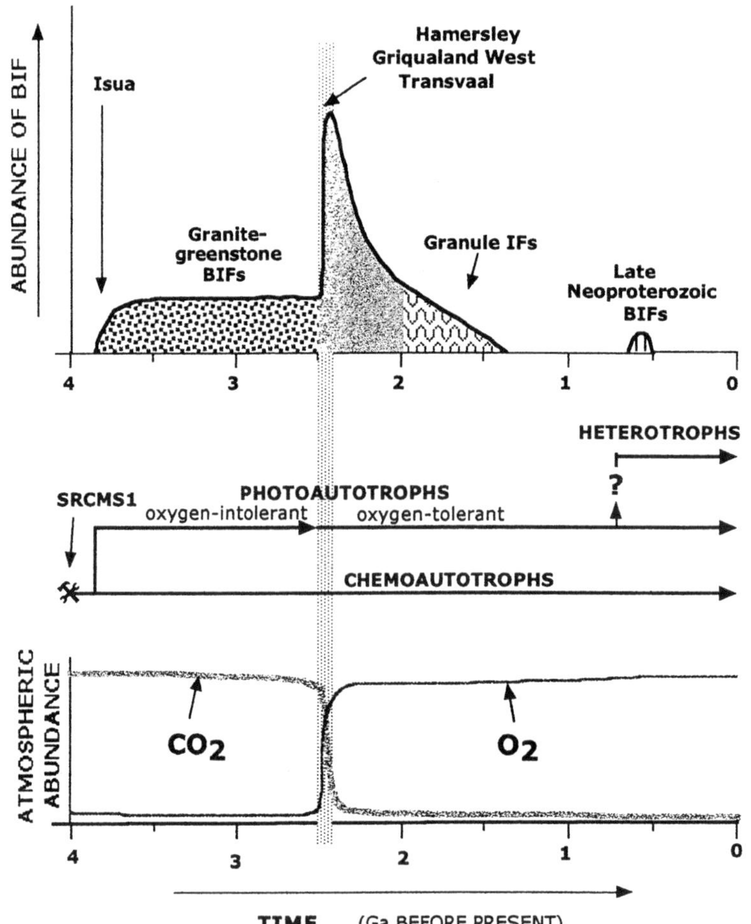


**Fig. 7.** Graphic summary of the Cloud hypothesis. The top diagram shows the time-distribution of BIF predicted by the Cloud hypothesis, the central diagram shows the biochemical evolution on which the hypothesis is based and the lower graph shows the related evolution of atmospheric carbon dioxide and oxygen levels (unquantified) proposed by the hypothesis. All three diagrams are linked by a vertical shaded line marking the 'Cloud event' (see text). The upper diagram essentially reproduces Fig. 6, modified to accord with Preston Cloud's concept: BIFs start abruptly with the appearance of the first (oxygen-intolerant) photoautotrophs, continue at an abundance limited by the efficiency with which oxygen can be disposed of, increase dramatically when an oxygen-mediating mechanism is evolved and then decrease until the oceans are flushed free of dissolved iron. 'SRCMS1' stands for 'self-replicating complex molecular system 1'—the common ancestor of all life on Earth.

to take up the oxygen resulting from their photosynthesizing activity. This was conveniently done by oxidation of oceanic ferrous iron, with resultant precipitation of IFs.

During the early Precambrian the size and scale of IF deposition would have been proportionate to the rate at which this chemical disposal could have been achieved. An oxygen production rate by the first photosynthesizers faster than the disposal rate would have led to their own demise, so that the process was self-limiting. Hence the common presence of IF in early Precambrian rocks, combined with an absence of giant-sized ones.

The next step of the Cloud hypothesis was a critical one. At some stage after the development of photosynthesis, it was supposed, photoautotrophs did develop a system of oxygen mediation that allowed them to survive and flourish in an oxidizing environment. Freed of their earlier biochemical restraints there was a massive population explosion; this caused the Great Gondwana BIFs and the peak in IF deposition.

It followed from the Cloud hypothesis that this great step in biochemical evolution would have resulted in a geologically almost instantaneous depletion of atmospheric $CO_2$, and its replacement by oxygen. This great quadruple event might be termed the Cloud event: its four interrelated components were the contemporaneous flushing of iron from the oceans, its deposition as IF, the change in atmospheric composition and the advent of photoautotrophs able to survive with high atmospheric (and oceanic) oxygen levels.

## The hypothesis reappraised

In 1992 the low precision with which the ages of the Great Gondwana BIFs were known permitted a belief in their synchronous deposition: a prediction of the Cloud hypothesis. In that year a project was initiated at Curtin University to test their synchroneity by using SHRIMP zircon geochronology. This work is still in progress, and the results have not been fully published, but it is now clear that the Great Gondwana BIFs were not laid down at the same time. For example, the Carajás Formation of Brazil (Trendall *et al.*, 1998) and the Mulaingiri Formation of India (Trendall *et al.*, 1997) were each deposited during separate narrow time intervals between 2.7 and 2.8 Ga, and their deposition was completed about $10^7$ years before the BIF of the the Hamersley Basin: BIF deposition began in the Hamersley Basin prior to 2.6 Ga, but the main period of deposition was (Fig. 5) between about 2.5 and 2.4 Ga. Deposition of the Transvaal and Griqualand West basins is less well controlled, but began at *c.*2.5 Ga, so that it is in part coeval with the Hamersley Basin (Nelson *et al.*, 1999). This time range of the Great Gondwana BIFs implies that the Cloud event as an abrupt four-component milestone in Earth evolution does not, therefore, exist.

Cloud's acceptance of a biological role in the precipitation of IF has also been challenged, and alternative abiogenic precipitation mechanisms have been suggested (e.g. Braterman *et al.*, 1984; Francois, 1986; Braterman and Cairns-Smith, 1987a, b; Draganic *et al.*, 1991). More generally, Klein & Beukes (1989, p. 1768) assert that 'the lack of organic carbon in iron-formation is a serious problem in any model that couples the deposition of BIF to microbial activity, especially photosynthetic activity'.

Drawing largely on the results of their detailed joint studies of the Transvaal Supergroup (Klein & Beukes, 1989; Beukes & Klein, 1990; Beukes *et al.*, 1990), Beukes & Klein (1992, p. 149) list three specific lines of evidence for the decoupling of IF deposition and primary organic productivity: (i) very low organic carbon content of microbanded BIF, despite its high depositional rate, which would have favoured the preservation of organic matter; (ii) very low P and low Ba contents in most IFs; and (iii) little preservation of organic C in siderite-rich IFs, which must have been deposited under reducing conditions. They therefore favour an abiogenic precipitation mechanism controlled by variation in Eh–pH conditions at a chemocline between somewhat oxic surface waters and deeper waters that were anoxic and enriched in dissolved ferrous iron (Beukes & Klein, 1992, p. 150). Klemm (2000) has more recently described an abiogenic model in greater chemical detail.

Beukes & Klein's (1992) cogent case against the biogenic control of IF deposition must be set against the well documented presence of isotopically light C in IFs, whether in carbonate or organic matter (Becker & Clayton, 1972; Perry *et al.*, 1973; Goodwin *et al.*, 1976; Schidlowski *et al.*, 1983; Baur *et al.*, 1985; Beukes *et al.*, 1990). As an alternative to acceptance of this feature as indicative of biogenicity, the case for which is well argued by Baur *et al.* (1985), Beukes & Klein (1992) suggest that it represents a primary signature of the Early Proterozoic deep ocean. My own preferred belief in biological control of IF deposition is influenced by the situation in the depository of the Hamersley Group, which provided the microbanded BIF samples with light carbon reported by Baur *et al.* (1985). In the Hamersley Basin the uppermost unit (Roy Hill Shale Member) within the formation (Jeerinah Formation) underlying the lowermost BIF (Marra Mamba Iron Formation) of the Hamersley Group typically contains microbanded chert bands which are texturally similar to microbanded chert mesobands of Hamersley Group BIFs (Davy, 1985; Thorne & Trendall, 2001). They differ in being thicker (commonly *c.*10 cm), in having thicker (*c.*1 cm) microbands, and in the definition of these microbands by thin dark carbonaceous layers. Trendall & Blockley (1970, p. 149) reported a carbon content of about 17% from a sample of this shale, indicating abundant organic productivity in the basin water immediately before BIF deposition. Brocks *et al.* (1999) have now succeeded in extracting indigenous hydrocarbon markers from the Roy Hill Shale Member, and in particular 2α-methylhopanes, which they interpret as firm evidence for the presence of photosynthesizing cyanobacteria. They also detected the same hydrocarbon suite in black chert samples from the Roy Hill Shale Member, and from shale within the Marra Mamba Iron Formation. Given the presence of abundant photosynthesizing cyanobacteria in the Hamersley Basin immediately before, and during earliest, BIF deposition it seems reasonable to accept their involvement in the precipitation process.

While it remains true that non-carbonate C in any form is essentially absent from BIF of the Hamersley Basin, I find the simple reaction

$$6Fe_2O_3 + C \rightarrow 4Fe_3O_4 + CO_2$$

during BIF diagenesis a simple explanation both for this absence and, in part, for the post-depositional growth of magnetite and ferrous carbonates (siderite, ankerite) in Hamersley Group BIF; Perry *et al.* (1973), Baur (1985) and Walker (1984, 1987) have previously accepted this mechanism. Beukes & Klein (1992) refer to magnetite as a common primary precipitate in IFs, but no BIF in the sense in which the term is used in this paper has clear evidence for primary depositional mineralogy.

Alternative biochemical controls of IF deposition, such as anoxygenic bacterial photosynthesis, have also been proposed (e.g. Kump, 1993; Widdell *et al.*, 1993), and these seem equally consistent with both the C and S (Goodwin *et al.*, 1976; Arnold *et al.*, 1991) isotopic evidence. Although the biogenic/abiogenic debate on IF deposition has not been finally resolved, a promising new approach, using Fe isotopes, has been attempted by Beard *et al.* (1999). My personal belief is that wider application of this method, especially to appropriately selected samples of the microbanded BIFs, will show that life, throughout the early Precambrian, was as closely associated with both Fe and Si deposition as it still locally remains on the present Earth (Konhauser & Ferris, 1996). That belief can only be strengthened by growing evidence for organic involvement in sedimentation, including BIF deposition, from the earliest geological record (Mojzsis *et al.*, 1996; Nutman *et al.*, 1997; Schidlowski, 1998; Rösing, 1999); as well as in association with Archaean BIF deposition (Venkatachala *et al.*, 1986, 1990); although the microfossils reported by La Berge (1973) were discredited by Heaney & Veblen (1991), they left open the possibility of some organic connection because of the presence of apatite. Megascopic algae have also been described from the Negaunee Iron-Formation (Han & Runnegar, 1992), suggesting the need for a reappraisal of the possible biogenic structures reported earlier from the same locality and formation by Mancuso *et al.* (1971). Traditional concepts of the Precambrian Earth as relatively lifeless and inhospitable appear less and less secure, as the evidence for biological activity throughout the stratigraphic record steadily increases.

The link between the late IF depositional peak and atmospheric oxygen and carbon dioxide levels, another key component of the Cloud hypothesis, appears less secure than it did a decade ago. The concentration of oxygen in Earth's early atmosphere is still the subject of robust debate (Ohmoto, 1997; Holland, 1999); this debate does not directly affect the model for IF deposition preferred here. However, it may be noted that two separate lines of evidence for a sharp rise in atmospheric oxygen level (Holland & Beukes, 1990; Collerson & Kamber, 1999) suggest that it was substantially later than deposition of the main contributors (Hamersley and Transvaal) to the peak of IF deposition of Fig. 6.

The existence of Hamersley-style microbanding in many BIFs older than those of the Hamersley Basin suggests that the rates of BIF deposition throughout the early Precambrian were controlled, like those of the Hamersley Group, by mean annual insolation rates mediated by biological productivity within the basin water. The reality of the *c.*2.5 Ga peak, despite this continuity of depositional control, and even though it does not include all the Great Gondwana BIFs, makes it necessary to find some alternative explanation for the enormous size and excellent preservation of the Hamersley and Transvaal BIFs, in particular. It is suggested that one of the main parameters controlling Precambrian IF deposition, and one which exerted critical control on the size, extent and preservation of individual IFs, was the changing nature of sedimentary basins during early continental growth. The secular evolution of sedimentary basins is thus the most important factor to consider in answering the question 'Why was IF deposited?', and this question is therefore addressed immediately after comment on the Neoproterozoic IFs.

## Neoproterozoic IFs: a different problem

The Neoproterozoic IFs were not discussed in detail in the address on which this paper is based, and are therefore also treated briefly here. They have been described from Australia (Braemar Iron-Formation: Whitten, 1970; Holowilena Iron-Formation: Dalgarno & Johnson, 1965), north-west Canada (Rapitan Group: Young, 1976; Klein & Beukes, 1993), Brazil (Jacadigo Formation: Dorr, 1973; Urucum: Walde *et al.*, 1993) and Namibia (Damara Supergroup: Martin, 1965); other occurrences are listed by Yeo (1986). They have ages in the approximate range 800–600 Ma, and most have some evidence of glacial association. The Rapitan IF is one of the most closely studied, and may be taken as representative. Like other Neoproterozoic IFs it is essentially a fine-grained, thin-bedded, haematite-quartz rock lacking either the sharply defined mesobanding of early Precambrian BIF or the coarser banding of GIF. Its sedimentology was described by Young (1976), who also provided a review of its possible significance, including the proposal of Williams (1972) based on the secular variation in the inclination of the rotational axis of the Earth; Young concluded that there was no definitive explanation for the presence of these Neoproterozoic IFs. The later detailed study by Klein & Beukes (1993) showed that the simple mineralogy and major element chemistry are distinctly different from those of most early Precambrian

IFs. They concluded that the Rapitan IF was deposited during a major trangression after a glacial event, and drew attention to the suggestion of Kirschvink (1992) that it could be related to development of anoxic, and iron-rich, ocean-bottom water consequent upon an almost completely ice covered ('snowball') Earth.

Hoffman *et al.* (1998) have subsequently developed Kirschvink's (1992) idea of a snowball Earth, which was based on an earlier idea of Harland (1964), into an integrated hypothesis in which the IFs are only one geological outcome of rapid and extreme oscillations of global climate. During cold excursions ice covered much of the land, and the oceans were also insulated from the atmosphere by thick ice. There was a consequent fall in dissolved oxygen level, resulting in an increase in dissolved iron. Meanwhile, continuing volcanism built up atmospheric carbon dioxide levels in the atmosphere, to a point where greenhouse warming led to rapid melting of both terrestrial and oceanic ice. The resultant melt-related deposition of glaciogene sediments at the start of the warm period was coeval with precipitation of dissolved ferrous iron from the re-oxygenated oceans, leading to a global association of IF with glacial material.

While more work is needed to test the validity of this hypothesis, it provides an elegant working model. Schmidt & Williams (1995), for example, have demonstrated that the glaciogene deposits of South Australia were formed in association with grounded glaciers near sea level at near-equatorial latitude, but have pointed out that explanations other than greenhouse oscillation need to be considered. Whatever the outcome, it seems well established that the Neoproterozoic IFs are a separate and unrelated phenomenon from the long early Precambrian (*c.*3.8 to *c.*1.6 Ga) period of IF deposition; the controlling factors of this are now examined.

## WHY WAS IF DEPOSITED?

### Necessary and sufficient conditions for IF deposition

It was argued above that the BIFs of the Hamersley Basin share so many features characteristic of early Precambrian (say >2.0 Ga) BIFs in general that all must have a common set of depositional parameters: there must be a set of necessary and sufficient conditions for BIF deposition. I suggest, building on earlier discussion of the depositional conditions of the Hamersley Basin, that there were three basic requirements for BIF depositions during this period, such that if, and only if, all three were met BIF deposition would inevitably follow. They were:

**1** The development of a depository that remained tectonically stable for periods approaching $10^6$ years.

**2** The depository had sufficient depth of water.

**3** The shape of the depository was such that ocean water with dissolved ferrous iron was able to circulate freely into and out of it.

A fourth requirement of BIF deposition, in my opinion, was the existence of phytoplankton in the water of the depository, but the distribution of such organisms in the marine environment throughout the early Precambrian was probably such that whenever the first three simple conditions existed, BIF would be laid down. However, the presence of the appropriate phytoplankton population without the concurrent existence of the three critical conditions listed could not achieve BIF deposition. Some additional comments on each of these three critical requirements follow.

The tectonic stability of the depository is required by the depositional rate of BIF, which is best known for the Hamersley Basin, and is assumed to be of the same order for all early BIFs. On that basis, a 100 m thick BIF unit requires a depository which was stable for a period approaching $10^6$ years. The depositional rate is not known with sufficient precision to preclude the possibility that $10^5$ years, or even less, may be adequate for a thin BIF unit, and the figure of $10^6$ years is only an order-of-magnitude estimate based on the probable mean rate of BIF deposition in the Hamersley Basin.

Three arguments for the deep-water origin of BIFs have already been outlined for those of the Hamersley Group: absence of an epiclastic component, absence of current-generated structures and extrapolation of the basin-floor sinking line defined by Fortescue Group deposition into Hamersley Group time. Simonson & Hassler (1996) have argued the need for deep water for early Precambrian BIFs more generally, and have related their deposition to global sea-level highstands. Their argument hinges on 'the widely accepted idea . . . that the deposition of large iron formations was made possible by a reservoir of dissolved iron in the deep oceans, while surface waters contained relatively low concentrations of dissolved iron' (Simonson & Hassler, 1996, p. 666). The earliest authority cited by Simonson & Hassler (1996) for this idea was Button (1982), reporting a concensus acceptance of a two-layer ocean at the 1980 Dahlem Conference on 'Mineral Deposits and the Evolution of the Biosphere'. But the suggestion that the 'deeper waters of the early oceans' were richer in iron than surface water was first made by Holland (1973), who saw the upwelling of deep water as the source of iron for BIF deposition in 'shallow marine areas'. In a later discussion Holland (1984) drew attention to the low iron content of early Precambrian shallow-water carbonates to support the restriction of iron-rich water to the deep oceans; Holland still saw BIF as the product of a near-shore shallow-water setting at that time. The significance of a stratified ocean in relation to the deposition of the Transvaal Supergroup was discussed in detail by Klein & Beukes (1989), and this concept is now widely accepted, as Simonson & Hassler (1996) state: for example, Eriksson *et al.* (1997, p. 49) accept that 'Archaean sea-water also was enriched in $Fe^{++}$ but only below the pycnocline' without supporting argument. It is a view which is also accepted here.

The continuous deposition of BIF in a depository such as the Hamersley Basin requires a basin geometry such that deep iron-rich water can circulate into and out of it fast enough to provide iron for the known depositional rate. Holland (1973) addressed this point, and found his calculations 'not discouraging'; it is revisited in subsequent discussion.

Discussion so far has focused on the deposition of BIF. There is general consensus in published work (e.g. Morey, 1983; Simonson, 1985; Beukes & Klein, 1990) that GIF formed in relatively shallow water, and that its characteristic lithology is caused by the reworking, whether by storm-generated or other current activity, of iron-rich sediments whose initial deposition was caused by the same factors that led to the deposition of non-granular iron-formation: more briefly put, GIF is BIF deposited or reworked in shallower water. I accept this interpretation here, so that the continuing discussion below is relevant to the deposition of IF, not just of BIF. It is the tectonic stability and geometry of sedimentary basins that, given the continuous abundance of both deep oceanic iron and suitable phytoplankton populations, constitute the critical controls on IF deposition; and secular changes in basin stability and geometry have been the critical influences on the time-distribution of IF. The reasons for these changes are now examined in the context of a model for the origin and early growth of continents.

## The secular evolution of sedimentary basins

### *The pre-geological Earth (>4.3 Ga) and the origin of continents*

'An understanding of the origin and growth of continental crust stands as one of the major goals in the study of the Earth's evolution' (Stein & Hofmann, 1994). The implication of this statement, which is as valid in 2002 as it was in 1994, is that the origin and growth of continental crust are essentially unknown. But the early development of sedimentary basins was controlled by the development of continents, and in the absence of firm knowledge of how this happened some speculative model must be used. While the model to be outlined here is indeed speculative, it is one which provides a potential explanation of BIF distribution in time; it has been previously published only in abstract (Trendall, 1990b).

If geological time is taken to be the period within which it is possible to make deductions about the condition of the Earth directly from the rock record, its beginning can be taken as *c.*4.0 Ga, the approximate age of the oldest dated *rock* (Bowring *et al.*, 1989; Bowring & Williams, 1999), and one commonly taken as the start of the Archaean. The condition of the pre-geological Earth cannot be reconstructed using evidence from contemporaneously formed rocks, and must rely instead on conceptual modelling, working either forwards in time from concepts of early planetary accretion, or backwards in time from constraints based on beliefs about the Archaean Earth.

The most generally preferred view for the pre-geological development of the Earth involves early collision with a Mars-sized body which led to the separation of the Moon (Stevenson, 1987). Allègre *et al.* (1995) have clearly summarized the evidence that puts these earliest events, including core formation, into a time-frame, and McCulloch & Bennett (1998) have more recently reviewed the evidence from an isotopic perspective. For pre-geological events relevant to the evolution of sedimentary basins it will suffice to begin with a view of the Earth at 4.3 Ga. The speculative model to be presented here starts arbitrarily at that time, and follows Pollack (1997), with the Earth totally molten, with a differentiated core and whole-mantle convection (Davies, 1992). The essential features of the model, which relates the deposition of BIF to continental evolution, are summarized in Figs 8–12.

It is suggested that convection took place in a pattern of very large polygonal cells, possibly as few as 12, although their number is speculative. Mantle convection on the modern Earth is still incompletely understood (see Albarède & van der Hilst, 1999, for a recent overview); an important reason for this is that the depth-dependence of important parameters remains unclear (e.g. Bunge *et al.*, 1996). Given this uncertainty concerning the modern Earth, it is hardly surprising that the nature of early Precambrian mantle

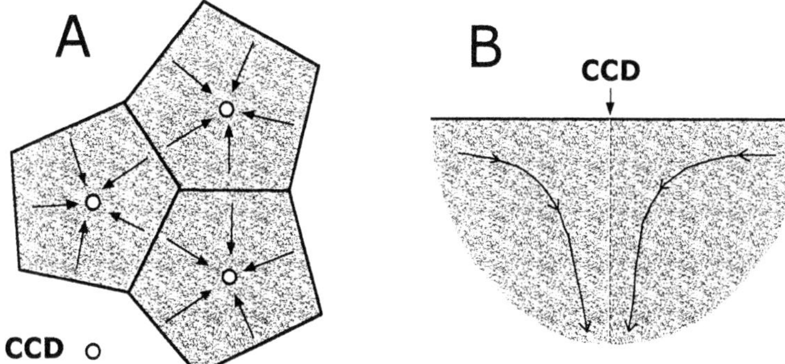


**Fig. 8.** Centres of convective descent (CCDs) in plan and cross-section at *c*.4.3 Ga.

convection is model-dependent, although there are significant constraints from chemical and isotopic data (Allègre, 1997; McCulloch & Bennett, 1998).

The whole-mantle large-cell convective system envisaged is shown diagrammatically in Fig. 8; globally, the pattern of polygons would have resembled that of a football. It is supposed that material rose at the sides of the polygons and moved inwards towards their centres, which are called centres of convective descent (CCDs). There appears to be no published analysis of the convection pattern to be expected in an Earth-sized, cooling, molten, rotating sphere with an internal heat source, due to radioactive decay. It is therefore possible that the main upward flow would have been at the centres of the polygons, and that flow would then have been outwards to descending polygon sides. In that case, equivalent CCDs would have been formed by the faster downflow at the triple-junctions of the polygons. Downward flow at the centres is here preferred, not only because it leads to a more realistic number of CCDs, but because it provides for ultimate evolution of the polygon sides into the present mid-ocean ridges.

The possible evolution of CCDs between 4.3 and 4.0 Ga is shown in Fig. 9. Cooling would initially have been rapid, so that a crust of devolatilized mantle composition would have begun to form, always thinnest at the 'ocean' ridge analogues (there were no oceans then) and thickening towards the CCDs. McKenzie & Weiss (1975) note that convective velocities in the early Precambrian must have been more than an order of magnitude greater because of the greater rate of radioactive heat generation. Thus, all ultramafic crust so formed was carried downwards at the CCDs, remelted and recirculated.

As cooling continued, and convection became slower, the cold ultramafic crust within the CCDs

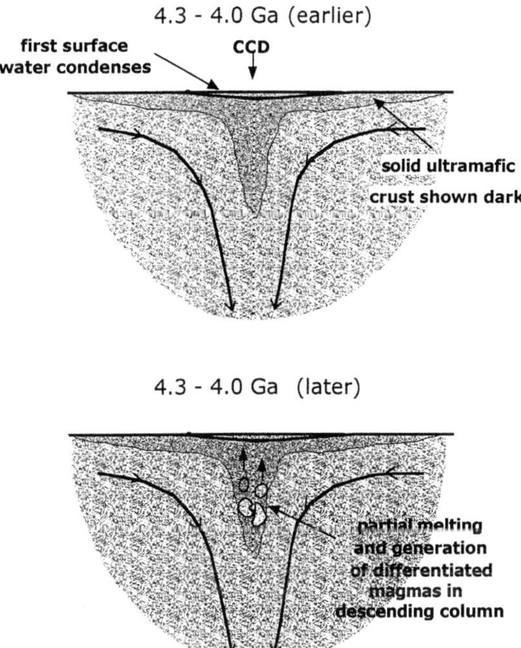


**Fig. 9.** The evolution of CCDs in the interval *c*.4.3–4.0 Ga.

would have undergone partial melting as it reheated during descent. This would have resulted first in the generation of basaltic magmas by partial melting, and later of silicic (granitoid) differentiates, whose buoyancy would have caused them to rise diapirically in the descending column. However, the rate of descent was initially much faster than the diapiric rise, so that all these differentiates were initially dragged back down and recirculated to the deep mantle.

However, in the latter part of this period (Fig. 9, lower diagram) substantial plugs of differentiates would have accumulated at the top of the CCDs, where they would be subject to a smaller downward pull than those lower in the column. Conditions in these plugs are speculative, and have not been modelled by any published theory for the origin of continents, since their formation over convective sinks is the converse of most popular models of early magma generation. For example, Kroner & Layer (1992, Fig. 3) illustrate early sial generation over the rising currents of convection cells; the model suggested here is closest to that of Davies (1992, 1993) The expectation seems reasonable that such plugs would have undergone intense tectonism, and that recycling and remelting of earlier differentiates would have been commonplace. I speculate that the earliest zircons so far isolated from Archaean terranes originated in such an environment. The >4.0 Ga zircons described by Nelson *et al.* (2000) from the Yilgarn Craton of Western Australia show repeated high-grade thermal events between 4186 and 2945 Ma, which may be the result of recycling of this kind.

The first surface water probably condensed in the course of this period, possibly to form separate pools over each CCD, which may initially have been slightly depressed. As described below, the rectilinear system of polygon sides in the model described here, at which mantle-composition material rose and spread, is regarded as the precursor of the later mid-ocean ridges; in that context it is worth noting that de Wit & Hynes (1995) suggest that mid-oceanic ridges were above sea-level on the 'Hadean Earth'.

*The c.4.0 Ga event*

Continued cooling would have led to an exponential slowing of the rate of mantle convection, and of the descent rate at CCDs. However, the rate of rise of differentiates within them is likely to have decreased more slowly, since this would have been controlled by Stokes's Law, and would have been dependent throughout on the density difference between the wall rock and the rising magma.

Under Stokes's Law the rate of rise would also have been inversely dependent on the viscosity of the surrounding rock, but this cannot be meaningfully estimated for the speculative model described. For this reason the *y*-axis of Fig. 10, which provides a diagrammatic representation of the cross-over concept, is unquantified. At the cross-over point, which is speculated to be at *c*.4.0 Ga, the rise of the lightest (granitoid) products of the petrogenetic processes taking place below the CCDs was in exact balance with their convective descent, so that the deep complex

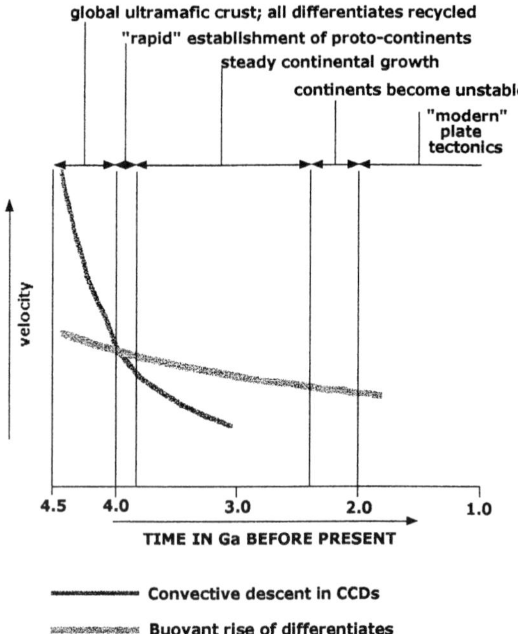


**Fig. 10.** Diagrammatic relationship between rate of descent in CCDs and buoyant rise of differentiated magmas, illustrating the proposed 4.0 Ga cross-over point.

plugs of mixed sialic material, which, as already noted, probably began to develop late in the Earth's pre-geological history, were steadily enlarged from below.

*The Archaean (*c.*4.0 to* c.*2.5 Ga)*

Immediately following the cross-over event, the sialic plugs within CCDs rose isostatically to become the 'ancient gneiss complexes' of the earliest continents (Fig. 11, upper diagram). The term protocontinent is used in that figure to indicate that these initial centres were very different from the continents of the modern Earth. Not only were they very small, but they were also relatively hot and ductile, more prone to respond to stress by flow than by fracture. These sialic islands, now with strong positive topography, would have displaced the pools of surface water over the CCDs to form encircling moats which later extended and coalesced to form the global ocean. Chemical weathering in the $CO_2$-rich atmosphere would have been intense, rapidly breaking down feldspar and ferromagnesian minerals; this in turn assisted equally rapid mechanical erosion, so that the earliest continents were a potential source of quartz-rich sediments, incidentally likely to contain zircons with >4.0 Ga ages. As noted earlier, concepts of the Precambrian Earth as relatively

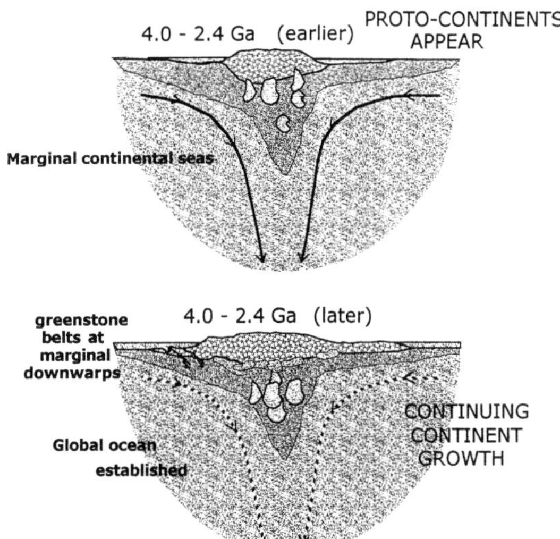


**Fig. 11.** Evolution of CCDs after the 4.0 Ga cross-over point.

lifeless and inhospitable now appear less secure, and biological activity may have contributed to Archaean weathering in much the same way as on the modern Earth.

At the same time, continuing mantle convection below the protocontinents (see Davies, 1993) would have established a process similar in many respects to modern subduction. Partial melting below the protocontinents would have continued to generate a range of magmas which added to their volume, both basally and then moving outwards towards the margins. Whether silicic magma could have been derived directly from mantle material remains a contentious issue (e.g. Evans & Hanson, 1997; Wyllie *et al.*, 1997; and references in both papers). Martin (1994) has presented detailed argument that the distinctive composition of Archaean TTGs could not have been achieved directly by partial melting of the mantle. For present purposes it is immaterial whether the early TTGs were derived directly from partial melting of descending peridotite, from differentiation of basaltic magmas themselves so derived or from the secondary partial melting of mafic rocks during recycling of earlier plug material. Once generated, the rise of silicic magmas would have taken place by processes essentially mirrored on the modern Earth (Clemens, 1998).

In the course of the Archaean, and probably gradationally rather than at a specific time, adiabatic partial melting of rising mantle at the boundaries of the convecting cells began to take place; and by the end of the period the compositional and physical structure of those boundaries resembled those of modern mid-ocean ridges, and the oceans would have been floored by mafic, rather than ultramafic, material.

The protocontinents would have grown steadily throughout Archaean time, effectively evolving into continents. The mechanisms by which this growth took place are speculative. Very early in their growth, the protocontinents would probably have risen so fast that gravitation collapse and spread (e.g. Sandiford, 1989) may have played an important role, and the full tectonic implications of this process have yet to be analysed fully. In the later Archaean (Fig. 11, lower diagram) conditions would not have been radically different from the 'plate tectonic' concepts of Kröner (1985). However, there is now a swing away from an earlier tendency to force the early Precambrian Earth into a plate tectonic mould (e.g. Windley, 1993), and mechanisms of the kind suggested by Bailey (1999) may well have played an important part. Whatever the detail of the processes contributing to the growth of continents during the Archaean they are likely to have become steadily thicker and more extensive. Throughout the period, continuing but slowing convection variously buckled the early continental margins, or caused the the formation of inward underthrusts (Fig. 11, lower), to form relatively small, often crudely linear, basins, variously along or just within protocontinent margins. These near-marginal troughs were the depositories of the greenstone belt successions.

Despite the recent appearance of a major monograph on greenstone belt geology (de Wit & Ashwal, 1997a) it is clear both from the thoughtful preface to the volume (de Wit & Ashwal, 1997b) and from many of the contributions within it that there is a long way to go before their structure, stratigraphy, sedimentology, chronology and genesis are fully understood. Greenstone belts are not the subject of this paper, and I am concerned here only to make sufficient general suppositions about their formation to suggest a reason for the common presence of BIF within them.

My own perceptions of greenstone belt successions is that they are highly variable, so that generalization is dangerous. However that may be, a common pattern is the early eruption of thick mafic and/or ultramafic volcanics, followed by mixed sedimentary rocks often with BIF, and capped by thick epiclastic, often arenaceous, sequences; felsic volcanics seem to occur irregularly throughout; sequences may be disproportionately thick in relation to the widths of belts. In noting this point, Lowe (1980) concludes that 'the initial volcanic history of greenstone belts cannot be

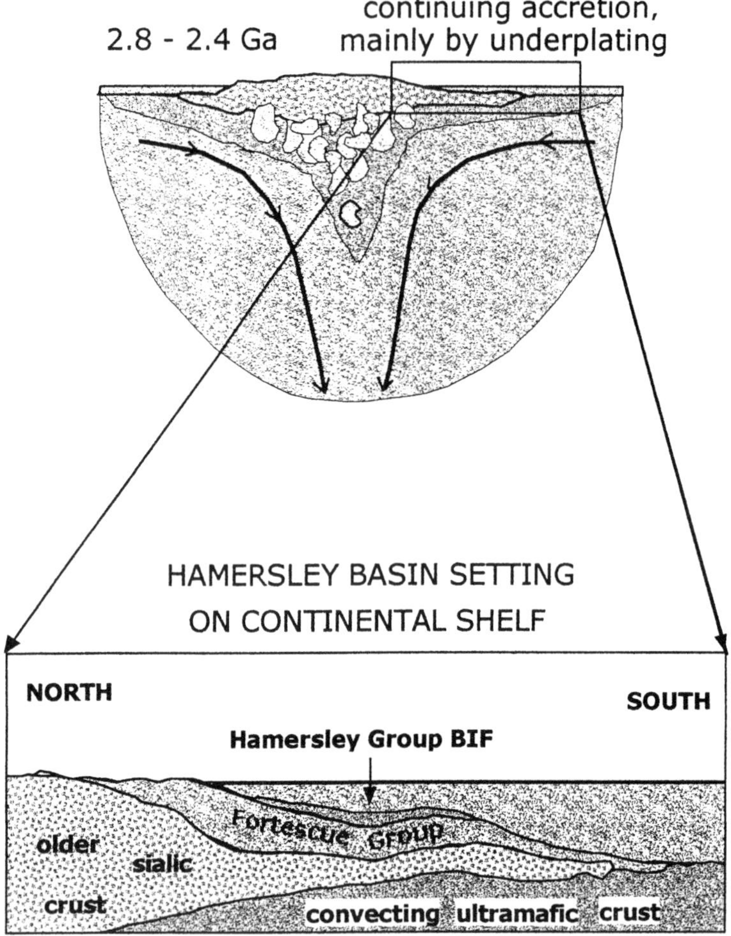


**Fig. 12.** Continuing evolution of CCDs, to *c*.2.4 Ga.

interpreted in terms of orogenic events'. The generalizations put forward by Groves *et al.* (1978) also seem to me to have stood the test of time well, and are consistent with the development of greenstone belts either at or slightly within, as well as parallel with, the margins of protocontinents (Fig. 11). In the context of the necessary and sufficent conditions for BIF deposition already discussed, a very simple model follows from these general points.

The protocontinents must have been in a mechanically critical state from their earliest inception. Initially very hot and ductile, they quickly became, at least locally, sufficiently brittle during greenstone belt development to allow passage of mafic and felsic magma through fractures; this occurred at least >3.7 Ga at Isua. Because greenstone basin floors were ductile (Choukrune *et al.*, 1995) they could not support the weight of significant mafic volcanism, so that they sank quickly, to increase water depth in the basin. I postulate, without a clear understanding of the mechanism, that this process led to a sufficient pause in the further evolution of the depository for the resultant BIF to reach its characteristic thickness.

In the later Archaean the same sequence of basin-forming events was continually repeated, but BIF deposition was enhanced because of the evolving mechanical properties of the continents. They had grown even more extensive, largely by continuing underplating from the initial CCDs, but also possibly by 'modern' marginal subduction melting and petrogenesis. As sialic growth and stability increased, the continental crust became increasingly brittle, so that dyking was periodically capable of releasing increasingly large volumes of mafic magma to the surface. Its increasing thickness and strength also meant that response to mafic volcanism was slower. But where the bulk of volcanics was sufficiently great, especially near continental margins, there were the same consequences: the accumulated flood basalts led to crustal depression, and the creation of deep marginal seas where BIF could accumulate. Their stabilizing effect was also greater, so that these depositories—those of the Great Gondwana BIFs—had sufficient stability for large BIF thicknesses to accumulate undisturbed. A very simple cross-section showing the continental margin structure during deposition of a Great Gondwana BIF, as exemplified by the Hamersley Basin, shows the equally simple position at which our arguments have arrived.

Although the late peak of BIF deposition (Fig. 6) may suggest some well defined 'event' in Earth history it is a corollary of the model outlined above that no fundamental difference exists between greenstone belt depositories and presently well preserved basins of the Hamersley type, as far as BIF deposition is concerned. The latter are simply potential greenstone belts which escaped the intense deformation experience by their generally earlier cousins. This point was emphasized by Arndt *et al.* (1991) from a geochronological viewpoint.

### *Post-Archaean developments*

The address on which this paper is based dealt with later continental evolution in only a cursory way, and this paper consequently follows that format. In summary, it is suggested that the steady areal growth and thickening of the first continental masses reached a point where they became mechanically unstable, and were liable to break-up and dispersal. The slowing of the large-cell whole-mantle convection pattern, as a result of effective cooling, probably led ultimately to a change from single-layer to two-layer convection. This dual event—initial continental break-up and dispersal, and the change in mantle convection structure—probably underlies the 'Archaean–Proterozoic' transition, whose existence many geologists have perceived and debated without clear resolution. The relatively shallow shelf seas on which the GIFs formed were still able to tap the reservoir of iron-rich water remaining in the deep oceans. But as the oceans became thoroughly oxygenated, and a blanket of abyssal sediments covered the earlier source of iron, BIF deposition essentially ceased, and iron-rich sedimentary rocks were deposited only in special, and often local, circumstances.

## Some paradoxes resolved?

The model of continental development summarized above, although presented here to account for the distribution of BIF in the Precambrian stratigraphic record, also fits comfortably with a number of issues in early Precambrian geology which appear paradoxical if a modern plate-tectonic model is used to interpret early Earth development. Some of these are identified and discussed below, but in brief outline only since they are tangential to the main subject of the paper.

### *Greenstones over sialic crust*

Arndt (1999) has recently drawn attention to the abundant evidence that the mafic lavas of greenstone belts were erupted through, or on, a basement of continental crust; and he has noted that this points to their emplacement 'in a tectonic setting that was rare or absent throughout the Phanerozoic' (*ibid.*, p. 155). The formation of greenstone belts over continental crust is a requirement of the model described in this paper, and should not be a matter for puzzlement. In this context Burke (1997, p. vi), in a Foreword commenting on a major problem of Archaean greenstone belts, suggests that because thick accumulations of subaqueous basalt are 'in later Earth history confined to the oceans and their contained arcs, plateaux, and oceanic islands', it is improbable that those of greenstone belts were laid down over older continental crust; and he adds that if they were indeed deposited in such an environment the 'world would have had to be astonishingly different in the Archaean'. I emphasize in the conclusions below that it was.

### *The global distribution of ancient crustal rocks*

Many papers on early Earth evolution and crustal growth include global maps showing the distribution of patches of 'older' rocks within the Precambrian cratons, using varying cut-off values to define 'older'. Developments in isotope analytical techniques, recognition of the unique role of zircon in recording early Earth events and the increasing output of more geochronological laboratories have all combined to raise the quality of such maps, but the validity of some published even as little as five years ago (e.g. Rogers, 1996) has been overtaken by a flood of new data. Although there is probably at least another decade to go before a reliable map showing the global distribution of all areas of Precambrian crust older than, say, 3.0 Ga becomes available, the present indication is that only a few exist (e.g. Kaapvaal, Zimbabwe (Horstwood *et al.*, 1999, confirm this status), Pilbara,

North-west Yilgarn, Greenland, Slave, West Africa, North China, Karelia–Kola, Karnataka, Amazon and the recent addition of North-west Superior: Böhm *et al.*, 2000). No general hypothesis for the small number and scattered distribution of these areas exists in the literature, and it is suggested here that the number of such areas is directly related to the number of CCDs in the pre-geological convection pattern of the Earth, of which they remain as witnesses. This is in contrast to a common current view of gradual aggregation of intra-oceanically formed differentiated rocks, variously termed crustal fragments or terranes, into continents by a more or less random process of amalgamation (e.g. de Wit *et al.*, 1992; Myers, 1995).

Relating these ancient cratonic cores to large convection cells also predicts that there was substantial angular separation between neighbouring cratons through their early history. This contrasts with hypotheses proposing contiguity of individual cratons on the grounds of their similar early development; palaeomagnetic and geochronological data should ultimately provide a secure test between these two mutually exclusive expectations. An example is provided by the Vaalbara hypothesis of Cheney *et al.* (1988), which proposes contiguity of the Pilbara and Kaapvaal cratons on the basis of sequence stratigraphic reconstruction (Cheney, 1996). Zegers *et al.* (1998) have used (*inter alia*) selected palaeomagnetic results to support early contiguous development of these two cratons. On the other hand, Wingate's (1998) interpretation of new palaeomagnetic evidence requires the two cratons to be widely separated at 2.78 Ga. More evidence is clearly needed for a definitive conclusion; meanwhile, care should be taken to avoid interpreting closely similar geological history as a necessary indication of contiguity (e.g. Nelson *et al.*, 1999).

*Cool Archaean cratons*

It has been common knowledge for decades among geologists mapping Archaean greenstone sequences that metamorphic grades appear similar to those of later orogens, and that the rocks are not obviously the products of processes on an Earth which must have been substantially hotter (Pollack, 1997). Sclater *et al.* (1980) showed that heat flow in Precambrian terranes decreases with tectonic age. Nyblade & Pollack (1993) later extended that analysis to show specifically that Archaean cratons have anomalously low heat flows; they attributed this to the existence of thick cratonic lithosphere below them. Low Archaean heat flow has never been satisfactorily incorporated into mainstream hypotheses of continental growth, but the existence of deep, and relatively cool, roots below older Precambrian areas is consistent with their formation over CCDs.

*Deep craton roots*

Polet & Anderson (1995) have recently reviewed and analysed the available seismic evidence for the depth extent of cratons, and have found that older cratons have high-velocity anomalies to depths up to 450 km. They conclude that 'The absence of a consistent high-velocity root beneath younger crust probably means that there was some process operating in early Earth history that has not been duplicated since.' This process is suggested to be the formation and activity of CCDs, whose remnants continue to define craton roots.

*Non-existent primordial crust, and the Late Heavy Bombardment (LHB)*

The concept of a 'primitive' or 'primordial' global sialic crust was popular not long ago (Hargraves, 1976; Lowman, 1989), though not among geologists with extensive experience of Archaean terranes; it arose largely from lunar models (O'Hara, 1970). The failure to find remnants of such material has been attributed to destruction by the LHB (Grieve, 1980). In the model described here the the bulk of the LHB would have been absorbed by the magma ocean (or its quenched crust) prior to the cross-over point that immediately preceded the appearance of sialic crust at the surface of the Earth.

*The abrupt start of the geological record at* c.*4.0 Ga*

The final paradox of early crustal development is the problem of why the geological record starts at *c.*4.0 Ga. The model for the origin and early development of continents proposed here predicts that no material older than the remnants of the first differentiated plugs above CCDs will ever be recovered from continental rocks. It predicts further that detrital, or otherwise isolated, zircons are likely to be found in rocks of all the oldest continental nucleii.

## QUANTITATIVE QUESTIONS

The model so far described for BIF distribution is a simple one. It envisages that all early Precambrian BIF was precipitated, through some biogenic agency, out of deep ocean water with dissolved ferrous iron, in basins which became tectonically overdeepened, and stabilized for long periods; and it envisages also that the principal control over the evidently irregular distribution of BIF abundance in time is due to tectonic changes associated with a particular model of early continental evolution. But there are still some important questions about BIF deposition that have not yet been directly addressed here: where did the iron come from, and how was it brought into the basin? Those questions, and another, are discussed below in the context of the simple model so far presented, and using a quantitative approach.

### The 'many oceans' problem

The 'many oceans' problem, as put to me by a number of people in discussion of the Hamersley Basin, can be stated simply as follows: 'There is far too much iron in the Hamersley BIFs to have come simply from ocean water. Even if there were 10 ppm Fe in the oceans at *c.*2.8–2.4 Ga it would have taken all the iron in the global oceans (*c.*$1.2 \times 10^{13}$ t) simply to make the Dales Gorge Member, and many oceans to make the Hamersley Group, which has eight times the iron of the Dales Gorge Member. It is inconceivable that the world's oceans can keep flowing into the Hamersley Basin and dumping their dissolved iron year after year.'

This problem is easily resolved by looking quantitatively at the annual deposition of microbands. The microbands-are-varves hypothesis yields an annual Fe precipitation rate of 225 t $km^{-2}$, so that, with a basin area of $10^5$ $km^2$, $22.5 \times 10^6$ t of iron are precipitated each year. The water volume of the Hamersley Basin was $5 \times 10^4$ $km^3$, assuming a mean depth of 0.5 km. If it is assumed that incoming water has 10 ppm Fe and that outgoing water has 7 ppm Fe, the annual water turnover (in equals out) would then have to be about $0.75 \times 10^4$ $km^3$, of the order of 15% of basin volume, to provide the necessary amount of iron. The resultant decrease in the ocean Fe concentration after mixing would be negligible: for example, after the first year the ocean concentration would fall from 10 to 9.99998 ppm.

Morris & Horwitz (1983) have assessed and tabulated published estimates for the possible iron content of early Precambrian oceans, varying from the 100–400 ppm of Mel'nik (1973) to the <1–5 ppm of Eugster and Chou (1973). The 10 ppm value adopted for the model above seems a reasonable one, and is half that suggested by Ewers (1980).

### The iron supply problem

A key point in the continental evolution model described in this paper is that it places the early oceans in direct contact with (cooled) ultramafic/mafic crust. If this was so, an annual increment of 61 kg Fe $km^{-2}$ (61 μg Fe $m^{-2}$ yr) would have to move (be leached) from the ocean floor into the ocean water to maintain a 10 ppm Fe concentration. This seems easily credible, but presupposes abandonment of the modern Earth model, and the associated need to get all the iron from mid-ocean ridge activity (e.g. Isley, 1995). Archaean ocean floors would not have had an abyssal sediment cover like those of the modern Earth, and the iron supply problem for BIF deposition simply does not exist in the model put forward in this paper. Direct contact with cooled mantle for the water from which early Precambrian BIFs were precipitated is consistent with the significant bulk of BIF data for both the Sm–Nd isotopic system and REE content (Fryer, 1983; Miller & O'Nions, 1985; Klein & Beukes, 1989; Derry & Jacobsen, 1990; Bau & Möller, 1991, 1993; Alibert *et al.*, 1991; Danielson *et al.*, 1992; Alibert & McCulloch, 1993; Manikyamba *et al.*, 1993; Arora *et al.*, 1995; Gerlach *et al.*, 1998). Such contact is also consistent with the early domination of ocean chemistry by volcanogenic sources to which Fryer *et al.* (1979) drew attention.

### Effectiveness of photosynthesis

For the Dales Gorge Member we know that $22.5 \times 10^{12}$ g of iron were deposited annually. That needs about $32 \times 10^{11}$ g of oxygen to be produced by photosynthesis, equivalent to about $1.2 \times 10^{10}$ kWh of energy a year. Using the roughly 2 kWh $m^{-2}$ of insolation energy effectively used annually on the present Earth (Govindjee & Shopes, 1992), and taking the basin area as $10^{11}$ $m^2$, about $2 \times 10^{11}$ kWh would be used if mean global photosynthesis productivity is supposed. Thus, in very rough terms, even an order of magnitude less photosynthetic effectivness would have been enough to precipitate the iron of the Hamersley Group BIFs.

An alternative way of looking at this is by using Holland's (1984) estimate of 0.27 g $m^{-2}$ $day^{-1}$ of C as the productivity for modern oceans. Then for the Dales Gorge Member, 22.5 mg $cm^{-2}$ $yr^{-1}$ of iron translates to 225 g $m^{-2}$ $yr^{-1}$, or about 0.6 g $m^{-2}$ $day^{-1}$. Now 223.4 g of Fe require 32 g of oxygen to precipitate, and 32 g of oxygen require the fixation of 12 g of carbon, so 0.6 g of Fe are equivalent to about 0.032 g $m^{-2}$ $day^{-1}$ of C, again much smaller than modern productivity.

Finally, suppose that each annual increment of iron laid down during deposition of the Dales Gorge

Member was done biochemically (photosynthetically) by organisms which, having done their job, died and sank with the iron. How much carbon would then be in the Dales Gorge Member BIF? We have already seen that the precipitation of four Fe atoms would require one C atom, so that 223.4 g of Fe require 12 g of C. A BIF with 30% Fe would then contain 1.6% carbon. The carbon content of the Dales Gorge Member is (as carbonate) about 1.6%.

## CONCLUSIONS

BIFs have often been described as bizarre or unusual rocks, and correspondingly exceptional conditions have been advanced to explain their presence in the stratigraphic record; but this stance is only taken because they have no compelling analogues on the present Earth. A correct uniformitarian approach to the interpretation of the past, using the present as a key, involves acceptance that if a sedimentary phenomenon essentially non-existent on the modern Earth is both common in, and characteristic of, the Precambrian stratigraphic record then it follows that the ordinary depositional environments of those times differed significantly from those of the present. With this approach, it should not be asked what *strange* circumstances led to the deposition of BIF, but instead in what respects were the *ordinary* environments of the Precambrian Earth radically different from those now existing.

The significance of BIFs in the Precambrian stratigraphic record is thus that they are sensitive indicators of major differences in early Earth history. First, the oceanic environment was very different, in that (*inter alia*) for much of the Precambrian it was stratified, with the deeper water, in direct contact with cooled convecting ultramafic/mafic rocks, relatively rich in dissolved ferrous iron. Second, whenever that deep water was able to circulate freely through a suitable depository, some biological mechanism facilitated the precipitation of iron as BIF; throughout the early Precambrian such circulation was usually achieved through the deepening and stabilizing of basins after volcanism. Third, basins with the requisite architecture developed uniquely as a corollary of early Precambrian crustal evolution. The deposition of BIF ceased when the global ocean had been flushed of its iron content, at which time the supply of iron was negligible.

## ACKNOWLEDGEMENTS

Thoughtful and constructive reviews by Cornelis Klein, Dietrich Klemm and Bruce Simonson are acknowledged with gratitude. David Nelson and Martin van Kranendonk offered helpful comments on the model presented for the origin of continents, christened by David 'the plug-hole hypothesis'. Finally, I thank the Organizing Committee of the 19th Regional European Meeting of the International Association of Sedimentologists for its invitation to present the keynote address on which the paper is based.

## REFERENCES

Albarède, F. & van der Hilst, R.D. (1999) New mantle convection model may reconcile conflicting evidence. *Eos*, **80**(45), 535–539.

Alibert, C., Kinsley, L. & McCulloch, M.T. (1991) *Rare-earth Elements and Nd Isotope Systematics in the Banded Iron-formations from the Hamersley Basin (Western Australia) and Implications for the Composition of Early Proterozoic Seawater*. Australian National University, Research School of Earth Sciences, Annual Report for 1990.

Alibert, C. & McCulloch, M.T. (1993) Rare earth element and neodymium isotopic compositions of the banded iron-formations and associated shales from Hamersley, Western Australia. *Geochim. Cosmochim. Acta*, **57**, 187–204.

Allaart, J.H. (1976) The pre-3760 m.y. old supracrustal rocks of the Isua area, central west Greenland, and the associated occurrence of quartz-banded ironstone. In: *The Early History of the Earth* (Ed. Windley, B.W.), pp. 177–189. Wiley-Interscience, London.

Allègre, C.J. (1997) Limitations on the mass exchange between the upper and lower mantle: the evolving convection regimes of the Earth. *Earth planet. Sci. Lett.*, **150** (1/2), 1–6.

Allègre, C.J., Manhès, G. & Göpel, C. (1995) The age of the Earth. *Geochim. Cosmochim. Acta*, **59**, 1445–1456.

Appel, P.W.U. (1980) On the early Archaean Isua iron-formation, west Greenland. *Precam. Res.*, **11**, 73–87.

Appel, P.W.U. & La Berge, G.L. (Eds) (1987) *Precambrian Iron-formations*. Theophrastus Publications, Athens.

Arndt, N.T. (1999) Why was flood volcanism on submerged continental platforms so common in the Precambrian? *Precam. Res.*, **97**, 155.

Arndt, N.T., Nelson, D.R., Compston, W., Trendall, A.F. & Thorne, A.M. (1991) The age of the Fortescue Group, Hamersley Basin, Western Australia, from ion microprobe zircon U–Pb results. *Aust. J. Earth Sci.*, **38**, 261–281.

Arnold, M., Alibert, C. & Jacquier, B. (1991) $\delta^{34}S$ in Hamersley BIF, W Australia: the role of bacterial activity and thiosulphates as an alternative to the hydrothermal model. *Terra Abstr.*, **3**, 195.

Arora, M., Govil, P.K., Charan, S.N. *et al.* (1995) Geochemistry and origin of Archean banded iron formation from the Bababudan Schist Belt, India. *Econ. Geol.*, **90**, 2040–2057.

BAILEY, R.C. (1999) Gravity-driven continental overflow and Archaean tectonics. *Nature*, **398**, 413–415.

BARRETT, T.J. & FRALICK, P.W. (1985) Sediment redeposition in Archean iron formation: examples from the Beardmore–Geraldton greenstone belt, Ontario. *J. sediment. Petrol.*, **55**(2), 205–212.

BARRETT, T.J. & FRALICK, P.W. (1989) Turbidites and iron formations, Beardmore–Geraldton, Ontario: application of a combined ramp fan model to Archean clastic and chemical sedimentation. *Sedimentology*, **36**, 221–234.

BAU, M. & MÖLLER, P. (1991) Rare-earth element of Archean iron-formations: physico-chemical controls on Europium mobilization and precipitation. *Terra Abstr.*, **3**, 195.

BAU, M. & MÖLLER, P. (1993) Rare-earth element systematics of the chemically precipitated component in early Precambrian iron-formations and the evolution of the terrestrial atmosphere-hydrosphere-lithosphere system. *Geochim. Cosmochim. Acta*, **57**, 2239–2249.

BAUR, M.E., HAYES, J.M., STUDLEY, S.A. & WALTER, M.R. (1985) Millimeter-scale variations of stable isotope abundances in carbonates from banded iron-formations in the Hamersley Group of Western Australia. *Econ. Geol.*, **80**, 270–282.

BEARD, B.L., JOHNSON, C.M., COX, L., SUN, H., NEALSON, K.H. & AGUILAR, C. (1999) Iron isotope biosignatures. *Science*, **285**, 1889–1892.

BECKER, R.H. & CLAYTON, R.N. (1972) Carbon isotopic evidence for the origin of a banded iron-formation in Western Australia. *Geochim. Cosmochim. Acta*, **36**, 577–595.

BEUKES, N.J. (1973) Precambrian iron-formations of southern Africa. *Econ. Geol.*, **68**, 960–1004.

BEUKES, N.J. (1980) Suggestions towards a classification and nomenclature for iron-formation. *Trans. geol. Soc. S. Afr.*, **83**, 285–290.

BEUKES, N.J. & KLEIN, C. (1990) Geochemistry and sedimentology of a facies transition—from microbanded to granular iron-formation—in the early Proterozoic Transvaal Supergroup, South Africa. *Precam. Res.*, **47**, 99–139.

BEUKES, N.J. & KLEIN, C. (1992) Models for iron-formation deposition. In: *The Proterozoic Biosphere* (Eds Schopf, J.W. & Klein, C.), pp. 147–151. Cambridge University Press, Cambridge.

BEUKES, N.J., KLEIN, C., KAUFMAN, A.J. & HAYES, J.M. (1990) Carbonate petrography, kerogen distribution, and carbon and oxygen isotope variations in an early Proterozoic transition from limestone to iron-formation deposition, Transvaal Supergroup, South Africa. *Econ. Geol.*, **85**, 663–690.

BÖHM, C.O., HEAMAN, L.M., CREASER, R.A. & CORKERY, M.T. (2000) Discovery of pre-3.5 Ga exotic crust at the northwestern Superior Province margin, Manitoba. *Geology*, **28**, 75–78.

BOWRING, S.A. & WILLIAMS, I.S. (1999) Priscoan (4.00–4.03 Ga) orthogneisses from northwestern Canada. *Contrib. Miner. Petrol.*, **134**, 3–16.

BOWRING, S.A., WILLIAMS, I.S. & COMPSTON, W. (1989) 3.96 Ga gneisses from the Slave Province. *Geology*, **17**, 307–310.

BRATERMAN, P.S. & CAIRNS-SMITH, A.G. (1987a) Photoprecipitation and the banded iron-formations—some quantitative aspects. *Orig. Life*, **17**, 221–228.

BRATERMAN, P.S. & CAIRNS-SMITH, A.G. (1987b) Iron photoprecipitation and the genesis of the banded iron-formations. In: *Precambrian Iron-formations* (Eds Appel, P.W.U. & LaBerge, G.L.), pp. 215–245. Theophrastus Publications, Athens.

BRATERMAN, P.S., CAIRNS-SMITH, A.G., SLOPER, R.W., TRUSCOTT, T.G. & CRAW, M. (1984) Photo-oxidation of iron(II) in water between pH 7.5 and 4.0. *J. chem. Soc., Dalton Trans.*, 1441–1445.

BROCKS, J.J., LOGAN, G.A., BUICK, R. & SUMMONS, R.E. (1999) Archean molecular fossils and the early rise of eukaryotes. *Science*, **285**, 1033–1036.

BUNGE, H.-P., RICHARDS, M.A. & BAUMGARDNER, J.R. (1996) Effect of depth-dependent viscosity on the planform of mantle convection. *Nature*, **379**, 436–438.

BUNTING, J.A. (1986) Geology of the eastern part of the Nabberu basin. *W. Aust. Geol. Survey Bull.*, **131**, 129 pp.

BURKE, K.D. (1997) Foreword. In: *Greenstone Belts* (Eds de Wit, M.J. & Ashwal, L.D.), pp. v–vii. Oxford Science Publications, Oxford.

BUTTON, A. (1982) Sedimentary iron deposits, evaporites and phosphorites—state of the art report. In: *Mineral Deposits and the Evolution of the Biosphere* (Ed. Holland, H.D. & Schidlowski, M.), pp. 259–273. Springer-Verlag, Berlin.

CAREY, S.W. (1976) *The Expanding Earth*. Elsevier, Amsterdam, 488 pp.

CAREY, S.W. (1996) *Earth Universe Cosmos*. University of Tasmania Press, Hobart, 231 pp.

CASTRO, L.O. (1994) Genesis of banded iron-formations. *Econ. Geol.*, **89**, 1354–1397.

CHAKRABORTY, K.L. & MAJUMDER, T. (1992) An unusual diagenetic structure in the Precambrian banded iron formation (BIF) of Orissa, India, and its interpretation. *Miner. Deposita*, **27**, 55–57.

CHENEY, E.S., ROERING, C. & STETTLER, E. (1988) Vaalbara. *Geocongress '88, 22nd Earth Sciences Congress of the Geological Society of South Africa, Extended Abstracts Volume*, pp. 85–88.

CHENEY, E.S. (1996) Sequence stratigraphy and plate tectonic significance of the Transvaal succession of southern Africa and its equivalent in Western Australia. *Precam. Res.*, **79**, 3–34.

CHOUKROUNE, P., BOUHALLIER, H. & ARNDT, N.T. (1995) Soft lithosphere during periods of Archaean crustal growth or crustal reworking. In: *Early Precambrian Processes* (Eds Coward, M.P. & Ries, A.C.) Spec. Publs geol. Soc. London, No. 95, pp. 67–86. Geol. Soc. London, Bath.

CISNE, J.L. (1984) A basin model for massive banded iron-formation and its geophysical applications. *J. Geol.*, **5**, 471–489.

CLEMENS, J.D. (1998) Observations on the origins and ascent mechanisms of granitic magmas. *J. geol. Soc.*, **155**, 843–851.

CLOUD, P. (1968) Atmospheric and hydrospheric evolution on the primitive Earth. *Science*, **160**, 729–736.

CLOUD, P.E. (1972) A working model of the primitive Earth. *Am. J. Sci.*, **272**, 537–548.

CLOUD, P.E. (1973) Paleoecological significance of banded iron-formation. *Econ. Geol.*, **68**, 1135–1143.

CLOUD, P.E. (1976) Major features of crustal evolution. *Alex L. Du Toit Memorial Lecture* No. 14. Geol. Soc. S. Afr., Annexure to No. 79, 1–33.

Cloud, P.E. (1988) *Oasis in Space*. W.W. Norton & Company, New York & London.

Collerson, K.D. & Kamber, B.S. (1999) Evolution of the continents and the atmosphere inferred from Th–U–Nb systematics of the depleted mantle. *Science*, **283**, 1519–1522.

Compston, W., Williams, I.S., McCulloch, M.T., Foster, J.J., Arriens, P.A. & Trendall, A.F. (1981) A revised age for the Hamersley Group. *Geol. Soc. Aust. Abstr.*, **3**, 40.

Dalgarno, C.R. & Johnson, J.E. (1965) The Holowilena ironstone, a Sturtian glacigene unit. *Q. Notes geol. Surv. S. Aust.*, **13**, 2–4.

Danielson, A., Möller, P. & Dulski, P. (1992) The europium anomalies in banded iron formations and the thermal history of the oceanic crust. *Chem. Geol.*, **97**, 89–100.

Davies, G.F. (1992) Plates and plumes: dynamos of the Earth's mantle. *Science*, **257**, 493–494.

Davies, G.F. (1993) Conjectures on the thermal and tectonic evolution of the Earth. *Lithos*, **30**, 281–289.

Davy, R. (1983) A contribution on the chemical composition of the Precambrian iron-formations. In: *Iron-formation: Facts and Problems* (Ed. Trendall, A.F. & Morris, R.C.), pp. 325–343. Elsevier, Amsterdam.

Davy, R. (1985) The mineralogy and composition of a core which intersects the Marra Mamba Iron Formation and the Roy Hill Shale Member. W. Aust. Geol. Surv., Record 1985/6.

Derry, L.A. & Jacobsen, S.B. (1990) The chemical evolution of Precambrian seawater: evidence from REEs in banded iron formations. *Geochim. Cosmochim. Acta*, **54**, 2965–2977.

de Wit, M.J. & Ashwal, L.D. (Eds) (1997a) *Greenstone Belts*. Oxford Science Publications, Oxford.

de Wit, M.J. & Ashwal, L.D. (1997b) Preface. Convergence towards divergent models of greenstone belts. In: *Greenstone Belts* (Eds de Wit, M.J. & Ashwal, L.D.), pp. x–xvi. Oxford Science Publications, Oxford.

de Wit, M.J. & Hynes, A. (1995) The onset of interaction between the hydrosphere and oceanic crust, and the origin of the first continental lithosphere. In: *Early Precambrian Processes* (Eds Coward, M.P. & Ries, A.C.), Spec. Publs geol. Soc. London, No. 95, pp. 1–9. Geol. Soc. London, Bath.

de Wit, M.J., Roering, C., Hart, R.J. *et al.* (1992) Formation of an Archaean continent. *Nature*, **357**, 553–562.

Dimroth, E. (1975) Paleo-environment of iron-rich sedimentary rocks. *Geol. Rundsch.*, **64**, 751–767.

Dimroth, E. (1976) Aspects of the sedimentary petrology of cherty iron-formation. In: *Handbook of Strata-bound and Stratiform Ore Deposits*, Vol. 7 (Ed. Wolf, K.H.), pp. 203–254. Elsevier, Amsterdam.

Dimroth, E. & Chauvel, J.-J. (1974) Petrography of the Sokoman iron formation in part of the central Labrador Trough. *Geol. Soc. Am. Bull.*, **84**, 111–134.

Dorr, J.V. (1973) Iron-formation in South America. *Econ. Geol.*, **68**, 1005–1022.

Draganic, I.G., Bjergbakke, E., Draganic, Z.D. & Sehested, K. (1991) Decomposition of ocean waters by potassium-40 radiation 3800 Ma ago as a source of oxygen and oxidizing species. *Precam. Res.*, **52**, 321–336.

Dunbar, G.J. & McCall, G.J.H. (1971) Archaean turbidites and banded ironstones of the Mt.Belches area (Western Australia). *Sediment. Geol.*, **5**, 93–133.

Dymek, R.F. & Klein, C. (1988) Chemistry, petrology and origin of banded iron-formation lithologies from the 3800 Ma Isua supercrustal belt, west Greenland. *Precam. Res.*, **39**, 247–302.

Eriksson, K.A. (1983) Siliciclastic-hosted iron-formations in the early Archaean Barberton and Pilbara sequences. *J. geol. Soc. Aust.*, **30**, 473–482.

Eriksson, K.A., Krapez, B. & Fralick, P.W. (1997) Sedimentological aspects. In: *Greenstone Belts* (Eds de Wit, M.J. & Ashwal, L.D.), pp. 291–293. Oxford Science Publications, Oxford.

Eugster, H.P. & Chou, I.-M. (1973) The depositional environments of Precambrian banded iron-formations. *Econ. Geol.*, **68**, 1144–1168.

Evans, O.C. & Hanson, G.N. (1997) Late-to post-kinematic Archaean granitoids of the S.W. Superior Province: derivation through direct mantle melting. In: *Greenstone Belts* (Eds de Wit, M.J. & Ashwal, L.D.), pp. 280–295. Oxford Science Publications, Oxford.

Ewers, W.E. (1980) Chemical conditions for the precipitation of banded iron-formation. In: *Biogeochemistry of Ancient and Modern Environments* (Eds Trudinger, P.A., Walter, M.R. & Ralph, B.J.), pp. 83–92. Australian Academy of Science, Canberra.

Ewers, W.E. (1983) Chemical factors in the deposition and diagenesis of banded iron-formation. In: *Iron-formation: Facts and Problems* (Eds Trendall, A.F. & Morris, R.C.), pp. 491–512. Elsevier, Amsterdam.

Ewers, W.E. & Morris, R.C. (1981) Studies on the Dales Gorge Member of the Brockman Iron Formation. *Econ. Geol.*, **76**, 1929–1953.

Francois, L.M. (1986) Extensive deposition of banded iron formations was possible without photosynthesis. *Nature*, **320**, 352–354.

Fryer, B.J. (1983) Rare earth elements in iron-formation. In: *Iron-formation: Facts and Problems* (Eds Trendall, A.F. & Morris, R.C.), pp. 345–358. Elsevier, Amsterdam.

Fryer, B.J., Fyfe, W.S. & Kerrich, R. (1979) Archean volcanogenic oceans. *Chem. Geol.*, **24**, 25–33.

Gair, J.E. (1975) Bedrock geology and ore deposits of the Palmer quadrangle, Marquette County, Michigan. Prof. Paper US Geol. Surv., **769**, 159 pp.

Gerlach, D.C., Shirey, S.B. & Carlson, R.W. (1998) Nd isotopes in Proterozoic iron formations: evidence for mixed-age provenance and depositional variability. *Eos*, **69**, 1615.

Goldich, S.S. (1973) Ages of Precambrian iron-formations. *Econ. Geol.*, **68**, 1126–1135.

Gole, M. (1981) Archean banded iron-formations, Yilgarn Block, Western Australia. *Econ. Geol.*, **76**, 1954–1974.

Gole, M. & Klein, C. (1981) Banded iron-formations through much of Precambrian time. *J. Geol.*, **89**, 169–183.

Goodwin, A.M., Monster, J. & Thode, H.G. (1976) Carbon and sulfur isotope abundances in Archean iron-formations and Early Precambrian life. *Econ. Geol.*, **71**, 879–891.

Govindjee & Shopes, R.J. (1992) Photosynthesis. In: *McGraw-Hill Encyclopedia of Science and Technology*, Vol. 13, 7th edn, pp. 455–468. McGraw-Hill, New York.

Grieve, R.A.F. (1980) Impact bombardment and its role in proto-continental growth on the early Earth. *Precam. Res.*, **10**, 217–247.

Gross, G.A. (1972) Primary features in cherty iron-formations. *Sediment. Geol.*, **7**, 241–261.

Gross, G.A. (1980) A classification of iron formations based on depositional environments. *Can. Miner.*, **18**, 215–222.

Groves, D.I., Archibald, N.J., Bettenay, L.F. & Binns, R.A. (1978) Greenstone belts as ancient marginal basins or ensialic rift zones. *Nature*, **273**, 460–461.

Han, Tsu-Ming & Runnegar, B. (1992) Megascopic eukaryotic algae from the 2.1 billion-year-old Negaunee iron-formation, Michigan. *Science*, **257**, 232–235.

Hargraves, R.B. (1976) Precambrian geologic history. *Science*, **193**, 363–371.

Harland, W.B. (1964) Critical evidence for a great infra-Cambrian glaciation. *Geol. Rundsch.*, **54**, 45–61.

Heaney, P.J. & Veblen, D.R. (1991) An examination of spherulitic dubiomicrofossils in Precambrian banded iron formations using the transmission electron microscope. *Precam. Res.*, **49**, 355–372.

Hoffman, P.F., Kaufman, A.J., Halverson, G.P. & Schrag, D.P. (1998) A Neoproterozoic Snowball Earth. *Science*, **281**, 1342–1346.

Holland, H.D. (1973) The oceans: a possible source of iron in iron-formations. *Econ. Geol.*, **68**, 1169–1172.

Holland, H.D. (1984) *The Chemical Evolution of the Atmosphere and the Oceans*. Princeton University Press, Princeton, NJ, 582 pp.

Holland, H.D. (1999) When did the Earth's atmosphere become oxic? A reply. *Geochem. News*, **100**, 20–22.

Holland, H.D. & Beukes, N.J. (1990) A paleoweathering profile from Griqualand West, South Africa: evidence for a dramatic rise in atmospheric oxygen between 2.2 and 1.9 BY BP. *Am. J. Sci.*, **290A**, 1–34.

Horstwood, M.S.A., Nesbitt, R.W., Noble, S.R. & Wilson, J.F. (1999) U–Pb zircon evidence for an extensive early Archean craton in Zimbabwe: a reassessment of the timing of crayton formation, stabilization, and growth. *Geology*, **27**, 707–710.

Isley, A.E. (1995) Hydrothermal plumes and delivery of iron to banded iron formations. *J. Geol.*, **103**, 169–185.

Isley, A. E. & Abbott, D. H. (1999) Plume-related mafic volcanism and the deposition of banded iron formation. *J. geophys. Res. B*, **104**, 15461–15477.

James, H.L. (1954) Sedimentary facies of iron-formation. *Econ. Geol.*, **49**, 235–293.

James, H.L. (1983) Distribution of banded iron-formation in space and time. In: *Iron-formation: Facts and Problems* (Eds Trendall, A.F. & Morris, R.C.), pp. 471–490. Elsevier, Amsterdam.

James, H.L. & Sims, P.K. (Eds) (1973) Precambrian iron-formations of the world. *Econ. Geol.*, **68**, 913–1179.

James, H.L. & Trendall, A.F. (1982) Banded iron formation: distribution in time and paleoenvironmental significance. In: *Mineral Deposits and the Evolution of the Biosphere* (Eds Holland, H.D. & Schidlowski, M.), pp. 199–218. Springer-Verlag, New York.

Johnson, H.D. & Baldwin, C.T. (1996) Shallow clastic seas. In: *Sedimentary Environments: Processes, Facies and Stratigraphy* (Ed. Reading, H.G.), pp. 232–280. Blackwell Science, Oxford.

Kalugin, A.S. (1969) Formation of the volcanogenic-sedimentary banded iron ores in the Devonian deposits of the Altai. In *Problemy obrazovaniya zhelezistych porod dokembriya* (*Problems of Formation of the Precambrian Iron Formations*), pp. 89–104. Izd. Naukova Dumka, Kiev (in Russian).

Kalugin, A.S. (1973) Geology and genesis of the Devonian banded iron-formation in Altai, western Siberia and eastern Kazakhstan. In: *Genesis of Precambrian Iron and Manganese Deposits*. Unesco, Paris, Earth Sciences, **9**, 159–165.

Kimberley, M.M. (1974) Origin of iron ore by diagenetic replacement of calcareous oolite. *Nature*, **250**, 319–320.

Kimberley, M.M. (1978) Paleoenvironmental classification of iron formations. *Econ. Geol.*, **73**, 215–229.

Kimberley, M.M. (1990) Nomenclature for iron formations. *Ore Geol. Rev.*, **5**, 1–12.

Kirschvink, J.L. (1992) Late Proterozoic low-latitude global glaciation: the Snowball Earth. In: *The Proterozoic Biosphere* (Eds Schopf, J.W. & Klein, C.), pp. 51–52. Cambridge University Press, Cambridge.

Klein, C. & Beukes, N.J. (1989) Geochemistry and sedimentology of a facies transition from limestone to iron-formation deposition in the Early Proterozoic Transvaal Supergroup, South Africa. *Econ. Geol.*, **84**, 1733–1774.

Klein, C. & Beukes, N.J. (1992) Time distribution, stratigraphy, and sedimentologic setting, and geochemistry of Precambrian iron-formations. In: *The Proterozoic Biosphere* (Eds Schopf, J.W. & Klein, C.), pp. 139–146. Cambridge University Press, Cambridge.

Klein, C. & Beukes, N.J. (1993) Sedimentology and geochemistry of the glaciogenic late Proterozoic Rapitan iron-formation, in Canada. *Econ. Geol.*, **80**, 542–565.

Klein, C., Beukes, N.J., Holland, H.D., Kasting, J.F. & Lowe, D.R. (1992) Conclusions and unsolved problems. In: *The Proterozoic Biosphere* (Eds Schopf, J.W. & Klein, C.), pp. 173–174. Cambridge University Press, Cambridge.

Klemm, D.D. (2000) The formation of Palaeoproterozoic banded iron formations and their associated Fe and Mn deposits, with reference to the Griqualand West deposits, South Africa. *J. Afr. Earth Sci.*, **30**, 1–24.

Kocurek, G.A. (1996) Desert aeolian systems. In: *Sedimentary Environments and Fazies*, 3rd edn (Ed. Reading, H.G.), pp. 125–153. Blackwell Science, Oxford.

Konhauser, K.O. & Ferris, F.G. (1996) Diversity of iron and silica precipitation by microbial mats in hydrothermal waters, Iceland: implications for Precambrian iron formations. *Geology*, **24**, 323–326.

Kröner, A. (1985) Evolution of the Archean continental crust. *Ann. Rev. Earth planet. Sci.*, **13**, 49–74.

Kröner, A. & Layer, P.W. (1992) Crust formation and plate motion in the Early Archean. *Science*, **256**, 1405–1411.

Kump, L. (1993) Bacteria forge a new link. *Nature*, **362**, 790–791.

Kusky, T.M. & Vearncombe, J.R. (1997) Structural aspects. In: *Greenstone Belts* (Eds de Wit, M.J. & Ashwal, L.D.), pp. 91–123. Oxford Science Publications, Oxford.

La Berge, G.L. (1966) Altered pyroclastic rocks in iron-formations in the Hamersley Range, Western Australia. *Econ. Geol.*, **61**, 147–161.

La Berge, G.L. (1973) Possible biological origin of Precambrian iron-formations. *Econ. Geol.*, **68**, 1098–1109.

Lowe, D.R. (1980) Archean sedimentation. *Ann. Rev. Earth planet. Sci.*, **8**, 145–167.

Lowman, P.D. (1989) Comparative planetology and the origin of continental crust. *Precam. Res.*, **44**, 171–195.

McCulloch, M.T. & Bennett, V.C. (1998) Early differentiation of the Earth: an isotopic perspective. In: *The Earth's Mantle* (Ed. Jackson, I.), pp. 127–138. Cambridge University Press, Cambridge.

McKenzie, D. & Weiss, N. (1975) Speculations on the thermal and tectonic history of the Earth. *Geophys. J. R. astron. Soc.*, **42**, 131–174.

Majumder, T. & Chakraborty, K.L. (1977) Primary sedimentary structures in the banded iron-formation of Orissa, India. *Sediment. Geol.*, **19**, 287–300.

Mancuso, J.J., Lougheed, S.S. & Wygant, T. (1971) Possible biogenic structures from the Precambrian Negaunee (iron) formation, Marquette Range, Michigan. *Am. J. Sci.*, **271**, 181–186.

Manikyamba, C., Balaram, V. & Naqvi, S.M. (1993) Geochemical signatures of polygenetic origin of a banded iron formation (BIF) of the Archaean Sandur greenstone belt (schist belt) Karnataka nucleus, India. *Precam. Res.*, **61**, 137–164.

Martin, H. (1965) *The Precambrian Geology of South West Africa and Namaqualand.* University of Cape Town Precambrian Research Unit, 159 pp.

Martin, H. (1994) The Archean grey gneisses and the genesis of continental crust. In: *Archaean Crustal Evolution, Developments in Precambrian Geology*, Vol. 11 (Ed. Condie, K.C.), pp. 259–295. Elsevier, Amsterdam.

Matin, A. & Mukhopadhyay, D. (1992) Primary sedimentary structures in banded ferruginous quartzite of Sandur Schist Belt, Karnataka. *J. geol. Soc. India*, **40**, 403–413.

Mel'nik, Y.P. (1973) On the origin of the Precambrian ferruginous quartzites. *Geol. Zhurn. Akad. Nauk Ukr. SSR* (*Geol. J. Acad. Sci. Ukr. SSR*), **33**, 3–16 (in Russian).

Mel'nik, Y.P. (1982) *Precambrian Banded Iron-formations.* Elsevier, Amsterdam, 310 pp.

Mengel, J.T. (1965) Precambrian taconite iron formation: a special type of sandstone. *Geological Society of America, Annual Meeting, Kansas City, MO, Abstracts*, p. 106.

Miller, R.G. & O'Nions, R.K. (1985) Source of Precambrian and chemical clastic sediments. *Nature*, **314**, 325–330.

Mojzsis, S.J., Arrhenius, G., McKeegan, K.D., Harrison, T.M., Nutman, A.P. & Friend, C.R.L. (1996) Evidence for life on Earth before 3,800 million years ago. *Nature*, **384**, 55–59.

Morey, G.B. (1983) Animikie Basin, Lake Superior Region, USA. In: *Iron-formation: Facts and Problems* (Eds Trendall, A.F. & Morris, R.C.), pp. 13–68. Elsevier, Amsterdam.

Morey, G.B. & Southwick, D.L. (1995) Allostratigrphic relationships of Early Proterozoic iron-formations in the Lake Superior region. *Econ. Geol.*, **90**, 1983–1993.

Morris, R.C. (1993) Genetic modelling for banded iron-formation of the Hamersley Group, Pilbara Craton, Western Australia. *Precam. Res.*, **60**, 243–286.

Morris, R.C. & Horwitz, R.C. (1983) The origin of the iron-formation-rich Hanersley Group of Western Australia—deposition on a platform. *Precam. Res.*, **21**, 273–297.

Morris, R.C. & Trendall, A.F. (1988) Discussion: a model for the deposition of the microbanded Precambrian iron formations. *Am. J. Sci.*, **288**, 664–669.

Myers, J.S. (1995) The generation and assembly of an Archaean supercontinent: evidence from the Yilgarn craton, Western Australia. In: *Early Precambrian Processes* (Eds Coward, M.P. & Ries, A.C.), Spec. Publs geol. Soc. London, No. 95, pp. 143–154. Geol. Soc. London, Bath.

Nealson, K.H. & Myers, C.R. (1999) Iron reduction by bacteria: a potential role in the genesis of banded iron formations. *Am. J. Sci.*, **290A**, 35–45.

Nelson, D.R., Robinson, B.W. & Myers, J.S. (2000) Complex geological histories extending for >4.0 Ga deciphered from xenocryst zircon microstructures. *Earth planet. Sci. Lett.*, **181**, 89–102.

Nelson, D.R., Trendall, A.F. & Altermann, W. (1999) Chronological correlations between the Pilbara and Kaapvaal cratons. *Precam. Res.*, **97**, 165–189.

Nutman, A.P., McGregor, V.R., Friend, C.R.L., Bennett, V.C. & Kinny, P.D. (1996) The Itsaq Gneiss Complex of southern West Greenland: the world's most extensive record of early crustal evolution (3900–3600 Ma). *Precam. Res.*, **78**, 1–39.

Nutman, A.P., Mojzsis, S.J. & Friend, C.R.L. (1997) Recognition of >3850 Ma water-lain sediments in West Greenland and their significance for the early Archaean Earth. *Geochim. Cosmochim. Acta*, **61**, 2465–2484.

Nyblade, A.A. & Pollack, H.N. (1993) A global analysis of heat flow from Precambrian terrains: implications for the thermal structure of Archean and Proterozoic lithosphere. *J. geophys. Res.*, **98**, 12207–12218.

O'Hara, M.J. (1970) Selenology. *Nature*, **225**, 321–322.

Ohmoto, H. (1997) When did the Earth's atmosphere become oxic? *Geochem. News*, **93**, 12–13, 26–27.

Ohtake, M., Naraoka, H., Hayashi, K. & Ohmoto, H. (1992) Archean surface environments. Part IV: Geochemistry of ~3.8 Ga metasediments from the Isua district, West Greenland. *29th International Geological Congress, Kyoto, Japan, Abstracts*, **1**, 173.

Ojakangas, R.W. (1983) Tidal deposits in the early Proterozoic basin of the Lake Superior region. The Palms and Pokegama Formations: evidence for subtidal shelf deposition of Superior-type banded iron formation. In: *Early Proterozoic Geology of the Great Lakes Region* (Ed. Medaris, L.G.), Mem. geol. Soc. Am., Boulder, **60**, 49–66.

Perry, E.C., Tan, F.C. & Morey, G.B. (1973) Geology and stable isotope geochemistry of the Biwabik iron formation, northern Minnesota. *Econ. Geol.*, **68**, 1110–1125.

Polet, J. & Anderson, D.L. (1995) Depth extent of cratons as inferred from tomographic studies. *Geology*, **23**, 205–208.

Pollack, H.N. (1997) Thermal characteristics of the Archean. In: *Greenstone Belts* (Eds de Wit, M.J. & Ashwal, L.D.), pp. 223–293. Oxford Science Publications, Oxford.

Radhakrishna, B.P. (Ed.) (1986) Banded iron-formation of India. *Geol. Soc. India J.*, **28** (special issue), 270 pp.

Rapp, R.P. (1997) Heterogeneous source regions for archaean granitoids: experimental and geochemical evidence. In: Greenstone Belts (Eds de Wit, M.J. & Ashwal, L.D.), pp. 267–279. Oxford Science Publications, Oxford.

Rogers, J.J.W. (1996) A history of continents in the past three billion years. *J. Geol.*, **104**, 91–107.

Rösing, M. (1999) $^{13}$C-depleted carbon microparticles in >3700-Ma sea-floor sedimentary rocks from west Greenland. *Science*, **283**, 674–676.

Sandiford, M. (1989) Horizontal structures in granulite terrains: a record of mountain building or mountain collapse? *Geology*, **17**, 449–452.

Schidlowski, M. (1998) Beginnings of terrestrial life: problems of the early record and implications for extraterrestrial scenarios. *Int. Soc. Optical Eng. (SPIE), Proceedings Series, No. 3441*, pp. 149–157.

Schidlowski, M., Hayes, J.M. & Kaplan, I.R. (1983) Isotopic inferences of ancient biochemistries: carbon, sulfur, hydrogen, and nitrogen. In: *Earth's Earliest Biosphere*

(Ed. Schopf, J.W.), pp. 149–186. Princeton University Press, Princeton, NJ.

SCHMIDT, P.W. & WILLIAMS, G.E. (1995) The Neoproterozoic climatic paradox: equatorial palaeolatitude for Marinoan glaciation near sea level in South Australia. *Earth planet. Sci. Lett.*, **134**, 107–124.

SCLATER, J.G., JAUPART, C. & GALSON, D. (1980) The heat flow through oceanic and continental crust and the heat loss of the Earth. *Rev. Geophys.*, **18**, 269–311.

SIMONSON, B.M. (1985) Sedimentological constraints on the origins of Precambrian iron-formation. *Geol. Soc. Am. Bull.*, **96**, 244–252.

SIMONSON, B.M. (1987) Early silica cementation and subsequent diagnesis in arenites from four early Proterozoic iron formations of north America. *J. sediment. Petrol.*, **57**, 494–511.

SIMONSON, B.M. & GOODE, A.D.T. (1989) First discovery of ferruginous chert arenites in the early Precambrian Hamersley Group of Western Australia. *Geology*, **17**, 269–272.

SIMONSON, B.M. & HASSLER, S.W. (1996) Was the deposition of large Precambrian iron formations linked to major marine transgressions? *J. Geol.*, **104**, 665–676.

STEIN, M. & HOFMANN, A.W. (1994) Mantle plumes and episodic crustal growth. *Nature*, **372**, 63–68.

STEVENSON, D.J. (1987) Origin of the Moon—the collision hypothesis. *Ann. Rev. Earth planet. Sci.*, **15**, 271–315.

THORNE, A.M. & TRENDALL, A.F. (2001) The Fortescue Group. *W. Aust. geol. Surv. Bull.*, **144**, 249 pp.

TOWE, K.M. (1981) Environmental conditions surrounding the origin and early Archean evolution of life: a hypothesis. *Precam. Res.*, **16**, 1–10.

TRENDALL, A.F. (1965) Discussion of 'Origin of Precambrian Iron Formations'. *Econ. Geol.*, **60**, 1065–1070.

TRENDALL, A.F. (1972) Revolution in Earth history. *Geol. Soc. Aust. J.*, **19**, 287–311.

TRENDALL, A.F. (1973a) Varve cycles in the Weeli Wolli Formation of the Precambrian Hamersley Group, Western Australia. *Econ. Geol.*, **68**, 1089–1097.

TRENDALL, A.F. (1973b) Iron-formations of the Hamersley Group of Western Australia: type examples of varved Precambrian evaporites. In: *Genesis of Precambrian Iron and Manganese Deposits*. Unesco, Paris, Earth Sciences, **9**, 257–267.

TRENDALL, A.F. (1983a) Introduction. In: *Iron-formation: Facts and Problems* (Eds Trendall, A.F. & Morris, R.C.), pp. 1–11. Elsevier, Amsterdam.

TRENDALL, A.F. (1983b) The Hamersley Basin. In: *Iron-formation: Facts and Problems* (Eds Trendall, A.F. & Morris, R.C.), pp. 69–129. Elsevier, Amsterdam.

TRENDALL, A.F. (1990a) Hamersley Basin. In: *Geology and Mineral Resources of Western Australia*, Mem. W. Aust. geol. Surv., **3**, 163–169.

TRENDALL, A.F. (1990b) A tale of two cratons: speculations on the origin of continents (abstract). *R. Soc. W. Aust. J.*, **79**, 141.

TRENDALL, A.F. (1998) Precise measurements of the depositional rate of a Precambrian banded iron formation (BIF) by two independent methods. *Geoscience 98, Biennial meeting of the Geol. Soc. London, Keele, Abstract Volume*, p. 48.

TRENDALL, A.F., BASEI, M.A.S., DE LAETER, J.R. & NELSON, D.R. (1998) SHRIMP zircon U–Pb constraints on the age of the Carajás Formation, Grâo Pará Group, Amazon Craton. *J. S. Am. Earth Sci.*, **11**, 265–277.

TRENDALL, A.F. & BLOCKLEY, J.G. (1970) The iron formations of the Precambrian Hamersley Group, Western Australia, with special reference to the associated crocidolite. *W. Aust. geol. Surv. Bull.*, **119**, 365 pp.

TRENDALL, A.F. & DE LAETER, J.R. (1972) Apparent age, and origin, of black porcelanite of the Joffre Member. W. Aust. geol. Surv. Ann. Rep. 1971, pp. 68–74.

TRENDALL, A.F., DE LAETER, J.R., NELSON, D.R. & MUKHOPADHYAY, D. (1997) A precise zircon U–Pb age for the base of the BIF of the Mulaingiri Formation (Bababudan Group, Dharwar Supergroup) of the Karnataka Craton. *Geol. Soc. India J.*, **50**, 161–170.

TRENDALL, A.F. & MORRIS, R.C. (Eds) (1983) *Iron-formation: Facts and Problems*. Elsevier, Amsterdam, 558 pp.

UNESCO (1973) *Genesis of Precambrian Iron and Manganese Deposits*. Proceedings of the Kiev Symposium, August 1970. Unesco, Paris, Earth Sciences, **9**, 382 pp.

VAN HISE, C.R. & LEITH, C.K. (1911) *The Geology of the Lake Superior Region*. Monogr. US geol. Surv., **52**, 641 pp.

VENKATACHALA, B.S., SHARMA, M., SRINIVASAN, R., SHUKLA, M. & NAQVI, S.M. (1986) Bacteria from the Archaean banded iron-formation of Kudremukh region, Dharwar Craton, south India. *Palaeobotanist*, **35**, 200–203.

VENKATACHALA, B.S., SHUKLA, M., SHARMA, M., NAQVI, S.M., SRINIVASAN, R. & UDAIRAJ, B. (1990) Archaean microbiota from the Donimalai Formation, Dharwar Supergroup, India. *Precam. Res.*, **47**, 27–34.

WALDE, D.H.G., GIERTH, E. & LEONARDOS, O.H. (1981) Stratigraphy and mineralogy of the manganese ores of Urucum, Matto Grosso, Brazil. *Geol. Rundsch.*, **70**, 1077–1085.

WALKER, J.C.G. (1984) Suboxic diagenesis in banded iron-formations. *Nature*, **309**, 340–342.

WALKER, J.C.G. (1987) Was the Archean biosphere upside down? *Nature*, **329**, 710–712.

WALKER, J.G.C. & ZAHNLE, K.J. (1986) Lunar nodal tide and distance to the Moon during the Precambrian. *Nature*, **320**, 600–602.

WHITTEN, G.F. (1970) *The Investigation and Exploitation of the Razorback Ridge Iron Deposit*. Rep. Invest. S. Aust. geol. Surv., **33**, 165 pp.

WIDDELL, F., SCHNELL, S., HEISING, S., EHRENREICH, A., ASSMUSS, B. & SCHINK, B. (1993) Ferrous iron oxidation by anoxygenic phototrophic bacteria. *Nature*, **362**, 834–836.

WILLIAMS, G.E. (1972) Geological evidence relating to the origin and secular rotation of the Solar System. *Mod. Geol.*, **3**, 165–181.

WILLIAMS, G.E. (2000) Geological constraints on the Precambrian history of Earth's rotation, and the moon's orbit. *Rev. Geophys.*, **38**, 37–59.

WINDLEY, B.M. (1993) Uniformitarianism today: plate tectonics is the key to the past. *J. geol. Soc. London*, **150**, 7–19.

WINGATE, M.T.D. (1998) A palaeomagnetic test of the Kaapvaal–Pilbara (Vaalbara) connection at 2.78 Ga. *S. Afr. J. Geol.*, **10**(4), 257–274.

WYLLIE, P.J., WOLF, M.B. & VAN DER LAAN, S.R. (1997) Conditions for formation of tonalites and trondhjemites: magmatic sources and products. In: *Greenstone Belts* (Eds by de Wit, M.J. & Ashwal, L.D.), pp. 256–266. Oxford Science Publications, Oxford.

Yeo, G.M. (1986) *Iron-formation in the Late Proterozoic Rapitan Group, Yukon and Northwest Territories*. Spec. Publ. Can. Inst. Mining Metall., **37**, 142–153.

Young, G.M. (1976) Iron-formation and glaciogenic rocks of the Rapitan Group, Northwest Territories, Canada. *Precam. Res.*, **3**, 137–158.

Zajac, I.S. (1974) The stratigraphy and mineralogy of the Sokoman Formation in the King Lake area, Quebec and Newfoundland. *Geol. Surv. Can. Bull.*, **220**, 159 pp.

Zegers, T., de Wit, M.J., Dann, J. & White, S.H. (1998) Vaalbara, Earth's oldest assembled continent? A combined structural, geochronological, and palaeomagnetic test. *Terra Nova*, **10**, 250–259.

*Spec. Publs int. Ass. Sediment.* (2002) **33**, 67–105

# The isotopic composition of Precambrian sulphides—seawater chemistry and biological evolution

H. STRAUSS*

*Institut für Geologie, Ruhr-Universität Bochum, Universitätsstrasse 150, 44801 Bochum, Germany*

## ABSTRACT

The sulphur isotopic composition of sedimentary sulphides records the temporal evolution of the Precambrian sulphur cycle. Early Archaean sediments show $\delta^{34}S$ values for pyrite around 0‰ and are interpreted to reflect a predominantly magmatic origin for sedimentary pyrite. Pyrite in sediments across the Archaean–Proterozoic transition record a substantial variability in $\delta^{34}S$ that resembles modern style bacterial sulphate reduction. During Proterozoic time, bacterial sulphate reduction, although in principle organic carbon limited, also appears to be characterized by sulphate limitation as evident from frequently positive sulphur isotope values. This is attributed to effective sulphate turnover as a consequence of readily metabolizable organic matter and possibly a lower sulphate concentration in the ocean. Based on distinct distributions of $\delta^{34}S$ values for sedimentary pyrite, the Neoproterozoic sulphur cycle appears to have responded to perturbations of the global ocean/atmosphere system as a consequence of repeated glaciations and resulting biogeochemical changes. This is consistent with the proposed Snowball Earth concept. Interpretations of the Precambrian sedimentary sulphur cycle result from a systematic assessment of organic carbon, sulphide sulphur and reactive iron abundances, 334 new and several hundred previously published sulphur isotope values for sedimentary pyrite and a comparison to the isotopic composition of Precambrian seawater sulphate as evident from marine evaporites.

## INTRODUCTION

Sediments and sedimentary rocks and their geochemical and isotopic inventories act as proxies for geological evolution. The rocks record temporal variations in the isotopic composition, e.g. of carbon, oxygen, sulphur and strontium, and reflect the time-dependent variability in the importance of physical, chemical and/or biological processes which shape Earth's surface environments and affect the chemical inventory of the ocean/atmosphere system (e.g. Veizer *et al.*, 1999; Goddéris & Veizer, 2000).

The interdependence of these geochemical cycles has been documented by Veizer & Hoefs (1976), Veizer *et al.* (1980), Holser (1984) and most recently Veizer *et al.* (1999). It is expressed in a positive correlation between the sulphate sulphur and strontium isotopic compositions, and in a negative correlation between the carbonate carbon and sulphate sulphur isotopic compositions. While carbon, oxygen and strontium isotopes were measured on inorganic or biogenic carbonates, the sulphur isotope record was traditionally based on the study of marine evaporites (e.g. Holser & Kaplan, 1966; Claypool *et al.*, 1980). Correlations between the isotope records of sulphur and strontium and of sulphur and oxygen reflect the importance of basalt–sea water interaction, e.g. at mid-ocean ridges, whereas the correlation between carbon and sulphur isotopes documents the biological aspect of the sulphur cycle. While the former reflects the long-term tectonic control, the latter represents the shorter-term biological cycle (Goddéris & Veizer, 2000).

This contribution presents 334 new sulphur isotope measurements for Precambrian sedimentary pyrite, together with a thorough review of previously published data. Furthermore, a total of 1284 samples have

* Present address: Geologisch-Paläontologisches Institut und Museum, Westfalische Wilhelms-Universität Münster, Corrensstrasse 24, 48149 Münster, Germany.

been analysed for their elemental abundances of organic carbon, sulphide sulphur and reactive iron, representing the most comprehensive data set published to date for Precambrian sediments. Chemical and isotopic data are interpreted with respect to their significance as a proxy for seawater chemistry and biological evolution, thereby addressing current opinions about the Precambrian sulphur cycle.

## ISOTOPIC BIOGEOCHEMISTRY OF MODERN MARINE SEDIMENTS—INFERENCES FOR PRECAMBRIAN SEDIMENTARY SYSTEMS

Elemental abundances in solid phases, porewater profiles of dissolved constituents and the isotopic compositions of sulphur- and carbon-bearing dissolved, gaseous and solid compounds are studied to constrain the depositional and diagenetic history of modern marine sediments. This history is exemplified through a series of predominantly biologically driven redox processes, such as aerobic respiration, nitrate, iron and sulphate reduction and methanogenesis (e.g. Froelich *et al.*, 1979; Stumm & Morgan, 1996). Concentration profiles, cross-plots of elemental abundances and characteristic ranges of isotopic compositions provide evidence for the activity of these individual processes and document the actual redox sequence present in a given sedimentary system.

For sulphur, principal reactions were summarized by Berner (1970, 1984), notably the process of dissimilatory bacterial sulphate reduction and subsequent formation of iron sulphide:

$$8(SO_4)^{2-} + 2Fe_2O_3 + 8H_2O + 15C_{org} \rightarrow 4FeS_2 + 15CO_2 + 16(OH^-) \quad (1)$$

It has been shown that bacterial sulphate reduction represents the key process for the anaerobic mineralization of sedimentary organic matter (e.g. Jørgensen, 1982) and is second in overall magnitude to aerobic respiration (e.g. Canfield, 1991).

During the past 15 years, it has further been shown that a series of geochemical parameters can be utilized to constrain the pertinent geochemical environment of sulphate reduction and pyrite formation. These include the correlation of abundances of sulphide sulphur and organic carbon (e.g. Leventhal, 1983; Raiswell & Berner, 1986; Lyons & Berner, 1992; Lyons, 1997) and the determination of reactive iron and calculation of the degree of pyritization (DOP = pyrite-Fe/pyrite Fe + HCl-extractable Fe), including its correlation with organic carbon and/or sulphur abundances (Raiswell & Berner, 1985; Raiswell *et al.*, 1988, 1994; Raiswell & Canfield, 1998).

The conceptual approach to interpreting these data derives from the principles underlying equation (1). Sedimentary organic matter is recycled through a sequence of aerobic and anaerobic processes. During bacterial sulphate reduction, dissolved (sea water) sulphate is reduced to hydrogen sulphide, while metabolizable organic matter is oxidized. This results in an overall positive correlation between the availability of metabolizable sedimentary organic material and the activity of sulphate reduction, expressed in a positive correlation between pyrite sulphur and organic carbon abundances (subsequently abbreviated as sulphur/carbon or S/C ratio) in the sediment (Leventhal, 1983). An empirically derived regression line with an intercept through the origin and an S/C ratio of 0.36 characterizes Holocene normal marine siliciclastic sediments, i.e. those deposited under well oxygenated bottom water conditions (e.g. Berner, 1984). Sediments of post-Devonian age deposited under such normal marine conditions are believed to have a similar initial S/C ratio (Raiswell & Berner, 1986), which could increase during diagenesis through selective carbon consumption. Prior to the development of a terrestrial flora (i.e. broadly pre-Devonian) and the resulting deposition of terrestrial organic matter, marine sediments are characterized by a higher S/C ratio. This has been demonstrated by Raiswell & Berner (1986) and is thought to indicate that purely marine organic matter is more effectively metabolized by sulphate reducers. In contrast, euxinic sediments deposited under anoxic bottom water conditions display a positive intercept on the sulphur axis in a sulphide sulphur–organic carbon cross-plot. Furthermore, samples can show a positive correlation between both parameters or little variation in sulphide sulphur abundance at highly variable organic carbon contents (Raiswell & Berner, 1985; Lyons & Berner, 1992). This results from sulphate reduction within the water column as opposed to solely within the sediment as in the case of normal marine conditions (oxic bottom water).

Further important palaeoenvironmental information can be derived from the abundance of reactive iron, expressed through the parameter DOP (pyrite-bound iron/reactive iron: Raiswell *et al.*, 1988). Apart from metabolizable organic carbon as the principal limiting factor for bacterial sulphate reduction, pyrite formation can further be affected through limitation of reactive iron. Sedimentary environments characterized by iron deficiency include carbonate-dominated systems and, more importantly, euxinic settings. In the latter case, a cross-plot of sulphide sulphur and organic carbon would be likely to show a positive intercept on the sulphur axis, reflecting syngenetic pyrite formation in the water column. At the same time, however, no correlation between sulphur and carbon would be discernible due to the decoupling of Fe and organic carbon deposition (e.g. Lyons & Berner, 1992).

Finally, sulphate can become limited when the process of sulphate reduction exceeds sulphate replenishment. Under normal marine conditions, the latter is generally regulated through diffusion from the overlying water column. However, enhanced availability of easily metabolizable organic material, coupled to diffusion-limited sulphate replenishment, might result in progressive exhaustion of the sulphate supply until the final termination of sulphate reduction. This process would be even more severe in environments which are characterized by initial low-sulphate abundances. For example, modern freshwater settings display very low sulphide sulphur to organic carbon ratios (Berner & Raiswell, 1984).

In principle, different geochemical environments can be distinguished: normal marine, euxinic with or without iron limitation and sulphate-limited. They can be recognized through respective sulphur–carbon –iron relationships (for a more detailed discussion see, for example, Raiswell & Berner, 1985, 1986; Raiswell *et al.*, 1988; Lyons & Berner, 1992; Lyons, 1997, and references therein).

Application of these relationships with the objective of constraining the biogeochemical environment of a given sedimentary system has been quite successful for Phanerozoic examples, but only very limited in sediments of Precambrian age. The ability of these correlations to yield unequivocal characterization strongly depends on the availability of adequate samples. Ideally, these should be reasonably rich in organic carbon and devoid of any obvious sulphide enrichments (i.e. concretions, sulphide layers). Furthermore, metamorphic overprint should be minimal, as thermal alteration results in loss of organic matter and remobilization of sulphur. In addition, external addition of sulphides through magmatic/hydrothermal processes will similarly alter the sulphur/carbon ratio.

Following these precautions, what remains is a more general question regarding the applicability of these correlations to Precambrian environments. This pertains in particular to water-column oxygenation and the overall debate about oxygen in the Precambrian atmosphere, particularly during the Archaean. Two extremes have been considered: (i) no oxygen in the Archaean and an increase of oxygen in the atmosphere/ocean system during the Proterozoic (e.g. Kasting, 1993); or (ii) low to moderate concentrations of free oxygen in the atmosphere/ocean system as far back as the preserved (sedimentary) rock record (Ohmoto, 1997). What would this mean with respect to the application of a sulphur/carbon ratio? The importance of bacterial sulphate reduction could be discernible through a positive correlation between sulphide sulphur and organic carbon, while a positive intercept on the sulphur axis would suggest the presence of anoxic bottom waters. These palaeoenvironmental assessments will be further supported by characteristic sulphur isotope values pointing to the principal process of pyrite formation, as will be discussed below. It should be noted that oxygenation of the atmosphere/ocean system also relates to the sulphate content of sea water. Canfield (1998), for example, argued for the presence of a substantial seawater sulphate concentration not earlier than 2300 Ma.

Apart from elemental abundances and correlations, the isotopic composition of sulphur plays a critical role in the interpretation of modern and ancient marine sedimentary systems. As a key process, dissimilatory sulphate reduction is associated with a substantial isotope fractionation. The magnitude of this is dependent on various environmental conditions.

Some 30 years ago, the fundamentals of sulphur isotope geochemistry were described by Thode *et al.* (1961), Kaplan & Rittenberg (1964) and others. Recent reviews of this subject have been published, e.g. by Strauss (1997), and the reader is referred to this article for further details. Pertinent to the present discussion are the following observations.

(a) Various chemical sediments—calcium sulphate, barite and structurally substituted sulphate in carbonates and phosphate—can faithfully reflect the sulphur isotopic composition of sea water. Respective temporal sulphur isotope records have been measured for the Phanerozoic, or parts thereof, from marine evaporites (Claypool *et al.*, 1980), structurally substituted sulphate in carbonates (Burdett *et al.*, 1989; Kampschulte *et al.*, 2001; Kampschulte & Strauss, 2000) and barite (Paytan *et al.*, 1999). Data from Precambrian evaporites have been reviewed by Strauss (1993a). More recently, the sulphur isotopic composition for terminal Neoproterozoic and early Palaeozoic sea water has been further characterized by data from evaporites (e.g. Strauss *et al.*, 2000), phosphorites (Shields *et al.*, 1999b) and barite (Shields *et al.*,

1999a). However, the importance of sulphate sulphur isotope data as a globally representative signal has to be constrained, e.g. through multiple analyses of time-equivalent deposits from different depositional basins (e.g. Nielsen, 1989).

(b) Pyrite from marine sediments displays variable and frequently negative $\delta^{34}S$ values. This is largely a consequence of a kinetic isotope effect associated with bacterial sulphate reduction which favours the reduction of $^{32}S$ (e.g. Chambers & Trudinger, 1979). The magnitude of this isotope effect varies for different bacteria and strongly depends on parameters including the rate of sulphate reduction and the availability of sulphate (e.g. Harrison & Thode, 1958; Kaplan & Rittenberg, 1964; Ohmoto, 1992; Habicht & Canfield, 1996). In principle, the net isotope fractionation decreases with increasing reduction rate, although results by Habicht & Canfield (1996) provide contrasting information. Furthermore, sulphate availability affects the isotopic composition of the resulting hydrogen sulphide, with a decrease in net isotope fractionation at sulphate concentrations below 1 mM (e.g. Ohmoto, 1992).

In addition, sulphate can become limited if sulphate consumption exceeds the sulphate supply at the site of bacterial reduction (e.g. Zaback *et al.*, 1993), a situation commonly referred to as a closed system. As stated above, multiple causes can be responsible for this, including an originally lower sulphate concentration of Precambrian sea water (e.g. Kah *et al.*, 2000). For the sulphur isotope system, the continuous decrease in sulphate concentration is accompanied by increasingly more positive $\delta^{34}S$ values as a consequence of the preferential reduction of $^{32}S$, for both the remaining seawater sulphate and the later-formed sulphide (Rayleigh fractionation: Nakai & Jensen, 1964). For the apparent isotopic composition of bulk sedimentary pyrite, this results in a reduced overall fractionation between sulphate and sulphide, with the isotope value for the sum of all pyrites within a given environment attaining the isotopic composition of the initial seawater sulphate. Microanalysis with high spatial resolution (e.g. Crowe *et al.*, 1990), however, reveals that during the later stage of sulphate exhaustion, pyrite displays a $\delta^{34}S$ value which can be far more enriched in $^{34}S$ than the initial sulphate reservoir.

Further amplification of the net sulphur isotope fractionation, expressed through the apparent isotope difference between sulphate and sulphide that exceeds the experimentally determined maximum fractionation associated with bacterial sulphate reduction of 46‰ (Kaplan & Rittenberg, 1964), is attributed to the disproportionation of thiosulphate and/or elemental sulphur. The importance of these processes and the associated isotope effects has been determined for modern sediments (e.g. Jørgensen, 1990; Canfield & Thamdrup, 1994) and verified experimentally (e.g. Smock *et al.*, 1998). The temporal evolution of this biological recycling of sulphur and intermediate sulphur species has been evaluated by Canfield & Teske (1996).

(c) Pyrite in marine sediments can be of inorganic origin, either as a result of thermochemical sulphate reduction (TSR), a process frequently observed in hydrocarbon-bearing strata at temperatures between 100 and 200 °C, or as a product of magmatic/hydrothermal precipitation. Machel *et al.* (1995) reviewed criteria to distinguish bacterial from thermochemical sulphate reduction, among them quite narrow, temperature-dependent isotope fractionations between sulphate and sulphide during TSR. While this process cannot be ruled out in general for the Precambrian sediments studied, it is not further considered.

More important in the present context is the second alternative, notably the presence of magmatic/hydrothermal sulphides. Ohmoto & Goldhaber (1997) present a review of the pertinent isotope systematics associated with these processes. Overall, magmatic sulphides are characterized by a narrow range in $\delta^{34}S$, commonly close to 0‰, the value for mantle sulphur (von Gehlen, 1992). This suggests that the source of sulphur in the sulphide is similar to sulphur in submarine volcanic rocks. Alternatively, sulphide sulphur can be the result of high-temperature abiotic sulphate reduction, e.g. as observed in modern mid-ocean ridge settings (e.g. Bluth *et al.*, 1988). Depending on the temperature of the hydrothermal system and whether equilibrium or disequilibrium conditions prevail, the isotopic composition of the resulting hydrothermal sulphide will be more or less depleted with respect to sulphate (e.g. Zheng, 1991).

(d) The oxidation of sulphide does not result in a substantial isotope fractionation (e.g. Fry *et al.*, 1984).

(e) Isotope values characterizing organically bound sulphur vary as a function of the predominant mode of sulphur incorporation (Anderson & Pratt, 1995; Weise *et al.*, 1999).

An increasing suite of operationally defined sulphur-bearing compounds has been identified in recent sediments (e.g. Rice *et al.*, 1993; Canfield *et al.*, 1998). With respect to Precambrian sediments, however, only the thermodynamically stable end-members of the geochemical sulphur cycle—i.e. marine sulphate and sedimentary pyrite—have been studied in a systematic way with the aim of reconstructing their temporal evolution (e.g. Schidlowski *et al.*, 1983; Lambert & Donnelly, 1990; Hayes *et al.*, 1992; Strauss, 1993a,b). The sulphur isotopic composition of marine evaporites has been determined as a central reference point for the global sulphur cycle at any given time. However, due to times of non-deposition and the poor preservation potential of evaporites, only a very fragmentary temporal record exists for the sulphur isotopic composition of Precambrian sea water (Strauss, 1993a). In contrast, a substantial body of isotope data exists for sedimentary sulphides of Precambrian age, resulting from studies aimed at reconstructing the antiquity of bacterial sulphate reduction (e.g. Schidlowski *et al.*, 1983). By analogy with modern marine sediments, variable but predominantly negative $\delta^{34}S$ values have been interpreted as a clear indication of a biological fractionation of the sulphur isotopes. The absence of any notable variation in $\delta^{34}S$ for sedimentary sulphides, and a small difference between the known or assumed isotopic composition of the ancient global seawater sulphate and the sedimentary sulphide has been interpreted to indicate a magmatic/hydrothermal sulphide origin (e.g. Lambert & Donnelly, 1990; Strauss, 1993b, 1999b).

An alternative interpretation of the Archaean sulphide sulphur isotope record has been proposed by Ohmoto *et al.* (1993), Ohmoto (1997) and Watanabe *et al.* (1997). They attribute the small isotope fractionation between Archaean seawater sulphate and sedimentary sulphides to a high sulphate reduction rate (10–100 mol $l^{-1}$ $yr^{-1}$) at one-third of present day seawater sulphate concentration (Ohmoto *et al.*, 1993).

In summary, parameters that have been utilized to constrain the geochemical environment of marine sedimentary systems include the abundances of organic carbon, sulphide sulphur, reactive iron and the degree of pyritization. Absolute values and, moreover, cross-plots of these parameters in different combinations provide a crucial context for the interpretation of sulphur isotope values determined for sedimentary pyrite. The latter constrain the specifics of pyrite formation. Of particular interest is the identification of a biogenic pyrite and its temporal reconstruction through Earth's history.

## SULPHUR ISOTOPE SYSTEMATICS AND ANALYTICAL METHODS

The element sulphur has four stable isotopes: $^{32}S$, $^{33}S$, $^{34}S$ and $^{36}S$. Natural abundances are 95.02, 0.75, 4.21 and 0.02%, respectively (Hoefs, 1997). As a consequence of different inorganic and organic, high- and low-temperature processes, resulting sulphur-bearing compounds within the atmosphere, hydrosphere, lithosphere and biosphere display variations in the isotope ratio of $^{34}S/^{32}S$. Isotopic compositions are commonly expressed in the delta notation as:

$$\delta^{34}S\ [‰] = (Rsa/Rst - 1) \times 1000,$$

where Rsa and Rst = $^{34}S/^{32}S$ of the sample and standard, respectively. Isotope values are reported as per mil difference with respect to the Cañon Diablo troilite (CDT). Heterogeneities in the sulphur isotopic composition of this reference material (Beaudouin *et al.*, 1994) and the fact that CDT is no longer available has resulted in the recommendation to report sulphur isotope data against the new V-CDT standard (Coplen & Krouse, 1998). To simplify comparison to previously published results, CDT is used throughout the present study.

In general, samples (free of any surficial weathering and without fracture fillings) were pulverized prior to geochemical analyses. Sulphur abundances were determined via coulometric titration following combustion in an oxygen stream (Lange & Brumsack, 1977) at 1400 and 1000 °C, respectively. Combustion at 1400 °C yields the total sulphur fraction, while the lower temperature liberates all sulphides and organic-bound sulphur. Organically bound sulphur is considered to be of negligible importance in most Precambrian sediments. Thus, sulphur which is liberated at 1000 °C is considered here to reflect sulphide sulphur. Abundances of organic carbon resulted from a sealed tube combustion procedure performed to determine the organic carbon isotopic composition and described in detail by Strauss *et al.* (1992a). Thereby, $CO_2$ was generated at 850 °C and measured volumetrically. Total organic carbon (TOC) abundances were calculated from the dry weight. Reactive iron is defined as the sum of pyrite-bound iron and HCl-leachable iron (Berner, 1970; Leventhal & Taylor, 1990; Raiswell *et al.*, 1994). The latter parameter was quantified through atomic absorption spectroscopy (AAS) or induced coupled plasma spectrometry (ICP), following the leaching of sediment samples with 6 M HCl at room temperature for 5 h.

Wet chemical preparation for determining the sulphur isotopic composition of pyrite involved reaction of pulverized sediments with a chromium chloride solution following the method outlined by Canfield *et al.* (1986). Sulphur dioxide for isotope analyses was generated from the resulting silver sulphide precipitates through combustion with vanadium pentaoxide at 1100 °C under vacuum (Newton *et al.*, 1995), followed by cryogenic purification and transfer into a 6 mm pyrex tube. Isotope measurements were performed using a Finnigan MAT 251 mass spectrometer. Reproducibility, as determined through replicate measurements, was better than ±0.3‰. International reference materials have been measured at: +17.2‰ (NBS 123), +20.6‰ (NBS 127), −0.2‰ (IAEA-S-1, formerly NZ-1). Precision was further controlled through analyses of various internal laboratory standards.

## THE ARCHAEAN

Investigations of the biogeochemistry of Archaean sedimentary systems have played an integral part in unravelling the evolution of life on Earth, with valuable evidence stemming from studies of the isotopic composition of organic carbon and pyrite sulphur (e.g. Schidlowski *et al.*, 1983). The record of preserved sedimentary rock successions of Archaean age dates back to *c.*3800 Ma with the *Isua Metasediments* in south-west Greenland. These include clastic sediments, possibly of deep-water origin (Rosing, 1999), and banded iron-formations (e.g. Appel, 1980). The age of the Isua rocks is currently controversial (for a recent discussion, see Whitehouse, 2000). Radiometric dates vary between 3655 and ≥3850 Ma (Kamber & Moorbath, 1998; Nutman *et al.*, 1997, respectively). Independent of this discussion, the Isua Supracrustal Belt in south-west Greenland hosts the oldest evidence for presumably marine sedimentary rocks, although they are structurally deformed and have experienced high-grade metamorphism and metasomatic alteration. Sedimentary sulphides present in the Isua banded iron-formation were studied for their sulphur isotopic composition during this investigation.

The Kaapvaal Craton in southern Africa provides a long record of Archaean and Proterozoic sedimentary successions. Samples were collected from various stratigraphic levels within the Swaziland, Pongola, Witwatersrand and Ventersdorp Supergroups. The *Swaziland Supergroup* represents a volcano-sedimentary sequence with an age >3000 Ma. It can be further subdivided into the dominantly volcanogenic Onverwacht Group and the overlying, more clastic, Fig Tree and Moodies groups. The geology of the principal outcrop area, the Barberton Mountainland of South Africa and neighbouring Swaziland, was recently reviewed by Lowe & Byerly (1999). The Onverwacht Group has been dated at between 3548 Ma (Kröner *et al.*, 1996) and 3300 Ma (Byerly *et al.*, 1996). The unconformably overlying Fig Tree Group was deposited between 3260 and 3225 Ma (Byerly *et al.*, 1996). Largely clastic sediments of the *Pongola Supergroup* represent the oldest cratonic sequence in southern Africa, with an age between 3100 and 2900 Ma (Armstrong *et al.*, 1991; Hegner *et al.*, 1984, 1994). Variable depositional environments have been proposed for the Pongola sediments, ranging from fluvial to tidal (Watchorn, 1980), and including stromatolitic carbonates (e.g. Beukes & Lowe, 1989) and ferruginous shales to true banded iron-formation (Beukes & Cairncross, 1991). The *Witwatersrand Supergroup* has long been known for its economic importance, specifically the gold deposits of southern Africa (e.g. Pretorius, 1976). It represents a succession of clastic sediments, quartzites and shales, with intercalated 'reefs', the gold-bearing conglomeratic horizons of economic importance. Depositional environments are constantly debated, with interpretations ranging from fluvial to marine (tidal to shelf). The age of the Witwatersrand Supergroup can be placed between 3074 ± 6 Ma for the underlying Dominion Group and 2714 ± 8 Ma for the overlying Ventersdorp Supergroup (Armstrong *et al.*, 1991). The Witwatersrand Supergroup is overlain by volcanic and sedimentary rocks of the *Ventersdorp Supergroup*. Those clastic sediments, ranging from conglomerates to black shales and carbonates, were deposited in small graben structures, the latter under predominantly lacustrine conditions (Buck, 1980). Radiometric dates of 2721 ± 18 and 2709 ± 4 Ma were measured for volcanic rocks from the Klipriviersberg Group and the Makwassie Formation, respectively (Armstrong *et al.*, 1991), pointing to a late Archaean age for the Ventersdorp Supergroup. Nelson *et al.* (1999) provide further age constraints for the succession on the Kaapvaal Craton.

### Results

$\delta^{34}S$ values for sedimentary pyrite from banded iron-formation within the *Isua Supracrustal succession* display a narrow variation between −2.0 and +3.0‰ ($n = 20$). Data were generated from isolated pyrite grains as well as powdered samples subjected to the chromium reduction method. Isotope values are independent of grain size (<63 to >500 μm) or analytical procedure.

Siltstones and shales from the *Swaziland Supergroup* are largely characterized by low contents of sulphide sulphur (<0.2 wt%, $n = 89$) and total organic carbon abundances of up to 5.3 wt% ($n = 45$). Resulting S/C ratios for almost all samples are <0.1. HCl-leachable iron ranges from 3.5 to 18.9 wt% ($n = 89$). Low pyrite-bound iron contents (most samples were below 0.2 wt%) result in calculated DOP values of <0.1. The sulphur isotopic composition of pyrite varies between −0.2 and +1.4‰ ($n = 25$).

Sediments from the *Witwatersrand Supergroup* display sulphide sulphur abundances of up to 0.5 wt% ($n = 162$). A few samples with higher values up to 7.9 wt% probably contain massive sulphide aggregates. Organic carbon contents vary between <0.05 and 0.39 wt% ($n = 150$), with all but a few samples having <0.1 wt%. No clear differences exist between different lithologies, but organic carbon contents appear to be slightly higher in the shales, while the quartzites show more variable but higher sulphide sulphur abundances than the shales. HCl-leachable iron contents range between 0.7 and 46.2 wt% ($n = 63$), and DOP values are generally below 0.1. $\delta^{34}S$ values for sedimentary pyrite range between −1.0 and +5.5‰ ($n = 45$). Again, no systematic difference between the two lithologies is discernible.

Abundances of sulphide sulphur up to 1.3 wt% ($n = 31$) and organic carbon contents between <0.1 and 4.0 wt% ($n = 18$) characterize siltstones and shales from the Kameeldoorns Formation, *Ventersdorp Supergroup*. HCl-leachable iron contents range from 1.0 to 5.4% ($n = 13$) and DOP values are between 0.10 and 0.52 ($n = 13$, one sample at 0.97). The sulphur isotopic composition of sedimentary sulphide for the Ventersdorp Supergroup samples lies between −0.8 and +6.5‰ ($n = 6$).

## Discussion

Biogeochemical studies of Early Archaean sediments attract significant interest due to the potential to interpret the early evolution of life on Earth. The desired information, however, is frequently altered, and any interpretation favouring biogenesis should always be based on a set of unequivocal research results. In the present context, the isotopic composition for sedimentary pyrite, the difference between the isotopic composition of sulphate and sulphide, and the C–S–Fe relationships are considered.

Limited information is available with respect to the isotopic composition of Archaean seawater sulphate (for a review, see Strauss, 1993a). Barites from the Warrawoona Supergroup of Western Australia (ave. 3.9‰, $n = 13$), the Onverwacht and Fig Tree groups of southern Africa (ave. 3.3‰, $n = 17$; ave. 3.4‰, $n = 7$, respectively) and the Sargur Group in India (ave. 5.4‰, $n = 13$) are considered to be representative of the early Archaean seawater sulphate sulphur isotopic composition (e.g. Strauss, 1993a).

Sulphides associated with the banded iron-formation at Isua, south-west Greenland, occur within the quartz layers of quartz-magnetite alternations with pyrite, situated predominantly along the contact between these two layers. SEM-studies of isolated pyrite grains reveal an irregular morphology, suggesting the impregnation of pre-existing pore space within the quartz layers by a sulphur-bearing fluid. Whether this fluid also contained the iron or whether Fe was scavenged from the magnetite layers remains unclear. The close spatial relationship between magnetite and pyrite suggests the latter. Pyrite displays a narrow range of sulphur isotope values between −3.0 and +3.0‰, which is in good agreement with previously published values for these rocks (Monster *et al.*, 1979). Furthermore, these results underline the observation of a very limited range in $\delta^{34}S$ for sulphides in early Archaean sedimentary sequences (Fig. 1). In the absence of any major fractionation between the Isua sulphides and the assumed seawater composition, and considering the similarity in $\delta^{34}S$ to mantle sulphur (von Gehlen, 1992), these sedimentary sulphides are interpreted as magmatic/hydrothermal precipitates.

Elemental abundances for organic carbon, sulphide sulphur and reactive iron for clastic sediments from the Swaziland Supergroup allow no unequivocal characterization of the geochemical environment. Low sulphur abundances coincide with high organic carbon abundances (Fig. 2). By analogy with Phanerozoic examples, calculated S/C ratios below 0.1 (correlation between sulphur and carbon: $r^2 = 0.34$) would suggest a low-sulphate 'freshwater' environment. A narrow range of $\delta^{34}S$ values between −0.2 and +1.4‰ (ave. 0.5‰, $n = 13$) characterizes these sediments. These results are comparable to previously measured sulphur isotope data for the Swaziland Supergroup with a narrow range and an average value close to 0‰ (data summarized by Strauss & Moore, 1992). All values closely resemble the isotopic composition of mantle sulphur around 0‰ (von Gehlen, 1992). Compared to an assumed isotopic composition of seawater sulphate around +3‰, as exemplified by early Archaean barite from South Africa (e.g. Perry *et al.*, 1971; Reimer, 1980), no substantial isotope fractionation between sulphate and sulphide is discernible. Again, in the absence of unequivocal evidence for a biogenic origin, the sulphides from the Swaziland Supergroup are

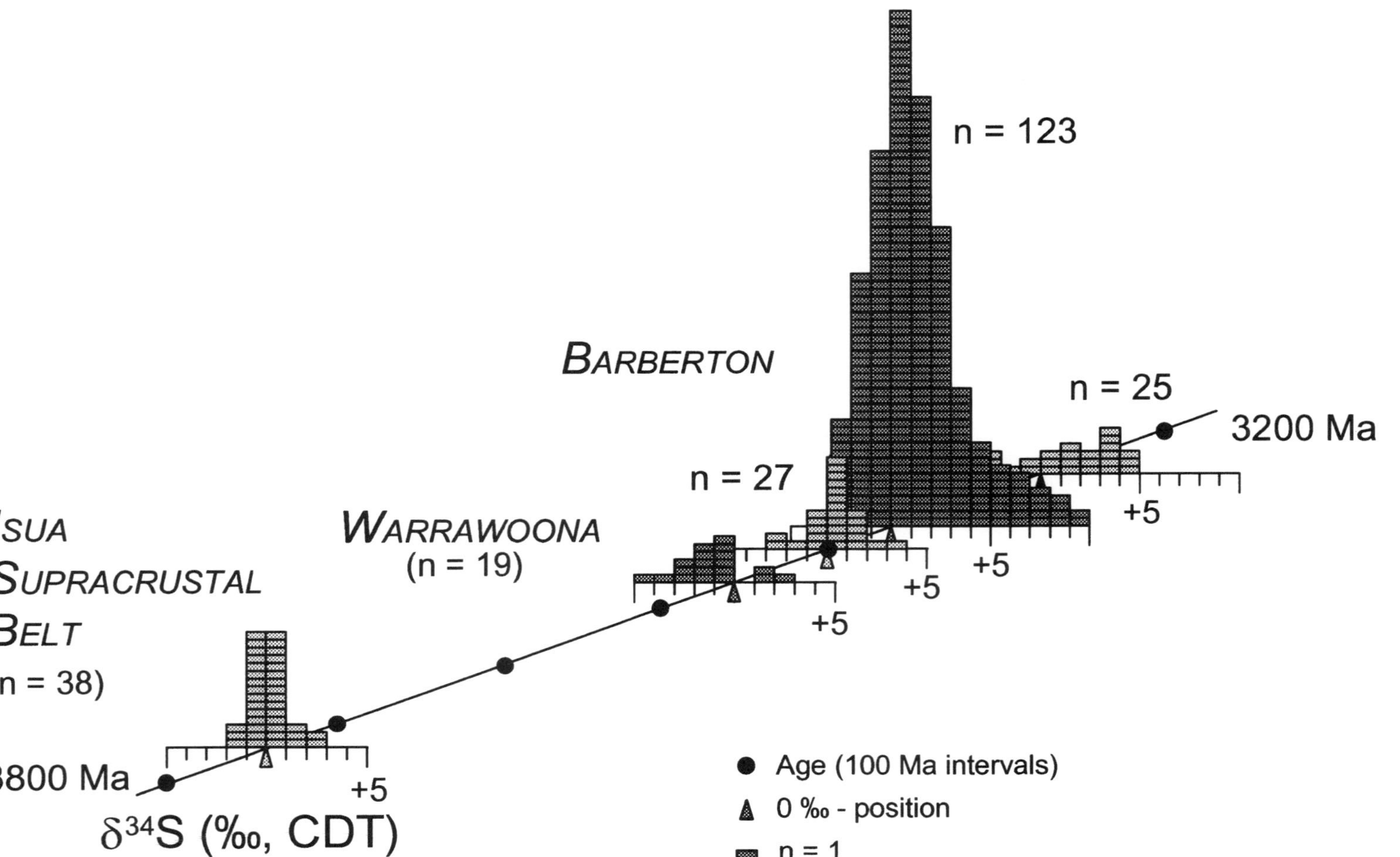


**Fig. 1.** The sulphur isotopic composition of early Archaean sedimentary sulphides (data from Lambert *et al.*, 1978; Strauss & Moore, 1992; Ohmoto *et al.*, 1993; Strauss, 1993b, 1999c).

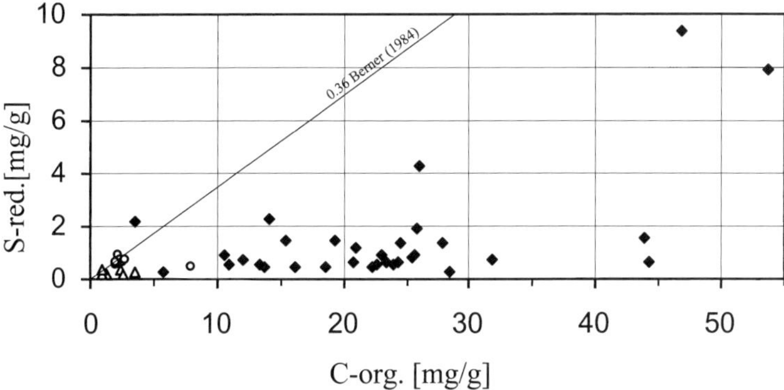


**Fig. 2.** Sulphur–carbon cross-plot for early Archaean sediments from South Africa.

interpreted as magmatic precipitates from a $H_2S$ rich fluid. Whether appropriate fluids discharged into the ocean where sulphides precipitated through reaction with dissolved iron, or whether such fluids percolated through the sediments and precipitation occurred in pore space, remains unresolved.

It should be noted that comparable isotope data have been published by Ohmoto (1992), Ohmoto *et al.* (1993) and most recently Kakegawa & Ohmoto (1999) for sulphides from the early Archaean sedimentary succession in the Barberton Mountain Land of southern Africa. Data were generated both by laser ablation in combination with mass spectrometry with high spatial resolution, and via conventional techniques on mineral separates. The former technique yielded a slightly larger range in $\delta^{34}S$, which still clustered around an average value close to 0‰. These authors favoured a biogenic origin for most of the sedimentary sulphides studied. Bacterial sulphate reduction is believed to have operated in generally warmer waters with a moderately high sulphate content and at high sulphate reduction rates. These conditions were thought to result in a minor sulphur isotopic fractionation. Habicht & Canfield (1996), however, argued against this by providing evidence from a combined microbiological and sulphur isotope study of modern microbial mats. Their results show that large isotope fractionation can be achieved even under very high sulphate reduction rates.

The Witwatersrand Supergroup represents a thick sedimentary succession (ave. 6000–7000 m) with three prominent sedimentary lithologies: shales, quartzites and the well known gold-bearing 'reefs' within coarse-grained conglomeratic sediments. Variable abundances of sulphide sulphur contrast with generally low organic carbon contents (Fig. 3). Differences exist with respect to the first two major lithologies. While shale samples exhibit a comparatively homogeneous distribution of sulphide sulphur below 0.15 wt%, numerous quartzite samples display substantially higher sulphur contents. Owing to the overall low organic carbon content (quartzites, however, show lower values than shales), the sulphur–carbon cross-plot exhibits highly variable sulphur values, independent of the carbon content, and S/C ratios significantly exceeding the normal marine threshold as determined for Holocene sediments (e.g. Berner, 1984). Two possible causes exist: sulphide mineralization of the entire succession, but preferentially within the more permeable quartzites, and/or a shift in the S/C ratio as a consequence of a preferred organic carbon loss during metamorphism (thermal overprint). The latter is consistent with a unidirectional shift in $\delta^{13}C$ of the organic matter towards more positive, thermally altered organic carbon (Beukes & Strauss, 1991). The abundance of reactive iron in all samples exceeds that of pyrite-bound iron by several orders of magnitude, clearly showing that no iron-limitation existed during formation of pyrite.

Sulphur isotope values between −1.0 and +5.5‰ have been measured during this study for various stratigraphic units within the Witwatersrand Supergroup. These data agree well with results from previous investigations (e.g. Hoefs *et al.*, 1968; Cameron, 1982; Hattori *et al.*, 1983; Förster, 1986), which generally focused on the auriferous reefs. In contrast, the present investigation with samples from the entire succession yielded similar $\delta^{34}S$ values around an average value of +3‰, with no systematic stratigraphic

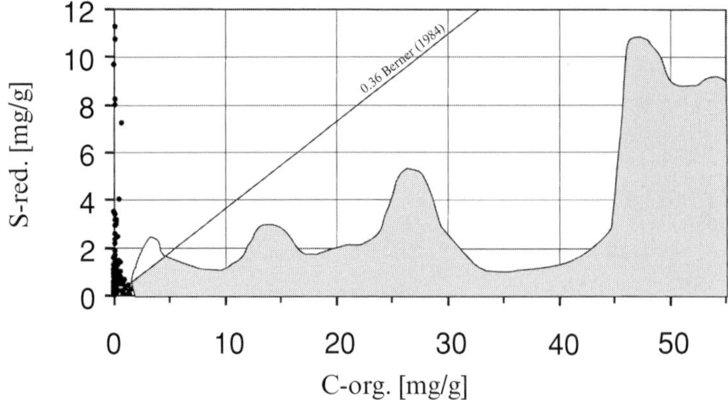


Moodies Gp., Swartkoppie Fm. und Sheba Fm. n = 45 • Witwatersrand Supergroup n = 151

**Fig. 3.** Sulphur–carbon cross-plot for sediments from the Witwatersrand Supergroup, South Africa (shaded area: data from Fig. 2 for comparison).

variation. Furthermore, no difference in the sulphur isotopic composition exists between shales and quartzites or between different pyrite morphologies, i.e. finely disseminated to coarse-grained, euhedral pyrite. This was already noted by Förster (1986). The homogeneity in $\delta^{34}S$ suggests a common origin with a homogeneous sulphur source and stable geochemical conditions during sulphide formation. Again, sulphur isotope values measured for the Witwatersrand sulphides overlap with the range for a magmatic sulphur source (von Gehlen, 1992) as well as a presumed seawater composition still around +3‰. Based on the sum of observations, the sulphides from the Witwatersrand Supergroup are regarded as abiotic precipitates.

Black shales from the Kameeldoorns Formation, Ventersdorp Supergroup, display variable, but on average low, S/C ratios, yet there is a clear positive correlation ($r^2 = 0.62$) between both parameters (Fig. 4). This would point to a coupling of the sulphur and carbon cycles analogous to younger sedimentary systems which results from bacterial sulphate reduction. Based on modern examples, most samples would be interpreted as indicative of a low-sulphate ('freshwater') environment. In fact, the interpretation of these sediments as having been deposited in small, possibly isolated, graben structures (Buck, 1980) might underline this conclusion. In addition, Veizer *et al.* (1989) studied carbonates from the Ventersdorp

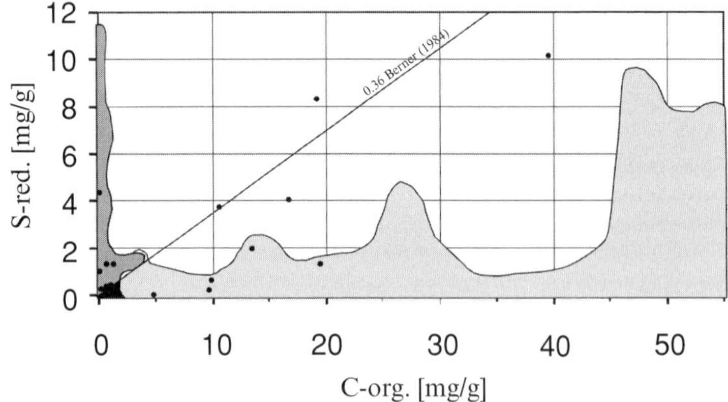


Moodies Gp., Swartkoppie Fm. und Sheba Fm. (n = 45)

Witwatersrand Spg. (n = 151)

• Ventersdorp Supergroup (n = 18)

**Fig. 4.** Sulphur–carbon cross-plot for sediments from the Ventersdorp Supergroup, South Africa (shaded areas: data from Fig. 3 for comparison).

Supergroup. Their conclusion, based on geochemical and isotopic data, favoured deposition of these carbonates in a lacustrine setting. Interestingly, samples from the present study with higher organic carbon and sulphide sulphur abundances display S/C ratios which approach the threshold for Phanerozoic 'normal marine shales'. DOP values between 0 and 0.52 indicate enhanced pyritization, yet no iron-limitation.

Carbon–sulphur–iron relationships, although not unequivocal, point towards bacterial sulphate reduction in a low-sulphate (?non-marine) environment. A limited number of $\delta^{34}S$ values range between −0.7 and +6.5‰ (mean 2.4‰). Interpretation of these isotope values remains difficult without a firm assessment of a truly marine depositional environment.

## Temporal trend in $\delta^{34}S_{sulphide}$: a synopsis

In addition to the new sulphur isotope results presented in this study, and the published data reported so far, a substantial body of data exists for sulphides associated with late Archaean greenstone belts and banded iron-formations from Australia and Canada (data summarized by Strauss & Moore, 1992) and Zimbabwe (Grassineau *et al.*, 1999). Reviewing the temporal evolution of the Archaean sulphur cycle (Fig. 5), a difference in the distribution of isotope values measured for sulphides in the individual stratigraphic units between the earlier part of the Archaean (i.e. >2750 Ma) and the late Archaean (<2750 Ma) is apparent. While sulphides from the former units display a relatively narrow range in $\delta^{34}S$ values, those from the latter units show a substantial spread in $\delta^{34}S$. Despite this difference in isotopic variability, most isotope values determined for sedimentary sulphides in late Archaean sediments lie close to 0‰.

In particular, the sulphur isotope results for sedimentary sulphides associated with the late Archaean banded iron-formations in Canada (Goodwin *et al.*, 1976, 1985; Shegelski, 1978; Thode & Goodwin, 1983) were initially interpreted as a biological signature, which started a long-standing debate about the antiquity of bacterial sulphate reduction. In contrast, a magmatic sulphur source for sulphides from these sedimentary environments was favoured by Cameron (1983a), Strauss (1986) and Bowins & Crocket (1994), although the latter authors proposed a possible addition of biogenic sulphur for some of the nodular pyrites.

More recently, quite variable sulphur isotope values between −18.3 and +16.7‰ for sedimentary sulphides from the late Archaean (2700 Ma) Belingwe Greenstone

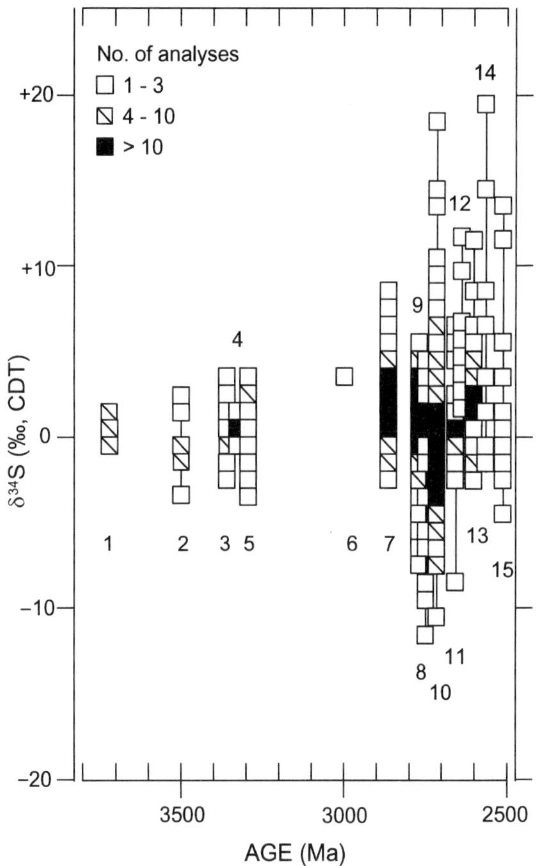


**Fig. 5.** The sulphur isotopic composition of Archaean sedimentary sulphides: 1, Isua Metasediments; 2, Warrawoona Supergroup; 3, Onverwacht and Fig Tree Group; 4, Swaziland Supergroup; 5, Sebakwian Group; 6, Mozaan Group; 7, Witwatersrand Supergroup; 8, Keewatin Group; 9, Uchi Greenstone Belt; 10, Late Archaean BIF; 11, Yilgarn Block and Ventersdorp Supergroup; 12, Jeerinah Formation/Lewin Shale, Marra Mamba IF; 13, Deer Lake Complex; 14, Wittenoom Dolomite/Carawine Dolomite; 15, Hamersley Group (data from Strauss & Moore, 1992; this study).

Belt in Zimbabwe were interpreted as a signal of bacterial sulphate reduction (Grassineau *et al.*, 1999). However, 60% of these sulphides define a substantially more narrow range of −0.4 ± 1.4‰, not unlike the $\delta^{34}S$ values of −0.7 ± 0.5‰ reported by these authors for sulphides from volcanic rocks within the same sequence.

Interpretations of the early Archaean sulphur cycle (>2700 Ma) based on the isotopic composition of sedimentary sulphides can be divided into two schools of thought. Acknowledging the absence of a sizeable isotope fractionation between seawater sulphate and sedimentary sulphide and an average isotopic composition for both close to 0‰, in particular for early Archaean successions, a magmatic/hydrothermal origin for these sulphides through discharge of hydrogen sulphide-bearing fluids into the seawater and precipitation due to reaction with dissolved iron is favoured by, for example, Schidlowski *et al.* (1983), Lambert & Donnelly (1990), Hayes *et al.* (1992) and Strauss (1999b). In contrast, the same record of isotope results has been interpreted as being indicative of a biological origin for these sedimentary sulphides (e.g. Ohmoto *et al.*, 1993; Watanabe *et al.*, 1997; Kakegawa & Ohmoto, 1999). These latter authors studied the observed sulphide sulphur isotope values and associated reduced isotope fractionation between sulphate and sulphide as a consequence of higher seawater temperature, moderate sulphate concentrations and high sulphate reduction rates. They inferred environmental conditions thought to result in a reduction in the magnitude of isotope fractionation. In particular, the biological prerequisite, namely a high reduction rate and a resulting reduced isotope effect, has recently been challenged (Habicht & Canfield, 1996). Canfield (1998) proposed that an increase in seawater sulphate concentration occurred not earlier than 2300 Ma, stimulating bacterial sulphate reduction.

A firm conclusion regarding the debate over biological versus magmatic sources of sulphur for the sulphides associated with sedimentary successions of Archaean age, particularly early Archaean, is difficult to attain based on the isotope evidence presently available. One major caveat is our extremely poor knowledge of the isotopic composition of late Archaean seawater sulphate. It provides the reference line against which any discussion of isotope fractionation as evidence for a biological origin has to be carried out. Furthermore, conclusive evidence is required with respect to the magnitude of sulphur isotope fractionation in relation to different parameters: environmental (e.g. sulphate concentrations, oxic versus anoxic water column) and microbiological (e.g. different types of sulphate-reducing bacteria, different reduction rates, effect of temperature). In particular, knowledge with respect to early microbial ecology is still very limited.

Considering the limitations in interpretation of the currently available isotope evidence, a definitive biogenic interpretation of the sedimentary sulphides in Archaean successions cannot be made. Instead, given the geotectonic settings for many of these successions, i.e. greenstone belts with prominent evidence for magmatic/hydrothermal activities, a sulphur cycle which is largely based on inorganic reactions is the favoured alternative. This interpretation is also supported by modelling results for the long-term isotope records of strontium, oxygen, carbon and sulphur (Goddéris & Veizer, 2000). These results clearly indicate that Earth's long-term evolution is driven by tectonics, which exert full control on the geochemical cycles and their temporal isotope trends. The early part of Earth's evolution, including seawater chemistry, is dominated by mantle processes. Followed by the growth of a sizeable continental crust, a substantial increase in the oxygen abundance of the ocean/atmosphere system and, thus, a corresponding increase in seawater sulphate concentration occurs at the Archaean–Proterozoic transition (2500 ± 250 Ma).

It should be pointed out that even the interpretation of a magmatic/hydrothermal origin for the early Archaean sedimentary sulphides does not preclude the operation of bacterial sulphate reduction early in Earth history. In fact, thermophilic sulphur bacteria are thought to be among the oldest organisms, based on phylogenetic work (e.g. Woese, 1987). In addition, the origin of life is thought to be related to the presence of sulphide (e.g. Wächtershäuser, 1990). The fact that early Archaean sedimentary sulphides on average do not display an isotope fractionation comparable in magnitude to modern marine sediments suggests that bacterial sulphate reduction might have been less important than in younger (and modern) sedimentary systems.

## THE ARCHAEAN–PROTEROZOIC TRANSITION

The Archaean–Proterozoic transition records some major rearrangements of Earth's geotectonic, environmental and possibly biological characteristics (Schopf, 1983; Schopf & Klein, 1992). It is thought to be a time in which the size of continental crust increased substantially (e.g. Condie, 1997) with evidence based on strontium isotope records (e.g. Veizer, 1988; Goddéris & Veizer, 2000). A major increase in oxygenation of the ocean/atmosphere system is suggested, based on the temporal distribution of banded iron-formations, red bed sequences, palaeosols and trends in organic carbon isotope ratios (discussions in Des Marais *et al.*, 1992; Holland, 1992; Kasting, 1993). As a consequence, the chemistry of redox-sensitive elements like sulphur would be affected. Finally, the appearance of eukaryotic life has recently been suggested for this time interval, based on biomarker data from the late Archaean Mount Bruce Supergroup (Brocks *et al.*, 1999).

Two prominent successions record this important transition in Earth's history, notably the Mount Bruce Supergroup on the Pilbara Craton, Western Australia, and the Transvaal Supergroup (and the time-equivalent Griqualand West Supergroup) on the Kaapvaal Craton, southern Africa. Both represent extensive sedimentary successions, hosting important mineral deposits, among them banded iron-formations (e.g. Beukes & Klein, 1992). Both successions contain volcanic units which have provided precise age constraints.

The *Mount Bruce Supergroup* comprises in ascending order the Fortescue, Hamersley and Turee Creek groups. These successions are present in the Hamersley Basin, Pilbara Craton, Western Australia. A comprehensive treatment of the geology and stratigraphy of the Pilbara Craton was provided by Hickman (1990). Recent geochronological work constrains the age of the Mount Bruce Supergroup between 2765 and 2209 Ma (for discussion see Nelson *et al.*, 1999).

Sedimentary units (largely shales) were sampled at various stratigraphic levels within the Fortescue and Hamersley Groups, including sediments associated with the Hamersley iron-formation. The Fortescue Group unconformably overlies the granite-greenstone terrane of the Pilbara Craton. Deposition of the Fortescue Group is constrained within ages of 2765 and 2687 Ma (Arndt *et al.*, 1991). The conformably overlying Hamersley Group was deposited between 2597 and 2449 Ma (Trendall *et al.*, 1998; Nelson *et al.*, 1999).

Beukes (1986), Altermann (1996) and Altermann & Nelson (1998) provided accounts of the geology, stratigraphy, sedimentology and depositional environments of the Griqualand West and Transvaal Supergroups. A maximum age of 2650 Ma was provided by Gutzmer and Beukes (1997) for the Vryburg Formation at the base of the Griqualand West Supergroup. The Ongeluk Lava in the upper part of the Griqualand West Supergroup has been dated at 2222 Ma (Cornell *et al.*, 1996), which was taken as the minimum age for the entire succession. However, a new Pb/Pb-age of 2394 ± 26 Ma for the Mooidrai Dolomite stratigraphically above the Ongeluk Lava (Bau *et al.*, 1999) suggests that this part of the Griqualand West Supergroup is at least 200 Myr older than previously thought. Samples for this sulphur isotope study were obtained from several stratigraphic units across the entire Griqualand West Supergroup and the overlying Mapedi Shale of the Olifantshoek Supergroup. The Olifantshoek Supergroup is a red bed sequence which includes some black shale units. These collectively overlie the Griqualand West Supergroup with angular unconformity. The age of the Mapedi Shale can be constrained by a depositional age of *c*.2400 Ma for the upper part of the Griqualand West Supergroup and an age of 1928 Ma for the Hartley Lava above (Cornell *et al.*, 1998). Results from a systematic study of the isotopic compositions of organic carbon and pyrite sulphur in this succession were previously published (Strauss & Beukes, 1996) and will only be briefly discussed here.

## Results

Sediments from the Mount Bruce Supergroup show abundances of organic carbon between <0.1 and 5.9 wt% ($n = 26$), with no systematic difference between the older Fortescue and the younger Hamersley Group. Sulphide sulphur contents are between <0.1 and 4.2 wt% ($n = 57$). Resulting S/C ratios for all but three samples vary between 0.0 and 1.0 ($n = 26$, see Fig. 6). A positive correlation between both parameters can be observed ($r^2 = 0.73$). HCl-leachable iron contents generally exceed pyrite-bound iron, commonly by substantial amounts. Notable exceptions include samples from the Jeerinah Formation/Lewin Shale, Fortescue Group, with DOP values larger than 0.50 and reaching 0.94. Somewhat higher DOP values (0.15–0.86) have been recorded for the Mount McRae Shale of the Hamersley Group. The sulphur isotopic composition of pyrite (Fig. 7) ranges from −2.2 to +19.2‰ ($n = 43$).

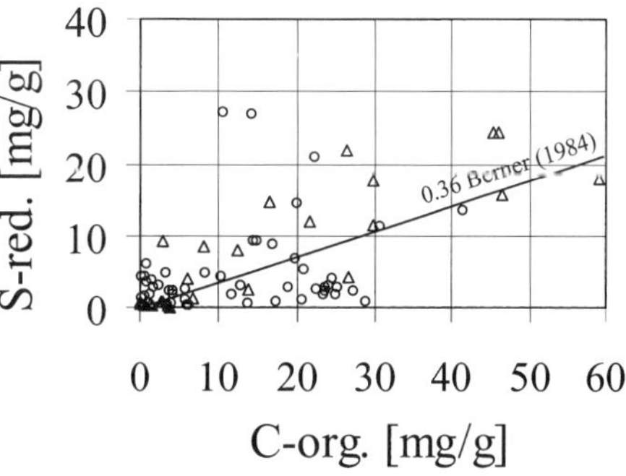


**Fig. 6.** Sulphur–carbon cross-plot for sediments from the Mount Bruce Supergroup, Western Australia, and the Transvaal/Griqualand West Supergroup, South Africa (data from Strauss & Beukes, 1996; this study).

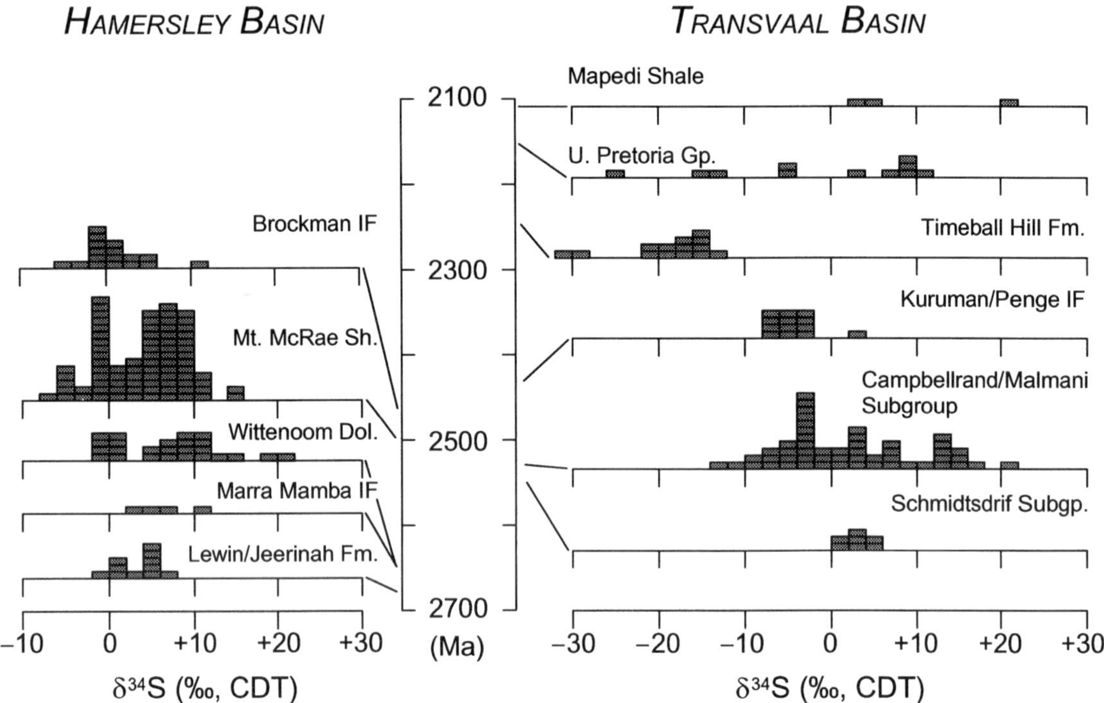


**Fig. 7.** The sulphur isotopic composition of sedimentary sulfides from the Mount Bruce Supergroup, Western Australia, and the Transvaal/Griqualand West Supergroup, South Africa (data from Strauss & Beukes, 1996; Kakegawa *et al.*, 1999; this study).

Organic carbon abundances range from <0.1 to 4.2 wt% ($n = 56$) for sediments from the Griqualand West Supergroup and sulphide sulphur abundances vary between <0.1 and 2.7 wt% ($n = 56$). Overall, both abundances display a concomitant increase, yet with substantial differences at the formation level. This is expressed in an only weak positive correlation ($r^2 < 0.2$). S/C ratios are between 0.1 and 2.4 ($n = 56$), with an average value of 0.4 (Fig. 6). Reactive iron contents (i.e. HCl-leachable plus pyrite-bound iron) vary between 0.3 and 10.9% ($n = 56$). DOP values for most samples from the Griqualand West Supergroup are below 0.5. Higher values between 0.8 and 0.9 were recorded for the Mapedi Shale. The sulphur isotopic composition of sedimentary pyrite (Fig. 7) displays a large range between −1.3 and +23.7‰ ($n = 30$). Individual analytical results have been presented by Strauss & Beukes (1996).

## Discussion

Both the late Archaean and the early Palaeoproterozoic have been proposed as the time in Earth's history when the sulphur isotopic composition of sedimentary sulphides provided the earliest evidence for bacterial sulphate reduction. The late Archaean occurrences from Canada (e.g. Goodwin *et al.*, 1976, 1985; Thode & Goodwin, 1983) or Zimbabwe (Grassineau *et al.*, 1999) quoted as examples for this interpretation were discussed in the previous section. Alternatively, the early Palaeoproterozoic was favoured by Cameron (1982, 1983b) as the oldest record for the activity of sulphate-reducing bacteria in marine sedimentary systems. The temporal evolution of the sulphide sulphur isotopic composition and, in particular, a major shift in $\delta^{34}S$ within the upper part of the Transvaal Supergroup were taken as clear evidence for biogenic processes.

Sulphur isotope results for pyrite and related sulphur–carbon–iron data from the Mount Bruce and the Transvaal/Griqualand West Supergroups will be utilized to assess whether these provide conclusive evidence for a biological origin of the pyrite. Consequently, these two sedimentary successions would then record a shift in the operational mode of the sedimentary sulphur cycle from largely inorganic to a biologically driven system.

As stated above, knowledge of the isotopic composition of ambient seawater sulphate is required if the isotopic fractionation between sulphate and sulphide ($\Delta\delta^{34}S$) can be considered as evidence for a biological origin of the sedimentary sulphides under study. In the absence of marine evaporites, the isotopic composition of early Palaeoproterozoic sea water is solely constrained by limited isotope results from trace quantities of sulphate within carbonate rocks. Bottomley *et al.* (1992) reported $\delta^{34}S$ values of +7.1 and +15.5‰ for two samples from the Carawine Dolomite, Hamersley Supergroup, Australia, although the authors suggested that this sulphate might in fact represent oxidized sulphide. Buchanan & Rouse (1982) suggested a $\delta^{34}S$ value around +15‰ for early Palaeoproterozoic sea water based on the analyses of carbonates from the Malmani Subgroup, Transvaal Supergroup, South Africa. Bottomley *et al.* (1992) measured an additional trace sulphate value of +11.8‰ for the *c.*2300 Ma Huronian Supergroup, Canada. Although poorly constrained, a range in $\delta^{34}S$ between approximately +10.0 and +15.0‰ could be characteristic of early Palaeoproterozoic seawater sulphate.

Sulphur/carbon ratios are quite variable for sediments from the Mount Bruce and Transvaal/Griqualand West Supergroups (Fig. 6). However, a significant positive correlation between both parameters for sediments from the Mount Bruce Supergroup points towards bacterial sulphate reduction as the likely origin of the sedimentary pyrite.

Sedimentary sulphides from various stratigraphic levels within the Fortescue and Hamersley groups (Fig. 7) define a range of $\delta^{34}S$ values between −6.0 and +21.0‰ ($n = 137$; data from this study; Strauss & Moore, 1992; Kakegawa *et al.*, 1999). Although a large spread in $\delta^{34}S$ values for sulphides was previously recorded in older sediments (e.g. late Archaean BIFs), the data represent for the first time in Earth's history an average $\delta^{34}S$ value that is well removed from 0‰ (typical for mantle sulphur; von Gehlen, 1992), in this case towards more positive values. The sulphur–carbon cross-plot (Fig. 6) suggests that pyrite formation was a result of bacterial reduction of seawater sulphate. Strongly $^{34}S$-depleted sulphides, however, are absent. Considering the S–C–Fe relationships and the $\delta^{34}S$ data, the overall picture is, nevertheless, clearly different from the Archaean examples discussed in the previous section. Although not fully comparable to younger sedimentary systems, available evidence points towards biogenic fractionation.

Assuming a biological origin for sedimentary sulphides from the Mount Bruce Supergroup, as has been recently proposed for the Mount McRae Shale, Hamersley Group (Kakegawa *et al.*, 1999), the maximum sulphur isotopic fractionation between a range in $\delta^{34}S$ for Palaeoproterozoic seawater sulphate of +10.0 to +15.0‰ (see above) and a minimum $\delta^{34}S$ value for pyrite of −6.0‰ would be between 16.0 and 21.0‰. While this is substantially less than that observed in modern marine or Phanerozoic sediments (for a recent review, see Strauss, 1997, 1999a), it is decidedly larger than fractionations observed for early Archaean sedimentary units. Isotope variability and in particular the positive $\delta^{34}S$ values would then be best explained in terms of limitations in sulphate availability. As a consequence of the preferred reduction of the $^{32}S$ isotope, progressive enrichment in $^{34}S$ of the residual sulphate occurs. This signal is subsequently transferred to the later-formed sedimentary sulphides.

An even larger spread in $\delta^{34}S$ values between −31.0 and +23.2‰ ($n = 100$) has been measured for sedimentary sulphides in the Transvaal/Griqualand West Supergroup (data from Strauss & Moore, 1992; Strauss & Beukes, 1996; Watanabe *et al.*, 1997). The temporal evolution of $\delta^{34}S$ records a clear difference in the sulphur isotopic composition of sedimentary sulphides between the lower and upper part of the succession (Fig. 7). Sediments from the Ghaap and Chuniespoort group (lateral equivalents in the Griqualand West and Eastern Transvaal areas, respectively) in the lower part of the Transvaal Supergroup display a range in $\delta^{34}S$ values between −12.0 and +22.0‰, with an average value close to 0‰. In contrast, sulphur isotope values ranging from −31.0 to +23.2‰ and an average $\delta^{34}S$ value displaced from 0‰ to clearly negative values characterize sulphides from the Pretoria and Olifantshoek Supergroups of the upper Transvaal and Griqualand West Supergroups.

Variable sulphur isotope values, but with an average value close to 0‰ for pyrite from the lower part of the succession, resemble examples from the late Archaean for which a largely inorganic origin has been suggested. The strongly negative sulphur isotope values for the upper part of the Transvaal/Griqualand West Supergroup are more clearly indicative of bacterial sulphate reduction as the predominant process of pyrite formation.

## Temporal trend in $\delta^{34}S_{sulphide}$: a synopsis

A comparison of $\delta^{34}S$ data for sedimentary sulphide in successions across the Archaean–Proterozoic transition with those of older Archaean data sets (Fig. 5) reveals differences in isotopic variability (i.e. total spread in $\delta^{34}S$ and average values), as noted by previous investigators (e.g. Schidlowski *et al.*, 1983; Lambert & Donnelly, 1990; Hayes *et al.*, 1992). An increase in the range of $\delta^{34}S$ values is quite obvious between early Archaean and late Archaean to early Palaeoproterozoic successions. More important are substantially negative $\delta^{34}S$ values like those measured for the early Palaeoproterozoic units in the Transvaal and Griqualand West Supergroup and an average sulphur isotopic composition well removed from 0‰. The resulting degree of isotopic fractionation resembles that associated with Phanerozoic/modern style bacterial sulphate reduction. Thus, the Archaean–Proterozoic transition records a fundamental change in the operational mode of the sedimentary sulphur cycle, namely a change from mantle-dominated to biologically driven. In addition, the substantial degree of sulphur isotope fractionation observed for pyrite from the upper Transvaal and Griqualand Supergroups would be consistent with a significant level of sulphate in the ocean.

It should be stated again that this interpretation is based on a conservative evaluation of the currently available evidence, considering geochemical and isotope data measured for relevant sedimentary successions. Our extremely poor knowledge of seawater isotopic composition for the early part of Earth's history and our currently limited understanding of changes in the magnitude of isotopic fractionation as a consequence of different environmental and/or microbiological parameters, like sulphate abundance, water temperature and reduction rate, have to be acknowledged. Historically, these caveats have resulted in two fundamentally opposite interpretations of the same isotope record.

Undoubtedly, the Archaean–Proterozoic transition hosts the oldest unequivocal sulphur isotope evidence for bacterial sulphate reduction, and thus demonstrates the emerging importance of this process for the anaerobic recycling of organic matter in sedimentary environments. Furthermore, this interpretation is consistent with independent evidence for the rate of crustal growth and geotectonic evolution on Earth (e.g. Goddéris & Veizer, 2000).

## THE PALAEOPROTEROZOIC

In addition to the results for the early Palaeoproterozoic units from the Mount Bruce and Transvaal/Griqualand West Supergroups presented in the previous section, several stratigraphic units within the McArthur Basin and Pine Creek Geosyncline, Northern Territory, Australia, and the Lawn Hill Platform and Georgetown Inlier, Queensland, Australia, have been studied for their sulphur isotope geochemistry.

The *McArthur Basin* contains a *c.*12 000 m thick sequence which is only weakly metamorphosed and structurally little deformed. In ascending order, the stratigraphic units are the Tawallah, McArthur, Nathan and Roper groups. Jackson *et al.* (1987) provided a comprehensive treatment of the geology of the McArthur Basin. Rough age control is given through disconformably underlying, *c.*1800 Ma crystalline basement and a suggested minimum age of *c.*1400 Ma for the Roper Group. Page & Sweet (1998), for example, provided geochronological constraints on the McArthur Group, with age data between 1638 ± 7 and 1642 ± 6 Ma for the Barney Creek Formation. A Rb–Sr-age of 1429 ± 31 Ma for illites from the McMinn Formation, Roper Group, was given by Kralik (1982) and is regarded as a minimum age for the entire succession.

Lithologically, the basal Tawallah Group consists of sandstones, some volcanics and a few carbonate horizons. It is disconformably overlain by the McArthur Group, which is dominated by partly stromatolitic carbonates and subordinate fine-grained clastics. Stromatolitic carbonates of the Nathan Group unconformably overlie the McArthur Group. Finally, fine-grained siliciclastics of the Roper Group are the uppermost part of the McArthur Basin sequence. Pb–Zn–Ag deposits within the McArthur Group (HYC deposit; Rye & Williams, 1981), as well as the hydrocarbon potential of the McArthur Basin (e.g. Powell *et al.*, 1987), are economically important.

The Lawn Hill Platform, Mount Isa Orogen and Georgetown Inlier are located to the south-east of the McArthur Basin. Plumb *et al.* (1981) described the geology of these regions. Lithostratigraphically, the deposits of the Lawn Hill Platform are similar to the McArthur Basin sequence, both being part of the North Australian Craton (Plumb, 1979). Comparable rock successions continue into the Mount Isa Orogen to the south-east, including the presence of important Pb–Zn–Ag deposits (McArthur River, Lady Loretta, Mount Isa). The sequence of the Georgetown Inlier is part of the North-east Queensland Precambrian Province.

On the *Lawn Hill Platform*, samples were collected from the McNamara Group. This unit comprises in ascending order the Torpedo Creek Quartzite, the Gunpowder Formation and the Paradise Creek

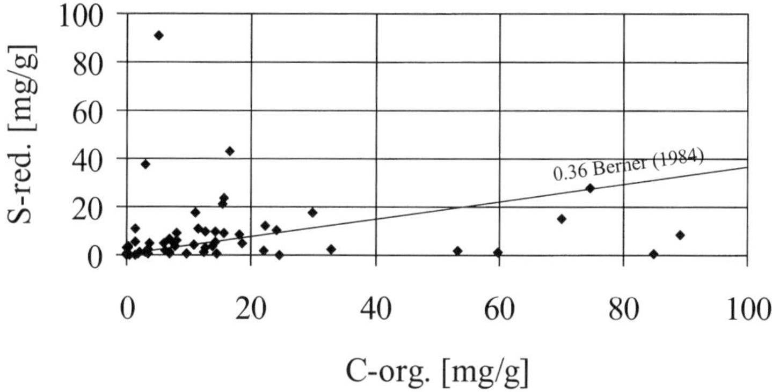


**Fig. 8.** Sulphur–carbon cross-plot for Palaeoproterozoic sediments from the McArthur Basin and Lawn Hill Platform, Australia.

Formation. The Gunpowder Formation consists of laminated mudstones rich in organic carbon and pyrite (subunit A), sandstones and algal dolostones (subunit B) and algal dolostone and intercalated sandstones and organic carbon- and pyrite-rich mudstones (subunit C). The overlying Paradise Creek Formation is composed of stromatolitic dolostones, dolomitic siltstones and chert layers. Hutton (1983) described the stratigraphy, sedimentology and depositional environment. Ages of 1694 ± 3 Ma and between 1653 ± 7 and 1659 ± 3 Ma were provided by Page & Sweet (1998) for the Gunpowder Creek and Paradise Creek formations, respectively.

The Etheridge Group, a 7500–11 000 m thick sedimentary succession with few intercalated volcanics, was deposited over large parts of the Georgetown Inlier. A minimum age of 1570 ± 20 Ma for this group is constrained by a younger folding/metamorphic event recorded in the correlative Einasleigh Metamorphics (Black *et al.*, 1979). The predominantly siliciclastic succession comprises in stratigraphically ascending order the Bernecker Creek Formation, the Robertson River Subgroup, the Townley Formation, the Heliman Formation, the Candlow Formation and the Langdon River Mudstone (Whitnall, 1984). The depositional environment is interpreted as shallow marine, shelf or epeiric sea. Metamorphism increases to the northeast. Samples were collected from the Stockyard Creek Mudstone Member of the Candlow Formation in the southwestern part of the Georgetown Inlier. Whitnall & Mackenzie (1980) and Whitnall (1984) described the geology.

## Results

The Palaeoproterozoic units within the McArthur Group which were studied during this investigation (Mallapunyah, Barney Creek, Lynott and Yalco Formations) have organic carbon contents between <0.1 and 2.4 wt% and sulphide sulphur contents from <0.1 to 4.3 wt%. Resulting S/C ratios (Fig. 8) vary between 0.1 and 12.0 (ave. 1.1 ± 2.3, $n = 24$). DOP values range from 0.10 to 0.62 with two samples showing values of 0.92 and 0.95. Sulphur isotope values for the McArthur Group are between –7.4 and +33.1‰ ($n = 24$). However, positive $\delta^{34}S$ values (+6.6 to +33.1‰) have been measured for all but one sample.

Two units from the Lawn Hill Platform were sampled for geochemical analyses. Low organic carbon abundances around <0.1 wt% and low to moderate sulphide sulphur contents between <0.1 and 0.3 wt% characterize the Paradise Creek Formation ($n = 4$). In contrast, high organic carbon values (0.1 to 2.5 wt%) and moderate sulphide sulphur contents (<0.1 to 1.1 wt%) were measured for the Gunpowder Creek Formation ($n = 11$). DOP values are quite variable between 0.10 and 0.92. A range of positive $\delta^{34}S$ values between 3.9 and 23.1‰ ($n = 7$) was determined for sedimentary pyrite from both formations.

Samples from the Candlow Formation, Georgetown Inlier ($n = 14$), display high organic carbon contents between 0.3 and 8.9 wt% and moderate to high sulphide sulphur values between 0.1 and 9.1 wt%. S/C ratios are generally below 0.1 with two exceptions (12.4 and 17.7). Quite variable DOP values between 0.04 and

0.94 characterize these sediments. Exclusively positive sulphur isotope values (+16.7 to +27.6‰, $n = 9$) were measured for sulphides from the Candlow Formation.

A few additional samples were studied from the Whites Formation, Pine Creek Geosyncline, Northern Australia (*c*.2000 Ma). They show moderate organic carbon contents (0.2 to 1.0 wt%), low sulphide sulphur abundances (<0.1 wt%) and DOP values that are <0.10 ($n = 4$). Two sulphur isotope values were determined for sedimentary pyrite: +21.5 and +23.7‰.

Moderate to high sulphide sulphur contents (0.2 to 2.2 wt%) and variable DOP values between 0.09 and 0.81 were measured for 12 samples from the *c*.1600 Ma Corunna Conglomerate, Adelaide Geosyncline, South Australia. The $\delta^{34}S$ values range from −4.2 to +31.1‰ ($n = 6$).

## Discussion

It was argued in the previous section that the Archaean–Proterozoic transition marks the earliest time in Earth's history with conclusive evidence that sedimentary pyrite is a result of bacterial sulphate reduction. This interpretation is based on the observation that the associated kinetic isotope effect is of comparable magnitude to Phanerozoic/modern systems. As a consequence, sulphur isotope results for pyrite from Proterozoic marine sediments should be discussed in reference to the isotopic composition of ambient seawater sulphate. Both the absolute values ($\delta^{34}S$) and the isotope differences ($\Delta^{34}S_{sulphate-sulphide}$) will be considered in subsequent sections.

The sulphur isotopic composition of Palaeoproterozoic sea water is poorly known, with data from trace sulphate within carbonates of early Palaeoproterozoic age yielding a range in $\delta^{34}S$ values between +10 and +15‰ (see above). Furthermore, barite occurrences in the McArthur Group of Northern Australia display a range in $\delta^{34}S$ between +18.4 and +24.7‰ (Muir *et al.*, 1985) and are interpreted as replacements of evaporitic calcium sulphate minerals, thus very broadly constraining the isotopic composition of late Palaeoproterozoic sea water.

A sulphur/carbon cross-plot (Fig. 8) reveals a positive correlation between both parameters (all samples, $r^2 = 0.32$; Barney Creek Formation, $r^2 = 0.42$, Lynott Formation, $r^2 = 0.82$) for sediments from the McArthur Group, but with quite variable S/C ratios. Thus, sedimentary pyrite is interpreted as biogenic. On average, S/C ratios are above the value for normal marine shales of post-Devonian age. Like Raiswell & Berner's (1986) interpretation for the early Palaeozoic, S/C ratios measured for samples from various stratigraphic levels within the McArthur Group are thought to reflect a more effective bacterial sulphate reduction owing to the highly metabolizable, purely marine organic material. Donnelly & Crick (1992) assume an additional hydrothermal sulphide component within the McArthur Group, based on their interpretation of related sulphur isotope values.

Sulphur isotope values determined during the present study are almost all positive, with a minimum $\delta^{34}S$ value at −7.4 and a maximum of +33.1‰. Compared to the range of $\delta^{34}S$ values determined for the barites from the McArthur Group and thought to represent seawater sulphate, the minimum $\delta^{34}S$ value for the sulphides would yield an overall isotope difference between 25 and 32‰. The magnitude of this $\Delta^{34}S$ value is quite comparable to Phanerozoic/modern sediments, without, however, the additional fractionation associated with disproportionation. The maximum $\delta^{34}S$ value for pyrite at +33.1‰ is more positive than the corresponding seawater composition. With respect to a biological origin for these pyrites, this points to a selective and strong $^{34}S$ enrichment owing to the preferential reduction of the $^{32}S$ isotope during bacterial sulphate reduction (e.g. Chambers & Trudinger, 1979) in a sulphate-limited environment—a process known as Rayleigh fractionation.

Sedimentary pyrite from units within the Lawn Hill Platform succession are regarded as biogenic. Although quite variable, S/C ratios for the Gunpowder Creek Formation are low, indicating an overabundance of organic carbon in a sulphate-limited environment (Fig. 8). Consistent with this interpretation are exclusively positive $\delta^{34}S$ values (+3.9 to +23.1‰), with some approaching the value for ambient seawater sulphate.

A similar interpretation is invoked for sedimentary pyrite from the Palaeoproterozoic Candlow Formation, Georgetown Inlier. Again, high organic carbon contents and moderate to high sulphide sulphur contents result in low S/C ratios. On a sulphur–carbon cross-plot, these samples define a field which is parallel to the carbon axis, suggesting a low sulphate environment. Positive $\delta^{34}S$ values for pyrite approaching (and some even exceeding) the assumed sulphur isotopic composition of seawater sulphate suggest Rayleigh fractionation processes linked to sulphate-limited conditions during pyrite formation.

Additional sulphur isotope data for sedimentary pyrite from the Whites Formation, Pine Creek Geosyncline, and from the Corunna Conglomerate, Adelaide Geosyncline, are equally variable, including strongly positive $\delta^{34}S$ values.

In summary, sulphur isotope data measured for pyrite from different late Palaeoproterozoic successions in Australia are quite similar, with numerous $\delta^{34}S$ values approaching and some even exceeding the sulphur isotopic composition of contemporaneous seawater sulphate. This pattern strongly suggests that bacterial sulphate reduction as the principal process of pyrite formation in these sediments was generally characterized by sulphate limitation. Whether this suggests a lower than present day concentration of sulphate in the seawater or whether it pertains largely to the site of bacterial sulphate reduction in the sediment remains unclear. High organic carbon contents at rather low sulphide sulphur abundances suggest a sulphate-limited environment, at least for some sediments.

### Temporal trend in $\delta^{34}S_{sulphide}$: a synopsis

Results obtained during this study supplement an existing set of sulphur isotope values for pyrite in Palaeoproterozoic sediments from Africa, Australia, North America and Europe (data compiled by Strauss & Moore, 1992; Strauss, 1993b). From the temporal trend of $\delta^{34}S$ values for sulphides (Fig. 9), some prominent features are discernible which clearly distinguish the Palaeoproterozoic from the Archaean record. These include:

**1** Highly variable sulphur isotope values for sedimentary pyrite within a single stratigraphic unit with a maximum spread in $\delta^{34}S$ of 60‰ (e.g. Gunflint Formation, Canada: Carrigan & Cameron, 1991).

**2** Negative sulphur isotope values in the range of −20 to −30‰ reflect a difference between the (poorly constrained) isotopic composition of ambient seawater sulphate and sedimentary pyrite of 30 to 40‰.

**3** Positive $\delta^{34}S$ values up to +40‰ exceed the sulphur isotopic composition of ambient seawater sulphate.

A significant increase in sulphur isotope variability for sulphides relative to the early Archaean has been recorded for the Archaean–Proterozoic transition and the continuation into the Palaeoproterozoic. In fact, this variability in $\delta^{34}S_{sulphide}$ appears to be a characteristic feature of the entire Proterozoic and also the Phanerozoic (for the latter see Strauss, 1997, 1999a). This variability in $\delta^{34}S$ values reflects a change in those parameters affecting the sulphur isotopic composition during progressing bacterial sulphate reduction under sulphate-limited conditions.

The difference between the isotopic composition of seawater sulphate and the most negative sulphur isotope value measured for pyrite in a given stratigraphic unit probably reflects the maximum isotope effect associated with the process of bacterial sulphate reduction and assumed to have occurred under optimal conditions (e.g. Strauss, 1997). Experimental work suggests a maximum net fractionation for bacterial sulphate reduction of 46‰ (Harrison & Thode, 1958). In modern marine sediments, however, the maximum overall sulphur isotope difference between seawater sulphate and sedimentary pyrite can be substantially larger as a consequence of isotopic fractionation associated with disproportionation reactions (e.g. Canfield & Thamdrup, 1994).

Although poorly constrained, the isotopic composition of seawater sulphate in the early Palaeoproterozoic is thought to be between +10 and +15‰ and closer to +20‰ in the later Palaeoproterozoic (see discussion above and Strauss, 1993a). Several sedimentary units of early Palaeoproterozoic age display $\delta^{34}S$ values for pyrite of approximately −20 to −30‰. This results in an isotope difference between sulphate and sulphide of 30 to 40‰. Isotope differences of similar magnitude result for the late Palaeoproterozoic. Both $\Delta^{34}S$ values are quite comparable to respective results for Phanerozoic/modern sediments.

In summary, sulphides from sediments of Palaeoproterozoic age were largely formed by bacterial sulphate reduction with environmental conditions ranging from optimal to sulphate-limited. The observed isotope differences between seawater sulphate and pyrite are attributed solely to bacterial sulphate reduction. No indication is given for disproportionation reactions, which would result in even larger fractionation effects.

## THE MESOPROTEROZOIC

The *Roper Group* represents the upper part of the succession in the McArthur Basin, Northern Territory, Australia. A minimum age of 1429 Ma (Kralik, 1982) has been suggested for this predominantly siliciclastic unit. The entire succession is structurally undeformed, has experienced little thermal overprint and is well known for its hydrocarbon potential (Jackson *et al.*, 1987). Several stratigraphic units within the Roper Group were sampled and analysed during this study (in ascending order: the Mainoru, Corcoran, McMinn and Velkerri formations), thus supplementing the results from the Palaeoproterozoic part of the McArthur Basin succession presented in the previous section.

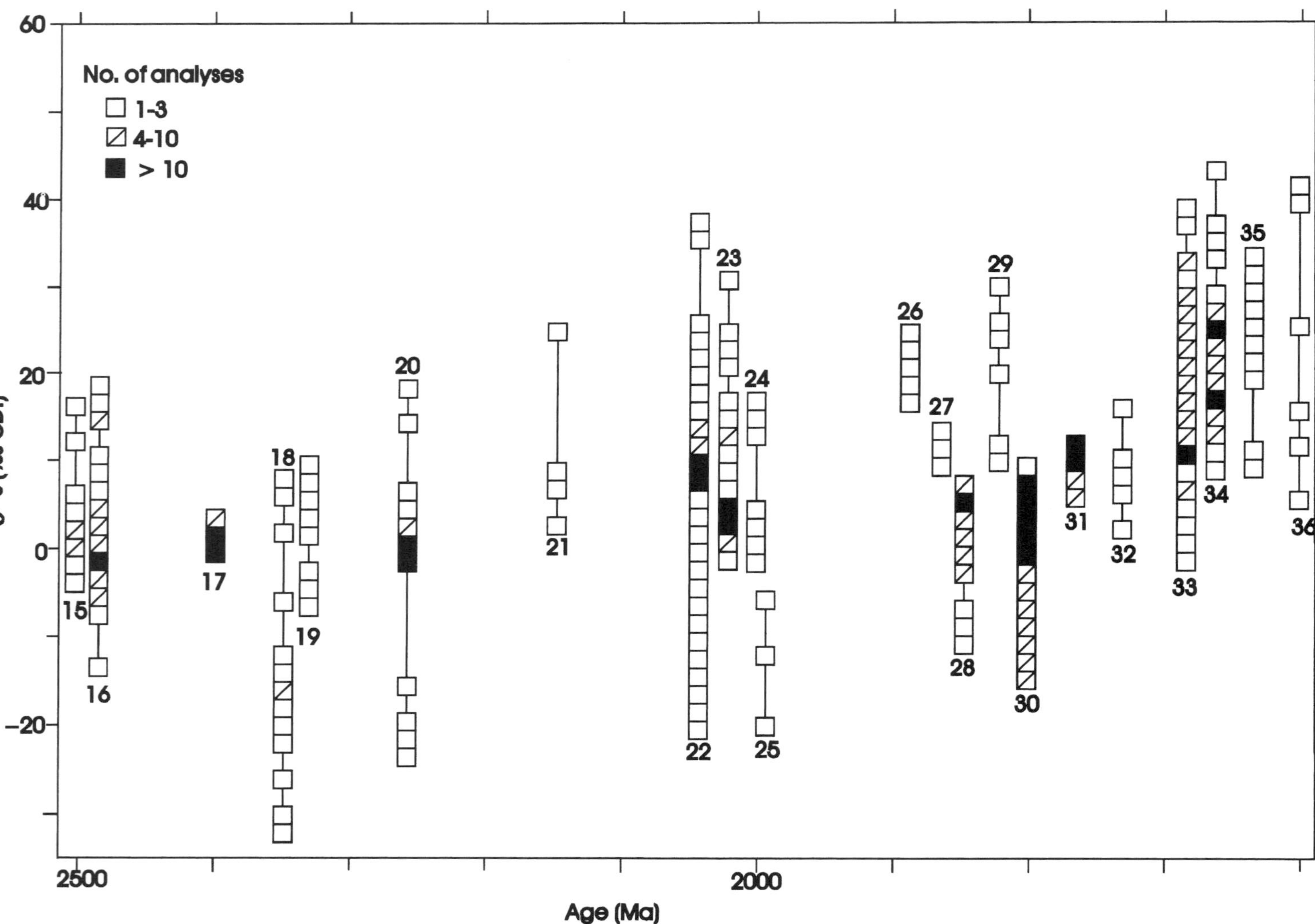


**Fig. 9.** The sulphur isotopic composition of Palaeoproterozoic sedimentary sulphides: 15, Hamersley Group; 16, Lower Transvaal Supergroup; 17, Lower Huronian Supergroup; 18, Aravalli Supergroup; 19, Upper Huronian Supergroup; 20, Upper Transvaal Supergroup; 21, Mapedi Shale; 22, Gunflint Formation; 23, Pine Creek Geosyncline; 24, Early Proterozoic BIF; 25, Udokan Series; 26, Rove Formation; 27, Virginia Formation; 28, Skellefte Field district; 29, Chelmsford and Onwatin Formation; 30, Outokumpu district; 31, Amisk Group; 32, Wollogorang Formation; 33, Lower McArthur Group; 34, Mt Isa Group; 35, Lawn Hill Platform and Georgetown Inlier; 36, Corunna Conglomerate (data from Strauss & Moore, 1992; this study).

## Results

Low to moderate abundances of organic carbon (<0.1 to 9.0 wt%, $n = 43$) and sulphide sulphur (<0.1 to 6.3 wt%, $n = 43$) abundances characterize the sedimentary units, although no clear stratigraphic variation is discernible. Furthermore, both parameters display a positive correlation ($r^2 = 0.76$), but with quite variable S/C ratios. With few exceptions, DOP values are <0.1 ($n = 42$), clearly indicating that no iron deficiency existed during pyrite formation. $\delta^{34}S$ values for sedimentary pyrite vary between −16.6 and +48.3‰ ($n = 27$), with a substantial number of samples that show positive to strongly positive sulphur isotopic compositions.

## Discussion

Data for organic carbon, sulphide sulphur and iron abundances obtained during this study supplement the results of a detailed previous study by Donnelly & Crick (1988). The positive correlation between organic carbon and sulphide sulphur abundances is consistent with a biological origin of the pyrite. Quite variable S/C ratios suggest that geochemical conditions might have been variable during sediment deposition and early diagenesis. Only a few samples—e.g. in the Velkerri Formation—exhibit moderate to high DOP values between 0.50 and 0.93, indicating some apparent iron deficiency during pyrite formation.

Sulphur isotope values measured for pyrite during this study and those previously published (Donnelly & Crick, 1988) display a total variation in $\delta^{34}S$ of more than 60‰, again clearly indicating variable conditions during pyrite genesis. For about half of these samples, $\delta^{34}S$ values higher than +20‰ have been measured. How does this compare to the isotopic composition of Mesoproterozoic sea water? Limited information is available in that respect (for discussion, see Strauss, 1993a). Relevant data have been obtained for barite occurrences within the lower Belt Supergroup, USA, ($\delta^{34}S$ between +13.3 and +18.3‰: Rye *et al.*, 1983; Strauss & Schieber, 1990; Zieg & Leitch, 1998) and in the Grenville of New York and Ontario (+14.5 to +28.6‰: Brown, 1973; Whelan *et al.*, 1990), which are thought to reflect original evaporitic calcium sulphates, and thus constrain the isotopic composition of ambient seawater sulphate. Gellatly *et al.* (2000) measured the sulphur isotopic composition of carbonate-associated sulphate in carbonates from the lower Belt Supergroup. Preliminary $\delta^{34}S$ values range from +13.7 to +23.2‰, with all but one value above +20‰. These results correspond to the sulphur isotope values measured for barite from this unit.

Kah *et al.* (2001) determined sulphur isotope data between +24 and +32‰ for gypsum from the *c.*1200 Ma (Kah and Marcantonio, personal communication) Society Cliffs Formation in the Canadian arctic. A detailed analysis of this evaporite section yielded a steady increase up-section. In summary, the sulphur isotopic composition of seawater sulphate appears to have experienced secular variations throughout the time interval between 1500 and 1000 Ma, some of which could be documented even within an individual evaporite unit (Kah *et al.*, 2001).

Given a range in $\delta^{34}S$ values for seawater sulphate at 1500 Ma between +13.3 and +18.3‰ (with evidence stemming from the lower Belt Supergroup), the observed minimum $\delta^{34}S$ value of −16.6‰ for sedimentary pyrite from the Roper Group reflects an isotopic difference between sulphate and sulphide of 30–35‰. While this is consistent with a biogenic origin for the pyrite through bacterial sulphate reduction, the calculated fractionation is less than what is often observed in Phanerozoic/modern sediments or late Proterozoic sedimentary units. It suggests the absence of disproportionation reactions (e.g. Canfield & Teske, 1996) and would also be consistent with a reduced magnitude of sulphur isotope fractionation as a consequence of a lower than present day sulphate concentration in seawater. Strongly positive $\delta^{34}S$ values for sedimentary sulphides as high as +48‰ exceed the value of Mesoproterozoic seawater sulphate. This is a feature which appears to be prominent in many Proterozoic and early Paleozoic units (Strauss, 1999a), and is consistent with the development of sulphate-limited conditions at the site of bacterial reduction and a corresponding Rayleigh type isotope fractionation. Ways to achieve this include turnover of available sulphate more rapidly than the sulphate is replenished. The effectiveness of sulphate utilization could be a consequence of easily metabolizable organic matter, but would also be consistent with a lower initial seawater sulphate concentration.

## Temporal trend in $\delta^{34}S_{sulphide}$: a synopsis

Apart from the data for the Roper Group, McArthur Basin, Australia, the temporal evolution of the sulphur isotopic composition of sedimentary pyrite in the Mesoproterozoic (Fig. 10) is further constrained by data sets for the Belt Supergroup, USA and Canada, and the Euralia Beds in southern Australia (data summarized by Strauss & Moore, 1992; Lyons *et al.*,

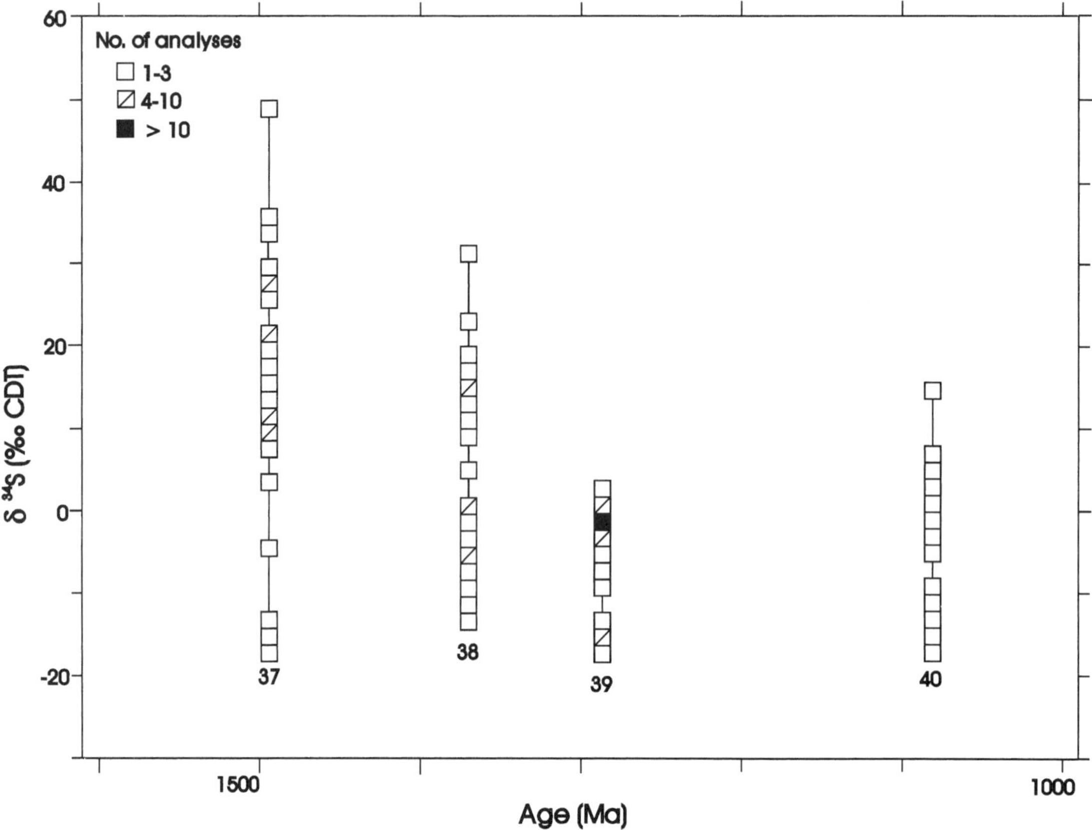


**Fig. 10.** The sulphur isotopic composition of Mesoproterozoic sedimentary sulphides: 37, Roper Group; 38, Lower Belt Supergroup; 39, Upper Belt Supergroup; 40, Eurelia Beds (data from Strauss & Moore, 1992; this study).

2000). Results imply pyrite genesis through bacterial sulphate reduction, partially under sulphate-limited conditions. For the Belt Supergroup, Lyons *et al.* (2000) suggest temporal variations in the supply of marine sulphate, given an initially lower concentration of sulphate in the Mesoproterozoic ocean. Considering minimum $\delta^{34}S$ values at −20‰ (Fig. 10) for sedimentary pyrite of Mesoproterozoic age and maximum $\delta^{34}S$ values at +32‰ for coeval seawater sulphate, a maximum difference in $\delta^{34}S$ between ambient seawater sulphate and biogenic pyrite of 52‰ can be calculated. This is quite comparable to values observed in Neoproterozoic and Phanerozoic sediments, which are interpreted to reflect bacterial sulphate reduction plus disproportionation. Future studies will have to reveal whether disproportionation was a significant process in the Mesoproterozoic versus the Neoproterozoic, as suggested by Canfield & Teske (1996). The large range in $\delta^{34}S$ values, including sedimentary pyrite with an isotopic composition more positive than the values for Mesoproterozoic seawater sulphate, suggests that this process frequently occurred under sulphate-limited conditions.

## THE NEOPROTEROZOIC AND ITS TRANSITION INTO THE EARLY PALAEOZOIC

The Neoproterozoic, in particular the terminal Neoproterozoic, represents a time in Earth's history that witnessed major rearrangements of the geotectonic regime (e.g. break-up of the supercontinent Rodinia; Hoffman, 1991), a set of global low-latitude glaciations including the proposed Snowball Earth (e.g. Hoffman *et al.*, 1998) and important changes in the evolution of life, including the appearance of metazoans (Glaessner, 1984) followed by a rapid radiation of all modern phyla during the so-called Cambrian Explosion. As a consequence, major perturbations of the global ocean/atmosphere system occurred. Sedimentary successions of Neoproterozoic age have attracted numerous geological, palaeobiological or geochemical studies. The latter have focused on chemical sediments as a proxy for seawater composition. Respective studies resulted in a major data set—e.g. for carbon and strontium isotopes (Kaufman *et al.*, 1993)—which has been utilized for chemostratigraphic correlations of Neoproterozoic sequences (e.g. Kaufman & Knoll, 1995) and the transition into the Phanerozoic. During this study, sedimentary successions have been analysed from the Amadeus Basin, central Australia; the Adelaide Geosyncline/Stuart Shelf, South Australia; the Varanger Peninsula, northern Norway; different locations in central and northern Spain; the East European Platform, Poland; and the stratotype for the Precambrian–Cambrian boundary, south-east Newfoundland, Canada. Sulphur isotope data for pyrite from the Newfoundland sequence have been published elsewhere (Strauss *et al.*, 1992b) and are discussed only briefly here.

The *Amadeus Basin* of central Australia represents an intracratonic basin which forms part of the Centralian Superbasin (Walter *et al.*, 1995). It comprises a *c.*14 km thick succession of predominantly shallow-water sediments which cover a stratigraphic range from the Neoproterozoic to the Devonian. Lithologically, the Neoproterozoic consists of terrigeneous siliciclastics, shallow-water carbonates, evaporites and two horizons of glacial sediments. Preiss *et al.* (1978), Lindsay (1987) and most recently Walter *et al.* (1995) have described the geology and stratigraphy of the Neoproterozoic and early Palaeozoic portions.

Age control for the largely sedimentary succession is rather poor and primarily based on lithostratigraphic correlations across the Centralian Superbasin, with successions in the Adelaide Geosyncline. Through correlation, a U–Pb zircon date of 802 ± 10 Ma for volcanics near the base of the Adelaide Geosyncline is regarded as a maximum age for the Centralian Superbasin sequence. The base of the Cambrian can be constrained biostratigraphically (e.g. Zang & Walter, 1992). An age of 543.9 ± 0.24 Ma has been determined elsewhere for the Precambrian–Cambrian boundary (Bowring *et al.*, 1993).

Lithologically, the succession rests unconformably on Archaean basement and begins with the fluvial to shallow marine clastics of the Heavitree Quartzite, followed up-section by partially stromatolitic carbonates and evaporites of the Bitter Springs Formation. These are overlain by two glacial units, the Areyonga and Olympic formations with the intercalated non-glacial siliciclastics of the Aralka Formation. The top part of the Neoproterozoic sequence is represented by predominantly clastic sediments of the Pertatataka and Julie formations. The Precambrian–Cambrian transition lies within the Arumbera Sandstone.

One of the most complete sedimentary successions of terminal Neoproterozoic age is preserved within the *Adelaide Geosyncline/Stuart Shelf*, South Australia. Stratigraphic units subdivide the sequence in ascending order into the Warrina Supergroup (Callana and Burra groups) and the Heysen Supergroup (Umberatana and Wilpena groups). A comprehensive treatment of the geology and stratigraphy is available in Preiss (1987). Volcanics in the basal Callana Group provide a maximum age of 802 ± 10 Ma (Fanning *et al.*, 1986) for the Adelaide Geosyncline/Stuart Shelf successions. Again, the Precambrian–Cambrian boundary provides the minimum age. Furthermore, lithostratigraphic correlations with the sequence in the Centralian Superbasin, including the two tillite horizons (Sturtian and Marinoan glacial units), provide additional stratigraphic control.

The depositional history of the Callana Group records stable platform sedimentation followed by rifting and basic volcanism. The sediments comprise siliciclastics, as well as carbonates, deposited in continental to marine environments. The overlying Burra Group begins with fluvial to marine clastic sediments, followed by transgressive–regressive fine-grained clastics and carbonates deposited in a nearshore to lagoonal setting. The unconformably overlying Umberatana Group contains predominantly clastic sediments, including the Sturt Tillite. Predominantly clastic marine sediments of the Wilpena Group represent the top of the Neoproterozoic sequence. The famous Ediacara fauna within the Pound Subgroup was first described by Glaessner & Wade (1966).

Virtually unmetamorphosed rocks of Neoproterozoic age are present on the *Varanger Peninsula*, northern Norway. They have been subdivided into two distinctive sequences of sedimentary rocks (Siedlecka, 1985) in the Tanafjord–Varangerfjord region to the west and south-west and in the Barents Sea region towards the north-east. Both depositional basins are separated by a major NW–SE trending fault system, the Trollfjord–Komagelv fault zone. Rock successions can be correlated using biostratigraphy (e.g. Vidal & Siedlecka, 1983).

In the Tanafjord–Varangerfjord region, the succession comprises in ascending order the Vadsø,

Tanafjorden and Vestertana groups. The Vadsø Group consists predominantly of terrigenous material and rests unconformably upon basement rocks. It is overlain by detrital rocks with subordinate carbonates of the Tanafjorden Group. Both sequences were deposited mainly in shallow marine environments. The Tanafjorden is overlain by rocks of the Vestertana Group. These include the Smalfjorden, Nyborg and Mortensnes formations, which represent two tillite levels of the Varanger glacials and an intercalated horizon of clastics and thin carbonates (Edwards, 1984), followed up-section by the siliciclastic Stappogiedde and Breivika formations. The Stappogiedde Formation hosts Ediacaran medusoid faunas (Farmer *et al.*, 1992). The Precambrian–Cambrian boundary is placed in the lower part of the Breivika Formation.

A whole rock Rb–Sr isochron from shales in the lower portion of the Vadsø Group yielding an age of 805 ± 7 Ma provides age control for this succession (Sturt *et al.*, 1975). Another Rb–Sr isochron age for the interglacial Nyborg Formation yielded an age of 653 ± 7 Ma. Further constraints are given by biostratigraphic data, notably acritarch occurrences which can be correlated across Baltoscania and the East European Platform (Vidal, 1981).

The succession within the Barents Sea region comprises the Barents Sea and Løkvikfjellet groups, which are considered to be older than 640 Ma. An age of 810 ± 19 Ma was determined for the Kongsfjord Formation at the base of this sequence (Siedlecka & Edwards, 1980). As stated before, the contact between both regions is fault-controlled, and correlation of both successions is difficult. Vidal & Moczydłowska (1995) provided a comprehensive treatment of the Neoproterozoic stratigraphy in northern Norway.

Neoproterozoic and Cambrian sedimentary rocks are exposed in *central and northern Spain*. Palacios & Vidal (1992) and Vidal *et al.* (1994) provided accounts on the geology and micropalaeontology and, most importantly, a reassessment of biostratigraphic correlations.

The Upper Vendian and Lower Cambrian succession in central Spain has been subdivided into the informal Domo Extremeño and Ibor groups (platformal sequence) and the Domo Extremeño and Rio Huso groups (basinal sequence). Sedimentary units studied here include in ascending order the Estenilla, Cijara and Pusa Shale formations. The Precambrian–Cambrian boundary probably occurs within the Pusa Shale (Vidal *et al.*, 1994). In northern Spain, a clear angular unconformity marks the Precambrian–Cambrian boundary. Siliciclastic rocks of the Neoproterozoic Narcea Group are unconformably overlain by the lower Cambrian Herreria Formation.

A thick succession of unmetamorphosed and structurally undeformed Neoproterozoic and Cambrian mudstones and siltstones rests unconformably upon crystalline basement of the *East European Platform*, Poland. Arén & Lendzion (1978) and Moczydłowska (1995) described the geology and stratigraphy.

During this study, siliciclastic sediments from the Lublin, Wlodawa and Mazowsze formations (in ascending order) were studied geochemically. Age control for these units is provided by a U–Pb SHRIMP-dating of single zircons from a tuff bed at the top of the Slawatycze Formation (551 ± 4 Ma; Compston *et al.*, 1995) below the Lublin Formation. Moczydłowska (1989, 1991) placed the Precambrian–Cambrian boundary within the Wlodawa Formation based on biostratigraphy.

## Results

Samples from the Pertatataka and Aralka formations, *Amadeus Basin*, central Australia, are characterized by organic carbon contents between <0.1 and 1.0 wt% ($n = 9$) and between 0.1 and 0.3 wt% ($n = 6$), respectively. Sulphide sulphur abundances range from <0.1 to 2.0 wt% ($n = 9$) and <0.1 to 0.4 wt% ($n = 6$), respectively. S/C ratios vary between 1.3 and 18.6 and between 0.2 and 2.5. DOP values between <0.10 and 0.56 for the Pertatataka Formation are generally higher than the values <0.15 for the Aralka Formation.

Quite variable $\delta^{34}S$ values between −19.5 and +22.2‰ ($n = 8$) have been measured for sulphides from the Pertatataka Formation, whereas sulphide samples from the Aralka Formation display strongly positive sulphur isotope values between +41.0 and +57.3‰ ($n = 5$).

Within the *Adelaide Geosyncline*, South Australia, samples from the Bunyeroo Formation, the Tarcowie Siltstone, the Tapley Hill Formation, the Sturt Tillite and the Burra Group were analysed during this study. Moderate to high organic carbon (0.3 to 2.2 wt%, $n = 10$) and sulphide sulphur (0.5 to 3.3 wt%, $n = 10$) abundances characterize the Bunyeroo Formation. Resulting S/C ratios vary between 0.7 and 7.0. Sulphide sulphur abundances between <0.1 and 1.1 wt% ($n = 17$) were measured for sedimentary pyrite from the Tapley Hill Formation. A few additional samples from the other sedimentary units display generally low to moderate sulphide sulphur contents. Similarly, low DOP values of <0.30 were determined for samples from this succession.

Quite variable and frequently strongly positive sulphur isotope values were measured for sulphides from these sediments. $\delta^{34}S$ values between −14.4 and +9.9‰ ($n = 6$) characterize pyrite from the Bunyeroo Formation. Ten samples from the Tapley Hill Formation yielded a range in $\delta^{34}S$ from +6.8 to +50.8‰ for sedimentary pyrite. A sulphur isotope value of +10.3‰ was measured for pyrite in the Sturt Tillite, and a range between +1.4 and +2.8‰ was obtained for three samples from the Burra Group.

Low to moderate organic carbon (<0.1 to 0.4 wt%, $n = 100$) and sulphide sulphur (<0.1 to 0.3 wt%, $n = 114$) contents were measured for different sedimentary units within the Vadsø, Tanafjorden and Vestertana groups, *Varanger Peninsula*, northern Norway. Calculated S/C ratios are highly variable (0.1 to 63.1, $n = 100$). DOP values are generally <0.1. Samples from the Barents Sea Group in the north-eastern part of the Varanger Peninsula show comparable abundances of organic carbon (<0.1 to 1.0 wt%, $n = 67$) and sulphide sulphur (<0.1 to 1.8 wt%, $n = 69$), with resulting S/C ratios between 0.1 and 24.4. Sulphur isotope values range from −19.4 to +36.7‰ for sedimentary sulphides within the Vadsø, Tanafjorden and Vestertana groups ($n = 17$) and from −9.4 to +28.7‰ ($n = 28$) for pyrites in sediments from the Barents Sea Group.

Sediments from central and northern *Spain* show organic carbon contents between <0.1 and 1.0 wt% ($n = 67$) and sulphide sulphur abundances between <0.1 and 5.9 wt% ($n = 83$). Resulting S/C ratios range from 0.1 to 7.1, with a few values substantially higher. DOP values are <0.40. $\delta^{34}S$ values between −17.4 and +49.3‰ ($n = 14$) characterize pyrite within these sediments; the strongly positive values between +32.4 and +49.3‰ derive from the Estenilla Formation in central Spain.

Terminal Neoproterozoic to early Cambrian sedimentary units from the *East European Platform*, Poland, display moderate contents of organic carbon (<0.1 to 0.3 wt%, $n = 60$) and sulphide sulphur (<0.1 to 0.9 wt%, $n = 60$, one value at 3.0 wt%). S/C ratios are variable between 0.5 and 14.0. DOP values are generally low (<0.40). Exclusively positive $\delta^{34}S$ values have been measured for sedimentary pyrite from the Mazowsze (+4.1 to +33.9‰, $n = 13$), Wlodawa (+11.4 to +33.9‰, $n = 7$) and Lublin (+16.8 to +40.5‰, $n = 8$) Formations.

## Discussion

A most prominent feature of Neoproterozoic sediments from Australia, Norway, Poland and Spain appears to be a positive correlation between organic carbon and sulphide sulphur (for all samples: $y = 1.49x + 1.43$; $r^2 = 0.2$; but see differences for individual groups of samples in Fig. 11), and high S/C ratios with extreme values of up to 20:1. The latter are in contrast to respective results from Palaeo- and Mesoproterozoic units and greatly exceed the range for normal marine shales of post-Devonian (Berner,

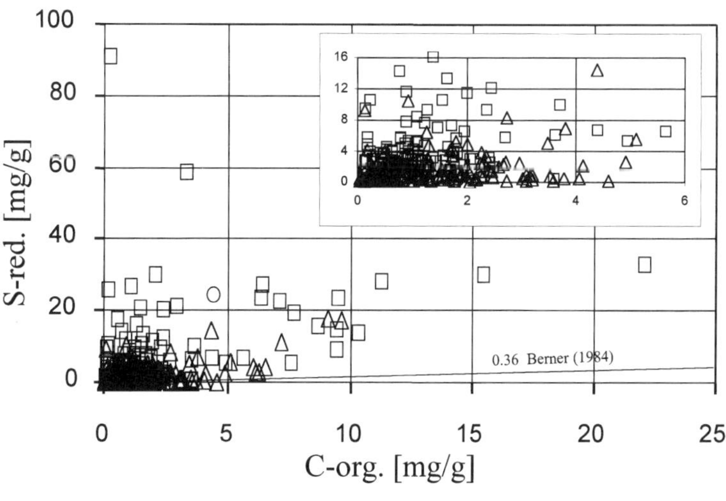


**Fig. 11.** Sulphur–carbon cross-plot for Neoproterozoic sediments from Australia, northern Norway, Poland and Spain.

1984), or even lower Palaeozoic (Raiswell & Berner, 1986) age. In principle, different options can be discussed in order to explain the overabundance of sulphur over organic carbon. Extremely high sulphur/carbon ratios can be a consequence of later diagenetic or epigenetic 'mineralization' (Leventhal, 1983, 1995). However, sulphides within these sediments generally appear as finely disseminated framboidal pyrites, but also idiomorphic pyrite cubes (5 mm) on bedding and cleavage planes. No concretions have been detected, nor do these sediments contain any major sulphide enrichment, pointing to hydrothermal activity. Thus, a biogenic origin of these pyrites is suggested. A relative enrichment of sulphur as a result of post-sedimentary metamorphic loss of organic matter can be excluded owing to the frequently excellent preservation of organic-walled microfossils within these successions. In summary, high S/C ratios are thought to reflect syngenetic to diagenetic values, warranting geochemical/depositional conditions which appear to be a prominent feature in Neoproterozoic sediments.

In their study of Phanerozoic marine shales, Raiswell & Berner (1986) noted a difference in S/C ratio between lower Palaeozoic and younger sediments. In particular, lower Palaeozoic sediments displayed higher S/C ratios, around an average of 2:1, as compared to a S/C ratio of 0.36 for Holocene sediments (Berner, 1984). Raiswell & Berner (1986) considered the bioavailability rather than the absolute abundance of sedimentary organic matter to be most important for the process of bacterial sulphate reduction. The late Palaeozoic (Devonian–Carboniferous) marks the rapid colonialization of terrestrial environments with land plants (e.g. Algeo *et al.*, 1995; Algeo & Scheckler, 1998). As a consequence, detrital terrigenous organic matter was transported into nearshore environments via rivers, resulting in a mixture of marine and terrestrial organic matter as substrates for sulphate reduction and other microbial processes. It is well known, however, that terrestrial organic material is less easily metabolized than marine organic material (Lyons & Gaudette, 1979). By analogy, S/C ratios for Neoproterozoic sediments, which are higher than post-Devonian Phanerozoic values, are interpreted as reflecting this source effect. An additional source of excess pyrite in the sediment results from its syngenetic formation as a consequence of bacterial sulphate reduction in an anoxic water column. This is also reflected by a positive intercept on the sulphur axis for the resulting regression line (Fig. 11). Such a scenario would be consistent with Logan *et al.* (1995), who proposed heterotrophic reworking of primary organic matter by sulphate reducers in the water column, based on their biomarker results.

Purely marine organic matter and anoxic water conditions would be equally likely in sedimentary systems older than Neoproterozoic, yet these older rocks do not show a high S/C ratio (Fig. 12). Thus, no firm conclusion can be drawn at this stage.

The sulphur isotopic composition of sedimentary pyrite from two sedimentary units within the *Amadeus Basin* exibit two distinct populations. Pyrite from the Aralka Formation displays $\delta^{34}S$ values between +41.0 and +57.3‰ ($n = 5$), whereas sedimentary pyrite from the Pertatataka Formation yielded a range in $\delta^{34}S$ between −19.5 and +22.2‰ ($n = 8$). The latter results are in good agreement with data published recently for the Pertatataka Formation by Logan *et al.* (1999). These results are interpreted to reflect a biogenic origin of pyrite through the process of bacterial sulphate reduction. No information exists with respect to the sulphur isotopic composition of sulphate, which acted as a sulphur source for the sedimentary pyrite in the Aralka Formation. The Aralka Formation was deposited in fault-bounded troughs, where the isotopic composition of a limited sulphate reservoir might have been altered significantly during bacterial sulphate reduction. Sediments of the Pertatataka Formation were deposited between the Marinoan glacial event(s) and the base of the Cambrian. During that time, the isotopic composition of global sea water was at +30 to +35‰, as constrained by numerous results from sedimentary evaporites and phosphorite-bound sulphate (e.g. Strauss, 1993a; Shields *et al.*, 1999b). Thus, maximum isotope fractionation between sulphate and sulphide as a result of bacterial sulphate reduction can be calculated at 50 to 55‰. This exceeds the magnitude that can be accounted for solely on the basis of bacterial sulphate reduction as determined through culture experiments (Kaplan & Rittenberg, 1964). Additional isotope fractionation, however, can be achieved through disproportionation of elemental sulphur and/or thiosulphate as has been discussed by, for example, Jørgensen (1990), Canfield & Thamdrup (1994) and Smock *et al.* (1998). This process is thought to have been operable over the past 800 Ma (Canfield & Teske, 1996) and related, among other things, to an increasing oxygen content of the Neoproterozoic atmosphere.

A second point relates to the observation that a spread of more than 40‰ has been recorded for sedimentary pyrite from the Pertatataka Formation, and that the Aralka and Pertatataka Formations show extremely $^{34}S$-enriched sulphur isotope values. These

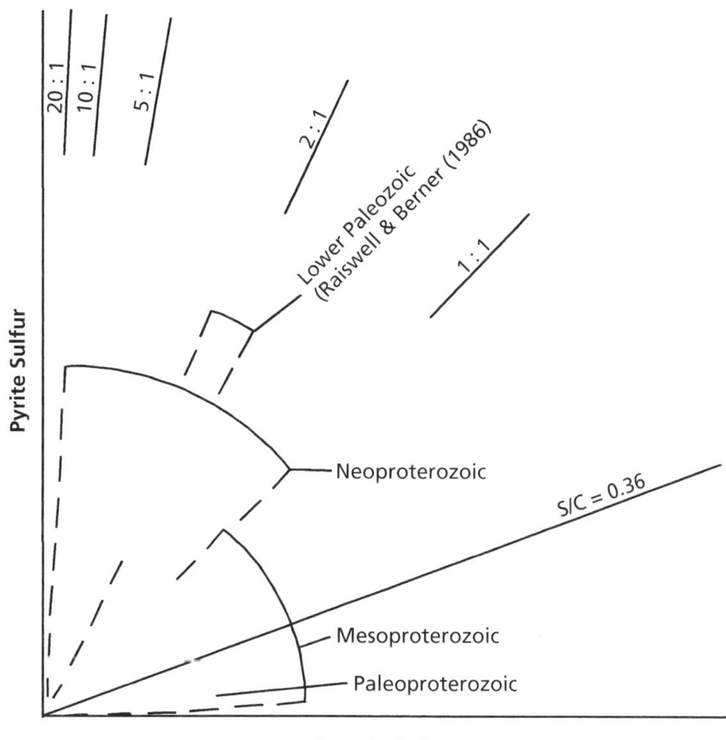


**Fig. 12.** Schematic sulphur–carbon cross-plot for Precambrian sediments.

approach or even exceed the $\delta^{34}S$ value of ambient seawater sulphate. A plausible explanation is the development of sulphate-limited conditions during the course of continuing bacterial sulphate reduction. As a consequence of the preferential reduction of $^{32}S$, a Rayleigh-type fractionation would result in progressively more $^{34}S$-enriched sulphide.

Sedimentary pyrite from different sedimentary units within the *Adelaide Geosyncline* yielded a total range in $\delta^{34}S$ between −14.4 and +50.8‰ with different distributions for the individual stratigraphic units. Sulphur isotope data are consistent with the basic assumption of a biogenic origin for pyrite. Positive to strongly positive $\delta^{34}S$ values were measured for sedimentary pyrite from the Tapley Hill Formation. These new results are in good agreement with sulphur isotope values measured previously for this stratigraphic unit (for compilation see Strauss & Moore, 1992), as well as those recently published by McKirdy *et al.* (2000). Maximum $\delta^{34}S$ values for pyrite greatly exceed the sulphur isotopic composition of ambient seawater sulphate, which was at *c.*+20‰ during this time interval (Strauss, 1993a). Again, this is consistent with bacterial sulphate reduction under sulphate-limiting conditions. In contrast, predominantly negative $\delta^{34}S$ values have been measured for sedimentary pyrite from the Bunyeroo Formation. Deposition of this unit occurred during terminal Neoproterozoic times (post-Marinoan). Given a sulphur isotopic composition of +30/+35‰ for seawater sulphate, maximum sulphur isotopic fractionation during bacterial sulphate reduction can be calculated at 45–50‰.

A thick succession of sedimentary units on the *Varanger Peninsula*, Northern Norway, has been studied for their sulphur isotopic composition. Sedimentary pyrites from different stratigraphic horizons in the Tanafjord–Varangerfjord region exhibit variable $\delta^{34}S$ values: Fugleberget Formation, +12.5‰, $n = 1$; Ekkerøy Formation, +22.9‰, $n = 1$; Dakovarre Formation, +6.3 to +8.5‰, $n = 2$; Smalfjord Formation, +9.0‰, $n = 1$; Stappogiedde Formation, −19.4 to +19.3‰, $n = 7$; Breivika Formation, +16.9 to +36.7‰, $n = 5$. Barents Sea region: Basnæring Formation, +12.3‰, $n = 1$; Båtsfjord Formation, −9.3 to +28.7‰, $n = 22$. Sedimentary pyrite is interpreted as biogenically based on the available elemental and isotopic data.

Sulphur isotope values between −17.4 and +49.3‰ ($n = 14$) were measured for samples from different stratigraphic horizons of terminal Neoproterozoic age in central and northern *Spain*. Results are consistent with a biogenic origin of the sedimentary pyrite. No clear stratigraphic (temporal) variation is discernible from these results.

Sedimentary pyrite in a succession across the Precambrian–Cambrian transition on the *East European Platform*, Poland, yielded exclusively positive sulphur isotope values between +4.1 and +40.5‰ ($n = 28$). These new results supplement a previously published range in $\delta^{34}S$ between +7.6 and +46.6‰ ($n = 13$; Bottomley *et al.*, 1992). Again, these sulphide sulphur isotope data are a result of bacterial reduction of seawater sulphate with a sulphur isotopic composition of +30 to +35‰. Thus, sulphide sulphur isotope values occasionally exceed the respective value for sea water, pointing to sulphate-limited conditions during bacterial sulphate reduction. No significant stratigraphic variations exist; however, pyrite samples from the Cambrian Mazowsze Formation with an average $\delta^{34}S$ value of +22.3‰ appear to be less $^{34}S$ enriched than samples from the Neoproterozoic Lublin Formation (average +32.6‰).

## Temporal trend in $\delta^{34}S_{sulphide}$: a synopsis

The Neoproterozoic is the earliest time in Earth's history for which a somewhat consistent data set documents the sulphur isotopic composition of global seawater sulphate and its temporal evolution. Strauss (1993a) reviewed the available information and provided a temporal trend for seawater sulphur isotopic composition. However, new sulphur isotope data and new age constraints warrant a renewed discussion.

Evaporites from the Society Cliffs Formation,Bylot Supergroup, arctic Canada, previously thought to constrain the sulphur isotopic composition of global sea water at 950 Ma, were probably deposited between *c.*1270 and 1000 Ma as constrained by palaeomagnetic, biostratigraphic and chemostratigraphic data, and supported by a recent Pb–Pb age of 1204 ± 22 Ma (cited in Lyons & Kah, 1999) for carbonates of the Society Cliffs and Victor Bay Formations. In addition, a recent sulphur isotope study of the evaporites yielded a range in $\delta^{34}S$ between +24 and +32‰, with a steady increase up-section (Lyons & Kah, 1999).

Evaporites within the Upper Roan Group, Zambian Copperbelt, yielded $\delta^{34}S$ values between +15.8 and +21.0‰ (Claypool *et al.*, 1980) for anhydrite. This evaporite unit was previously placed at 1100 Ma, but with large age uncertainty. New U–Pb data for the Katangan sequence constrain the deposition of these evaporites to between 880 and 830 Ma (Armstrong *et al.*, 1999).

Additional sulphur isotope data were recently published for evaporites from the Hanseran Formation, north-western India (Strauss *et al.*, 2001), and phosphorite-bound sulphate from the Precambrian–Cambrian transition at Meishucun and correlative sections, south China (Shields *et al.*, 1999b). Sulphate sulphur isotope values between +27.5 and +39.7‰ ($n = 26$) for north-western India and between +31.4 and +37.0‰ ($n = 40$) for south China corroborate the previously determined strongly positive $\delta^{34}S$ signature of terminal Neoproterozoic and early Cambrian sea water. It should be pointed out that all sulphates that display strongly enriched sulphur isotope values lie stratigraphically above the uppermost Marinoan glacial event. This has been dated at 590–600 Ma (Knoll & Walter, 1992) or 575–590 Ma, as suggested by Saylor *et al.* (1998). This age frame for the strongly positive $\delta^{34}S_{sulphate}$ values is further constrained through the biostratigraphic record of the Precambrian–Cambrian type section in south China (Shields *et al.*, 1999b), with the Precambrian–Cambrian boundary at 543.9 Ma (Bowring *et al.*, 1993). The strongly $^{34}S$-enriched signature of terminal Neoproterozoic and early Palaeozoic seawater sulphate is thought to be a result of enhanced bacterial sulphate reduction in a stratified water column and subsequent spill-over of these water masses on the shelf areas. Strauss *et al.* (2001) suggested a causal relationship to the proposed extreme palaeo-oceanographic conditions during a terminal Neoproterozoic Snowball Earth (e.g. Hoffman *et al.*, 1998).

Available sulphur isotope data can be used to construct a temporal trend for the isotopic composition of Neoproterozoic seawater sulphate (Fig. 13), with discussions provided by Shields *et al.* (1999b) and Strauss *et al.* (2001). This temporal trend represents the reference line for the following discussion of the sulphur isotopic composition of sedimentary pyrite.

Apart from the sulphur isotope results measured for pyrite during this study, approximately a dozen additional datasets exist for sedimentary units of Neoproterozoic age. Despite an irregular distribution and several gaps, the temporal record in $\delta^{34}S_{sulphide}$ (Fig. 14) exhibits some clearly discernible features. In order to provide an internally consistent stratigraphic reference, sulphur isotope results for sedimentary pyrite are being discussed as three consecutive sets, each separated in time by one of the two major Neoproterozoic glacial epochs, the Sturtian (*c.*750 Ma) and the Marinoan (*c.*600 Ma). This is also consist-

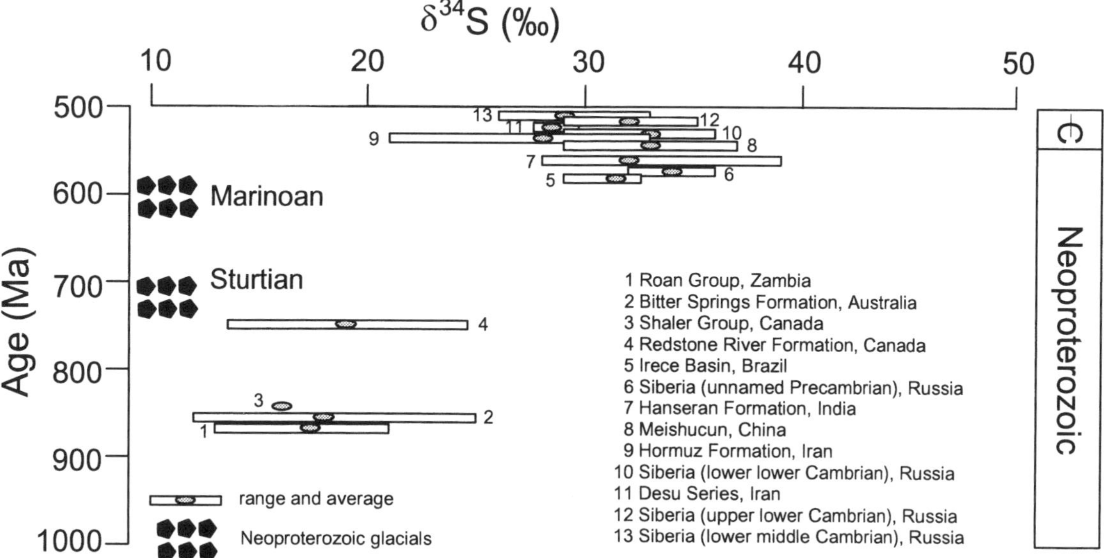


**Fig. 13.** The sulphur isotopic composition of Neoproterozoic sea water.

**Fig. 14.** The sulphur isotopic composition of Neoproterozoic sedimentary sulphides: 41, Nonesuch Shale; 42, Grenville Metasediments; 43, Bijaigarh Formation; 44, Lower Chuar Group; 45, Upper Chuar Group; 46, Little Dal Formation; 47, Burra Group; 48, Tapley Hill Formation; 49, Aralka Formation; 50, Vadsø, Tanafjord and Barents Sea Groups; 51, Ballachulish Slate; 52, Twitya Formation; 53, Sheepbed Formation; 54, Pertatataka Formation; 55, Vestertana Group; 56, Bunyeroo Formation; 57, post-Marinoan of Spain; 58, East European Platform; 59, Neufoundland; 60, Kaza and Isaac Formations; 61, Lesser Himalaya (data from Strauss & Moore, 1992; Ross *et al.*, 1995; Logan *et al.*, 1999; McKirdy *et al.*, 2000; Strauss, unpublished; this study).

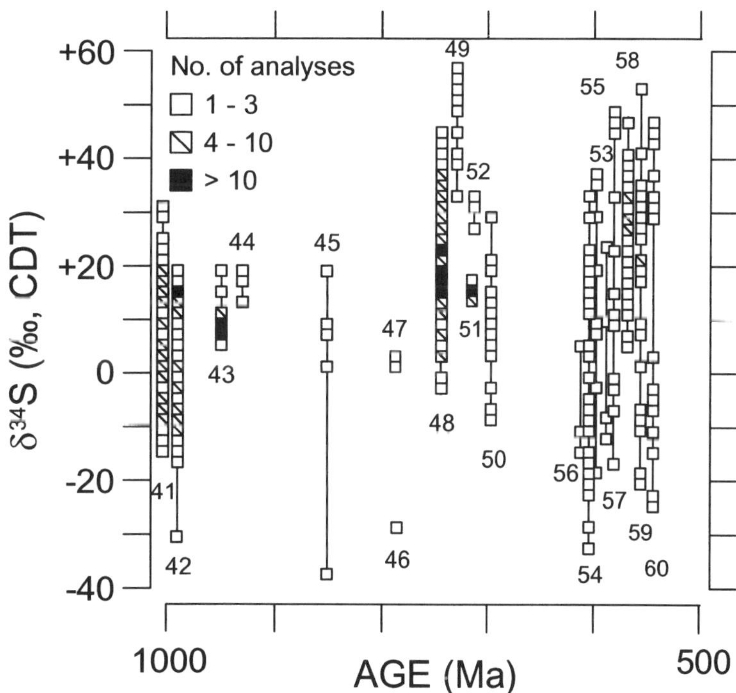


ent with the interpretation that the glacial horizons reflect important sequence boundaries (Christie-Blick *et al.*, 1995; Narbonne & Aitken, 1995; Walter *et al.*, 1995).

The *pre-Sturtian* record of $\delta^{34}S_{sulphide}$ comprises data compiled by Strauss & Moore (1992) that define a range of sulphur isotope values between −17 and +31‰, with sulphur isotope results for three individual samples being even more negative. Again, a substantial spread in $\delta^{34}S$ was demonstrated for the individual stratigraphic units, which probably reflects temporal changes in the geochemical conditions during bacterial sulphate reduction and subsequent pyrite formation, notably the development of sulphate limitation. As mentioned before, this is interpreted as reflecting pyrite genesis under evolving diagenetic conditions as a consequence of enhanced sulphate turnover owing to a more effective mineralization of highly reactive organic matter and/or a possibly lower sulphate concentration in sea water. Sulphide sulphur isotope values approach or sometimes even exceed the sulphur isotopic composition of ambient seawater sulphate, which has been constrained for this time interval at *c*.+20‰. Given this seawater sulphur isotopic composition, a maximum isotope fractionation between sulphate and sulphide of 37‰ would be consistent with values typically observed for bacterial sulphate reduction, yet less than the possible maximum of 46‰ (e.g. Kaplan & Rittenberg, 1964).

*Post-Sturtian, pre-Marinoan* sedimentary pyrites exhibit a total range in $\delta^{34}S$ of 66‰ (−9.4 to +57.3‰). Furthermore, individual stratigraphic units display comparably large ranges in $\delta^{34}S$, e.g. 49‰ for the Tapley Hill Formation, South Australia. Most prominent, however, are the extremely $^{34}S$-enriched values as measured, for example, for interglacial sediments from the Mackenzie Mountains of north-western Canada (Twitya Formation), the Amadeus Basin of central Australia (Aralka Formation) and the Adelaide Geosyncline/Stuart Shelf, South Australia (Tapley Hill Formation). $\delta^{34}S_{sulphide}$ values between −9.4 and +28.3‰ were measured for sediments from northern Norway, which are bound only by the Marinoan/Varanger glacial deposits at the top of this succession. The strongly positive $\delta^{34}S$ values must reflect geochemical conditions which are different from those characteristic of older Neoproterozoic settings. It is quite likely that these are related to the fact that this time interval is bracketed by two periods of intense global glaciations and different geochemical conditions in the water column.

In the absence of a $\delta^{34}S$ value for ambient seawater sulphate, interpretations of the unusual sulphur isotope signature for sedimentary pyrite can only be tentative. The sulphur isotopic composition of pre-Sturtian seawater sulphate was at *c*.+20‰. Post-Marinoan seawater sulphate had an isotopic composition of +30 to +35‰. Three options exist: (i) seawater sulphate during the interglacial interval had an isotopic composition at *c*.+20‰; (ii) the seawater isotopic composition varied between +20 and +30‰; or (iii) the seawater composition resembled that of post-Marinoan time at *c*.+30 to +35‰.

Given a sulphur isotopic composition for seawater sulphate at *c*.+20‰ (option i), the maximum sulphur isotope fractionation between ambient sea water and the sedimentary pyrite in these interglacial units would be approximately 30‰, while $^{34}S$-enriched values for sedimentary pyrite would exceed the seawater sulphate value by 37‰. A seawater sulphate sulphur isotopic composition at *c*.+30‰ (option iii), analogous to post-Marinoan time, would result in a maximum fractionation of 40‰, while the most positive $\delta^{34}S_{sulphide}$ value would exceed the isotopic composition of seawater sulphate by 27‰. This latter option is favoured here for the interglacial time (i.e. the post-Sturtian). Option ii, an intermediate sulphur isotopic composition for ambient seawater sulphate, or for that matter any other $\delta^{34}S$ value for seawater sulphate, cannot be constrained at all and is not considered here.

It should be noted that preliminary sulphur isotope results for trace sulphate in the cap carbonates above the glacial horizons in Namibia display strongly positive $\delta^{34}S$ values, reaching +40‰ (Hurtgen & Arthur, 2000).

*Post-Marinoan* sediments from the Mackenzie Mountains, Canada (Sheepbed Formation, Kaza and Isaac Formations), the Amadeus Basin, central Australia (Pertatataka Formation), the Adelaide Geosyncline/Stuart Shelf, South Australia (Bunyeroo Formation), the Varanger Peninsula, northern Norway (Vestertana Group), different stratigraphic units in central Spain, the East European Platform, the lesser Himalaya and the succession straddling the Precambrian–Cambrian transition in south-eastern Newfoundland have all yielded a total range in $\delta^{34}S_{sulphide}$ of 81‰ (−28 to +53‰). Again, individual data sets are equally characterized by major variations in $\delta^{34}S$. Terminal Neoproterozoic (and early Palaeozoic) seawater sulphate had a sulphur isotopic composition of +30 to +35‰, which defines the known maximum value for the temporal $\delta^{34}S$ record of global sea water through time (e.g. Strauss, 1993a; Shields *et al.*, 1999b; Strauss *et al.*, 2001).

Comparing the lowest $\delta^{34}S$ value for pyrite and this sulphur isotopic composition of seawater sulphate, a maximum isotope fractionation of approximately 60‰ results. This, however, greatly exceeds the maximum fractionation observed for bacterial sulphate reduction (as defined by experimental work; e.g. Kaplan & Rittenberg, 1964) but resembles modern marine sedimentary pyrite (for review, see Strauss,

1997). The large sulphur isotope fractionation suggests an additional (biologically driven) process that further amplifies the overall kinetic isotope fractionation between parent sulphate and resulting sulphide. As stated above, disproportionation of thiosulphate or elemental sulphur has been identified as a possible cause for enlarged apparent isotope fractionation (e.g. Jørgensen, 1990; Canfield & Thamdrup, 1994; Smock *et al.*, 1998). A combination of sulphate reduction and disproportionation was considered by Canfield & Teske (1996) to be responsible for the isotope pattern observed in marine sediments of Phanerozoic age and possibly as far back as 800 Ma.

## TEMPORAL EVOLUTION OF THE BIOLOGICAL SULPHUR CYCLE DURING PRECAMBRIAN TIMES

The sulphur isotopic composition of sedimentary pyrite as documented by 334 new results and a compilation of previously published data reflects the temporal evolution of the sulphur cycle during Precambrian time. Additional geochemical parameters, including the abundances of organic carbon, sulphide sulphur, reactive iron and the degree of pyritization, further constrain the biogeochemical environment at the time of sediment deposition. A final assessment of results discussed in previous sections provides unequivocal evidence for major changes in the temporal evolution of the sedimentary sulphur cycle during Precambrian time.

The Archaean sulphur cycle, in particular the *early Archaean*, appears to have been dominated by non-biological reactions. Seawater sulphate concentrations were lower than today and their ultimate sulphur source is thought to be magmatic, as evident from the isotopic composition of early Archaean barite. In the absence of any large-scale sulphur isotope fractionation between seawater sulphate and sedimentary pyrite, pyrite sulphur is interpreted as being of inorganic, hydrothermal/magmatic origin. This would be consistent with environmental conditions in the ocean/atmosphere system that were dominated by mantle processes, as suggested by independent geochemical data (e.g. Veizer *et al.*, 1989). In contrast, specific environmental and/or metabolic conditions, such as high rates of sulphate reduction at low to moderate seawater sulphate concentrations, have been postulated to account for a greatly reduced sulphur isotope fractionation during bacterial sulphate reduction (Ohmoto *et al.*, 1993), but have been challenged by microbiological and combined sulphur isotope data (Habicht & Canfield, 1996). At present, this interpretation, which favours a magmatic/hydrothermal origin of pyrite sulphur in the early Archaean, appears to be at variance with independent evidence from molecular phylogeny for the antiquity of sulphate reducers (Woese, 1987). However, no unequivocal isotopic evidence, i.e. large sulphur isotope fractionations between seawater sulphate and sedimentary pyrite, as observed in younger sedimentary systems, exists for a dominantly biologically controlled sedimentary sulphur cycle.

The *late Archaean* and its transition into the *early Palaeoproterozoic* marks the earliest appearance of major offsets between the sulphur isotopic composition of ambient seawater sulphate, although only poorly constrained at +10 to +15‰, and sedimentary pyrite, with values as negative as −20‰. This represents an isotope fractionation of some 30‰, which is well within the range observed for the process of bacterial sulphate reduction in modern marine sediments and laboratory experiments (e.g. Chambers & Trudinger, 1979; Strauss, 1997). Numerous $\delta^{34}S_{sulphide}$ values close to the magmatic value of 0‰ in many late Archaean sedimentary successions reflect the continued importance of magmatic sulphur input to the ocean. This would be consistent with geotectonic settings; for example, of late Archaean greenstone belts. On the other hand, platformal sequences (e.g. the Transvaal Supergroup in South Africa) display $^{34}S$-depleted pyrite and clearly reflect a sedimentary system in which sulphate reducers dominate the sulphur cycle and play an important role in the remineralization of organic matter. The temporal evolution of an inorganically dominated early Archaean sulphur cycle towards a biologically driven system starting around the Archaean–Proterozoic transition is consistent with ideas about the timing of crustal growth (e.g. Taylor & McLennan, 1985) and modelling results of temporal carbon, sulphur, oxygen and strontium isotope records that are entirely based on the decrease in mantle heat flow and the growth pattern of continental crust (Goddéris & Veizer, 2000).

At the onset of the Proterozoic, bacterial sulphate reduction was firmly established as a key process of sediment biogeochemistry. The isotopic composition of seawater sulphate increased during the *Palaeo- and Mesoproterozoic* from an initial +10 to +15‰ to a range between +15 and +25‰. However, this temporal evolution is only poorly constrained. The associated isotope fractionation between seawater sulphate and sedimentary pyrite attained maximum values of 30 to

40‰, which compare well with sulphur isotope fractionations observed for bacterial sulphate reduction in younger sediments. On the other hand, the temporal record of $\delta^{34}S_{pyrite}$ clearly exibits values that approach and even exceed the sulphur isotopic composition of seawater sulphate. These high $\delta^{34}S$ values reflect the development of sulphate-limited conditions during progressive bacterial sulphate reduction. This could be a consequence of effective sulphate turnover owing to highly reactive organic matter or low initial sulphate concentrations, or a combination of both. Either way, it appears to be a consistent signature which has been documented for nearly every sedimentary unit studied and, thus, characterizes the biogeochemical environment.

The *Neoproterozoic* records major changes in the sedimentary sulphur cycle as a consequence of environmental conditions forcing changes in biogeochemical patterns, including metabolic innovations. Prior to the lower Sturtian glacial interval, seawater sulphate had a sulphur isotopic composition of *c.*+20‰. Sedimentary pyrite was formed as a result of bacterial sulphate reduction, which was characterized by a maximum isotope fractionation of 37‰. Isotopic variability indicates sulphate limitation and progressive Rayleigh fractionation during sediment diagenesis. The interglacial interval—i.e. post-Sturtian, pre-Marinoan sediments—yielded $\delta^{34}S$ values for pyrite that are suggestive of bacterial sulphate reduction under sulphate-limited conditions. Although not documented, the range in $\delta^{34}S_{sulphide}$ would be consistent with a $^{34}S$-enriched isotopic composition of ambient seawater sulphate, analogous to post-Marinoan times. Seawater sulphate during post-Marinoan time was strongly $^{34}S$-enriched. An isotope mass balance for the global marine sulphur cycle would suggest that this reflects enhanced bacterial sulphate reduction and pyrite burial. This can be envisioned in the deeper parts of an anoxic glacial ocean, followed by ocean turnover during deglaciation and movement of $^{34}S$-enriched sulphate into the shelf water. Extremely high sedimentary S/C ratios suggest a high level of sulphate consumption while—at the same time—very positive $\delta^{34}S_{sulphide}$ values suggest that sulphate turnover via bacterial reduction was faster than sulphate replenishment, resulting in a Rayleigh fractionation for the sulphur isotope system. Maximum isotope fractionation between sulphate and sulphide, however, exceed the range typical solely for bacterial sulphate reduction. Amplification of the overall isotope effect can be achieved through disproportionation reactions involving elemental sulphur and/or thiosulphate, and the advent of this oxidative step in sulphur metabolism is placed in post-Marinoan time. It should be pointed out that the temporal evolution of the Neoproterozoic sulphur cycle is consistent with the implications derived from the carbon isotope record, notably an enhanced deposition of organic matter during the Neoproterozoic and a full rearrangement of carbon turnover and related isotopic perturbations during glacial and post-glacial times (e.g. Kaufman *et al.*, 1997; Hoffman *et al.*, 1998).

## CONCLUSIONS AND UNSOLVED PROBLEMS

The isotopic composition of redox-sensitive elements (C, S and N) can provide important information about the biogeochemistry of sedimentary systems. This is true for modern settings but can also be applied to ancient sediments and is particularly appealing for the Precambrian, where sediments frequently lack direct fossil evidence for these processes. In this paper, I reviewed new and previously published sulphur isotope data for sedimentary sulphides and sulphates and new geochemical results with the aim of reconstructing the temporal evolution of the biological sulphur cycle. In the course of this evaluation, I have tried to provide a conservative assessment of the currently available evidence for the advent of major biological innovations.

Certain aspects remain as unsolved problems to be studied in future research. Most important in that respect is the establishment of a firm understanding of the isotopic composition of Precambrian, and in particular Archaean and Palaeoproterozoic, sea water. This represents one of the most important aspects as it provides the reference line for the interpretation of sulphur isotope fractionations associated with different biological reactions. Equally important is a re-evaluation of isotope fractionation associated with bacterial sulphate reduction under different environmental conditions and via different sulphate-reducing micro-organisms. This can only be achieved through detailed laboratory experiments, and key aspects include the dependence of fractionation magnitudes on ambient temperature, sulphate concentration and reduction rate. Finally, a more detailed sulphur isotope record for sulphate and sulphide from Neoproterozoic sediments in the stratigraphic context of glacial horizons (analogous to the detailed carbon isotope work) would undoubtedly result in a better understanding of the suggested causal relationship between changes in the environmental conditions during the proposed Snowball Earth and resulting changes in sulphur biogeochemistry.

## ACKNOWLEDGEMENTS

This study presents results from several research projects over the past 10 years on the sulphur isotope geochemistry of Precambrian sedimentary systems. A substantial part of it represents research performed in the course of my habilitation, and special thanks go to J. Veizer for his continuous interest and support during this time. Financial support through the Deutsche Forschungsgemeinschaft (Grants: Ve 112/1-1/1-2; Str 281/7-1/7-2; Str 281/12-1) is gratefully acknowledged. Numerous colleagues have provided valuable support in the field, during sample acquisition and through stimulating discussions. These include in alphabetical order: P.W.U. Appel, D.M. Banerjee, N.J. Beukes, D. Buhl, M. Deb, J. Gutzmer, A.J. Kaufman, N. Lemon, B. Mayer, D.M. McKirdy, M. Moczydłowska, S.M. Moorbath, T. Palacios, D. Rawlings, G.A. Shields, R.E. Summons, G. Vidal. Thorough and most constructive reviews were provided by T.W. Lyons and B. Mayer. Work in Isua, south-west Greenland, was part of the Isua Multidisciplinary Research Project, funded by the Danish Research Council and the Geological Survey of Denmark and Greenland (GEUS). Finally, help from numerous students during lab work is acknowledged. Analytical data can be obtained from the author upon request.

## NOTE

Following the completion of the manuscript, the following articles, which are directly relevant to the discussion of the Precambrian sulphur isotope record, have been published.

Shen *et al.* (2001) present sulphur isotope results for barite and for microscopic sulphides contained within the barite from the *c.*3.47 Ga Dresser Formation, Warrawoona Supergroup, Pilbara Craton, Western Australia. The reported maximum sulphur isotope fractionation between sulphate and sulphide amounts to 21.1‰, indicating microbial sulphate reduction for this early Archaean environment.

Differences in magnitude of sulphur isotope fractionation during bacterial sulphate reduction as a consequence of changes in environmental parameters are presented by Canfield (2001) and Habicht and Canfield (2001), based on their studies of diverse natural populations. Similarly, new data are reported by Detmers *et al.* (2001), based on their cultural experiments.

## REFERENCES


ALGEO, T.J., BERNER, R.A., MAYNARD, J.B. & SCHECKLER, S.E. (1995) Late Devonian oceanic anoxic events and biotic crisis: rooted in the evolution of vascular land plants. *GSA Today*, **5**, 64–66.

ALGEO, T.J. & SCHECKLER, S.E. (1998) Terrestrial–marine teleconnections in the Devonian: links between the evolution of land plants, weathering processes, and marine anoxic events. *Phil. Trans. R. Soc. London B*, **353**, 113–130.

ALTERMANN, W. (1996) Sedimentology, geochemistry and palaeogeographic implications of volcanic rocks in the upper Archaean Campbell Group, western Kaapvaal Craton, South Africa. *Precam. Res.*, **79**, 73–100.

ALTERMANN, W. & NELSON, D.R. (1998) Sedimentation rates, basin analysis and regional correlations of three Neoarchean and Palaeoproterozoic sub-basins of the Kaapvaal craton as inferred from precise U–Pb zircon ages from volcaniclastic sediments. *Sediment. Geol.*, **120**, 225–256.

ANDERSON, T.F. & PRATT, L.M. (1995) Isotopic evidence for the origin of organic sulfur and elemental sulfur in marine sediments. In: *Geochemical Transformations of Sedimentary Sulfur* (Eds Vairavamurthy, M.A. & Schoonen, M.A.A.), pp. 378–396. American Chemistry Society, Washington, DC.

APPEL, P.W.U. (1980) On the early Archean Isua iron-formation, West Greenland. *Precam. Res.*, **11**, 73–87.

ARÉN, B. & LENDZION, K. (1978) Charakterystyka stratygraficzno–litologiczna wendu i kambru dolnego. *Prace Inst. Geol.*, **90**, 7–49 (English abstract).

ARMSTRONG, R.A., COMPSTON, W., RETIEF, E.A., WILLIAMS, I.S. & WELKE, H.J. (1991) Zircon ion microprobe studies bearing on the age of the Witwatersrand triad. *Precam. Res.*, **53**, 243–266.

ARMSTRONG, R.A., ROBB, L.J., MASTER, S., KRUGER, F.J. & MUMBA, P.A.C.C. (1999) New U–Pb age constraints on the Katangan Sequence, Central African Copperbelt. *J. Afr. Earth Sci.*, **28**, 6–7.

ARNDT, N.T., NELSON, D.R., COMPSTON, W., TRENDALL, A.F. & THORNE, A.M. (1991) The age of the Fortescue Group, Hamersley Basin, Western Australia, from ion-microprobe U–Pb zircon results. *Aust. J. Earth Sci.*, **38**, 261–281.

BAU, M., ROMER, R.L., LÜDERS, V. & BEUKES, N.J. (1999) Pb, O, and C isotopes in silicified Mooidraai dolomite (Transvaal Supergroup, South Africa): implications for the composition of Paleoproterozoic seawater and 'dating' the increase of oxygen in the Precambrian atmosphere. *Earth planet. Sci. Lett.*, **174**, 43–57.

BEAUDOUIN, G., TAYLOR, B.E., RUMBLE, D. III & THIEMENS, M.H. (1994) Variations in the sulfur isotope composition of troilite from Cañon Diablo iron meteorite. *Geochim. Cosmochim. Acta*, **58**, 4253–4255.

BERNER, R.A. (1970) Sedimentary pyrite formation. *Am. J. Sci*, **268**, 1–23.

BERNER, R.A. (1984) Sedimentary pyrite formation: an update. *Geochim. Cosmochim. Acta*, **48**, 605–615.

BERNER, R.A. & RAISWELL, R. (1984) C/S method for distinguishing freshwater from marine sedimentary rocks. *Geology*, **12**, 365–368.

BEUKES, N.J. (1986) The Transvaal Sequence in Griqualand West. In: *Mineral Deposits of Southern Africa* (Eds Anhaeusser, C.R. & Maske, S.), pp. 819–828. Geological Society of South Africa, Johannesburg.

BEUKES, N.J. & CAIRNCROSS, B. (1991) A lithostratigraphic–sedimentological reference profile for the late Mozaan Group, Pongola Sequence: application to sequence stratigraphy and correlation with the Witwatersrand Supergroup. *S. Afr. J. Geol.*, **94**, 44–69.

BEUKES, N.J. & KLEIN, C. (1992) Models for iron-formation deposition. In: *The Proterozoic Biosphere: a Multidisciplinary Study* (Eds Schopf, J.W. & Klein, C.), pp. 147–151. Cambridge University Press, Cambridge.

BEUKES, N.J. & LOWE, D.R. (1989) Environmental control on diverse stromatolite morphologies in the 3000 Myr Pongola Supergroup, South Africa. *Sedimentology*, **36**, 383–397.

BEUKES, N.J. & STRAUSS, H. (1991) A geochemical study of carbon and sulfur in sedimentary rocks from the Witwatersrand and Ventersdorp Supergroups and its bearing on the depositional environment. Unpublished report, 16 pp.

BLACK, L.P., BELL, T.H., RUBENACK, M.J. & WHITNALL, I.W. (1979) Geochronology of discrete structural–metamorphic events in a multiply-deformed Precambrian terrain. *Tectonophysics*, **54**, 103–138.

BLUTH, G.J., GRAHAM, U.H. & OHMOTO, H. (1988) Sulfide–sulfate chimneys on the East Pacific Rise, 11° and 13° N latitudes. Part 2: sulfur isotopes. *Can. Mineral.*, **26**, 505–515.

BOTTOMLEY, D.J., VEIZER, J., NIELSEN, H. & MOCZYDLOWSKA, M. (1992) Isotopic composition of disseminated sulfur in Precambrian sedimentary rocks. *Geochim. Cosmochim. Acta*, **56**, 3311–3322.

BOWINS, R.J. & CROCKET, J.H. (1994) Sulfur and carbon isotopes in Archean banded iron formations: implications for sulfur sources. *Chem. Geol.*, **111**, 307–323.

BOWRING, S., GROTZINGER, J., ISAACSON, G., KNOLL, A.H., PELECHATY, S. & KOLOSOV, P. (1993) Calibrating rates of Early Cambrian evolution. *Science*, **261**, 1293–1298.

BROCKS, J.J., LOGAN, G.A., BUICK, R. & SUMMONS, R.E. (1999) Archean molecular fossils and the early rise of eukaryotes. *Science*, **255**, 74–77.

BROWN, J.S. (1973) Sulfur isotopes of Precambrian sulfates and sulfides in the Grenville of New York and Ontario. *Econ. Geol.*, **68**, 362–370.

BUCHANAN, D.L. & ROUSE, J.E. (1982) Role of contamination in the precipitation of sulphides in the Platreef of the Bushveld complex. *Proc. IGCP Projects 91 and 161, 3rd Nickel Sulphide Field Conference, Perth, W. Aust.*, pp. 141–146.

BUCK, S.G. (1980) Stromatolite and ooid deposits within the fluvial and lacustrine sediments of the Precambrian Ventersdorp Supergroup of South Africa. *Precam. Res.*, **12**, 311–330.

BURDETT, J.W., ARTHUR, M.A. & RICHARDSON, M. (1989) A Neogene seawater sulfate isotope age curve from calcareous pelagic microfossils. *Earth planet. Sci. Lett.*, **94**, 189–198.

BYERLY, G.R., KRÖNER, A., LOWE, D.R., TODT, W. & WALSH, M.M. (1996) Prolonged magmatism and time constraints for sediment deposition in the early Archaean Barberton greenstone belt: evidence from the Upper Onverwacht and Fig Tree Groups. *Precam. Res.*, **78**, 125–138.

CAMERON, E.M. (1982) Sulphate and sulphate reduction in early Precambrian oceans. *Nature*, **296**, 145–148.

CAMERON, E.M. (1983a) Genesis of Proterozoic iron-formation: sulphur isotope evidence. *Geochim. Cosmochim. Acta*, **47**, 1069–1074.

CAMERON, E.M. (1983b) Evidence from early Proterozoic anhydrite for sulphur isotope partitioning in Precambrian oceans. *Nature*, **304**, 54–56.

CANFIELD, D.E. (1991) Sulfate reduction in deep-sea sediments. *Am. J. Sci.*, **291**, 177–188.

CANFIELD, D.E. (1998) A new model for Proterozoic ocean chemistry. *Nature*, **396**, 450–453.

CANFIELD, D.E. (2001) Isotope fractionation by natural populations of sulfate-reducing bacteria. *Geochim. Cosmochim. Acta*, **65**, 1117–1124.

CANFIELD, D.E. & TESKE, A. (1996) Late Proterozoic rise in atmospheric oxygen concentration inferred from phylogenetic and sulphur-isotope studies. *Nature*, **382**, 127–132.

CANFIELD, D.E. & THAMDRUP, B. (1994) The production of $^{34}S$-depleted sulfide during bacterial disproportionation of elemental sulfur. *Science*, **266**, 1973–1975.

CANFIELD, D.E., BOUDREAU, B.P., MUCCI, A. & GUNDERSEN, J.K. (1998) The early diagenetic formation of organic sulfur in the sediments of Mangrove Lake, Bermuda. *Geochim. Cosmochim. Acta*, **62**, 767–781.

CANFIELD, D.E., RAISWELL, R., WESTRICH, W.T., REAVES, C.M. & BERNER, R.A. (1986) The use of chromium reduction in the analysis of reduced inorganic sulfur in sediments and shales. *Chem. Geol.*, **54**, 149–155.

CARRIGAN, W.J. & CAMERON, E.M. (1991) Petrological and stable isotope studies of carbonate and sulfide minerals from the Gunflint Formation, Ontario: evidence for the origin of early Proterozoic iron-formation. *Precam. Res.*, **52**, 347–380.

CHAMBERS, L.A. & TRUDINGER, P.A. (1979) Microbiological fractionation of stable sulfur isotopes: a review and critique. *Geomicrobiol. J.*, **1**, 249–293.

CHRISTIE-BLICK, N., DYSON, I.A. & VON DER BORCH, C.C. (1995) Sequence stratigraphy and the interpretation of Neoproterozoic Earth history. *Precam. Res.*, **73**, 3–26.

CLAYPOOL, G.E., HOLSER, W.T., KAPLAN, I.R., SAKAI, H. & ZAK, I. (1980) The age curves of sulfur and oxygen isotopes in marine sulfate and their mutual interpretation. *Chem. Geol.*, **28**, 190–260.

CONDIE, K.C. (1997) *Plate Tectonics and Crustal Evolution*. Pergamon Press, Oxford, 288 pp.

COLEMAN, M.L. & RAISWELL, R. (1981) Carbon, oxygen and sulphur isotope variations in concretions from the Upper Lias of NE England. *Geochim. Cosmochim. Acta*, **45**, 329–340.

COMPSTON, W., SAMBRIDGE, M.S., REINFRANK, R.F., MOCZYDLOWSKA, M., VIDAL, G. & CLAESSON, S. (1995) Numerical ages of volcanics and the earliest faunal zone within the late Precambrian of East Poland. *J. geol. Soc. London*, **152**, 599–611.

COPLEN, T.B. & KROUSE, H.R. (1998) Sulphur isotope data consistency improved. *Nature*, **392**, 32.

Cornell, D.H., Armstrong, R.A. & Walraven, F. (1998) Geochronology of the Proterozoic Hartley Basalt Formation, South Africa: constraints on the Kheis tectogenesis and the Kaapvaal craton's earliest Wilson cycle. *J. Afr. Earth Sci.*, **26**, 5–27.

Cornell, D.H., Schütte, S.S. & Eglington, B.L. (1996) The Ongeluk Basaltic Andesite Formation in Griqualand West South Africa: submarine alteration in a 2222 Ma Proterozoic sea. *Precam. Res.*, **79**, 101–124.

Crowe, D.E., Valley, J.W. & Baker, K.L. (1990) Microanalysis of sulfur-isotope ratios and zonation by laser microporbe. *Geochim. Cosmochim. Acta*, **54**, 2075–2092.

Des Marais, D.J., Strauss, H., Summons, R.E. & Hayes, J.M. (1992) Carbon isotope evidence for the stepwise oxidation of the Proterozoic environment. *Nature*, **359**, 605–609.

Detmers, J., Brüchert, V., Habicht, K.S. & Kuever, J. (2001) Diversity of sulfur isotope fractionations by sulfate-reducing prokaryotes. *Appl. environ. Microbiol.*, **67**, 888–894.

Donnelly, T.H. & Crick, I.H. (1988) Depositional environment of the Middle Proterozoic Velkerri Formation in northern Australia: geochemical evidence. *Precam. Res.*, **42**, 165–172.

Donnelly, T.H. & Crick, I.H. (1992) Biological and abiological sulfate reduction in two Northern Australian Proterozoic basins. In: *Early Organic Evolution* (Eds Schidlowski, M., Golubic, S., Kimberley, M.M., McKirdy, D.M. & Trudinger, P.A.), pp. 398–407. Springer-Verlag, Berlin.

Edwards, M.B. (1984) Sedimentology of the Upper Proterozoic glacial record, Vestertana Group, Finnmark, North Norway. *Norsk geol. unders. Bull.*, **394**, 76 pp.

Fanning, C.M., Ludwig, K.R., Forbes, B.G. & Preiss, W.V. (1986) Single and multiple grain U–Pb zircon analyses for the early Adelaidean Rook Tuff, Willouran Ranges, South Australia. *Geol. Soc. Aust. Abstr.*, **15**, 71–72.

Farmer, J., Vidal, G., Strauss, H., Moczydłowska, M., Ahlberg, P. & Siedlecka, A. (1992) Ediacaran medusoid fossils from the Innerelva Member (late Proterozoic) of Digermul Peninsula, Tanafjorden area, northeastern Finnmark. *Geol. Mag.*, **129**, 181–195.

Förster, O. (1986) *Geochemische und isotopengeochemische Untersuchungen an Kohle- und Gold-führenden Konglomeraten des Proterozoikums (Central Rand Group, Südafrika).* Dissertation, TU Munich.

Froelich, P.N., Klinkhammer, G.P., Bender, M.L. *et al.* (1979) Early oxidation of organic matter in pelagic sediments of the eastern equatorial Atlantic; suboxic diagenesis. *Geochim. Cosmochim. Acta*, **43**, 1075–1090.

Fry, B., Gest, H. & Hayes, J.M. (1984) Isotope effects associated with anaerobic oxidation of sulfide by the purple synthetic bacterium *Chromatium vinosum. FEMS Microbiol. Lett.*, **22**, 282–287.

Gellatly, A.M., Lyons, T.W. & Kah, L.C. (2000) Trace sulfate within carbonates of the Mesoproterozoic Newland Formation, Montana: implications for global- and local-scale controls on sulfate availability. *Geol. Soc. Am., Rocky Mtn Sect., 52nd Annual Meeting Abstracts Volume, No. 32*, p. A-10.

Glaessner, M.F. (1984) *The Dawn of Animal Life: a Biohistorical Study*. Cambridge University Press, Cambridge, 244 pp.

Glaessner, M.F. & Wade, M. (1966) The late Precambrian fossils from Ediacara, South Australia. *Paleontology*, **9**, 599–628.

Goddéris, Y. & Veizer, J. (2000) Tectonic control of chemical and isotopic composition of ancient oceans: the impact of continental growth. *Am. J. Sci.*, **300**, 434–461.

Goodwin, A.M., Monster, J. & Thode, H. (1976) Carbon and sulfur isotope abundances in Archean iron-formations and early Precambrian life. *Econ. Geol.*, **71**, 870–891.

Goodwin, A.M., Thode, H.G., Chou, C.-L. & Karkhansis, S.N. (1985) Chemostratigraphy and origin of the late Archean siderite–pyrite-rich Helen Iron Formation, Michipicotan belt, Canada. *Can. J. Earth Sci.*, **22**, 72–84.

Grassineau, N.V., Lowry, D., Fowler, C.M.R., Nisbet, E.G. & Mattey, D.P. (1999) Sulphur and Carbon isotopic heterogeneities in the Manjeri Formation in the 2.7 Ga Belingwe Greenstone Belt, Zimbabwe: evidence of a biological signature. *J. Conf. Abstr.*, **4**, 261.

Gutzmer, J. & Beukes, N.J. (1997) High-grade manganese ores in the Kalahari manganese field: characterization and dating of ore-forming events. Fourth Quarterly Progress Report, Johannesburg, unpublished.

Habicht, K.S. & Canfield, D.E. (1996) Sulphur isotope fractionation in modern microbial mats and the evolution of the sulphur cycle. *Nature*, **382**, 342–343.

Habicht, K.S. & Canfield, D.E. (2001) Isotope fractionation by sulfate-reducing natural populations and the isotopic composition of sulfide in marine sediments. *Geology*, **29**, 555–558.

Harrison, A.G. & Thode, H.G. (1958) Mechanism of the bacterial reduction of sulfate from isotope fractionation studies. *Faraday Soc. Trans.*, **54**, 84–92.

Hattori, K., Campbell, F.A. & Krouse, H.R. (1983) Sulphur isotope abundances in Aphebian clastic rocks: implications for the coeval atmosphere. *Nature*, **302**, 323–326.

Hayes, J.M., Lambert, I.B. & Strauss, H. (1992) The sulfur-isotopic record. In: *The Proterozoic Biosphere: a Multidisciplinary Study* (Eds Schopf, J.W. & Klein, C.), pp. 129–131. Cambridge University Press, Cambridge.

Hegner, E., Kröner, A. & Hofmann, A.W. (1984) Age and isotope geochemistry of the Archaean Pongola and Ushushwana suites in Swaziland, Southern Africa: a case for crustal contamination of mantlederived magma. *Earth Planet. Sci. Lett.*, **70**, 267–279.

Hegner, E., Kröner, A. & Hunt, P. (1994) A precise U–Pb zircon age for the Archean Pongola Supergroup volcanics in Swaziland. *J. Afr. Earth Sci.*, **18**, 339–341.

Hickman, A.H. (1990) Excursion No. 5: Pilbara and Hamersley Basin. In: *Third International Archaean Symposium, Perth, 1990, Excursion Guidebook* (Eds Ho, S.E., Glover, J.E., Myers, J.S. & Muhling, J.R.), pp. 1–60. Geol. Dept. & Univ. Extension, Univ. W. Aust Publ. No. 21.

Hoefs, J. (1997) *Stable Isotope Geochemistry*. Springer-Verlag, Berlin, 201 pp.

Hoefs, J., Nielsen, H. & Schidlowski, M. (1968) Sulfur isotope abundances in pyrite from the Witwatersrand conglomerates. *Econ. Geol.*, **63**, 975–977.

Hoffman, P.F. (1991) Did the breakout of Laurentia turn Gondwanaland inside-out? *Science*, **252**, 1409–1412.

Hoffman, P.F., Kaufman, A.J., Halverson, G.P. & Schrag, D.P. (1998) A Neoproterozoic Snowball Earth. *Science*, **281**, 1342–1346.

HOLLAND, H.D. (1992) Distribution and Paleoenvironmental Interpretation of Proterozoic Paleosols. In: *The Proterozoic Biosphere: a Multidisciplinary Study* (Eds Schopf, J.W. & Klein, C.), pp. 153–155. Cambridge University Press, Cambridge.

HOLSER, W.T. (1984) Gradual and abrupt shifts in ocean chemistry during Phanerozoic time. In: *Patterns of Change in Earth Evolution* (Eds Holland, H.D. & Trendall, A.F.), pp. 123–143. Springer-Verlag, Berlin.

HOLSER, W.T. & KAPLAN, I.R. (1966) Isotope geochemistry of sedimentary sulfates. *Chem. Geol.*, **1**, 93–135.

HURTGEN, M.T. & ARTHUR, M.A. (2000) The sulfur isotopic composition of Neoproterozoic seawater sulfate: implications for a 'Snowball Earth'. *Geol. Soc. Am. Abstr. Progr.*, **32**, 1380.

HUTTON, J.L. (1983) Stratigraphic drilling report—GSQ Lawn Hill 1-4. *Queensland Gov. Mining J.*, **84**, 228–240.

JACKSON, M.J., MUIR, M.D. & PLUMB, K.A. (1987) Geology of the southern McArthur Basin, Northern Territory. *Bur. Mines Res. Bull.*, **220**, 173 pp.

JØRGENSEN, B.B. (1982) Mineralization of organic matter in the sea bed—the role of sulphate reduction. *Nature*, **296**, 643–645.

JØRGENSEN, B.B. (1990) A thiosulfate shunt in the sulfur cycle of marine sediments. *Science*, **249**, 152–154.

KAH, L.C., LYONS, T.W. & CHESLEY, J.T. (2001) Geochemistry of a 1.2 Ga carbonate–evaporite succession, northern Baffin and Bylot islands: implications for Mesoproterozoic marine evolution. *Precam. Res.*, **111**, 203–234.

KAKEGAWA, T. & OHMOTO, H. (1999) Sulfur isotope evidence for the origin of 3.4 to 3.1 Ga pyrite at the Princeton gold mine, Barberton Greenstone Belt, South Africa. *Precam. Res.*, **96**, 209–224.

KAKEGAWA, T., KAWAI, H. & OHMOTO, H. (1999) Origins of pyrites in the ~2.5 Ga Mt McRae Shale, the Hamersley District, Western Australia. *Geochim. Cosmochim. Acta*, **62**, 3205–3220.

KAMBER, B.S. & MOORBATH, S. (1998) Initial Pb of the Amîtsoq gneiss revisited: implication for the timing of early Archaean crustal evolution in West Greenland. *Chem. Geol.*, **150**, 19–41.

KAMPSCHULTE, A. & STRAUSS, H. (2000) The isotopic evolution of Phanerozoic seawater based on structurally substituted sulfate in carbonates. *Abstract GeoCanada 2000*.

KAMPSCHULTE, A., BRUCKSCHEN, P. & STRAUSS, H. (2001) The sulphur isotopic composition of trace sulphates in Carboniferous brachiopods: implications for coeval seawater, correlation with other geochemical cycles and isotope stratigraphy. *Chem. Geol.*, **175**, 149–173.

KAPLAN, I.R. & RITTENBERG, S.C. (1964) Microbiological fractionation of sulfur isotopes. *J. gen. Microbiol.*, **34**, 195–212.

KASTING, J.-F. (1993) Earth's early atmosphere. *Science*, **259**, 920–926.

KAUFMAN, A.J. & KNOLL, A.H. (1995) Neoproterozoic variations in the C-isotopic composition of seawater: stratigraphic and biogeochemical implications. *Precam. Res.*, **73**, 27–49.

KAUFMAN, A.J., JACOBSEN, S.B. & KNOLL, A. (1993) The Vendian record of Sr and C isotopic variations in seawater: implications for tectonics and paleoclimate. *Earth planet. Sci. Lett.*, **120**, 409–430.

KAUFMAN, A.J., KNOLL, A.H. & NARBONNE, G.M. (1997) Isotopes, ice ages, and terminal Proterozoic Earth history. *Proc. Natl Acad. Sci.*, **94**, 6600–6605.

KNOLL, A.H. & WALTER, M.R. (1992) Latest Proterozoic stratigraphy and Earth history. *Nature*, **356**, 673–678.

KRALIK, M. (1982) Rb–Sr age determinations on Precambrian carbonate rocks of the Carpentarian McArthur basin, Northern Territories, Australia. *Precam. Res.*, **18**, 157–170.

KRÖNER, A., HEGNER, E., WENDT, J.I. & BYERLY, G.R. (1996) The oldest part of the Barberton granite–greenstone terrain, South Africa: evidence for crust formation between 3.5 and 3.7 Ga. *Precam. Res.*, **78**, 105–124.

LAMBERT, I.B. & DONNELLY, T.H. (1990) The paleoenvironmental significance of trends in sulphur isotope compositions in the Precambrian: a critical review. In: *Stable Isotopes and Fluid Processes in Mineralization* (Eds Herbert, H.K. & Ho, S.E.), Spec. Publ. Univ. W. Aust., **23**, 260–268.

LAMBERT, I.B., DONNELLY, T.H., DUNLOP, J.S.R. & GROVES, D.I. (1978) Stable isotopic compositions of early Archean sulphate deposits of probable evaporitic and volcanogenic origins. *Nature*, **276**, 808–811.

LANGE, J. & BRUMSACK, H.-J. (1977) Total sulphur analysis in geological and biological materials by coulometric titration following combustion. *Fresenius Z. Anal. Chem.*, **286**, 361–366.

LEVENTHAL, J.S. (1983) An interpretation of carbon and sulfur relationships in Black Sea sediments as indicators of environments of deposition. *Geochim. Cosmochim. Acta*, **47**, 133–138.

LEVENTHAL, J.S. (1995) Carbon–sulfur plots to show diagenetic and epigenetic sulfidation in sediments. *Geochim. Cosmochim. Acta*, **59**, 1207–1211.

LEVENTHAL, J.S. & TAYLOR, C. (1990) Comparison of methods to determine degree of pyritization. *Geochim. Cosmochim. Acta*, **54**, 2621–2625.

LINDSAY, J.F. (1987) Upper Proterozoic evaporites in the Amadeus basin, central Australia, and their role in basin tectonics. *Geol. Soc. Am. Bull.*, **99**, 852–865.

LOGAN, G.A., CALVER, C.R., GORJAN, P., SUMMONS, R.E., HAYES, J.M. & WALTER, M.R. (1999) Terminal Proterozoic mid-shelf benthic microbial mats in the Centralian Superbasin and their environmental significance. *Geochim. Cosmochim. Acta*, **63**, 1345–1358.

LOGAN, G.A., HAYES, J.M., HIESHIMA, G.B. & SUMMONS, R.E. (1995) Terminal Proterozoic reorganization of biogeochemical cycles. *Nature*, **376**, 53–56.

LOGAN, G.A., SUMMONS, R.E. & HAYES, J.M. (1997) An isotopic biogeochemical study of Neoproterozoic and Early Cambrian sediments from the Centralian Superbasin, Australia. *Geochim. Cosmochim. Acta*, **61**, 5391–5409.

LOWE, D.R. & BYERLY, G.R. (1999) *Geologic Evolution of the Barberton Greenstone Belt, South Africa.* Spec. Publ. Geol. Soc. Am., **329**, 319 pp.

LYONS, T.W. (1997) Sulfur isotopic trends and pathways of iron sulfide formation in upper Holocene sediments of the anoxic Black Sea. *Geochim. Cosmochim. Acta*, **61**, 3367–3382.

LYONS, T.W. & BERNER, R.A. (1992) Carbon–sulfur–iron systematics of the uppermost deep-water sediments of the Black Sea. *Chem. Geol.*, **99**, 1–27.

LYONS, T.W. & KAH, L.C. (1999) Sulfur geochemistry of the Mesoproterozoic Bylot Supergroup, northern Baffin and

Bylot Islands, Canada: local and global implications. In: *Ninth Annual V.M. Goldschmidt Conference*, pp. 179–180. LPI Contribution No. 971, Lunar and Planetary Institute, Houston.

Lyons, T.W., Luepke, J.J., Schreiber, M.E. & Zieg, G.A. (2000) Sulfur geochemical constraints on Mesoproterozoic restricted marine deposition: lower Belt Supergroup, northwestern United States. *Geochim. Cosmochim. Acta*, **64**, 427–437.

Lyons, W.B. & Gaudette, M.E. (1979) Sulfate reduction and the nature of organic matter in estuarine sediments. *Org. Geochem.*, **1**, 151–155.

Machel, H.G., Krouse, H.R. & Sassen, R. (1995) Products and distinguishing criteria of bacterial and thermochemical sulfate reduction. *Appl. Geochem.*, **10**, 373–389.

McKirdy, D.M., Burgess, J.M., Lemon, N.M. *et al.* (2000) A chemostratigraphic overview of the late Cryogenian interglacial sequence in the Adelaide fold-thrust belt, South Australia. *Precam. Res.*, **106**, 149–186.

Moczydłowska, M. (1989) Upper Proterozoic and lower Cambrian acritarchs from Poland—micropaleontology, biostratigraphy, and thermal study. *Lund Publs Geol.*, **75**, 1–30.

Moczydłowska, M. (1991) Acritarch biostratigraphy of the lower Cambrian and the Precambrian–Cambrian boundary in southeastern Poland. *Fossils and Strata*, **29**, 1–127.

Moczydłowska, M. (1995) Neoproterozoic and Cambrian successions deposited on the East European Platform and Cadomian basement area in Poland. *Stud. geoph. geod.*, **39**, 276–285.

Monster, J., Appel, P.W.U., Thode, H.G., Schidlowski, M., Carmichael, C.M. & Bridgewater, D. (1979) Sulfur isotope studies in Early Archaean sediments from Isua, West Greenland: implications for the antiquity of bacterial sulfate reduction. *Geochim. Cosmochim. Acta*, **43**, 405–413.

Muir, M.D., Donnelly, T.H., Wilkins, R.W.T. & Armstrong, K.J. (1985) Stable isotope, petrological, and fluid inclusions studies of minor mineral deposits from the McArthur Basin: implications for the genesis of some sediment-hosted base metal mineralization from the Northern Territory. *Aust. J. Earth Sci.*, **32**, 239–260.

Nakai, N. & Jensen, M.L. (1964) The kinetic isotope effect in the bacterial reduction and oxidation of sulfur. *Geochim. Cosmochim. Acta*, **28**, 1893–1912.

Narbonne, G.M. & Aitken, J.D. (1995) Neoproterozoic of the Mackenzie Mountains, northwestern Canada. *Precam. Res.*, **73**, 101–121.

Nelson, D.R., Trendall, A.F. & Altermann, W. (1999) Chronological correlations between the Pilbara and Kaapvaal cratons. *Precam. Res.*, **97**, 165–189.

Newton, R.J., Bottrell, S.H., Dean, S.P., Hatfield, D. & Raiswell, R. (1995) An evaluation of the use of the chromous chloride reduction method for isotopic analyses of pyrite in rocks and sediment. *Chem. Geol.*, **125**, 317–320.

Nielsen, H. (1989) Local and global aspects of the sulphur isotope age curve of oceanic sulphate. In: *Evolution of the Global Biogeochemical Sulphur Cycle* (Eds Brimblecombe, P. and Lein, A.Yu.), pp. 57–64. Wiley, New York.

Nutman, A.P., Bennett, V.C., Friend, C.R.L. & Rosing, M.T. (1997) ~3710 and ≥3790 Ma volcanic sequences in the Isua (Greenland) supracrustal belt; structural and Nd isotope implications. *Chem. Geol.*, **141**, 271–287.

Ohmoto, H. (1992) Biogeochemistry of sulfur and the mechanisms of sulfide–sulfate mineralization in Archean oceans. In: *Early Organic Evolution* (Eds Schidlowski, M., Golubic, S., Kimberley, M.M., McKirdy, D.M. & Trudinger, P.A.), pp. 378–397. Springer-Verlag, Berlin.

Ohmoto, H. (1997) When did the Earth's atmosphere become oxic? *Geochem. News*, **93** (Fall), 12–28.

Ohmoto, H. & Goldhaber, M.B. (1997) Sulfur and carbon isotopes. In: *Geochemistry of Hydrothermal Ore Deposits*. (Ed. Barnes, H.L.), pp. 517–611. Wiley, New York.

Ohmoto, H., Kakegawa, T. & Lowe, D.R. (1993) 3.4-billion-year-old biogenic pyrites from Barberton, South Africa: sulfur isotope evidence. *Science*, **267**, 555–557.

Page, R.W. & Sweet, I.P. (1998) Geochronology of basin phases in the western Mount Isa Inlier, and correlation with the McArthur Basin. *Aust. J. Earth Sci.*, **45**, 219–232.

Palacios, T. & Vidal, G. (1992) Lower Cambrian acritarchs from northern Spain: the Precambrian–Cambrian boundary and biostratigraphic implications. *Geol. Mag.*, **129**, 421–436.

Paytan, A., Kastner, M., Campbell, D. & Thiemens, M.H. (1999) Sulfur isotopic composition of Cenozoic seawater sulfate. *Science*, **282**, 1459–1462.

Perry, E.C., Monster, J. & Reimer, T.O. (1971) Sulfur isotopes in Swaziland System baryte and the evolution of the Earth's atmosphere. *Science*, **171**, 1015–1016.

Plumb, K.A. (1979) The tectonic evolution of Australia. *Earth Sci. Rev.*, **14**, 205–249.

Plumb, K.A., Derrick, G.M., Needham, R.S. & Shaw, R.D. (1981) The Proterozoic of northern Australia. In: *Precambrian of the Southern Hemisphere* (Ed. Hunter, D.R.), pp. 205–307. Elsevier, Amsterdam.

Powell, T.G., Jackson, M.J., Sweet, I.P., Crick, I.H., Boreham, C.J. & Summons, R.E. (1987) Petroleum geology and geochemistry, Middle Proterozoic McArthur Basin. *Bur. Mines Res. Record*, **48**, 286 pp.

Preiss, W.V. (1987) The Adelaide Geosyncline—late Proterozoic stratigraphy, sedimentation, palaeontology, and tectonics. *Geol. Surv. S. Aust. Bull.*, **53**, 438 pp.

Preiss, W.V., Walter, M.R., Coats, R.P. & Wells, A.T. (1978) Lithological correlations of Adelaidean glaciogenic rocks in parts of the Amadeus, Ngalia, and Georgina Basins. *BMR J. Aust. Geol. Geophys.*, **3**, 45–53.

Pretorius, D.A. (1976) The nature of the Witwatersrand gold–uranium deposits. In: *Handbook of Strata-bound and Stratiform Ore Deposits*, Vol. 7 (Ed. Wolf, K.H.), pp. 29–88. Elsevier, Amsterdam.

Raiswell, R. & Berner, R.A. (1985) Pyrite formation in euxinic and semi-euxinic sediments. *Am. J. Sci.*, **285**, 710–724.

Raiswell, R. & Berner, R.A. (1986) Pyrite and organic matter in Phanerozoic normal marine shales. *Geochim. Cosmochim. Acta*, **50**, 1967–1976.

Raiswell, R. & Canfield, D.E. (1998) Sources of iron for pyrite formation in marine sediments. *Am. J. Sci.*, **289**, 219–245.

Raiswell, R., Canfield, D.E. & Berner, R.A. (1994) A comparison of iron extraction methods for the determination of degree of pyritisation and the recognition of iron-limited pyrite formation. *Chem. Geol.*, **111**, 101–111.

Raiswell, R., Buckley, F., Berner, R.A. & Anderson, T.F. (1988) Degree of pyritization of iron as a paleoenvironmental indicator of bottom-water oxygenation. *J. sediment. Petrol.*, **58**, 812–819.

REIMER, T.O. (1980) Archaean sedimentary baryte deposits of the Swaziland Supergroup (Barberton Mountain Land, South Africa). *Precam. Res.*, **12**, 393–410.

RICE, C.A., TUTTLE, M.L. & REYNOLDS, R.L. (1993) The analysis of forms of sulfur in ancient sediments and sedimentary rocks: comments and cautions. *Chem. Geol.*, **107**, 83–95.

ROSING, M.T. (1999) $^{13}$C-depleted carbon microparticles in >3700-Ma sea-floor sedimentary rocks from West Greenland. *Science*, **283**, 674–676.

ROSS, G.M., BLOCH, J.D. & KROUSE, H.R. (1995) Neoproterozoic strata of the southern Cordillera and the isotopic evolution of seawater sulfate. *Precam. Res.*, **73**, 71–99.

RYE, R.O. & WILLIAMS, N. (1981) Studies of the base metal deposits at McArthur river, Northern Territory: III. The stable isotope geochemistry of the HYC, Ridge and Cooley deposits. *Econ. Geol.*, **76**, 1–26.

RYE, R.O., WHELAN, J.F., HARRISON, J.E. & HAYES, T.S. (1983) The origin of copper–silver mineralization in the Ravalli Group as indicated by preliminary stable isotope studies. Spec. Publ. Montana Bur. Mines Geol., **90**, 104–111.

SAYLOR, B.Z., KAUFMAN, A.J., GROTZINGER, J.P. & URBAN, F. (1998) A composite reference section for terminal Neoproterozoic strata of southern Namibia. *J. sediment. Res.*, **68**, 1223–1235.

SCHIDLOWSKI, M., HAYES, J.M. & KAPLAN, I.R. (1983) Isotopic inferences of ancient biochemistries. Carbon, sulfur, hydrogen, and nitrogen. In: *Earth's Earliest Biosphere. Its Origin and Evolution* (Ed. Schopf, J.W.), pp. 149–186. Princeton University Press, Princeton, NJ.

SCHOPF, J.W. (1983) *Earth's Earliest Biosphere: Its Origin and Evolution*. Princeton University Press, Princeton, NJ, 543 pp.

SCHOPF, J.W. & KLEIN, C. (1992) *The Proterozoic Biosphere: a Multidisciplinary Study*. Cambridge University Press, Cambridge, 1348 pp.

SHEGELSKI, R.J. (1978) *Stratigraphy and geochemistry of Archean iron formations in the Sturgeon Lake–Savant Lake Greenstone Terrain, Northwestern Ontario*. PhD thesis, University of Toronto.

SHEN, Y., BUICK, R. & CANFIELD, D.E. (2001) Isotopic evidence for microbial sulphate reduction in the early Archaean era. *Nature*, **410**, 77–81.

SHIELDS, G.A., STILLE, P., DEYNOUX, M., CLAUER, N., STRAUSS, H. & BRASIER, M. (1999a) Divining the origin of the barite-bearing, post-glacial, Neoproterozoic–Cambrian cap carbonates of Jbéliat, near Atar, Mauritania: a petrographic and isotopic (Nd–Sr–S–C–O) study. *J. Conf. Abstr.*, **4**, 732.

SHIELDS, G.A., STRAUSS, H., HOWE, S.S. & SIEGMUND, H. (1999b) Sulphur isotope compositions of sedimentary phosphorites from the basal Cambrian of China: implications for Neoproterozoic–Cambrian biogeochemical cycling. *J. geol. Soc.*, **156**, 943–955.

SIEDLECKA, A. (1985) Development of the Upper Proterozoic sedimentary basins of the Varanger Peninsula, East Finnmark, North Norway. *Geol. Surv. Finl. Bull.*, **331**, 175–185.

SIEDLECKA, A. & EDWARDS, M.B. (1980) Lithostratigraphy and sedimentation of the Riphean Båsnæring Formation, Varanger Peninsula, North Norway. *Norges geol. unders.*, **355**, 27–47.

SMOCK, A.M., BÖTTCHER, M.E. & CYPIONKA, H. (1998) Fractionation of sulfur isotopes during thiosulfate reduction by *Desulfovibrio desulfuricans*. *Arch. Microbiol.*, **169**, 460–463.

STRAUSS, H. (1986) Carbon and sulfur isotopes in Precambrian sediments from the Canadian Shield. *Geochim. Cosmochim. Acta*, **50**, 2653–2662.

STRAUSS, H. (1993a) The sulfur isotopic record of Precambrian sulfates: new data and a critical evaluation of the existing record. *Precam. Res.*, **63**, 225–246.

STRAUSS, H. (1993b) *Geochemische und isotopengeochemische Untersuchungen zum Schwefelkreislauf in präkambrischen Sedimenten*. Habilitationsschrift, Ruhr-Universität, Bochum.

STRAUSS, H. (1997) The isotopic composition of sedimentary sulfur through time. *Palaeogeogr. Palaeoclimatol. Palaeoecol.*, **132**, 97–118.

STRAUSS, H. (1999a) Geological evolution from isotope proxy signals—sulfur. *Chem. Geol.*, **161**, 89–101.

STRAUSS, H. (1999b) The sulfur-isotopic composition of Precambrian sediments: seawater chemistry and biological evolution. In: *Ninth Annual V.M. Goldschmidt Conference*, p. 286. LPI Contribution No. 971, Lunar and Planetary Institute, Houston.

STRAUSS, H. (1999c) A sulfur isotope investigation of the Isua metasediments—evidence for biological activity? *J. Conf. Abstr.*, **4**, 259.

STRAUSS, H., BANERJEE, D.M. & KUMAR, V. (2001) The sulfur isotopic composition of Neoproterozoic to Early Cambrian seawater—evidence from the cyclic Hanseran evaporites, NW India. *Chem. Geol.*, **175**, 17–28.

STRAUSS, H., BENGTSON, S., MYROW, P.M. & VIDAL, G. (1992b) Stable isotope geochemistry and palynology of the late Precambrian to early Cambrian sequence in Newfoundland. *Can. J. Earth Sci.*, **29**, 1662–1673.

STRAUSS, H. & BEUKES, N.J. (1996) Carbon and sulfur isotopic compositions of organic carbon and pyrite in sediments from the Transvaal Supergroup, South Africa. *Precam. Res.*, **79**, 57–71.

STRAUSS, H., DES MARAIS, D.J., HAYES, J.M., LAMBERT, I.B. & SUMMONS, R.E. (1992a) Procedures of whole rock and kerogen analysis. In: *The Proterozoic Biosphere: a Multidisciplinary Study* (Eds Schopf, J.W. & Klein, C.), pp. 699–707. Cambridge University Press, Cambridge.

STRAUSS, H. & MOORE, T.B. (1992) Abundances and isotopic compositions of carbon and sulfur species in whole rock and kerogen. In: *The Proterozoic Biosphere: a Multidisciplinary Study* (Eds Schopf, J.W. & Klein, C.), pp. 709–797. Cambridge University Press, Cambridge.

STRAUSS, H. & SCHIEBER, J. (1990) A sulfur isotope study of pyrite genesis: the Mid-Proterozoic Newland Formation, Belt Supergroup, Montana. *Geochim. Cosmochim. Acta*, **54**, 197–204.

STUMM, W. & MORGAN, J.J. (1996) *Aquatic Chemistry*. John Wiley & Sons, New York, 1022 pp.

STURT, B.A., PRINGLE, I.R. & ROBERTS, D. (1975) Caledonian nappe sequence of Finnmark, northern Norway, and the timing of the orogenic deformation and metamorphism. *Geol. Soc. Am. Bull.*, **86**, 710–718.

TAYLOR, S.R. & MCLENNAN, S.M. (1985) *The Continental Crust: Its Composition and Evolution*. Blackwell Scientific Publications, Oxford, 312 pp.

THODE, H.G. & GOODWIN, A.M. (1983) Further sulfur and carbon isotope studies of late Archean iron-formations of the Canadian Shield and the rise of sulfate-reducing bacteria. *Precam. Res.*, **20**, 337–356.

THODE, H.G., MONSTER, J. & DUNFORD, H.B. (1961) Sulphur isotope geochemistry. *Geochim. Cosmochim. Acta*, **25**, 159–174.

TRENDALL, A.F., NELSON, D.R., DE LAETER, J.R. & HASSLER, S. (1998) Precise U–Pb ages from the Marra Mamba Iron Formation and the Wittenoom Formation, Hamersley Group, Western Australia. *Aust. J. Earth Sci.*, **45**, 137–142.

VEIZER, J. (1988) The Earth and its life: system perspective. *Origins of Life*, 18, 13–39.

VEIZER, J., ALA, D., AZMY, K. *et al.* (1999) $^{87}Sr/^{86}Sr$, $\delta^{13}C$ and $\delta^{18}O$ evolution of Phanerozoic seawater. *Chem. Geol.*, **161**, 59–88.

VEIZER, J. & HOEFS, J. (1976) The nature of $^{18}O/^{16}O$ and $^{13}C/^{12}C$ secular trends in sedimentary carbonate rocks. *Geochim. Cosmochim. Acta*, **40**, 1387–1395.

VEIZER, J., HOEFS, J., LOWE, D.R. & THURSTON, P.C. (1989) Geochemistry of Precambrian carbonates 2. Archean greenstone belts and Archean sea water. *Geochim. Cosmochim. Acta*, **53**, 859–871.

VEIZER, J., HOLSER, W.T. & WILGUS, C.K. (1980) Correlation of $^{13}C/^{12}C$ and $^{34}S/^{32}S$ secular variations. *Geochim. Cosmochim. Acta*, **44**, 579–587.

VIDAL, G. (1981) Micropalaeontology and biostratigraphy of the Upper Proterozoic and Lower Cambrian sequences in East Finnmark, northern Norway. *Norsk geol. unders. Bull.*, **362**, 53 pp.

VIDAL, G. & MOCZYDŁOWSKA, M. (1995) The Neoproterozoic of Baltica—stratigraphy, palaeobiology and general geological evolution. *Precam. Res.*, **73**, 197–216.

VIDAL, G., PALACIOS, T., GÁMEZ-VINTANED, J.A., DIÉZ BALDA, M.A. & GRANT, S.W.F. (1994) Neoproterozoic—early Cambrian geology and paleontology of Iberia. *Geol. Mag.*, **131**, 729–765.

VIDAL, G. & SIEDLECKA, A. (1983) Planktonic, acid-resistant micro-fossils from the Upper Proterozoic strata of the Barents Sea region of Varanger Peninsula, East Finnmark, Northern Norway. *Norsk geol. unders. Bull.*, **382**, 45–79.

VON GEHLEN, K. (1992) Sulfur in the Earth's mantle a review. In: *Early Organic Evolution* (Eds Schidlowski, M., Golubic, S., Kimberley, M.M., McKirdy, D.M. & Trudinger, P.A.), pp. 359–366. Springer-Verlag, Berlin.

WÄCHTERSHÄUSER, G. (1990) The case for the chemoautotrophic origin of life in an iron–sulfur world. *Orig. Life*, **20**, 173–176.

WALTER, M.R., VEEVERS, J.J., CALVER, C.R. & GREY, K. (1995) Neoproterozoic stratigraphy of the Centralian Superbasin, Australia. *Precam. Res.*, **73**, 173–195.

WATANABE, Y., NARAOKA, H., WRONKIEWICZ, D.J., CONDIE, K.C. & OHMOTO, H. (1997) Carbon, nitrogen, and sulfur geochemistry of Archean and Proterozoic shales from the Kaapvaal Craton, South Africa. *Geochim. Cosmochim. Acta*, **61**, 3441–3459.

WATCHORN, M.B. (1980) Fluvial and tidal sedimentation in the 3000 Ma Mozaan basin, South Africa. *Precam. Res.*, **13**, 27–42.

WEISE, A., LÜCKGE, A. & STRAUSS, H. (1999) Untersuchungen gelöster und sedimentärer Schwefelspezies im Arabischen Meer (Sonne-130 Fahrt). *Terra Nostra*, **4**, 285–287.

WHELAN, J.F., RYE, R.O., DELORRAINE, W. & OHMOTO, H. (1990) Isotopic geochemistry of a mid-Proterozoic evaporite basin: Balmat, New York. *Am. J. Sci.*, **290**, 396–424.

WHITEHOUSE, M. (2000) Time constraints on when life began: the oldest record of life on Earth? *Geochem. News*, **103** (April), 10–14.

WHITNALL, I.W. (1984) Stratigraphy, structure, and metamorphism of the Proterozoic Etheridge and Langlovale Groups, central Georgetown Inlier, North Queensland. *Queensland geol. Surv. Rec.*, **59**, 66 pp.

WHITNALL, I.W. & MACKENZIE, D.E. (1980) New and revised stratigraphic units in the Proterozoic Georgetown Inlier, North Queensland. *Queensland Gov. Mining J.*, **81**, 28–43.

WOESE, C.R. (1987) Bacterial Evolution. *Microbiol. Rev.*, **51**, 221–271.

ZABACK, D.A., PRATT, L.M. & HAYES, J.M. (1993) Transport and reduction of sulfate and immobilization of sulfide in marine black shales. *Geology*, **21**, 141–144.

ZANG, W. & WALTER, M.R. (1992) Late Proterozoic and Cambrian microfossils and biostratigraphy, Amadeus Basin, central Australia. *Ass. Australas. Palaeontol. Mem.*, **12**, 132.

ZHENG, Y.-F. (1991) Sulphur isotopic fractionation between sulphate and sulphide in hydrothermal ore deposits: disequilibrium *vs* equilibrium processes. *Terra Nova*, **3**, 510–516.

ZIEG, G.A. & LEITCH, C.H.B. (1998) The geology of the Sheep Creek copper deposit, *Meager County, Montana. Belt Symposium III Abstr., Montana Bur. Mines Geol. Open-File Report MBMG*, No. 381.

*Spec. Publs int. Ass. Sediment.* (2002) **33**, 107–120

# Decimetre-thick encrustations of calcite and aragonite on the sea-floor and implications for Neoarchaean and Neoproterozoic ocean chemistry

D. Y. SUMNER

*Geology Department, University of California, Davis, CA 95616, USA*

## ABSTRACT

Centimetre- to metre-tall aragonite pseudomorphs are abundant in Neoarchaean marine carbonates and less common, but present, in some Proterozoic carbonates. Neoarchaean carbonates also contain substantial proportions of decimetre-thick calcite encrustations on the sea-floor, whereas their abundance declines through time. In contrast, the relative proportion of micritic sediment increases during Proterozoic time. The precipitation of large aragonite crystals and thick layers of calcite on the sea-floor implies that crystal growth rates were very high relative to sedimentation rates. For metre-tall Neoarchaean fans in shallow subtidal depositional environments, crystal growth rates must have substantially exceeded estimates of modern aragonite cement precipitation rates. High aragonite saturation states in sea water could produce rapid precipitation rates. The paucity of micrite suggests that nucleation of carbonate in the water column was limited. These conditions need to be maintained for millions of years and may be due to the presence of chemical inhibitors that slow crystal nucleation and precipitation rates. Rapid local crystal growth may reflect the local absence of inhibitors and globally low carbonate accumulation rates relative to calcium influx to the oceans. Some Neoproterozoic 'cap carbonates', which immediately overly glacial deposits, contain large aragonite pseudomorphs, similar to those common in Neoarchaean carbonates. Unlike Neoarchaean carbonates, however, sea-floor calcite encrustations are rare, and micrite precipitation was abundant. These differences suggest that the circumstances leading to the growth of Neoproterozoic large aragonite fans were different. The limited stratigraphic distribution of aragonite pseudomorphs also suggests that changes in ocean dynamics may have produced a temporary increase in carbonate saturation states, in contrast to the long-term maintenance of high supersaturation required by the distribution of Neoarchaean aragonite pseudomorphs.

## INTRODUCTION

Carbonate mineral precipitation is one of the key components of the global carbon cycle, and secular variation in the mineralogy of both abiotic and biotic carbonates reflects changes in seawater chemistry through time (e.g. Sandberg, 1985b; Wilkinson & Given, 1986; Grotzinger, 1990). Evaluating the carbonate precipitation processes that led to formation of Precambrian carbonate platforms is one of the best ways to constrain the carbonate chemistry of early sea water (Grotzinger & Kasting, 1993; Sami & James, 1993; 1994; Kah & Knoll, 1996; Sumner & Grotzinger, 1996b, 2000; Sumner, 1997a; Turner *et al.*, 2000). As better constraints on carbonate chemistry are developed, they will provide new insights into a variety of other studies, including environments for the early evolution of life, models of atmospheric chemistry and the dynamics of carbon and other elemental cycling.

Analyses of numerous Precambrian carbonate platforms have demonstrated that many of the fundamental architectural properties of Precambrian platforms are identical to those of their Phanerozoic counterparts despite the lack of shell-secreting metazoans (see summary in Grotzinger, 1989; Knoll & Swett, 1990; Sami & James, 1993; Saylor *et al.*, 1995; Pelechaty *et al.*, 1996; Turner *et al.*, 1997). However, some of the styles of carbonate precipitation are significantly different, and they vary within Archaean and Proterozoic time (e.g. Grotzinger, 1990; Sami & James, 1994; Kah & Knoll, 1996; Sumner & Grotzinger, 1996a,b; Grotzinger & Knoll, 1999).

One particularly significant change in carbonate precipitation is a decline in the abundance of thick encrustations of calcite and aragonite crystals on the sea-floor. Neoarchaean carbonates commonly contain decimetre- to metre-thick coatings of calcite and pseudomorphed aragonite that grew directly on the sea-floor from ambient marine water (Sumner, 1997a; Sumner & Grotzinger, 2000). These encrustations are abundant in all Neoarchaean carbonates studied (Sumner & Grotzinger, 2000). In contrast, such thick coatings of carbonate 'cements' are much less abundant in Palaeoproterozoic carbonates (Grotzinger & Read, 1983; Grotzinger, 1990; Sami & James, 1994), are limited to shallow, inner-platform depositional environments in Mesoproterozoic carbonates (Kah & Knoll, 1996) and are generally absent from Neoproterozoic carbonates, with the exception of carbonates associated with glacial deposits (Knoll & Swett, 1990; Peryt *et al.*, 1990; Narbonne *et al.*, 1994; Grotzinger & Knoll, 1995; Kennedy, 1996; Soffer & Hoffman, 1998; James, 1999; Turner *et al.*, 2000). Decimetre-thick coatings of calcite or aragonite are present only in unusual Phanerozoic carbonates (e.g. Grotzinger, 1990; Grotzinger & Kasting, 1993; Grotzinger & Knoll, 1995). The abundance of these textures in diverse depositional environments in Neoarchaean carbonates, in contrast to their relative paucity and limited distribution in Proterozoic carbonates, suggests an overall decline in sea-floor carbonate encrustation and an increase in micrite precipitation (Grotzinger, 1990; Kah & Knoll, 1996; Sumner & Grotzinger, 1996b; Grotzinger & Knoll, 1999). However, the presence of large Neoproterozoic aragonite pseudomorphs invites comparisons with Neoarchaean precipitation conditions.

Most of the thick Neoproterozoic coatings of sea-floor cements directly overlie glacial diamictites. Some of these carbonates contain decimetre-high aragonite fans (Peryt *et al.*, 1990; Narbonne *et al.*, 1994; Grotzinger & Knoll, 1995; Kennedy, 1996; Soffer & Hoffman, 1998; James, 1999). The close association of the carbonates with glacial diamictites implies substantial accumulation of carbonates in cold climates or extremely rapid deglaciation (e.g. Fairchild *et al.*, 1989; Kennedy, 1996; Hoffman *et al.*, 1998). Although carbonates have been observed in some modern glacial environments, such as rare deep fjords and glacial lakes, these carbonates do not form thick accumulations and do not provide a suitable analogue for the Neoproterozoic cap carbonate–glaciation associations (e.g. Fairchild & Spiro, 1990; Fairchild, 1993; Fairchild *et al.*, 1999). In fact, the presence of large aragonite pseudomorphs is inconsistent with deposition in cold water, which normally inhibits aragonite precipitation (e.g. Burton & Walter, 1987). Thus, substantial fluctuations in climate and carbonate chemistry are needed to explain the cap carbonate facies and their glacial associations.

The first step in understanding the changes in ocean carbonate chemistry through Archaean and Proterozoic time is to characterize carbonate precipitation styles and their distribution with depositional environment. The following is a summary of the Precambrian distribution of decimetre-thick sea-floor encrustations of calcite and aragonite that lack stromatolitic textures. The abundance of these coatings generally declines through time, suggesting secular variation in seawater chemistry.

## SEA-FLOOR ENCRUSTING CALCITE AND ARAGONITE CRYSTALS

Three features distinguish Neoarchaean carbonates from their younger counterparts: (i) Neoarchaean carbonates contain abundant aragonite pseudomorphs that radiated up to several metres above the sea floor (Sumner & Grotzinger, 2000); (ii) calcite encrustations, up to decimetres thick, commonly coated sea-floor surfaces, including sedimentary structures such as wave ripples and channel margins (Sumner, 1997a); and (iii) micrite is rare in subwave base depositional facies (Sumner, 1997a). Although some aspects of each can be found in Proterozoic and Phanerozoic carbonates, these three features are strikingly characteristic of Neoarchaean carbonates. Through Proterozoic time, there is a transition to fewer precipitated coatings and more abundant micrite.

### Precipitation styles

#### *Aragonite pseudomorphs*

Large limestone and dolomite crystal pseudomorphs that form centimetre- to metre-tall fans are known from every well preserved Neoarchaean carbonate platform on Earth (e.g. Hofmann, 1971; Bertrand-Sarfati, 1976; Bertrand-Sarfati & Eriksson, 1977; Martin *et al.*, 1980; Walter, 1983; Holland, 1984; Abell *et al.*, 1985; Hofmann *et al.*, 1985; Wilks, 1986; Simonson *et al.*, 1993; Sumner & Grotzinger, 2000). Fans always grew upwards or outwards from depositional surfaces into the overlying water column. Many fans are draped by sediment (Fig. 1); however, some

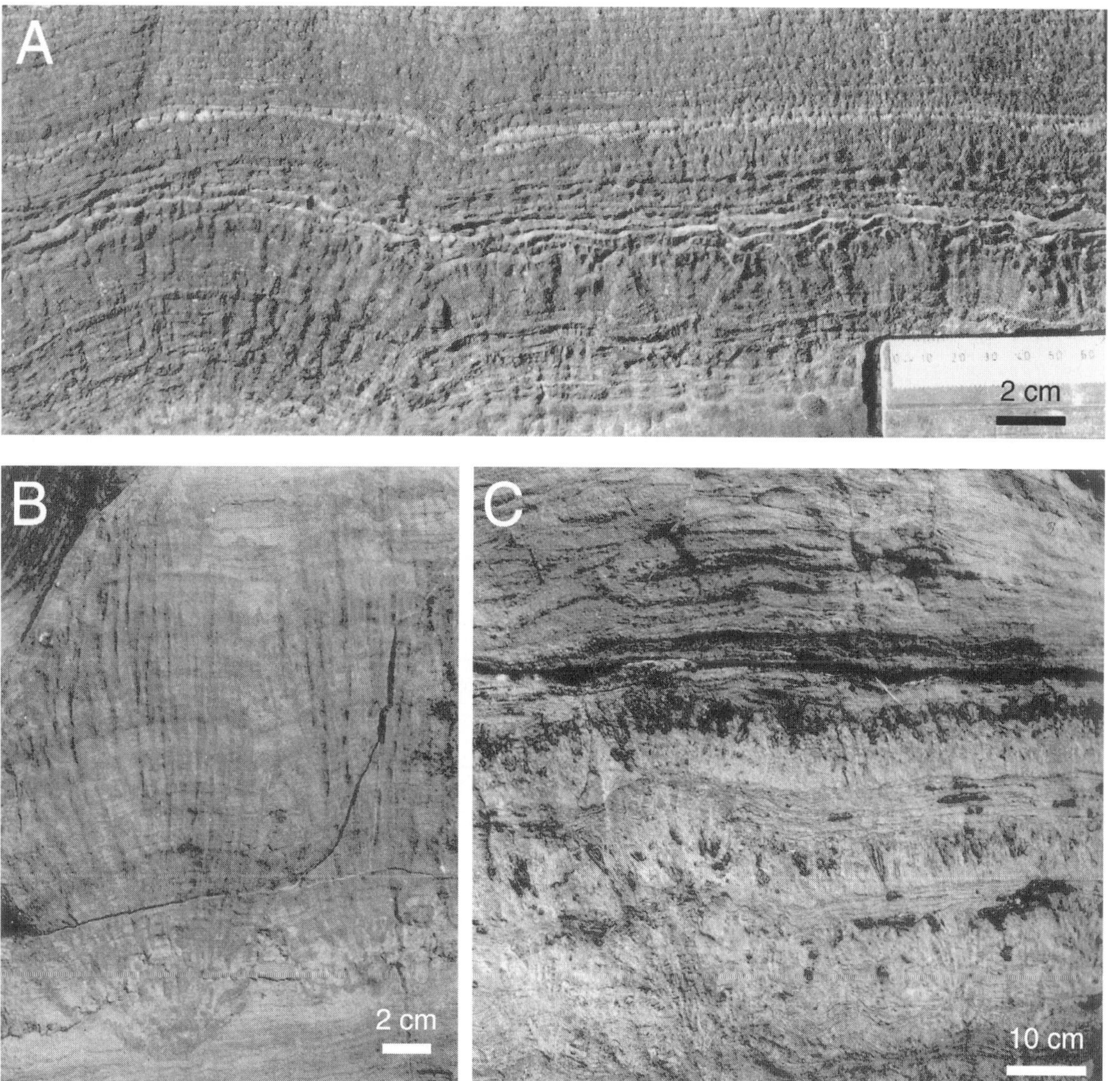


**Fig. 1.** Late Archaean aragonite pseudomorphs. (A) A layer of aragonite pseudomorphs overlain by laminated sediment from the Campbellrand–Malmani carbonate platform, South Africa. (B) Decimetre-scale fans of aragonite pseudomorphs draped by sediment from the Campbellrand–Malmani carbonate platform. (C) Fans interbedded with cross-stratified grainstones from the Cheshire Formation, Zimbabwe.

fans, up to 160 cm tall, lack evidence for detrital carbonate between the crystals, demonstrating that they did not grow within the sediment (Sumner & Grotzinger, 2000).

The fans have been interpreted as either gypsum or aragonite pseudomorphs, but Sumner & Grotzinger (2000) reinterpret all the pseudomorphs as replacing aragonite with the exception of morphologically distinct gypsum pseudomorphs from the 2.6 Ga Carawine Dolomite, Australia (Simonson *et al.*, 1993). The fans have flat to feathery crystal terminations and hexagonal cross-sections, and trains of inclusions within the pseudomorphs define a fibrous texture for the precursor mineral. Some well preserved pseudomorphs are replaced by optically unoriented 50–250 μm calcite crystals that are equant to rarely elongate and can contain up to several thousand ppm Sr (Sumner & Grotzinger, 2000). These properties are all characteristic

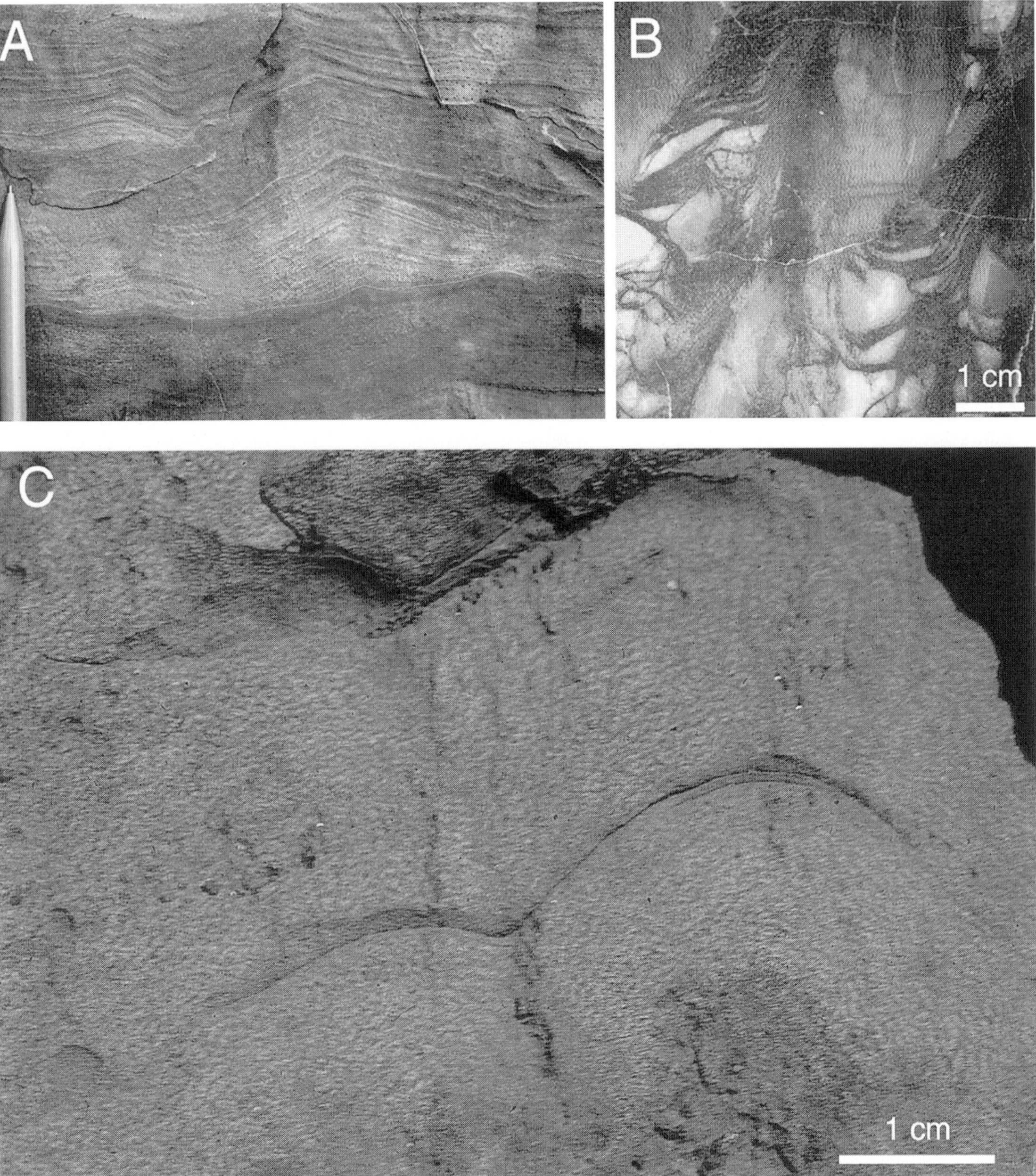


**Fig. 2.** Calcite encrustations of the sea-floor. (A) Fibrous marine calcite coating a wave ripple. Grainstones at the base of the bed are light coloured, and the calcite encrustations at the tops of beds are dark. Note that the crest of the ripple acted as a high during deposition of the next bed. (B) Microbialite from the Gamohaan Formation consisting of laminated mat encased in a serrate, fibrous marine cement (herringbone calcite of Sumner & Grotzinger, 1996a). Note the abundant voids created by the microbial structures. None of these contains any micrite. (C) A decimetre-thick coating of herringbone calcite that can be laterally traced for over 7000 km$^2$ (Sumner, 1997a). It is coating a microbial structure.

of an aragonitic precursor mineralogy (e.g. Loucks & Folk, 1976; Assereto & Folk, 1980; Mazzullo, 1980; Sandberg, 1985a; Peryt *et al.*, 1990) and strongly support an aragonitic precursor mineralogy for the crystal fans.

The environments in which the pseudomorphed aragonite precipitated are diverse, ranging from agitated subtidal to evaporitic facies. Depositional environments are best constrained in the 2580 to 2520 Ma Campbellrand–Malmani carbonate platform, South Africa, where preservation of the platform for 800 km across strike allows evaluation of the extent of restriction during growth of the fans (Sumner & Grotzinger, 2000). Fans are abundant throughout the platform and are prominent in wave-swept, subtidal giant stromatolite facies deposited in open marine environments (Sumner & Grotzinger, 2000). In addition, the crystal pseudomorph fans are abundant along the platform margin in lower intertidal reef equivalent facies. The presence of fans in these agitated, open marine environments implies that they precipitated from normal sea water and not solely from evaporitically concentrated waters (Sumner & Grotzinger, 2000).

Aragonite pseudomorphs from other Archaean carbonates, ranging in age from 2.9 to 2.6 Ga, are found in peritidal to below wave-base subtidal depositional environments (Hofmann *et al.*, 1985; Simonson *et al.*, 1993; Sumner & Grotzinger, 2000). For example, in the *c.*2.7 Ga Cheshire Formation, Belingwe Greenstone Belt, Zimbabwe, fans are an important facies, constituting as much as 50–100% of the observed volume of carbonate rock for decimetre- to decametre-thick sections (Martin *et al.*, 1980; Sumner & Grotzinger, 2000). Commonly, decimetre-tall fans are interbedded with fine quartz sands and silts containing cross-stratification most consistent with wave- or storm-dominated open marine shelf settings and water depths of 50 m and shallower (Fig. 1C; Sumner & Grotzinger, 2000). Lithoherms consisting solely of crystal fans are surrounded by siltstone, grainstone and intraclasts, demonstrating that the pseudomorph precursor crystals could form current-resistant highs on the sea-floor (Sumner & Grotzinger, 2000).

The diverse depositional environments in which aragonite fans precipitated as well as their local dominance implies abundant precipitation of aragonite crystals on the sea-floor during Neoarchaean time. Rare centimetre- to decimetre-tall aragonite fans are present in younger carbonates, but they are generally only present in localized environments (e.g. Mazzullo & Cys, 1977; Grotzinger & Read, 1983; Tucker, 1986; Babcock & Yurewicz, 1989; Peryt *et al.*, 1990; Grotzinger, 1994; Grotzinger & Knoll, 1995; Soffer & Hoffman, 1998; James, 1999). The ubiquity of large aragonite pseudomorphs in diverse depositional environment distinguishes Neoarchaean carbonate facies from younger counterparts.

### *Calcite encrustations*

Calcite encrustations on the sea-floor also are common in most, possibly all Neoarchaean carbonates (Fig. 2). The encrustations coat various sedimentary features, including the tops of pseudomorphed aragonite fans, columnar stromatolites, wave ripples, intraclasts and channel walls (Sumner & Grotzinger, 1996a). In deep subtidal depositional environments, calcite encrustation of microbial structures was a particularly important depositional process in the 2520 Ma Gamohaan Formation, South Africa (Sumner, 1997a,b). More than 90% of the rock in a 40 m section through the Gamohaan Formation consists of calcite that precipitated on *in situ* microbial mats or as sea-floor encrustations (Sumner, 1997a). Carbonate grains and micrite are absent. This stratigraphic interval contains similar concentrations of calcite cement over 7000 km$^2$ of exposed platform, demonstrating the significance of *in situ* calcite precipitation as a depositional process.

Sea-floor calcite encrustations are rare in Palaeoproterozoic carbonate platforms. Centimetre-thick layers of calcite cement usually only occur in stromatolites or neptunian dikes (Grotzinger, 1986; Sami & James, 1996; Pope & Grotzinger, 1997). The paucity of thick calcite coatings documents a significant decrease in the importance of precipitation of encrustations on the sea-floor as a platform-building process after Neoarchaean time. Although calcite encrustations are rare, many Proterozoic stromatolites contain cement-like crystal textures, particularly those that grew during Palaeoproterozoic time (Grotzinger, 1990; Kah & Knoll, 1996; Sami & James, 1996; Grotzinger & Knoll, 1999). However, most Neoproterozoic stromatolites contain micritic or microsparitic microtextures, due either to precipitation of micrite within a microbial mat or to trapping-and-binding of micrite by the mat (Grotzinger, 1990; Kah & Knoll, 1996; Knoll & Semikhatov, 1998; Grotzinger & Knoll, 1999; Turner *et al.*, 2000). This secular variation in stromatolite textures mimics the decline in thick calcite encrustations on the sea-floor, although it occurred later during Proterozoic time.

### *Micrite precipitation*

In contrast to the abundance of *in situ* precipitated calcite and aragonite, micrite precipitation was probably not widespread during Neoarchaean time. Recrystallization can mask micritic deposits in some circumstances, but micritic sediment is absent from the well preserved Gamohaan Formation (Sumner, 1997a). In 40 m of section, logged and described on a 5 cm scale, not a single micritic drape was observed. Since the microbialite facies of the Gamohaan Formation were deposited below wave base (Sumner, 1997a), any micrite that precipitated in the overlying water column should have accumulated on the microbialites. Thus, the absence of micrite is an indication that significant micrite precipitation was not occurring in the overlying water column. Similar deep subtidal facies in the 2.6 Ga Bulawayo greenstone belt, Zimbabwe, and the 2.6 Ga Carawine Formation, Australia, also lack micritic drapes (Sumner, 2000; unpublished data). In contrast, near-slope deep water facies of the Wittenoom Formation, which correlates to the Carawine Formation, contain a significant component of possible periplatform ooze deposited with fine-grained turbidites sourced from a shallow shelf (Simonson *et al.*, 1993). In the Campbellrand–Malmani carbonate platform, rare thin beds of micrite are present only in restricted peritidal environments. Thus, it appears that micrite did precipitate on shallow platforms during Neoarchaean time, but it may not have precipitated in the water column above more basinal facies. Additional characterizations of Neoarchaean carbonate platforms are necessary to test this hypothesis.

In contrast to the apparent paucity of micrite in many Neoarchaean carbonates, a number of Proterozoic carbonate platforms contain abundant microspar derived from micrite (e.g. Grotzinger, 1986, 1989; Knoll & Swett, 1990; Sami & James, 1994, 1996; Turner *et al.*, 1997, 2000). Although much of this micrite may have precipitated within benthic microbial mats (Fairchild, 1991; Kah & Knoll, 1996; Sami & James, 1996; Turner *et al.*, 2000), in several cases, significant micrite precipitation from the water column is the most likely source of some of the fine-grained carbonate (Knoll & Swett, 1990; Sami & James, 1994; Turner *et al.*, 1997). In particular, the 1.8 Ga Pethei Group, north-west Canada, contains a thick package of slope to basinal limestone rhythmites composed of fine grainstone and micrite (Sami & James, 1993, 1994). In contrast to the large volume of slope to basinal micrite that was deposited as loose sediment, micrite on the platform is less abundant and is almost exclusively associated with stromatolites, which produced clasts rather than loose micrite during erosion (Sami & James, 1994, 1996). The lack of evidence for loose micritic sediment on the platform led Sami & James (1994) to propose that the micrite deposited in slope and basinal rythmites precipitated in the water column rather than having been transported off the platform as periplatform ooze. The abundance of micrite in at least one Palaeoproterozoic platform and numerous younger ones suggests a significant change in carbonate precipitation near the Archaean–Proterozoic boundary.

#### Geochemical constraints

The substantial differences between cement-rich Neoarchaean carbonate textures and micrite-rich younger carbonates suggest secular variation in the dynamics of carbonate precipitation (Grotzinger, 1990; Kah & Knoll, 1996; Sumner & Grotzinger, 1996b, 2000; Knoll & Semikhatov, 1998; Grotzinger & Knoll, 1999). The decimetre to metre heights of many Neoarchaean aragonite pseudomorphs in shallow subtidal depositional environments would have required average sedimentation rates to be very low, crystal growth rates to be locally very high or a combination of both. If crystal growth rates were equivalent to estimates of modern aragonite cement precipitation rates of up to 29 mm 100 $yr^{-1}$ (Grammer *et al.*, 1993, 1996), intervals of hundreds to thousands of years with no sediment influx would be required for growth of Neoarchaean aragonite fans. In many cases, the crystal fans and calcite cement coatings precipitated in agitated shallow subtidal depositional environments with interbedded grainstones and siliciclastic siltstones (Simonson *et al.*, 1993; Sumner & Grotzinger, 2000) where it is unrealistic to expect no sediment influx for hundreds to thousands of years. Average sedimentation rates may have been lower owing to early cementation of carbonate sediments and a lack of significant micrite precipitation (see below), which could allow larger crystals to grow on the sea-floor. However, no evidence supports the interpretation that most depositional environments in all Neoarchaean carbonates had the extremely low sedimentation rates required to allow precipitation of decimetre and thicker carbonate encrustations at modern growth rates. Thus, localized crystal precipitation rates were probably higher in Neoarchaean oceans than the estimates of modern cement growth rates, possibly owing to higher aragonite supersaturation of surface waters.

The saturation state of sea water, i.e. the calcium and carbonate ion concentrations, is controlled by the net flux of calcium to the oceans, the carbonate chemistry of sea water, the total shelf area on which carbonate can accumulate and carbonate precipitation rates. The global area of carbonate deposition during Neoarchaean time is unconstrained. However, a high saturation state for aragonite could have been maintained for millions to hundreds of millions of years by either high calcium fluxes to the oceans or limited areas of carbonate accumulation owing to globally limited shelf area. Either could result in rapid growth of individual aragonite and calcite crystals on the sea-floor in the few environments where carbonate accumulated.

Even though crystal growth rates suggest that Neoarchaean sea water was highly supersaturated with respect to aragonite, micrite precipitation was rare. This requires low nucleation rates from open marine sea water (Sumner & Grotzinger, 1996b), which may have been caused either by a low pelagic biomass, or by the chemistry of sea water. If the main source of micrite prior to the evolution of calcifying algae was microbially induced precipitation, a low abundance of pelagic organisms in Neoarchaean oceans could have limited micrite precipitation. Others have suggested that the flux of organic matter to the sea-floor may have been low based on the paucity of organic matter in basinal iron-formations (Klein & Beukes, 1989; Beukes & Klein, 1990; Kaufman *et al.*, 1990), which is consistent with a low pelagic biomass. However, organic-rich shales of similar age suggest that Neoarchaean pelagic biomass was similar to that of younger oceans (e.g. Watanabe *et al.*, 1997). Furthermore, both cyanobacteria and heterotrophic bacteria provide a significant component of the modern pelagic biomass (Fogg, 1995). Similar species are likely to have been pelagic during Neoarchaean time, and some are capable of inducing micrite precipitation in the water column (Robbins & Blackwelder, 1992; Yates & Robbins, 1995), suggesting that pelagic biomass may not have been the most important factor in the paucity of Neoarchaean micrite. Alternatively, both biotic and abiotic micrite precipitation could have been kinetically inhibited by the chemical properties of sea water. In modern sea water, magnesium and sulphate slow the rates of calcite precipitation and nucleation, allowing the maintenance of supersaturated sea water (e.g. Berner, 1975). Numerous other ions in addition to various organic compounds also slow calcite and aragonite precipitation rates (e.g. Mucci & Morse, 1983; Meyer, 1984; Mucci, 1986; Burton & Walter, 1990; Dromgoole & Walter, 1990a,b; Zuddas & Mucci, 1994). If one or more of these were abundant in Neoarchaean sea water, micrite precipitation may have been geochemically limited even if a substantial pelagic biomass was present.

Two redox-sensitive elements, iron and manganese, were present in Archaean sea water and were less abundant after the oceans became more oxidizing at roughly 2 Ga (e.g. Ewers, 1983; Holland, 1984; Beukes & Klein, 1992; Kasting, 1992; Holland *et al.*, 1994; Simonson & Hassler, 1996). Both iron and manganese, in their reduced states, substantially slow the rate of calcite precipitation (Meyer, 1984; Dromgoole & Walter, 1990a,b; Katz *et al.*, 1993; Sumner, unpublished data), and either may have significantly reduced calcite nucleation rates (Sumner & Grotzinger, 1996b). Reduced iron is a particularly good candidate for a calcite precipitation inhibitor in Neoarchaean sea water for two reasons: (i) it is the most effective inhibitor to calcite precipitation studied to date (Meyer, 1984); and (ii) extensive Neoarchaean and Palaeoproterozoic banded iron-formations demonstrate that $Fe^{2+}$ was abundant in deep sea water (e.g. Ewers, 1983; Beukes & Klein, 1992; Simonson & Hassler, 1996). If even trace amounts of $Fe^{2+}$ were present in shallow ocean waters, it may have reduced the rate of calcite nucleation from sea water, because it limits spontaneous calcite nucleation in experimental NaCl solutions, consistent with the theoretical effects of inhibitors on nucleation (Sumner & Grotzinger, 1996b; Sumner, unpublished data). On extensive carbonate platforms, photosynthetic $O_2$ production and photolitic oxidation of $Fe^{2+}$ to $Fe^{3+}$ may have removed the inhibiting effects of $Fe^{2+}$ on calcite nucleation, allowing platformal precipitation of micrite. A similar effect may have occurred as oxygen concentrations rose in the surface oceans and atmosphere. With even small amounts of $O_2$, $Fe^{2+}$ would have been completely absent from surface waters, leading to more rapid calcite precipitation and an overall decrease in the saturation state of sea water. If aragonitic micrite was also rare, a similar inhibitor to aragonite nucleation may also have been present, although a specific candidate has yet to be identified.

In addition to reducing nucleation rates, precipitation inhibitors slow overall precipitation rates, thereby sustaining higher supersaturation states. If inhibitors to aragonite and calcite nucleation were present in Neoarchaean sea water, they could have allowed Neoarchaean sea water to remain more highly supersaturated with respect to aragonite and calcite than in their absence (Sumner & Grotzinger, 1996b; Sumner, 1997a). If local conditions favoured carbonate precipitation owing to low concentrations of inhibitors, precipitation could have occurred extremely rapidly. For example, high photosynthetic production of $O_2$ could have oxidized $Fe^{2+}$, thereby removing it as an inhibitor to calcite precipitation. Alternatively, if $Fe^{2+}$ only affected calcite precipitation, aragonite precipitation may have occurred very rapidly where temperatures and pressures were favourable.

Although the $Fe^{2+}$ inhibitory model is speculative, it does satisfy the need for both high saturation states and little micrite precipitation during Neoarchaean time. To test this model of redox-sensitive nucleation inhibitors, carbonate platforms between 2.5 and 1.8 Ga need to be studied from a petrographic perspective. The isotopic characteristics of carbonates of this age have been reported, and a large positive $\delta^{13}C$ excursion has been attributed to dramatic change in carbon cycling related to a rise in atmospheric $O_2$ (e.g. DesMarais *et al.*, 1992; Karhu & Holland, 1996; Melezhik *et al.*, 1999). Petrographic and textural studies of these units may be able to identify whether or not the change from Neoarchaean carbonate precipitation styles to more micritic facies does correlate to changes in atmospheric $O_2$. Additionally, experimental precipitation studies to characterize the dynamics and morphological characteristics of carbonate that precipitates from anoxic solutions with various chemical properties may provide kinetic constraints for the $Fe^{2+}$ inhibitory model, and identify specific characteristics that could correlate to natural cement textures.

## NEOPROTEROZOIC CAP CARBONATES

Some of the rare younger carbonate facies containing large aragonite pseudomorph fans similar to those of Neoarchaean age are associated with extensive Neoproterozoic glacial deposits (Peryt *et al.*, 1990; Narbonne *et al.*, 1994; Grotzinger & Knoll, 1995; Kennedy, 1996; Soffer & Hoffman, 1998; James, 1999). The Neoproterozoic glacial intervals represent extreme fluctuations in climate with significant accumulations of ice at equatorial latitudes (e.g. Harland, 1965; Williams *et al.*, 1998; Sohl *et al.*, 1999; Li, 2000). The glacially derived diamictites commonly are overlain by carbonates, called 'cap carbonates', with textures that suggest abundant carbonate precipitation from ambient sea water. The association of cold water glacial deposits with extensive carbonates is unusual and has elicited numerous explanations, including rapid climate change and unusual oceanographic properties during deglaciation (Grotzinger & Knoll, 1995; Hoffman *et al.*, 1998; Myrow & Kaufman, 1999). Developing a consistent, testable model is difficult because the cap carbonates require an unusual combination of climatic and oceanographic parameters that promote abundant carbonate precipitation during deglaciation.

### Stratigraphic setting

Neoproterozoic cap carbonates typically overly glacial deposits including tillites, debris flows, and siliciclastic or carbonate mudstones with abundant ice rafted debris (see papers in Hambrey & Harland, 1981). Cap carbonates vary in thickness from a veneer to more than tens of metres, and occasionally post-glacial carbonate platform deposition resumes, leading to accumulations of hundreds of metres of carbonate (e.g. Hoffman *et al.*, 1998; Prave, 1999). Stratigraphic data suggest that many cap carbonates were deposited during the rise in sea level during melting of extensive ice sheets and were occasionally deposited directly on erosional surfaces interpreted as having experienced terrestrial glaciation (e.g. Bertrand-Sarfati & Moussine-Pouchkine, 1983; Aitken, 1991; Kennedy, 1996). Depositional environments range from shallow water platforms (e.g. James, 1999) to middle to lower submarine fan environments (e.g. Myrow & Kaufman, 1999).

### Cap carbonate precipitated facies

Cap carbonates typically contain a combination of clastic carbonate, sheet cracks filled with calcite or aragonite cements and/or fibrous aragonite pseudomorphs that form fans similar to those seen in Neoarchaean carbonates. Other components, such as barite crystals, iron minerals and siliciclastic mud, commonly are present, but will not be discussed here.

#### *Aragonite pseudomorphs*

Centimetre- to metre-tall aragonite pseudomorphs have been documented in cap carbonates from the Bambuí Group, Brazil (Fig. 3a; Peryt *et al.*, 1990), the Ravensthroat Cap Carbonate (James, 1999) and 'Tepee Dolostone' (Narbonne *et al.*, 1994), north-west Canada, the Maieberg (Soffer & Hoffman, 1998) and Bushmansklippe (Grotzinger & Knoll, 1995) formations, Namibia, and the Nuccaleena Cap Carbonate, Australia (Kennedy *et al.*, 1997). Fans in the Bambuí Group, Ravensthroat Cap Carbonate and the Maieberg Formation all have well preserved crystal structures characteristic of an aragonitic precursor (Peryt *et al.*, 1990; Soffer & Hoffman, 1998; James, 1999). In

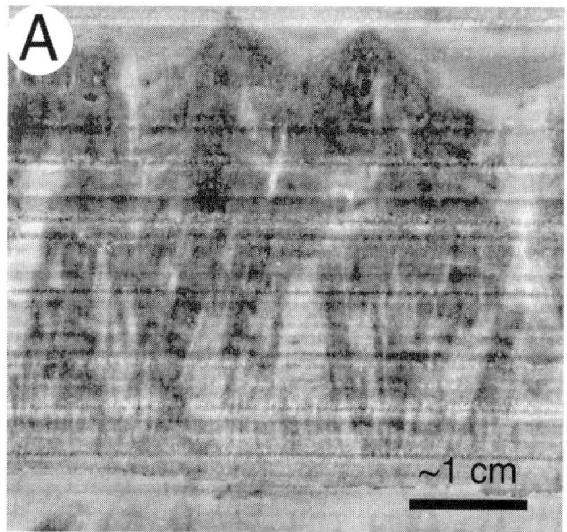

**Fig. 3.** (A) Aragonite pseudomorphs from the Pedro Leopoldo facies of the Bambuí Group, Brazil. They are interpreted as part of a cap carbonate (A.J. Kaufman, 2000, personal communication). Photo courtesy of A.J. Kaufman. (B) Cement-filled sheet cracks from the cap carbonate immediately overlying the Ghaub diamictite, Namibia. The sheet cracks are lighter grey. Note the complex synsedimentary folding of the surrounding carbonate.

addition, those from the Bambuí Group and the Ravensthroat Cap Carbonate have elevated strontium concentrations, which support an aragonitic interpretation (Peryt *et al.*, 1990; James, 1999).

The aragonite pseudomorphs are generally present in the deeper water or platform edge facies of cap carbonates (Kennedy, 1996; Soffer & Hoffman, 1998; James, 1999). Pseudomorphed aragonite fans in the Maieberg Formation were deposited near the platform margin, possibly along a 5 km topographic high (Soffer & Hoffman, 1998). Fans in the Ravensthroat Cap Carbonate are most abundant along a shallow water shelf edge where they built 'reef-like structures', sometimes including elongate stromatolites composed of aragonite pseudomorphs (James, 1999). Shoreward and basinward, micrite dominates and aragonite pseudomorphs are rare (James, 1999). In contrast, Kennedy (1996) reports that the Nuccaleena fans are associated with the deepest water cap carbonate facies in the Adelaide basin. More proximal facies contain abundant carbonate turbidites sourced off nearby slopes (Kennedy, 1996). Aragonite fans may have only precipitated in environments beyond the average flow distance for the turbidites. For shelf edge aragonite fans, a bathymetric high on the shelf margin may have had low local sedimentation rates, allowing significant aragonite fan growth, or high wave energy could have swept competing micrite out of the depositional environment, also allowing significant aragonite fan growth. In either case, the growth of aragonite fans may correlate with locally low sedimentation rates. No aragonite pseudomorphs have been documented from depositional environments with demonstrably high sedimentation rates.

*Sheet cracks*

The sheet crack facies of cap carbonates consists of abundant fibrous marine cements that infill neptunian dikes and sills in clastic carbonate. This facies is abundant in the Adelaide and Amadeus basins, Australia (Kennedy, 1996), and is present as a 1.3 m bed immediately overlying the Ghaub diamictite, Fransfontein, Namibia (Fig. 3b; Sumner & Prave, unpublished data). In both areas, the sediment and cements have been dolomitized. Similar facies may also be present in glacial limestones in western Hoggar, Algeria, where the cements consist of calcite (Caby & Fabre, 1981).

Many of the sheet cracks appear to have formed during synsedimentary deformation of carbonate turbidites (Kennedy, 1996) or in centimetre- to decimetre-thick beds of unknown origin. The sheet cracks are commonly concentrated into specific horizons and can be truncated by overlying layers (Kennedy, 1996), suggesting a synsedimentary origin. In the Adelaide and Amadeus basins, Australia, the fibrous cements are overlain by internal sediment, also demonstrating a synsedimentary origin for the cements (Kennedy, 1996). Kennedy (1996) also reports dissolution surfaces with up to 20 cm of relief that may have formed contemporaneously with cement precipitation within the sheet cracks. He interprets this as evidence that sea water was undersaturated with respect to carbonate, whereas pore waters were supersaturated (Kennedy, 1996).

The abundance of cements in the sheet crack facies suggests that pore waters had the potential for substantial carbonate precipitation. However, if Kennedy (1996) is correct, and contemporaneous sea water was undersaturated with respect to the carbonate sediment, the abundance of cement precipitation may be due to unstable chemistry or abundant crystal defects in the carbonate sediment. If much of the sediment was derived from glacial grinding of underlying carbonate platforms, the resulting carbonate may have contained abundant dislocations and defects, making it particularly susceptible to dissolution (Fairchild *et al.*, 1999). More stable phases, such as low-defect calcite, may have become supersaturated in the poor waters and precipitated in the sheet cracks as the high-defect carbonate sediment dissolved. Alternatively, if the relief on surfaces observed by Kennedy (1996) is not due to dissolution, the abundance of carbonate cements may reflect highly supersaturated sea water.

This second interpretation is consistent with the presence of aragonite fans in deeper depositional environments, which requires high saturation states. In contrast, if dissolution surfaces are present, the saturation state of end-glacial sea water was probably highly variable; locally conditions were supersaturated sufficiently to precipitate centimetre-tall aragonite fans, whereas contemporaneous carbonate in more proximal depositional environments was dissolving. Additional field-work in diverse areas is needed to distinguish between these interpretations.

### *Micrite*

Micrite is abundant in Neoproterozoic carbonates in general (e.g. Knoll & Swett, 1990; Turner *et al.*, 1997) and is a significant component of many cap carbonates. A particularly striking micritic cap carbonate is present in the Gaskiers Formation, Newfoundland (Myrow & Kaufman, 1999). It consists of 50 cm of micrite with rare millimetre thick green mudstone laminae and immediately overlies glacial diamictites and mudstones with drop stones (Myrow & Kaufman, 1999). It was deposited in a mid to outer fan environment and is the only bedded carbonate within a 15 km thick sequence of volcaniclastic flysch and molasse deposits (Myrow & Kaufman, 1999). The absence of sedimentary structures in the bed and the interbedding of green mudstone and micrite suggest that the bed formed from the settling of suspended micrite (Myrow & Kaufman, 1999). The accumulation of micrite in a volcaniclastic-dominated basin suggests rapid micrite precipitation in the water column and reduced siliciclastic influxes due to end-glacial transgression (Myrow and Kaufman, 1999). The presence of micrite in such a deep water, siliciclastic environment supports the theory that carbonate saturation states were very high in Neoproterozoic end-glacial sea water and is consistent with the growth of large aragonite fans in other depositional environments.

The abundance of micrite in cap carbonates is critical to developing models of carbonate precipitation during deglaciation. Substantial micrite accumulation demonstrates that carbonate nucleation, either calcite or aragonite, was not significantly inhibited in the water column. Carbonate nucleation, whether it was abiotic or biogenically induced, occurred in abundance in open sea water, unlike during Neoarchaean time. If micrite precipitation was abiotic, it may have been a response to highly supersaturated surface water conditions due to rapid deglacial changes in ocean chemistry (Grotzinger & Knoll, 1995; Kennedy, 1996; Hoffman *et al.*, 1998; Myrow & Kaufman, 1999). Within a geologically short time, rapid precipitation would have slowed to more typical Neoproterozoic rates as calcite and aragonite were removed from sea water and the extent of supersaturation declined.

Alternatively, if micrite precipitation was biologically driven, the precipitation of abundant micrite might reflect an increase in planktonic biomass response to deglaciation. Very rapid growth of the planktonic biomass, owing to increased open ocean surface, sunlight, nutrients etc. (e.g. Kimura *et al.*, 1997), could have led to rapid removal of $CO_2$ from surface waters promoting micrite precipitation. Severe stress on the Neoproterozoic biosphere owing to extensive glaciations has been proposed as a significant evolutionary force (Kimura *et al.*, 1997; Hoffman *et al.*, 1998). The potentially high pelagic productivity during end-glacial times may have affected the sedimentary record by promoting abundant micrite precipitation.

## Neoarchaean–Neoproterozoic comparisons

The presence of pseudomorphed aragonite fans in Neoproterozoic and Neoarchaean carbonates requires sea water that was supersaturated with respect to aragonite and high crystal growth rates. For Neoproterozoic oceans, the high saturation states need to be maintained for a short time and in relatively few environments. In contrast, Neoarchaean oceans must have remained supersaturated for millions of years. This difference in time-scales places significantly different constraints on the geochemical models required. For Neoproterozoic aragonite precipitation, changes in ocean circulation, ocean temperature and atmospheric $CO_2$ could all produce rapid increases in aragonite saturation states. In contrast, the highly supersaturated state of Neoarchaean sea water had to have been maintained in a dynamic equilibrium controlled by continental run-off, sea-floor hydrothermal alteration and the global accumulation and burial of carbonate sediment.

The abundance of Neoproterozoic micrite and its potential paucity in Neoarchaean carbonates also suggest that the mechanisms for producing the high saturation states were different. The nucleation of carbonate in Neoarchaean sea water may have been chemically or biologically limited, forcing most carbonate precipitation to occur directly on the sea-floor (Sumner & Grotzinger, 1996b). In contrast, nucleation was not a substantial limitation during Neoproterozoic deglaciation. Instead, high saturation states may have been produced by the dramatic imbalance and mixing of highly alkaline waters with waters charged with inorganic carbon (e.g. Kaufman *et al.*, 1991; Grotzinger & Knoll, 1995; Hoffman *et al.*, 1998).

## CONCLUSIONS

The distribution of Neoarchaean and Proterozoic carbonate facies shows a shift from the precipitation of calcite and aragonite directly on the sea-floor to more abundant micrite precipitation. The general decline in the abundance of decimetre-thick coatings of aragonite and calcite on the sea-floor reflects a change in the dynamics of carbonate precipitation through Precambrian time, particularly near the Archaean–Proterozoic boundary. This change could be due to changes in the level of seawater aragonite supersaturation and the nucleation rate of calcite and aragonite crystals in sea water. Changes in the concentrations of trace elements that slow precipitation and nucleation rates may have caused this secular variation.

Models for precipitation of Neoproterozoic aragonite fans can be fundamentally different from those needed to explain Neoarchaean aragonite fans. Neoproterozoic fans precipitated over short intervals of time in environments with low sediment accumulation rates. Thus, elevated aragonite supersaturation states could have been due to temporary changes in ocean dynamics related to the dramatic changes associated with the end of the extensive glaciations.

## ACKNOWLEDGEMENTS

I thank Paul Hoffman, Linda Kah and Bruce Wilkinson for helpful reviews that led to significant improvements in the manuscript. Field work in Neoarchaean carbonates was supported by NASA grant NAGW-2795 to John Grotzinger. Paul Hoffman provided field support in Namibia.

## REFERENCES

Abell, P.I., McClory, J., Martin, A. & Nisbet, E.G. (1985) Archean stromatolites from the Ngezi Group, Belingwe Greenstone Belt, Zimbabwe; preservation and stable isotopes—preliminary results. *Precam. Res.*, **27**, 357–383.

Aitken, J.D. (1991) Two late Proterozoic glaciations, Mackenzie Mountains, northwestern Canada. *Geology*, **19**, 445–448.

Assereto, R. & Folk, R.L. (1980) Diagenetic fabrics of aragonite, calcite, and dolomite in an ancient peritidal–spelean environment: Triassic Calcare Rosso, Lombardia, Italy. *J. sediment. Petrol.*, **50**, 371–394.

Babcock, J.A. & Yurewicz, D.A. (1989) The massive facies of the Capitan limestone, Guadalupe mountains, Texas and New Mexico. In: *Subsurface and Outcrop Examination of the Capitan Shelf Margin, Northern Delaware Basin* (Eds Harris, P.M. & Grover, G.A.). Soc. econ. Paleontol. Miner., Tulsa, Core Workshop 13, pp. 365–372.

Berner, R.A. (1975) The role of magnesium in the crystal growth of calcite and aragonite from seawater. *Geochim. Cosmochim. Acta*, **39**, 489–504.

Bertrand-Sarfati, J. (1976) Psedomorphoses de gypse en rosettes dans un calcaire cryptalgo-laminaire du Précambrien inférieur (système du Transvaal. Afrique du Sud). *Bull. Soc. géol. France*, **18**, 99–102.

Bertrand-Sarfati, J. & Eriksson, K.A. (1977) Columnar stromatolites from the Early Proterozoic Schmidtsdrift Formation, northern Cape Province, South Africa. Part 1: Systmatic and diagnostic features. *Palaeontol. Afr.*, **20**, 1–26.

Bertrand-Sarfati, J. & Moussine-Pouchkine, A. (1983) Platform-to-basin facies evolution: the carbonates of late Proterozoic (Vendian) Gourma (west Africa). *J. sediment. Petrol.*, **53**, 275–293.

Beukes, N.J. & Klein, C. (1990) Geochemistry and sedimentology of a facies transition—from microbanded to granular iron-formation—in the early Proterozoic Transvaal Supergroup, South Africa. *Precam. Res.*, **47**, 99–139.

Beukes, N.J. & Klein, C. (1992) Models for iron-formation deposition. In: *The Proterozoic Biosphere* (Eds Schopf, J.W. & Klein, C.), pp. 147–151. Cambridge University Press, Cambridge.

Burton, E.A. & Walter, L.M. (1987) Relative precipitation rates of aragonite and Mg calcite from seawater: temperature or carbonate ion control? *Geology*, **15**, 111–114.

Burton, E.A. & Walter, L.M. (1990) The role of pH in phosphate inhibition of calcite and aragonite precipitation rates in seawater. *Geochim. Cosmochim. Acta*, **54**, 797–808.

Caby, R. & Fabre, J. (1981) Late Proterozoic to Early Palaeozoic diamictites, tillites and associated glacigenic sediments in the Série Pourprée of western Hoggar, Algeria. In: *Earth's Pre-Pleistocene Glacial Record* (Eds Hambrey, M.J. & Harland, W.B.), pp. 140–145. Cambridge University Press, Cambridge.

DesMarais, D.J., Strauss, H., Summons, R.E. & Hayes, J.M. (1992) Carbon isotope evidence for the stepwise oxidation of the Proterozoic environment. *Nature*, **359**, 605–609.

Dromgoole, E.L. & Walter, L.M. (1990a) Inhibition of calcite growth rates by $Mn^{2+}$ in $CaCl_2$ solutions at 10, 25, and 50 °C. *Geochim. Cosmochim. Acta*, **54**, 2991–3000.

Dromgoole, E.L. & Walter, L.M. (1990b) Iron and manganese incorporation into calcite: effects of growth kinetics, temperature and solution chemistry. *Chem. Geol.*, **81**, 311–336.

Ewers, W.E. (1983) Chemical factors in the deposition and diagenesis of banded iron-formation In: *Iron-formation: Facts and Problems* (Eds Trendall, A.F. & Morris, R.C.), pp. 491–512. Elsevier, Amsterdam.

Fairchild, I.J. (1991). Origins of carbonate in Neoproterozoic stromatolites and the identification of modern analogues. *Precam. Res.*, **53**, 281–299.

Fairchild, I.J. (1993) Balmy shores and icy wastes; the paradox of carbonates associated with glacial deposits in Neoproterozoic times. *Sediment. Rev.*, **1**, 1–16.

Fairchild, I.J., Hambrey, M.J., Spiro, B. & Jefferson, T.H. (1989) Late Proterozoic glacial carbonates in Northeast Spitsbergen; new insights into the carbonate–tillite association. *Geol. Mag.*, **126**, 469–490.

Fairchild, I.J., Killawee, J.A., Sharp, M.J. *et al.* (1999) Solute generation and transfer from a chemically reactive alpine glacial–proglacial system. *Earth Surf. Process. Landf.*, **24**, 1189–1211.

Fairchild, I.J. & Spiro, B. (1990) Carbonate minerals in glacial sediments; geochemical clues to palaeoenvironment In: *Glacimarine Environments: Processes and Sediments* (Ed. Dowdeswell, J.A.), Spec. Publs geol. Soc. London, No. 53, pp. 201–216.

Fogg, G.E. (1995) Some comments on picoplankton and its importance in the pelagic ecosystem. *Aquat. microb. Ecol.*, **9**, 33–39.

Grammer, G.M., Ginsburg, R.N., Swart, P.K., McNeill, D.F., Jull, A.J.T. & Prezbindowski, D.R. (1993) Rapid growth rates of syndepositional marine aragonite cements in steep marginal slope deposits, Bahamas and Belize. *J. sediment. Petrol.*, **63**, 983–989.

Grammer, G.M., McNeill, D.F. & Crescini, C.M. (1996) Quantifying rates of syndepositional marine cementation across a carbonate platform and margin, Bahamas (abstract). *Geol. Soc. Am.*, **28**, 337.

Grotzinger, J.P. (1986) Cyclicity and paleoenvironmental dynamics, Rocknest platform, northwest Canada. *Geol. Soc. Am. Bull.*, **97**, 1208–1231.

Grotzinger, J.P. (1989) Facies and evolution of Precambrian carbonate depositional systems: Emergence of the modern platform archetype In: *Controls on Carbonate Platform and Basin Development* (Eds Crevello, P.D., Wilson, J.L., Sarg, J.F. & Read, J.F.). Spec. Publ. Soc. econ. Paleont. Miner., Tulsa, **44**, 79–106.

Grotzinger, J.P. (1990) Geochemical model for Proterozoic stromatolite decline. *Am. J. Sci.*, **290A**, 80–103.

Grotzinger, J.P. (1994) Trends in Precambrian carbonate sediments and their implication for understanding evolution. In: *Early Life on Earth* (Ed. Bengtson, S.), pp. 245–258. Columbia University Press, New York.

Grotzinger, J.P. & Kasting, J.F. (1993) New constraints on Precambrian ocean composition. *J. Geol.*, **101**, 235–243.

Grotzinger, J.P. & Knoll, A.H. (1995) Anomalous carbonate precipitates: is the Precambrian the key to the Permian? *Palaios*, **10**, 578–596.

Grotzinger, J.P. & Knoll, A.H. (1999) Stromatolites in Precambrian carbonates: evolutionary mileposts or environmental dipsticks? *Ann. Rev. Earth planet. Sci.*, **27**, 313–358.

Grotzinger, J.P. & Read, J.F. (1983) Evidence for primary aragonite precipitation, lower Proterozoic (1.9 Ga) dolomite, Wopmay orogen, northwest Canada. *Geology*, **11**, 710–713.

Hambrey, M.J. & Harland, W.B. (Eds) (1981) *Earth's Pre-Pleistocene Glacial Record.* Cambridge University Press, Cambridge, 1004 pp.

Harland, W.B. (1965) Critical evidence for a great Infra-Cambrian glaciation. *Geol. Rundsch.*, **54**, 45–61.

Hoffman, P.F., Kaufman, A.J., Halverson, G.P. & Schrag, D.P. (1998) A Neoproterozoic snowball earth. *Science*, **281**, 1342–1346.

Hofmann, H.J. (1971) Precambrian fossils, pseudofossils, and problematica in Canada. *Geol. Surv. Can. Bull.*, **189**, 1–146.

Hofmann, H.J., Thurston, P.C. & Wallace, H. (1985) Archean Stromatolites from Uchi greenstone belt, Northwestern Ontario In: *Evolution of Archean Supracrustal Sequences* (Eds Ayres, L.D., Thurston, P.C., Card, K.D. & Weber, W.), Spec. Publ. geol. Ass. Can., **28**, 1125–1132.

Holland, H.D. (1984) *The Chemical Evolution of the Atmosphere and Oceans.* Princeton University Press, Princeton, NJ, 582 pp.

Holland, H.D., Kuo, P.H. & Rye, R.O. (1994) $O_2$ and $CO_2$ in the late Archaean and early Proterozoic atmosphere (abstract). *M. Goldschmidt Conf.*, **58A**(A-K), 424–425.

James, N.P. (1999) Neoproterozoic cap carbonate facies; Mackenzie Mountains, NW Canada: abiotic precipitation and global glacial meltdown (abstract). *Geol. Soc. Am. Bull.*, **31**, A-487.

Kah, L.C. & Knoll, A.H. (1996) Microbenthic distribution in Proterozoic tidal flats: Environmental and taphanomic considerations. *Geology*, **24**, 79–82.

Karhu, J.A. & Holland, H.D. (1996) Carbon isotopes and the rise of atmospheric oxygen. *Geology*, **24**, 867–870.

Kasting, J.F. (1992) Models relating to Proterozoic atmospheric and oceanic chemistry In: *The Proterozoic Biosphere* (Eds Schopf, W.J. & Klein, C.), pp. 1185–1187. Cambridge University Press, Cambridge.

Katz, J.L., Reick, M.R., Herzog, R.E. & Parsiegla, K.I. (1993) Calcite growth-inhibition by iron. *Langmuir*, **9**, 1423–1430.

Kaufman, A.J., Hayes, J.M. & Klein, C. (1990) Primary and diagenetic controls of isotopic compositions of iron-formation carbonates. *Geochim. Cosmochim. Acta*, **54**, 3461–3473.

Kaufman, A.J., Hayes, J.M., Knoll, A.H. & Germs, G.B. (1991) Isotopic compositions of carbonates and organic carbon from upper Proterozoic succession in Namibia: stratigraphic variation and the effects of diagenesis and metamorphism. *Precam. Res.*, **73**, 301–327.

KENNEDY, M.J. (1996) Stratigraphy, sedimentology, and isotopic geochemistry of Australian Neoproterozoic post-glacial cap dolostones; deglaciation, $\delta^{13}C$ excursions, and carbonate precipitation. *J. sediment. Res.*, **66**, 1050–1064.

KENNEDY, M.J., PRAVE, A.R. & HOFFMANN, K.H. (1997) A carbon isotopic record through a Neoproterozoic glacial cycle (abstract). *Geol. Soc. Am. Bull.*, **29**, 196.

KIMURA, H., MATSUMOTO, R., KAKUWA, Y., HAMDI, B. & ZIBASERESHT, H. (1997) The Vendian–Cambrian $^{13}C$ record, North Iran; evidence for overturning of the ocean before the Cambrian explosion. *Earth planet. Sci. Lett.*, **147**, E1–E7.

KLEIN, C. & BEUKES, N.J. (1989) Geochemistry and sedimentology of a facies transition from limestone to iron-formation deposition in the Early Proterozoic Transvaal Supergroup, South Africa. *Econ. Geol.*, **84**, 1733–1742.

KNOLL, A.H. & SEMIKHATOV, M.A. (1998) The genesis and time distribution of two distinctive Proterozoic stromatolite microstructures. *Palaios*, **13**, 408–422.

KNOLL, A.H. & SWETT, K. (1990) Carbonate deposition during the later Proterozoic Era: an example from Spitsbergen. *Am. J. Sci.*, **290A**, 104–132.

LI, Z.X. (2000) New palaeomagnetic results from the 'cap dolomite' of the Neoproterozoic Walsh Tillite, northwestern Australia. *Precam. Res.*, **100**, 359–370.

LOUCKS, R.G. & FOLK, R.L. (1976) Fanlike rays of former aragonite in Permian Capitan Reef pisolite. *J. sediment. Petrol.*, **46**, 483–485.

MARTIN, A., NISBET, E.G. & BICKLE, M.J. (1980) Archean stromatolites of the Belingwe Greenstone Belt, Zimbabwe (Rhodesia). *Precam. Res.*, **13**, 337–362.

MAZZULLO, S.J. (1980) Calcite pseudospar replacive of marine acicular aragonite, and implications for aragonite cement diagenesis. *J. sediment. Petrol.*, **50**, 409–422.

MAZZULLO, S.J. & CYS, J.M. (1977) Submarine cements in Permian boundstones and reef-associated rocks, Guadalupe Mountains, west Texas and southeastern New Mexico. In: *Upper Guadalupian Facies, Permian Reef Complex, Guadalupe Mountains, New Mexico and West Texas* (Eds Hileman, M.E. & Mazzullo, S.J.), pp. 151–200. Permian Basin Section, Soc. econ. Paleont. Miner., Tulsa.

MELEZHIK, V.A., FALLICK, A.E., MEDVEDEV, P.V. & MAKARIKHIN, V.V. (1999) Extreme C-13(carb) enrichment in *ca* 2.0 Ga magnesite–stromatolite–dolomite–'red beds' association in a global context: a case for the world-wide signal enhanced by a local environment. *Earth Sci. Rev.*, **48**, 71–120.

MEYER, H.J. (1984) The influence of impurities on the growth rate of calcite. *J. Crystal Growth*, **66**, 639–646.

MUCCI, A. (1986) Growth kinetics and composition of magnesium calcite overgrowths precipitated from seawater: quantitative influence of orthosphosphate ions. *Geochim. Cosmochim. Acta*, **50**, 2255–2265.

MUCCI, A. & MORSE, J.W. (1983) The incorporation of $Mg^{2+}$ and $Sr^{2+}$ into calcite overgrowths: influences of growth rate and solution composition. *Geochim. Cosmochim. Acta*, **47**, 217–233.

MYROW, P.M. & KAUFMAN, A.J. (1999) A newly discovered cap carbonate above Varanger-age glacial deposits in Newfoundland, Canada. *J. sediment. Res.*, **69**, 784–793.

NARBONNE, G.M., KAUFMAN, A.J. & KNOLL, A.H. (1994) Integrated chemostratigraphy and biostratigraphy of the Windermere Supergroup, northwestern Canada; implications for Neoproterozoic correlations and the early evolution of animals. *Geol. Soc. Am. Bull.*, **106**, 1281–1292.

PELECHATY, S.M., GROTZINGER, J.P., KASHIRTSEV, V.A. & ZHERNOVSKY, V.P. (1996) Chemostratigraphic and sequence stratigraphic constraints on Vendian–Cambrian basin dynamics, Northeast Siberian Craton. *J. Geol.*, **104**, 543–563.

PERYT, T.M., HOPPE, A., BECHSTAEDT, T., KOESTER, J., PIERRE, C. & RICHTER, D.K. (1990) Late Proterozoic aragonitic cement crusts, Bambui Group, Minas Gerais, Brazil. *Sedimentology*, **37**, 279–286.

POPE, M.C. & GROTZINGER, J.P. (1997) The 1.8 Ga Stark Megabreccia; facies association and reconstruction of a major Paleoproterozoic evaporite (abstract). *Am. Ass. Petrol. Geol. 1997 Annual Convention*, **6**, 94.

PRAVE, A.R. (1999) Two diamictites, two cap carbonates, two delta (super 13) C excursions, two rifts; the Neoproterozoic Kingston Peak Formation, Death Valley, California. *Geology*, **27**, 339–342.

ROBBINS, L.L. & BLACKWELDER, P.L. (1992) Biochemical and ultrastructural evidence for the origin of whitings; a biologically induced calcium carbonate precipitation mechanism. *Geology*, **20**, 464–468.

SAMI, T.T. & JAMES, N.P. (1993) Evolution of an early Proterozoic foreland basin carbonate platform, lower Pethei Group, Great Slave Lake, north-west Canada. *Sedimentology*, **40**, 403–430.

SAMI, T.T. & JAMES, N.P. (1994) Peritidal carbonate platform growth and cyclicity in an early Proterozoic foreland basin, Upper Pethei Group, northwest Canada. *J. sediment. Res.*, **B64**, 111–131.

SAMI, T.T. & JAMES, N.P. (1996) Synsedimentary cements as Paleoproterozoic platform building blocks, Pethei Group, Northwestern Canada. *J. sediment. Res.*, **66**, 209–222.

SANDBERG, P.A. (1985a) Aragonite cements and their occurrence in ancient limestone. In: *Carbonate Cements* (Eds Schneidermann, N. & Harris, P.M.). Spec. Publ. Soc. econ. Paleont. Miner, Tulsa, **36**, 33–57.

SANDBERG, P.A. (1985b) Nonskeletal aragonite and $pCO_2$ in the Phanerozoic and Proterozoic. In: *The Carbon Cycle and Atmospheric $CO_2$: Natural Variations Archean to Present* (Eds Sundquist, E.T. & Broecker, W.S.), pp. 585–594. American Geophysical Union, Washington, DC.

SAYLOR, B.Z., GROTZINGER, J.P. & GERMS, G.J.B. (1995) Sequence stratigraphy and sedimentology of the Neoproterozoic Kuibis and Schwarzrand subgroups (Nama Group), southwestern Namibia. *Precam. Res.*, **73**, 153–171.

SIMONSON, B.M. & HASSLER, S.W. (1996) Was the deposition of large Precambrian iron formations linked to major marine transgressions? *J. Geol.*, **104**, 665–676.

SIMONSON, B.M., SCHUBEL, K.A. & HASSLER, S.W. (1993) Carbonate sedimentology of the early Precambrian Hamersley Group of Western Australia. *Precam. Res.*, **60**, 287–335.

SOFFER, G. & HOFFMAN, P.F. (1998) Sea floor precipitates pseudomorphic after aragonite in a Neoproterozoic 'cap carbonate' from northwestern Namibia (abstract). *Geol. Soc. Am. Annual Meeting, Abstracts with Programs*, p. 30.

Sohl, L.E., Christie-Blick, N. & Kent, D.V. (1999) Paleomagnetic polarity reversals in Marinoan (*ca* 600 Ma) glacial deposits of Australia: implications for the duration of low-latitude glaciation in Neoproterozoic time. *Geol. Soc. Am. Bull.*, **111**, 1120–1139.

Sumner, D.Y. (1997a) Carbonate precipitation and oxygen stratification in late Archean seawater as deduced from facies and stratigraphy of the Gamohaan and Frisco formations, Transvaal Supergroup, South Africa. *Am. J. Sci.*, **297**, 455–487.

Sumner, D.Y. (1997b) Late Archean calcite–microbe interactions: two morphologically distinct microbial communities that affected calcite nucleation differently. *Palaios*, **12**, 300–316.

Sumner, D.Y. (2000) Microbial versus environmental influences on the morphology of Late Archean fenestrate microbialites. In: *Microbial Sediments* (Eds Riding, R.E. & Awramik, S.M.), pp. 307–314. Springer-Verlag, Berlin.

Sumner, D.Y. & Grotzinger, J.P. (1996a) Herringbone calcite: petrography and environmental significance. *J. sediment. Res.*, **66**, 419–429.

Sumner, D.Y. & Grotzinger, J.P. (1996b) Were kinetics of Archean calcium carbonate precipitation related to oxygen concentration? *Geology*, **24**, 119–122.

Sumner, D.Y. & Grotzinger, J.P. (2000) Late Archean aragonite precipitation: Petrography, facies associations, and environmental significance. In: *Carbonate Sedimentation and Diagenesis in the Evolving Precambrian World* (Eds Grotzinger, J.P. & James, N.P.). Spec. Publ. Soc. econ. Paleont. Miner.,Tulsa, **65**.

Tucker, M.E. (1986) Formerly aragonitic limestones associated with tillites in the late Proterozoic of Death Valley, California. *J. sediment. Petrol.*, **56**, 818–830.

Turner, E.C., James, N.P. & Narbonne, G.M. (1997) Growth dynamics of Neopreoterozoic calcimicrobial reefs, Mackenzie Mountains, northwestern Canada. *J. sediment. Res.*, **67**, 437–450.

Turner, E.C., James, N.P. & Narbonne, G.M. (2000) Taphonomic control on microstructure in early Neoproterozoic reefal stromatolites and thrombolites. *Palaios*, **15**, 87–111.

Walter, M.R. (1983) Archean stromatolites: evidence of the Earth's earliest benthos. In: *Earth's Earliest Biosphere* (Ed. Schopf, J.W.), pp. 187–213. Princeton University Press, Princeton, NJ.

Watanabe, Y., Naraoka, H., Wronkiewicz, D.J., Condie, K.C. & Ohmoto, H. (1997) Carbon, nitrogen, and sulfur geochemistry of Archean and Proterozoic shales from the Kaapvaal Craton, South Africa. *Geochim. Cosmochim. Acta*, **61**, 3441–3459.

Wilkinson, B.H. & Given, R.K. (1986) Secular variation in abiotic marine carbonates: Constraints on Phanerozoic atmospheric carbon dioxide contents and oceanic Mg/Ca ratios. *J. Geol.*, **94**, 321–333.

Wilks, M.E. (1986) *The geology of the Steep Rock Group, NW Ontario: a major Archean unconformity and Archean stromatolites*. MSc thesis, University of Saskatchewan, Saskatoon.

Williams, D.M., Kasting, J.F. & Frakes, L.A. (1998) Low-latitude glaciation and rapid changes in the Earth's obliquity explained by obliquity–oblateness feedback. *Nature*, **396**, 453–455.

Yates, K.K. & Robbins, L.L. (1995) Experimental evidence for a $CaCO_3$ precipitation mechanism for marine Synechocystis. *Bull. Inst. Océanograph. Monaco*, **14**(2), 51–59.

Zuddas, P. & Mucci, A. (1994) Kinetics of calcite precipitation from seawater: I. A classical chemical kinetics description for strong electrolyte solutions. *Geochim. Cosmochim. Acta*, **58**, 4353–4362.

# Case Studies

*Spec. Publs int. Ass. Sediment.* (2002) **33**, 123–152

# The 2.7–2.63 Ga Indin Lake supracrustal belt: an Archaean marginal basin–foredeep succession preserved in the western Slave Province, Canada

S. J. PEHRSSON

*Continental Geoscience Division, Geological Survey of Canada, 601 Booth Street, Ottawa, Ontario, K1A 0E8, Canada*

## ABSTRACT

The 2.7–2.63 Ga Indin Lake supracrustal belt contains the sedimentary record of a relatively juvenile, Neoarchaean tectonostratigraphic sequence unique to the south-western part of the Slave Province, Northwest Territories, Canada. Yellowknife Supergroup rocks of the belt comprise three distinct lithostratigraphic groups. The basal, >2.67 Ga, Hewitt Lake group consists of homogeneous, pillowed mafic tholeiitic and lesser felsic calc-alkaline volcanic rocks, generated from effusive eruptions and syneruptive mass and sediment gravity flows on the flanks of moderately deep subaqueous shield volcanoes. The overlying *c.*2.67 Ga Leta Arm group includes pillowed and massive mafic to felsic flows and diverse intermediate to felsic reworked pyroclastic and volcanogenic epiclastic rocks, all characterized by abrupt lateral and vertical facies changes. It is interpreted to have largely formed from effusive eruptions and syneruptive mass flows on the medial flanks of moderate to shallow, originally locally emergent, marine stratovolcanoes. Both groups are interpreted to have formed on thinned, somewhat older, continental crust, based on the variety of interbedded siliciclastic facies, isotopic and geochemical characteristics and inherited zircon component. The unconformably overlying, <2.65 Ga Chalco Lake group, consisting of conglomerates, graded sandstone–mudstones and minor felsic and mafic volcanic rocks including peperites, was largely deposited from sediment gravity flows in a deep marine basin. The basin formed during compressional orogenesis in the southern Slave Province and contains detritus eroded from both its volcanic substrate and Mesoarchaean sialic crust. The lithostratigraphic record of the Indin Lake supracrustal belt is interpreted to reflect early establishment of a marginal basin–arc sequence on thinned Mesoarchaean crust and subsequent development of a marine, syncollisional foredeep basin. Textures, associations and facies of the various lithostratigraphic units are strikingly similar to those of similar modern day environments. The Hewitt Lake and Leta Arm groups, like modern arc–marginal basin systems, are dominated by products of submarine effusive eruptions, debris and sediment gravity flows. The deep water facies of the syncollisional Chalco Lake group are typical of modern, marine orogenic basins such as foredeeps or underfilled peripheral foreland basins. Association of bimodal magmatism with foredeep sedimentation, however, is a feature unique to the Precambrian. The quartz-rich nature of first-cycle sandstones derived largely from volcanic sources is thought to reflect aggressive chemical weathering in the humid, $CO_2$-rich Archaean atmosphere. The predominance of mass and sediment gravity flows in these Neoarchaean sedimentary environments may be linked to high rates of erosion in a high relief, unvegetated terrain.

## INTRODUCTION

Establishing the depositional environments of supracrustal sequences is fundamental to the reconstruction of tectonic settings, both modern and ancient. This becomes even more difficult for the Archaean, where the understanding of lateral variations of lithostratigraphy, inferred palaeogeography and tectonothermal history is hampered by poor chronostratigraphic control.

The implications of lithostratigraphic variations to the tectonic history of the Archaean Slave Province (Fig. 1) have recently been the focus of the Slave Natmap project, which identified a number of distinct Meso- and Neoarchaean supracrustal sequences (Villeneuve *et al.*, 1994; Henderson *et al.*, 1995; Hrabi *et al.*, 1995; Bleeker *et al.*, 1999; Pehrsson & Villeneuve,

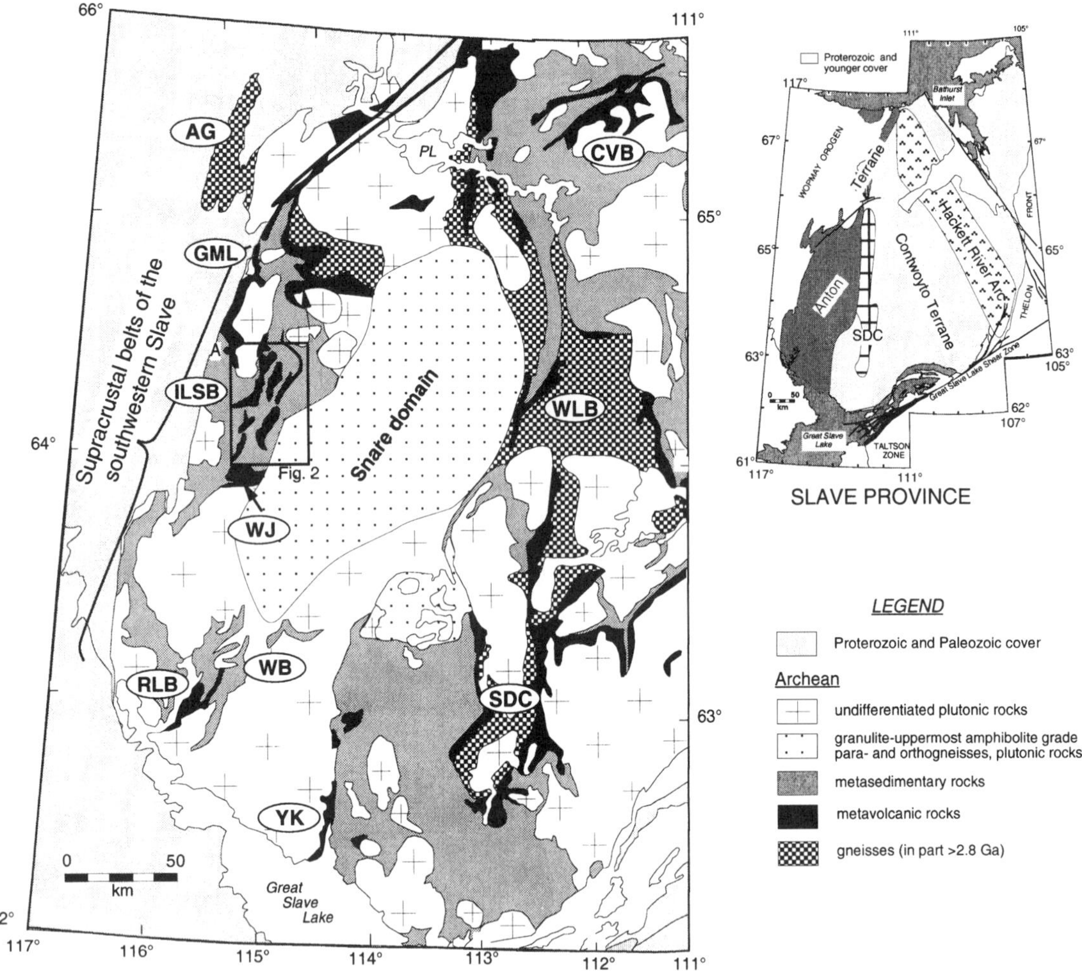


**Fig. 1.** Simplified geology of the western Slave Province. Inset: terrane subdivisions of the Slave Province modified after Kusky (1989). Abbreviations as follows: CVB, Central Volcanic belt; AG, Acasta Gneisses; GML, Grenville Mesa Lake belts; PL, Point Lake; RLB, Russell Lake belt; SDC, Sleepy Dragon Complex; WB, Wheeler Lake belt; WLB, Winter Lake belt; WJ, Wijinnedi belt; YK, Yellowknife belt.

1999). The relationship between a western basement complex, the Anton terrane (Fig. 1 inset; Kusky, 1989; Bleeker *et al.*, 1999) and overlying Neoarchaean supracrustal belts of the north- and south-central Slave Province has been a central and controversial element in models espousing ensialic rift (Henderson, 1981) versus collisional/accretionary settings (Fyson & Helmstaedt, 1988; Kusky, 1989; Davis & Hegner, 1992; King *et al.*, 1994). The lithostratigraphic record of this Neoarchaean supracrustal 'cover' in the poorly studied south-western part of the Province (Fig. 1) is critical to tectonic models of the Province as a whole, as this region also structurally overlies Mesoarchaean crust (Kusky, 1989; Frith, 1993; Pehrsson & Chacko, 1997).

Supracrustal belts of the south-western Slave Province have been characterized as dominantly submarine mafic volcanic sequences, *c.*2700–2600 Ma in age, deposited entirely upon older basement and overlain by coeval felsic volcanic rocks and related turbidites (Padgham, 1985; Frith, 1993; Isachsen & Bowring, 1994; King & Helmstaedt, 1997). This succession is superficially similar to that of the Yellowknife Supergroup in its type area, the south-central Slave Province

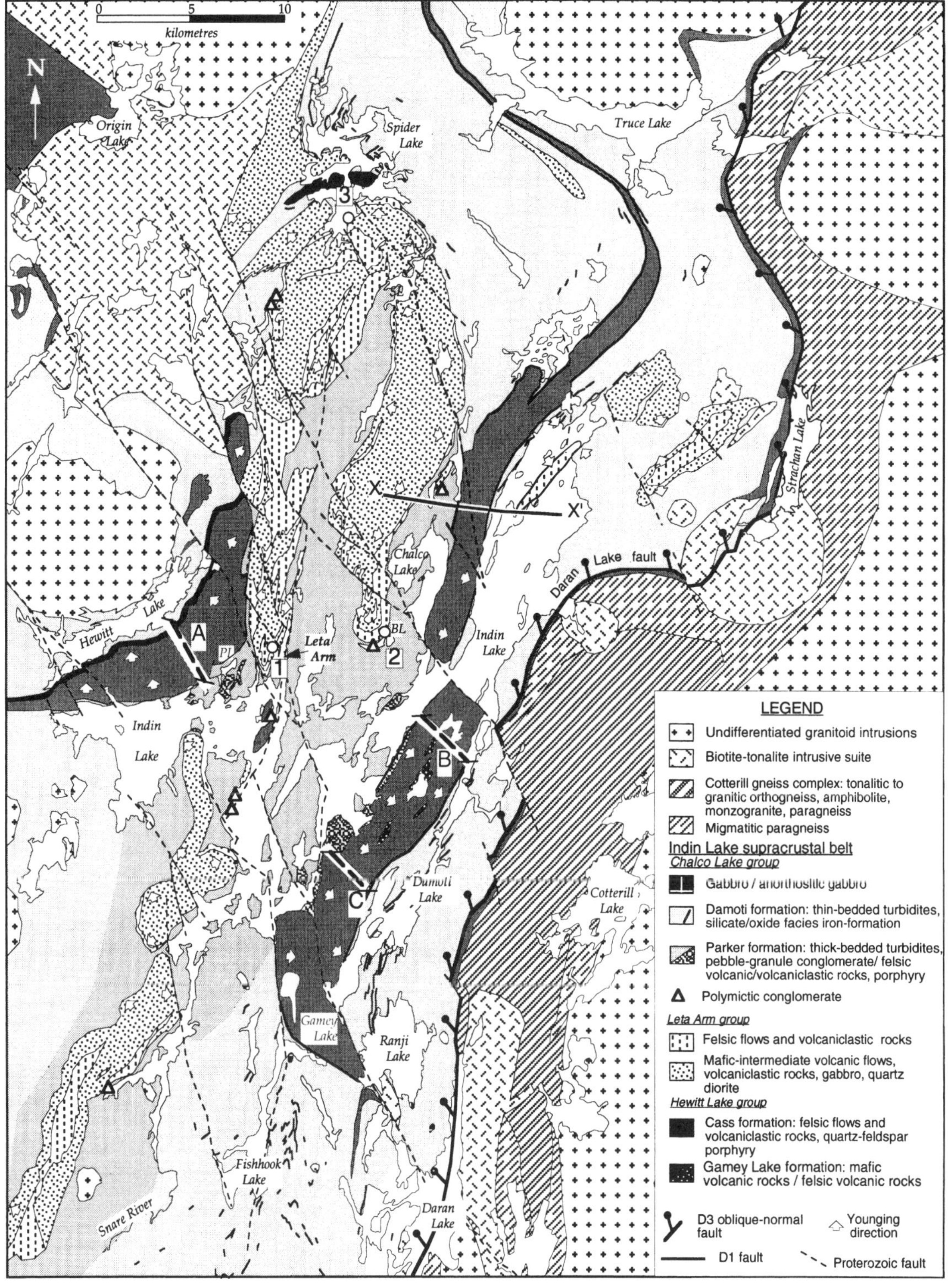


**Fig. 2.** Simplified geology of the Indin Lake supracrustal belt and surrounding area. Section line X–X′ refers to Fig. 3B. Type and representative sections A–C and 1–3 of the Hewitt Lake and Leta Arm groups, respectively, are noted.

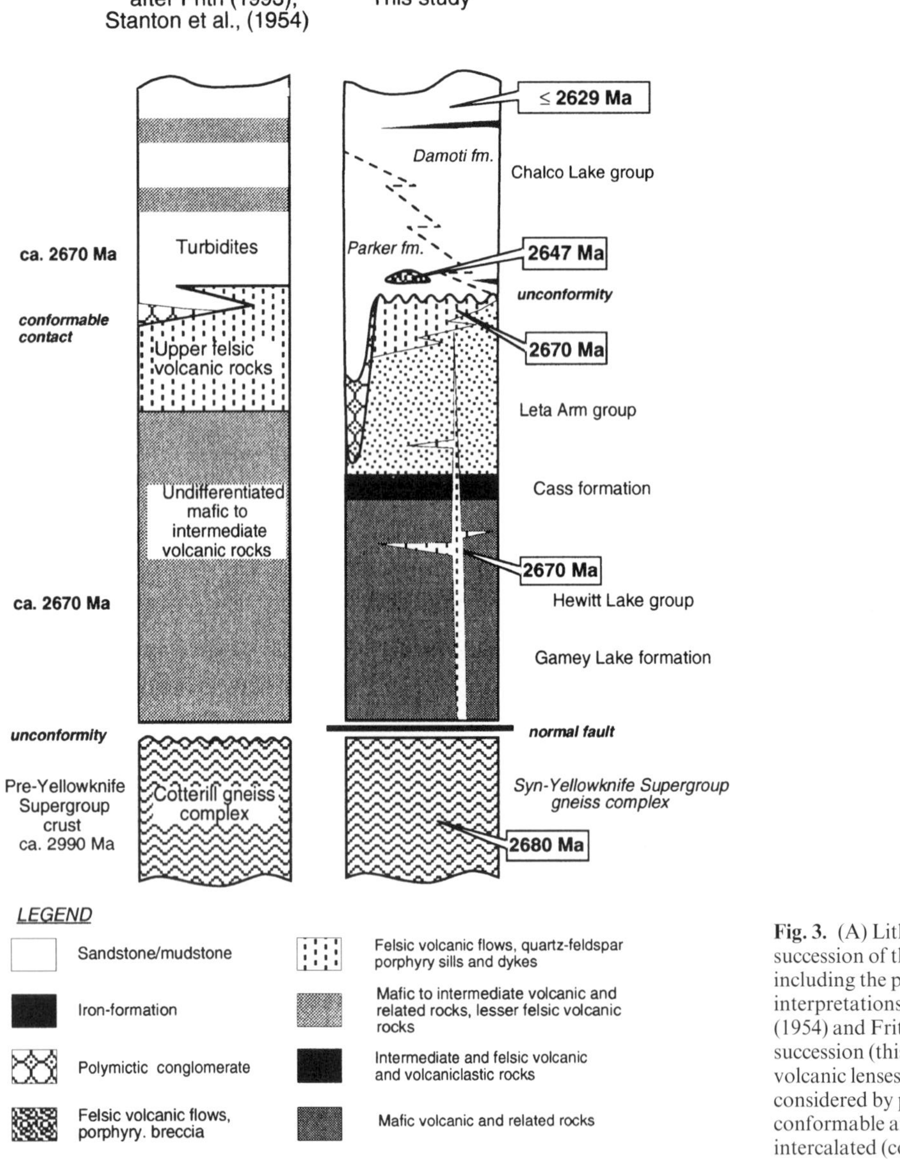


**Fig. 3.** (A) Lithostratigraphic succession of the Indin Lake area, including the previous stratigraphic interpretations of Stanton *et al.* (1954) and Frith (1993) and a revised succession (this study). Thick volcanic lenses in turbidites considered by previous studies to be conformable are structurally intercalated (compare with B). Approximate thickness of the entire succession is >10 km. (*continued*)

(Helmstaedt & Padgham, 1986), lending support to early models of a simple east–west tectonostratigraphic subdivision in the craton (Davis *et al.*, 1996).

A recent structural and stratigraphic study of the Indin Lake supracrustal belt of the south-western Slave Province (Figs 1 & 2) identified an imbricated, anomalously young, 2670–2630 Ma, lithostratigraphic succession (Fig. 3; Pehrsson, 1998; Pehrsson & Villeneuve, 1999). Similar young ages have been reported from the Wheeler Lake supracrustal belt,

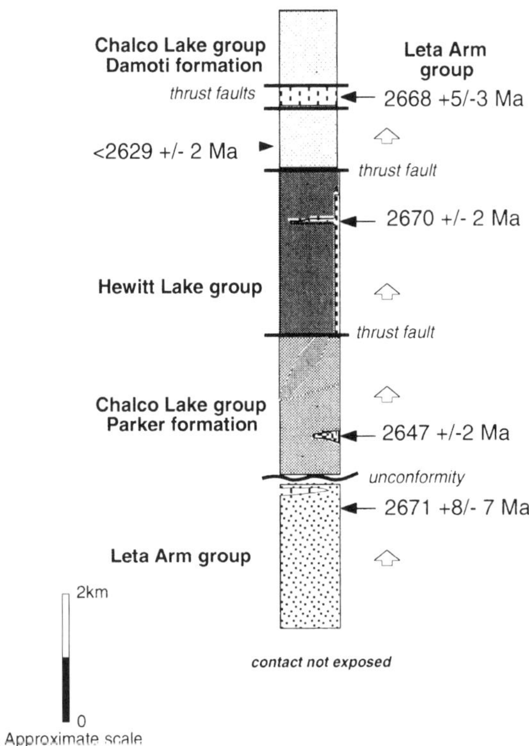


**Fig. 3.** (*cont'd*) (B) $D_1$ tectonostratigraphic section of the east half of the Indin Lake supracrustal belt between line X–X′. Age dating of units in the upward-facing section, on the limb of a regional $F_2$ fold, indicate that older and younger units are structurally imbricated and are not repeated due to cyclic volcanism. White arrows indicate the direction of stratigraphic younging.

also situated in the south-western part of the Province (Isachsen & Bowring, 1994). Isotopic studies have noted less involvement of Mesoarchaean crust in the generation of post-tectonic plutons of the south-west (10–30%; Davis *et al.*, 1996; Yamashita *et al.*, 1998, 1999) relative to the central and northern Slave Province (>50%; Davis & Hegner, 1992; Davis *et al.*, 1996). These data collectively point to a distinct tectonostratigraphic history for this region. Thus, establishment of the depositional setting of this supracrustal succession is important for understanding the relevance of this terrane to the tectonic evolution of the Slave Province.

The purpose of this paper is threefold: (i) to present the stratigraphy and physical volcanology of the Indin Lake supracrustal belt in order to establish a local depositional model; (ii) to use these data as first-order constraints on the tectonic setting of this distinct south-western terrane; and (iii) to compare the interpreted depositional environment with modern settings.

## REGIONAL SETTING

The Slave Province of the north-western Canadian Shield is a composite granite-greenstone terrane consisting of volcano-sedimentary belts (collectively termed the Yellowknife Supergroup; Helmstaedt & Padgham, 1986), predominantly 2720–2600 Ma in age, intruded by voluminous 2700–2550 Ma syn- to post-volcanic plutons (Fig. 1; van Breemen *et al.*, 1992; Isachsen & Bowring, 1994; King & Helmstaedt, 1997 and references therein). All supracrustal rocks are metamorphosed to greenschist to upper-amphibolite facies.

On the basis of field, geochronologic and isotopic studies, the Yellowknife Supergroup (Helmstaedt & Padgham, 1986) is interpreted to overlie older, >2750 Ma sialic crust only in the western part of the province (Thorpe, 1972; Baragar & McGlynn, 1976; Padgham, 1985; Kusky, 1989; Davis & Hegner, 1992; Davis *et al.*, 1996). This older crust consists of 2850–4000 Ma plutonic and supracrustal units including banded tonalite-amphibolite and biotite gneisses, massive to foliated granitic rocks, felsic volcanic rocks and quartzite–iron-formation–conglomerate sequences (Bowring *et al.*, 1989; Roscoe *et al.*, 1989; Hrabi *et al.*, 1995; Isachsen & Bowring, 1997; Henderson, 1998; Bleeker *et al.*, 1999). Contacts between Mesoarchaean rocks and overlying supracrustal sequences are variously interpreted to be unconformities (Henderson, 1981), thrust decollements (Kusky, 1989; Hrabi *et al.*, 1995), extensional detachments (James & Mortensen 1992) or some combination of all three (Bleeker *et al.*, 1999). Widespread deposition of sedimentary rocks commenced during volcanism from 2700 to 2650 Ma and overlapped with 2640–2580 Ma compressional deformation (van Breemen *et al.*, 1992; Isachsen & Bowring, 1994; Davis & Bleeker, 1998; Pehrsson & Villeneuve, 1999).

### The Indin Lake supracrustal belt

The Indin Lake supracrustal belt is part of a 300 km long, N–NE trending volcano-sedimentary belt that extends from Grenville Lake to Russell Lake, in the

**Table 1.** Summary of events in the Indin Lake area.

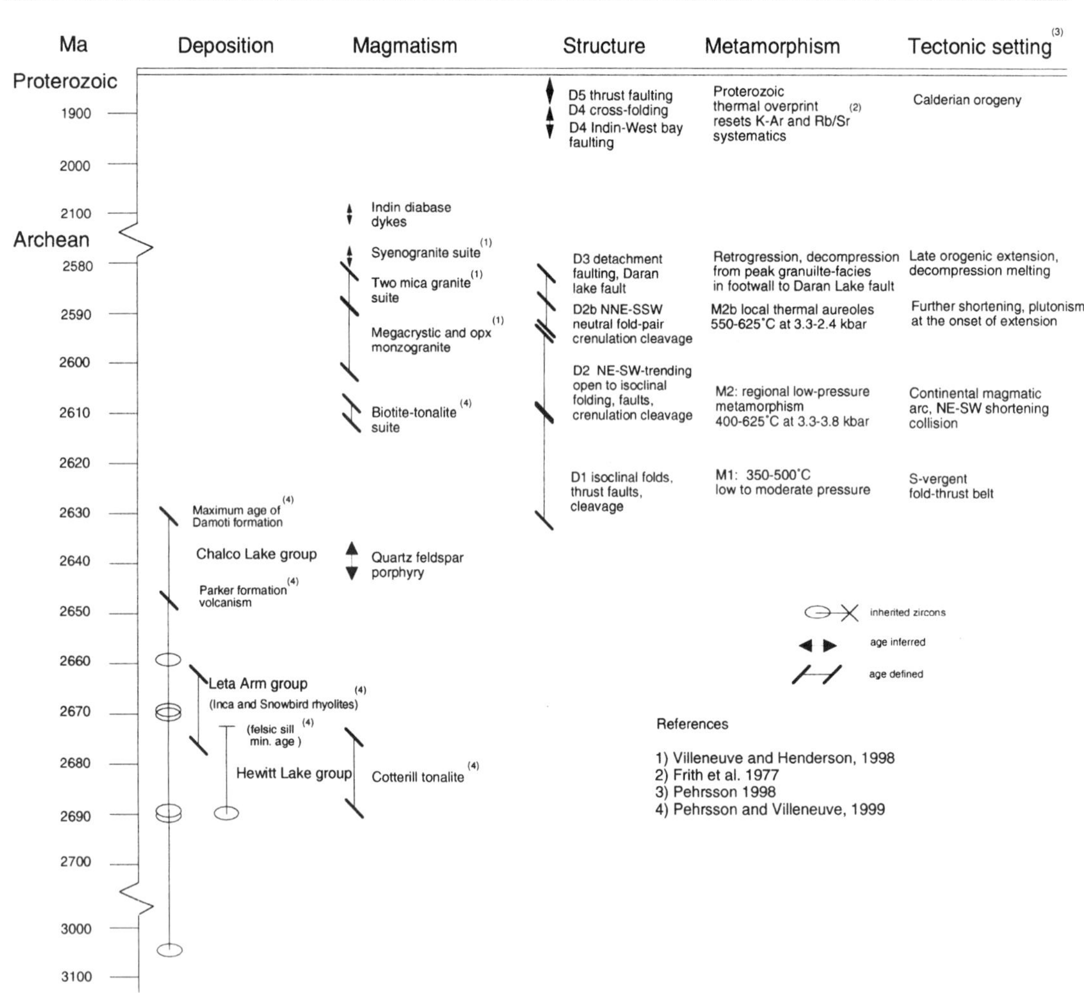


south-western Slave Province (Fig. 1; Pehrsson, 1998). It is in the central, widest portion of this tract and is bounded by Palaeoproterozoic rocks of Wopmay Orogen on the west and plutonic and high-grade gneissic rocks of the Snare domain on the east (Pehrsson & Chacko, 1997). The Indin Lake supracrustal belt is composed of intercalated Neoarchaean volcanic and sedimentary rocks of the Yellowknife Supergroup and is structurally underlain by Mesoarchaean crust in the Grenville Lake area (Fig. 1; Frith *et al.*, 1986; Villeneuve, unpublished data). The supracrustal rocks are metamorphosed to low-pressure, greenschist to upper-amphibolite facies conditions (Pehrsson & Chacko, 1997). The belt is intruded by suites of syn- to post-tectonic granitoid rocks (Table 1) and Proterozoic diabase dykes that have been described in detail elsewhere (Frith, 1993; Pehrsson *et al.*, 1995).

Previous mapping of the Indin Lake supracrustal belt delineated ten individual volcanic lenses intercalated with sedimentary rocks (Fig. 2; Fortier, 1949; Tremblay *et al.*, 1953; Stanton *et al.*, 1954; Frith, 1993). A simple stratigraphic succession was proposed wherein mafic and felsic volcanic rocks were conformably overlain by, and stratigraphically intercalated with, coeval sedimentary rocks (Fig. 3A; Stanton *et al.*, 1954; Frith, 1993). Volcanic and sedimentary rocks were interpreted to have been erupted through, and deposited on, proposed basement gneisses, represented by the high-grade Cotterill gneiss complex to the east (Fig. 2; Frith, 1993).

A revised, informal lithostratigraphic succession for the Indin Lake supracrustal belt, based on recent regional mapping and new U–Pb age data (Pehrsson & Villeneuve, 1999), is presented in Fig. 3A. The oldest rocks of the Indin Lake supracrustal belt are minimum 2670 Ma mafic volcanic rocks of the Hewitt Lake group. Mafic to felsic volcanic rocks of the 2670 Ma Leta Arm group, coeval in age with felsic intrusions that cut the Hewitt Lake group, are inferred to have originally overlain the Hewitt Lake group (Fig. 3A). The Leta Arm group is unconformably overlain by 2647–2629 Ma sedimentary and felsic volcanic rocks of the Chalco Lake group. The Indin Lake supracrustal belt is underlain by 2680 Ma tonalitic orthogneisses of the moderate-pressure Cotterill gneiss complex. The mid-crustal orthogneisses do not represent *in situ* older basement as proposed by Frith (1993), as they do not pre-date the Yellowknife Supergroup. Isotopic data, however, suggest that the complex may have assimilated older crust (Fig. 3A; cf. Pehrsson & Villeneuve, 1999).

Deformation of the Indin Lake supracrustal belt consists of three Archaean and two Proterozoic deformation events (Table 1; Pehrsson, 1998). The earliest deformation ($D_1$) event, bracketed between 2630 and 2609 Ma, formed an E–NE trending, weakly south-vergent fold-fault belt (Pehrsson & Chacko, 1997). U–Pb dating of units in an upward-facing sequence (Fig. 3B) confirms local 'older over younger' relationships resulting from $D_1$ thrust faulting. Subsequent $D_2$ deformation, broadly coeval with the regional low-pressure metamorphism at *c.*2609 Ma, produced a N–NE trending, upright fold-belt. Interference between regional-scale $D_2$ and $D_1$ folds further repeated the intercalated units. Late Archaean oblique–normal faulting ($D_3$, Daran Lake fault, Fig. 2) juxtaposed high-grade, moderate-pressure rocks of the Cotterill gneiss complex with lower-grade, low-pressure rocks of the Indin Lake supracrustal belt (Pehrsson & Chacko, 1997). Subsequent Proterozoic deformation included open folding, strike-slip and minor thrust faulting ($D_4$–$D_5$) (Table 1; Pehrsson, 1998).

## STRATIGRAPHY AND PHYSICAL VOLCANOLOGY

The Indin Lake supracrustal belt is subdivided into three lithostratigraphic groups on the basis of character and homogeneity of facies and primary volcanic structures (Fig. 2, Table 1; Pehrsson, 1998; Pehrsson & Villeneuve, 1999). The following descriptions are based primarily on 1 : 50 000 scale mapping, although local 1 : 1000 scale mapping was completed on parts of the Hewitt and Chalco Lake groups.

Despite the localized strong penetrative strain and regional greenschist to amphibolite-facies metamorphism, delicate primary textures are commonly preserved and the prefix 'meta' is omitted for simplicity. Sections and figures presented are from the best exposed parts of the map area.

Volcanic rocks were informally classified in the field based on texture, colour index and phenocryst populations. Whole rock and trace element geochemistry on reference samples supports this compositional classification (Pehrsson, 1998), and is summarized in Table 2. Volcaniclastic rocks are described following the terminology of Fisher & Schminke (1984), using standard granulometric classification (blocks ≥64 mm, lapilli 2–64 mm, ash ≤2 mm) and rock descriptions (tuff has ≥75% ash; lapilli tuff has ≥25% lapilli, ≥25% ash and ≤25% blocks; tuff breccia has ≥25% blocks and ≥25% ash and/or lapilli). Pyroclastic rocks are considered to be composed of fragments that originated from or as a direct consequence of a volcanic eruption. This definition includes pyroclastic debris remobilized by water, including pumice, glass shards, euhedral crystals, angular juvenile or other delicate fragments like fiamme. Epiclastic fragments are those derived by penecontemporaneous or later erosion and weathering of volcanic rocks.

### The Hewitt Lake group

The Hewitt Lake group is a sequence of predominantly mafic volcanic rocks that occurs in three N–NE trending lenses within the Indin Lake supracrustal belt and a fourth along its eastern margin (Fig. 2; Pehrsson, 1998). These lenses, 1–3.5 km wide by 10–60 km long, are steeply dipping homoclines with faulted upper and lower contacts. The present maximum thickness of the group is 3.5 km. It is subdivided into two formations, a lower, mafic Gamey Lake formation and an upper, intermediate-felsic Cass formation (Fig. 3A). Lithofacies and representative sections

**Table 2.** Summary of representative geochemical data from the Indin Lake supracrustal belt (after Pehrsson, 1998).

| Group/fm<br>sample<br>composition | Gamey Lake<br>G203<br>mafic | Gamey Lake<br>G204<br>mafic | Cass<br>D54<br>intermediate | Leta Arm<br>G72<br>intermediate | Leta Arm<br>C268<br>intermediate | Leta Arm<br>55<br>felsic | Chalco Lake<br>215<br>felsic |
|---|---|---|---|---|---|---|---|
| $SiO_2$ | 49.48 | 52 | 64.4 | 55.58 | 64.75 | 69.05 | 71.95 |
| $TiO_2$ | 0.98 | 1.36 | 0.5 | 1.19 | 0.88 | 0.51 | 0.14 |
| $Al_2O_3$ | 12.99 | 11.76 | 15.89 | 15 | 15 | 15.51 | 16.78 |
| $Fe_2O_3$ | 13.9 | 14.23 | 3.96 | 8.1 | 4.25 | 2.49 | 1.09 |
| MnO | 0.27 | 0.27 | 0.18 | 0.24 | 0.11 | 0.05 | 0.05 |
| MgO | 8.91 | 6.27 | 2.24 | 6 | 3.07 | 2.54 | 1.15 |
| CaO | 10.96 | 10.99 | 8.09 | 8.83 | 6.31 | 1.35 | 1.13 |
| $Na_2O$ | 2.33 | 1.87 | 2.76 | 4.37 | 4.38 | 0.31 | 6.19 |
| $K_2O$ | 0.11 | 0.12 | 1.85 | 0.2 | 1 | 8.09 | 1.47 |
| $P_2O_5$ | 0.06 | 0.12 | 0.13 | 0.49 | 0.26 | 0.09 | 0.03 |
| Total | 100 | 100 | 100 | 100 | 100 | 100 | 100 |
| Mg | 55.94 | 46.58 | 52.82 | 59.47 | 58.85 | 66.93 | 67.65 |
| Cr | 233 | 9 | 60 | 673 | 232 | 103 | 11 |
| Ni | 77 | 97 | 30 | 121 | 69 | 43 | 2 |
| Sc | 56 | 34 | 16 | 33 | 11 | 12 | 3 |
| V | 339 | 337 | 88 | 199 | 111 | 52 | 18 |
| Cu | 97 | 89 | 15 | 4 | 15 | 15 | 7 |
| Pb | 3 | | 5 | 6 | 11 | 22 | 1 |
| Zn | 82 | 92 | 1 | 68 | 27 | 39 | 2 |
| K | 945 | 982 | 15 367 | 1675 | 8318 | 67 138 | 12 209 |
| Rb | 2 | 1 | 43 | 4 | 26 | 94 | 35 |
| Ba | 38 | 51 | 770 | 200 | 338 | 2621 | 316 |
| Sr | 114 | 163 | 280 | 480 | 307 | 69 | 265 |
| Ga | 15 | 18 | 14 | 19 | 18 | 17 | 19 |
| Ta | 0.1 | 0.21 | | 0.33 | 0.68 | 0.8 | 0.06 |
| Nb | 2.7 | 4 | 5.8 | 6.6 | 10.2 | 7 | 0.5 |
| Hf | 1.55 | 2.61 | | 5.38 | 7.1 | 4.6 | 3.01 |
| Zr | 54 | 101 | 148 | 232 | 302 | 193 | 105 |
| Ti | 5861 | 8182 | 2972 | 7112 | 5249 | 3084 | 857 |
| Y | 19 | 24 | 14 | 29 | 23 | 12 | 3 |
| Th | 0.21 | 1.14 | 2.02 | 4.24 | 9.21 | 9.86 | 2.29 |
| U | 1.62 | 0.27 | | | 3.91 | 2.35 | |

*Lava compositions (%)*

| Group | Basalt | Bas. andesite | Andesite | Dacite | Rhyolite |
|---|---|---|---|---|---|
| Hewitt Lake | 47 | 31 | | 7 | 15 |
| Leta Arm | 32 | 22 | 13 | 17 | 16 |
| Chalco | 5 | | | 40 | 55 |

*Relative compositional trends (%)*

| Group | Tholeiite–mafic | Calc–alkaline–mafic | Calc–alkaline–int.–felsic |
|---|---|---|---|
| Hewitt Lake | 78 | 0 | 22 |
| Leta Arm | 28 | 22 | 50 |
| Chalco | 5 | | 95 |

*Nd isotopic signature**

| Group | Mafic | Intermediate | Felsic |
|---|---|---|---|
| Hewitt Lake† | 1.87–1.21 | | 1.42 |
| Leta Arm‡ | 0.02 | 1.08 | 1.45–0.46 |
| Chalco§ | | | 2.86 |

* Theriault & Pehrsson, unpublished data.
† EpNd calculated at 2680 Ma.
‡ EpNd calculated at 2670 Ma.
§ EpNd calculated at 2650 Ma.

from both formations are presented in Fig. 4 and Table 3.

*Gamey Lake formation*

The Gamey Lake formation is a homogeneous package of mafic volcanic and related rocks that occurs in all belts of the Hewitt Lake group (Fig. 2). It comprises over 80% of the group and locally attains a thickness of nearly 3200 m. The lower contact is everywhere faulted, and the upper contact is only exposed south-east of Hewitt Lake.

Mafic lava flows make up almost 50% of the Gamey Lake formation and are classified as tholeiitic basalts (60%) and basaltic-andesites (40%, Table 2; Pehrsson, 1998). They are composed of pillowed and lesser non-pillowed lava flows in continuous sequences up to several hundred metres thick and are traceable along strike for several kilometres (Fig. 4). The thickness of individual flows (Table 3) is comparable to that of modern examples (Cas, 1992), although the pillows themselves are somewhat smaller (Hargreaves & Ayres, 1979; G. Walker, 1992). No systematic variation in the overall ratio of pillowed to non-pillowed flows (4 : 1) has been noted through the formation, and pillowed lava flows are present throughout (Fig. 4).

Mafic lapilli tuffs and tuff breccia occur as discontinuous thin units, <5 m thick, between and along strike of individual mafic flows. Clasts in these facies are typically of similar vesicularity or phenocryst content as adjacent flows and are dominantly composed of pillow fragments. Contacts with underlying flows are locally gradational, with transitions over a metre from intact or partial pillows surrounded by jigsaw-fit tuff breccia to blocky or irregular breccia and lapilli tuff. Epiclastic rocks are also locally interbedded on a metre scale with the mafic volcanic flows, and consist of normally graded, mafic wackes in beds <2 m thick that drape pillow tops, and structureless matrix-supported volcanic conglomerates in beds <5 m thick, with mafic, felsic and rare granodiorite clasts (Fig. 5A; Table 3).

Gabbro sills and dykes comprise up to 25% of the formation. The sills have lateral gradational contacts with massive flows and lack well developed chill margins. Dykes locally follow pillow outlines or terminate in pillowed flows (Table 3). They are interpreted as synvolcanic intrusions that injected while the volcanic complex was still warm (see Walker, 1993).

Felsic volcanic and volcaniclastic rocks are a minor but widespread component of the Gamey Lake formation, occuring as units, 2–1000 m thick by up to 2500 m long, interbedded with mafic lavas (Fig. 4). This facies is absent in thinner belts of the Hewitt Lake group in the Strachan, Spider and Chalco Lake areas (Fig. 2). The Gamey and Hewitt Lake belts contain several thinner units too small to represent in Fig. 2.

Compositionally, the felsic volcanic rocks are calc-alkaline rhyodacites to rhyolites (Table 2). Felsic lapilli tuffs and tuff breccias dominate thinner, <200 m thick, volcaniclastic lenses that are rarely continuous for more than a few hundred metres along strike. Larger sequences consisting of felsic tuff breccia, flows and porphyritic stocks, and lesser lapilli tuff, are 50–2000 m thick by up to 2.5 km long. They are characterized by the interstratification of lobe-like, non-brecciated and locally flow-banded felsic lavas, clast-supported, angular felsic tuff breccia and cross-cutting porphyry sills that locally terminate in the lavas. Swarms of felsic porphyry dykes cross-cut the mafic flows throughout the formation.

Several rare but distinctive lithofacies are associated with the thicker felsic volcanic packages. Thinly bedded to laminated, sulphidic or graphitic argillite and siltstone locally occur as minor interbeds (<5 m thick) within felsic lapilli tuffs and tuff breccias. Massive beds (<10 m) of compositionally bimodal lapilli tuff are locally present in the felsic tuff breccias, as are rare beds of massive to planar-laminated dolostone. The tuff breccias adjacent to the latter facies typically have a dolomite matrix between the clasts.

*Cass formation*

The Cass formation is a thin but laterally continuous sequence, 750 m thick by 50 km long, of intermediate to felsic volcanic rocks that overlies the Gamey Lake formation at Hewitt Lake (Fig. 2). It occurs only on the west side of the Indin Lake supracrustal belt and is interpreted to have been structurally cut out against $D_1$ thrust faults in other belts of the Hewitt Lake group (Fig. 2). The formation, steeply NW-dipping and upright, is associated with a prominent aeromagnetic high on regional total field magnetic maps (Pehrsson, 1998).

The basal contact of the Cass formation is marked by a heterolithic breccia, found in thin beds and narrow, 5–10 cm wide, fractures that cut down into the underlying Gamey Lake formation by up to half a metre (Figs 4 & 5B; Table 3). The breccia is dominated by clasts with the same concentration of amygdules in pillow selvages as subjacent pillowed lava flows (Fig. 4). A progressive decrease in the proportion of mafic clasts is noted upsection away from the basal contact. The upper contact of the formation with the overlying Chalco Lake group is faulted. Cross-cutting gabbro

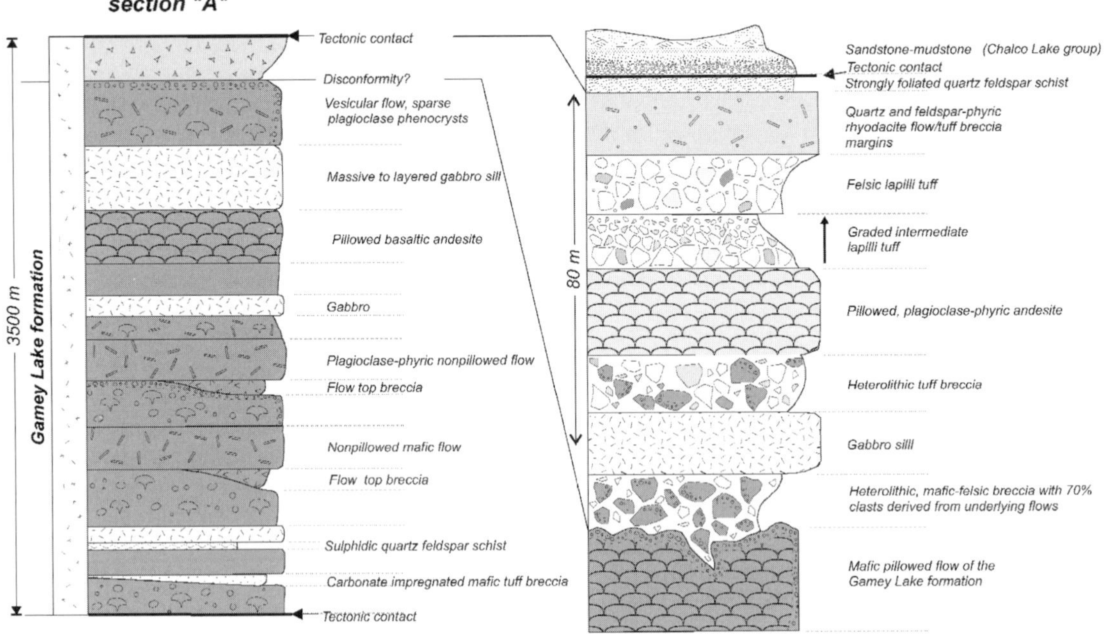


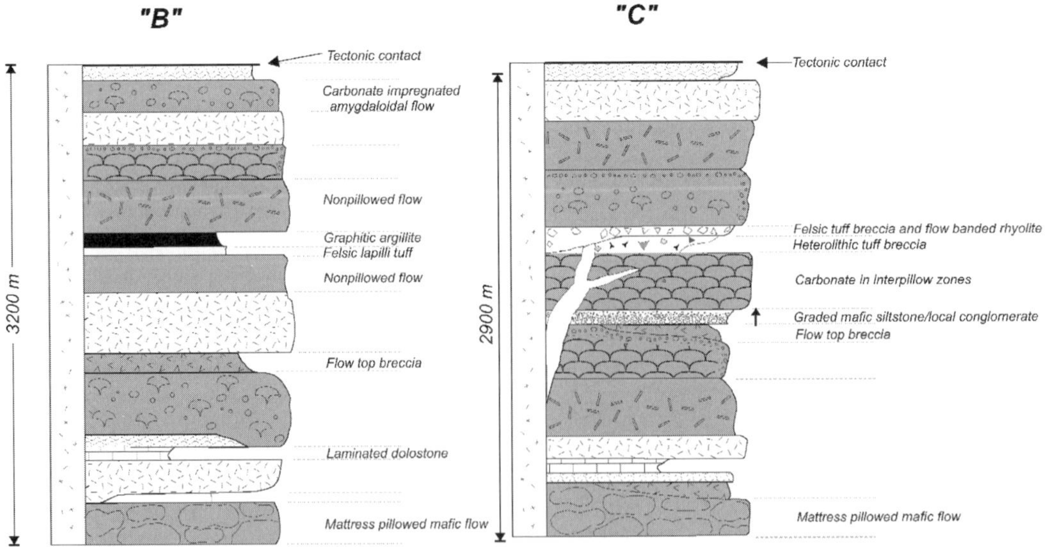


**Fig. 4.** Representative sections of the Hewitt Lake group. The sections are located near Gamey and Cass Lakes, respectively (Fig. 2), in areas with the best preserved volcanic features owing to their relatively low metamorphic grade and internal strain state. The sections are composite, based on 1 : 50 000 scale mapping, and do not depict minor covered intervals. Thicknesses represent aggregate thickness of units, not individual beds. Thickness of felsic units is exaggerated for clarity. Note the dominance of mafic pillowed and massive flows in the Gamey Lake formation and the change to more felsic composition and volcaniclastic nature in the upper Cass formation.

**LEGEND**

Late gabbro dykes and sills

Layered to massive gabbro

Basalt and basaltic andesite

Andesite

Dacite

Rhyolite / rhyodacite

Fine-grained carbonate, bedded

Carbonate-cemented pillow breccia

Graphitic or sulphidic argillite

Sulphidic quartz feldspar schist

Textures

Medium size pillows

Small pillows

Large mattress pillows and tubes

Plagioclase-phyric flows

Vesiculated rims

Amygdules

Breccia, blocks, lapilli

Pillow breccia

Gradation in volcaniclastic rocks fining upwards in clast size and increase in matrix proportion

sills and dykes inflate the thickness of the formation by up to 20%, but because they have chilled margins and locally occur in the overlying Chalco Lake group, they are considered to be post-volcanic.

The Cass formation is dominated by intermediate-felsic volcanic rocks with a much higher proportion of volcaniclastic facies than the Gamey Lake formation and only minor pillowed lava flows. The various facies of the formation are characteristically discontinuous, with individual units rarely traceable for more than 250–500 m and abrupt facies changes both across and along strike.

Intermediate tuff breccias and lapilli tuffs are the predominant facies and occur throughout the formation as tens of metres scale units of interlayered structureless and normally graded beds (Table 3). Gradational changes in lithofacies, including proportions of coarse clasts and clast- to matrix-support, were noted over a strike length of <250 m. Heterolithic breccias that are interbedded with intermediate lapilli tuffs higher in the section typically have more rounded clasts and few or no mafic clasts in contrast to the basal breccias (Fig. 4).

Thin, pillowed intermediate flows are restricted to the lower part of the formation and are generally overlain by intermediate lapilli tuffs and tuff breccia (Table 3). Non-brecciated felsic lavas are restricted to the top of the formation, where they occur as lobes surrounded by thick units of clast-supported felsic tuff breccia. Felsic porphyry sills and dykes that cut across bedding locally terminate at the contact of felsic flows. The intermediate-felsic lava flows and porphyritic sills are classified as calc-alkaline andesites, rhyodacite and rhyolites, the latter being similar to flows and sills in the underlying Gamey Lake formation (Table 2; Pehrsson 1998).

*Depositional processes*

The predominance of pillowed flows in the Hewitt Lake group and presence of partial selvages on many clasts in mafic lapilli tuffs and breccia indicate a subaqueous eruptive setting (Table 3; Ballard & Moore, 1977; Walker, 1993). The lateral continuity of flows, widespread carbonate alteration and interbedded dolomite (possibly from $CO_2$-bearing sea water) support a marine environment.

The jigsaw fit, blocky to irregular morphology, clast textures and contact relationships of the mafic lapilli tuffs and tuff breccias are consistent with an interpretation as hyaloclastite breccias, derived by quench fragmentation of adjacent lava flows rather than explosive fragmentation (Dimroth *et al.*, 1978;

**Table 3.** Lithofacies of the Hewitt Lake group.

| Facies | Proportion (%)* | Description† |
|---|---|---|
| *Cass formation* | | |
| Tuff breccia and lapilli tuff | 55 | Intermediate tuff breccia and lapilli tuffs: poor size sorting; blocky plagioclase–phyric, pumiceous (>30% amygdules) and rare aphyric clasts, angular to subrounded, 1–20 cm in diameter; euhedral/broken quartz + plagioclase crystal-rich matrix; in (i) sharp-based normally graded, tabular beds (<1 m thick), clast- to matrix-supported, showing an upward decrease in clast size and concomitant increase in matrix; and (ii) matrix-supported structureless beds, 2–10 m thick |
| Heterolithic lapilli tuff and breccia | 20 | Lapilli tuff and breccia: 0.5–10 m thick structureless beds, clast- to matrix-supported, poorly sorted, angular to rounded, 1–20 cm clasts; clast composition: basal units, 70% mafic pillow, 20% intermediate, 10% felsic, phenocryst and amygdule proportions match underlying mafic pillowed flows; upper units, 0–20% mafic, 40–70% andesite, 30–40% rhyolite |
| Felsic lapilli tuff | 10 | Felsic lapilli tuff: matrix- to clast-supported, poorly sorted, sharp-based, 1–3 m beds; angular quartz and plagioclase–phyric rhyolite clasts in a crystal-rich matrix with euhedral bipyramidal quartz and broken/euhedral feldspar crystals |
| Intermediate volcanic flows | 10 | Intermediate flows: pillowed, plagioclase–phyric (1–5 mm phenocrysts), 5–10 m thick, amygdules‡ 0.2–1 cm, ave. 0.5 cm, range 0–30%, ave. 25%, pillows range 10–70 cm, ave. §30 × 50 cm, ave. selvage thickness 2.5 cm |
| Felsic volcanic flows, breccia and porphyry | 5 | Felsic flows: quartz and feldspar–phyric, non-brecciated lobes (up to 2 m thick) surrounded by tuff breccias (4–10 m thick) with angular 6–20 cm blocky clasts; dominantly non-vesicular, locally 5% or less vesicules; breccias are clast-supported locally to jigsaw fit, both clasts and lobes locally have mm-scale flow banding; quartz– and feldspar–phyric porphyry as <10 m thick sills and swarms of narrow (<1 m wide) dykes; dykes locally terminate at the lower contact of sills and flow/breccia units |
| *Gamey Lake formation* | | |
| Mafic flows and lapilli tuff/tuff breccia | 60 | Pillowed (42%) and non-pillowed flows (8%), 0.01–5 m thick, aphyric to sparsely plagioclase-phyric (1–4 mm phenocrysts), amygdules 0.2–3.0 cm, ave. 0.4 cm, 0–30%, ave. 20%; pillows range 10–200 cm, ave. 25 × 40 cm; lava shelves, varioles and radial cooling cracks rare; selvage thickness ranges 0.5–2 cm, ave. 1 cm; basaltic andesite pillows on ave. larger (30 × 60 cm) with longer aspect-ratios (3 : 1); pervasive carbonate alteration of groundmass; amygdules filled with carbonate/quartz, concentrated in pillow rims/cores. Mafic lapilli tuff and breccias (10%): in discontinuous beds <5 m thick by 100 m long, clast-supported (locally jigsaw fit), angular, blocky to irregular, aphyric to plagioclase–phyric clasts, clasts with pillow selvages where adjacent to pillowed flows, phenocryst and amygdule proportions similar to adjacent flows, local highly amygdaloidal (30–40%) clasts (<20% of population); commonly with a fine-grained dolomite or chlorite schist matrix |
| Gabbro sills and dykes | 25 | Gabbro sills, 3–400 m thick, in massive to layered sill complexes up to 1000s of metres long, tabular to irregular in outline, locally glomerocrystic; lack well developed chilled margins, local gradational contacts with overlying massive flows, sills at a low angle to bedding and flow contacts; iregular and discontinuous dykes that cross-cut but locally follow pillow outlines and terminate in pillowed flows |
| Felsic lapilli tuff/tuff | 5 | Felsic lapilli tuffs (4%): poorly sorted, matrix- to clast-supported, angular to subrounded, aphyric and quartz + plagioclase-phyric clasts (rare mafic clasts comprise <20% of population); in featureless beds, typically 2–30 m, bedded units up to 200 m thick; matrix with <1.5 cm lapilli, euhedral quartz plagioclase crystals; locally with fine-grained dolomite matrix. Tuffs (1%): with rare angular shards, broken and euhedral quartz and plagioclase crystals, typically as cm-scale interbeds in lapilli tuffs; both tuffs and lapilli tuffs with 1–5% sulphide (pyrite + pyrrhotite) in matrix |
| Felsic flows/tuff breccia and porphyry | 5 | Felsic flows: as unbrecciated lobes 2–40 m thick by 500 m long, quartz– + feldspar–phyric (1–3 mm phenocrysts, bipyramidal and resorbed quartz). Tuff breccia: poorly sorted, clast-supported, in featureless beds, typically 2–40 m, locally 200 m thick; angular, blocky aphyric and quartz– + plagioclase–phyric clasts, clasts 6–15 cm, ave. 10 cm. Porphyry intrusions: quartz– and plagioclase–phyric sills 40–500 m thick by 2500 m long, cut across bedding at <30°; dykes, <2 m thick, truncate at sills and flow contacts |
| Bimodal lapilli tuff | <5 | Lapilli tuff: matrix-supported, poorly sorted, structureless beds, <10 m thick; with angular aphyric and quartz– and/or plagioclase–phyric felsic clasts (60%) and aphyric mafic clasts (40%), 1–20 cm, ave. 4–6 cm; in a fine-grained, intermediate matrix with <50% bipyramidal quartz crystals |
| Volcanic conglomerate/argillite/carbonate | <5 | Boulder to cobble conglomerate, matrix-supported <5 m thick beds, rounded to subrounded mafic, felsic and rare granodiorite clasts in a siltstone matrix with sulphide and magnetite. Mafic wacke: normally graded beds <2 m thick, basal coarse-grained sand component with rounded 0.5–2 cm clasts drapes the tops of pillowed flows. Sulphidic or graphitic argillite and quartzofeldspathic siltstone: thin-bedded (0.5–2 cm) units up to 5 m thick; rare massive and planar-laminated, dolomite and carbonate, <1 m thick, with silicic or magnetite-rich seams <5 mm thick |

* Represents volume percentage of facies in formation based on field mapping.

† Maximum thicknesses of beds are likely minimum figures due to local cover on facies contacts; tectonized or schistose units in Fig. 2 not described here.

‡ Visual estimates on two-dimensional outcrops.

§ Pillow dimensions averaged from visual estimates of two-dimensional outcrops in pillows with 2 : 1 or less aspect ratio, effects of stretching in third dimension not included, but stretch averages 3 : 1 down-dip.

**Fig. 5.** (A) Pillowed basalt of the Gamey Lake formation overlain by graded mafic siltstone. (B) Contact between vesicular pillows of the Gamey Lake formation (left) and heterolithic breccia of the Cass formation (right). Similarity between the texture of the mafic clasts and underlying pillows suggests the breccia represents debris reworked from the pillowed flow. Hammer for scale is 40 cm long.

Staudigel & Schminke, 1984; Walker, 1993; Doucet *et al.*, 1994). The jigsaw fit of the clasts and limited extent of this facies suggests minimal clast transport.

The composite felsic volcanic flow/tuff breccia units in both the Gamey Lake and Cass formations, although thin compared to modern felsic lava flows (e.g. De Rosen-Spence *et al.*, 1980), have similar facies characteristics (jigsaw-fit, clast-supported breccias) and geometric relationships (lobes of lava enveloped by breccia). The associated clast-supported tuff breccias are interpreted as autoclastic breccias.

The poor sorting, subrounded clasts, local mafic clasts and massive to graded nature of matrix-supported, felsic or intermediate lapilli tuff–tuff breccia–tuff sequences are consistent with deposition from sediment gravity flows such as turbidites (Lowe, 1982). The predominantly monolithologic composition, angular clasts and euhedral or broken single crystals in these facies suggest that they are largely composed of pyroclastic debris, which has not been significantly texturally modified before subaqueous deposition (Dolozoi & Ayres, 1991). Such syneruptive, resedimented deposits can move downslope from the site of eruption by hundreds of kilometres (Staudigel & Schminke, 1984; Cas, 1992).

The bimodal lapilli tuffs of the Gamey Lake formation and heterolithic breccias of the Cass formation, characterized by poor sorting, texturally modified clasts and massive, matrix-supported structure, are best interpreted as resedimented units deposited from debris flows (Car & Ayres, 1991; Dolozoi & Ayres, 1991). Because they are interbedded with the syneruptive sediment gravity flows and thin argillite units, they may have resulted from periods of mass-wasting between volcanic eruptions (see Fisher & Schminke, 1984). The matrix-supported, mixed composition

**Table 4.** Lithofacies of the Leta Arm group.

| Facies | Proportion (%)* | Description† |
|---|---|---|
| Mafic–intermediate flows/breccia | 40 | Mafic (18%) and intermediate (12%) pillowed and non-pillowed flows (ratio p/np‡: 65/35), <10 m thick × 500 m long, lava shelves, varioles and radial cooling cracks noted. Mafic flows: plagioclase–phyric (1–5 mm phenocrysts), amygdules §0.3–3.0 cm, ave. 0.5 cm, 10–30%, ave. 25%; pillows range 15–150 cm, ave.‖ 30 × 60 cm; selvage thickness 0.5–3 cm, ave. 1.5 cm. Intermediate flows: plagioclase–, clinopyroxene– or rarely biotite–phyric (1–5 mm phenocrysts), pillows on ave. larger (35 × 70 cm), amygdules 0.3–3.0 cm, ave. 0.6 cm, 10–40%, ave. 30%. Mafic lapilli tuff and breccias (10%): clast- to matrix-supported, in discontinuous beds, 0.2–50 m thick by 300 m long, clast-supported (locally jigsaw fit), angular blocky, locally fluidal, plagioclase–phyric and scoriaceous clasts, 1–40 cm, ave. 10 cm, clasts with pillow selvages where adjacent to pillowed flows, phenocryst and amygdule proportions similar to adjacent flows, local highly amygdaloidal (40–50%) clasts (10–80% of population); matrix-supported units with 1–5 mm shards and vesicular lapilli in matrix, commonly with fine-grained dolomite or chlorite schist matrix |
| Heterolithic lapilli tuff and tuff breccia | 19 | Lapilli tuff and tuff breccia (ratio 2.5 : 1): units 0.5–20 m thick by up to 500 m long, structureless to normally graded beds up to 1 m thick, clast- to matrix-supported, poorly sorted, 1–20 cm lithic and pumiaceous clasts; angular to rounded, clast composition >50% intermediate with lesser mafic, felsic, diorite clasts, proportions can vary within individual beds from >50% of a single composition at the base to less than 10% at the top, mafic clasts most abundant where beds overlie mafic flows; matrix composed of angular to subrounded lithic fragments and broken to euhedral quartz and plagioclase crystals |
| Gabbro/diorite/ quartz diorite | 10 | Gabbro, diorite and quartz–diorite sills: massive to layered, 10–300 m thick, up to 3000 m long, tabular to irregular in outline, local glomerocrystic or quartz–feldspar porphyry phases; sills typically lack well developed chilled margins, subparallel to bedding and flow contacts; diorite dykes crosscut bedding, show chilled margins but are restricted to this group |
| Intermediate lapilli tuff and tuff breccia | 9 | Lapilli tuff and tuff breccia (ratio 2.5 : 1): structureless to normally graded beds, 2–10 m thick, clast- to matrix-supported, poorly sorted, angular to subrounded, 1–20 cm clasts; in a crystal (euhedral/broken plagioclase and bipyramidal quartz) and lithic-rich matrix; locally with pumiaceous, irregular or fluidal intermediate clasts and fiamme, rare scoriaceous mafic clasts |
| Felsic lapilli and crystal tuff, tuff breccia | 14 | Felsic lapilli tuff (7%): poorly sorted, matrix-clast supported (5–30% clasts); angular to subrounded clasts, 0.5–5 cm, in poorly sorted and bedded units, 10–300 m thick. Clast composition: porphyritic and massive rhyolite, dacite, rarely amygdaloidal basalt (<5% of population), pink cherty tuffaceous clasts; matrix contains 5%, 0.5–1 cm lapilli, 5% quartz and 30% euhedral plagioclase crystals. Crystal tuffs (3%): well bedded (beds 3–15 cm thick) units up to 10 m thick, locally normally graded, matrix-supported, pumiaceous felsic clasts with amoeboid shapes, diffuse boundaries and fiamme (<10%), matrix has angular bubble wall shards, 20% felsic lithic clasts <1 cm, 5% amygdaloidal basalt clasts, <5 mm, quartz (1–2 mm) and euhedral or broken plagioclase crystals (1–5 mm). Tuff breccia (2%): normally graded (clast-supported at base, matrix-supported at top) and structureless beds (<1 m thick) in units 5–30 m thick, angular to subrounded clasts (6–30 cm, ave. 10 cm) of plagioclase + quartz–phyric, amygdaloidal and aphyric felsic lithic clasts, in a crystal and lithic-rich matrix similar to above. Cherty tuffs (2%): pink and pale buff, laminated tuffs with 0.5–1 cm beds in units to 2 m thick; fine-grained dolomite between clasts, tuffs and lapilli tuffs with 1–5% sulphide (pyrite + pyrrhotite) in matrix |
| Felsic flows and porphyry intrusions | 6 | Felsic flows: as unbrecciated lobes 2–20 m thick by 300 m long; quartz– and feldspar–phyric (1–5 mm phenocrysts, bipyramidal and resorbed quartz), locally flow-banded, non-brecciated flows; bounded by clast-supported felsic breccia with angular, blocky quartz– and plagioclase–phyric clasts, in featureless beds up to 20 m thick; quartz– and feldspar–phyric sills cut across bedding at <30°; 10 m thick by 500 m long, haematitic alteration common |
| Conglomerate/ sandstone/mudstone | 2 | Granule to boulder conglomerate and sandstone: in discontinuous beds, 1–5 m thick by 10–50 m long, clast- to matrix-supported, rounded and locally subangular clasts of porphyritic felsic, mafic and intermediate volcanic, and intraformational sandstone in medium-grained sandstone matrix, thicker units (>3 m) tend to be polymictic, thinner units (<3 m) oligomictic. Sandstone and argillite: as 2–10 cm thick interbeds in cherty tuffs or 10–50 cm thick, interbedded or normally graded beds in sections up to 10 m thick, fine–medium-grained sandstone with <10% euhedral to subrounded and broken, 1–5 mm, plagioclase crystals |

* Represents volume percentage of facies in formation based on field mapping.
† Maximum thicknesses of beds are likely minimum figures due to local cover on facies contacts; tectonized or schistose units in Fig. 2 not described here.
‡ p/np: pillowed/non-pillowed.
§ Visual estimates on two-dimensional outcrops.
‖ Pillow dimensions averaged from visual estimates of two-dimensional outcrops in pillows with 2 : 1 or less aspect ratio, effects of stretching in third dimension not included, but stretch averages 3 : 1 down-dip.

volcanic conglomerates interbedded with the mafic pillow lavas are possibly products of similar processes.

The finely layered dolomite interbeds with magnetite and silica-rich lenses do not preserve unequivocal sedimentary structures and are typically strongly tectonized. They could be hydrothermal carbonate precipitates or carbonate bank deposits that fringed the volcanic edifice.

The depth of eruption and deposition for the group is constrained by a number of features. The abundance and relatively large average size of amygdules in the mafic flows (Table 3), considered a reflection of the water depth for tholeiitic volatile-poor basalts, support moderate water depths (<1000 m; McBirney, 1963; Moore & Schilling, 1973; Moore, 1979; Dolozoi & Ayres 1991; see also Gill *et al.*, 1990). The paucity of mafic fragmental rocks relative to flows is also consistent with this interpretation (Dimroth *et al.*, 1978; Staudigel & Schminke, 1984). Similarly, the unusual thinness of the rhyolite flows (<40 m) has been interpreted by other workers as a result of highly fluidal rhyolite eruptions in fairly deep water, below the volatile suppression depth (Yamagishi & Dimroth, 1985; Mueller & White, 1992). Taken together, and considering the lack of wave-influenced or tidal structures in the epiclastic and resedimented volcaniclastic units, eruption and deposition must have been below storm wave base (>200 m; Leckie, 1988), but above 1000 m.

## The Leta Arm group

The Leta Arm group is a heterogeneous package of mafic to felsic volcanic rocks (Fig. 2), that forms seven N–NE trending belts within the Indin Lake supracrustal belt (Fig. 2, Table 4). These belts, 1–4 km wide by 5–30 km long, are interpreted as antiformal fold interference structures (Pehrsson, 1998). The upper contact of the group is sharp and marked by the first appearance of polymictic conglomerate and sandstone-mudstones of the Chalco Lake group. The present maximum exposed thickness of the Leta Arm group is approximately 1.5 km (half the width of the widest fold structure), although its base is not exposed.

The Leta Arm group consists of intimately interlayered mafic to felsic lava flows (46%), heterolithic and monolithic volcaniclastic rocks (42%), mafic–intermediate sills and dykes (10%) and rare epiclastic rocks (2%; Table 4). Lithofacies and their distribution are characteristically heterogeneous, with units of varying composition and facies interlayered on a 10–1000 m scale and displaying complex interfingering relationships both along and across strike (Figs 2 & 6, Table 4). Consequently, relative proportions of individual lithofacies are highly variable between the major belts (Fig. 2).

The Leta Arm group is subdivided into mafic–intermediate and felsic volcanic map units, respectively (Fig. 2). As individual beds are rarely traceable in outcrop beyond a few tens or hundreds of metres, the specific internal stratigraphic order of the group is unclear. An overall change upsection to more felsic composition and volcaniclastic texture is nevertheless noted (Fig. 6, Table 4).

Nearly half the group is composed of mafic to intermediate, pillowed and non-pillowed lavas that are characterized as tholeiitic and calc-alkaline, basalt and basaltic andesite and calc-alkaline andesites and dacites, in proportions of roughly 3/2 (Tables 2 & 4). Mafic lava flows of the Leta Arm group tend to be slightly more vesicular than those of the Hewitt Lake group and include a greater proportion of non-pillowed flows (Table 4). Individual mafic–intermediate pillows are also slightly larger than those of the Hewitt Lake group, with thicker selvages. Pillowed lava flows are found throughout the group, except in the uppermost 250–300 m in the Bow Lake and Leta Arm areas (Fig. 2). Mafic lapilli tuff and breccias occur between and along strike of individual lava flows and locally contain a significant proportion of highly amygdaloidal (>40% amygdules) scoriaceous or fluidal clasts. Thin interbeds, <5 m thick, of mafic or intermediate volcanic conglomerate with rounded clasts locally have scoured bases.

Heterolithic lapilli tuffs and tuff breccias are the most abundant volcaniclastic units and typically occur as thin (<10 m), clast-supported units between individual mafic–intermediate lava flows (Fig. 7A). The thicker (>10 m), more laterally continuous (>300 m), matrix-supported units near the top of the group tend to be normally graded, have a more heterogeneous composition, have local interbeds of mudstone and are overlain by 3–5 m thick, structureless beds of matrix-supported, polymictic volcanic conglomerate (Fig. 6).

Intermediate lapilli tuffs and tuff breccias occur predominantly between the mafic–intermediate flows, but also form thicker, normally graded units (>30 m) where interlayered with intermediate flows or felsic lapilli tuffs. The felsic lapilli and crystal tuffs and subordinate tuff breccias are found as individual beds between flows, but most commonly in continuous felsic sections hundreds of metres thick (Fig. 7B). At the top of the group these facies contain thin interbeds of

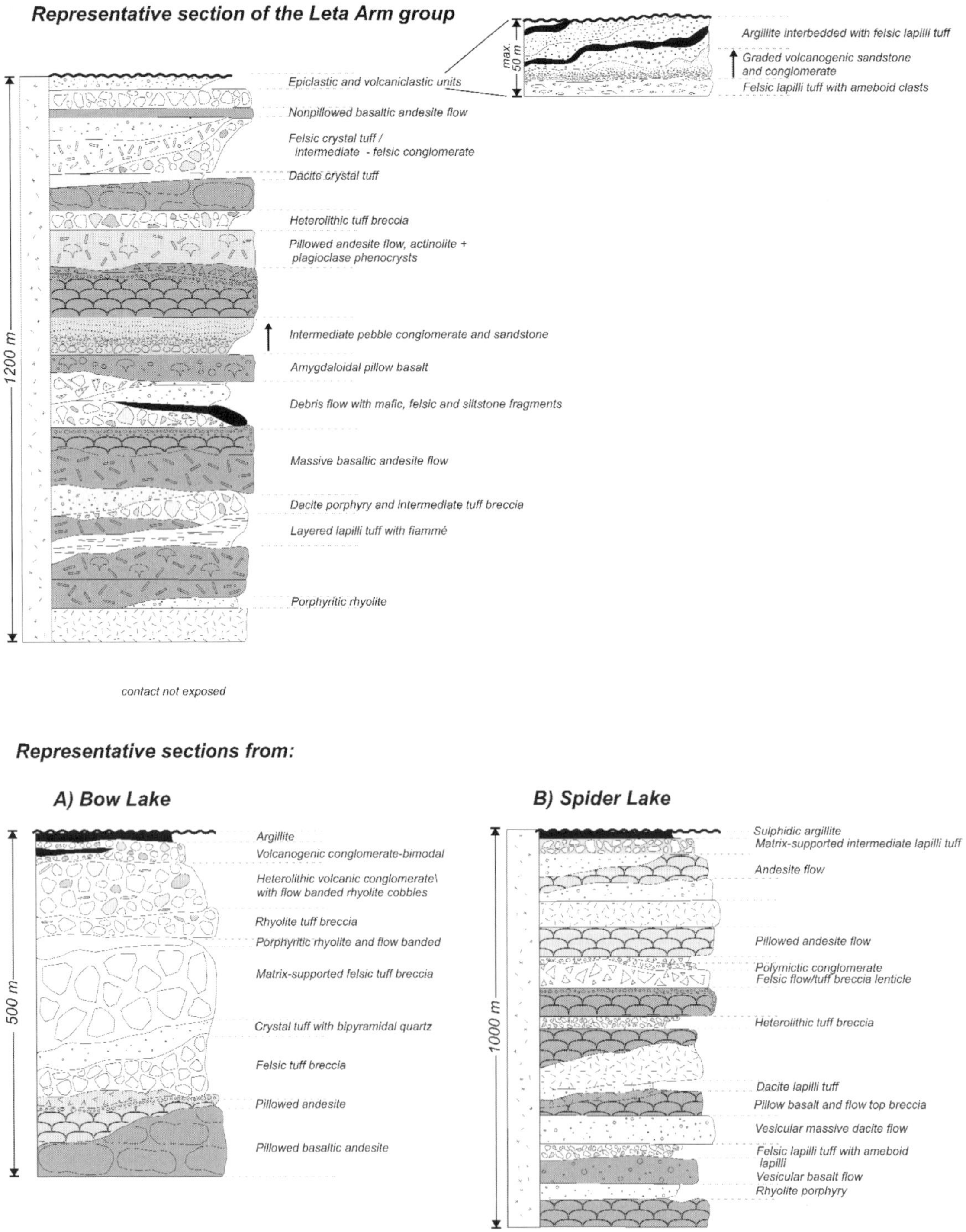

Representative section of the Leta Arm group
max. 50 m
Argillite interbedded with felsic lapilli tuff
Graded volcanogenic sandstone and conglomerate
Felsic lapilli tuff with ameboid clasts
Epiclastic and volcaniclastic units
Nonpillowed basaltic andesite flow
Felsic crystal tuff / intermediate - felsic conglomerate
Dacite crystal tuff
Heterolithic tuff breccia
Pillowed andesite flow, actinolite + plagioclase phenocrysts
Intermediate pebble conglomerate and sandstone
Amygdaloidal pillow basalt
Debris flow with mafic, felsic and siltstone fragments
Massive basaltic andesite flow
Dacite porphyry and intermediate tuff breccia
Layered lapilli tuff with fiammé
Porphyritic rhyolite
1200 m
contact not exposed
Representative sections from:
A) Bow Lake
Argillite
Volcanogenic conglomerate-bimodal
Heterolithic volcanic conglomerate\ with flow banded rhyolite cobbles
Rhyolite tuff breccia
Porphyritic rhyolite and flow banded
Matrix-supported felsic tuff breccia
Crystal tuff with bipyramidal quartz
Felsic tuff breccia
Pillowed andesite
Pillowed basaltic andesite
500 m
B) Spider Lake
Sulphidic argillite
Matrix-supported intermediate lapilli tuff
Andesite flow
Pillowed andesite flow
Polymictic conglomerate
Felsic flow/tuff breccia lenticle
Heterolithic tuff breccia
Dacite lapilli tuff
Pillow basalt and flow top breccia
Vesicular massive dacite flow
Felsic lapilli tuff with ameboid lapilli
Vesicular basalt flow
Rhyolite porphyry
1000 m


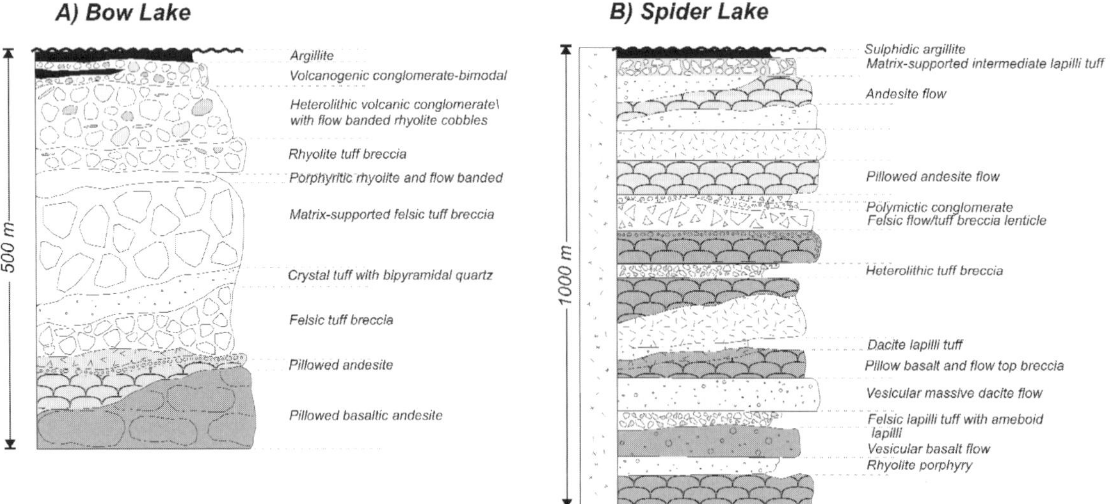

**Fig. 6.** Representative sections of the Leta Arm group. Note the variability of facies between and throughout the various sections. Legend and thickness of units as per Fig. 4.

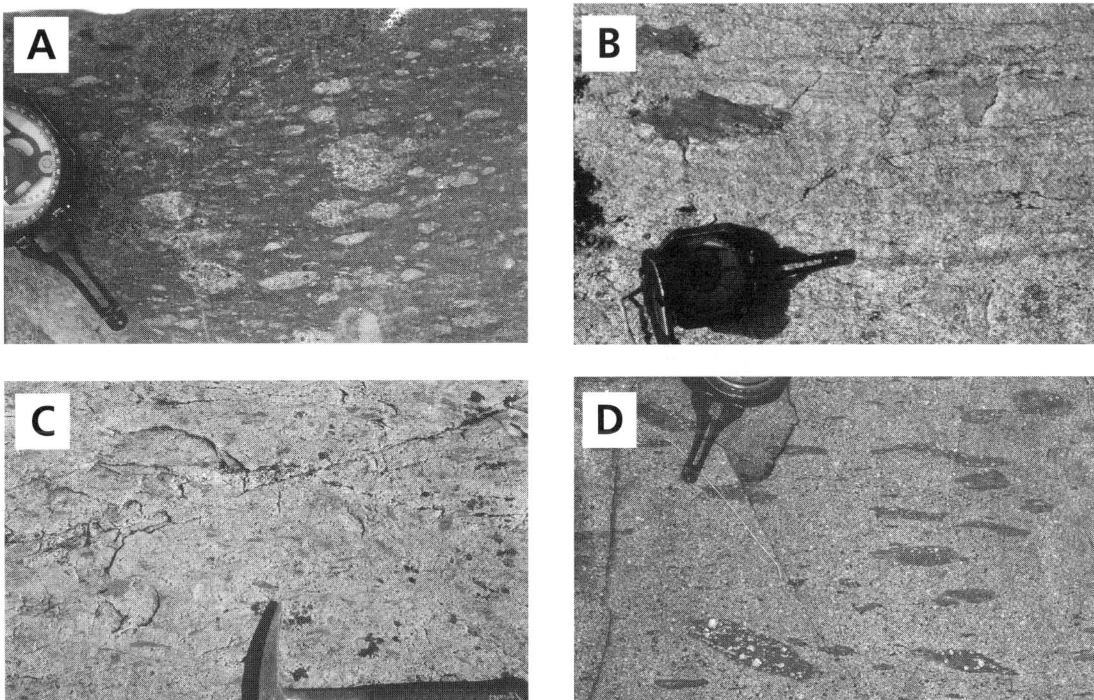

**Fig. 7.** Representative textures Leta Arm group facies. (A) Heterolithic lapilli tuff composed of rhyolite, dacite and diorite clasts in a chloritic matrix. (B) Normally graded, bedded felsic crystal-lithic tuff from west of Spider Lake. (C) Fiamme (above hammer) in rhyolite lapilli tuff, west of Leta Arm. (D) Highly vesicular pumice lapilli in crystal-rich rhyolite tuff. Sample from the culmination west of Strachan Lake. Scales: coin, 2.5 cm long; hammer, 40 cm long; brunton, 22 cm long.

argillite or heterolithic breccia. Although dominantly matrix-supported, both facies can pass along tens of metres of strike into clast-supported units. The cherty felsic tuffs are restricted to the upper 100 m of the group, where they are interbedded with lapilli tuffs, sandstone and argillite.

Felsic lava flows and associated breccias are a minor component of the Leta Arm group and are restricted to the thicker, continuous felsic sections in the uppermost part of the group. They are interbedded with thick felsic lapilli and crystal tuffs, but also with thin (<5 m) non-pillowed andesite lavas. The felsic porphyries occur as sill and dyke complexes cutting the lower mafic–intermediate flows and terminating in the upper felsic sections.

*Depositional processes*

Many facies of the Leta Arm group are strikingly similar to those of the Hewitt Lake group, particularly the Cass formation, and are inferred to result from similar depositional processes (compare Tables 3 & 4). Like the Gamey Lake formation, the Leta Arm group is dominated by pillowed lavas indicative of subaqueous effusive eruption. The mafic lapilli tuff and tuff breccias largely resemble hyaloclastite breccias derived from quench fragmentation. The monolithologic intermediate and felsic volcaniclastic deposits (lapilli tuffs/tuffs/tuff breccia) with abundant angular clasts and broken or euhedral crystals, like similar units in the Cass formation, are interpreted as pyroclastic debris deposited from debris and other sediment gravity flows. The facies characteristics of the heterolithic volcaniclastic and epiclastic rocks (lapilli tuffs/tuff breccia, conglomerate) are also consistent with deposition from debris or other mass flows.

Differences in characteristics and abundance of certain facies signal important differences in depositional process and environment between the two groups. The larger pillows with thicker selvages of mafic–intermediate lavas in the group may be due to the greater viscosity of the calc-alkaline basaltic andesite and andesite flows (see Griffiths & Fink,

**Table 5.** Lithofacies of the Chalco Lake group.

| Facies | Proportion (%)* | Description |
|---|---|---|
| *Parker formation* | | |
| Sandstone–mudstone | 80 | Sharp-based, normally graded, sandstone–mudstone beds, 0.15–3 m thick, typically >0.15 m; laterally continuous beds for tens of m; beds of basal massive sandstone, middle graded sandstone–mudstone and upper parallel-laminated silty mudstone; proportion of siltstone/mudstone to sandstone 1 : 2, with channel scours up to 20 cm deep by 40–60 cm long, load casts, ball-and-pillow and flame structures, mud rip-up clasts; thick sandstone beds (10–60 cm) with heavy mineral banding, planar cross-beds, isolated mud chips, occurring singly or in discrete channels 10–30 m thick that excise underlying strata; 1 m to tens of m-scale irregular, disharmonically folded and disrupted sandstone layers bounded by planar beds, faults with cm to 10 m offset locally sole into bedding surfaces; sandstone comprises coarse- to fine-grained, poorly sorted wacke consisting of angular to subangular grains of quartz, feldspar and lithic fragments in a finer-grained matrix of quartz, feldspar, sericite and chlorite or biotite (metamorphic) |
| Felsic volcanic breccia, bimodal breccia and porphyry | 10 | Felsic breccia (3%): sharp-based, structureless units 0.25–300 m thick by up to 1000s m long, poorly sorted, clast-supported, angular blocky, locally jigsaw fit, polyhedral, clasts of aphyric and quartz + feldspar–phyric rhyolite in lobe-shaped or lenticular masses; basal contact (5 cm thick) with sandstone–mudstone consists of clast to matrix-supported, jigsaw fit, felsic fragments in a siltstone matrix, matrix grades up section to felsic lithic tuff with 2–4 mm lapilli; local m-scale xenoliths of sandstone–mudstone, basal contacts truncate bedding in adjacent sandstone–mudstones. Bimodal breccia (2%): unbedded, poorly sorted, matrix-supported with 30–50% blocky to irregular clasts (10–200 cm) of bedded sandstone–mudstone or aphryic to quartz + feldspar–phryic rhyolite in a siltstone or ash matrix with 2–20%, <2 cm rhyolite lapilli; occurs as irregular lobes 2–20 m thick by 100 m long, poorly-size sorted. Rhyolite porphyry (5%): quartz + feldspar–phyric, in tabular to lobe-shaped units, 0.25–150 m thick × 10s–1000s m long |
| Heterolithic and mafic volcaniclastic rocks | 5 | Heterolithic lapilli tuffs: poorly sorted, matrix- to clast-supported units 10–300 m thick by <1 km long, angular to subrounded mafic and felsic volcanic clasts in an intermediate matrix. Felsic lapilli tuffs: poorly sorted, matrix-supported beds, 5–25 cm thick by 10s m long, angular aphyric rhyolite clasts (0.5–15 cm, ave. 2 cm) in a sandstone or mudstone matrix, 10–40% clasts, ave. 20%. Mafic tuff breccia, as a single occurrence within the Parker formation, 50 m thick by 100 m long; clast-supported, angular blocky to irregular clasts with selvages, 5 cm to 0.5 m long, unit isolated in sandstone–mudstone |
| Polymictic conglomerate | 3 | Polymictic cobble–boulder conglomerate: discontinuous, massive beds, 2–30 m thick, 30–1000 m long, clast-supported, well rounded to subangular 2–20 cm clasts, ave. 4–8 cm, of intermediate, felsic and mafic volcanic rocks, pink cherty tuffs, diorite, quartz–feldspar porphryry, sandstone, mudstone and granite in a mudstone to siltstone matrix |
| Bimodal conglomerate | 2 | Bimodal conglomerate: discontinuous, massive beds, 2–10 cm by min. 10 m long of granule conglomerate; matrix-supported, rounded to subangular clasts of aphyric-rhyolite (90%) and rare sandstone (10%) in a poorly sorted, coarse sand matrix; occurs between individual and massive sandstone beds and as basal layers (10–50 cm thick) in some graded sandstone–mudstone beds |
| *Damoti formation* | | |
| Sandstone–mudstone | 95 | Sandstone–mudstone: in sharp, flat-based, normally graded beds, 1–25 cm thick, typically >15 cm, laterally continuous beds for 10s m, proportion of siltstone/mudstone to sandstone 1 : 1, lower graded, upper laminated bedforms, upper parts of beds with climbing ripples, few scours or flames, sandstone comprises medium- to fine-grained, poorly sorted wacke consisting of angular to subangular grains of quartz, feldspar and lithic fragments in a finer-grained matrix of quartz, feldspar, sericite and chlorite or biotite (metamorphic) |
| Iron-formation/ mudstone | 5 | Silicate- and minor oxide-facies iron-formation: beds 2–20 m thick, ave. 1–2 m, by up to 2000 m long, with sandstone–mudstone units, contain layers of white microquartz, 1–5 cm thick by 10s m long with magnetite-rich seams. Sulphidic and graphitic mudstone: laminated 0.5–1 m beds in units 10–200 cm thick by up to 500 m long, mudstones typically as interbeds within iron-formation |
| Mafic intrusive rocks: common to both formations | | Gabbro to anorthositic gabbro: medium- to coarse-grained, locally layered, equant stocks, 250–500 m across; rare mafic dykes, <1 m wide, that cross-cut bedding but are themselves folded and deformed, noted particularly in the Daran/Fishhook/N Damoti Lake areas; intrusions are recrystallized, metamorphosed and deformed to same degree as host sandstone–mudstones, unlike phases of younger intrusive suites |

* Represents volume percentage of facies in formation based on field observation.

1992). Systematic changes in the vesicularity, proportion of pillowed/sheet flows and a greater abundance and thickness of mafic pillow breccias and flow-top breccias (Table 4), have been interpreted by some authors to reflect shallower depths of eruption (Dimroth *et al.*, 1978; Staudigel & Schminke, 1984; Busby-Spera, 1987), although the lack of wave- or tidal-structures in epiclastic deposits still suggests below storm wave base conditions. A depth of eruption and deposition between 1000 and 200 m is considered consistent with the facies characteristics of the group.

The highly amygdaloidal or fluidal clasts characterizing the mafic lapilli tuffs and tuff breccias have been interpreted in terms of redeposition from shallow subaqueous or subaerial eruptive environments (Dolozoi & Ayres, 1991) or explosive fragmentation in sub-wave base subaqueous fire-fountains (Mueller & White, 1992). Cas (1992) suggests, however, that the latter process can only operate in relatively shallow water. Overall, these lithofacies suggest that downslope remobilization and mixing of material from possibly different water depths was an important depositional process for the group.

A number of the felsic and intermediate volcaniclastic facies comprise pyroclastic rocks. The fluidal, pumiceous or amoeboid clasts and fiamme are considered evidence of hot emplacement of pyroclastic deposits (Fig. 7C & D; Cas & Wright, 1991). As the felsic-dominated units in the upper few hundred metres of the Leta Arm group lack pillow lavas and contain a significant proportion of the clast-supported volcanic conglomerates, it is possible that the volcanic complex became emergent during the final stages of its growth. However, pyroclastic material, including that generated by hot explosive fragmentation, can be present in subaqueous environments. This can occur directly, as a result of subaqueous pyroclastic eruption (Fouquet *et al.*, 1998), or indirectly through passage of subaerial pyroclastic flows into the ocean or subaqueous redeposition of subaerial unconsolidated material. Given the large volume of pyroclastic material, its presence in sediment gravity flows interbedded with dominantly subaqueous lavas (Fig. 7) and that not all entrained debris contains evidence of hot emplacement, the simplest interpretation is that the pyroclasts are redeposited, possibly from a shallow subaqueous or subaerial eruptive environment.

## The Chalco Lake group

The Chalco Lake group is the highest and most widespread lithostratigraphic unit of the Indin Lake supracrustal belt, forming approximately 60% of the belt. Its lower contact with the Leta Arm group is interpreted to be an unconformity, based on isotopic age relationships, stratigraphic truncations and the local presence of a basal polymictic boulder conglomerate (Fig. 3A; Pehrsson, 1998). The upper contact is not exposed. The group contains steeply dipping, graded sandstone–mudstones that are isoclinally folded. Its maximum preserved stratigraphic thickness is no more than 10 km, half the width of the thickest section, taking into account the isoclinal folding (Fig. 2).

The Chalco Lake group has been further subdivided into the Parker and Damoti formations, distinguished by differences in mean bed thickness, sedimentary structures, sand content of the sandstone–mudstones and associated minor facies (Table 5). Where the sandstone-mudstones are medium-bedded and lack subordinate interbedded facies, the contact between the two formations is placed between known exposures of these units. Where exposed, the contact is gradational, with a gradual decrease in mean bed thickness towards the Damoti formation. Both formations are intruded by numerous gabbro to anorthositic gabbro stocks and sills and dolerite dykes (Table 5).

### *Parker formation*

The Parker formation occurs in the central part of the Indin Lake supracrustal belt, where it surrounds and overlies the Leta Arm group (Fig. 2). Its estimated thickness ranges from 2 to 5 km and it makes up approximately 30–40% of the Chalco Lake group.

The lowermost unit of the formation is a polymictic, cobble to boulder conglomerate that forms discontinuous beds no more than 1000 m long by up to 30 m thick (Fig. 8A). Eight localities of this facies are identified in the map area, along nearly 100 km strike length (Fig. 2). It is characteristically clast-supported and dominated by lithological units derived from the underlying Leta Arm group (Table 5). Exotic clasts including granite and tonalite are locally abundant near Chartrand Lake to the north-west (Locality A, Fig. 2). In this area, this lithofacies forms 2–50 m thick beds in a sandstone/conglomerate sequence up to 1500 m thick (McGlynn & Ross, 1963). Lateral progressive thinning and cut-out of thick felsic units at the top of Leta Arm group just west of Chalco Lake (Fig. 2) and a 15 Myr age gap between the formations support the interpretation of an unconformity below this facies (Pehrsson & Villeneuve, 1999).

The basal conglomerate is abruptly succeeded both upsection and along strike by a thick package of sandstone-mudstones. This facies comprises 80% of

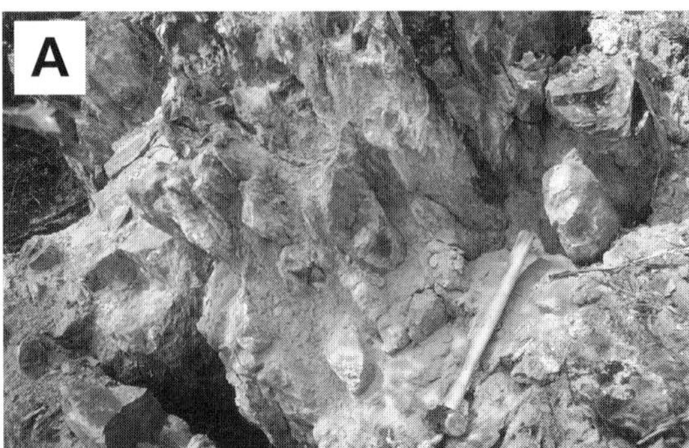

**Fig. 8.** Features of the Parker formation. (A) Basal conglomerate between the Leta Arm group and the Chalco Lake group. Note the well rounded nature of cobbles weathering out of the sulphidic sandstone. (B) Thick, massive sandstone–mudstone basal layers grading upward to siltstone, overlain by laminated silty-mudstone. Notebook 25 cm long; hammer 80 cm long.

the formation and is characterized by medium to thick beds (0.15–3 m) of basal massive sandstone overlain by graded sandstone–mudstone divisions and upper parallel-laminated, non-graded, silty mudstone (Bouma AD divisions; Table 5, Fig. 8b; Bouma, 1962; Pehrsson, 1998). Load-structures, including flames, load casts and ball and pillow structures, are common throughout the sandstone–mudstone sequences (Table 5). Thick planar sandstone beds with isolated mud chips are interpreted as amalgamated beds (see Henderson, 1972), and occur predominantly in the central Indin Lake area in proximity to felsic volcanic rocks of the formation (Table 5). Rare, matrix-supported, bimodal conglomerates are present throughout the formation as individual, discontinuous interbeds within the sandstone–mudstones and as thin units at the base of graded beds (Table 5). Heterolithic, mafic–felsic lapilli tuffs form minor, discontinuous volcaniclastic lenses within the sandstone–mudstones and are associated with a single occurrence of mafic tuff breccia on central Indin Lake. The poorly sorted tuff breccia contains texturally unmodified, blocky to irregular, mafic pillow fragments, up to 0.5 m long.

Felsic volcanic rocks and porphyry form a minor but distinctive unit near the base and middle of the Parker formation (Fig. 9 inset & Table 5). Rhyolite porphyry is the most abundant subtype in this facies, forming lobe-shaped to irregular intrusive masses, 0.25–150 m thick by 10–1000 m long, surrounded by thick units (up to 300 m) of porphyritic rhyolite breccia (Figs 2 & 9). Locally, the lower and lateral contacts of the porphyry with adjacent sandstone–mudstone consists of 5 cm of jigsaw-fit, blocky and polyhedral rhyolite clasts in a matrix of siltstone (Fig. 9A). The porphyry can be unbrecciated within the larger >20 m thick lobes but thinner lobes consist entirely of blocky, angular rhyolite clasts supported by a felsic tuff matrix with abundant 2–4 mm rhyolite lapilli. Bedding in adjacent sandstone–mudstones is disrupted or truncated.

Compositionally bimodal breccias are present as metre-thick layers along the margins of some porphyry bodies, as metre-scale lobes enveloped by felsic breccia and as tens of metres irregular bodies within the sandstone–mudstones (Fig. 9). The breccias consist of poorly sorted, massive, up to 2 m wide blocky fragments of sandstone–mudstone and porphyritic rhyolite in a siltstone or felsic tuff matrix (Fig. 9B). Beds in adjacent sandstone-mudstones are disrupted, faulted or display centimetre to metre scale disharmonic (locally isoclinal) folds of bedding that are bounded by undisturbed, planar layers (Fig. 9C). Thin beds of felsic lapilli tuff, consisting of rhyolite lapilli and blocks with a felsic, wacke or mudstone matrix (Fig. 9D), are found within the sandstone–mudstones several metres to tens of metres above and below the porphyries.

**Fig. 9.** (*opposite*) Detailed geology of the Parker formation south-east of Parker Lake. The inset map depicts the location of Fig. 9 and other occurrences of rhyolite. Letters refer to the localities of inset photos. $F_2$ and $F_1$ fold axes not shown for clarity. Jigsaw-fit rhyolite breccia with a silt matrix (A) occurs at the lower contact of brecciated rhyolite porphyry. Beds of rhyolite lapilli and blocks in a silt to mudstone matrix (B) occur above and below the porphyries. In places, rhyolite blocks, massive rhyolite, mudstone and sandstone clasts are chaotically mixed (C), resembling peperites formed from the mixing of unconsolidated sediment and magma. Intrabedded disrupted and disharmonic folds of bedding (D) in the turbidites below individual flows suggest synsedimentary disruption. Pen for scale is 13 cm long.

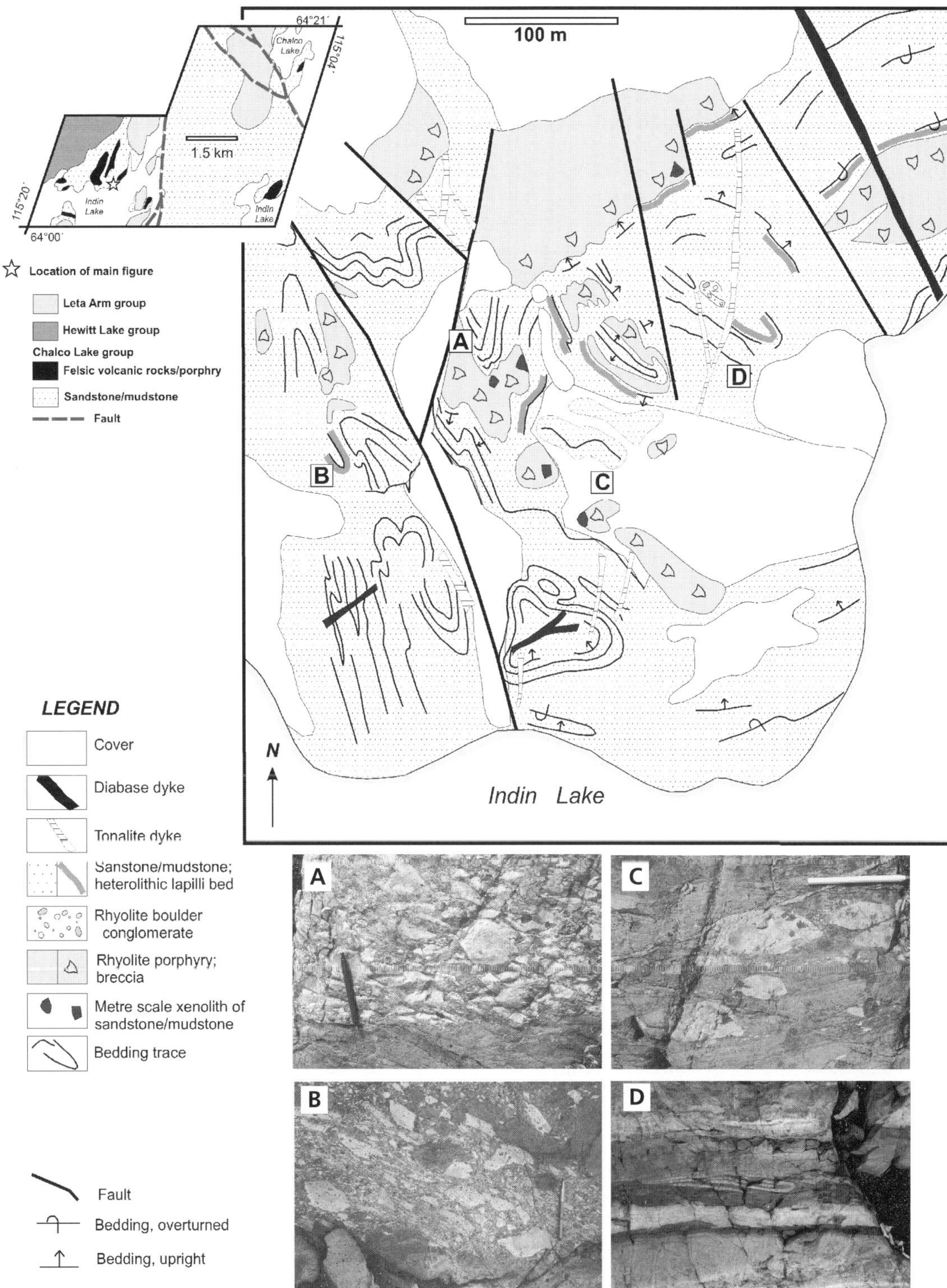
64°21′
Chalco Lake
115°04′
1.5 km
115°20′
Indin Lake
Indin Lake
64°00′
Location of main figure
Leta Arm group
Hewitt Lake group
Chalco Lake group
Felsic volcanic rocks/porphyry
Sandstone/mudstone
Fault
100 m
A
B
C
D
N
Indin Lake
LEGEND
Cover
Diabase dyke
Tonalite dyke
Sanstone/mudstone; heterolithic lapilli bed
Rhyolite boulder conglomerate
Rhyolite porphyry; breccia
Metre scale xenolith of sandstone/mudstone
Bedding trace
Fault
Bedding, overturned
Bedding, upright

*Damoti formation*

The Damoti formation underlies the centre and margins of the Indin Lake supracrustal belt and comprises nearly 60% of the Chalco Lake group. Its approximate thickness ranges from 3 to 5 km, although the upper contact is not exposed. Regional facing directions and isotopic age data suggest that the Damoti formation both overlies and occurs along strike of the Parker formation above the Leta Arm group (Pehrsson, 1998; Pehrsson & Villeneuve, 1999). This is consistent with the present interpretation that the two formations are lateral facies equivalents (Fig. 3A), although discontinuous exposure obscures contact relationships in detail.

The Damoti formation is dominated by thin- to medium-bedded, graded sandstone-mudstones, which tend to be thinner-bedded (1–15 cm) and less sand-rich and have fewer load-structures than those of the Parker formation (Table 5). Beds are characterized by basal units of sandstone or siltstone grading to mudstone and upper, plane-laminated, mudstone (Bouma divisions AE). Locally, individual beds contain ripple cross-laminated siltstone (Bouma division C). This facies forms extremely monotonous, thick units with individual beds continuous for up to 100 m.

Thin units of silicate and oxide facies iron-formation and sulphidic to graphitic mudstone are the characteristic associated facies of the Damoti formation (Table 5 & Fig. 2). The iron-formations occur in clusters at three distinct stratigraphic levels within the sandstone-mudstones to the south, south-east and north of Indin Lake (Fig. 2).

*Depositional processes*

The basal conglomerates of the Chalco Lake group in the Indin Lake area are not sufficiently well exposed to yield unequivocal information about their depositional setting. Clast- to matrix-supported, unstructured, poorly sorted conglomerates can be deposited from debris flows in a variety of settings, including subaerial alluvial fan and submarine environments. Polymictic boulder conglomerates of the Beniah Formation in the central Slave Province, typified by similar facies characteristics and also interstratified with sandstone, have been interpreted by Corcoran *et al.* (1999) as alluvial fanglomerates. Whatever their setting, the basal conglomerates at Indin Lake are clearly situated above a significant hiatus in the stratigraphy, during which rocks of the Leta Arm group were uplifted and eroded.

The scale and continuity of units and lack of wave structures indicate that sandstone-mudstones of the Chalco Lake group were deposited below storm wave base in a marine setting (Leckie, 1988; R. Walker, 1992). The bedforms and facies distributions of these units are characteristic of deposition from turbidity currents (Bouma, 1962; Mutti & Ricchi-Luchi, 1978), an intepretation pioneered for rocks of the Slave Province by Henderson (1972). The tightness of folding in the group precludes any meaningful interpretation of fan or apron settings.

The Parker formation sandstone–mudstones resemble deposits of high-density turbidity current deposits and submarine slides, with local conglomerates deposited from debris flows (Lowe, 1982). Deposition was below storm wave base, >200 m, possibly in deep water in a seismically active or unstable area. The sandstone–mudstones of the Damoti formation are similar in many respects to deposits of low-density, levee-forming turbidity currents of modern submarine fans (Table 4; Piper *et al.*, 1995). Periodic deposition of iron-formation and mudstone among the turbidites suggests varying oxygen conditions and periods of restricted sediment input in the deep marine basin (Kimberley, 1989). The texture, clast-supported nature and clast size of the mafic tuff breccia interbedded with the Parker formation suggest that it is a hyaloclastite breccia derived from a subaqueous mafic lava flow. The mafic dykes and stocks in both formations, which clearly predate major plutonism and deformation of the region, could be feeders and subvolcanic intrusions related to the mafic tuff breccia.

The felsic and heterolithic breccia facies of the Parker formation have the diagnostic characteristics of blocky peperites (Hanson & Schweickert, 1982; Busby-Spera & White, 1987; Hanson & Wilson, 1993; Hanson & Hargrove, 1999): the margins of the rhyolite porphyries have massive, chaotically mixed zones of magma and unconsolidated host-sediment; the breccias are poorly sorted and ungraded, and have jigsaw fit or polyhedral, blocky clasts; and adjacent bedding in the sediments is disrupted or truncated. A blocky texture is typical of peperites formed by intrusion into poorly sorted, coarse-grained sediments like the wackes of the Parker formation (Busby-Spera & White, 1987). The clasts are formed by quenching and phreatomagmatic fragmentation of the magma, typically with steam-driven mixing and displacement of clasts (Hanson & Wilson, 1993; Hanson & Hargrove, 1999). Entrainment of the silt and wacke into the breccia at the immediate contact of the porphyry suggests that there was *in situ* fluidization of the sandstones. The rhyolite porphyries are probably hypabyssal sills rather than effusive flows because of their overall morphology and the presence of peperites on both the sides and base of the bodies, uncommon for lava flows (Hanson & Schweickert, 1982).

Irregular units of magmatic clasts supported by host-sediments several metres away from the site of intrusion have been interpreted to form as a result of *in situ* clast-dispersal processes (Busby-Spera & White, 1987; Hanson & Hargrove, 1999). The felsic lapilli tuff beds within the sandstone–mudstones are superficially similar to units formed by this process. However, given their planar bedded nature and the fact that they are present up to tens of metres above or below the rhyolite porphyries, a distance much greater than that described by other workers, they are more likely to result from syneruptive redeposition of fragmental material from effusive flows.

## DISCUSSION

### Depositional and tectonic setting of the Indin Lake supracrustal belt

The Indin Lake supracrustal belt, like many Archaean greenstone belts, is composite, representing the product of distinct depositional episodes spanning at least 40 Myr. It formed in several environments, with depositional episodes separated by as much as 20 Myr, and represents different water depths, compositions and character of volcanism. The Indin Lake supracrustal belt has similar stratigraphy, lithofacies, ages and structural style to the other contiguous supracrustal belts of the south-western Slave Province, which also include fault-bounded, submarine mafic volcanic successions (Mesa, Chartrand, Russell Lake, Fig. 1), antiformal culminations of 2660–2670 Ma intermediate–felsic volcanic and volcaniclastic rocks (Mesa, Russell Lake, Wijinnedi) and basal conglomerates (Chartrand) overlain by <2.65 Ga turbidites (Wheeler, Fig. 2; McGlynn & Ross, 1963; Ross, 1966; Jackson, 1988; Villeneuve & Henderson, 1998). Consequently the depositional model and interpreted settings for the ISLB summarized below and depicted in Fig. 10 are considered applicable to the south-west Slave Province as a whole.

#### *Pre-2670 Ma: the Hewitt Lake group*

The predominance of extensive, homogeneous pillowed and massive sheet flows in the preserved, albeit incomplete, sections of the Hewitt Lake group is indicative of eruption on the flanks of a subaqueous mafic shield volcano or seamount (Fig. 10A; cf. Cas, 1992; Walker, 1993). Relative proportions of pillowed to massive lava flows are thought to be influenced by supply rate and slope, with sheet flows favoured by high rates of eruption (Dimroth *et al.*, 1978; Griffiths & Fink, 1992). The greater proportion of pillowed lava in the Gamey Lake formation and the presence of epiclastic interbeds suggest relatively low rates of eruption and periodic cessation of active volcanism, during which dominantly volcanic highlands were eroded.

Minor felsic centres are interpreted to have been intermittently built on the shield volcano surface (Fig. 10A). Association of epiclastic facies such as mudstones and heterolithic breccias with the felsic volcanic units and rare, scoriaceous clasts in the mafic lapilli tuffs, suggest local reworking and erosion of a shallower volcanic pile outside the final depositional site. Periodic pauses between eruptions, such as those marked by conglomerate and mudstone deposition, often coincided with episodes of felsic edifice construction (e.g. Fisher & Schminke, 1984).

The transition to more intermediate and felsic calc-alkaline volcanism at the top of the Hewitt Lake group indicates a significant change in the character and composition of volcanism with time. The dominance of pyroclastic debris in resedimented syneruptive units, the composition, abrupt vertical and lateral facies changes and overall poorly bedded nature are consistent with the construction of steep-sided stratovolcanoes, prone to mass-wasting and reworking of pyroclastic debris (Staudigel & Schminke, 1984). The formation represents the subaqueous part of the volcano where effusive eruptions occurred and volcaniclastic debris, formed higher on the volcano, accumulated (Fig. 10A).

Subaqueous mafic shield volcanoes can form in a variety of settings, including oceanic or marginal rifts, island arcs, oceanic seamounts and backarc basins (Cas, 1992). Many features of the Gamey Lake formation, including the incorporation of terrigenous or epiclastic material, interpreted low rate of eruption, older inherited zircon component (Table 1) and geochemistry (Pehrsson, 1998), are inconsistent with oceanic rifts or seamounts. The subaqueous setting and abundance of tholeiitic mafic volcanic rocks do not compare well with modern continental arcs, although immature island arcs can be dominantly tholeiitic (Ewart & Hawkesworth, 1987). Marginal or backarc basin volcanoes typically have compositions,

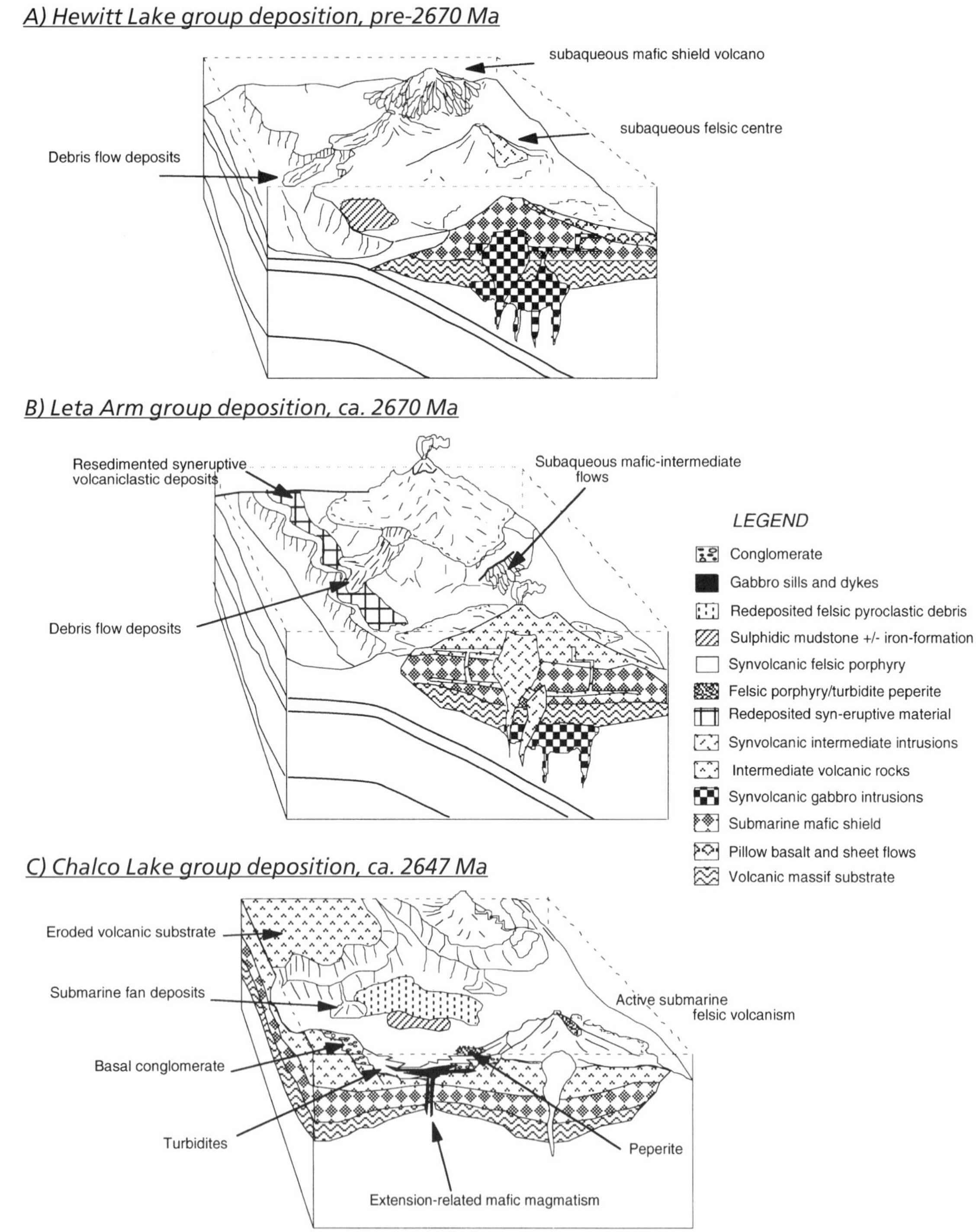

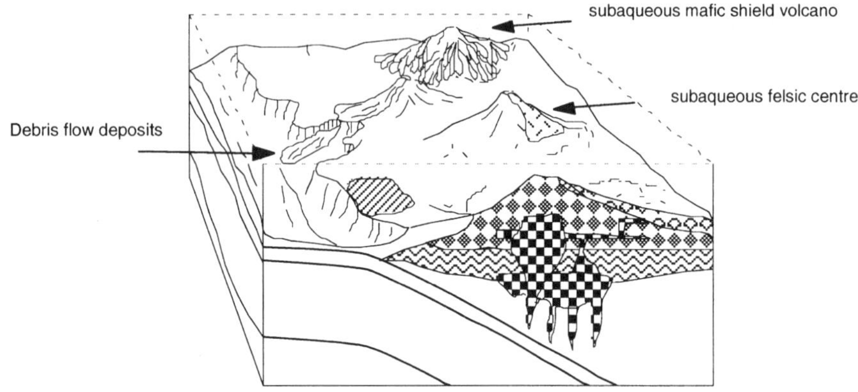






**Fig. 10.** Depositional model for the Yellowknife Supergroup in the Indin Lake area. (A) Deposition of mafic volcanic rocks of the Hewitt Lake group in a moderately deep, submarine environment. Lesser felsic–intermediate volcanic rocks of the group form shallower centres on the mafic sheet flows. Material from these centres is dominantly deposited as debris and sediment gravity flows. (B) Deposition of mafic, intermediate and felsic volcanic rocks of the Leta Arm group in a moderately deep submarine environment from locally emergent stratovolcanoes environment built upon the Hewitt Lake group. (C) Later deposition of the Chalco Lake group sedimentary and felsic volcanic rocks in a marginal, possibly foredeep environment. Erosion of the Leta Arm arc and older sialic crust contributes to the detritus.

bathymetry and lithofacies that are similar to those inferred for the Gamey Lake formation (Tamaki, 1988; Marsaglia, 1995). A marginal volcanic-dominated rift setting has been proposed for the 2700 Ma Kam Group of the Yellowknife greenstone belt, the uppermost formation of which has nearly identical geochemistry and lithofacies to the Gamey Lake formation (Helmstaedt & Padgham, 1986; Cousens, 2000).

Differentiating between these settings hinges on the extent of interaction of the Hewitt Lake group with older crust. There is no known *in situ* basement to the Hewitt Lake group. Mesoarchaean crust is present 100 km north of Indin Lake (Fig. 1), but mafic volcanic rocks structurally overlying the old orthogneisses appear to be more iron-rich than those of the Gamey Lake formation, lack felsic volcanic interbeds and are exclusively aphyric basalts (Pehrsson, unpublished data), suggesting they are not part of the Hewitt Lake group. Their relationship to older crust cannot be used to infer the nature of the basal contact. Indirect evidence, such as the lack of a depleted mantle Nd signature in the basalts and rhyodacites (+1.9 to +1.5, versus +2.5 to +3.0 for depleted mantle at 2700 Ma, Table 2), favours some assimilation of slightly older crust, a feature inconsistent with an island arc setting, but noted in modern backarc basins built on attenuated continental crust such as the Sea of Japan (Tamaki, 1988).

### *Syn-2670 Ma: the Leta Arm group*

Deposition of the Leta Arm group commenced by 2670 Ma, possibly on the submarine mafic substrate represented by the Hewitt Lake group (Fig. 10B). The abrupt and highly varied facies changes, both along and across strike, abundance of intermediate–felsic subaqueous effusive flows and large volume of resedimented syneruptive and epiclastic sediment-gravity flows are typical of the construction and mass-wasting of an active stratovolcano (Fig. 10B; cf. Fisher & Schminke, 1984). The proportions of lava to reworked or resedimented volcaniclastic facies in the Leta Arm group are characteristic of the medial facies of Cenozoic stratovolcanoes, in which there is continuous redeposition of shallow subaqueous to subaerial pyroclastic and epiclastic debris into moderately deep water on the flanks of the edifice (Staudigel & Schminke, 1984). The Leta Arm group may represent the apron and margin of a stratovolcano in a marine setting.

The Cass formation at the top of the Hewitt Lake group greatly resembles the thick, laterally continuous intermediate–felsic sections of the Leta Arm group. Both are dominated by calc-alkaline intermediate to felsic volcanic rocks and heterolithic volcaniclastic facies, and have characteristically abrupt facies changes both along and across strike. Present isotopic age constraints allow that the Hewitt Lake group is either coeval with or older than the Leta Arm group, raising the possibility that the Cass formation and Leta Arm group are facies equivalents. If there is no significant hiatus between deposition of the Leta Arm group and Cass formation and the underlying Gamey Lake formation, the noted changes in facies, composition and depth of eruption may simply represent the natural progression from shield to cone building in a single magmatic system.

Calc-alkaline basalt–andesite–dacite volcanic sequences are common associations of both oceanic and continental arcs, although submarine calc-alkaline stratovolcanoes are more common in oceanic than continental environments (Miyashiro, 1974). This setting is consistent with the facies, compositions and geochemistry of the group, but problematic given the somewhat more evolved Nd signatures of lavas with a range of composition (+1.45 to +0.02, Table 4). A nascent arc built upon previously stretched, subaqueous older crust most simply explains the features of the Leta Arm group.

### *2647–2629 Ma: the Chalco Lake group*

The Chalco Lake group basin became active some 20 Myr later, with deposition of the basal conglomerate on the eroded flanks of the Leta Arm group. The dominant facies of the group, the sandstone–mudstone turbidites, represent a deeper marine, sediment-dominated environment. High density turbidity current deposits of the Parker formation are interpreted to have been deposited in an unstable zone near the shelf-slope break, where earthquakes triggered slumping and sediment gravity flows (Fig. 10C). Turbidites of the Damoti formation possibly represent a more distal environment, although interbedding with chemogenic sediments (restricted detrital sediment input) is also consistent with channel levee complexes on upper fan slopes (Lowe, 1982). The abundance of 2660–2700 Ma detrital zircons in the Damoti sandstone–mudstones and Parker rhyolites substantiate sedimentological evidence of erosion of the underlying Leta Arm group, but a lesser mode of 2760–3048 Ma zircons also suggests recycling of older Mesoarchaean crust (Pehrsson & Villeneuve, 1999).

Magmatism did not cease with the Leta Arm group. Calc-alkaline felsic magmas were emplaced into unconsolidated Parker formation sandstone-mudstones at *c*.2647 Ma (Fig. 3A), resulting in the formation of blocky peperites. The dominance of felsic porphyry magmatism over effusive eruption is thought to be a consequence of the buoyancy factors of magma emplaced into thick sequences of unconsolidated sediment, favouring shallow intrusions over extrusive eruption (McBirney, 1963). The rhyolite lapilli and blocks in heterolithic lapilli interbeds could be derived from felsic lavas fed from a subaqueous fissure on the basin margin (Busby-Spera, 1988).

The presence of felsic peperites, numerous minor mafic intrusions and rare mafic hyalcolastite breccia in the sandstone–mudstones suggests that Chalco Lake group magmatism was bimodal (Fig. 10C). Association of bimodal magmatism with turbidite deposition is strikingly similar to the proximal parts of modern and ancient subaqueous arcs, backarc and rift basins, which are typical sites for peperite sill-sediment complexes (Tamaki, 1988; Wilson, 1991; Smith & Landis, 1995).

Deposition of the Chalco Lake group occurred between at least 2647 and 2629 Ma, and possibly as late as 2609 Ma. During this time, early compressional deformation of the Yellowknife–Hearne basin had begun to the south-east (Davis & Bleeker, 1998; Pehrsson & Villeneuve, 1999). The basin thus developed in a tectonically active, marine orogenic environment contemporaneous with ongoing deformation elsewhere in the orogen. Shortly thereafter the basin was itself subjected to compressional orogenesis in a fold–thrust belt ($D_1$ deformation, Table 3).

Underfilled foredeeps and marginal basins near an active orogenic system are also characterized by rapid build-up of thick turbidite successions like the Chalco Lake group, (Marsaglia, 1995; Miall, 1995; Sinclair, 1997). Contemporaneous magmatism, as noted above, is a common feature of modern and ancient backarcs or marginal seas, but is rare or absent in Phanerozoic foredeeps (Miall, 1995), and characteristic of Early Proterozoic foredeeps (Hoffman, 1987) and Archaean successions interpreted as foreland- or foredeep-basins, such as the Fig Tree Group of the Barberton, South Africa (Jackson *et al.*, 1987). Foredeep magmatism is commonly mantle-derived (Hoffman, 1987), consistent with the significantly more juvenile isotopic signature of the Parker formation rhyolites (Table 2).

A foredeep environment reconciles many of the features of the Chalco Lake group and explains the sudden increase in influx of older Neoarchaean to Mesoarchaean detritus due to uplift and erosion of varied crustal levels and components of the orogen. Although such a hypothesis is speculative, especially given the long hiatus between deposition of the Chalco Lake and Leta Arm groups, it can be tested through integrated structural/sedimentologic/isotopic studies. Diachroneity in the timing of compressional deformation and deposition is predicted; continued turbidite deposition should overlap with *c*.2.61–2.59 Ga deformation of the Chalco Lake group. If a foredeep setting is correct, the unconformity between the Leta Arm and Chalco Lake groups would be a fingerprint of collision in the orogen, and have more regional tectonic significance.

### Comparison with modern environments

The volumetrically largest component of the Neoarchaean greenstone belts of the south-west Slave Province is the 2.7–2.63 Ga volcano-sedimentary succession represented by the three lithostratigraphic groups described herein. In all three groups, sediment gravity flows or mass flows are the dominant depositional medium for clastic material, be it epiclastic debris deposited in the Chalco Lake group or resedimented, syneruptive debris in the marginal basin–arc-like environment of the Hewitt Lake and Leta Arm groups. These inferred basinal environments are strikingly similar to modern ones. The subaqueous parts of modern marginal or backarc basins are also often dominated by volcaniclastic debris deposited from fan- and debris-flow systems, a function of high rates of tectonic uplift (Klein, 1985). These basins can also lack shallower, shoreface deposits, especially where the basin margins are fault-bounded (Tamaki, 1988).

The subaqueous flanks of marine stratovolcanoes are relatively poorly documented compared to associated sedimentary basins (Soh *et al.*, 1991; Marsaglia, 1995; Sowerbutts & Underhill, 1998), but are thought to be dominated by mixed flows and primary and secondary volcaniclastic rocks (Cas, 1992), facies strikingly similar to those of the Leta Arm group. Phanerozoic calc-alkaline stratovolcanoes are generally linked to subduction processes, an interpretation equally favoured (Bickle *et al.*, 1993; Erikkson *et al.*, 1994) and disfavoured (Hamilton, 1993) for Archaean sequences. The facies associations, composition and geochemistry of the Leta Arm group are arc-like, although its Nd isotopic signature suggests some recycling of older continental crust. High rates of erosion are favoured in unvegetated, subaerial Archaean terrains (Erikkson *et al.*, 1994) and probably contributed to redeposition of volcanic material by mass and sediment gravity flows.

The similarity between the sedimentary record of modern foredeeps and Precambrian deep marine turbidite sequences has been noted by a number of workers (Hoffman, 1991; Krapez, 1993; Eriksson *et al.*, 1994). The major difference is the associated magmatism, in Early Proterozoic and Archaean examples (see Hoffman, 1987), where the magmatism has been linked to anomalous thermal regimes caused by slab break-off (Hildebrand *et al.*, 1991). The magmatic load may also have contributed to basin subsidence (Ingersoll & Busby, 1995), favouring underfilled foredeeps in the Archean.

The $CO_2$-rich Archaean atmosphere (Kasting, 1993) is thought to have had a strong influence on sandstone composition, preferentially breaking down feldspars and generating anomalously quartz-rich first-cycle sandstones (Eriksson *et al.*, 1998). This climatic influence has been interpreted by Corcoran *et al.* (1998, 1999) as the cause for the quartz-rich, shoreface sandstones in the Keskarrah and Beaulieu Rapids formations of the central Slave Province. The sandstones of the Chalco Lake group (Pehrsson, 1998), and other Slave turbidites show similar compositions, which were interpreted to reflect a preponderance of sialic detritus (Davidson, 1967; Henderson, 1998) inconsistent with the dominance of volcanic detrital zircon modes (Pehrsson & Villeneuve, 1999). Although the long hiatus between deposition of the Leta Arm and Chalco Lake groups allows for multi-cycle reworking, the complete lack of intervening sedimentary rocks suggests that the sandstone composition is more probably a general effect of Archaean climatic conditions.

## ACKNOWLEDGEMENTS

Herb Helmstaedt, Robert Rainbird and John Henderson are thanked for discussions on the stratigraphy and physical volcanology at Indin Lake. Deborah Lemkow is thanked for digital drafting of some figures. This paper represents part of the author's PhD dissertation at Queen's University, supported by the Geological Survey of Canada, the Canada–Northwest Territories 1991–1996 Mineral Development Agreement and an NSERC post-graduate scholarship. Additional logistical support from Royal Oak Mines Ltd and SouthernEra Resources is gratefully acknowledged. Robert Rainbird, Bob Baragar, Wulf Mueller and L.D. Ayres are thanked for helpful reviews on an earlier draft. Larry Aspler, James White and Mike Easton are thanked for critical reviews. This is GSC Contribution 1998203.

## REFERENCES

Ballard, R.D. & Moore, J.G. (1977) *Photographic Atlas of the Mid-Atlantic Ridge Rift Valley*. Springer-Verlag, New York, 114 pp.

Baragar, W.R.A. & McGlynn, J.C. (1976) Early Archean basement in the Canadian Shield: a review of the evidence. *Paper 76-14*. Geological Survey of Canada, Ottawa.

Bickle, M.J., Bettenay, L.F., Barley, M.E., Chapman, H.J., Groves, D.J., Champbell, I.H. & De Laeter, J.R. (1993) A 3500 Ma plutonic and volcanic calc-alkaline province in the Archean east Pilbara Block. *Contrib. Mineral. Petrol.*, **84**, 25–35.

Bleeker, W., Ketchum, J., Jackson, V. & Villeneuve, M. (1999) The Central Slave Basement Complex, Part 1: Its structural topology and authochtonous cover. *Can. J. Earth Sci.*, **36**, 1083–1109.

Bouma, A.H. (1962) *Sedimentology of Some Flysch Deposits*. Elsevier, Amsterdam, 168 pp.

Bowring, S.A., Williams, I.S. & Compston, W. (1989) 3.96 Ga gneisses from the Slave Province, Northwest Territories. *Geology*, **17**, 969–1064.

Busby-Spera, C.J. (1987) Lithofacies of deep marine basalts emplaced on a Jurassic back-arc apron, Baja California (Mexico). *J. Geol.*, **95**, 671–686.

Busby-Spera C.J. (1988) Evolution of a Middle Jurassic back-arc basin, Cedros Island, Baja California: evidence from a marine volcaniclastic apron. *Bull. geol. Soc. Am.*, **100**, 218–233.

Busby-Spera, C.J. & White, J.D.L. (1987) Variation in peperite textures associated with differing host-sediment properties. *Bull. Volcanol.*, **49**, 65–775.

Car, D. & Ayres, L.D. (1991) A thick dacitic debris flow sequence, Lake of the Woods greenstone terrane, central Canada: resedimented products of Archean vulcanian, plinian and dome-building eruption. *Precam. Res.*, **50**, 239–260.

Cas, R.A.F. (1992) Submarine volcanism: eruption styles, products and relevance to understanding the host-rock successions to volcanic-hosted massive sulphide deposits. *Econ. Geol.*, **87**, 511–541.

Cas, R.A.F. & Wright, J.V. (1991) Subaqueous pyroclastic flows and ignimbrites: an assessment. *Bull. Volcanol.*, **53**, 357–380.

Corcoran, P.L., Mueller, W.U. & Chown, E.H. (1998) Climatic and tectonic influences on fan deltas and wave- to tide-controlled shoreface deposits: evidence from the Archaean Keskarrah Formation, Slave Province, Canada. *Sediment. Geol.*, **120**, 125–152.

Corcoran, P.L., Mueller, W.U. & Padgham, W.A. (1999) Influence of tectonism and climate on lithofacies distribution and sandstone and conglomerate composition in the Archean Beaulieu Rapids Formation Northwest Territories, Canada. *Precam. Res.*, **94**, 175–204.

Cousens, B.L. (2000) Geochemistry of the Archean Kam Group, Yellowknife Greenstone Belt, Slave Province, Canada. *J. Geol.*, **108**, 181–197.

Davidson, A. (1967) *Metamorphism and intrusion in the Benjamin Lake map area, NWT*. PhD thesis, University of British Columbia, Vancouver.

Davis, W.J. & Bleeker, W. (1998) Timing of plutonism and regional deformation in the Yellowknife–Sleepy Dragon area, southern Slave Province. Twenty-fifth annual NWT

Geoscience Forum program with abstracts, pp. 34–36. Ind. N. Aff. Can., Yellowknife, NT.

Davis, W.J., Gariepy, C. & van Breemen, O. (1996) Pb isotopic composition of late Archaean granites and the extent of recycling of early Archaean crust in the Slave Province, northwest Canada. *Chem. Geol.*, **130**, 255–269.

Davis, W.J. & Hegner, E. (1992) Neodymium isotopic evidence for the accretionary development of the Late Archean Slave Province. *Contrib. Mineral. Petrol.*, **111**, 493–504.

De Rosen-Spence, A.F., Provost, G., Dimroth, E., Gochnauer, K. & Owen, V. (1980) Archean subaqueous felsic flows, Rouyn–Noranda, Quebec, Canada and their Quaternary equivalents. *Precam. Res.*, **12**, 43–77.

Dimroth, E., Cousineau, P., Leduc, M. & Sanschagrin, Y. (1978) Structure and organization of Archean subaqueous basalt flows, Rouyn–Noranda area, Quebec, Canada. *Can. J. Earth Sci.*, **15**, 902–918.

Dolozoi, M.B. & Ayres, L.D. (1991) Early Proterozoic, basaltic andesite tuff–breccia: downslope, subaqueous mass transport of phreatomagmatically-generated tephra. *Bull. Volcanol.*, **53**, 477–495.

Doucet, P., Mueller, W. & Chartrand, F. (1994) Archean, deep-marine, volcanic eruptive products associated with the Conigas massive sulphide deposit, Quebec, Canada. *Can. J. Earth Sci.*, **31**, 1569–1584.

Eriksson, K.A., Krapez, B. & Fralick, P.W. (1994) Sedimentology of Archean greenstone belts; signatures of tectonic evolution. *Earth Sci. Rev.*, **37**(1/2), 1–88.

Erikkson, P.A., Condie, K.C., Tirsgaard, H. *et al.* (1998) Precambrian clastic sedimentation systems. *Sediment. Geol.*, **120**, 5–53.

Ewart, A. & Hawkesworth, C.J. (1987) The Pleistocene–Recent Tonga–Kermadec arc lavas: interpretation of new isotopic and rare earth data in terms of a depleted mantle source. *J. Petrol.*, **28**, 495–530.

Fisher, R.V. & Schminke, H.-U. (1984) *Pyroclastic Rocks*. Springer-Verlag, New York.

Fortier, Y.O. (1949) Preliminary map 49-10a, Indin Lake, E1/2, District of Mackenzie, scale 1 : 125 000. Geological Survey of Canada, Ottawa.

Fouquet, Y., Eissen, J.-P., Ondreas, H., Barriga, F., Batiza, R. & Danyushevsky, L. (1998) Extensive volcaniclastic deposits at the Mid-Atlantic Ridge axis: results of deep-water basaltic explosive volcanic activity? *Terra Nova*, **10**, 280–286.

Frith, R., Frith, R.A. & Doig, R. (1977) The geochronology of the granitic rocks along the Bear–Slave Structural Province boundary, northwest Canadian Shield. *Can. J. Earth Sci.*, **14**, 1356–1373.

Frith, R.A. (1993) *Precambrian Geology of the Indin Lake Map Area, District of Mackenzie, Northwest Territories.* Mem. geol. Surv. Can., Ottawa, **424**.

Frith, R.A., Loveridge, W.D. & van Breemen, O. (1986) U–Pb ages on zircon from basement granitoids of the western Slave Structural Province, northern Canadian Shield. In: *Current Research, Part A, Paper 86-1A*, pp. 113–119. Geological Survey of Canada, Ottawa.

Fyson, W.K. & Helmstaedt, H. (1988) Structural patterns and tectonic evolution of supracrustal domains in the Archean Slave Province, Canada. *Can. J. Earth Sci.*, **25**, 301–315.

Gill, J., Torssander, H., Lapierre, R. *et al.* (1990) Explosive deep water basalt in the Sumisu Backarc Rift. *Science*, **248**, 1214–1217.

Griffiths, R.W. & Fink, J.H. (1992) Solidification and morphology of submarine lavas: a dependance on extrusion rate. *J. geophys. Res.*, **97**(19), 729–19, 737.

Hamilton, W.B. (1993) Evolution of Archean mantle and crust. In: *The Geology of North America* (Eds Reed, J.C., Ball, T.T., Farmer, G.L. & Hamilton, W.B.), pp. 597–636. Geological Society of America, Boulder.

Hanson, R.E. & Hargrove, U.S. (1999) Processes of magma/wet sediment interaction in a large-scale Jurassic andesitic peperite complex, northern Sierra Nevada, California. *Bull. Volcanol.*, **60**, 610–626.

Hanson, R.E. & Schweickert, R.A. (1982) Chilling and brecciation of a Devonian rhyolite sill intruded into wet sediments, Northern Sierra Nevada, California. *J. Geol.*, **90**, 717–724.

Hanson, R.E. & Wilson, T.J. (1993) Large-scale rhyolite peperites (Jurassic, southern Chile). *J. Volcanol. geothermal. Res.*, **54**(3/4), 247–264.

Hargreaves, R. & Ayres L.D. (1979) Morphology of Archean metabasalt flows, Utik Lake, Manitoba. *Can. J. Earth Sci.*, **16**, 1452–1466.

Helmstaedt, H. & Padgham, W.A. (1986) A new look at the stratigraphy of the Yellowknife Supergroup at Yellowknife, NWT—implications for the age of gold-bearing shear zones and Archean basin evolution. *Can. J. Earth Sci.*, **23**, 454–475.

Henderson, J.B. (1972) Sedimentology of Archean turbidites at Yellowknife, Northwest Territories. *Can. J. Earth Sci.*, **9**, 882–902.

Henderson, J.B. (1981) Archean basin evolution in the Slave Province, Canada. In: *Plate Tectonics in the Precambrian* (Ed. Kroner, A.), pp. 213–235. Elsevier, Amsterdam.

Henderson, J.B. (1998) Geology of the Keskarrah Bay area, District of Mackenzie, Northwest Territories. *Geol. Surv. Can. Bull.*, **527**.

Henderson, J.R., Kerswill, J.A., Henderson, M.N. *et al.* (1995) Geology, geochronology, and metallogeny of High Lake greenstone belt, Archean Slave Structural Province, Northwest Territories. In: *Current Research, Paper 95-1C*, pp. 97–106. Geological Survey of Canada, Ottawa.

Hildebrand, R.S., Hoffman, P.F. & Bowring, S.A. (1991) Magmatic evidence for roll-back and failure of the subducting slab during arc-continent collision in Wopmay Orogen, Northwest, Canada. *Eos*, **72**(17), 287.

Hoffman, P.F. (1987) Early Proterozoic foredeeps, foredeep magmatism, and superior-type iron-formation of the Canadian Shield. In: *Proterozoic Lithospheric Evolution* (Ed. Kroner, A.). Am. Geophys. Un. Geody. Ser., **17**, 85–98.

Hoffman, P.F. (1991) On accretion of granite–greenstone terranes. *NUNA Conference on Greenstone Gold and Crustal Evolution, Val D'Or, Quebec*, pp. 32–45.

Hrabi, R.B., Nelson, M.D. & Helmstaedt, H. (1995) Diverse metavolcanic sequences and late polymictic conglomerate-associated metasedimentary rocks in the Winter Lake supracrustal belt, Slave Province, Northwest Territories. In: *Current Research, Paper 95-1E*, pp. 137–148. Geological Survey of Canada, Ottawa.

Ingersoll, R.V. & Busby, C.J. (1995) Tectonics of sedimentary basins. In: *Tectonics of Sedimentary Basins* (Eds Busby, C.J. & Ingersoll, R.V.), pp. 1–52. Blackwell Science, Oxford.

Isachsen, C.E. & Bowring, S.A. (1994) Evolution of the Slave craton. *Geology*, **22**, 917–920.

Isachsen, C.E. & Bowring, S.A. (1997) The Bell Lake group and Anton Complex; a basement–cover sequence beneath the Archean Yellowknife greenstone belt revealed and implicated in greenstone belt formation. *Can. J. Earth Sci.*, **34**(2), 169–189.

Jackson, M.P.A., Eriksson, K.A. & Harris, C.W. (1987) Early Archean foredeep sedimentation related to crustal shortening: a reinterpretation of the Barberton sequence, southern Africa. *Tectonophysics*, **136**, 197–221.

Jackson, V.A. (1988) Geology of the Russell–Slemon Lakes area, NTS 85O/4. NWT Geology Division, DIAND, EGS 1988–11.

James, D.T. & Mortensen, J.K. (1992) An Archean metamorphic core complex in the southern Slave Province: basement–cover relations between the Sleepy Dragon complex and the Yellowknife Supergroup. *Can. J. Earth Sci.*, **29**, 2133–2145.

Kasting, J.F. (1993) Earth's early atmosphere. *Science*, **259**, 920–926.

Kimberley, K.M. (1989) Exhalative origins of iron formations. *Ore Geol. Rev.*, **5**, 13–145.

King, J.E., Pehrsson, S.J. & Helmstaedt, H. (1994) Application of terrane analysis to the Archean Slave Province. *Geol. Soc. Am. Annual Meeting, Program with Abstracts*, pp. A339.

King, J.E. & Helmstaedt, H. (1997) The Slave Province, North-West Territories, Canada. In: *Greenstone Belts* (Eds de Wit M.J. & Ashwal, L.D.), pp. 459–479. Oxford Science, Oxford.

Klein, G.D. (1985) The control of depositional depth, tectonic uplift, and volcanism on sedimentation processes in the back-arc basins of the western Pacific Ocean. *J. Geol.*, **93**, 8–25.

Krapez, B. (1993) Sequence stratigraphy of the Archaean supracrustal belts of the Pilbara Block, Western Australia. *Precam. Res.*, **60**, 1–45.

Kusky, T.M. (1989) Accretion of the Archean Slave Province. *Geology*, **17**, 63–67.

Leckie, D. (1988) Wave-formed coarse-grained ripples and their relationship to hummocky cross-stratification. *J. sediment. Petrol.*, **58**, 607–622.

Lowe, D.R. (1982) Sediment gravity flows II. Depositional models with special reference to the deposits of high-density turbidity currents. *J. sediment. Petrol.*, **52**, 279–297.

McBirney, A.R. (1963) Factors governing the nature of submarine volcanism. *Bull. Volcanol.*, **26**, 455–469.

McGlynn, J.C. & Ross, J.V. (1963) Arseno Lake map area, District of Mackenzie. *Paper 63-26*. Geological Survey of Canada, Ottawa.

Marsaglia, K.M. (1995) Interarc and backarc basins. In: *Tectonics of Sedimentary Basins* (Eds Busby, C.J. & Ingersoll, R.V.), pp. 299–330. Blackwell Science, Oxford.

Miall, A.D. (1995) Collision-related foreland basins. In: *Tectonics of Sedimentary Basins* (Eds Busby, C.J. & Ingersoll, R.V.), pp. 393–424. Blackwell Science, Oxford.

Miyashiro, A. (1974) Volcanic rock series in island arcs and active continental margins. *Am. J. Sci.*, **274**, 321–355.

Moore, J.G. (1979) Vesicularity and $CO_2$ in mid-ocean ridge basalt. *Nature*, **282**, 250–253.

Moore, J.G. & Schilling, J.G. (1973) Vesicles, water, and sulpher in Reykjanes Ridge basalt. *Contrib. Mineral. Petrol.*, **41**, 105–118.

Mueller, W. & White, J.D.L. (1992) Felsic fire-fountaining beneath Archean seas: pyroclastic deposits of the 2730 Ma Hunter Mine Group, Quebec, Canada. *J. Volcanol. geothermal Res.*, **54**, 117–134.

Mutti, E. & Ricci-Lucchi, F. (1978) Turbidites of the northern Apennines: introduction to facies analysis (English translation). *Int. geol. Rev.*, **20**, 125–166.

Padgham, W.A. (1985) Observations and speculations on supracrustal successions in the Slave Structural Province. In: *Evolution of Archean Supracrustal Sequences* (Eds Ayres, L.D., Thurston, P.C., Card, K.D. & Weber, W.), Spec. Publ. geol. Ass. Can., St John's, **28**, 133–151.

Pehrsson, S.J. (1998) *Deposition, deformation and preservation of the Indin Lake supracrustal belt, Slave Province, Northwest Territories*. PhD thesis, Queen's University, Kingston, Ontario.

Pehrsson, S.J. & Chacko, T. (1997) Contrasting styles of deformation and metamorphism between mid- and upper-crustal rocks of the western Slave Province, Northwest Territories. In: *Current Research, Paper 97-1C*, pp. 15–25. Geological Survey of Canada, Ottawa.

Pehrsson, S.J., Grant, J.A., Dorval, A.C. & Lewis, M. (1995) Stratigraphy and structure of the Indin Lake area, western Slave Province, Northwest Territories. In: *Current Research, Paper 95-1E*, pp. 149–159. Geological Survey of Canada, Ottawa.

Pehrsson, S.J. & Villeneuve, M.E. (1999) Deposition and imbrication of a 2680–2629 Ma supracrustal sequence in the Indin Lake area, southwestern Slave Province, Canada. *Can. J. Earth Sci.*, **36**, 1149–1168.

Piper, D.J.W. & Leg 155 Shipboard Party (1995) ODP Leg 155 on the Amazon deep-sea fan: Amazingly like the Meguma group. *Atl. Geol. Soc. Collo. Ann. Gen. Mtng Program and Abstracts*, 25 pp.

Roscoe, S., Stubley, M. & Roach, D. (1989) Archean quartz arenites and pyritic paleoplacers in the Beaulieu River supracrustal belt, Slave structural province, NWT. In: *Current Research, Paper 89-1C*, pp. 199–214. Geological Survey of Canada, Ottawa.

Ross, J.V. (1966) The structure and metamorphism of the Mesa Lake map area, District of Mackenzie (86B/14,west half). *Geol. Surv. Can. Bull.*, **124**.

Smith, G.A. & Landis, C.A. (1995) Intra-arc basins. In: *Tectonics of Sedimentary Basins* (Eds Busby, C.J. & Ingersoll, R.V.), pp. 263–298, Blackwell Science, Oxford.

Sinclair, H.D. (1997) Tectonostratigraphic model for underfilled peripheral foreland basins: an Alpine perspective. *Bull. Geol. Soc. Am.*, **109**, 324–346.

Soh, W., Pickering, K.T., Taira, A. & Tokuyama, H. (1991) Basin evolution in the arc–arc Izu Collision Zone, Mio–Pliocene Miura Group, central Japan. *J. geol. Soc. Lon.*, **148**, 317–330.

Sowerbutts, A.A. & Underhill, J.R. (1998) Sedimentary response to intra-oceanic extension: controls on Oligo–Miocene depositon, Sarcidano sub-basin, Sardinia. *J. geol. Soc. London*, **155**, 491–508.

Stanton, M.S., Tremblay, L.P. & Yardley, D.H. (1954) Geology Chalco Lake, District of Mackenzie, Northwest Territories, scale 1:63,360. Geological Survey of Canada, Ottawa.

Staudigel, H. & Schminke, H.-U. (1984) The Pliocene seamount series of LaPalma/Canary Islands. *J. geophys. Res.*, **89**(11), 195–11, 215.

Tamaki, K. (1988) Geological structure of the Sea of Japan and its tectonic implications. *Bull. geol. Surv. Japan*, **39**, 269–365.

THORPE, R.I. (1972) Interpretation of lead isotope data for base metal and gold deposits, Slave Province, Northwest Territories. In: *Report of Activities, Paper 72-1B*, pp. 72–77. Geological Survey of Canada, Ottawa.

TREMBLAY, L.P., WRIGHT, G.M. & MILLER, M.L. (1953) Geology Ranji Lake, District of Mackenzie, Northwest Territories 86B/3, scale 1:63,360. Geological Survey of Canada, Ottawa.

VAN BREEMEN, O., DAVIS, W.J. & KING, J.E. (1992) Temporal distribution of granitoid plutonic rocks in the Archean Slave Province, northwest Canadian Shield. *Can. J. Earth Sci.*, **29**, 2186–2199.

VILLENEUVE, M.E. (1994) Correlating within and between Slave Province greenstone belts using U–Pb dating. In: *Exploration Overview 1994*. NWT Geol. Div., Ind. N. Aff. Can., 60 pp.

VILLENEUVE, M.E. & HENDERSON, J.B. (1998) U–Pb geochronology of Wijineddi Lake area, Slave Province, District of Mackenzie, NWT. In: *Radiogenic Age and Isotopic Studies: Report 11, Paper 1998-F*. Geological Survey of Canada, Ottawa.

VILLENEUVE, M.E., RELF, C., HRABI, B. & JACKSON, V. (1994) Geochronology of supracrustal sequences in the Slave Province, NWT, Canada: implications for age of basement. In: *Abstracts of the Eighth International Conference on Geochronology, Cosmochronology and Isotope Geology*, p. 342.

WALKER, G.P.L. (1992) A morphometric study of pillow-size spectrum among pillow lavas. *Bull. Volcanol.*, **54**, 459–474.

WALKER, G.P.L. (1993) Basaltic-volcano systems. In: *Magmatic Processes and Plate Tectonics* (Eds Pritchard, H.M., Alabaster, T., Harris, N.B.W. & Neary, C.R.), Spec. Publs geol. Soc. London, No. 76, pp. 3–38. Geol. Soc. London, Bath.

WALKER, R.G. (1992) Turbidites and submarine fans. In: *Facies Models: Response to Sea Level Change* (Eds Walker, R.G. & James, N.P.), pp. 239–264. Geological Association of Canada, St John's.

WILSON, T.J. (1991) Transition from back-arc to foreland basin development in the southernmost Andes: stratigraphic record from the Ultima Esperanza District, Chile. *Bull. geol. Soc. Am.*, **103**, 98–111.

YAMAGISHI, H. & DIMROTH, E. (1985) A comparison of Miocene and Archean rhyolite hyaloclastites: evidence for a hot and fluid rhyolite lava. *J. Volcanol. geothermal Res.*, **23**, 337–355.

YAMASHITA, K., CREASER, R.A. & HEAMAN, L.M. (1998) Geochemical and isotopic constraints for tectonic evolution of the Slave Province. In: *Slave–Northern Cordillera Lithospheric Experiment (SNORCLE), Transect Meeting 1998*, pp. 18–19.

YAMASHITA, K., CREASER, R.A., STEMLER, J.U. & ZIMARO, T.W. (1999) Geochemical and Nd–Pb isotopic systematics of late Archean granitoids, southwestern Slave Province, Canada: constraints for granitoid origin and crustal isotopic structure. *Can. J. Earth Sci.*, **36**, 1131–1137.

*Spec. Publs int. Ass. Sediment.* (2002) **33**, 153–182

# Sedimentology of a tide- and wave-influenced high-energy Archaean coastline: the Jackson Lake Formation, Slave Province, Canada

W. U. MUELLER*, P. L. CORCORAN† *and* J. A. DONALDSON‡

**Sciences de la terre, Université du Québec à Chicoutimi, G7H 2B1, Canada;*
*†Department of Earth Sciences, Dalhousie University, Nova Scotia, BAH 3J5, Canada; and*
*‡Department of Earth Sciences, Carleton University, Ottawa, K1S 5B6, Canada*

## ABSTRACT

The 50–300 m thick 2.6 Ga Jackson Lake Formation in the Slave Province is a coarse clastic N-trending sedimentary sequence that formed during the terminal stage of Archaean cratonization. The western basin margin displays an unconformable contact with older volcanic rocks and the eastern margin is fault-bounded. The conglomerate–sandstone, sandstone–argillite and argillite–sandstone are the three principal lithofacies that collectively form a large-scale, fining-upward sequence. The conglomerate–sandstone lithofacies is interpreted as coalescing, coarse, clastic fan-deltas that prograded either into a shallow marine setting below fair weather wave base or on to a high-energy tidal shelf. The sandstone–argillite lithofacies is divided into: (i) a tidal channel sublithofacies which is composed of tabular to wedge-shaped cross-beds, interpreted as subtidal in-channel straight to sinuous-crested sandwaves with abundant argillite drapes forming sigmoidal tidal bundles; and (ii) a tidal shoal sublithofacies with prevalent medium-scale cross-beds grading into laminated argillite-dominated units, suggestive of sandwaves at the margin of a subtidal channel and its gradual abandonment, resulting in flat-topped suspension-dominated shoals. Palaeocurrent data generally display a bimodal–bipolar distribution with a prominent NNE-trend, consistent with a tide interpretation. The argillite–sandstone lithofacies, composed of finely laminated argillite and graded sandstone interbeds, is characteristic of a shallow-marine subtidal setting with local storm influence. Clast compositions and pebble to boulder clast-sizes in the conglomerate are consistent with local provenance and significant topographic relief, signifying tectonic influence on sedimentation, in contrast to the sandstone and argillite controlled by tides and waves. Tectonism associated with tide- and wave-influenced sedimentation may be the result of distinct Archaean Earth–Moon dynamics. The envisaged model for the Jackson Lake Formation is a high-relief, fault- and unconformity-bounded basin with a well developed, N-trending embayment with access to the open ocean. A modern depositional counterpart would combine the Bay of Fundy with its embayed macrotidal basin geometry, and Baja California with coarse-grained fan-deltas that demonstrate a significant sediment influx.

## INTRODUCTION

Detailed sedimentary facies analyses of >2.5 Ga Archaean supracrustal sequences have been problematic because of the fragmentary record of sedimentary units, regional deformation and high metamorphic grade. Notwithstanding this, Eriksson *et al.* (1988) demonstrated that sedimentary structures in metamorphosed rocks can be readily recognized and that a modern sedimentological evaluation may be conducted. In comparison with Phanerozoic sedimentary sequences, little is known about Archaean transport processes, dispersal systems, facies sequences and architecture. Emphasis has generally been placed on identifying sedimentary successions and their geodynamic settings (Eriksson, 1980; Eriksson *et al.*, 1994; Lowe, 1994), rather than depositional processes and facies architecture. Only recently has there been a shift to process-oriented sedimentary studies (Eriksson *et al.*, 1998) which take into account climatic (Corcoran *et al.*, 1998), volcanic (Mueller & Corcoran, 1998) and tectonic (Mueller *et al.*, 1994) controls.

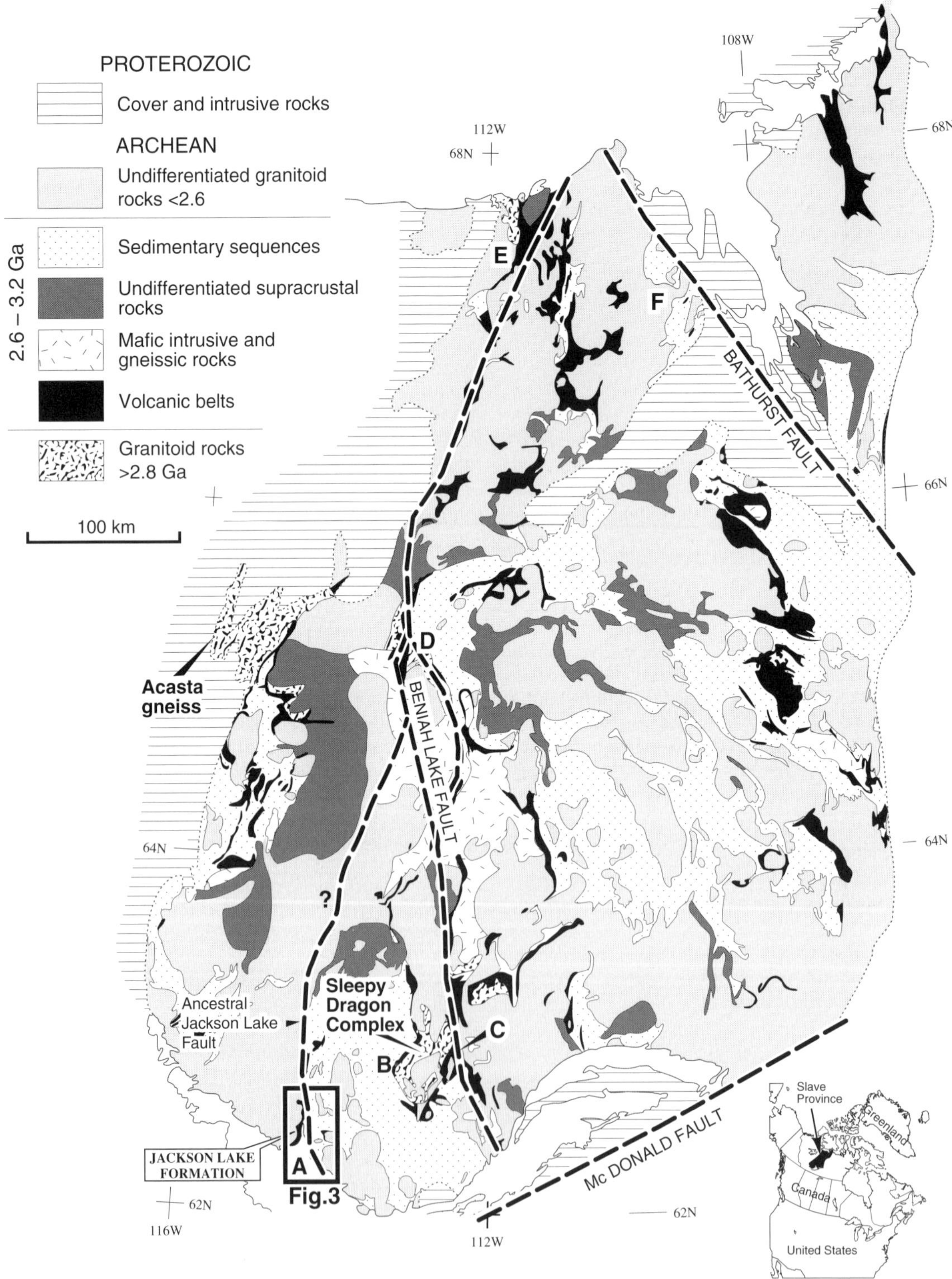


**Fig. 1.** General geology of the Archaean Slave Province with Yellowknife area outlined (Fig. 3). Locations in which a stratigraphy has been established (see Fig. 2): (A) Yellowknife volcanic belt; (B) Cameron River volcanic belt; (C) Beaulieu River volcanic belt; (D) Point Lake volcanic belt; (E) Anialik River volcanic belt; (F) Hood River volcanic belt (modified from Corcoran *et al.*, 1998).

In the absence of vegetation, an inferred oxygen-poor atmosphere (Des Marais, 1994) and different Earth–Moon dynamics (Williams, 1998), Eriksson *et al.* (1998) argued that Precambrian sedimentary depositional systems may have differed significantly from their modern analogues. The coastal regions at the subaerial to subaqueous interface would be particularly affected because of significant detrital input into the sea. Coarse, clastic braid- and fan-deltas are important loci where fluvial traction currents and subaerial mass flows interact with oscillatory wave movement and tide-induced currents. Collectively, the deposits and setting are a function of first-order tectonic controls and second-order eustatic sea-level changes. Recognition of such a critical ancient setting allows for direct comparison with well documented ancient and modern fan-delta and alluvial–fluvial settings (e.g. Koster & Steel, 1984; Colella & Prior, 1990; Marzo & Puigdefábregas, 1993) or shallow-water tidal environments (Nio & Yang, 1991; Johnson & Levell, 1995).

This study presents the physical sedimentology and depositional history of the Archaean Jackson Lake Formation, emphasizing tide and wave influence in a high-energy, tectonically controlled depositional environment. Tide influence appears to be a recurring theme in many late-Archaean, fault-bounded dispersal systems (Corcoran *et al.*, 1998) and synorogenic arc basins (Mueller and Dimroth, 1987), suggesting that tidal cycles in tectonically active areas are possibly linked to a more dynamic Archaean Earth–Moon system (Williams, 1998). The objectives of this paper are to: (i) discuss the prominent sedimentary lithofacies; (ii) evaluate the influence of tectonism on sedimentation; (iii) assess the possibility of tidal influence in a high-relief system; and (iv) propose a depositional model for the Jackson Lake Formation.

## ARCHAEAN SLAVE CRATON

The Archaean Slave Province is a complex amalgamation of N-trending sedimentary sequences and small but widely dispersed mafic to felsic volcanic belts, referred to as supracrustal rocks or as greenstone belts. Plutonic rocks either intrude supracrustal sequences or represent older crust (Fig. 1). The 210 000 km$^2$ Archaean Slave Province, which includes the 4.00–4.03 Ga Acasta gneisses (Bowring & Williams, 1998) to *c.*2.52–2.64 Ga late- to syn-tectonic, Archaean plutonic rocks (Isachsen & Bowring, 1997), spans *c.*1.5 Gyr. Sedimentary assemblages are important pan-craton markers because they facilitate identification of specific basin-forming events, such as arc formation, unroofing, collision or fragmentation (Mueller & Donaldson, 1992; Mueller *et al.*, 1996), which in turn help to delineate large-scale tectonic processes (Ingersoll & Busby, 1995). Henderson (1981) proposed a rifted continental tectonic model for the Yellowknife belt based on a detailed sedimentary and petrographic study of the Burwash Formation turbidites (Henderson, 1975) and previous mapping (Henderson & Brown, 1966). Numerous models have since been proposed, but all represent a variation on a plate tectonic theme (Kusky, 1989; MacLachlan & Helmstaedt, 1995; Kusky & Polat, 1999). Plate tectonics, occurring as early as 3.8 Ga (Komiya *et al.*, 1999), require horizontal lithospheric plate movement to produce late-orogenic strike-slip basins (Corcoran *et al.*, 1999).

The Yellowknife volcanic belt, unconformably overlying a gneissic tonalitic basement (Henderson, 1981; Helmstaedt & Padgham, 1986; Isachsen & Bowring, 1994, 1997), has been extrapolated on a pan-Slave scale for comparative purposes (Henderson, 1998). The Slave craton is different from most Archaean cratons, displaying (i) more sedimentary than volcanic rocks, (ii) a higher felsic/mafic ratio in the volcanic portions of greenstone belts, (iii) an exposed sialic basement, and (iv) more evolved potassium-rich granites (Henderson, 1981; Padgham & Fyson, 1992; Fig. 1). In addition, numerous unconformities between basement and volcanic or sedimentary successions (Corcoran *et al.*, 1998; Henderson, 1998; Bleeker *et al.*, 1999), and volcanic and sedimentary successions (Mueller *et al.*, 1998) have been recorded, and in conjunction with radiometric age determinations, helped to define pan-Slave stratigraphic tie lines (Fig. 2; Mueller *et al.*, 1998; Corcoran *et al.*, 1999). The Yellowknife stratigraphic succession, established by Henderson & Brown (1966), modified by Helmstaedt & Padgham (1986) and subsequently extrapolated to include other volcano-sedimentary assemblages (MacLachlan & Helmstaedt 1995), contains mafic to felsic volcanic cycles interstratified with extensive turbiditic sequences. These are unconformably overlain by late-stage clastic sedimentary rocks.

### Yellowknife stratigraphy

The basal Kam Group of the Yellowknife volcanic belt (section A of Fig. 2; Helmstaedt & Padgham, 1986) is a thick mafic volcanic sequence that formed between 2722 and 2701 Ma (Isachsen & Bowring,

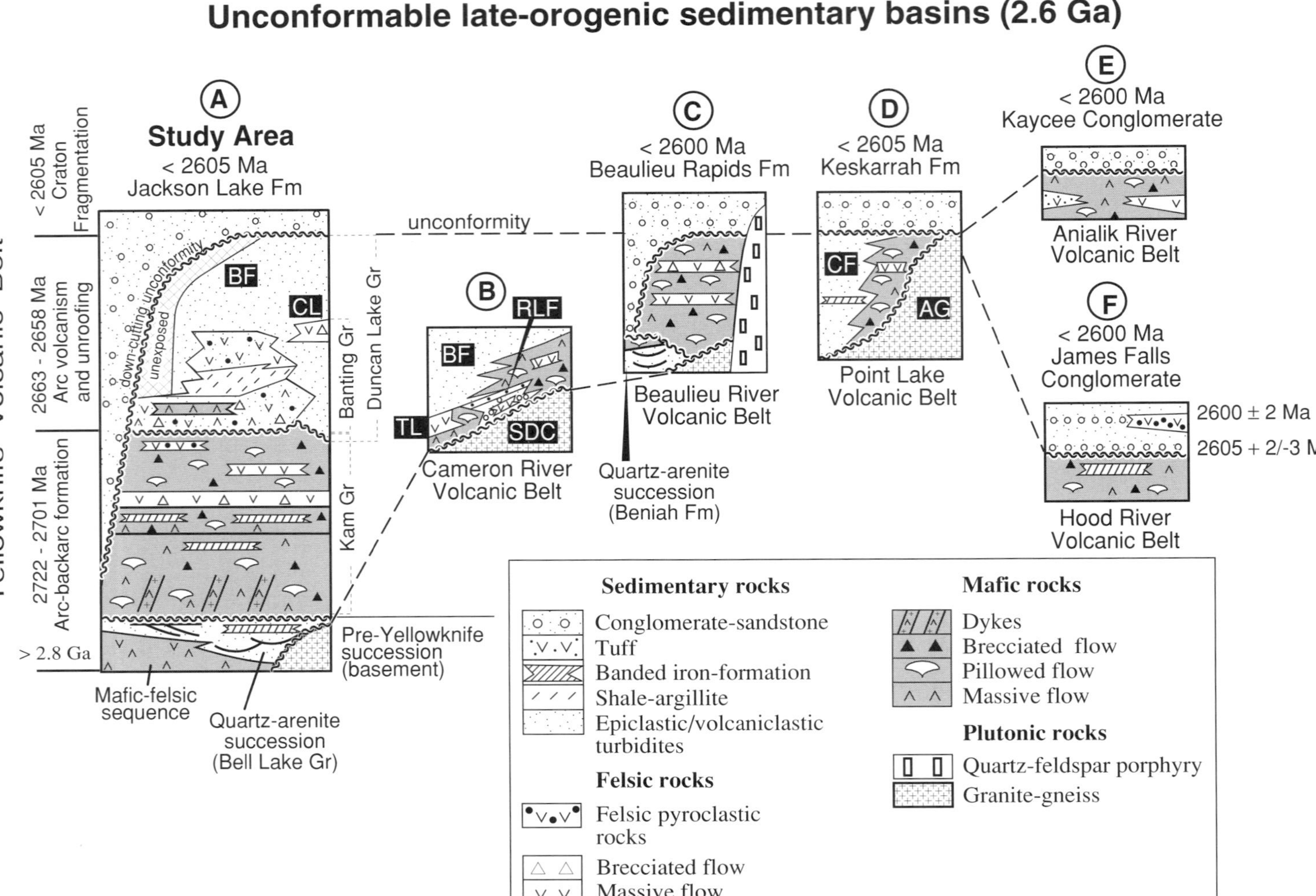


**Fig. 2.** General stratigraphy of the Yellowknife volcanic belt and proposed pan-Slave correlation of 2.6 Ga late-stage, orogenic basins based on precise U–Pb age determinations. Modified from Corcoran *et al.* (1998). AG, Augustus Granite; BF, Burwash Formation; CL, Clan Lake felsic volcanic complex; CF, Contwoyto Formation; RLF, Raquette Lake Formation; SDC, Sleepy Dragon Complex; TL, Turnback Lake.

1994, 1997). Similar to a modern ocean- or sea-floor with sporadic subaqueous felsic centres, it locally contains volcanic edifices that breached the ocean surface, as suggested by felsic tuff and tuffaceous beds with wave-formed cross-beds and significant up-section increase in brecciated flows and volcaniclastic deposits (Falck, 1990; Mueller, unpublished data). Development of the Kam Group over 20 Myr compares favourably to composite shield volcano construction locally capped by small felsic centres. Cousens (2000) advocated a rifted continental margin setting based on the tholeiitic geochemical signature of mafic–felsic volcanic rocks and isotope geochemistry ($\varepsilon_{Nd}$ values).

In contrast, the 2663–2658 Ma arc-related Banting Group (Isachsen & Bowring, 1994), overlying the Kam Group unconformably (Helmstaedt & Padgham, 1986) with a pronounced 35–40 Myr hiatus (Isachsen & Bowring, 1994), is composed of abundant turbiditic volcaniclastic and pyroclastic rocks as well as andesitic to felsic flows (Padgham, 1987a, b). The Banting rocks are interstratified with the turbiditic deposits of the Burwash Formation (Duncan Lake Group; Fig. 2A) and age dates from felsic volcanic centres in the Burwash Formation at Clan (2661 ± 2 Ma; Mortensen *et al.*, 1992) and Turnback (2663 ± 7 Ma; Henderson *et al.*, 1987; Fig. 2) Lakes, as well as felsic tuffs in the Burwash turbidites yielding ages of 2661 ± 2 Ma (Bleeker & Villeneuve, 1995), support contemporaneity. In the east, the Beaulieu River volcanic belt and a magmatic arc sequence (Mueller & Corcoran, 1997), represented by the *c.*2690–2670 Ma volcano-sedimentary Raquette Lake Formation (Mueller *et al.*, 1998) of the Cameron River volcanic belt (Lambert 1988), developed unconformably on top of the N-trending Sleepy Dragon basement complex (Fig. 1 & Fig. 2B; Henderson, 1985; Corcoran *et al.*, 1999). This eastern arc sequence forming between 2690 and 2660 Ma constitutes the missing link between Kam and Banting Group volcanism (Figs 1 & 2).

### *Jackson Lake Formation*

The N-trending, 50–300 m thick Jackson Lake Formation, traceable intermittently for 35 km along strike, is the youngest Archaean depositional unit (Figs 2 & 3; Henderson, 1975), as indicated by a U–Pb age of 2.6 Ga from a granitic clast (Isachsen & Bowring, 1994, 1997). The sedimentary rocks represent a remnant basin that unconformably overlies the Kam Group (Figs 4 & 5A & B), and locally displays an inferred regolith (Fig. 4). The formation is delineated on the east by the ancestral Jackson Lake fault (Helmstaedt & Bailey, 1987), which was subsequently reactivated during the Proterozoic (Hay–Duck fault). The Jackson Lake sequence contains N–NNE striking (0–30°), steeply dipping (70–90°) strata with locally overturned beds that young to the east. Late folding and faulting disrupted the N-trending sedimentary rocks. Greenschist facies metamorphism is prominent, but the prefix 'meta' is omitted to simplify rock description. In addition to the pioneer work of Henderson (1975), geological maps of Henderson & Brown (1966) and Helmstaedt *et al.* (1980) served as a basis. Four localities—(A) Yellowknife Bay, (B) Shot Lake, ($C_1$, $C_2$) Walsh Lake and (D) Jackson Lake (Fig. 3)—were mapped at scales of 1 : 20, 1 : 100 and 1 : 400.

## SEDIMENTOLOGY

Characteristic lithofacies, identified on the basis of lithology and sedimentary structures, include: (i) conglomerate–sandstone; (ii) sandstone–argillite; and (iii) argillite–sandstone (Table 1). Argillite-dominated beds, <150 cm thick, and medium- to very coarse-grained sandstone characterize the sandstone–argillite lithofacies, whereas thick argillite-dominated beds (>150 cm) and very fine- to fine-grained sandstone comprise the argillite–sandstone lithofacies. The petrography, mineral composition and geochemistry of the sedimentary rocks are provided in Corcoran & Mueller (this volume, pp. 183–212).

### Conglomerate–sandstone lithofacies

The 2–70 m-thick conglomerate–sandstone lithofacies (Table 1), best exposed at Yellowknife Bay (Figs 3 & 6A), is prominent along the unconformable contact with Kam Group rocks and is exposed higher up-section above erosional contacts with the sandstone–argillite lithofacies (Fig. 6B). Amalgamated, clast-supported conglomerate (Fig. 5C) up to 30 m thick, matrix-supported conglomerate in a quartz-rich, coarse-grained sandstone matrix (Figs 5D & 6A, section I) and angular clast-dominated conglomerate (Fig. 5E) define this lithofacies. Both matrix- and angular clast-supported conglomerates are predominant at the unconformity. Conglomerates are massive to stratified and well stratified at the interface with the argillite–sandstone lithofacies (Fig. 5F). Locally, clasts fill erosional channels or flat scours. Individual beds in amalgamated sequences are identified by a change in clast composition and size, or are delineated by lenticular sandstone interbeds. Clasts are subrounded to angular and range

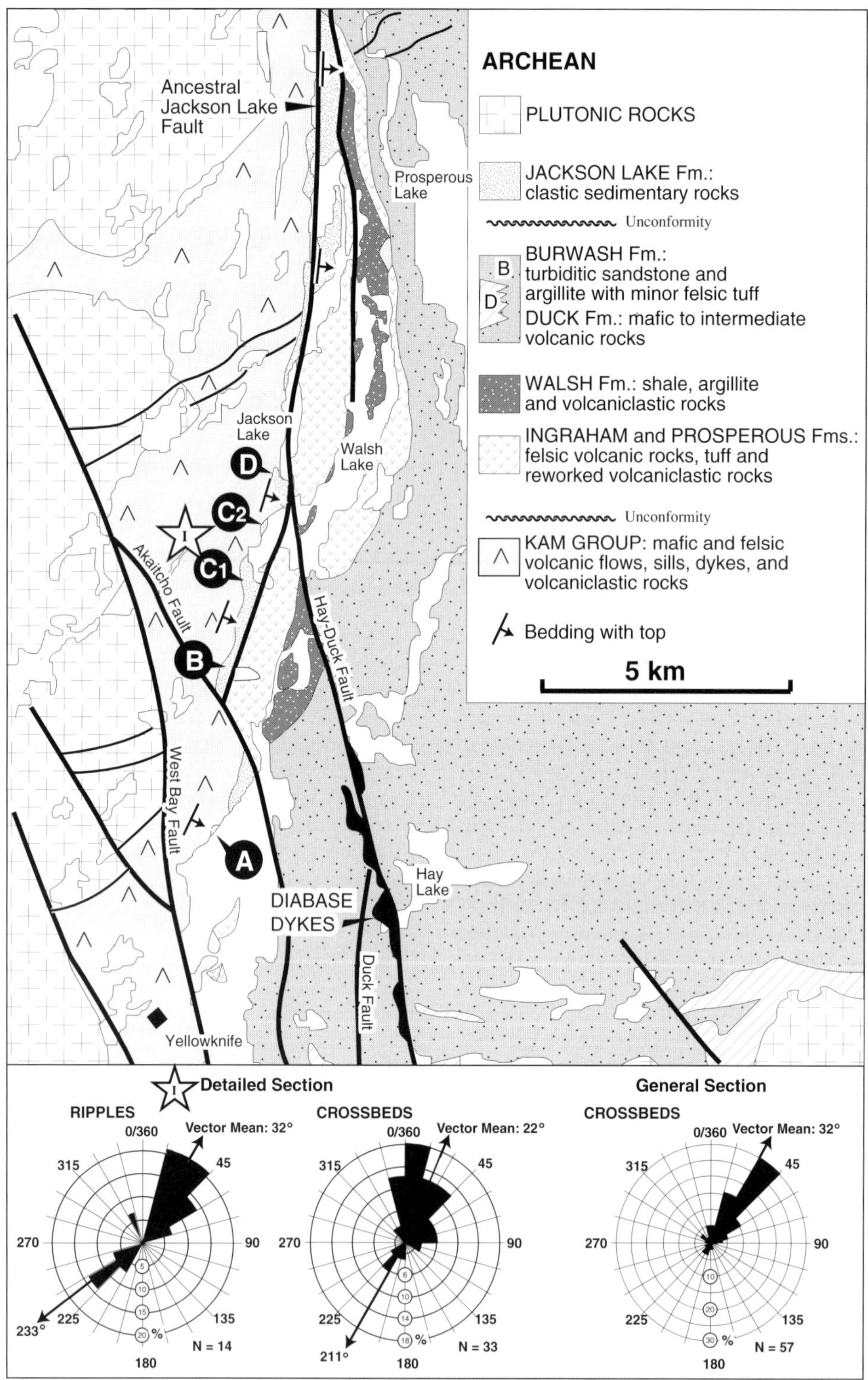


**Fig. 3.** Geology of the Yellowknife area and distribution of stratigraphic sections in the Jackson Lake Formation: (A) Yellowknife Bay; (B) Shot Lake; (C) Walsh Lake ($C_1$, south; $C_2$, north); and (D) Jackson Lake. The western contact of the Jackson Lake Formation is characterized by an unconformity, whereas the eastern contact is fault-bounded. Palaeocurrent data are from section $C_1$ from stratigraphic column I (Fig. 7A) with a detailed study (Fig. 10C) and a general palaeocurrent analysis of remaining column I of Fig. 7A. Map modified from Henderson & Brown (1966) and Helmstaedt *et al.* (1980).

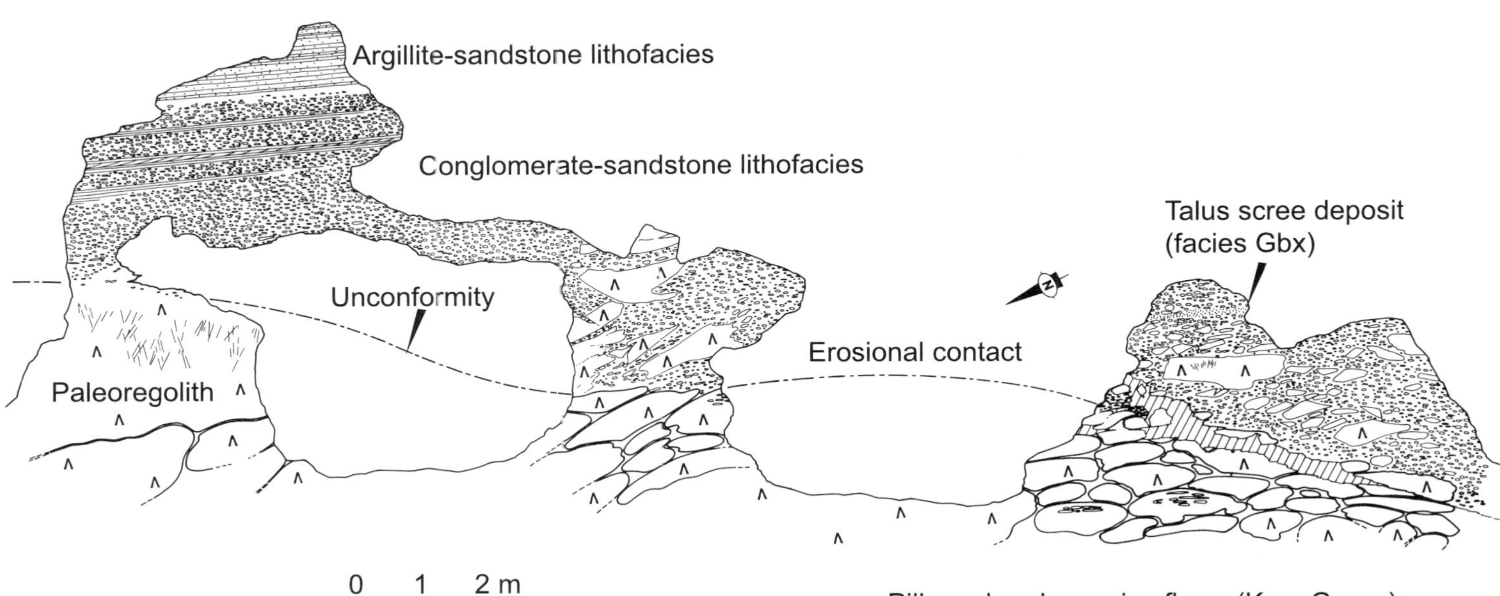


**Fig. 4.** Detailed sketch of unconformity with an inferred palaeoregolith between pillowed flows of the Kam Group and Jackson Lake Formation. Note the angular–clast conglomerate of the conglomerate–sandstone lithofacies, interpreted as a talus scree deposit (facies Gbx) and composed exclusively of mafic clasts (Walsh Lake, section $C_1$, Fig. 3; section I of Fig. 6A). Argillite–sandstone lithofacies overlies the angular–clast conglomerate.

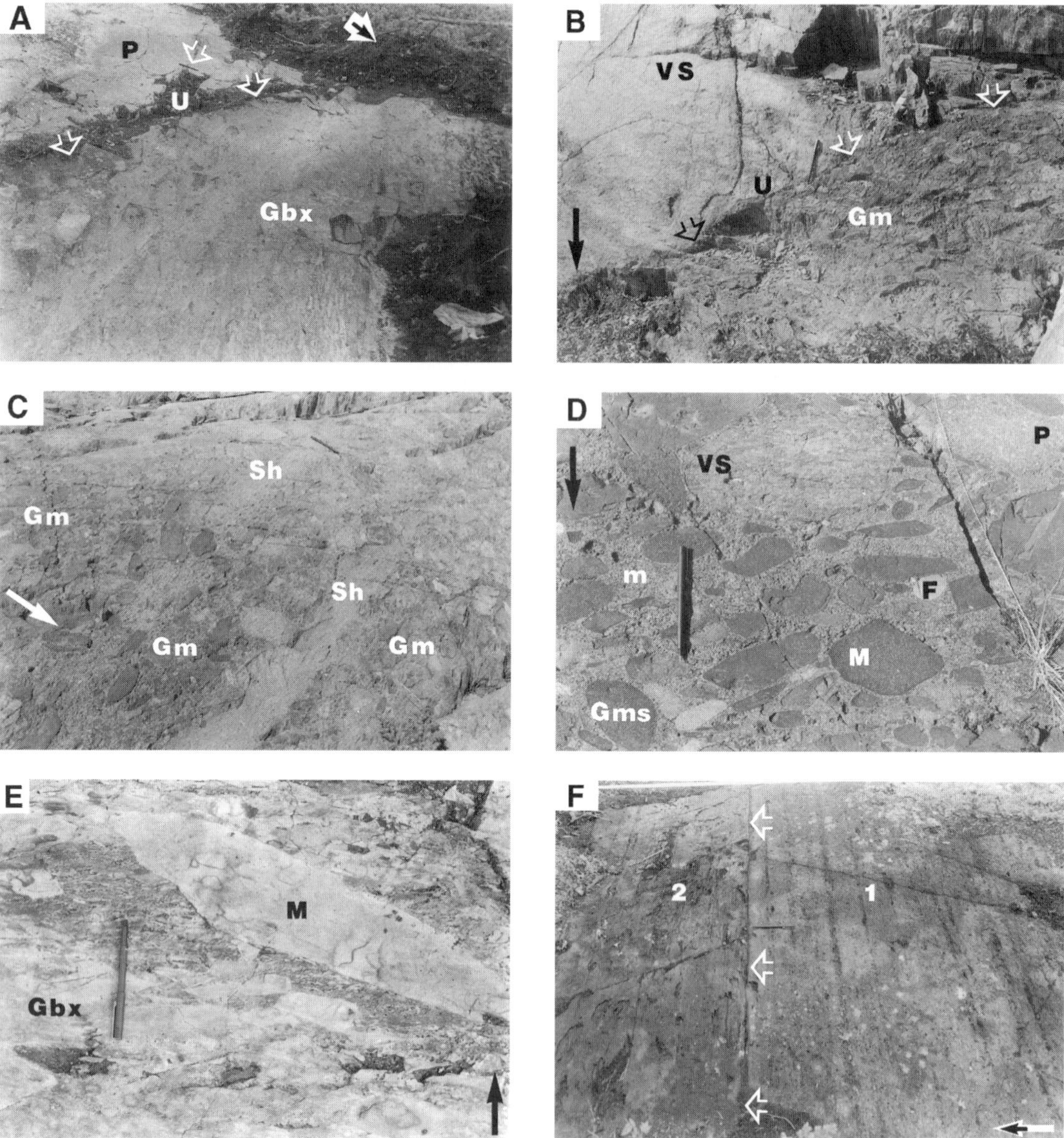


**Fig. 5.** Characteristics of the conglomerate–sandstone lithofacies. Large black or white arrow indicates tops in all photographs. Pen length is 15 cm. (A) Unconformity (U and arrows) between pillowed flows (P) of the Kam Group and angular clast conglomerate (sedimentary breccia; facies Gbx) of the Jackson Lake Formation (see Fig. 4). (B) Unconformity (U and arrows) between felsic volcaniclastic sedimentary rocks (VS) of the upper Kam Group and Jackson Lake clast-supported conglomerates (facies Gm; section D of Fig. 3 and section II of Fig. 6A). (C) Interstratified sandstone beds (facies Sh) between amalgamated, clast-supported conglomerates (facies Gm; column III of Fig. 6A). (D) Matrix-supported conglomerate (facies Gms) with abundant mafic (M), local felsic volcaniclastic sedimentary (VS), felsic volcanic (F) and plutonic (P) clasts in a quartz-rich, coarse-grained sandstone matrix (m). (E) Angular clast conglomerate (facies Gbx) composed of mafic constituents (M), just above unconformity (see Fig. 4). (F) Well stratified subaqueous pebble conglomerate of conglomerate–sandstone lithofacies (1) with a sharp contact with the argillite–sandstone lithofacies (2) indicated by arrows and pen (column IIc of Fig. 6).

**Table 1.** Characteristics of the Jackson Lake sedimentary deposits.

| Lithofacies, thickness | Bedform, characteristics | Process, interpretation | Setting |
|---|---|---|---|
| *Conglomerate–sandstone*<br>2–70 m thick (15–25% of formation)<br>Conglomerate: pebble- to boulder-sized<br>Sandstone: granule to coarse-grained | Angular-clast conglomerate or sedimentary breccia.<br>Amalgamated matrix- and clast-supported conglomerate; clasts are subangular to subrounded.<br>Conglomerate is massive, poorly to well stratified with erosive bases. Beds are channel-shaped or show flat scours.<br>Lenticular sandstone, trough cross-bedded and planar to low-angle bedded; massive sandstone. | Rock fall: talus scree (Gbx). Matrix- and clast-supported conglomerate: mass flow deposits (Gms), traction current structures related to floods or high fluvial run-off gravel bars (Gm), channel-shaped scours (Gt). Well stratified beds are reworked by waves and tides.<br>Sandstone: traction current structures; in-channel dunes (St) and bar top sands (Sh, Sl); hyperconcentrated flood flow (Sm).<br>Complete facies sequence: Gbx–Gms–Gm–Gt–Sm–Sh–Sl–St. | Fan-delta or braid-delta with prominent stream-dominated sediments and local mass flow deposits.<br>*Location*: rugged, high-energy coastline with tide and wave action affecting subaqueous portion of fans. Coarse clastic margin of tidal embayment. Baja California and Bay of Fundy are analogues. |
| *Sandstone–argillite*<br>10–152 m thick (60–70% of formation) with two sublithofacies:<br>**1** *Tidal channel sublithofacies*<br>**2** *Tidal shoal sublithofacies*<br>Conglomerate: pebble- to cobble-sized<br>Sandstone: medium- to very coarse-grained<br>Argillite: very fine-grained sandstone, siltstone, mudstone | **1** *Tidal channel sublithofacies.*<br>Tabular crossbeds, 20–120 cm thick, with argillite drapes on tangential to sigmoidal-shaped foresets forming 5–20 m thick cross-bed cosets; abundant reactivation surfaces; local pebble trains with rip-ups on top of bedforms or incised as pebble–cobble-filled channels. Argillite 5–20 cm thick between larger bedforms with graded and laminated siltstone-mudstone beds and rippled horizons.<br>**2** *Tidal shoal sublithofacies*<br>Composite strata 50–200 cm thick with planar to low-angle planar beds and cross-beds forming packages 5–10 m thick.<br>Argillite drapes on foresets, on reactivation surfaces and between cross-beds. Capped by 20–150 cm thick, argillite units with ripples, millimetre mudstone–siltstone couplets, and planar beds. Form upward-fining sequences. | **1** Straight- to sinuous-crested sandwaves forming in subtidal high-energy channels.<br>Highly fluctuating tidal action causes deposition of sandstone during dominant current action followed by deposition of slack-water argillite-drapes. Reactivation surfaces reflect high-energy stages. Rip currents during storms transported conglomerate debris into subtidal channels.<br>**2** Complex strata with cross-beds and abundant slack-water argillite laminae grading into low-energy argillite beds.<br>Deposits reflect change from marginal in-channel to subtidal low-energy flat-topped shoal deposits. Argillite units indicate suspension deposition with minor current action; millimetre mudstone–siltstone couplets similar to vertical tidal bundles. | Shallow-water subtidal zone composed of tidal channels and tidal shoals in contact with coarse clastic fan-deltas.<br>*Location*: part of tidal embayment surrounded by fault-bounded topographic highs.<br>**1** High-energy tidal channels in a macrotidal estuarine complex. Local floods and storms transport conglomerate debris.<br>**2** Low-energy, marginal subtidal channel grading into argillite-dominated shoals. Abandonment of channel and formation of subtidal shoals. Results: upward-fining sequence. |
| *Argillite–sandstone*<br>5–30 m thick (5–10% of formation)<br>Argillite with minor coarse-grained sandstone beds | Graded and laminated couplets of fine-grained sandstone and mudstone–siltstone; horizons of thin-rippled intervals.<br>Coarse-grained graded sandstone. | Suspension deposits with weak tide/wave influence produced current ripples and graded laminated beds.<br>Storm influence with possible rip currents. | Shallow-water subtidal zone in direct contact with subaqueous portion of fan-deltas. Close to fault-bounded margin.<br>*Location*: part of sandy tidal embayment. |

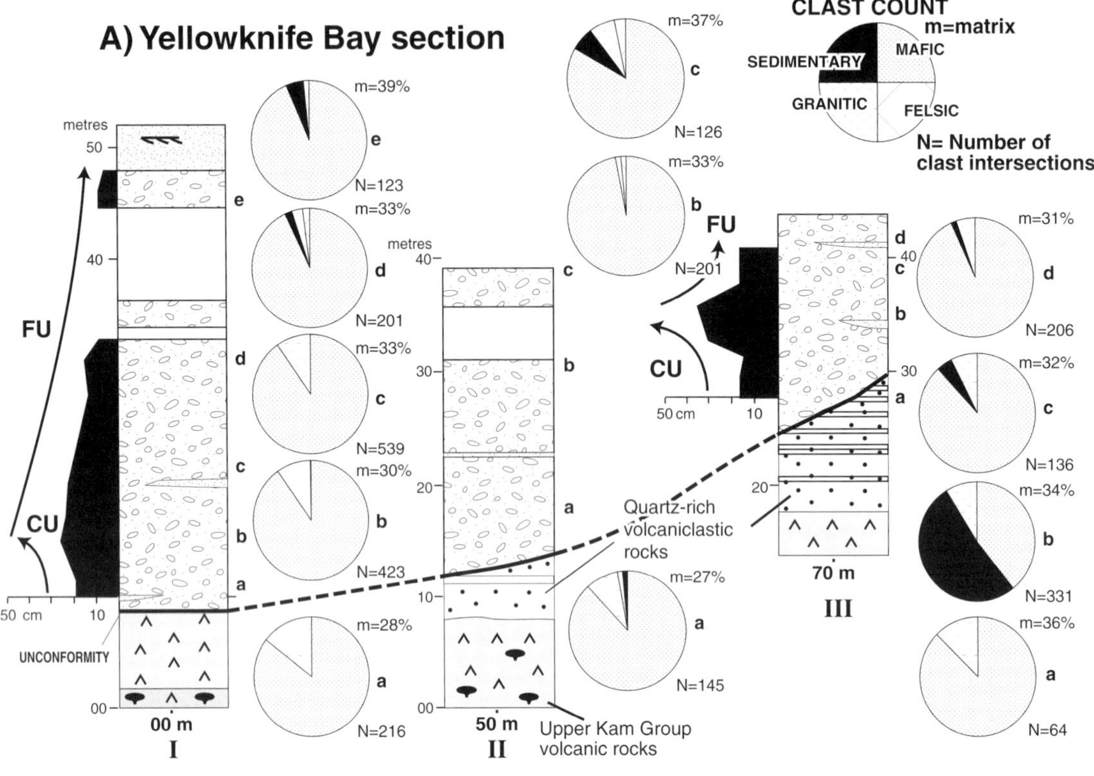


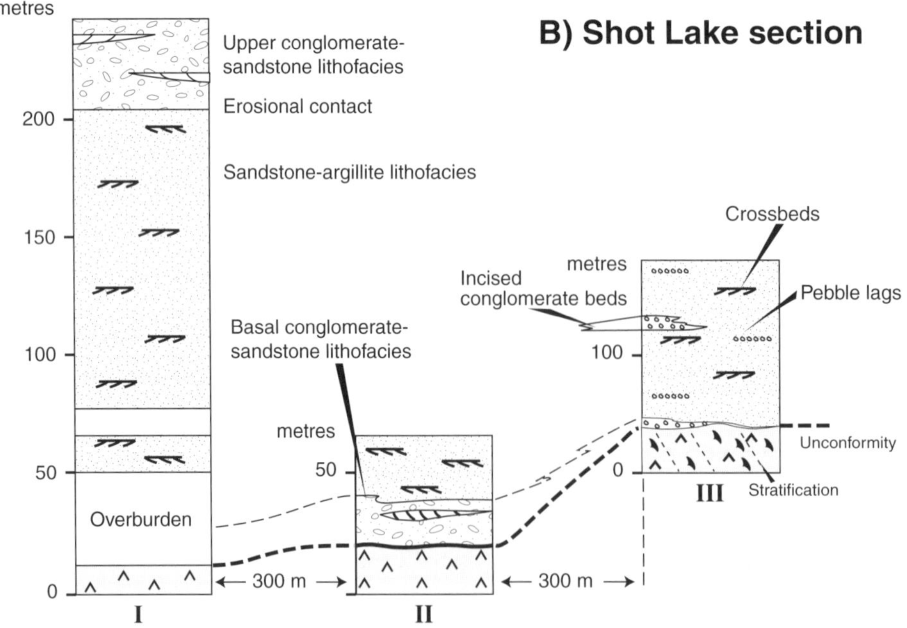


**Fig. 6.** Stratigraphic sections in the Yellowknife Bay and Shot Lake areas (sections A and B, Fig. 3). (A) Columns I to III show conglomerates cutting down into the felsic volcaniclastic deposits of the Kam Group (Yellowknife Bay Formation). A local prevalence of sedimentary and mafic clasts derived from the quartz-rich volcaniclastic rocks and mafic volcanic rocks is observed. Upward-coarsening (CU) and fining (FU) sequences are common in conglomerate beds. (B) Stratigraphy at Shot Lake with a basal and an upper conglomerate–sandstone lithofacies. Note angular unconformity between pillowed Kam Group flows and clastic sedimentary rocks.

from pebble to boulder in size, with pebble-dominated conglomerate displaying a pronounced stratification, in contrast to beds with cobble- to boulder-sized clasts. Laminated sandstone and argillite are interstratified with the pebble conglomerate close to the argillite–siltstone lithofacies. Boulder-size clasts constitute <10% of the clast population. Clast sizes in composite conglomerate beds display coarsening- and fining-upward sequences.

The 5–50 cm-thick, coarse- to very coarse-grained sandstone interstratified with conglomerate comprises less than 10% of the lithofacies (Fig. 5C). Laterally discontinuous planar beds and trough cross-beds are intercalated with rare massive, very coarse-grained to granular sandstone beds. Planar to low-angle planar beds generally cap conglomerate beds and are commonly incised by cross-beds. Cosets of truncating trough cross-beds are locally prominent. The upper contact between sandstone and conglomerate is erosional, whereas the lower contact from conglomerate to sandstone is gradational but abrupt.

Upper contacts between the conglomerate-sandstone and other lithofacies are generally sharp but non-erosional (Fig. 5F), or are gradational, but lower contacts are erosional where the conglomerate–sandstone overlies the sandstone–argillite or argillite–sandstone lithofacies. Stacking of sedimentary bedforms is variable, but generally, conglomerate changes up-section or laterally into sandstone. A general succession of beds is observed and includes matrix-supported, angular clast-dominated or clast-supported conglomerate at the base, with incised channel conglomerate or stratified conglomerate forming higher up-section. Interstratified planar bedded, trough cross-bedded and massive sandstone beds are best developed in the clast-supported conglomerate and increase up-section.

### Interpretation

The conglomerate–sandstone lithofacies displays the salient characteristics of traction current and mass flow processes that operate on alluvial fans (Ridgway & DeCelles, 1993; Blair & McPherson, 1994; Chamyal *et al.*, 1997), or on subaerial portions of marine fan-deltas (Wescott & Ethridge, 1980; Hwang *et al.*, 1995) and lacustrine fan-deltas (Laird, 1995; Horton & Schmitt, 1996), as well as proximal braided streams (Rust, 1984). The sedimentary facies code for fluvial–alluvial deposits as defined by Miall (1978, 1992), with modifications for Archaean sedimentary rocks (Mueller *et al.*, 1994, 1998), is employed (Table 1).

The matrix-supported and angular clast-dominated conglomerates are interpreted as cohesive debris flows (facies Gms; Laird, 1995) and talus scree (facies Gbx; Mueller & Corcoran, 1998) or rock fall deposits (Blair & McPherson, 1994), respectively. Both are common to proximal fan reaches (Mack & Rasmussen, 1984; Blair & McPherson, 1994). The massive to poorly stratified, clast-supported conglomerates (facies Gm) represent longitudinal gravel bars (Galloway & Hobday, 1983), whereas shallow conglomerate-filled scours (facies Gt) developed as high-energy, channelized flood deposits (Rust & Koster, 1984). The well stratified pebble conglomerate, locally accentuated by planar sandstone interbeds, can be explained as gravel sheets formed by high-energy, wave-induced processes, such as those operating at the shoreface in the breaker zone (Hart & Plint, 1995). The contact with the subaqueous argillite–sandstone lithofacies (Fig. 5E) and laminated sandstone and argillite interbeds supports this notion.

Interstratified lenticular sandstone associated with the boulder- to cobble-size amalgamated conglomerate represents the product of bedload processes. The planar-bedded to low-angle planar sandstone (facies Sh and Sl) is consistent with bar top sand deposits that developed during incipient waning flood stages (Eriksson, 1978; Best & Bridge, 1992). Trough cross-beds (facies St) are diagnostic of in-channel lunate dune migration (Eriksson, 1978). The massive sandstone (Sm) is considered a product of hyperconcentrated flood flows (Simpson & Eriksson, 1989; Horton & Schmitt, 1996). A hierarchy of sedimentary structures concomitant with an up-section grain size decrease is recognized in the conglomerate–sandstone lithofacies: a complete facies sequence of Gbx–Gms–Gm–Gt–Sm–Sh–Sl–St is consistent with subaerial fans.

### Sandstone–argillite lithofacies

The 10–152 m thick sandstone–argillite lithofacies (Table 1; Fig. 7A & B), composed of medium- to very coarse-grained quartz-rich sandstone with local pebble–cobble conglomerate interbeds (<2%), constitutes 90–95%, whereas very fine-grained sandstone, siltstone and mudstone, collectively referred to as argillite, represents 5–10% of the lithofacies. Contacts with the underlying argillite–siltstone and overlying conglomerate–sandstone are sharp and erosional, respectively.

Coarse-grained sandstone and ubiquitous argillite drapes, in addition to interstratified 5–150 cm thick argillite-dominated units, characterize the lithofacies.

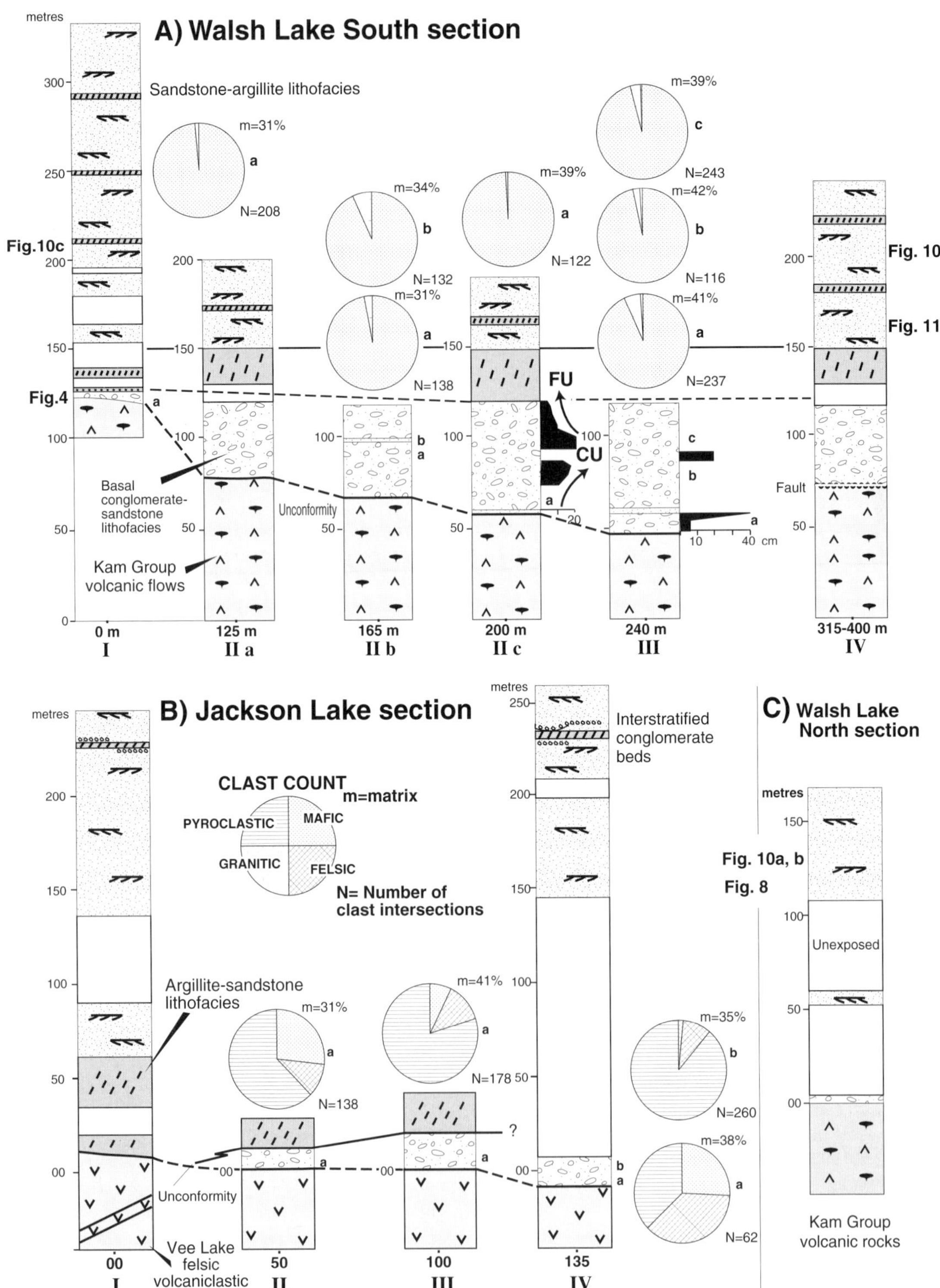


**Fig. 7.** Stratigraphic sections at Walsh (south A and north C) and Jackson Lakes (B) represent sections $C_1$, $C_2$ and D of Fig. 3, respectively. Argillite–sandstone lithofacies, prominent adjacent to the unconformity, is used as a marker horizon and accentuates the irregular unconformable contact between the conglomerate–sandstone lithofacies and Kam Group volcanic rocks.

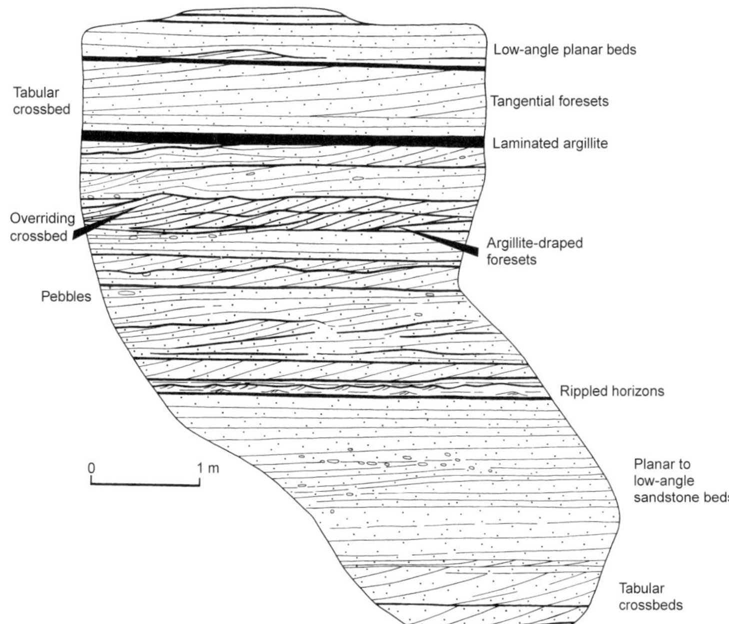


**Fig. 8.** Outcrop characteristics of the tidal channel sublithofacies of the sandstone–argillite lithofacies. Argillite beds and drapes delineate individual cross-beds, occur on foresets or accentuate reactivation surfaces. Note the presence of pebble trains or isolated pebbles on bedding planes.

The medium- to large-scale sedimentary structures are divided into (i) 20–120 cm thick, predominantly tabular-shaped cross-beds with tangential foresets (Figs 8 & 9A) and minor argillite units, and (ii) 50–200 cm thick complex strata composed of tangential cross-beds as well as low-angle to planar beds (Fig. 9B) with abundant argillite interbeds. Tabular cross-beds have been designated as the *tidal channel sublithofacies* and complex strata as the *tidal shoal sublithofacies*, in order to emphasize the inferred depositional process and setting.

*Tidal channel sublithofacies*

Internally complex tabular cross-beds (Fig. 10A & B), traceable for up to 30 m along strike, are present as 5–20 m thick cosets and separated by 5–50 cm thick argillite-dominated units. The overall geometry is a tabular to planar shape with ubiquitous tangential foresets (Figs 8 & 9C) that becomes wedge-shaped at cross-bed terminations. Foreset toes generally flatten out at the base of the beds (Fig. 9A). Contacts between large-scale cross-beds are sharp and planar (Fig. 9A & C) to locally wavy, commonly accentuated by thin beds or laminae of argillite (Fig. 9C). In addition, superposed cross-beds locally display erosive contacts in which the upper cross-beds override and erode the underlying cross-bedded unit (Figs 8 & 9C). Cross-beds have unidirectional, generally low-angle tangential-shaped foresets (Fig. 10A & B) with abundant internal erosive and reactivation surfaces. Reactivation surfaces, either convex-upward or planar angular, are readily recognized where draped by argillite (Fig. 9C & D). In addition to foresets, underlying beds are also eroded (Fig. 9E). Low-angle foresets, in part graded, are common (Fig. 9C & D) and may be capped by 1–10 mm thick, discontinuous, argillite drapes. The sandstone–argillite also forms centimetre-scale couplets that locally exhibit sigmoidal foreset shapes (Fig. 9F & G). The spacing of these couplets varies at the centimetre scale, becoming thicker and subsequently thinner (Fig. 10A). The argillite drape is commonly partially eroded by the subsequent sandstone bed and may be completely absent. Evidence of erosion is also documented by the presence of thin argillite rip-up clasts in the overlapping sandstone-dominated foreset or on foreset toes. The sandstone part of the couplet is generally laminated to massive and the argillite is finely laminated. Rippled laminae were not observed.

Isolated granite and felsic volcanic pebbles (Fig. 9G), minor 2–10 cm thick pebble trains (Fig. 10A) and 20–50 cm thick lenticular, channelized, matrix-supported

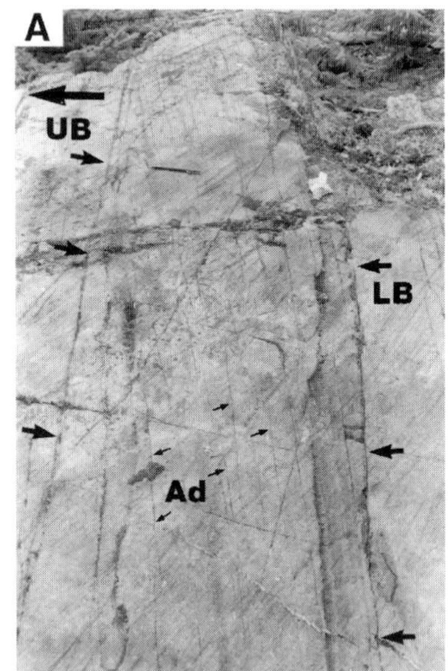


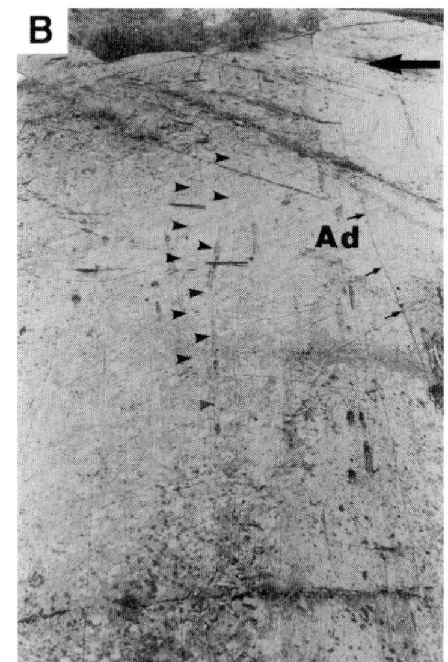


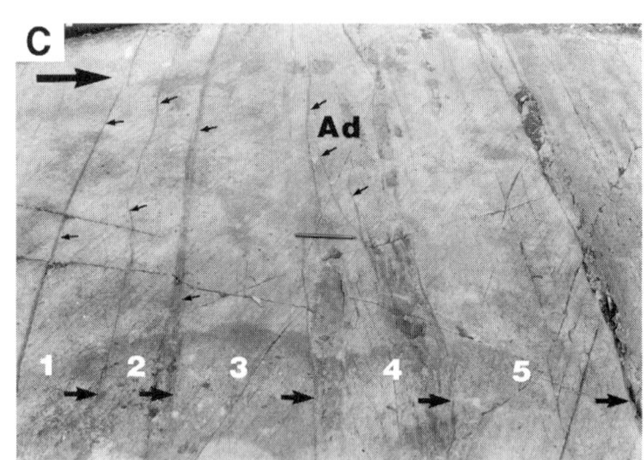

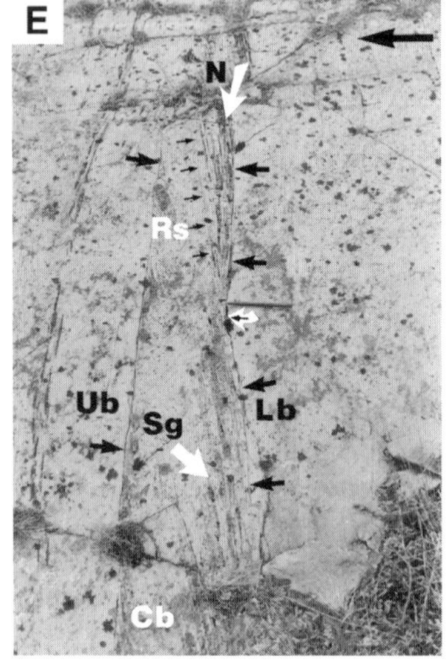


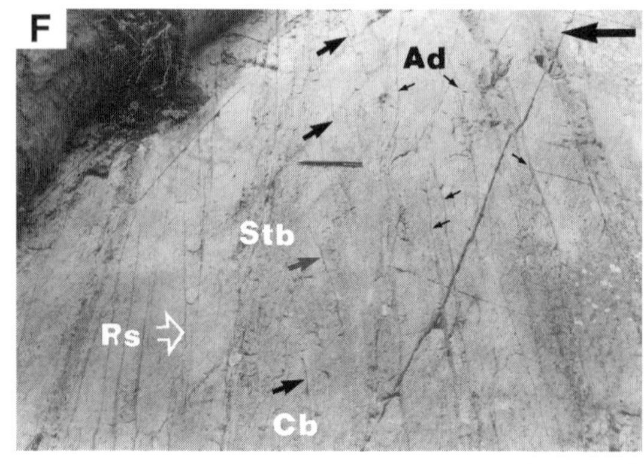


(*continued*)

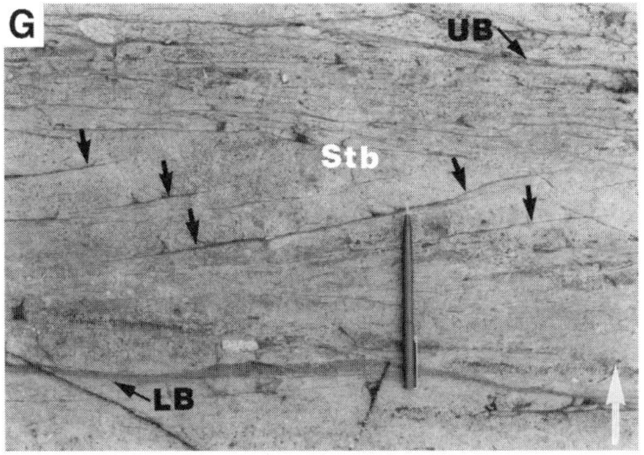


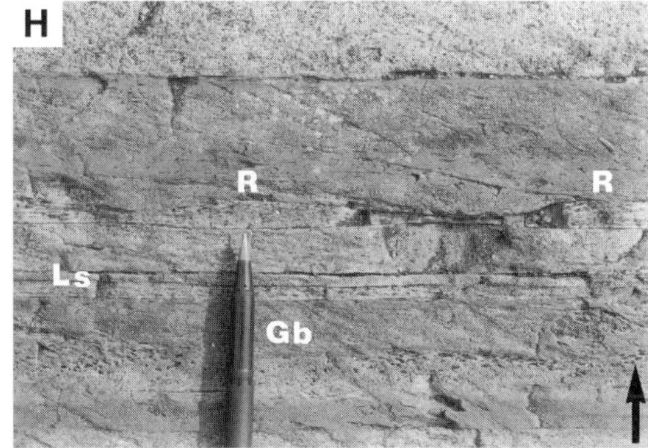


**Fig. 9.** Characteristics of the tidal channel sublithofacies of the sandstone–argillite lithofacies (except for photograph B). Large black or white arrow indicates tops in all photographs. Pen length is 15 cm. (A–F *opposite*) (A) A large-scale cross-bed with an upper (UB) and lower bounding (LB) argillite-draped cross-bed (arrows). The tabular cross-bed shows lateral migration of tangential foresets with subtle low-angle reactivation surfaces. Numerous foresets feature discontinuous argillite drapes (Ad; small arrows). (B) Tidal shoal sublithofacies composed of small- to medium-scale cross-beds. Argillite drapes (Ad; small arrows) are prominent between complex cross-beds and preserved on foresets. Arrowheads indicate migrating spur of sandwave. See Fig. 11 for details. (C) Outcrop displaying cross-beds 1, 2, 3, 4 and 5 with argillite drapes (Ad; small arows). Argillite laminae accentuate complex bedding units and form on foresets of cross-beds. (D) Close-up of complex cross-bed 4 with argillite drapes on foresets and reactivation surfaces (Rs). Sandstone displays faint lamination (L). (E) Cross-bed (Cb) with a well developed reactivation surface (Rs; small arrows) that scours into underlying bed. The cross-bed is interpreted as part of a neap (N)–spring cycle (SG). Ub, upper bounding surface; Lb, lower bounding surface. (F) Tabular to trough-shaped cross-beds (Cb) with abundant argillite drapes (Ad; small arrows) on foresets and separating cross-beds. Convex-upward shaped foresets with argillite drapes (near pen), forming sigmoidal tidal bundles (Stb), and drape on planar reactivation surface (Rs). (G) Detail of F illustrating delicate argillite-draped foresets on inferred sigmoidal tidal bundles (Stb). Isolated pebbles are located at top and bottom of cross-bed. UB, upper bounding surface; LB, lower bounding surface. (H) Argillite-dominated units with graded bedded siltstone–mudstone couplets (Gb), ripples (R) and laminated sandstone (Ls). Note sharp sandstone contact.

conglomerate beds are locally preserved in flat scours. Isolated pebbles were found at foreset toes, whereas sporadic pebble trains formed on the tops and bottoms of complex cross-beds. Argillite rip-up clasts are locally abundant (Fig. 10A & B) and are common in scoured conglomerate-filled channels. Argillite beds, 5–50 cm thick, delineate thicker cosets of tabular cross-bed sets and display prominent fine-grained sandstone to mudstone grading, parallel sandstone–siltstone–mudstone laminae and rare rippled intervals (Fig. 9H). Bed contacts between argillite and sandstone are sharp (Fig. 9H) to erosive and can be followed laterally along strike for 20 m.

*Interpretation*

The lateral continuity of tabular sets of cross-beds with ubiquitous argillite on foresets and between cross-beds reflects highly fluctuating energy conditions that are restricted to distinct settings and are formed by specific transport processes. Prominent bedload-dominated sandy fluvial systems may exhibit similar sedimentary structures (Rainbird, 1992; Røe & Hermansen, 1993), but this setting is discounted based on (i) ubiquitous argillite drapes formed at many scales and their sequential repetition, (ii) rapid and distinct stratigraphic changes into conglomerate- or argillite-dominated units over tens of metres and (iii) the geological context showing an adjacent basal unconformity and fault proximity. A high-energy clastic shoreline influenced by tides, waves and storms best accommodates the size and variety of crossbeds, in addition to the systematic interaction of argillite with sandstone (Dalrymple, 1992; Orton & Reading, 1993; Richards, 1994; Reading & Collinson, 1996). Internal cross-bed characteristics, size and coset arrangement favour comparison with ancient (Johnson & Levell, 1995) and modern (Nio & Yang, 1991) subtidal in-channel sandwave deposits.

The constant interplay between argillite and sandstone requires suspension-fallout processes alternating with high- to low-energy bedload movement from wave- and tide-action. The large-scale tabular to wedge-shape cross-beds are suggestive of lateral migration of straight-crested sandwaves (Houthuys & Gullentops,

A
Sandstone-argillite lithofacies
metres
10
9
8
7
6
5
4
3
2
1
0
Pebble trains
Unidirectional cross-beds
Rip-up clasts
Argillite-draped foresets
Reactivation surfaces
Argillite separating cross-beds
metres
Spring
Neap
Spring
Sigmoidal tidal bundles
Planar bed
Pebble train
6
5
4
3
2
1
0

B
Sandstone-argillite lithofacies
Low-angle cross-beds
Rip-up clasts
Planar beds
Reactivation surface
Laminated argillite between bedforms
Cross-beds
Argillite-draped foresets
Pebble train
Argillite separating cross-beds
Cross-beds
Low-angle cross-beds
metres
0
1
2
3
4
5
6
7
8
9
10
11
12
13
14
15
16
17
18
19
20
21
22
27
28
29
30
31
32
33

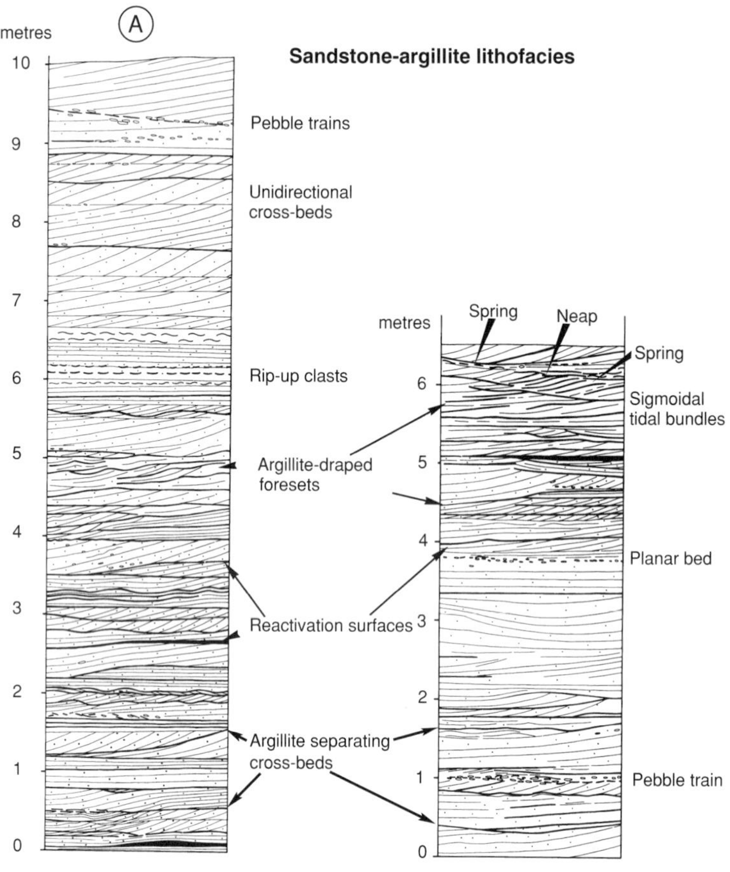

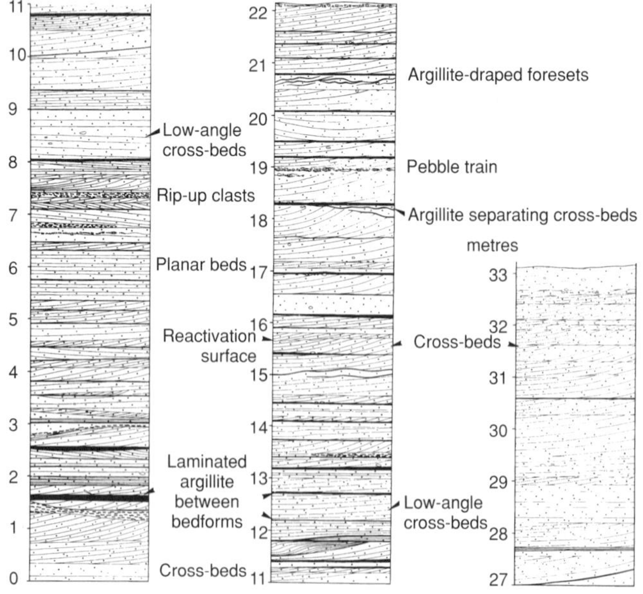

*(continued)*

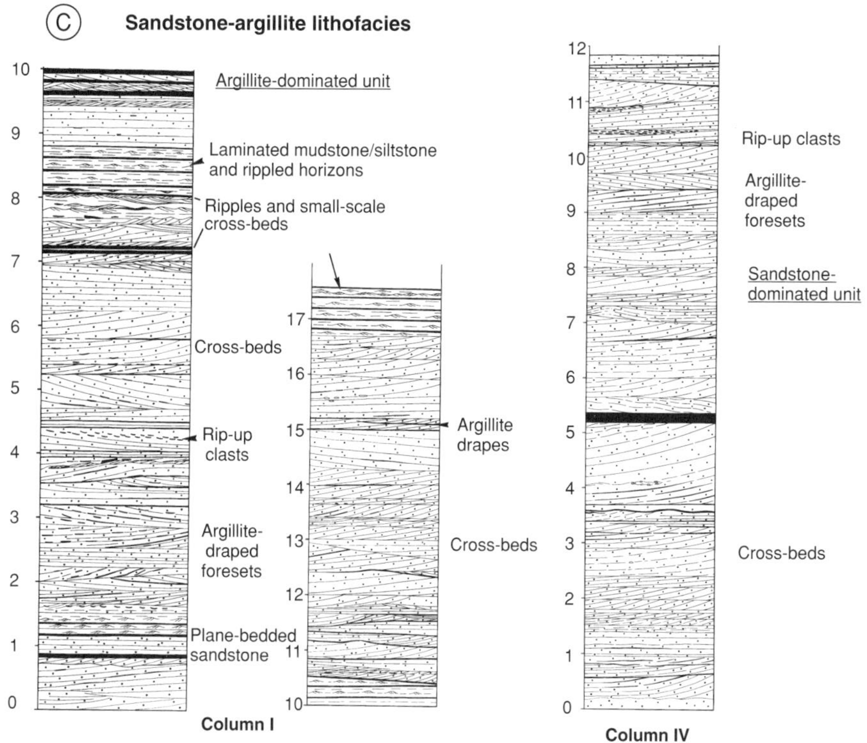


**Fig. 10.** Detailed stratigraphic columns of the sandstone–argillite lithofacies. (A) and (B) (*opposite*) Prominent cross-bedded structures with argillite drapes and abundant surfaces define the tidal channel sublithofacies (Figs 3, section $C_2$, & 7C). An inferred spring–neap–spring cycle is displayed in A. (C) Abundant medium- and small-scale cross-beds, ripples, argillite beds and planar beds characterize the tidal shoal sublithofacies (Figs 3, section $C_1$, & 7A). Note distinct argillite- and sandstone-dominated units.

1988; George, 1994; Richards, 1994), whereas wedge- to almost trough-shaped crossbeds indicate slightly sinuous-crested bedforms (George, 1994). The repetition of lateral accreting, sigmoidal-shaped sandstone–argillite beds in composite cross-strata (Fig. 9F & G) are considered prime Archaean candidates for tide-generated processes, commonly referred to as tidal bundles (Visser, 1980; Kreisa & Moiola, 1986; Nio & Yang, 1991). The graded and laminated sandstone on foresets represents sediment avalanching during maximal tidal transport of the dominant current. The argillite represents a pause from high-velocity sedimentation (Kreisa & Moiola, 1986; Simpson & Eriksson, 1991), and reflects suspension-fallout from high sediment concentrations (Johnson, 1978, p. 236) during the slack-water stage after the dominant tidal current. Discontinuous argillite drapes attest to the erosive power of either the subordinate current or new dominant current. An absence of argillite drapes implies (i) wave action inhibiting suspension deposition (Dalrymple, 1992), (ii) rotary tides with no slack water periods (Dalrymple, 1992) or (iii) strong dominant tidal currents eroding slack water and weak subordinant current deposits.

Overriding cross-beds, shown in Fig. 10A, indicate different rates of cross-bed growth due to variable current velocities under unidirectional flow (compare with Figs 17 & 18 of Nio & Yang, 1991). This feature is common to tidal regimes. In addition, reactivation surfaces draped by argillite indicate highly variable tidal velocities and bedload transport during cross-bed growth (Richards, 1986, 1994). The Jackson Lake deposits, containing reactivation surfaces with a pronounced channel geometry cutting down into underlying beds (Fig. 9E), are interpreted as the products of the dominant tidal current during spring tide (possibly enhanced by storm surge). Additionally, this down-cutting reactivation surface may coincide with a neap–spring cycle change. Observed convex-upward reactivation surfaces (type B of Nio & Yang, 1991) are explained by Houthuys & Gullentops (1988) as vortex erosion in front of the cross-bed and related to

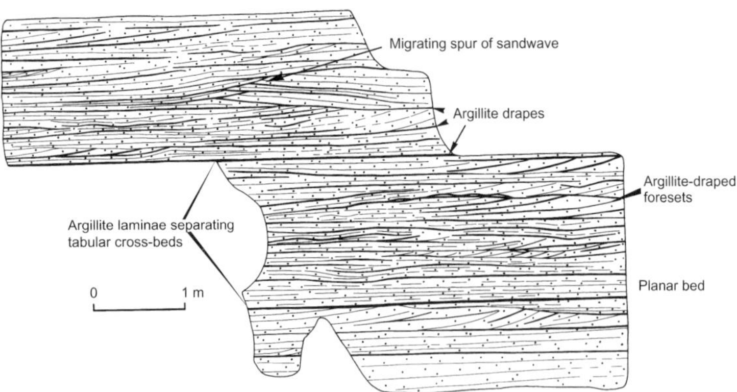


**Fig. 11.** Detailed field sketch of the tidal shoal sublithofacies, composed of planar beds and cross-beds. Individual beds are separated by argillite laminae and cross-beds feature an abundance of argillite drapes. The spur of the sandwave is characterized by interstratification of cross-beds with opposing dips.

the strength of the subordinate current. The numerous identified planar angular erosion surfaces are similar to type C of Nio & Yang (1991).

Interchannel argillites deposited between tabular cross-beds record channel abandonment and vertical aggradation commensurate with low-energy suspension deposits. The presence of ripples as well as laminated and graded siltstone–mudstone couplets supports this notion (Fig. 9H). A low-energy subtidal setting is envisaged, whereas a tidal flat setting for the argillite is considered improbable because of the absence of flaser bedding and desiccation cracks in the sequence (Deynoux *et al.*, 1993). The introduction of pebbly debris into the subtidal setting is reconciled with debris originating from the adjacent conglomerate–siltstone lithofacies during high wave surge, floods or storms (Mueller & Dimroth, 1987; Duke & Prave, 1991). These channelized conglomerate beds can be explained as storm-generated rip currents (Mueller & Dimroth, 1987), in which mass flow processes operated. The pebble trains are considered top-surface bedload deposits that formed during storm surge (Richards, 1986).

### *Tidal shoal sublithofacies*

The 5–10 m thick tidal ridge sublithofacies (Figs 10C & 11), the lateral and vertical equivalent of the tidal channel sublithofacies, is composed of: (i) cosets of small- to medium-scale (5–50 cm thick) tangential cross-beds that are tabular to trough-shaped; (ii) 10–50 cm thick low-angle to planar beds; and (iii) 5–150 cm thick argillite-dominated units. Collectively, these three units define 1–10 m thick upward-fining sequences commencing with a sandstone overlain by an argillite-dominated unit (Fig. 10C; e.g. 2–9 m of column I). The characteristic feature is the ubiquitous continuous and discontinuous argillite drapes on tangential foresets or between large cross-beds. Composite strata are commonly tabular (Fig. 11) and exhibit ripples and cross-beds with bidirectional flow on well exposed outcrop surfaces (Fig. 3).

The cross-beds with tangential foresets and local convex-upward reactivation surfaces have argillite drapes on foresets and contain abundant argillite rip-up clasts (Fig. 12A). Small centimetre-scale sand wedges, encased by argillite laminae (Fig. 12A), were locally observed on cross-bed foresets. Some superposed cross-beds display opposing foreset directions and prominent erosion surfaces cutting down into lower beds. Small-scale cross-beds and ripples occasionally cap larger cross-beds. The planar to low-angle beds generally erode cross-bed sets, but also conformably overlap thin beds of argillite. Cross-beds with opposing dip directions on two-dimensional outcrop surfaces are locally observed in these complex strata (Figs 11 & 12B). They are subsequently overlain by wavy to low-angle beds with abundant argillite laminae or covered by thicker argillite beds. Conspicuous continuous argillite veneers are locally preserved between individual beds (Fig. 12B).

The laterally continuous argillite beds (20–150 cm thick), representing a combination of mudstone, siltstone and very fine-grained sandstone, contain abundant parallel laminated beds usually capped by symmetrical and asymmetrical rippled horizons (Fig. 12C), or 1–2 cm thick graded beds. In addition to 5–10 cm thick cross-beds, local thin contorted beds and sandstone scours, rich in argillite rip-up clasts, are present. Of special interest are individual millimetre-scale planar beds displaying alternating mudstone-siltstone couplets

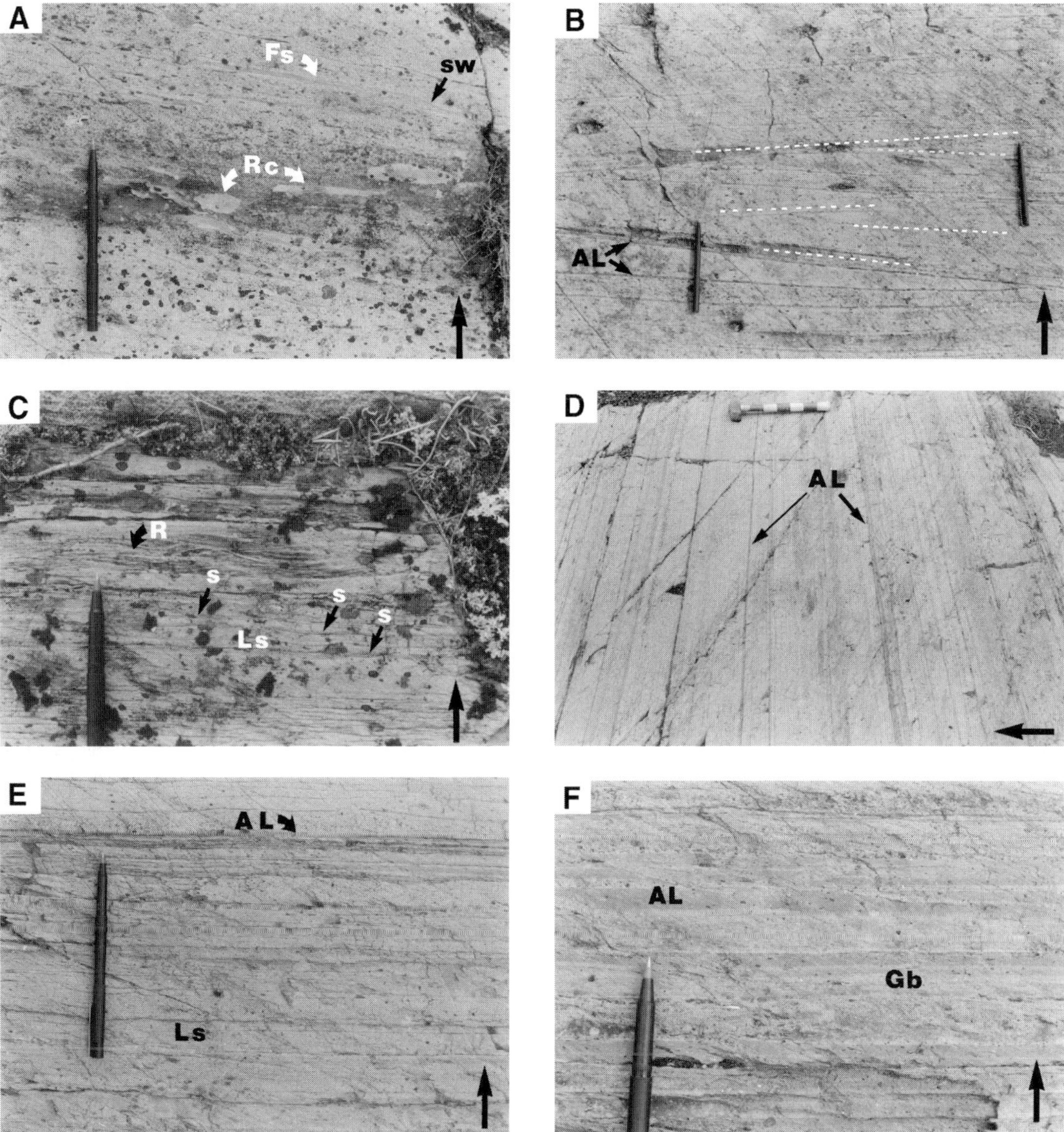


**Fig. 12.** Characteristics of the tidal shoal sublithofacies (A to C) and argillite–sandstone lithofacies (D to F). Pen length is 15 cm and tip of pen indicates younging direction. (A) Rip-ups (Rc) in clast-rich cross-bed with argillite-draped foreset (Fs). Note small sandstone wedge (sw) between argillite drapes. (B) Migrating spur of sandwave (left to right as indicated by pens and dashed line) with abundant argillite laminae (AL) accentuating individual beds. Detail of Fig. 8. (C) Argillite-dominated unit with laminated very fine-grained sandstone (Ls) and siltstone, as well as rippled horizons (R). Millimetre-scale argillite may indicate subordinate (s) tidal current deposits in vertical tidal bundles. (D) Planar-bedded argillite–sandstone lithofacies. Note lateral continuity of thin beds and argillite laminae (AL). Scale, hammer, 30 cm. (E) Finely laminated fine-grained sandstone, siltstone (Ls) and mudstone (argillite laminae, AL). (F) Close-up of argillite–sandstone lithofacies with graded beds rich in rip-up clasts (Gb), laminated sandstone beds and distinct mudstone cappings (argillite laminae, AL).

(Fig. 12C) that appear to change thickness at the millimetre scale in a repetitious manner. The 1–5 cm thick rippled horizons commonly contain superposed ripples with opposing foreset laminae and small-scale erosion surfaces.

*Interpretation*

The tidal shoal sublithofacies, with a significant increase in argillite and absence of pebble trains, represents a low-energy equivalent of the tidal channel sublithofacies. Abundant argillite drapes and reactivation surfaces in cross-beds argue for highly fluctuating energy conditions attributable to tide- and wave-induced processes (Reading & Collinson, 1996). The cross-bedded sandstone-dominated units reflect in-channel bed migration, whereas argillite-dominated units are consistent with channel abandonment and a low-energy, interchannel setting. The tide-controlled channels, represented by the tabular cross-strata and trough-shaped cross-beds, are not as large and well defined as in the tidal channel sublithofacies, so that a marginal channel setting is envisaged. An overall setting would be flat-topped subtidal shoals (Simpson & Eriksson, 1989; Richards, 1994; Johnson & Levell, 1995) or subtidal flats (Dalrymple, 1992) fringing subordinate tidal channels or zones off-axis from the principal tidal channel. The poorly defined upward-fining sequences, which are consistent with lateral channel migration, support such an interpretation.

The tabular to trough-shaped cross-bed sets are consistent with lateral migration of straight- to sinuous-crested sandwaves (Houthuys & Gullentops, 1988; George, 1994; Richards, 1994). Superposed argillite drapes indicate slack-water tidal periods and subordinate current deposition (Kreisa & Moiola, 1986; Nio & Yang, 1991). Argillite rip-up clasts on sandy foresets illustrate the erosive power of tidal currents (Fig. 12A) and reactivation surfaces reflect variable tidal velocities during sandbar growth. The sandstone wedges between argillite laminae (Fig. 12A) are of particular interest because they are sound indicators of tidal influence and are commonly employed to demonstrate semidiurnal ebb- and flood-dominated tidal cycles (Nio & Yang, 1991; Dalrymple, 1992). The rarely preserved sandstone wedges preserved on argillite drapes are inferred to represent the subordinate tidal current, whereas the overlying argillite drape indicates the subsequent slack water stage.

Laterally interstratified cross-beds with opposing dip directions and an up-section asymmetric growth compare favourably with migrating sandwaves (Figs 11 & 12C). The low-angle to planar beds without argillite drapes are interpreted as upper flow regime sands that formed during high wave-surge and maximal tidal current activity (Mueller & Dimroth, 1987; Richards, 1994).

The argillite-dominated units (Fig. 12C) represent a significantly lower tide- and wave-influenced energy regime. The combined wave- and current-influenced ripples support a subtidal setting where wave action and tidal currents are capable of generating complex combinations of small-scale structures (see Simpson & Eriksson, 1991). Individual millimetre-scale planar beds composed of alternating mudstone–siltstone couplets (Fig. 12C) may be an Archaean analogue of vertically stacked tidal bundles (e.g. Tessier & Gigot, 1989; Deynoux *et al.*, 1993). Parallel laminated graded beds, in conjunction with argillite beds, reflect dissipating lower flow energy regimes on an unconfined sheet-like topography rather than a confined channel structure. Large flat-topped subtidal shoals would be an appropriate locus for the low-energy sedimentary structures (see Deynoux *et al.*, 1993). Locally incised cross-beds and graded beds were probably generated during storms or indicate secondary tidal channel fills.

## Argillite–sandstone lithofacies

The 5–30 m thick argillite–sandstone lithofacies (Table 1; Fig. 12D), the predominant lithofacies adjacent to the unconformity, has (i) an abrupt to gradational lower contact with the conglomerate–sandstone lithofacies, (ii) a sharp, erosive upper contact with the sandstone–argillite lithofacies and (iii) an unconformable contact with the volcanic and volcaniclastic rocks of the Kam Group. The consistent stratigraphic position (Fig. 7A & B), sedimentary structures and thickness enable distinction from argillite-dominated units of the sandstone–argillite lithofacies. Parallel-laminated sandstone and siltstone–mudstone couplets (Fig. 12E), unidirectional rippled intervals and very fine-grained to mud-size graded sandstone beds (Fig. 12G) are characteristic. Argillite rip-up clasts are observed in graded beds. The lateral continuity of beds (Fig. 12D) and the absence of flaser-type bedding, cross-beds or well defined channelized scours is characteristic. Coarse-grained, 10–20 cm thick graded beds with erosional bases are locally intercalated.

### Interpretation

The argillite–sandstone lithofacies is best explained in relation to its association with the other lithofacies. A low-energy environment dominated by suspension sedimentation and wave- and tide-induced current structures can account for the formation of ripples, parallel lamination as well as small-scale graded beds. Sporadic storms are an appealing process to explain the development of graded sandstone beds with rip-up clasts (Howard & Reineck, 1981; Kreisa, 1981). Laterally continuous rippled horizons may be the product of wave-induced currents or wave-action (de Raaf *et al.*, 1977). The lack of complex bedforms as observed in the sandstone–argillite lithofacies, thickness of lithofacies and a prevalence of fine-grained sedimentary material argue for a proximal offshore setting below fair weather wave base in which local storms interrupted the calm suspension-dominated sedimentation pattern (Mueller & Dimroth, 1987). Subtidal, vertically accreted tidal rhythmites also appear to be present, but additional detailed work is required.

## CONGLOMERATE COMPOSITION AND CLAST SIZE

The conglomerate clast population provides insight into provenance and dispersal patterns, whereas clast and grain size variations provide an indication of distance from source. The importance of clast populations identifying source, dispersal patterns, tectonic influence and basin migration was illustrated by Crowell & Link (1982) for the Ridge basin. Extensive clast counts to determine provenance and tectonic influence have been conducted in the Archaean for late-orogenic strike-slip basins (Mueller *et al.*, 1991, 1994; Corcoran *et al.*, 1998, 1999) and thrust tip or piggyback basins (Eriksson, 1978; Heubeck & Lowe, 1994). In addition, clast and grain size distribution is commonly used to discern proximal and distal settings on alluvial fans, fan-deltas or gravelly braided streams. Dimensions of kilometres for small fans to tens of kilometres for large fans (Heward, 1978; Rust & Koster, 1984) can be indirectly monitored by clast size.

The clast population in the Jackson Lake Formation was measured using an orthogonal grid with 10 cm intervals on outcrops that permitted proper evaluation of clast constituents. Stations were chosen according to stratigraphic position so that critical vertical and lateral changes in clast composition could be registered (Figs 6A, 7A & B). A total of 4545 clast intersection points (from a total of 6900) counted at Yellowknife Bay, Walsh and Jackson Lakes (Fig. 3, sections A, $C_1$, D, respectively) yield a reasonable estimate of prevalent and subordinate clast population, and help to establish provenance. Mean total rock matrix accounts for 34.13% of the conglomerate beds.

Pie diagrams (Figs 6A, 7A & B) show that: (i) the clasts were derived from a volcanic-dominated hinterland; (ii) sources are local; and (iii) plutonic, felsic volcanic and volcaniclastic (sedimentary) clasts are subordinate. The Yellowknife Bay section (Fig. 6A), with 4000 intersection points and a 33% matrix component, exhibits a predominance of mafic volcanic clasts (84%) and a subordinate population of felsic volcanic (1%), granitic and quartz–feldspar porphyry (7.8 and 0.2%) and sedimentary (7%) components. The sedimentary clasts were derived from the volcaniclastic turbiditic rocks exposed at the unconformity (Fig. 5A) and are locally prevalent at certain stratigraphic levels (Fig. 6A, column III). Mafic and felsic volcanic clasts originated from the adjacent mafic-dominated Yellowknife Bay Formation, and granitic fragments were derived from the western Plutonic Complex (Fig. 3). The Walsh Lake section (Fig. 3, $C_1$), based on 1900 intersection points and a 37% matrix component, predominantly contains mafic clasts (96%), and minor granitic (4%) and felsic (<1%) clasts, all of which were derived from the Kam Group volcanic rocks and the western Plutonic Complex. The Jackson Lake section (Fig. 3, D), with 1000 intersection points and a 36% matrix component, is characterized by a strong local felsic component with 75.5% pyroclastic fragments derived from the adjacent Vee Lake Lenticle (Fig. 7B) and 13.5% quartz–feldspar phyric clasts originating from the adjacent dykes/flows, whereas mafic constituents are minor with 10.5%.

Clast sizes in conglomerates are especially useful because size variations can indicate the influence of tectonism on sedimentation (Steel *et al.*, 1977; Steel & Gloppen, 1980; Rust & Koster, 1984). Clast measurements were standardized by averaging the apparent long 'a' and short 'b' axes on outcrop surfaces, minimizing the effects of deformation. The average of the ten largest clasts in each bed was taken. The size of boulders ranges up to 73 × 17 cm (average 45 cm) and 80 × 19 cm (average 49.5). Local boulder-dominated beds up to 2.5 m thick average 41.6 cm for the ten largest clasts (Fig. 7A, column III). The best exposed outcrop zone for clast measurements (Fig. 6A) exhibits overall size averages of 16.55 cm (Fig. 6A, column I) and 20.4 cm (Fig. 6A, column III), indicating a cobble- to boulder-size range. Abundance of large clasts requires significant topographic relief, and a proximity to source in the order of kilometres can be inferred. Upward-coarsening then fining (CU–FU) sequences in the conglomerate are sound indicators of tectonic influence on sedimentation. CU–FU sequences can be explained by uplift and basin subsidence along faults where tectonic activity initially exceeded the rate of erosion, but as the effects of tectonism deteriorated, fault scarp retreat occurred (Heward, 1978; Mack & Rasmussen, 1984; Rust & Koster, 1984).

## PALAEOCURRENT DATA

Measuring viable palaeocurrent directions in Archaean rocks is difficult due to the degree of deformation, metamorphic grade, and scarcity of exposures suitable for direct palaeocurrent measurement. None the less, the importance of current or wave directions in the rock record to establish source and dispersal patterns is paramount in determining the evolutionary history of a basin. Work by Heubeck & Lowe (1994) and Eriksson (1977) in the Archaean Barberton greenstone belt demonstrates the applicability and usefulness of palaeocurrent data in constructing a basin model. In addition to determining source and basin evolution, palaeocurrent measurements are essential for recognizing sedimentary processes. Tidal processes dominated by ebb and flood currents are revealed in the ancient record through tidal bundles, flaser beds and herringbone crossbeds. The palaeocurrent data should ideally exhibit a bimodal and bipolar current-flow distribution.

A total of 104 measurements (Fig. 3) were conducted in the best exposed area of the sandstone–argillite lithofacies (Fig. 7A, column I) to determine if palaeocurrent measurements support tide influence, as suggested by sedimentary structures. Measurements were conducted on tabular to slightly wedge-shaped cross-beds with well defined foresets, which are generally argillite-draped. The vector mean provides an accurate assessment of the flow direction because bedforms are constructional, formed by accretion or sandwave growth. All readings were rotated into the horizontal position to correct for dip of bedding. In addition to large-scale structures, small-scale cross-beds and ripples were measured to determine possible differences in flow direction.

The measurement density was conducted at two scales with 47 readings in a detailed mapped section (Fig. 7A, column I, 224–242 m, and in detail Fig. 10C, column I) and 57 readings for the segment related to column I (Fig. 7A). Results display a bimodal, generally bipolar distribution where the NNE–NE and SSW–SW flow components are dominant and subordinate, respectively (Fig. 3). The apparent flow direction of current-induced sedimentary structures is scale independent: both large- and small-scale sedimentary structures show similar distributions on rose diagrams. The palaeocurrent data support inferred tide influence.

## DISCUSSION

The Archaean Jackson Lake Formation displays numerous characteristics that are uncommon to modern depositional systems and shed light on the possibly unique conditions that operated during the Archaean (>2.5 Ga). To construct a viable depositional model, several relevant aspects require attention: (i) tectonic influence on sedimentation; (ii) Archaean environmental parameters; and (iii) Archaean tidal deposit characteristics and conditions.

### Tectonic influence on sedimentation

The up to 300 m thick Jackson Lake Formation displays lithofacies stacking that can be reconciled with first order tectonic control. The lithofacies succession of conglomerate–sandstone to argillite–sandstone and/or sandstone–argillite (Fig. 6A) collectively defines a 150–200 m thick upward-fining sequence. A large-scale recurrence of lithofacies (Fig. 6B) in constrained basins is a prime indicator of tectonic uplift and basin subsidence (Steel *et al.*, 1977; Heward, 1978; Rust & Koster, 1984; Hartley, 1993). A climate change may also affect lithofacies stacking. Coarse clastic upward-fining sequences develop during fault deterioration and/or fault back-stepping (Heward, 1978; Mack & Rasmussen, 1984), whereby each upward-fining sequence suggests fault rejuvenation with renewed introduction of coarse clastic material. Smaller-scale 10–25 m thick upward-coarsening then fining cycles in the amalgamated conglomerate beds (Fig. 6A) suggest fault activation and rapid deterioration, and are generally considered to represent the immediate clastic response of tectonically induced sedimentation. In addition to a prominent angular unconformity (Figs 4 & 5A) at the base of the Jackson Lake Formation and a bounding fault to the east, abundant angular to subrounded boulder-size clasts are locally up to 80 cm in diameter, suggesting significant topographic relief and proximity to an escarpment.

Alternatively, Blair & Bilodeau (1988) contended that the response to fault-dominated tectonism was flooding, accompanied by subaqueous deposition of fine-grained sediments, whereas relative tectonic quiescence is characterized by coarse clastic debris. Similarly, Hartley (1993) interpreted upward-coarsening megasequences as the products of reduced source area uplift/subsidence after tectonic activity and upward-fining sequences as increased tectonic activity. In the case of the Jackson Lake Formation, which is an open marine system, this is a tenable explanation for loci where the argillite–sandstone overlies the conglomerate–sandstone lithofacies (Fig. 7A & B). Fault scarp segments directly adjacent to a marine setting with abrupt and significant tectonic activity would favour marine flooding with large upward-coarsening sequences representing increased sedimentation, whereas fault sections slightly removed from the coast with gradational fault displacement and deterioration would display upward-fining sequences. Both case scenarios probably occurred along the ancestral Jackson Lake fault because differential rates of movement occur along the strike of a fault plane or splay.

The lithofacies architecture represented by a complete facies sequence Gbx–Gms–Gm–Gt–Sm–Sh–Sl–St (Table 1) is consistent with the proximal to medial reaches of coarse clastic fans, and is supported by talus scree (facies Gbx) and mass flow deposits (facies Gms; see Mack & Rasmussen, 1984). Fan deposits of this nature coincide with active tectonic settings (Lonergan & Scheiber, 1993). Accumulation rates in fault-bounded basins are generally very high, and if tectonically active for 1–10 Myr, depositional rates can be in excess of 1000 m $Myr^{-1}$. Gravel-dominated alluvial fans are generally adjacent to active fault-bounded mountain fronts (Heward, 1978), or tectonically active settings with high relief, such as intermontain basins (Long, 1981). Similarly, fan-deltas are controlled by tectonics (Wescott & Ethridge, 1980; Nemec & Postma, 1993; Laird, 1995) and the interpreted coarse clastic fan-deltas of the Jackson Lake Formation display the salient attributes of tectonically controlled sedimentation (e.g. Frostick & Steel, 1993).

## Archaean environmental parameters

In the absence of vegetation, Archaean subaerial alluvial–fluvial dispersal patterns would have favoured low-sinuosity streams and promoted unconfined, hyperconcentrated flood flows (Mueller & Corcoran, 1998). In addition, high $CO_2$ levels (Young, 1991), elevated surface temperatures of up to 85°C (Kasting, 1993) and a prevalence of humid atmospheric conditions (Des Marais, 1994) were highly efficient weathering parameters that influenced sandstone composition even in tectonically active areas (Corcoran *et al.*, 1998). Corcoran *et al.* (1998) suggested that first cycle quartz-arenites are possible even in high-relief systems if (i) a prominent granitic source was nearby and (ii) high-energy tide- and wave-influenced systems were present to facilitate sorting and labile mineral breakdown. Clast compositions in conglomerates from the Jackson Lake Formation display a preponderance of volcanic debris derived from volcaniclastic sediments or flows, and basaltic clasts constitute >90% of the volcanic clast population. The source of the volcanic clasts is the underlying Kam Group, and therefore local. Plutonic clasts constitute <5%, so that generating quartz-arenites via tide and wave sorting and attrition processes is improbable. The relevance of conglomerate composition, sandstone petrography and geochemistry with respect to climate, source and weathering in constrained high-relief depositional settings is treated separately by Corcoran and Mueller (this volume, pp. 183–212).

The coalescing fans along the western margin display local proximal talus breccia and mass flow deposits (<10%) that are characteristic of small catchment areas. Preservation of talus scree deposits implies steep slopes, but the interpreted fan-deltas are primarily sheetflood and bedload deposits, indicating stream-dominated processes. Steep sloped fans dominated by fluvial processes develop preferentially in humid hot or temperate climates (Galloway & Hobday, 1983). Modern environments favouring unconfined flow are encountered in high-relief settings with sparse vegetation, including arid to semi-arid (e.g. western United States and Atacama desert, Chile) and some humid temperate settings (e.g. New Zealand). The fan-deltas of the Jackson Lake Formation developed along steep slopes under humid and hot conditions, in which unconfined sheetflows were common.

## Archaean tidal deposit charateristics and conditions

Eriksson (1977) initially documented Archaean tidal flat and channel deposits in an estuarine to back-barrier setting based on the presence of flaser bedding, herringbone cross-bedding and clay/siltstone-draped beds. Detailed case studies in ancient sequences are constrained to stable intracratonic settings with extensive epeiric seas (Simpson & Eriksson, 1991) and passive margin environments with tide-influenced barrier island systems (Hamberg, 1991; George, 1994) or estuarine complexes (Dalrymple, 1992; Johnson & Levell, 1995; Ehlers & Chan, 1999). The tidal channel and tidal shoal deposits of the Jackson Lake Formation display sigmoidal tidal bundles (Fig. 9G), bimodal–bipolar palaeocurrent trends (Fig. 3) and prominent argillite drapes on sedimentary structures that formed in a very constrained and protected, but high relief, setting. These distinctive Archaean deposits may be the results of vastly different atmospheric conditions coupled with a more influential Earth–Moon system.

The development of tidal sequences or clay-draped sedimentary bedforms in tectonically active high relief zones is striking and evident in the Slave Province (Corcoran *et al.*, 1998, 1999; this study) as well as in the Abitibi greenstone belt (Mueller & Dimroth, 1987; Mueller *et al.*, 1991). The Jackson Lake strata display tidal sequences akin to modern tide-dominated shelf deposits on stable platforms, but the lithofacies relationships indicate a tectonically active setting. Were tidal processes different during the Archaean? Tidal range is problematic and can only be inferred based on modern analogues. The Jackson Lake in-channel sandwaves, composed of laterally accreted tidal bundles and possible vertically accreted tidal bundles on tidal shoals, coupled with coarse- to very coarse-grained immature sandstone, are consistent with a macrotidal regime (Richards, 1994; Ehlers & Chan, 1999). Currents in this regime are strong enough to transport the coarsest grain sizes and allow for suspension deposition during slack water periods. In modern settings, estuarine complexes (e.g. Bay of Fundy) commonly display macrotidal conditions, whereas barrier shoreline and tidal inlets are more akin to micro- to mesotidal settings. The stratigraphic sequence in the Jackson Lake Formation favours fan-deltas associated with an estuarine-type dispersal system.

### *Earth–Moon conditions through time*

The combination of coarse clastic fan-deltas prograding on to a macrotidal shelf can be reasonably explained by postulating a greater tidal range during the Archaean. Although Archaean tidal bundles have not been specifically identified, the recognition of Proterozoic and Cambrian tidal rhythmites (Williams, 1998) and laterally accreted tidal bundles (Deynoux *et al.*, 1993; Bose *et al.*, 1997) provides a basis for determining lunar retreat rate and ancient lunar cyclicity (Sonett *et al.*, 1996a,b; Williams, 1998; Kvale *et al.*, 1999), and may be extrapolated to infer stronger Archaean tidal conditions. A calculated lunar retreat rate of 2.19 cm $yr^{-1}$ for the *c.*620 Ma Elatina-Reynella rhythmites (Williams, 1998) and 2.1 cm $yr^{-1}$ for the 900 Ma Big Cottonwood Formation (Sonett *et al.*, 1996b), compared with a present rate of 3.82 cm $yr^{-1}$ (Sonett *et al.*, 1996b), shows a temporal increase in rate of lunar retreat. The lunar semimajor axis (mean Earth–Moon distance) has increased from 900 Ma to Present from $3.45 \times 10^{10}$ to $3.844 \times 10^{10}$ cm or 13.4%. The lunar orbital period is today 27.32 (Kvale *et al.*, 1999) compared to 25.89 (620 Ma) and 25.24 days during the Proterozoic (900 Ma) based on tidalite-derived lunar parameters (Sonett *et al.*, 1996b). As determined by Williams (1998) and Sonett *et al.* (1996a), the terrestrial day had 21.9 (Cambrian age) and 18.2 h (Proterozoic age), although the calculations of Williams (1998) imply 20.9 h for the latter. A more dynamic Earth–Moon tidal system is probable because (i) shorter days equate with a faster rate of Earth rotation, (ii) a shorter lunar period implies a shorter lunar orbit distance (and hence distance from Earth), (iii) a shorter Earth–Moon distance translates into stronger gravitation forces and (iv) faster Earth rotations increase the neap–spring cycles per year. Collectively, these data suggest a stronger tidal regime during the Neoproterozoic, and when extrapolated to the Archaean, tidal range should have been even higher. The association of tectonically controlled sedimentation in a high relief zone with sedimentological evidence for significant tidal influence can thus be adequately explained by Archaean Earth–Moon dynamics. One of the major inferences of this study is that a macrotidal range in a high relief zone can compensate for tectonic influence on sedimentation during periods of tectonic quiescence, but is subordinate during active tectonism. The tidal deposits at Jackson Lake, as well as those from the Keskarrah Formation at Point Lake (Fig. 2; Corcoran *et al.*, 1998), appear to have been significantly influenced by the dynamics of the Archaean Earth–Moon system.

## Depositional model

A depositional model for the Jackson Lake Formation must combine (i) the interstratification, distribution and stacking of the three lithofacies, (ii) the internal sedimentary characteristics of the lithofacies, (iii) the upward-coarsening and fining sequences in the conglomerate, (iv) the presence of a bounding unconformity and local derivation of clasts, (v) basin geometry and (vi) tidal influence (Fig. 13). Modern-day tectonic settings with similar characteristics include fault-bounded strike-slip basins that prograde from coarse clastic fan-deltas into sandy–gravelly braidplains and finally into a standing body of water (Crowell & Link,

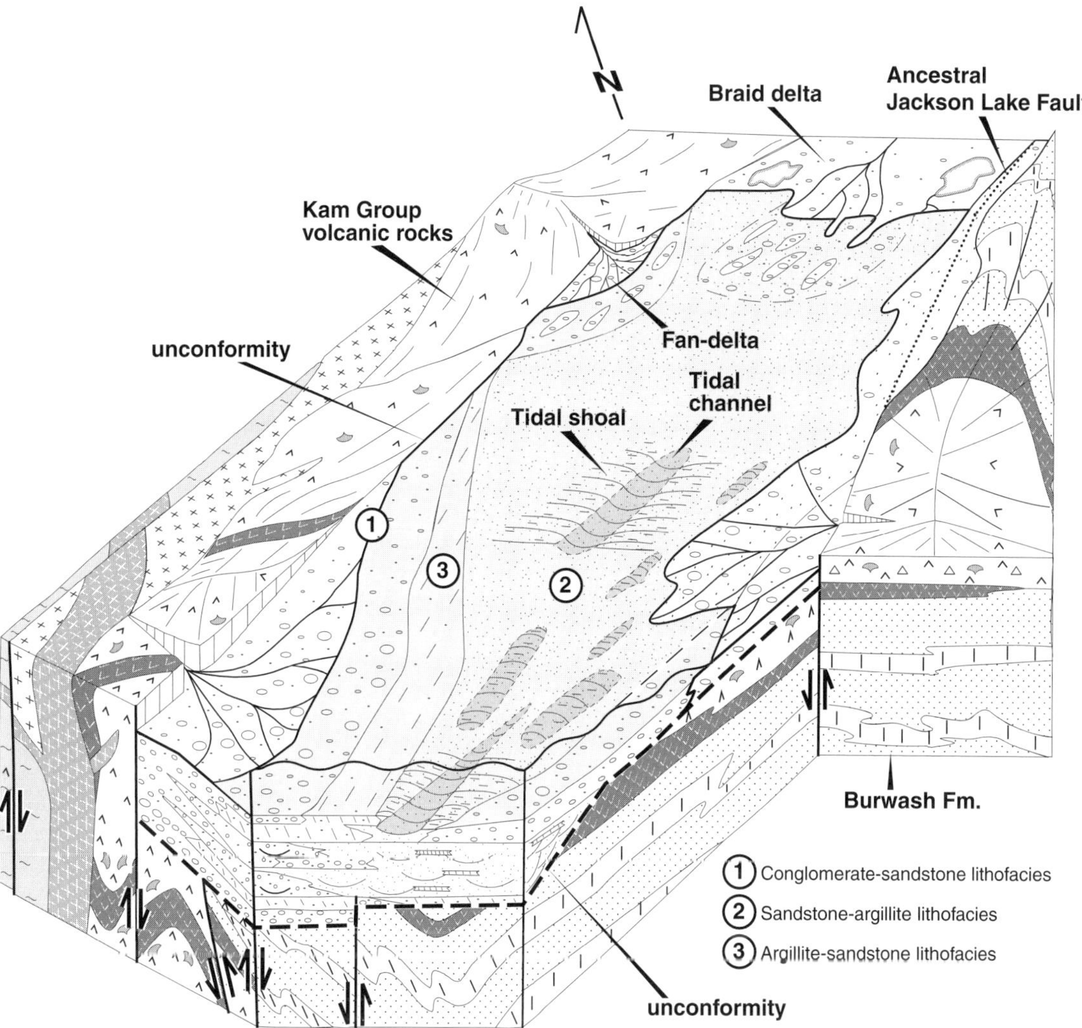


**Fig. 13.** Palaeogeographic reconstruction of the depositional environment for the Jackson Lake Formation. A high relief setting with bounding faults and coastal fan-deltas prograding into a N-trending, tide-dominated embayment is envisaged. The prominent large-scale sedimentary structures in the embayment are tidal channels and tide-influenced shoals which are juxtaposed and interstratified with prograding fan-deltas. The Jackson Lake sediments received significant volcanic detritus from parts of the underlying Kam Group that were elevated in fault-bounded blocks. The lithofacies are numbered and setting is indicated.

1982; Hempton *et al.*, 1983; Nilsen & McLaughlin, 1985). Active fault systems ranging from 15 to 100 km in length in modern settings are generally responsible for alluvial fan formation (Nilsen & McLaughlin, 1985). Local clast provenance, as shown for the Jackson Lake Formation, is common to tectonically controlled basins (e.g. Christie-Blick & Biddle, 1985). Modern fan-deltas around volcanic islands display abundant wave- and minor tide-influence on shallow platforms (Wescott & Ethridge, 1980; Orton & Reading, 1993). Extensive continental-type platforms with macrotidal conditions, similar to that of the Bay of Fundy (maximum tide range of 16 m), are generally required to produce tidal bundles.

The unconformity and ancestral Jackson Lake fault with adjacent fan-deltas, represented by the conglomerate–sandstone lithofacies, and the shallow-water argillite–sandstone lithofacies emphasize tectonic control on sedimentation and represent the margins of an embayment. The sandstone–argillite lithofacies, composed of the tidal channel and tidal shoal sublithofacies, developed in a NNE–SSW trending embayment protected from the open ocean, amplifying the tidal range (Fig. 13). An asymmetric ebb- and flood-dominated tidal current regime can be suggested, based on bimodal palaeocurrent data and inferred tidal bundles. A central estuarine-type morphology that developed under macrotidal conditions (George, 1994; Ehlers & Chan, 1999) with a narrow connection to the ocean, but limited by faults causing high relief, is the proposed palaeoenvironment (Fig. 13).

## CONCLUSIONS

The Archaean Jackson Lake Formation is preserved in a N-trending, remnant, clastic sedimentary basin. The three principal lithofacies, represented by the conglomerate–sandstone, sandstone–argillite and argillite–sandstone, display lithofacies repetition and upward-fining sequences that are consistent with tectonically controlled sedimentation. The bounding Jackson Lake fault was one of the principal tectonic structures influencing sedimentation and lithofacies distribution. In addition, the presence of an angular unconformity and lithofacies stacking exhibiting a basin margin conglomerate-dominated sequence passing into a basin-central, argillite–sandstone succession are sound indicators of tectonic influence on sedimentation. High relief intermountain basins or tectonically active, rugged, high relief coastlines are considered modern counterparts. The conglomerate-dominated lithofacies with the up-section transition into argillite-dominated lithofacies is consistent with a fan-delta environment.

The sandstone–argillite is inferred to be a tide- and wave-influenced lithofacies that developed under macrotidal conditions. The tidal channel and tidal shoal sublithofacies display the lateral change from in-tidal channel sandwave migration to channel margin or secondary tidal channel deposits grading into subtidal flat-topped shoals during channel abandonment. Clay-draped cross-beds are considered the Archaean analogue of modern sigmoidal tidal bundles. The millimetre-thick mudstone–siltstone couplets in argillite-dominated units may reflect preserved vertically accreted tidal bundles.

The unusual combination of a high relief, tectonically controlled basin influenced by macrotidal conditions requires special consideration. The results of this study, based on a detailed sedimentological facies analysis, are consistent with a more dynamic Earth–Moon system during the Archaean. Although much work is still required to quantify Archaean tidal cyclicities, a higher and more pronounced tidal regime, owing to a smaller semi-major axis (mean Earth–Moon distance) and stronger gravitational and rotational forces, would have affected sedimentary basins and restricted depositional settings far more effectively than today, even in highly tectonically active environments. Modern counterparts such as Baja California, envisaged to be morphologically and tectonically similar to the proposed Jackson Lake setting, would have been affected far more efficiently by tides during the Archaean.

Future identification and evaluation of well preserved vertically and laterally accreted tidal bundles in Archaean sequences are required to test and confirm or refute the proposed hypothesis. In the light of this study, the best preserved sections of the Jackson Lake Formation need to be re-evaluated. In addition, the well preserved 2.8–2.9 Ga platformal quartz–arenite sequences elsewhere in the Slave Province (P. Pickett, MSc thesis in progress at Chicoutimi) are prime candidates for the preservation of vertically stacked tidal bundles.

## ACKNOWLEDGEMENTS

This research project was principally supported by the Geology Division of Indian Affairs and Northern Development in Yellowknife (DIAND CONTRIBUTION #00-003), Northwest Territories, headed at that time by Dr W.A. Padgham. The authors thank Dr Padgham for his logistical and financial support and continued interest. P. Doucet and S. Belley were able field assistants. This study benefited from LITHOPROBE (LITHOPROBE CONTRIBUTION No. 1159), and NSERC grants to W.U.M. and J.A.D. Thought provoking and rigorous reviews by E. Simpson and H. Stollhofen were greatly appreciated and significantly improved the manuscript.

## REFERENCES

Best, J. & Bridge, J. (1992) The morphology and dynamics of low amplitude waves upon upper stage plane beds and the preservation of planar laminae. *Sedimentology*, **39**, 737–752.

Blair, T.C. (1987) Sedimentary processes, vertical stratification sequences, and geomorphology of the Roaring

River alluvial fan, Rocky Mountain National Park, Colorado. *J. sediment. Petrol.*, **57**, 1–18.

BLAIR, T.C. & BILODEAU, W.L. (1988) Development of tectonic cyclothems in rift, pull-apart, and foreland basins: Sedimentary response to episodic tectonism. *Geology*, **16**, 517–520.

BLAIR, T.C. & MCPHERSON, J.G. (1994) Alluvial fans and their natural distinction from rivers based on morphology, hydraulic processes, sedimentary processes, and facies assemblages. *J. sediment. Res.*, **A64**, 450–489.

BLEEKER, W., KETCHUM, J.W.F., JACKSON, V.A. & VILLENEUVE, M.E. (1999) The central Slave basement complex, part I: its structural topology and autochthonous cover. *Can. J. Earth Sci.*, **36**, 1083–1109.

BLEEKER, W. & VILLENEUVE, M. (1995) Structural studies along the Slave portion of the SNORCLE transect. *Slave–Northern Cordillera Lithospheric Evolution Transect Workshop Meeting, 8–9 April, University of Calgary*, 8–13.

BOOTHROYD, J.C. & NUMMEDAL, D. (1978) Proglacial braided outwash: a model for humid alluvial-fan deposits. In: *Fluvial Sedimentology* (Ed. Miall, A.D.). Mem. Can. Soc. petrol. Geol., Calgary, **5**, 641–668.

BOSE, P.K., MAZUMDER, R. & SARKER, S. (1997) Tidal sandwaves and related storm deposits in the transgressive Protoproterozoic Chaibas Formation, India. *Precam. Res.*, **84**, 63–81.

BOWRING, S.A. & WILLIAMS, I.S. (1998) Priscoan (4.00–4.03 Ga) orthogneisses from northwestern Canada. *Contrib. Mineral. Petrol.*, **134**, 3–16.

CHAMYAL, L.S., KHADKIKAR, A.S., MALIK, J.N. & MAURYA, D.M. (1997) Sedimentology of the Narmada alluvial fan, western India. *Sediment. Geol.*, **107**, 263–279.

CHRISTIE-BLICK, N. & BIDDLE, K.T. (1985) Deformation and basin formation along strike-slip faults. In: *Strike-slip Deformation, Basin Formation, and Sedimentation* (Eds Biddle, K.T. & Christie-Blick, N.). Spec. Publ. Soc. econ. Paleont. Miner. Tulsa, **37**, 1–34.

COLELLA, A. & PRIOR, D.B. (1990) *Coarse-grained Deltas*. Spec. Publs int. Ass. Sediment., No. 10. Blackwell Scientific Publications, Oxford.

CORCORAN, P.L., MUELLER, W.U. & CHOWN, E.H. (1998) Climatic and tectonic influences on fan deltas and wave-to tide-controlled shoreface deposits: evidence from the Archean Keskarrah Formation, Slave Province, Canada. *Sediment. Geol.*, **120**, 125–152.

CORCORAN, P.L., MUELLER, W.U. & PADGHAM, W. (1999) Influence of tectonism and climate on lithofacies distribution and sandstone and conglomerate composition in the Archean Beaulieu Rapids Formation, Northwest Territories, Canada. *Precam. Res.*, **94**, 175–204.

COUSENS, B.L. (2000) Geochemistry of the Archean Kam Group, Yellowknife Greenstone Belt, Slave Province, Canada. *J. Geol.*, **108**, 181–197.

CROWELL, J.C. & LINK, M.H. (1982) *Geologic History of the Ridge Basin, Southern California*. Society of Economic Paleontologists and Mineralogists, Tulsa, 304 pp.

DALRYMPLE, R.W. (1992) Tidal depositional systems. In: *Facies Models: Response to Sea Level Change* (Eds Walker, R.G. & James, N.P.), pp. 195–218. Geological Association of Canada.

DECELLES, P.G., GRAY, M.B., RIDGWAY, K.D. *et al.* (1991) Kinematic history of a foreland uplift Paleocene synorogenic conglomerate, Beartooth Range, Wyoming and Montana. *Geol. Soc. Am. Bull.*, **103**, 1458–1475.

DE RAAF, J.F.M., BOERSMA, J.R. & VAN GELDER, A. (1977) Wave-generated structures and sequences from a shallow marine succession, Lower Carboniferous, County Cork, Ireland. *Sedimentology*, **24**, 451–483.

DES MARAIS, D.J. (1994) The Archean atmosphere: its composition and fate. In: *Archean Crustal Evolution: Developments in Precambrian Geology 11* (Ed. Condie, K.C.). Elsevier, Amsterdam, pp. 505–519.

DEYNOUX, M., DURINGER, P., KHATIB, R. & VILLENEUVE, M. (1993) Laterally and vertically accreted tidal deposits in the Upper Proterozoic Madina–Kouta basin, southeastern Senegal, West Africa. *Sediment. Geol.*, **84**, 179–188.

DUKE, W.L. & PRAVE, A.R. (1991) Storm- and tide-influenced prograding shoreline sequences in the Middle Devonian Mahantango Formation, Pennsylvania. In: *Clastic Tidal Sedimentology* (Eds Smith, D.G., Reinson, G.E., Zaitlin, B.A. & Rahmani, R.A.). Mem. Can. Soc. petrol. Geol., Calgary, **16**, 349–370.

EHLERS, T.A. & CHAN, M.A. (1999) Tidal sedimentology and estuarine deposition of the Proterozoic Big Cottonwood Formation, Utah. *J. sediment. Res.*, **69**, 1169–1180.

ERIKSSON, K.A. (1977) Tidal deposits from the Archaean Moodies Group, Barberton Mountain Land, South Africa. *Sediment. Geol.*, **18**, 257–281.

ERIKSSON, K.A. (1978) Alluvial and destructive beach facies from the Archaean Moodies Group, Barberton Mountain Land, South Africa and Swaziland. In: *Fluvial Sedimentology* (Ed. Miall, A.D.), Mem. Can. Soc. petrol. Geol., Calgary, 5, 287–311.

ERIKSSON, K.A. (1980) Transitional sedimentation styles in the Moodies and Fig Tree Groups, Barberton Mountain Land, South Africa: evidence favouring an Archean continental margin. *Precam. Res.*, **12**, 141–160.

ERIKSSON, K.A., KIDD, W.S.F. & KRAPEZ, B. (1988) Basin analysis in regionally metamorphosed and deformed Early Archean terrains: examples from southern Africa and western Australia. In: *New Perspectives in Basin Analysis* (Eds Kleinspehn, K.L. & Paola, C.), pp. 371–404. Springer-Verlag, New York.

ERIKSSON, K.A., KRAPEZ, B. & FRALICK, P.W. (1994) Sedimentology of Archean greenstone belts: signatures of tectonic evolution. *Earth Sci. Rev.*, **37**, 1–88.

ERIKSSON, P.G., CONDIE, K.C., TIRSGAARD, H. *et al.* (1998) Precambrian clastic sedimentation systems. *Sediment. Geol.*, **120**, 5–53.

FALCK, H. (1990) *Volcanic and sedimentary rocks of the Yellowknife Bay Formation, Giant section, Yellowknife greenstone belt, NWT.* Unpblished MSc thesis, Carleton University, Canada.

FROSTICK, L.E. & STEEL, R.J. (1993) Tectonic signatures in sedimentary basins fills: an overview. In: *Tectonic Controls and Signatures in Sedimentary Successions* (Eds Frostick, L.E. & Steel, R.J.), Spec. Publs int. Ass. Sediment. No. 20, pp. 1–9. Blackwell Scientific Publication, Oxford.

GALLOWAY, W.E. & HOBDAY D.K. (1983) *Terrigenous Clastic Depositional Systems.* Springer-Verlag, Berlin, 423 pp.

GEORGE, A.D. (1994) Tidal sedimentation in part of the late Silurian Grampians Basin, southeastern Australia. *J. sediment. Res.*, **B64**, 311–325.

HAMBERG, L. (1991) Tidal and seasonal cycles in a Lower Cambrian shallow marine sandstone (Hardeberga Fm.)

Scania, southern Sweden. In: *Clastic Tidal Sedimentology* (Eds Smith, D.G., Reinson, G.E., Zaitlin, B.A. & Rahmani, R.A.), Mem. Can. Soc. petrol. Geol., **16**, 255–274.

Hart, B.S. & Plint, A.G. (1995) Gravelly shoreface and beachface deposits. In: *Sedimentary Facies Analysis* (Ed. Plint, A.G.), Spec. Publs int. Ass. Sediment., No. 22, pp. 75–99. Blackwell Science, Oxford.

Hartley, A.J. (1993) Sedimentological response of an alluvial system to source area tectonism: the Seilao Member of the Late Cretaceous to Eocene Purilactic Formation of northern Chile. In: *Tectonic Controls and Signatures in Sedimentary Successions* (Eds Frostick, L.E. & Steel, R.J.), Spec. Publs int. Ass. Sediment., No. 20, pp. 489–500. Blackwell Scientific Publications, Oxford.

Helmstaedt, H. & Bailey, H. (1987) Problems of structural geology in the Yellowknife greenstone belt. In: *Yellowknife Guide Book* (Ed. Padgham, W.A.), pp. 33–39. Min. Deposits Div., Geolological Survey of Canada, Ottawa.

Helmstaedt, H., King, J. & Boodle, R. (1980) Geology of the Banting and Walsh Lakes map area, NTS 85J/9. Geology Division, Indian Affairs and Northern Development, EGS 1980-5.

Helmstaedt, H. & Padgham, W.A. (1986) A new look at the stratigraphy of the Yellowknife Supergroup at Yellowknife, NWT—implications for the age of gold-bearing shear zones and Archean basin evolution. *Can. J. Earth Sci.*, **23**, 454–475.

Hempton, M.R., Dunne, L.A. & Dewey, J.F. (1983) Sedimentation in an active strike-slip basin, southeastern Turkey. *J. Geol.*, **91**, 401–412.

Henderson, J.B. (1975) Sedimentology of the Archean Yellowknife Supergroup at Yellowknife, District of Mackenzie. *Geol. Surv. Can. Bull.*, **246**, 62 pp.

Henderson, J.B. (1981) Archean basin evolution in the Slave Province, Canada. In: *Precambrian Plate Tectonics* (Ed. Kroner, A.). Elsevier, Amsterdam, pp. 213–236.

Henderson, J.B. (1985) Geology of the Yellowknife–Hearne Lake area, District of Mackenzie: a segment across an Archean basin. Mem. Geol. Surv. Can., Ottawa, **414**.

Henderson, J.B. (1998) Geology, Keskarrah Bay Area, District of Mackenzie, Northwest Territories. Map 1679A, scale 1 : 50 000. Geological Survey of Canada, Ottawa.

Henderson J.B. & Brown, I.C. (1966) Geology and structure of the Yellowknife greenstone belt, District of Mackenzie. *Geol. Surv. Can. Bull.*, **14**, 87 pp.

Henderson, J.B., van Breemen, O. & Loveridge, W.D. (1987) Some U–Pb zircon ages from Archean basement, supracrustal and intrusive rocks, Yellowknife–Hearne Lake area, District of Mackenzie. In: *Radiogenic Age and Isotopic Studies Report 1, Paper 87-2*, pp. 111–121. Geological Survey of Canada, Ottawa.

Heubeck, C. & Lowe, D.R. (1994) Depositional and tectonic setting of the Archean Moodies Group, Barberton greenstone belt, South Africa. *Precam. Res.*, **68**, 257–290.

Heward, A.P. (1978) Alluvial fan sequence and megasequence models: with examples from Westphalian D–Stefanian B coalfields, northern Spain. In: *Fluvial Sedimentology* (Ed. Miall, A.D.), Mem. Can. Soc. petrol. Geol., Calgary, **5**, 669–702.

Horton, B.K. & Schmitt, J. (1996) Sedimentology of a lacustrine fan-delta system, Miocene Horse Camp Formation, Nevada, USA. *Sedimentology*, **43**, 133–156.

Houthuys, R. & Gullentops, F. (1988) Tidal transverse bars building up a longitudinal sand body (Middle Eocene, Belgium). In: *Tide-influenced Sedimentary Environments and Facies* (Eds de Boer, P.L., Van Gelder, A. & Nio, S.D.), pp. 153–166. Reidel, Dordrecht.

Howard, J.D. & Reineck, H.-E. (1981) Depositional facies of high-energy beach to offshore sequence: a comparison with a low-energy sequence. *Am. Ass. petrol. Geol. Bull.*, **65**, 807–830.

Hwang, I.G., Chough, S.K., Hong, S.W. & Choe, M.Y. (1995) Controls and evolution of fan delta systems in the Miocene Pohang Basin, SE Korea. *Sediment. Geol.*, **98**, 147–179.

Ingersoll, R.V. & Busby, C.J. (1995) Tectonics of sedimentary basins. In: *Tectonics of Sedimentary Basins* (Eds Busby, C.J. & Ingersoll, R.V.), pp. 1–51. Blackwell Science, Oxford.

Isachsen, C.E. & Bowring, S.A. (1994) Evolution of the Slave Craton. *Geology*, **22**, 917–920.

Isachsen, C.E. & Bowring, S.A. (1997) The Bell Lake group and Anton Complex: a basement–cover sequence beneath the Archean Yellowknife greenstone belt revealed and implicated in greenstone belt formation. *Can. J. Earth Sci.*, **34**, 169–189.

Johnson, H.D. (1978) Shallow siliciclastic seas. In: *Sedimentary Environments and Facies* (Ed. Reading, H.G.), pp. 207–258. Blackwell Scientific Publications, Oxford.

Johnson, H.D. & Levell, B.K. (1995) Sedimentology of a transgressive, estaurine sand complex: the Lower Cretaceous Woburn Sands (Lower Greensand), southern England. In: *Sedimentary Facies Analysis* (Ed. Plint, G.), Spec. Publs Int. Ass. Sediment., No. 22, pp. 17–46. Blackwell Science, Oxford.

Kasting, J.F. (1993) Earth's early atmosphere. *Science*, **259**, 920–926.

Komiya, T., Maruyama, S., Masuda, T., Nohda, S., Hayashi, M. & Okamoto, K. (1999) Plate tectonics at 3.8–3.7 Ga: field evidence from the Isua accretionary complex, southern West Greenland. *J. Geol.*, **107**, 515–554.

Koster, E. H. & Steel, R.J. (1984) Sedimentology of gravels and conglomerates. *Can. Soc. petrol. Geol. Mem.*, **10**, 441 pp.

Kreisa, R.D. (1981) Storm-generated sedimentary structures in subtidal marine facies with examples from the Middle and Upper Ordovician of southwestern Virginia. *J. sediment. Petrol.*, **51**, 823–848.

Kreisa, R.D. & Moiola, R.J. (1986) Sigmoidal tidal bundles and other tide-generated sedimentary structures of the Curtis Formation, Utah. *Geol. Soc. Am. Bull.*, **83**, 381–387.

Kreisa, R.D., Moiola, R.J. & Nøttvedt, A. (1986) Tidal sand wave facies, Rancho Rojo sandstone (Permian), Arizona. In: *Shelf Sands and Sandstones* (Eds Knight, R.J. & McLean. J.R.). Mem. Can. Soc. petrol. Geol., Calgary, **2**, 277–291.

Kusky, T.M. (1989) Accretion of the Archean Slave province. *Geology*, **17**, 63–67.

Kusky, T.M. & Polat, A. (1999) Growth of granite–greenstone terranes at convergent margins, and stabilization of Archean Cratons. *Tectonophysics*, **305**, 43–73.

Kvale, E.P., Johnson, H.W., Sonett, C.P., Archer, A.W. & Zawistoski, A. (1999) Calculating lunar retreat rates using tidal rhythmites. *J. sediment. Res.*, **69**, 1154–1168.

Laird, M.G. (1995) Coarse-grained lacustrine fan-delta deposits (Pororari Group) of the northwestern South Island, New Zealand: evidence for Mid-Cretaceous rifting. In: *Sedimentary Facies Analysis* (Ed Plint, A.G.), Spec. Publs int. Ass. Sediment., No. 22, pp. 197–217. Blackwell Science, Oxford.

Lambert, M.B. (1988) The Cameron and Beaulieu River volcanic belts, District of Mackenzie, Northwest Territories. *Geol. Surv. Can. Bull.*, **382**, 145 pp.

Lonegran, L. & Scheiber, B.C. (1993) Proximal deposits a fault-controlled basin margin, Upper Miocene, SE Spain. *J. geol. Soc. London*, **150**, 719–727.

Long, D.G.F. (1981) Dextral strike slip faults in the Canadian Cordillera and depositional environments of related fresh-water intermontane coal basins. In: *Sedimentation and Tectonics in Alluvial Basins* (Ed. Miall, A.D.), Spec. Publ. Geol. Ass. Can., St John's, **23**, 153–186.

Lowe, D.R. (1994) Archean greenstone-related sedimentary rocks. In: *Archean Crustal Evolution* (Ed. Condie, K.C.), pp. 121–169. Developments in Precambrian Geology, 11. Elsevier, Amsterdam.

Mack, G.H. & Rasmussen, K.A. (1984) Alluvial-fan sedimentation of the Cutler Formation (Permo-Pennsylvanian) near Gateway, Colorado. *Geol. Soc. Am. Bull.*, **95**, 109–116.

MacLachlan, K. & Helmstaedt, H. (1995) Geology and geochemistry of an Archean mafic dike complex in the Chan Formation: basis for a revised plate-tectonic model of the Yellowknife greenstone belt. *Can. J. Earth Sci.*, **32**, 614–630.

Marzo, M. & Puigdefábregas, C. (Eds) (1993) *Alluvial Sedimentation*. Spec. Publs int. Ass. Sediment., No. 17. Blackwell Scientific Publications, Oxford.

Miall, A.D. (1978) Lithofacies types and vertical profile models in braided river deposits: a summary. In: *Fluvial Sedimentology* (Ed. Miall, A.D.), Mem. Can. Soc. petrol. Geol., Calgary, **5**, 801–829.

Miall, A.D. (1992) Alluvial deposits. In: *Facies Models: Response to Sea Level Change* (Eds Walker, R.G. & James, N.P.), pp. 119–142. Geological Association of Canada.

Mortensen, J.K., Henderson, J.B., Jackson, V.A. & Padgham, W.A. (1992) U–Pb geochronology of Yellowknife Supergroup felsic volcanic rocks in the Russell Lake and Clan Lake areas, southwestern Slave Province, NWT In: *Radiogenic Age and Isotopic Studies 5, Paper 91-2*, pp. 1–7. Geological Survey of Canada, Ottawa.

Mueller, W.U., Bowring, S.A., Corcoran, P.L. & Pickett, C. (1998) Unconformities, major faults and the evolution of volcano-sedimentary basins on the Slave craton. *Slave–Northern Cordillera Lithospheric Evolution Transect and Cordilleran Tectonics Workshop Meeting, 6–8 March, Simon Fraser University*, pp. 15–16.

Mueller, W.U. & Corcoran, P.L. (1997) Volcanology and sedimentology of the Raquette Lake Formation: a remnant dissected arc sequence. NWT Geology Division, Indian and Northern Affairs Canada, EGS-1997-11.

Mueller W.U. & Corcoran, P.L. (1998) Late-orogenic basins in the Archean Superior Province, Canada: characteristics and inferences. *Sediment. Geol.*, **120**, 177–203.

Mueller, W.U., Daigneault, R., Mortensen, J.K. & Chown, E.H. (1996) Archean terrane docking: upper crust collision tectonics, Abitibi greenstone belt, Quebec, Canada. *Tectonophysics*, **265**, 127–150.

Mueller, W. & Dimroth, E. (1987) A terrestrial–shallow marine transition in the Archean Opemisca Group east of Chapais, Quebec. *Precam. Res.*, **37**, 29–55.

Mueller, W. & Donaldson, J.A. (1992) Development of sedimentary basins in the Abitibi belt: an overview. *Can. J. Earth Sci.*, **29**, 2249–2265.

Mueller, W., Donaldson, J.A. & Doucet, P. (1994) Volcanic and tectono-plutonic influences on sedimentation in the Archean Kirkland Basin Abitibi greenstone belt, Canada. *Precam. Res.*, **68**, 201–230.

Mueller, W., Donaldson, J.A., Dufresne, D. & Rocheleau, M. (1991) The Duparquet Formation: sedimentation in a late Archean successor basin, Abitibi greenstone belt, Quebec, Canada. *Can. J. Earth Sci.*, **28**, 1394–1406.

Nemec, W. & Postma, G. (1993) Quaternary alluvial fans in southwestern Crete: sedimentation processes and geomorphic evolution. In: *Alluvial Sedimentation* (Eds Marzo, M. & Puigdefábregas, C.), Spec. Publs int. Ass. Sediment., No. 17, pp. 235–276. Blackwell Scientific Publications, Oxford.

Nilsen, T.H. & McLaughlin, J. (1985) Comparison of tectonic framework and depositional patterns of the Hornelen strike-slip basin of Norway and the Ridge and Little Sulphur Creek strike-slip basins of California. In: *Strike-slip Deformation, Basin Formation, and Sedimentation* (Eds Biddle, K.T. & Christie-Blick, N.), Spec. Publs Soc. econ. Paleont. Miner., Tulsa, **37**, 79–103.

Nio, S.-D. & Yang, C.-S. (1991) Diagnostic attributes of clastic tidal deposits: a review. In: *Clastic Tidal Sedimentology* (Eds Smith, D.G., Reinson, G.E., Zaitlin, B.A. & Rahmani, R.A.). Mem. Can. Soc. petrol. Geol., Calgary, **16**, 3–28.

Orton, G.J. & Reading, H.G. (1993) Variability of deltaic processes in terms of sediment supply, with particular emphasis on grain size. *Sedimentology*, **40**, 475–512.

Padgham, W.A. (1987a) The Yellowknife volcanic belt: setting and stratigraphy. In: *Yellowknife Guide Book* (Ed. Padgham, W.A.), pp. 11–19. Min. Dep. Div., Geological Survey of Canada, Ottawa.

Padgham, W.A. (1987b) Guide to parts of the Crestaurum, Townsite, and Yellowknife Bay Formations and the Banting Group. In: *Yellowknife Guide Book* (Ed. Padgham, W.A.), pp. 55–79. Min. Dep. Div., Geological Survey of Canada, Ottawa.

Padgham, W.A. & Fyson, W.K. (1992) The Slave Province: a distinct Archean craton. *Can. J. Earth Sci.*, **29**, 2072–2086.

Rainbird, R.H. (1992) Anatomy of a large-scale braid-plain quartzarenite from the Neoprotero: Shaler Group, Victoria Island, Northwest Territories, Canada. *Can. J. Earth Sci.*, **29**, 2537–2550.

Reading, H.G. & Collinson, J.D. (1996) Clastic coasts. In: *Sedimentary Environments: Processes, Facies and Stratigraphy*, 3rd edn (Ed. Reading, H.G.), pp. 154–231. Blackwell Science, Oxford.

Reineck, H.E. & Singh, I.B. (1980) *Depositional Sedimentary Environments*. Springer-Verlag, Heidelberg, 549 pp.

Relf, C., Chouinard, A., Sandeman, H. & Villeneuve, M. (1994) Contact relationships between the Anialik River volcanic belt and the Kangguyak gneiss belt, northwestern Slave Province, Northwest Territories. In: *Current Research 1994-C*, pp. 49–59. Geological Survey of Canada, Ottawa.

Richards, M.T. (1986) Tidal bed form migration in shallow marine environments: evidence from the lower Triassic, western Alps, France. In: *Shelf Sands and Sandstones* (Eds Knight, R.J. & McLean, J.R.), Mem. Can. Soc. Petr. Geol., Calgary, **11**, 257–276.

Richards, M.T. (1994) Transgression of an estuarine channel and tidal flat complex: the lower Triassic of Barles, Alpes de Haute Provence, France. *Sedimentology*, **41**, 55–82.

Ridgway, K.D. & DeCelles, P.G. (1993) Stream-dominated alluvial fan and lacustrine depositional systems in Cenozoic strike-slip basins, Denali fault system, Yukon Territory, Canada. *Sedimentology*, **40**, 645–666.

Røe, S.-L. & Hermansen, M. (1993) Processes and products of large, Late Precambrian sandy rivers in northern Norway. In: *Alluvial Sedimentation* (Eds Marzo, M. & Puigdefabregas, C.), Spec. Publs int. Ass. Sediment., No. 17, pp. 151–166. Blackwell Scientific Publications, Oxford.

Rust, B.R. (1984) Proximal braidplain deposits in the Middle Devonian Malbaie Formation of eastern Gaspé, Canada: *Sedimentology*, **31**, 675–695.

Rust, B.R. & Koster, E.H. (1984) Coarse clastic deposits. In: *Facies Models*, 2nd edn (Ed. Walker, R.G.). Geosci. Can. Reprint Ser. 1, 53–69.

Simpson, E.L. & Eriksson, K.A. (1989) Sedimentology of the Unicoi Formation in southern and central Virginia: evidence for late Proterozoic to early Cambrian rift-to-passive margin transition. *Geol. Soc. Am. Bull.*, **101**, 42–54.

Simpson, E.L. & Eriksson, K.A. (1991) Depositional facies and controls on parasequence development in siliciclastic tidal deposits from the Lower Proterozoic, upper Mount Guide Quartzite, Mount Isa Inlier, Australia. In: *Clastic Tidal Sedimentology* (Eds Smith, D.G., Reinson, G.E., Zaitlin, B.A. & Rahmani, R.A.), Mem. Can. Soc. petrol. Geol., Calgary, **16**, 371–387.

Sonett, C.P., Kvale, E.P., Zakharian, A., Chan, M.A. & Demko, T.M. (1996a) Late Proterozoic and Paleozoic tides, retreat of the Moon, and rotation of the Earth. *Science*, **274**, 100–104.

Sonett, C.P., Zakharian, A. & Kvale, E.P. (1996b) Ancient tides and length of day: correction. *Science*, **274**, 1068–1069.

Steel, R.J. & Gloppen, T.G. (1980) Late Caledonian (Devonian) basin formation, western Norway: signs of strike-slip tectonics during basin infilling. In: *Sedimentation in Oblique-slip Mobile Zones* (Eds Ballance, P.F. & Reading, H.G.), Spec. publs int. Ass. Sediment., No. 4, pp. 79–103. Blackwell Scientific Publications, Oxford.

Steel, R.J., Maehle, S., Nielsen, H., Røe, S.L. & Spinnagr, A. (1977) Coarsening-upward cycles in the alluvium of Hornelen Basin (Devonian), Norway: sedimentary response to tectonic events. *Geol. Soc. Am. Bull.*, **88**, 1124–1134.

Tessier, B. & Gigot, P. (1989) A vertical record of different tidal cyclicities: an example from the Miocene Marine Molasse of Digne (Haute Provence, France). *Sedimentology*, **36**, 767–776.

Visser, M.J. (1980) Neap–spring cycles reflected in Holocene subtidal large-scale bedform deposits: a preliminary note. *Geology*, **8**, 543–546.

Wescott, W.A. & Ethridge, F.G. (1980) Fan-delta sedimentology and tectonic setting—Yallahs fan delta, southeast Jamaica. *Am. Ass. petrol. Geol. Bull.*, **64**, 374–399.

Williams, G.E. (1998) Precambrian tidal and glacial clastic deposits: implications for Precambrian earth–Moon dynamics and paleoclimate. *Sediment. Geol.*, **120**, 55–74.

Young, G.M. (1991) The geologic record of glaciation: relevance to the climatic history of the Earth. *Geosci. Can.*, **18**, 100–108.

*Spec. Publs int. Ass. Sediment.* (2002) **33**, 183–211

# The effects of weathering, sorting and source composition in Archaean high-relief basins: examples from the Slave Province, Northwest Territories, Canada

P. L. CORCORAN* *and* W. U. MUELLER†

**Department of Earth Sciences, Dalhousie University, Halifax, Nova Scotia, B3H 3J5, Canada; and †Sciences de la terre, Université du Québec a Chicoutimi, Québec, G7H 2B1, Canada*

## ABSTRACT

The Archaean Keskarrah, Beaulieu Rapids and Jackson Lake formations are late-orogenic, tectonically controlled sedimentary sequences characterized by major bounding unconformities and crustal-scale faults. Representing the youngest depostional event in the Slave Province, the successions contain similar lithofacies, rapid lateral and vertical lithofacies changes, and upward-fining and upward-coarsening then fining sequences, which, in association with faults and unconformities, are convincing indicators of tectonic influence. The Keskarrah and Jackson Lake formations were deposited as alluvial fans and fan-deltas prograding into shallow-water shelf settings, whereas the Beaulieu Rapids Formation represents ancient alluvial fan and fan-delta deposits passing into fluvial and minor lacustrine environments. Petrography, clast counts and geochemistry indicate that the compositions of the sedimentary sequences were influenced by source rock composition, sorting and chemical weathering. The Keskarrah Formation, containing quartz-rich sandstones and quartz arenites, received detritus from a plutonic-dominated source, as indicated by 48% plutonic clasts in conglomerate, high proportions of quartz in sandstones, high Th values and $La_n/Yb_n$ and $Sm_n/Yb_n$ ratios and low Ti, Sc and Ni values. In addition, CIA values of 68–95 demonstrate that the Keskarrah sediments underwent significant chemical weathering while being subjected to high degrees of sorting by tides and waves along a shelf setting. Mafic volcanic rocks were the main source for the Jackson Lake sedimentary sequence, represented by 72% mafic volcanic clasts in conglomerate, high proportions of volcanic rock fragments in the sandstones, in addition to low Th values and $La_n/Yb_n$ and $Sm_n/Yb_n$ ratios, and higher Ti, Sc and Ni values compared with those of the Keskarrah Formation. CIA values ranging from 67 to 83 indicate substantial chemical weathering, and additional diminution of labile minerals occurred in the macrotidal shelf depositional setting. Major source rocks for the Beaulieu Rapids Formation were plutonic and mafic volcanic, as indicated by 38 and 43% plutonic and mafic volcanic clast components, respectively, high proportions of polycrystalline quartz and Th, Ti, Sc and Ni values and $La_n/Yb_n$ and $Sm_n/Yb_n$ ratios intermediate between the Keskarrah and Jackson Lake formations. CIA values for the Beaulieu Rapids sedimentary rocks vary from 69 to 78. The quartz-rich nature of the sedimentary sequences is suggestive of the complex interplay between chemical weathering and sorting, two factors that were enhanced in the Archaean owing to a humid, $CO_2$-rich atmosphere and a greater tidal influence.

## INTRODUCTION

The chemical and mineralogical composition of sandstones and siltstones, combined with conglomerate clast populations, provide significant information concerning source areas, in addition to geomorphological and atmospheric factors involved in the evolution of a particular rock suite. In combination with sedimentary facies analyses and stratigraphy, compositional data are an invaluable tool for determining the most crucial allocyclic factor(s) controlling sedimentary rock formation. The composition of a clastic sequence, as determined from point- and clast-counting procedures, is influenced by source area (Basu, 1976; Mack, 1981; Picard & McBride, 1993), geodynamic setting (Dickinson & Suczek, 1979; Schwab, 1981; Dickinson *et al.*, 1983), relief (Johnsson & Stallard, 1989), climate (Suttner & Dutta, 1986; Nesbitt *et al.*, 1996; Corcoran *et al.*, 1998), transport (Franzinelli & Potter, 1983; Nesbitt & Young, 1996) and depositional

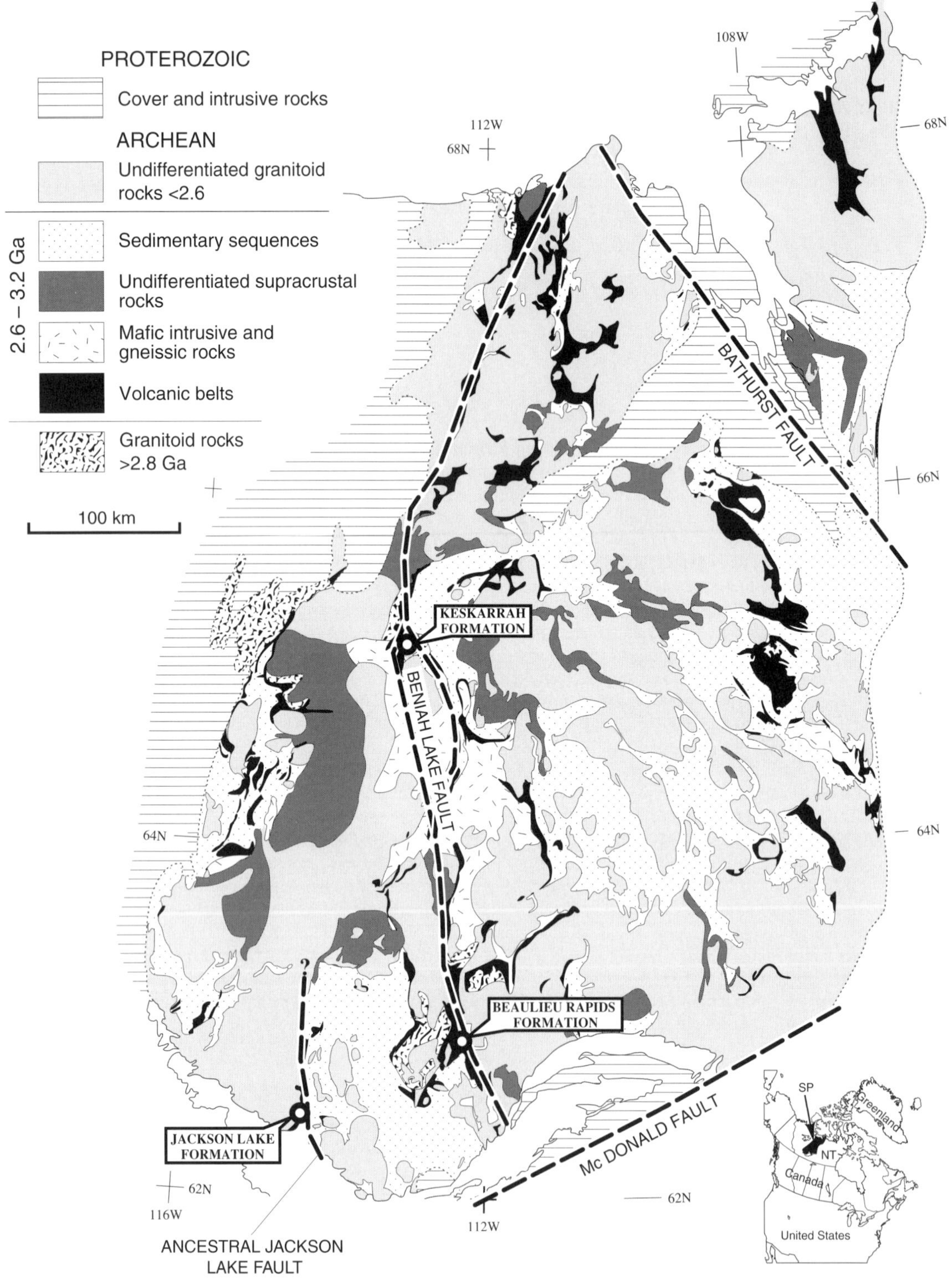


**Fig. 1.** Lithological map of the Slave Province (SP) in the Northwest Territories (NT), Canada, illustrating the location of the Keskarrah and Beaulieu Rapids formations adjacent to the Beniah Lake fault, and the Jackson Lake Formation adjacent to the ancestral Jackson Lake fault. Modified from Corcoran *et al.* (1998).

environment (McBride *et al.*, 1996; Corcoran *et al.*, 1998). Similarly, the geochemistry of sedimentary rocks and possible sources has been employed to deduce provenance (Roser & Korsch, 1988; Fralick & Kronberg, 1997; Holail & Moghazi, 1998), tectonic setting (Bhatia, 1983; Bhatia & Crook, 1986; Feng & Kerrich, 1990) and the degree of chemical weathering (Chemical Index of Alteration, CIA; Nesbitt & Young, 1982; Fedo *et al.*, 1997a,b). Incorporating data obtained from geochemical analyses and point- and clast-counting allows comparison of the results provided from two independent methods. Studies of this nature are significant in Archaean terranes because much of the evidence supporting source rock composition and tectonic setting has been affected by structural deformation and metamorphism. In addition, the composition of ancient sedimentary rocks reflects unique Archaean atmospheric conditions wherein a humid (Des Marais, 1994), $CO_2$-rich (Kasting, 1993) atmosphere would have promoted greater precipitation and increased chemical weathering of source rocks and sediments (Corcoran *et al.*, 1998). Supplementary modifications in sediment composition would have occurred in ancient shelf-type settings where mechanical abrasion and hydraulic sorting during wave- and tide-action could have been augmented owing to a more dynamic Earth–Moon system during the Archaean (Williams, 1998; Mueller *et al.*, this volume, pp. 153–182).

This study concerns three *c.*2.6 Ga (Isachsen *et al.*, 1991; Isachsen & Bowring, 1994) sedimentary basins in the Slave Province, Northwest Territories, Canada, that are located along two crustal-scale fault zones: (i) the Keskarrah Formation and (ii) the Beaulieu Rapids Formation along the Beniah Lake fault, and (iii) the Jackson Lake Formation along the ancestral Jackson Lake fault. Corcoran *et al.* (1998, 1999) previously emphasized the quartz-rich nature of sandstones in the Keskarrah and Beaulieu Rapids formations, which were tectonically active basins during their evolution. High quartz percentages (up to 98%) in first-cycle sedimentary rocks that were deposited in high relief settings were interpreted as being influenced by source rock composition and sorting, but were mainly attributed to intense chemical weathering. A combination of geochemistry and petrography of sandstones and siltstone–sandstones, in addition to clast counting in conglomerates, is used to discern the effects of Archaean chemical weathering. This study also evaluates additional factors that control the composition of sedimentary rocks, while demonstrating that each basin is governed by these factors to varying degrees.

## SLAVE PROVINCE GEOLOGY

The Archaean Keskarrah, Beaulieu Rapids and Jackson Lake formations are located in the approximately 500 × 700 km Slave Province in the Northwest Territories, Canada (Fig. 1). The Slave Craton is composed of basement gneisses, narrowly exposed quartz arenite sequences, linear, N–S trending volcanic belts, extensive turbiditic successions, conglomeratic sequences preserved along faults and younger granitoids (Fig. 1). Characteristics distinguishing the Slave Province from most Archaean cratons include a sialic basement exposed in the west, an elevated felsic/mafic volcanic rock ratio and abundant sedimentary sequences (Henderson, 1981; Padgham & Fyson, 1992). Crustal-scale, north-trending lineaments characterize the Slave craton and are locally associated with linear, mafic-dominated volcanic belts, porphyry stocks and most of the late-orogenic conglomeratic sequences. The Keskarrah Formation of the Point Lake belt, the Beaulieu Rapids Formation of the Beaulieu River volcanic belt and the Jackson Lake Formation of the Yellowknife volcanic belt are considered part of the Yellowknife Supergroup (Henderson, 1970, 1988), in addition to other late-orogenic, coarse clastic successions, such as the <2600 Ma James Falls and Kaycee conglomerates of the Hood River and Anialik belts, respectively (Villeneuve *et al.*, 1993; Relf *et al.*, 1994).

Models proposed for the evolution of the Slave Province are based on either the development of individual greenstone belts or the entire Slave Craton. Henderson (1981, 1985) and Easton (1985) suggested that sedimentary rocks in the Yellowknife and Point Lake belts formed as rift basin deposits that formed on pre-existing continental crust. MacLachlan & Helmstaedt (1995) and Cousens (2000) proposed that the Yellowknife belt formed during rifting of a continental marginal arc. Kusky (1989, 1991) discounted the ensialic rift model and extrapolated the arc-related formation of greenstone belts craton-wide. According to Kusky (1989, 1991), the Slave Province evolved through accretion of distinct terranes, some containing sialic basement. The *c.*2.6 Ga sedimentary sequences unconformably overlying volcanic rocks, flysch-type sedimentary sequences and gneissic basement developed during the late stages of Slave Province cratonization and resemble deposits comprising late-orogenic basins of the Superior Province (Mueller & Corcoran, 1998) and modern fault-bound, pull-apart or strike-slip basins (Corcoran *et al.*, 1999; Mueller *et al.*, this volume, pp. 153–182).

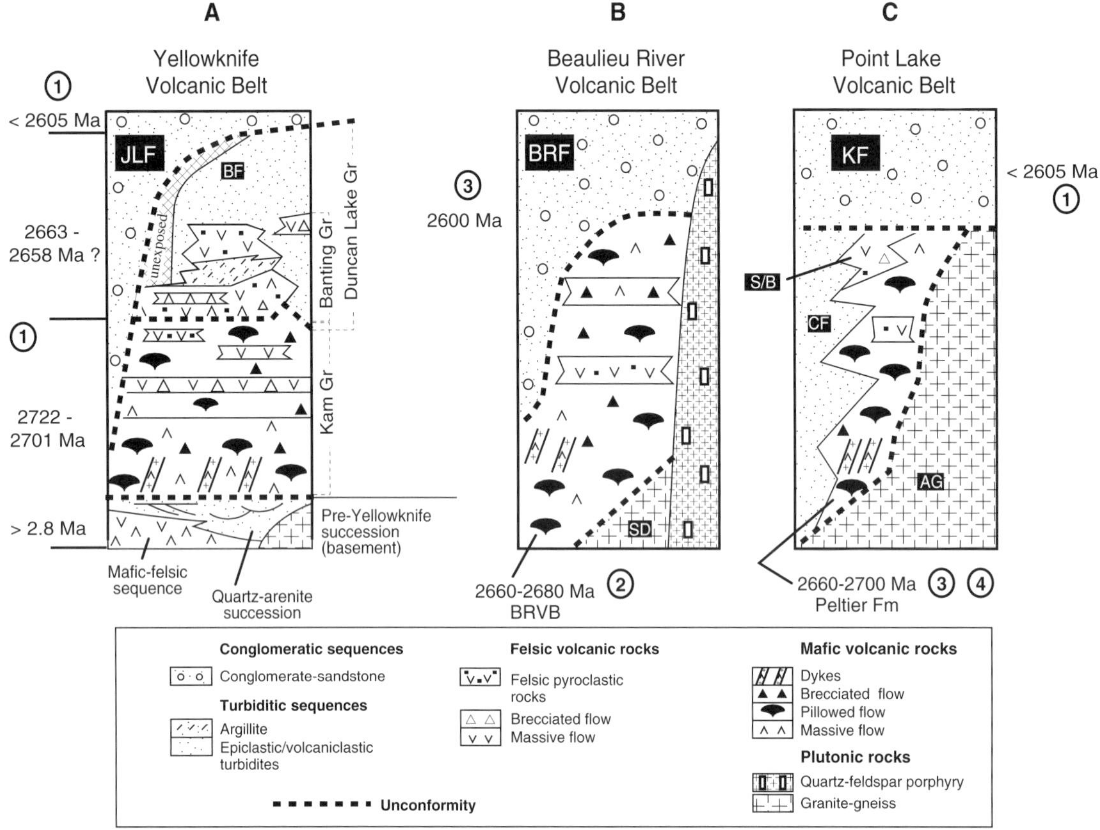


**Fig. 2.** Stratigraphy of the Yellowknife volcanic belt, Slave Province, and correlations with the Beaulieu River and Point Lake volcanic belts. Note that the Jackson Lake, Beaulieu Rapids and Keskarrah formations represent the youngest depositional event. Age dates from: (1) Isachsen *et al.* (1991), Isachsen & Bowring (1994, 1997); (2) Henderson *et al.* (1987); (3) Mueller *et al.* (1998); (4) Northrup *et al.* (1999). JLF, Jackson Lake Formation; BF, Burwash Formation; BRF, Beaulieu Rapids Formation; BRVB, Beaulieu River volcanic belt; SD, Sleepy Dragon Complex; KF, Keskarrah Formation; S/B, Samandre and Beauparlant formations; CF, Contwoyto Formation; AG, Augustus Granite. Modified from Corcoran *et al.* (1998).

## Local geology and unconformable relationships

The Keskarrah, Beaulieu Rapids and Jackson Lake formations represent the youngest depositional event in the Slave Province and thus form a significant pan-Slave tie-line (Mueller *et al.*, 1998; Fig. 2). These successions provide an excellent data set with which to compare factors controlling the composition of conglomerates, sandstones, and siltstones. All rocks in the three study areas have undergone greenschist grade metamorphism, but the prefix 'meta' is omitted for simplicity.

The Keskarrah Formation, in the north-central Slave Province overlies the 3.15–3.22 Ga Augustus Granite (Krogh & Gibbins, 1978; Henderson *et al.*, 1982; Northrup *et al.*, 1999) unconformably at several well exposed localities (Fig. 3, localities A–D; Henderson, 1988; Corcoran *et al.*, 1998). The Keskarrah conglomerate contains large granitic boulders, up to 4 m in size, adjacent to the unconformity (Fig. 3, locality A), whereas cobble-size angular fragments are locally predominant (Fig. 3, locality C; Corcoran *et al.*, 1998). The unconformity between the pillowed and massive mafic volcanic flows of the 2.67–2.69 Ga Point Lake belt (Northrup *et al.*, 1999) and the conglomerate, sandstone and siltstone–sandstone of the Keskarrah Formation is well exposed on Cyclops Peninsula (Fig. 3, localities E, F; Corcoran *et al.*, 1998). In addition, the Keskarrah Formation unconformably overlies the turbiditic Contwoyto Formation, which is interstratified with the volcanic sequence (Fig. 3, locality G; Corcoran *et al.*, 1998; Henderson, 1998; Corcoran,

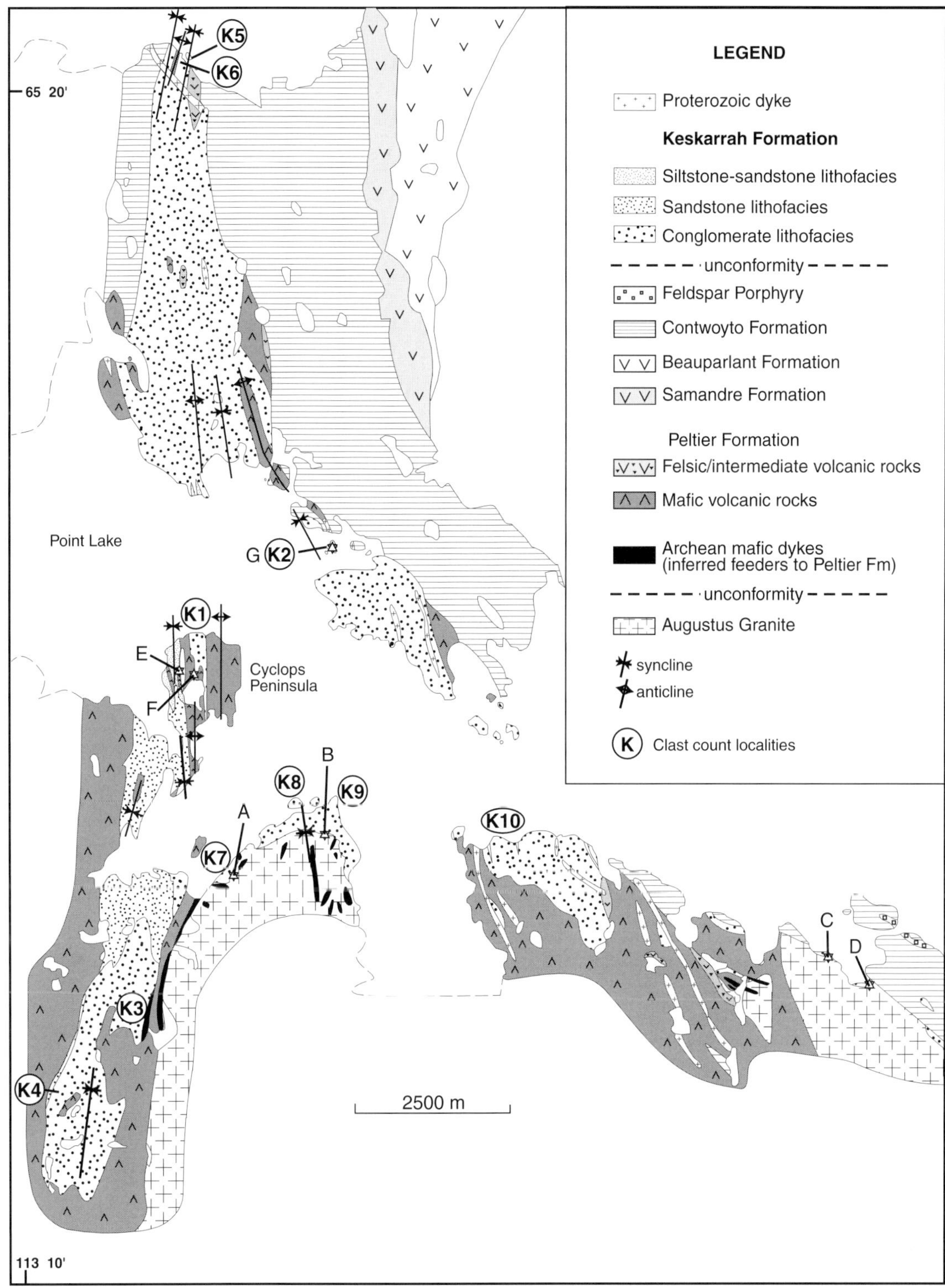


**Fig. 3.** Lithological map illustrating the distribution of lithofacies in the Keskarrah Formation and the well exposed unconformable contacts with the underlying Augustus Granite (localities A–D), Peltier Formation (localities E, F) and Contwoyto Formation (locality G). Clast count stations are indicated K1–K10. Modified from Corcoran *et al.* (1998).

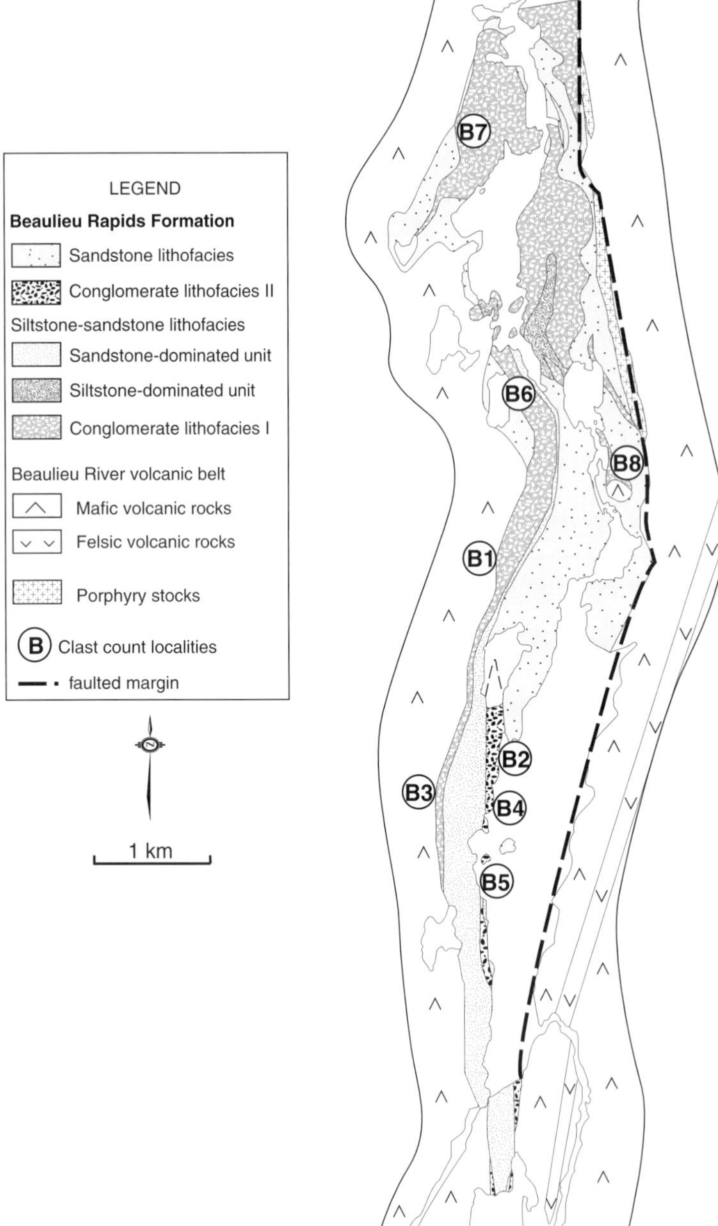


**Fig. 4.** Lithological map of the Beaulieu Rapids Formation which unconformably overlies the Beaulieu River volcanic belt on the west. Note the location of porphyry stocks along the eastern faulted margin. Clast count stations are indicated B1–B8. Modified from Corcoran *et al.* (1999).

2000). The 600 km long, N-trending, crustal-scale Beniah Lake Fault, located approximately 2 km west of the Keskarrah Formation, probably played a crucial role in the formation of the tectonically influenced sedimentary basin (Corcoran *et al.*, 1998). Numerous faults are intimately associated with basin evolution (Henderson, 1988) and local porphyry stocks (Fig. 3).

The Beaulieu Rapids Formation, located in the south-central Slave Province (Fig. 1), unconformably overlies the Beaulieu River volcanic belt (Figs 2 & 4), which is inferred to be time-equivalent with the 2663 Ma Cameron River belt (Henderson *et al.*, 1987; Lambert *et al.*, 1992). The >2.8 Ga (Henderson *et al.*, 1987) Sleepy Dragon Complex is overlain unconformably by the Beaulieu River volcanic belt. Beaulieu Rapids conglomerate and pebbly sandstone are in contact with mafic massive, pillowed and pillow brecciated flows and hyaloclastite (Corcoran, 2000). Felsic tuffs and breccias and quartz–feldspar porphyry stocks are confined to the east (Fig. 4). The stocks are located along the Beniah Lake fault which marks the eastern margin of the basin. Conglomerate deposited along this margin locally contains quartz–feldspar porphyry clasts (Corcoran *et al.*, 1999).

The Jackson Lake Formation, located in the southern Slave Province near Yellowknife (Fig. 1), unconformably overlies the 2658–2722 Ma (Isachsen *et al.*, 1991; van Breemen *et al.*, 1992; Isachsen & Bowring, 1994) Kam, Banting and Duncan Lake Groups of the Yellowknife volcanic belt (Figs 2 & 5). A 35–40 Myr hiatus is inferred between the mafic flow-dominated Kam and felsic flow- and volcaniclastic-dominated Banting Groups (Isachsen & Bowring, 1994). The turbiditic flysch-type Burwash Formation of the Duncan Lake Group is interstratified with age-equivalent Banting Group felsic volcanic centres at Turnback (2663 + 7 or −5 Ma; Henderson *et al.*, 1987) and Clan (2661 ± 2 Ma; Mortensen *et al.*, 1992) Lakes. The Yellowknife volcanic belt rests unconformably on an older gneissic basement (Henderson, 1981; Helmstaedt & Padgham, 1986; Isachsen & Bowring, 1997). The Archaean Jackson Lake fault, later rejuvenated during the Proterozoic as the Hay–Duck fault, constrains the sedimentary basin on the east (Fig. 5).

## Sedimentology

Lithofacies in the Keskarrah, Beaulieu Rapids and Jackson Lake Formations were studied in detail (Corcoran *et al.*, 1998, 1999; Mueller *et al.*, this volume, pp. 153–182). Table 1 lists the sedimentary structures inherent to each of the remnant basins in addition to the interpreted depositional environments. A general review of the sedimentology is provided in this section to place the sedimentary sequences into a geodynamic context. The three Archaean basins demonstrate strikingly similar features, including: (i) association with crustal-scale faults; (ii) unconformable contacts with older volcanic, sedimentary and gneissic rocks; (iii) comparable sedimentary lithofacies; (iv) rapid lateral and vertical lithofacies changes; and (v) well defined, large-scale upward-fining or upward-coarsening then fining sequences. These characteristics are all consistent with deposition in fault-controlled, tectonically influenced basins.

### *Keskarrah Formation*

The 5–700 m thick Keskarrah Formation is composed of conglomerate, sandstone and siltstone–sandstone lithofacies (Fig. 3). The basal conglomerate lithofacies, 75–700 m thick, and located above unconformities with the Augustus Granite, mafic Peltier Formation and sedimentary Contwoyto Formation, grades upsection into the sandstone and, locally, siltstone–sandstone lithofacies. Matrix-supported conglomerate was deposited during debris flow episodes, whereas clast-supported varieties with rounded components represent coalescing gravel bars or sheets. Clast-supported conglomerate with prominent angular clasts formed from talus scree deposition. Sandstone interbeds developed as bar top sand and in-channel dune deposits during waning flood conditions. The Keskarrah conglomerate lithofacies represents alluvial fan and fan delta deposits that developed in a high relief setting.

The sandstone lithofacies, up to 625 m thick, locally overlies mafic volcanic flows unconformably. Contacts between the underlying conglomerate and overlying siltstone–sandstone are generally gradational. Complex tabular cross-strata intimately associated with argillite drapes and laminae are interpreted as subtidal shoreface deposits. Tidal- and wave-influence is inferred, based on tabular structures, reactivation surfaces and argillite-draped foresets. Planar sandstone beds alternating with thinner argillite laminae developed during fluctuating wave and tide energy conditions during storm activity.

The siltstone–sandstone lithofacies, 5–60 m thick, overlies the sandstone and conglomerate lithofacies gradationally. Tabular structures are ubiquitous and represent dunes or sandwaves that were generated during wave and tide action. Abundant argillite drapes and laminae, in addition to reactivation surfaces, further support tidal influence. The relative abundance of argillite indicates elevated levels of suspension sedimentation and a decrease in current and wave energy. The characteristics of the siltstone–sandstone lithofacies indicate deposition in a lower shoreface to proximal offshore setting.

### *Beaulieu Rapids Formation*

The 0.2–1 km thick Beaulieu Rapids Formation is composed of two large-scale depositional sequences:

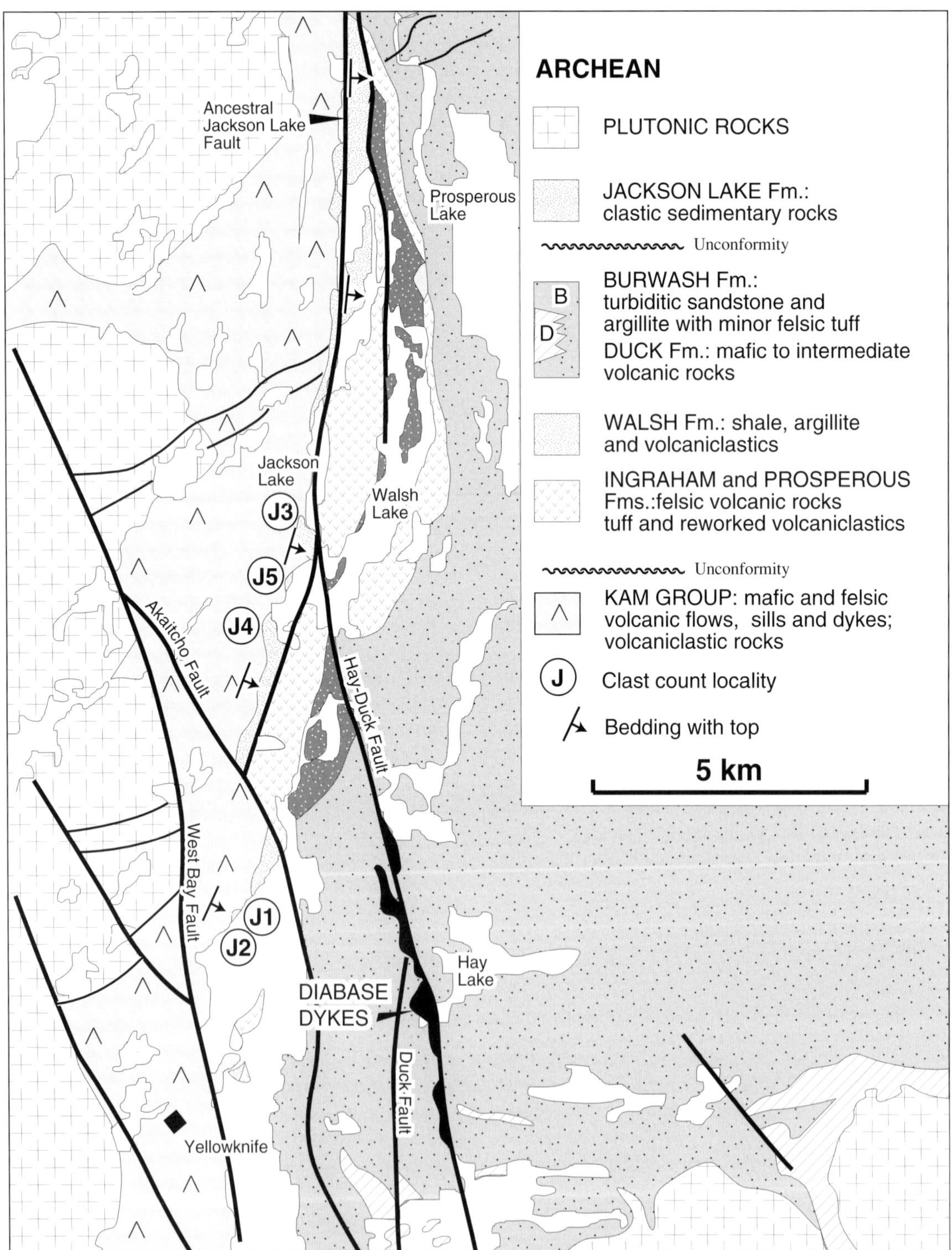


**Fig. 5.** Lithological map of the Jackson Lake Formation demonstrating the western unconformity with the underlying Kam Group, and the eastern fault-bound margin. Clast count stations are indicated J1–J5. Modified from Mueller *et al.* (this volume, pp. 153–182).

**Table 1.** Characteristics of the sedimentary lithofacies in the Keskarrah, Beaulieu Rapids and Jackson Lake formations.

| | | | | |
|---|---|---|---|---|
| *Keskarrah Formation* | | | | |
| Lithofacies | Conglomerate | Sandstone | Siltstone–sandstone | |
| Sedimentary structures | Massive crudely to well stratified, clast-supported conglomerate; massive, matrix-supported conglomerate with local angular clasts; planar bedded sandstone; trough cross-bedded sandstone | Tangential cross-strata; planar beds; planar cross-beds; reactivation surfaces; argillite | Tangential cross-strata; planar beds; planar cross-beds; small scale trough cross-beds; ripples; reactivation surfaces; abundant argillite | |
| Inferred setting | Coalescing streamflow-dominated fan-deltas and alluvial fans | Shallow-water shoreface deposits influenced by waves and tides | Lower shoreface to proximal offshore dominated by waves and tides | |
| *Beaulieu Rapids Formation* | | | | |
| Lithofacies | Conglomerate I | Siltstone–sandstone | Conglomerate II | Sandstone |
| Sedimentary structures | Massive crudely bedded, matrix-supported conglomerate; trough cross-bedded, clast-supported conglomerate; planar cross-bedded conglomerate; massive pebbly sandstone; trough cross-bedded sandstone; planar bedded sandstone; argillite | (a) Sandstone-dominated unit: pebble trains; trough cross-beds; planar cross-beds; planar beds; argillite<br>(b) Siltstone-dominated unit: asymmetrical wave ripples; graded beds; dish and pillar structures; clastic dykes | Massive crudely bedded, clast-supported conglomerate; trough cross-bedded, clast-supported conglomerate; planar bedded sandstone; trough cross-bedded sandstone | High-angle trough cross-beds; planar beds; low-angle planar cross-beds; pebbles on foresets; argillite |
| Inferred setting | Proximal to medial reaches of streamflow-dominated alluvial fans and fan deltas with suspended fines | (a) Subaerial to subaqueous, wave- to fluvial-influenced fan- or braid-deltas<br>(b) Ephemeral lakes or ponds with flash flooding | Distal streamflow-dominated regions of alluvial fans or proximal braided stream | Sandy braided stream |
| *Jackson Lake Formation* | | | | |
| Lithofacies | Conglomerate–sandstone | Sandstone–argillite | Argillite–sandstone | |
| Sedimentary structures | Massive crudely to well stratified, clast-supported conglomerate; matrix-supported conglomerate with local angular clasts; trough cross-bedded sandstone; planar bedded sandstone | (a) Tidal channel sublithofacies: tabular cross-beds with tangential foresets; reactivation surfaces; pebble trains; rip-ups; graded beds; ripples; argillite<br>(b) Tidal ridge sublithofacies: composite bedforms; planar beds; bidirectional cross-beds; reactivation surfaces; graded beds, ripples; abundant argillite | Graded sandstone–mudstone couplets; graded sandstone; ripples | |
| Inferred setting | Streamflow-dominated fan- or braid-deltas with local mass flow deposits | (a) Tidal channels in a shallow-water subtidal zone<br>(b) Tidal ridges in a shallow-water sub- to intertidal zone | Subaqeous portion of fan-deltas in lower shoreface to proximal offshore setting | |

(i) the 300–375 m thick conglomerate I and siltstone–sandstone lithofacies, and (ii) the 300–400 m thick conglomerate II and sandstone lithofacies (Fig. 4). Conglomerate I, overlying the unconformity with the Beaulieu River volcanic belt, is divided into matrix- to clast-supported conglomerate, and massive to stratified pebble–cobble conglomerate. Matrix- to clast-supported conglomerate, <120 m thick, represents debris flow deposits where massive, and longitudinal gravel bars or sheets where cross-bedded and interstratified with sandstone. Laminated argillite within the conglomerate was deposited during overbank deposition. The pebble–cobble conglomerate, 50–200 m thick, contains erosive channels and well developed stratification that formed during confined stream deposition. Stream incision is represented by interstratified trough cross-bedded and planar bedded sandstone, whereas metre-thick upward-fining sequences capped by argillite are products of autocyclic floods. The sedimentology of conglomerate I (Table 1) is consistent with deposition on the proximal to distal portions of alluvial fans and fan deltas.

The siltstone–sandstone lithofacies, overlying conglomerate I gradationally and overlain by conglomerate II and the quartz-rich sandstone erosionally, is divided into sandstone- and siltstone-dominated units. The sandstone-dominated unit, up to 305 m thick, contains truncating sets of trough cross-beds and planar cross-beds consistent with in-channel dunes and transverse sand bars, respectively, in an alluvial–fluvial setting. Argillite drapes on and between bedforms represent fluctuating flow energy conditions. Sedimentary structures (Table 1) are consistent with a fluvial-influenced sandy braid delta or braidplain. The siltstone-dominated unit, 10–40 m thick, contains structures that develop during suspension sedimentation and weak wave action. Wave ripples, siltstone–mudstone couplets, graded beds, clastic dykes and dish and pillar structures are common features of the siltstone-dominated unit, and indicate that deposition occurred in shallow ephemeral lakes or ponds.

The conglomerate II lithofacies, <120 m thick, erosionally overlies the siltstone–sandstone lithofacies. Sedimentary structures are similar to those of conglomerate I (Table 1). The clast-supported conglomerate is consistent with deposition as gravel bars on proximal braided streams or distal parts of alluvial fans. The interstratification of trough cross-bedded and planar bedded sandstone represents migration of sinuous-crested dunes within stream channels and upper flow regime sand deposition along gravel bars, respectively.

The sandstone lithofacies, up to 320 m thick, is dominated by granular to pebbly, high-angle, truncating trough cross-beds, consistent with in-channel migration of sinuous-crested dunes. Vertical aggradation of dunes is inferred from ubiquitous truncation of cross-beds. Pebbles on foresets attest to high flow energy, whereas minor argillite between bedforms and on foresets was deposited during waning energy conditions. Sedimentary structures (Table 1) are suggestive of a high-energy sandy braidplain setting.

### *Jackson Lake Formation*

The Jackson Lake Formation is composed of conglomerate–sandstone, sandstone–argillite and argillite–sandstone lithofacies that most commonly form a large-scale upward-fining sequence, although stratigraphic positions are locally variable. The conglomerate–sandstone lithofacies, 2–70 m thick, is well exposed at the unconformity with the mafic volcanic Kam Group. Contacts with overlying lithofacies are sharp or gradational, but where conglomerate overlies the sandstone–argillite or argillite–sandstone, the contact is erosive. Matrix-supported and angular conglomerate represent debris flow and talus scree deposits, respectively, whereas clast-supported varieties are interpreted as longitudinal gravel bars. Lenticular, planar bedded and trough cross-bedded sandstone interstratified with the clast-supported conglomerate developed as bar top sands and in-channel dunes, respectively. The sedimentary structures of the conglomerate–sandstone lithofacies (Table 1), and its association with the two other lithofacies, support an interpretation of coalescing fan-deltas.

The sandstone–argillite lithofacies, 10–152 m thick, normally overlies the conglomerate–sandstone gradationally, but locally overlies the argillite–sandstone erosionally. Tidal channel and tidal ridge sublithofacies are distinguished by a predominance of tabular cross-bed sets in the former and complex bedforms in the latter. The tidal channel lithofacies, characterized by tabular–planar cross-beds with tangential foresets, and argillite between bedforms and on foresets, represents in-channel subtidal dune migration in a shelf-shoreline setting (Table 1). Local sandstone–argillite couplets are analogous to modern tidal bundles. Additional tidal influence is inferred from reactivation surfaces with argillite drapes. Minor pebble and conglomerate beds in the tidal channel deposits indicate rapid erosion of the conglomerate–sandstone lithofacies during high storm and high wave-surge activity combined with maximum tidal current action.

The tidal ridge lithofacies is composed of small- to medium-scale cross-beds, low-angle to planar beds, argillite beds, drapes on foresets and laminae between bedforms. Argillite-dominated units characterized by ripples and parallel lamination support a shallow water, tide-influenced shoreline, where current energy is highly variable (Table 1).

The argillite–sandstone lithofacies, 5–30 m thick, overlies the conglomerate–sandstone sharply to gradationally and is eroded by the sandstone–argillite lithofacies. Very fine-grained sandstone, siltstone and mudstone, which are parallel laminated, graded and rippled, are associated with medium-bedded, coarse-grained graded sandstone deposits. The sedimentary structures inherent in the abundant fine-grained deposits are consistent with low-energy wave action and tides in a proximal offshore setting that was affected periodically by storms, as indicated by local coarse-grained graded beds (Table 1).

### Petrography and clast compositions

A total of 53 quartz-rich sandstones and siltstone–sandstones from the Keskarrah and Beaulieu Rapids formations, in addition to 18 sandstone–argillites from the Jackson Lake Formation, were point counted using 350–400 grains. Data sets from the Keskarrah and Beaulieu Rapids formations are provided in Corcoran *et al.* (1998, 1999), whereas significant new data from the Jackson Lake Formation are presented here (Table 2). Eight grain parameters were applied to determine composition and possible source rocks: (i) total quartz (Q); (ii) polycrystalline quartz over total quartz ratio (Qp/Q); (iii) total feldspar (F); (iv) plagioclase feldspar over total feldspar ratio (P/F); (v) rock fragments (R); (vi) plutonic rock fragments and/or polycrystalline quartz (Rp); (vii) volcanic rock fragments (Rv); and (viii) sedimentary rock fragments (Rs). The results are illustrated on QFR (Folk, 1974) and RpRvRs diagrams (Fig. 6A & B).

Clast components and percentage matrix in conglomerates were determined by counting intersection points (clasts and matrix) using a 100-point orthogonal grid with 10 cm spacing. Totals of 3524 clast intersection points from 10 stations in the Keskarrah Formation, 5353 clast intersection points from eight stations in the Beaulieu Rapids Formation and 4545 clast intersection points from five main stations in the

**Table 2.** Point count results of 22 samples from the Keskarrah Formation, 31 samples from the Beaulieu Rapids Formation, and 18 samples from the Jackson Lake Formation.

| Sample | Q (%) | Qp/Q | F (%) | P/F | R (%) | Rp (%) | Rv (%) | Rs (%) |
|---|---|---|---|---|---|---|---|---|
| *Keskarrah Formation* | | | | | | | | |
| 93-1 | 95 | 0.75 | 1 | 1 | 4 | 94 | 3 | 3 |
| 93-2 | 94 | 0.72 | 2 | 0.5 | 4 | 94 | 3 | 3 |
| 93-4 | 83 | 0.64 | 10 | 0.8 | 7 | 88 | 7 | 5 |
| 95-81 | 81 | 0.41 | 9 | 0.67 | 10 | 77 | 2 | 21 |
| 95-82 | 79 | 0.38 | 9 | 0.56 | 12 | 71 | 8 | 21 |
| 95-108 | 80 | 0.5 | 11 | 0.73 | 9 | 82 | 4 | 14 |
| 95-110 | 82 | 0.37 | 9 | 0.67 | 9 | 77 | 5 | 17 |
| 95-111 | 95 | 0.62 | 0 | | 5 | 92 | 1 | 7 |
| 95-135 | 68 | 0.32 | 9 | 0.78 | 23 | 49 | 9 | 42 |
| 96-54 | 94 | 0.52 | 1.5 | 0.67 | 4.5 | 92 | 0 | 8 |
| 96-55 | 81 | 0.6 | 5.5 | 0.91 | 13.5 | 78 | 0 | 22 |
| 96-56 | 95 | 0.69 | 1.5 | 0.67 | 3.5 | 95 | 1 | 4 |
| 96-56a | 98 | 0.87 | 1 | 0.5 | 1 | 99 | 0.5 | 0.5 |
| 96-68 | 87 | 0.62 | 6 | 1 | 7 | 89 | 4 | 7 |
| 96-83 | 73 | 0.13 | 7 | 0.57 | 20 | 33 | 37 | 30 |
| 96-2 | 78 | 0.46 | 15 | 0.73 | 7 | 84 | 12 | 4 |
| 96-7 | 73 | 0.71 | 19 | 0.74 | 8 | 87 | 6.5 | 6.5 |
| 96-8 | 90 | 0.54 | 6 | 0.83 | 4 | 92 | 4 | 4 |
| **Sst mean** | **85** | **0.56** | **7** | **0.71** | **8** | **82** | **6** | **12** |
| 95-83 | 71 | 0.42 | 17 | 0.53 | 12 | 71 | 5 | 24 |
| 95-107 | 72 | 0.32 | 14 | 0.57 | 14 | 62 | 5 | 33 |
| 95-138 | 74 | 0.19 | 8 | 0.5 | 18 | 44 | 0 | 56 |
| 95-140 | 76 | 0.25 | 6 | 0.5 | 18 | 51 | 3 | 46 |
| **Slst-sst mean** | **73** | **0.29** | **11** | **0.55** | **16** | **57** | **3** | **40** |

(*continued on p. 194*)

**Table 2.** (*cont'd*)

| Sample | Q (%) | Qp/Q | F (%) | P/F | R (%) | Rp (%) | Rv (%) | Rs (%) |
|---|---|---|---|---|---|---|---|---|
| *Beaulieu Rapids Formation* | | | | | | | | |
| 94-03 | 69 | 0.86 | 4 | 0.63 | 27 | 86 | 1 | 13 |
| 94-05 | 73 | 0.79 | 12 | 0.67 | 15 | 94 | 1 | 5 |
| 94-08 | 75 | 0.93 | 9 | 0.93 | 16 | 90 | 4 | 6 |
| 94-09 | 76 | 0.88 | 12 | 0.89 | 12 | 95 | 2 | 3 |
| 94-10 | 80 | 0.79 | 13 | 0.77 | 7 | 94 | 2 | 4 |
| 94-11 | 71 | 0.82 | 17 | 0.67 | 12 | 93 | 2 | 5 |
| 94-13 | 65 | 0.80 | 19 | 0.76 | 16 | 93 | 2 | 5 |
| 94-15 | 80 | 0.76 | 7 | 1.0 | 13 | 96 | 4 | 0 |
| 94-16 | 75 | 0.81 | 19 | 0.80 | 6 | 96 | 4 | 0 |
| 94-17 | 73 | 0.86 | 12 | 0.74 | 15 | 93 | 3 | 4 |
| 94-18 | 62 | 0.90 | 28 | 1.0 | 12 | 95 | 0 | 5 |
| 94-24 | 72 | 0.83 | 12 | 0.65 | 16 | 92 | 4 | 4 |
| 94-25 | 81 | 0.83 | 5 | 0.46 | 14 | 89 | 6 | 5 |
| 94-26 | 80 | 0.91 | 11 | 0.71 | 9 | 94 | 3 | 3 |
| 94-27 | 72 | 0.86 | 13 | 0.82 | 14 | 96 | 4 | 0 |
| 94-31 | 86 | 0.74 | 8 | 1.0 | 5 | 100 | 0 | 0 |
| 94-33 | 83 | 0.78 | 11 | 0.96 | 6 | 100 | 0 | 0 |
| 94-38 | 78 | 0.90 | 14 | 0.77 | 9 | 99 | 1 | 0 |
| 94-43 | 81 | 0.90 | 6 | 0.93 | 13 | 95 | 5 | 0 |
| 94-44 | 80 | 0.89 | 18 | 0.76 | 2 | 99 | 0.5 | 0.5 |
| 94-48 | 77 | 0.91 | 14 | 0.65 | 9 | 94 | 3 | 3 |
| 94-49 | 73 | 0.84 | 18 | 0.69 | 9 | 95 | 2.5 | 2.5 |
| 95-04 | 90 | 0.83 | 4 | 0.75 | 5 | 94 | 0 | 6 |
| 95-19 | 86 | 0.92 | 1 | 0.75 | 12 | 92 | 1 | 7 |
| 95-34 | 66 | 0.86 | 27 | 0.50 | 7 | 92 | 4 | 4 |
| **Sst mean** | **76** | **0.85** | **13** | **0.76** | **11** | **95** | **2** | **3** |
| 94-22 | 63 | 0.83 | 21 | 0.86 | 16 | 100 | 0 | 0 |
| 94-23 | 74 | 0.69 | 8 | 1.0 | 18 | 98 | 2 | 0 |
| 94-30 | 69 | 0.78 | 24 | 0.75 | 7 | 88 | 2 | 0 |
| 94-32 | 83 | 0.78 | 13 | 0.97 | 4 | 100 | 0 | 0 |
| 94-40 | 73 | 0.81 | 23 | 0.87 | 4 | 98 | 1 | 1 |
| 94-42 | 64 | 0.92 | 28 | 0.71 | 8 | 100 | 0 | 0 |
| **Slst-sst mean** | **73** | **0.80** | **18** | **0.87** | **9** | **99** | **1** | **0** |
| *Jackson Lake Formation* | | | | | | | | |
| 2-93 | 75 | 0.36 | 10 | 0.68 | 15 | 27 | 72 | 28 |
| 4a-93 | 81 | 0.37 | 10 | 0.60 | 9 | 30 | 66 | 34 |
| 6-93 | 79 | 0.29 | 7 | 0.56 | 14 | 23 | 80 | 20 |
| 7-93 | 67 | 0.50 | 15 | 0.79 | 18 | 34 | 82 | 18 |
| 8-93 | 83 | 0.40 | 5 | 0.75 | 12 | 33 | 59 | 41 |
| 10-93 | 78 | 0.44 | 12 | 0.89 | 10 | 34 | 78 | 22 |
| 12-93 | 62 | 0.40 | 27 | 0.88 | 11 | 25 | 75 | 25 |
| 15-93 | 70 | 0.38 | 21 | 0.85 | 9 | 27 | 61 | 39 |
| 16-93 | 71 | 0.38 | 9 | 0.72 | 20 | 27 | 71 | 29 |
| 19-93 | 60 | 0.33 | 29 | 0.86 | 11 | 20 | 68 | 32 |
| 20-93 | 57 | 0.32 | 29 | 0.83 | 14 | 18 | 75 | 25 |
| 21-93 | 53 | 0.29 | 30 | 0.81 | 17 | 15 | 79 | 21 |
| 24-93 | 61 | 0.33 | 29 | 0.78 | 10 | 20 | 69 | 31 |
| 26-93 | 57 | 0.40 | 14 | 0.78 | 29 | 23 | 71 | 29 |
| 28-93 | 69 | 0.26 | 20 | 0.76 | 11 | 18 | 86 | 14 |
| 35-93 | 79 | 0.23 | 10 | 0.67 | 11 | 18 | 68 | 32 |
| 36-93 | 77 | 0.29 | 2 | 0.38 | 21 | 22 | 79 | 21 |
| 37-93 | 52 | 0.26 | 30 | 0.87 | 18 | 14 | 79 | 21 |
| **Sst mean** | **69** | **0.35** | **17** | **0.74** | **14** | **24** | **73** | **27** |

Grain parameters: Qp/Q, where Q is total quartz, and Qp is polycrystalline quartz; P/F, where F is total feldspar (plagioclase + potassium) and P is plagioclase feldspar; R, total rock fragments; Rp, plutonic rock fragments (including polycrystalline quartz); Rv, volcanic rock fragments; and Rs, sedimentary rock fragments.

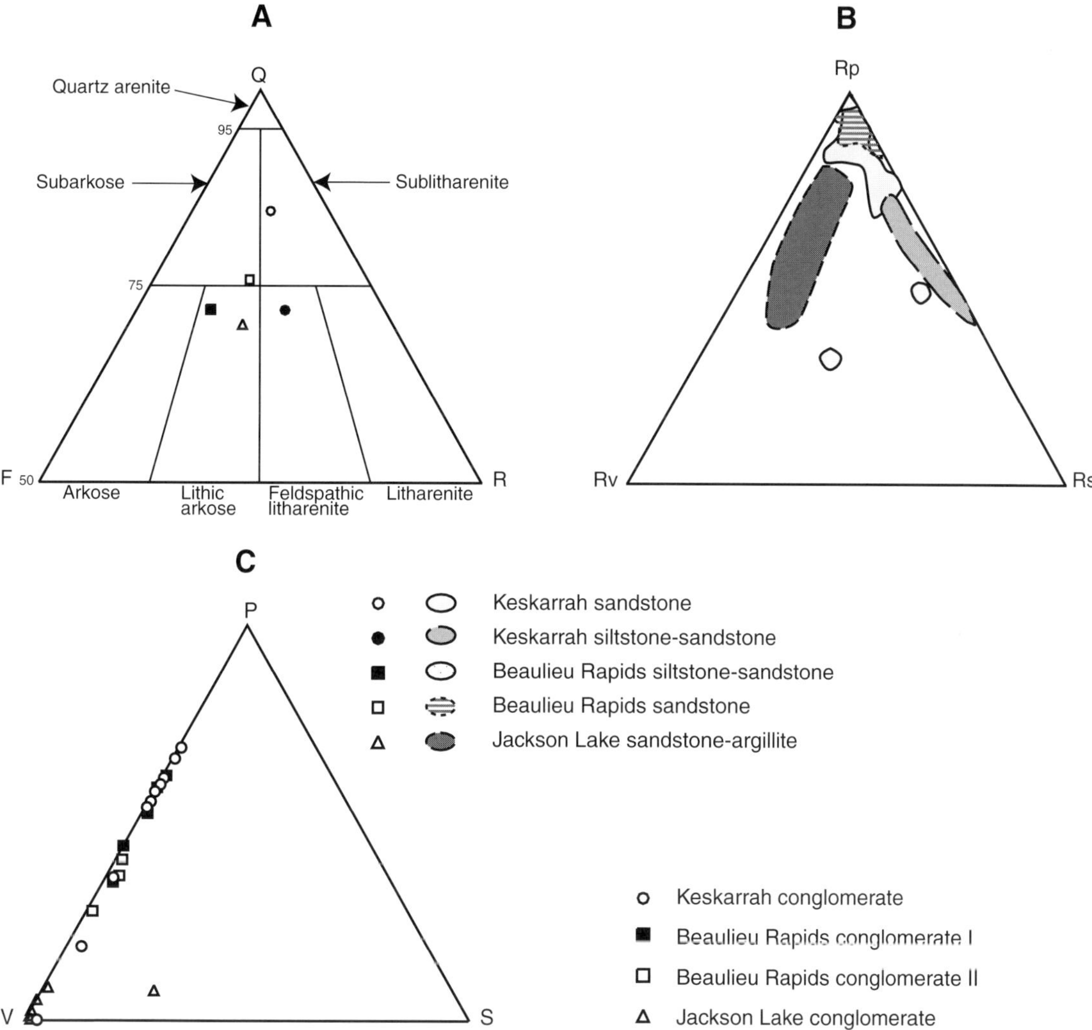


**Fig. 6.** (A) Ternary diagram illustrating the fields and average compositions of sandstones and siltstone–sandstones of the Keskarrah and Beaulieu Rapids formations, and sandstone–argillites of the Jackson Lake Formation on a QFR diagram (Folk, 1974), truncated at 50% quartz. (B) RpRvRs diagram illustrating the fields of rock fragment compositions. Rp, plutonic fragments; Rv, volcanic fragments; Rs, sedimentary fragments. (C) PVS diagram illustrating the composition of clasts from 10 stations in the Keskarrah Formation, eight stations in the Beaulieu Rapids Formation and five stations in the Jackson Lake Formation. P, plutonic clasts; V, volcanic (mafic and felsic) clasts; S, sedimentary clasts.

Jackson Lake Formation were plotted on a plutonic (excluding porphyries), volcanic and sedimentary clast ternary diagram (PVC; Fig. 6C), and on compositional pie charts (Fig. 7). Porphyry clasts were excluded from the classification because the percentage of porphyry components is minimal compared with volcanic, gneisso-plutonic and sedimentary clasts.

### *Keskarrah Formation*

Keskarrah Formation sandstones have an average QFR ratio of 85 : 7 : 8, plotting as sublitharenites, whereas a mean QFR ratio of 73 : 11 : 16 defines the siltstone-sandstones as feldspathic litharenites (Folk, 1974; Fig. 6A & Table 2). The sandstones are

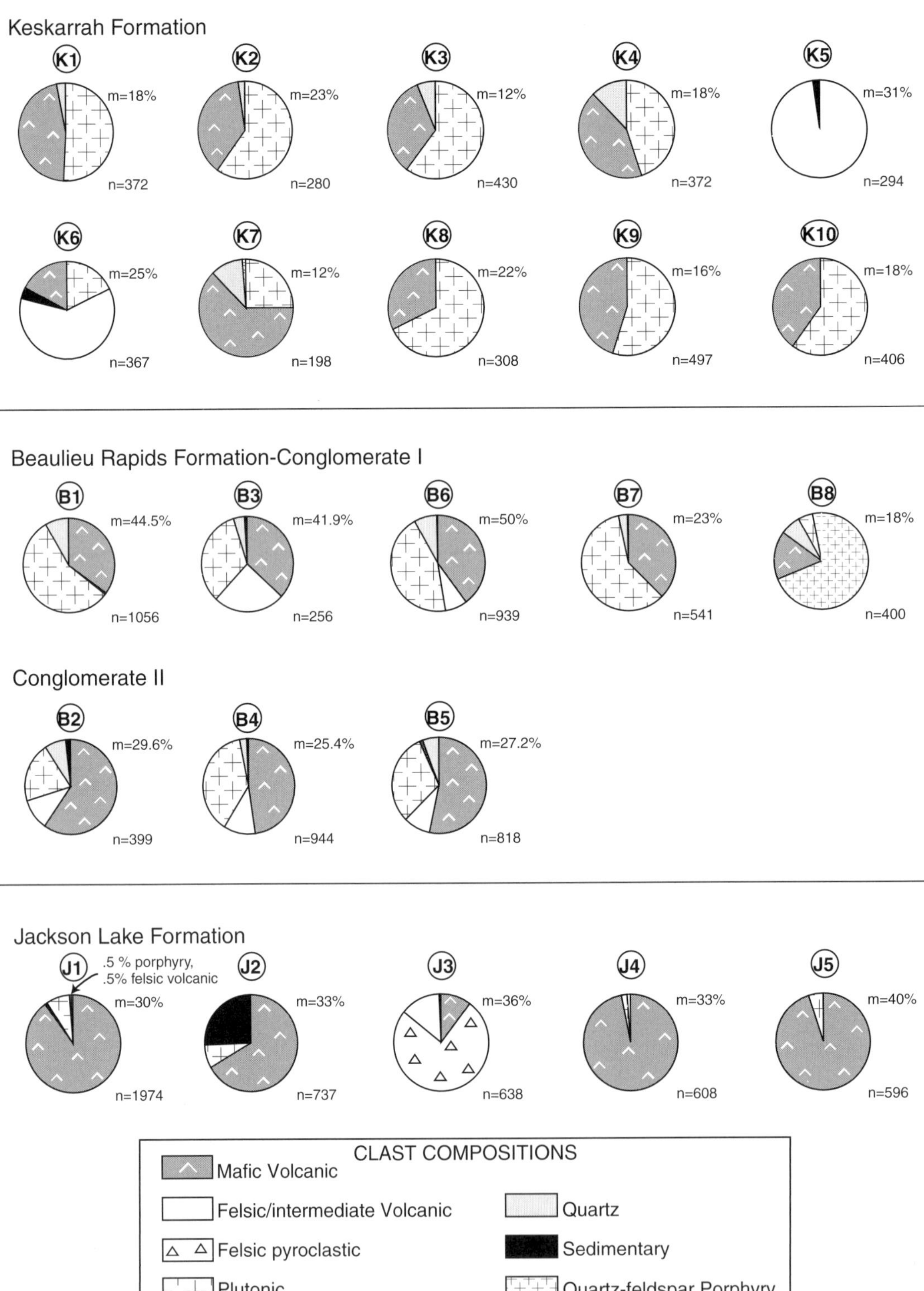


**Fig. 7.** Compositional pie diagrams illustrating the relative proportions of clast types in conglomerates of the Keskarrah, Beaulieu Rapids and Jackson Lake formations.

quartz-rich as indicated by quartz percentages >90% for seven samples, and >95% for four samples. An average Qp/Q ratio of 0.56 indicates a mixture of mono- and polycrystalline quartz varieties. Samples classified as quartz arenites are chiefly composed of subangular to subrounded mono- and polycrystalline quartz with minor plagioclase and potassium feldspar (ave. 1%), and rock fragments (ave. 3.5%). The matrix is clay-rich, but constitutes <10% of the sandstone samples. Higher feldspar percentages (up to 19%) are normally associated with an increase in clay-rich matrix. A P/F mean ratio of 0.71 indicates that the sandstones predominantly contain plagioclase feldspar. Rock fragments are mainly plutonic (82%) and contain minimal sedimentary (12%) and volcanic (6%) varieties (Fig. 6B). The sandstones are medium- to very coarse-grained and are moderately to well sorted.

Siltstone–sandstones generally contain more feldspar and rock fragments than the sandstones, but are relatively quartz-rich with percentages between 71 and 76% (Qp/Q mean ratio of 0.29). The samples are poorly sorted and contain subrounded quartz and rock fragments, and angular to subangular feldspar (P/F mean ratio of 0.55) in a quartzose–clay matrix. The clay, mainly a result of feldspar alteration, is now composed of sericite and chlorite. Rock fragments are predominantly plutonic (57%) and sedimentary (40%), with subsidiary volcanic clasts (3%; Fig. 6B).

Conglomerates in the Keskarrah Formation contain a variety of clast types, including plutonic (granite, granodiorite, tonalite), volcanic (felsic and mafic aphyric to porphyritic), sedimentary (siltstone, chert) and quartz veins. Predominant compositions vary with respect to subjacent source rocks. Plutonic clasts are prevalent, accounting for 46% of the conglomerate lithofacies (Fig. 7). Mafic volcanic clasts comprise 36% of the conglomerate (Fig. 7), whereas felsic and intermediate volcanic components (14.5%) are found near contacts with felsic and intermediate volcanic source rocks (Corcoran *et al.*, 1998). Similarly, sedimentary fragments, constituting 0.5% of the clasts (Fig. 7), are locally predominant where the conglomerate overlies the turbiditic Contwoyto Formation unconformably (Corcoran *et al.*, 1998). Quartz clasts account for 3% of the conglomerate and matrix percentage varies between 12 and 31% (Fig. 7). Clast counts from most of the localities in the Keskarrah Formation plot within the upper 50% plutonic portion on the PVS diagram (Fig. 6C), with a substantially greater plutonic (including quartz) component than the Jackson Lake conglomerate and conglomerate II of the Beaulieu Rapids Formation (Fig. 6C).

### *Beaulieu Rapids Formation*

Sandstones of the Beaulieu Rapids Formation have an average QFR ratio of 76 : 13 : 11, plotting in the subarkose field, whereas the siltstone–sandstone average plots in the lithic arkose field with a mean QFR ratio of 73 : 18 : 9 (Fig. 6A & Table 2). The sandstones are moderately to well sorted, coarse- to very coarse-grained, and contain abundant subangular to subrounded polycrystalline quartz, as indicated by an average Qp/Q ratio of 0.85. A mean P/F ratio of 0.76 demonstrates that plagioclase is more common than potassium feldspar (Table 2). Rock fragments are predominantly plutonic (61%), with a moderate amount of sedimentary (22%) and volcanic (17%) components (Fig. 6B). The sandstone matrix is composed of quartz, chlorite, actinolite and sericite.

Siltstone–sandstones are poorly to moderately sorted and are composed of very fine- to fine-grained sandstone. Subangular to subrounded quartz is the predominant mineral (63–83%), with a pronounced polycrystalline component indicated by a mean Qp/Q ratio of 0.80 (Table 2). The P/F ratio of 0.87 indicates that plagioclase is the dominant feldspar. Rock fragments are mainly plutonic, with an average of 83%, whereas volcanic and sedimentary rock fragments account for only 14 and 3%, respectively (Fig. 6B). The matrix of siltstone–sandstones is quartzose and clay-rich, similar to that of sandstones.

A wide variety of clast types comprise the conglomerates of the Beaulieu Rapids Formation, including plutonic (granite, granodiorite, tonalite, magmatic breccia), volcanic (felsic and mafic aphyric and porphyritic), porphyry (quartz–feldspar, hornblende), sedimentary (chert) and quartz. Conglomerate I predominantly contains plutonic (40%) and mafic volcanic (28%) clasts with subordinate porphyry (15%), felsic volcanic (11%), quartz (6%) and sedimentary (<0.5%) components. Porphyry clasts were found at only one locality (B8), where they account for 73% of the components (Figs 4 & 7). Omitting clast count B8 results in notable increases in plutonic (49%) and mafic volcanic (31%) clasts in conglomerate I. Matrix percentage varies from 18 to 50%, in contrast with conglomerate II where the matrix component changes slightly from approximately 25 to 30% (Fig. 7). Clasts in conglomerate II are mafic volcanic (55%), plutonic (29%), felsic volcanic (10%), quartz (5%) and sedimentary (1%). There is a marked increase in mafic volcanic clasts and decrease in plutonic clasts from conglomerate I to conglomerate II (Fig. 6C).

**Table 3.** Major and trace element compositions of the Keskarrah, Beaulieu Rapids and Jackson Lake sandstones and siltstone–sandstones.

| | Keskarrah Formation | | | | Beaulieu Rapids Formation | | | | | Jackson Lake Formation | | | |
|---|---|---|---|---|---|---|---|---|---|---|---|---|---|
| Sample | KS-2-93 sst | KS-4-93 sst | PLC-96-56 sst | PLC-95-138 slst-sst | PLC-94-10 sst | PLC-94-12 sst | PLC-94-16 sst | PLC-94-23 slst-sst | PLC-94-44 sst | JL-28-93 sst-arg | JL-29-93 arg-sst | JL-30-93 arg-sst | JL-35-93 sst-arg |
| SiO2 | 85.71 | 76.73 | 90.95 | 76.92 | 83.76 | 72.12 | 75.84 | 80.56 | 73.7 | 75.38 | 61.79 | 65.52 | 71.18 |
| TiO2 | 0.23 | 0.46 | 0.13 | 0.69 | 0.28 | 0.58 | 0.44 | 0.36 | 0.53 | 0.42 | 0.82 | 0.63 | 0.58 |
| Al2O3 | 8.75 | 10.39 | 6.48 | 13.39 | 7.83 | 13.7 | 11.06 | 10.3 | 13.25 | 12.97 | 19.3 | 13.82 | 10.91 |
| Fe2O3 | 0.64 | 3.01 | 0.54 | 5.4 | 2.61 | 5.01 | 4.59 | 3.19 | 4.78 | 2.81 | 7.76 | 6.9 | 5.56 |
| MnO | | 0.04 | | 0.08 | 0.02 | 0.04 | 0.05 | 0.03 | 0.04 | 0.01 | 0.01 | 0.04 | 0.05 |
| MgO | 0.3 | 1.03 | 0.18 | 0.68 | 0.95 | 2.08 | 2.41 | 1.24 | 2.08 | 1.15 | 3.07 | 3.56 | 2.45 |
| CaO | 0.14 | 1.32 | 0.04 | 0.11 | 0.39 | 0.21 | 1.66 | 0.61 | 0.19 | 0.82 | 0.2 | 1.55 | 2.86 |
| Na2O | 0.4 | 1.26 | 0.09 | 0.11 | 0.88 | 1.12 | 0.32 | 0.35 | 0.14 | 2.09 | 0.82 | 0.51 | 1 |
| K2O | 2.7 | 2.39 | 1.89 | 0.5 | 2.21 | 3.73 | 1.6 | 2.31 | 3.48 | 2.27 | 3 | 2.17 | 1.59 |
| P2O5 | 0.05 | 0.07 | 0.04 | 0.09 | 0.06 | 0.1 | 0.08 | 0.08 | 0.14 | 0.09 | 0.15 | 0.13 | 0.11 |
| LOI | 1.23 | 2.88 | 0.82 | 2.53 | 0.99 | 2.03 | 2.05 | 1.67 | 2.47 | 2.5 | 3.7 | 4.81 | 4.21 |
| Total | 100.15 | 99.58 | 101.16 | 100.5 | 99.98 | 100.72 | 100.1 | 100.7 | 100.8 | 100.51 | 100.62 | 99.64 | 100.5 |
| CIA Values | 73.53 | 67.64 | 76.24 | 94.9 | 69.23 | 73.03 | 75.55 | 75.9 | 77.67 | 71.46 | 82.76 | 76.62 | 66.69 |
| | | | | | | | | | | | | | |
| Ni | 8 | 30 | 92 | 72 | 31 | 60 | 51 | 37 | 50 | 50 | 75 | 73 | 54 |
| Sc | 3 | 7 | 2 | 8 | 4 | 9 | 8 | 6 | 8 | 6 | 16 | 12 | 11 |
| V | 21 | 62 | 15 | 83 | 41 | 86 | 70 | 50 | 74 | 62 | 153 | 107 | 93 |
| Cu | 15 | 32 | 9 | 14 | 30 | 35 | 58 | 37 | 37 | 14 | | 54 | 11 |
| Zn | 12 | 41 | 110 | 95 | 62 | 115 | 96 | 55 | 63 | 40 | 121 | 100 | 63 |
| Sr | 25 | 152 | 35 | 65 | 54 | 46 | 106 | 72 | 31 | 202 | 188 | 123 | 127 |
| Ta | 0.28 | 0.4 | 0.14 | 0.3 | 0.38 | 0.57 | 0.38 | 0.43 | 0.56 | 0.27 | 0.37 | 0.27 | 0.28 |
| Nb | 2.66 | 4 | 1.26 | 3.71 | 4.46 | 7.87 | 4.24 | 5.12 | 7.65 | 3.26 | 4.79 | 3.28 | 3.35 |
| Hf | 2.63 | 2.97 | 1.49 | 3.71 | 2.11 | 3.66 | 2.9 | 3.35 | 5.19 | 2.63 | 3.74 | 4.23 | 3.02 |
| Zr | 99.18 | 109.23 | 53.39 | 156.36 | 79.25 | 141.1 | 111.93 | 128.75 | 205.06 | 102.64 | 146.7 | 169.24 | 123.49 |
| Ti | 1380 | 2760 | 780 | 4140 | 3480 | 2640 | 2160 | 3180 | 2520 | 4920 | 3780 | 3480 | 1380 |
| Y | 7.12 | 11.17 | 4.39 | 11.15 | 9.96 | 13.3 | 13.4 | 12.49 | 13.06 | 9.25 | 18.01 | 16.11 | 14.58 |
| Th | 10.02 | 12.55 | 7.58 | 10.72 | 6.31 | 8.08 | 6.84 | 9.1 | 9.38 | 7.5 | 9.16 | 8.24 | 8.84 |
| Th/Sc | 3.34 | 1.79 | 3.79 | 1.34 | 1.58 | 0.9 | 0.86 | 1.52 | 1.17 | 1.25 | 0.57 | 0.69 | 0.8 |
| | | | | | | | | | | | | | |
| La | 25.86 | 35.04 | 20.77 | 35.98 | 24.98 | 19.86 | 29.39 | 43.11 | 45.88 | 25.92 | 36 | 28.87 | 31.21 |
| Ce | 41.72 | 59.39 | 33.18 | 58.8 | 43.37 | 39.93 | 51.93 | 76.34 | 86.38 | 44.81 | 69.08 | 51.96 | 53.78 |
| Pr | 4.88 | 6.97 | 3.67 | 7.59 | 5.5 | 5.08 | 6.85 | 9.65 | 11.18 | 5.66 | 8.86 | 6.64 | 6.85 |
| Nd | 16.91 | 23.66 | 12.16 | 26.26 | 19.58 | 17.52 | 24.93 | 34.02 | 38.58 | 20.7 | 32.53 | 24.54 | 24.41 |
| Sm | 2.75 | 4.07 | 2.03 | 4.21 | 3.22 | 3.07 | 4.12 | 4.93 | 6.09 | 3.1 | 5.66 | 4.21 | 4.16 |
| Eu | 0.63 | 0.94 | 0.46 | 0.98 | 0.83 | 0.79 | 0.94 | 1.19 | 1.13 | 1.01 | 1.54 | 1.16 | 1.21 |
| Gd | 2.38 | 3.22 | 1.67 | 3.12 | 2.54 | 2.4 | 3.4 | 3.84 | 4.52 | 2.62 | 4.59 | 3.57 | 3.62 |
| Tb | 0.33 | 0.48 | 0.23 | 0.42 | 0.37 | 0.38 | 0.48 | 0.54 | 0.63 | 0.35 | 0.61 | 0.53 | 0.53 |
| Dy | 1.44 | 2.08 | 0.87 | 1.85 | 1.62 | 1.97 | 2.22 | 2.29 | 2.54 | 1.56 | 2.82 | 2.63 | 2.46 |
| Ho | 0.27 | 0.41 | 0.17 | 0.41 | 0.33 | 0.46 | 0.48 | 0.43 | 0.51 | 0.32 | 0.64 | 0.56 | 0.5 |
| Er | 0.61 | 1.12 | 0.38 | 1.05 | 0.94 | 1.2 | 1.23 | 1.07 | 1.36 | 0.85 | 1.53 | 1.49 | 1.26 |
| Tm | 0.08 | 0.15 | 0.06 | 0.17 | 0.13 | 0.19 | 0.18 | 0.17 | 0.21 | 0.13 | 0.25 | 0.23 | 0.17 |
| Yb | 0.64 | 1.1 | 0.42 | 1.11 | 0.92 | 1.3 | 1.23 | 1.13 | 1.55 | 0.92 | 1.58 | 1.46 | 1.12 |
| Lu | 0.09 | 0.16 | 0.06 | 0.17 | 0.13 | 0.2 | 0.17 | 0.17 | 0.23 | 0.12 | 0.25 | 0.22 | 0.19 |
| Lan/Ybn | 34.63 | 25 | 37.14 | 22.64 | 20.58 | 11.18 | 17.48 | 27.13 | 23.36 | 21.35 | 15.41 | 13.44 | 19.64 |
| Smn/Ybn | 5.97 | 4.71 | 5.88 | 4.29 | 4.3 | 2.8 | 3.97 | 5.03 | 5.03 | 4.14 | 3.93 | 3.18 | 4.24 |

*Jackson Lake Formation*

The Jackson Lake sandstone-argillite samples plot as lithic arkoses with an average QFR ratio of 69 : 17 : 14 (Fig. 6A & Table 2). The samples are not as quartz-rich as the sandstones and siltstone–sandstones of the Keskarrah and Beaulieu Rapids formations (Fig. 6A). In addition, a Qp/Q ratio of 0.35 for the Jackson Lake samples indicates that monocrystalline quartz predominates over polycrystalline varieties, in contrast to the relatively high Qp/Q ratios of the samples from the other sequences (Table 2). The fine- to coarse-grained sandstone–argillites are moderately to well sorted. Quartz grains are subrounded and are generally larger (up to 1 mm) than the subangular to subrounded feldspar grains and rock fragments (up to 0.75 mm). The feldspar, predominantly plagioclase as indicated by the P/F ratio of 0.74 (Table 2), is more common in finer-grained samples with higher matrix percentages. The matrix is quartzo-feldspathic, displaying sericite and carbonate alteration. Rock fragments are mainly volcanic (73%), with subordinate sedimentary (27%) and plutonic (24%) components, as opposed to the Keskarrah and Beaulieu Rapids sandstones and siltstone–sandstones which are dominated by plutonic fragments (Table 2 & Fig. 6B). Felsic volcanic fragments (66%) are more common than their mafic volcanic counterparts (34%).

Clast compositions in the Jackson Lake conglomerate, with matrix percentages between 30 and 40%, are highly dependent on proximity to specific source rocks (Mueller *et al.*, this volume, pp. 153–182). Several types of clasts comprise the conglomerate, including plutonic (granite, granodiorite, tonalite, pegmatite), volcanic (felsic and mafic aphyric to porphyritic; felsic pyroclastic), quartz–feldspar porphyry and sedimentary (volcaniclastic sandstone, siltstone, chert). Mafic volcanic clasts (72%) predominate, whereas the sedimentary (5.5%), plutonic (5%) and felsic volcanic (3%) components are minor (Fig. 7). Felsic pyroclastic and quartz–feldspar porphyry clasts contribute 15 and <0.1%, respectively, but both components were each identified at one locality (J1, J3; Fig. 7). All five clast count stations plot in the lower 25% plutonic portion on the PVS diagram, emphasizing a significant volcanic input (Fig. 6C).

## Geochemistry

Nine representative sandstone and sandstone–argillite, and four siltstone– and argillite–sandstone samples were selected from the Keskarrah, Beaulieu Rapids and Jackson Lake formations for geochemical analysis to: (i) permit comparison with point and clast counting results; (ii) determine the intensity of chemical weathering using CIA values; (iii) elucidate possible source areas based on trace element patterns; and (iv) distinguish tectonic setting using specific trace elements. Geochemical data and results from Jenner *et al.* (1981), Easton (1985) and Cousens (2000) are incorporated with the 13 samples used for this study.

The samples were analysed at the Geoscience Laboratories of the Ontario Geological Survey. Major elements were determined using fused-disc X-ray fluorescence (XRF), and loss on ignition (LOI) values, as determined by gravimetric methods, are <2.50 for most of the samples with no values >4.81 (Table 3). Co, Cu, Ni, Sc, V and Zn were analysed by inductively coupled plasma-optical emission spectrometry (ICP-OES), whereas the rare earth elements (REE) Nb, Cs, Hf, Ta, Th, U, Y and Zr were determined by inductively coupled plasma-mass spectrometry (ICP-MS). The accuracy and precision of the methods are discussed in Longerich *et al.* (1990).

*Keskarrah Formation*

Three sandstones and one siltstone–sandstone that were previously point counted (Corcoran *et al.*, 1998) were analysed geochemically (Table 3). All four samples were collected from Cyclops Peninsula, where the greatest proportion of the siltstone–sandstone lithofacies and quartz arenites of the sandstone lithofacies is located (Fig. 3). Sandstones KS-2-93 and PLC-96-56 contain 86 and 91 wt% $SiO_2$, respectively, whereas sample KS-4-93 contains 77% $SiO_2$. This discrepancy is also evident in the point counting results, where PLC-96-56 is represented by 95% and KS-4-93 by 83% total quartz (Table 2). Sample PLC-95-138, a siltstone–sandstone, contains 77% $SiO_2$, and is similarly represented by a lower quartz component (74%) than the sandstones. There is a marked difference in MgO content, wherein the values are much lower for the Keskarrah Formation samples (0.18–1.03%) than those of the Beaulieu Rapids (0.95–2.41%) and Jackson Lake (1.15–3.56%) samples (Fig. 8A). In addition, the Keskarrah Formation has lower Ti, Sc and Ni, and higher Th and Th/Sc than the other sedimentary sequences (Fig. 8B & Table 3). The degree of chemical alteration to which the sedimentary rocks were subjected is determined by calculating CIA values (Nesbitt & Young, 1982), which measures and compares the molecular proportions of alumina and alkalis in a particular sample. Chemically weathered

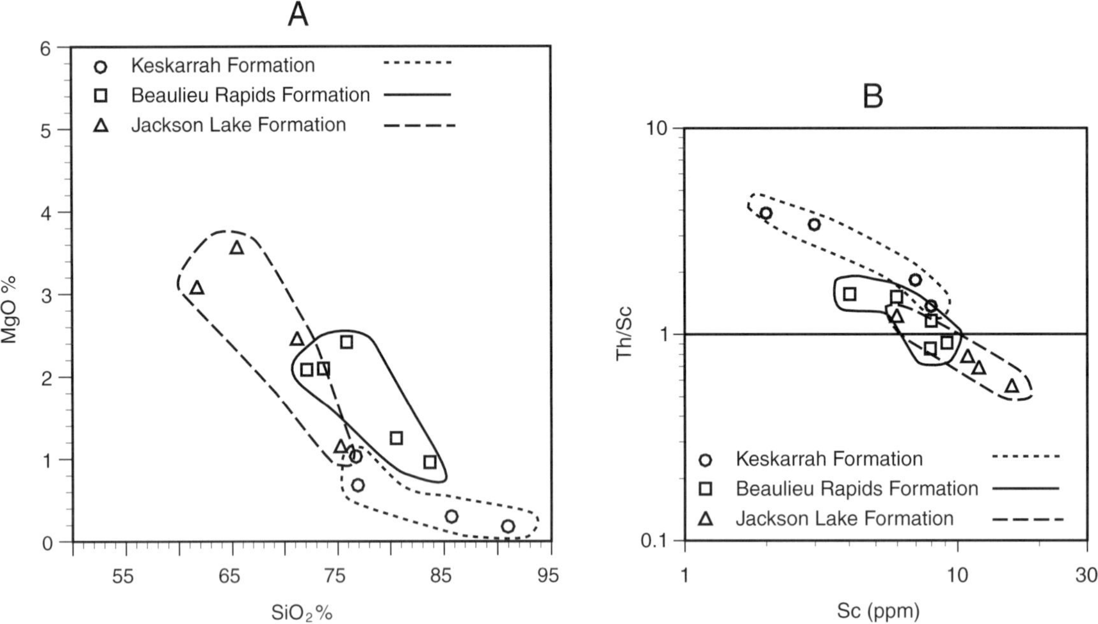


**Fig. 8.** Harker variation diagrams of (A) MgO wt% versus $SiO_2$ wt% and (B) Th/Sc (ppm) versus Sc (ppm) demonstrating clear distinctions between the Keskarrah, Beaulieu Rapids and Jackson Lake samples.

material will produce high values due to the breakdown of feldspars into clay, whereas low values indicate minimal chemical alteration. CIA values for the Keskarrah Formation samples range from 74 to 95, with the highest value represented by the siltstone–sandstone sample (Table 3). Higher values for most finer-grained rocks are expected because of the greater quantity of aluminous minerals produced during chemical weathering (Nesbitt & Young, 1982). Similarly, the three sandstones and one siltstone–sandstone have $Al_2O_3/Na_2O$ ratios of 8, 72 and 122, respectively, which are indicative of an essential weathering component because average sandstones produce $Al_2O_3/Na_2O$ ratios between 5 and 11 (Pettijohn, 1975). The samples follow a weak trend that intersects with the ($CaO + Na_2O$)–$Al_2O_3$ boundary on the alkali-alumina ternary diagram (Fig. 9). This trend is related to the 'initial' stages of weathering, whereas the siltstone–sandstone situated along the $K_2O$–$Al_2O_3$ line represents an advanced weathering stage with a significant loss in $K_2O$.

In order to determine the tectonic setting with which ancient sedimentary rocks are affiliated, specific elements that are immobile must be considered. The least mobile elements include the REE, Th, Zr, Nb, Ti and Sc, as they are relatively insensitive to weathering and metamorphism (Holland, 1978). Bhatia & Crook (1986) developed tectonic classification diagrams based on the proportion of immobile trace elements in modern graywackes. The Keskarrah Formation samples lie in the active continental margin (ACM)/passive margin (PM) field on the La–Th–Sc diagram, but are concentrated around the ACM field on the Th–Sc–Zr/10 classification (Fig. 10A & B). This supports the earlier work of Corcoran *et al.* (1998) which illustrated that the Keskarrah sedimentary sequence was deposited in a tectonically active environment rather than a passive margin setting as would normally be associated with quartz arenites. In addition to tectonic setting, the REE are indicators of possible source rocks because they often retain their source rock chemistry. The Keskarrah Formation samples are HREE depleted with $La_n/Yb_n = 23$–$37$, $Sm_n/Yb_n = 4$–$6$ and slight negative Eu anomalies (Fig. 11A). The REE patterns of some possible source rocks are provided in Fig. 11B & C. $La_n/Yb_n$ ratios of 28 and 27 for PD-96-12 (intermediate volcanic rock interstratified with Peltier Formation basalts) and Aug-97-2a (Augustus granite), respectively, are comparable to the sedimentary rocks. The only negative Eu anomaly produced by a possible source rock is illustrated by sample S-3, a dacite from the upper part of the Point Lake belt (Easton, 1985).

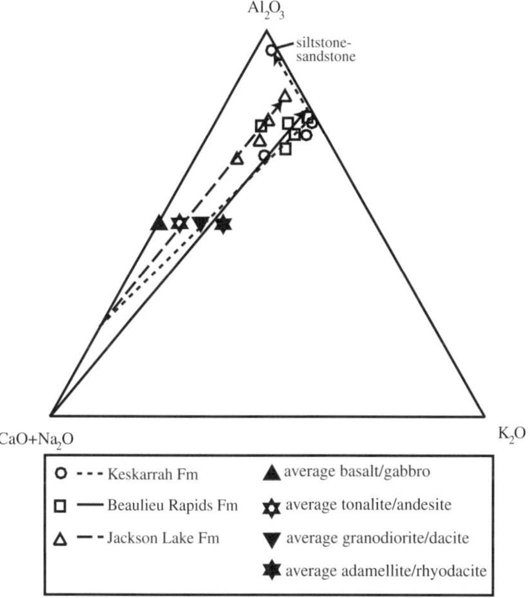


**Fig. 9.** $Al_2O_3$-($CaO + Na_2O$)-$K_2O$ ternary diagram based on CIA (chemical index of alteration; Nesbitt & Young, 1982, 1984) values produced by samples from the Keskarrah, Beaulieu Rapids and Jackson Lake formations. Note the relatively high values for all samples, especially the Keskarrah siltstone–sandstone. Arrows indicate general trends which intersect specific source rock compositions. Note how the Keskarrah and Jackson Lake trends intersect the $Al_2O_3$-($CaO + Na_2O$) boundary, indicating that chemical weathering was not the only factor controlling the composition of the samples.

### *Beaulieu Rapids Formation*

Four sandstones and one siltstone–sandstone were selected from the Beaulieu Rapids Formation for geochemical analysis. Point counting results from four of the samples (PLC-94-10, 94-16, 94-23, 94-44) are published in Corcoran *et al.* (1999), whereas sample PLC-94-12 was collected from a matrix-rich, fine-grained bed in the sandstone lithofacies. $SiO_2$ wt% varies considerably from 72 to 84 and there appears to be a weak correlation between point counted quartz component (74–80%) and $SiO_2$ wt% (Tables 2 & 3). However, sample PLC-94-23, a siltstone–sandstone, contains the least amount of total quartz (74%) compared to the sandstones, but $SiO_2$ wt% is significantly high (approximately 81%; Table 3). This discrepancy is explained by increased quartz (<0.03 mm) in the matrix of the moderately sorted siltstone–sandstones as opposed to the generally well sorted sandstones, which are characterized by a sericitic matrix. MgO contents range from 0.95 to 2.14% (Fig. 8A) and CIA values are between 69 and 78, following a general trend subparallel to and eventually intersecting with the ($CaO + Na_2O$)–$Al_2O_3$ boundary, excluding one sample which plots with less $K_2O$ than the other samples (Fig. 9). Ti, Sc and Ni are higher, and Th and the Th/Sc ratio are lower, compared with the Keskarrah Formation (Fig. 8B & Table 3). $Al_2O_3/Na_2O$ ratios of 9–95 are consistent with strong chemical weathering.

In terms of tectonic setting, the samples plot in the ACM/PM and continental volcanic arc (CVA) fields on the La–Th–Sc diagram (Bhatia & Crook, 1986), but are more closely clustered in the CVA field on the Th–Sc–Zr/10 triangle (Fig. 10A & B). Because the samples are removed from the PM field (Fig. 10B), a passive margin is discounted, which is consistent with the tectonically active, strike-slip interpretation of Corcoran *et al.* (1999). Samples from the Beaulieu Rapids Formation have $La_n/Yb_n = 11–27$ and $Sm_n/Yb_n = 3–5$, with PLC-94-16, 23 and 44 demonstrating very slight Eu anomalies (Fig. 12). REE patterns of potential source rocks are unavailable, but the relatively high $La_n/Yb_n$ ratios and the presence of Eu anomalies support a predominant felsic extrusive or intrusive source.

### *Jackson Lake Formation*

Two sandstone–argillites and two argillite–sandstones were selected from the Jackson Lake Formation. Point counting results from the sandstone–argillites (JL-28-93, JL-35-93) are provided in Table 2. Samples JL-29-93 and JL-30-93 were not point counted because of their high (>15%) matrix component. $SiO_2$ weight percentages are relatively low (approximately 62–75%) in comparison with samples from the Keskarrah and Beaulieu Rapids formations. Similarly, the total quartz component for the sandstone–argillites (69%, 79%) is normally lower than total quartz in sandstones from the other sedimentary sequences (Table 2). MgO contents range from 1.15 to 3.56% (Fig. 8A) and CIA values vary greatly between 67 and 83, but follow a distinct trend on the alkali–alumina ternary diagram, markedly closer to the ($CaO + Na_2O$)/$Al_2O_3$ line than trends produced by the two other formations (Fig. 9). Ti, Sc and Ni are greater than in the Keskarrah and Beaulieu Rapids formations, and Th and the Th/Sc ratio are lower compared with the Keskarrah Formation (Fig. 8B & Table 3). $Al_2O_3/Na_2O$ ratios range from 6 to 27, representative of chemical maturity.

On both the La–Th–Sc and Th–Sc–Zr/10 discrimination diagrams, the Jackson Lake Formation

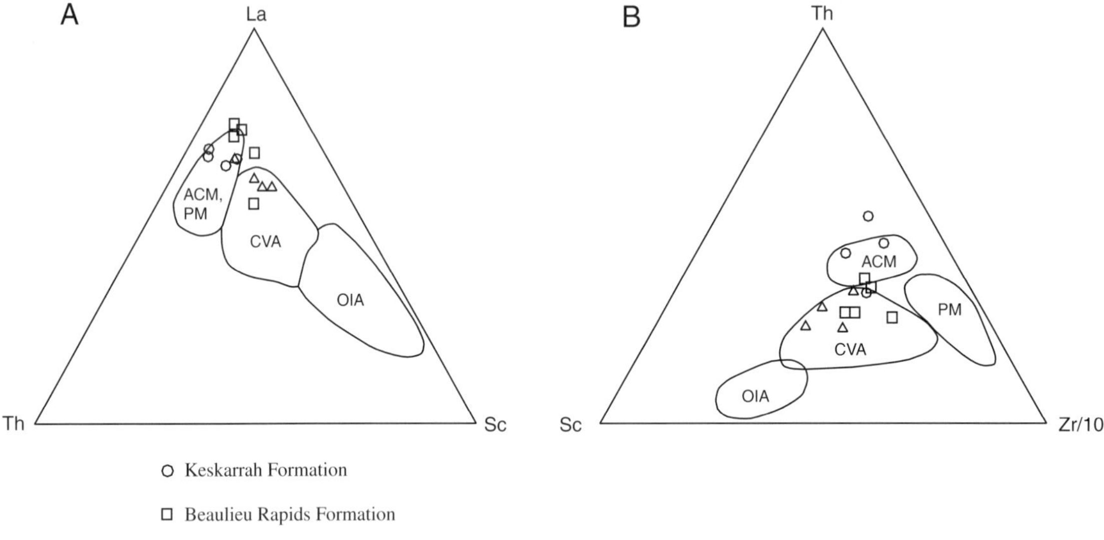


**Fig. 10.** La–Th–Sc and Th–Sc–Zr/10 plots (after Bhatia & Crook, 1986) for discriminating the tectonic settings of the Keskarrah, Beaulieu Rapids and Jackson Lake formations. ACM, active continental margin; PM, passive margin; CVA, continental volcanic arc; OIA, oceanic island arc. Note how the Keskarrah samples plot consistently in the ACM field; the Beaulieu Rapids samples in the CVA and ACM fields; and the Jackson Lake samples in the CVA field.

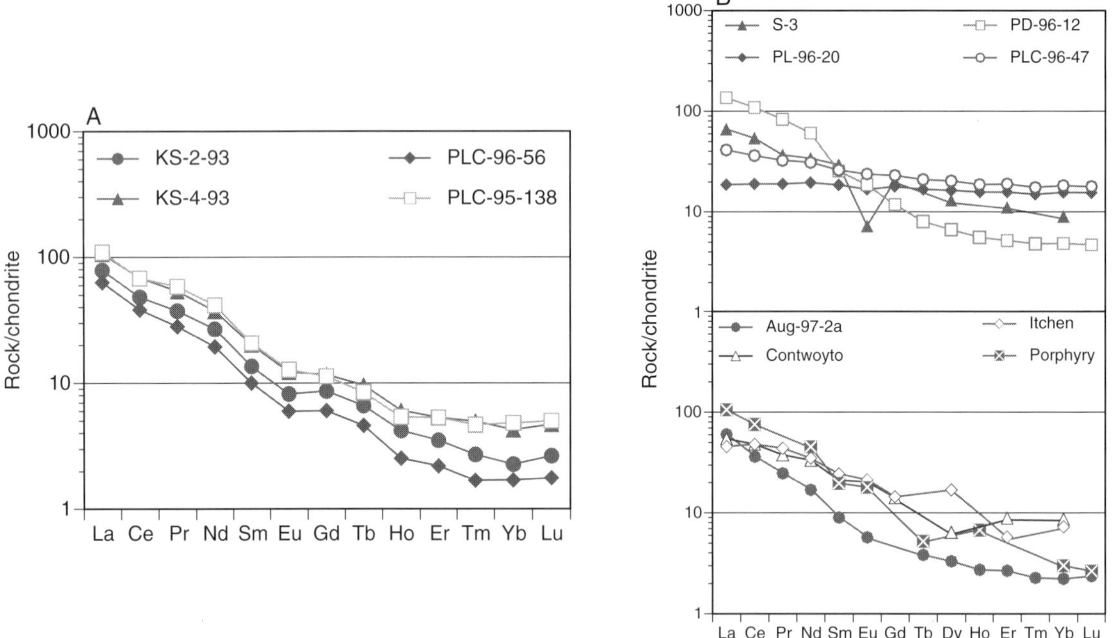


**Fig. 11.** Chondrite-normalized REE abundances for samples from the (A) Keskarrah Formation (KS-2-93, KS-4-93, PLC-96-56, PLC-95-138), and (B) volcanic Peltier Formation (PL-96-20, PLC-96-47, PD-96-12), dacite from the Point Lake belt (S-3; Easton, 1985), Augustus granite (Aug-97-2a), Contwoyto and Itchen formations (Easton, 1985) and a feldspar porphyry. Normalizing values after Sun (1982).

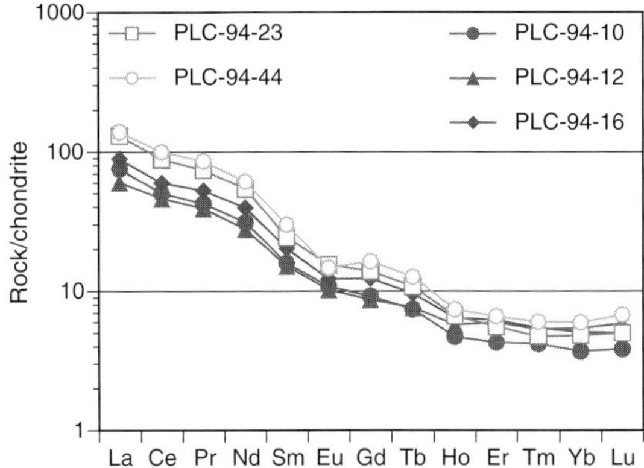


**Fig. 12.** Chondrite-normalized REE abundances for samples from the Beaulieu Rapids Formation (PLC-94-10, 12, 16, 23, 44). Normalizing values after Sun (1982).

samples plot in the CVA field with only one sample lying near the boundary of the ACM/PM field on the La–Th–Sc diagram (Fig. 10A & B). $La_n/Yb_n = 13–21$ and $Sm_n/Yb_n = 3–4$ for the Jackson Lake samples produce slightly flatter REE patterns than those of the other sedimentary sequences (Fig. 13A). There are no Eu anomalies, as previously noted by Jenner *et al.* (1981). REE patterns of potential source rocks are illustrated in Fig. 13B. $La_n/Yb_n$ ratios of 62 and 49 for samples from the felsic-dominated Banting Group and the granitic basement, respectively, suggest a minimal source contribution to the Jackson Lake Formation. In addition, negative Eu anomalies demonstrated by the Townsite and basement samples preclude these as major contributors of detritus. The REE patterns of the Jackson Lake samples resemble the pattern produced by the turbiditic Burwash sample, indicating that the two sedimentary sequences received material from similar source rocks.

## DISCUSSION

The mineralogy and chemical composition of sedimentary rocks are sound indicators of tectonic setting, source rocks and the intensity of chemical weathering. Combined with sedimentary facies analyses, rock composition can contribute to determination of palaeoenvironments and the allocyclic processes that occurred during transport and at the site of deposition. These factors are part of a complex interplay that varies depending on age, atmospheric conditions, available source materials and sedimentary basin type. Although the Keskarrah, Beaulieu Rapids and Jackson Lake formations all developed during the late Archaean, differences in composition indicate that the factors operating during the deposition of each of the sequences were not identical. The conventional petrographic discrimination diagrams of Dickinson *et al.* (1983) are often difficult to apply to Archaean sedimentary sequences, especially to small, strike-slip or late-orogenic basins, because a good database is lacking. As a result, the tectonic settings of the Keskarrah,

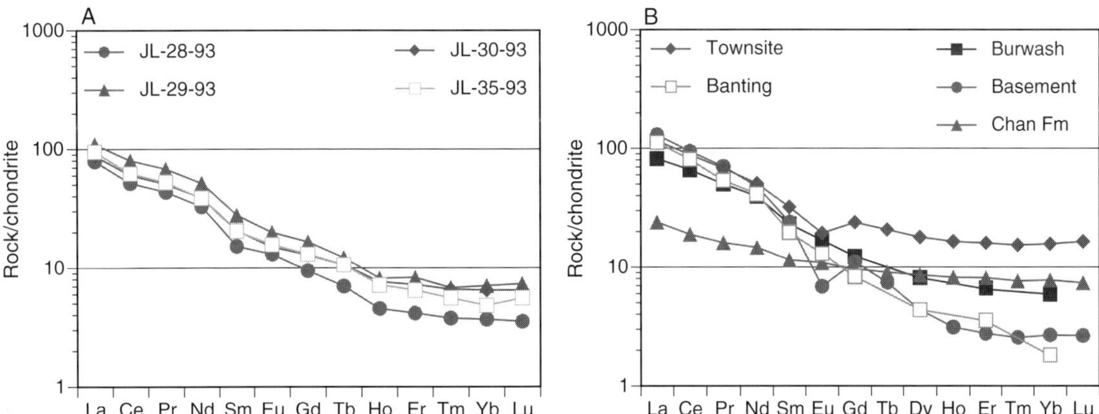


**Fig. 13.** Chondrite-normalized REE abundances for samples from the (A) Jackson Lake Formation (JL-28-93, 29-93, 30-93, 35-93), and (B) volcanic Townsite and Chan formations of the Kam Group, and older plutonic rocks (Cousens, 1999), in addition to the sedimentary Burwash and felsic volcanic-dominated Banting Group (Jenner *et al.*, 1981). Normalizing values after Sun (1982).

Beaulieu Rapids and Jackson Lake formations are based on geochemical data, whereas the petrography plays a more important role in determining source rocks.

### Tectonic setting

The chemical composition of sedimentary rocks is strongly influenced by the nature of the source rocks and specific tectonic settings. Generally, the fingerprint of a tectonic setting is represented by various lithofacies and their distinct distribution in a sedimentary basin. The Keskarrah and Jackson Lake formations, which contain alluvial fan–fan delta conglomeratic deposits that prograde into wave- and tide-influenced shoreline-shelf sandstone and argillite, are consistent with high relief, suggestive of an active tectonic setting interpretation (Hempton *et al.*, 1983; Corcoran *et al.*, 1998; Mueller *et al.*, this volume, pp. 153–182). Similarly, the Beaulieu Rapids Formation with alluvial fan–fan delta conglomerates prograding into braided stream sandstones and braidplain–lacustrine siltstone–sandstones, which are organized into two large-scale, well defined upward-fining sequences, indicate tectonic influence on sedimentation (Rust & Koster, 1984; Corcoran *et al.*, 1999). Bounding faults and unconformities, characteristic of the three sedimentary successions, lend further support for basin development in an active tectonic setting (Johnson, 1985; Krapez & Barley, 1987).

The chemical composition of sandstones can further elaborate on the setting in which tectonically influenced sedimentary sequences are deposited (Bhatia, 1983; Bhatia & Crook, 1986; Roser & Korsch, 1988; Girty *et al.*, 1993). Samples from the Keskarrah Formation are affiliated with the active continental margin (ACM) setting on tectonic discrimination diagrams (Fig. 10A & B). The ACM field represents sedimentary rocks associated with plate convergence, strike-slip margins and thick continental crust (Bhatia & Crook, 1986). This implies elevated proportions of La and Th which are normally associated with a high quartz component, suggestive of a granitic source. The Beaulieu Rapids Formation is affiliated with the ACM field in addition to the continental volcanic arc setting (CVA; Fig. 10A & B). The strike-slip margin associated with the ACM field (Bhatia & Crook, 1986) is in good agreement with the strike-slip basin interpretation of Corcoran *et al.* (1999), which was based on lithofacies, petrography and structural geology. The Jackson Lake samples, which plot in the CVA field (Fig. 10A & B), are representative of plate convergence. A strong relationship with volcanic source rocks is indicated by lower La, Th and Th/Sc ratio, and higher Sc values (Bhatia & Crook, 1986; Fedo *et al.*, 1997a,b). The geochemistry and related tectonic discrimination diagrams are consistent with the interpreted depositional settings of the respective remnant basins.

### Source rocks

Preliminary interpretations of the source rocks as determined from tectonic discrimination diagrams indicate that the Keskarrah Formation received detritus primarily from a granitic source; the Beaulieu Rapids formation from a mixed granitic–mafic volcanic source; and the Jackson Lake Formation from a mainly mafic volcanic source. Relatively high Th values, low Ti, Sc and Ni values and higher $La_n/Yb_n$ and $Sm_n/Yb_n$ ratios from the Keskarrah samples further support a predominantly granitic source, as opposed to lower Th, Th/Sc ratios, $La_n/Yb_n$ and $Sm_n/Yb_n$, and higher Ti, Sc and Ni produced by the Jackson Lake rocks. These elemental components suggest a predominant mafic source for the Jackson Lake Formation, whereas intermediate values and ratios for the Beaulieu Rapids Formation indicate a mixed mafic–granitic source, with granitic rocks slightly dominant. Negative Eu anomalies noted for the Keskarrah and Beaulieu Rapids formations imply that felsic volcanic rocks played at least a minimal role in contributing detritus to the sedimentary basins. Although Jackson Lake samples lack negative Eu anomalies, most of the potential source rocks in the Yellowknife belt, including some felsic units, are not characterized by the anomalies (Jenner *et al.*, 1981); thus a source characterization based on Eu is not attempted. Trends on the CIA plot demonstrate that the Keskarrah and Beaulieu Rapids formations contain detritus resembling weathered average granodioritic material, whereas the Jackson Lake Formation follows a trend associated with derivation from a more mafic average tonalitic/andesitic source.

Point and clast counting results are generally consistent with geochemical data with respect to source rocks. The abundance of quartz and plutonic rock fragments in the Keskarrah Formation supports a granitic source. Minimal sedimentary and volcanic rocks fragments indicate that detritus from the Point Lake volcanic belt and the turbiditic Contwoyto Formation were incorporated into the sedimentary basin. In addition, clast counts in conglomerates reflect predominant plutonic and mafic volcanic sources. The low proportion of volcanic rock fragments in the sandstones is attributed to breakdown of sand-sized feldspar during chemical weathering, abrasion and sorting because feldspars are considerably more labile than quartz (Potter, 1986; Heins, 1995; Corcoran *et al.*, 1999).

The Beaulieu Rapids Formation is also rich in quartz, especially polycrystalline varieties, suggesting a proximal granitic source (Basu, 1976; Johnsson, 1993). Plutonic rock fragments are predominant, but sedimentary and volcanic fragments additionally contribute to the sandstone framework. The majority of the sedimentary fragments are in the sandstone lithofacies of the second large-scale upward-fining sequence. The increase is attributed to cannibalization of the older siltstone–sandstones of the first upward-fining sequence (Corcoran *et al.*, 1999). The predominant plutonic and mafic volcanic clasts in the conglomerate indicate that the plutonic Sleepy Dragon Complex and Beaulieu River volcanic belt were the major source areas.

The Jackson Lake Formation is characterized by a moderate quartz percentage, but unlike sandstones in the Keskarrah and Beaulieu Rapids formations which have Qp/Q ratios of 0.56 and 0.85, respectively, the Jackson Lake samples are dominated by monocrystalline varieties (Qp/Q = 0.35). Although most quartz comprising sandstones is generally attributed to a plutonic source (Pettijohn, 1975), felsic volcanic sources also contribute to the monocrystalline proportion. A felsic source is supported by a large proportion of felsic volcanic rock fragments in the Jackson Lake samples. Mafic volcanic clasts are abundant in conglomerates as opposed to felsic volcanic, sedimentary, plutonic and porphyry counterparts, indicating that the major contributor of detritus was the mafic-dominated Kam Group underlying the sedimentary rocks.

Geochemical, petrographical and clast counting results demonstrate that the major source rocks for the sedimentary sequences include: (i) primarily the Augustus granite and subordinate mafic volcanic Peltier Formation for the Keskarrah Formation; (ii) a mixture of the plutonic Sleepy Dragon Complex and the mafic-dominated Beaulieu River volcanic belt for the Beaulieu Rapids Formation; and (iii) primarily the mafic-dominated Kam Group, and subordinate plutonic and felsic volcanic rocks for the Jackson Lake Formation.

## Chemical weathering

The effects of chemical weathering on sedimentary rock composition are dependent on climatic environment, absence or abundance of vegetation and relief with related sediment residence time. These effects, well documented by several authors (Basu, 1976; Grantham & Velbel, 1988; Johnsson & Stallard, 1989; Johnsson, 1993; Corcoran *et al.*, 1998, 1999), would have been greatly enhanced during the Archaean because of inferred higher atmospheric $CO_2$ levels (Young, 1991; Kasting, 1993), elevated temperatures (Kasting, 1993) and more humid conditions (Des Marais, 1994). Chemical weathering is maximal in humid climates where breakdown of feldspars and clay minerals is enhanced (Basu, 1976; Potter, 1986; Fedo *et al.*, 1995, 1997a,b). Consequently, with all other factors being equal, Archaean sedimentary rocks should contain greater proportions of quartz, a relatively stable mineral during chemical weathering, compared to younger sequences. The quartz-rich sandstones and quartz arenites reported from the Keskarrah Formation have been attributed to intense *in situ* chemical weathering of the plutonic and volcanic source rocks, and of sediments during deposition and transport (Corcoran *et al.*, 1998). This assumption is further supported by the general trends for the Keskarrah and Jackson Lake formations that intersect the ($CaO + Na_2O$)–$Al_2O_3$ line on the CIA diagram (Fig. 9). Because the trends do not terminate at the base of the ternary diagram where natural groundwater compositions lie, this implies that potassium metasomatism was an important factor governing the proportion of feldspar in the samples (McLennan *et al.*, 1993; Fedo *et al.*, 1997a,b). High CIA values for the Keskarrah siltstone–sandstone and the Jackson Lake argillite–sandstones indicate an elevated amount of clay derived from feldspar decomposition. However, CIA values from the Keskarrah Formation indicate that one sample (PLC-95-138) underwent intense chemical weathering, whereas the other three, in addition to the samples from Easton (1985), were highly but not intensely weathered. Similar results for samples from the Beaulieu Rapids and Jackson Lake formations imply that chemical weathering played a significant role in the composition of the sandstones, but was not the only factor.

## Depositional setting and sorting

The moderately high CIA values for all three sequences support a significant weathering component during the Archaean. Notwithstanding this, the lack of intense weathering, represented by CIA values generally <85, indicates that weathering alone cannot account for quartz arenites in a high relief, tectonically active setting (e.g. Keskarrah Formation), and significantly high (69%) quartz and silica (62–75%) components in a sedimentary sequence that received most of its detritus from a mafic volcanic source (e.g. Jackson Lake Formation).

The presence of quartz-rich sedimentary rocks that formed in high relief palaeoenvironments can also, in part, be the result of processes operating at the site of deposition. Hydraulic sorting and mechanical abrasion during transport and in the depositional environment are possibly two of the most significant factors controlling the composition of sedimentary rocks (Morton & Hallsworth, 1999). Feldspars are preferentially sorted and removed from sediment along beach-shelf type environments (Picard & McBride, 1993; Kairo *et al.*, 1993; McBride *et al.*, 1996). The shoreface to proximal offshore setting of the Keskarrah Formation would have been a favourable site for extensive diminution of feldspar in the sandstones (Fig. 14). The combination of a predominantly granitic source, high levels of chemical weathering, associated K-metasomatism and increased winnowing and mechanical breakdown of feldspar through swash/backwash, tidal currents, longshore drift and storm transport was sufficient to counterbalance the features of a high relief, tectonically active setting. This setting normally favours low sediment residence times with an increase in rock fragments and feldspars (Johnsson *et al.*, 1988; Grantham & Velbel, 1988).

Similarly, macrotidal currents played a prominent role in the Jackson Lake palaeoenvironmental setting. Trace element geochemistry and clast counting indicate a predominantly mafic volcanic source, but the sandstone–argillites are rich in quartz and silica. The low feldspar and rock fragment percentages in the sandstone–argillites could be attributed in part to chemical weathering with CIA values between 67 and 71. K-metasomatism was also a factor in feldspar diminution as indicated by relatively high CIA values

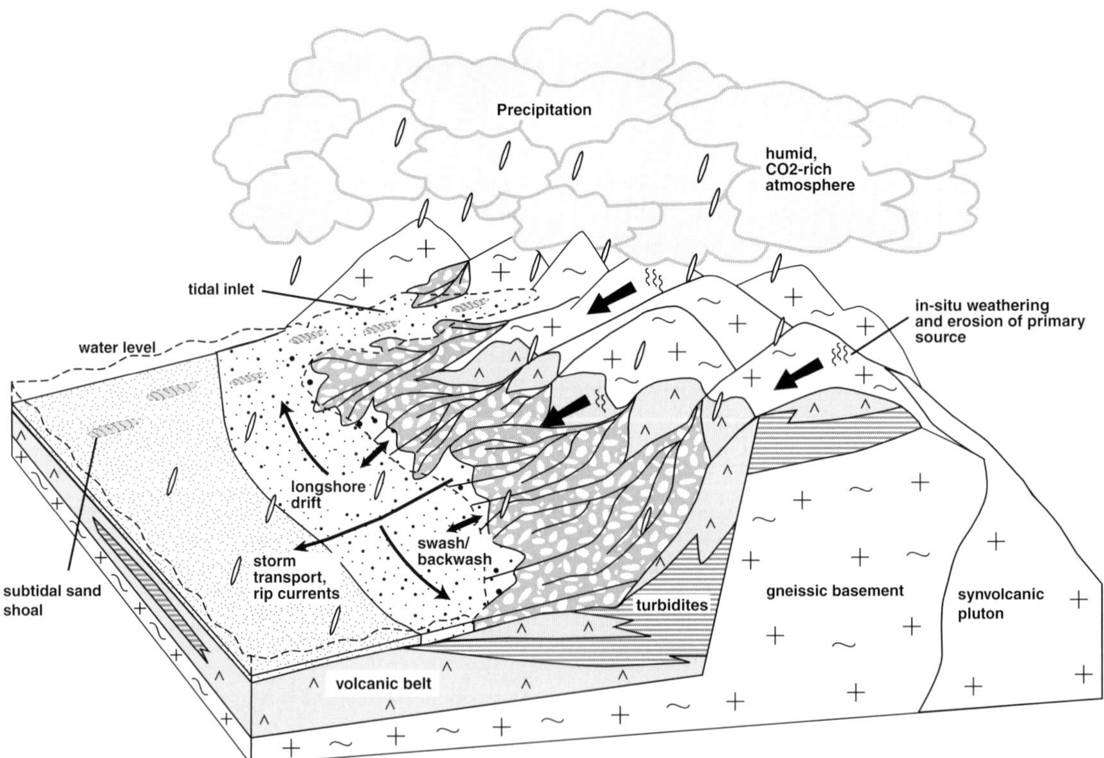


**Fig. 14.** Palaeogeographic model illustrating various factors controlling the composition of the Keskarrah Formation sediments at the time of deposition. Plutonic and volcanic source rocks were weathered *in situ* under humid, warm, $CO_2$-rich Archaean atmospheric conditions. The material was eroded while constantly being subjected to chemical weathering during transport. Coalescing alluvial fans and fan-deltas developed in the tectonically active, high-relief setting. Fan-deltas prograded directly on to a marine shelf and into a tidal inlet where the sediments were subjected to longshore drift, swash/backwash processes, storm transport and subtidal waves.

of 77 and 83 for the argillite–sandstones. The high, but not extreme, CIA values lead to the assumption that the Jackson Lake Formation tide-dominated shoreline would have been conducive to sorting and further breakdown of labile minerals.

The Beaulieu Rapids Formation demonstrates predominant plutonic and mafic volcanic source components, as indicated from clast counts and chemistry. Although quartz and silica percentages are high, the paucity of quartz arenites, similar to those characterizing the Keskarrah Formation, could be related to the fluvial to lacustrine settings of the sandstone and siltstone–sandstone lithofacies, which are not environments that promote high degrees of sorting.

The results from facies analyses, petrography, clast counts and geochemistry of the Keskarrah, Beaulieu Rapids and Jackson Lake formations indicate that source rock composition, chemical weathering under unique Archaean atmospheric conditions and sorting played crucial roles in generating quartz-rich sedimentary rocks. The combination of a plutonic-dominated source, high levels of chemical weathering and good sorting produced quartz-rich sandstones and quartz arenites in the Keskarrah Formation. Although the Beaulieu Rapids Formation received detritus from a plutonic source and underwent similar degrees of chemical weathering, the lack of quartz arenites in this sequence may be attributed to the alluvial–fluvial setting, which is not as favourable to sorting as a shelf-type environment. The Jackson Lake Formation was strongly influenced by tides and by chemical weathering, but does not contain as much quartz as the other sequences because plutonic and felsic volcanic sources were minor.

### Comparisons with Phanerozoic and modern counterparts

The effects of chemical weathering, provenance and sorting on sandstone composition have been well established in previous studies concerning Phanerozoic and modern sedimentary deposits. The results from the Archaean Keskarrah, Beaulieu Rapids and Jackson Lake formations are consistent with those of younger examples from fluvial and shoreline settings (Table 4). Basu (1976) and Mack (1981) underlined the importance of source rock composition on framework mineral percentages in Holocene and modern sands, respectively, whereas Nesbitt *et al.* (1996) demonstrated from Recent sediments in the Mallacoota Basin, Australia, that under humid climatic conditions, sediment composition reflects the nature of the souce rock weathering profiles. Basu (1976), Franzinelli & Potter (1983) and Suttner & Dutta (1986) convincingly documented an increase in quartz and decrease in plagioclase in modern and ancient humid or tropical settings compared with sediments deposited in arid or semiarid environments. The effects of chemical weathering are dependent on duration and intensity according to Grantham & Velbel (1988) and Johnsson & Stallard (1989). A decrease in labile minerals (e.g. plagioclase) in modern sands from North Carolina and Panama resulted from weathering under humid conditions where intensity was greatest, and on low slopes where duration was increased. Kairo *et al.* (1993) documented an abundance of quartz in Palaeozoic foreshore sandstones compared with offshore/transition sandstones in a wave-dominated delta, and attributed this abundance to shoreline reworking. Similarly, the influence of sorting on sediment composition was proven by McBride *et al.* (1996) where quartz percentage increases and feldspar and rock fragments decrease as sediments move northward along a South Texas Barrier Island beach. Although chemical weathering, source rocks with associated weathering profiles and sorting control sediment composition to some degree, only a combination of these factors significantly alters the proportion of framework minerals. Nesbitt & Young (1996) determined that without chemical weathering, abrasion and sorting had a minimal effect on the composition of Recent sediments from Baffin Island.

## CONCLUSIONS

The late-orogenic Keskarrah, Beaulieu Rapids and Jackson Lake formations display petrographical and geochemical characteristics that, combined with the physical sedimentology, reflect unique Archaean conditions. The remnant sedimentary basins were primarily controlled by the nature of source rocks, high degrees of chemical weathering in a humid, $CO_2$-rich Archaean atmosphere and sorting during transport and at the site of deposition. The Keskarrah and Beaulieu Rapids formations, adjacent to the Beniah Lake fault, and the Jackson Lake Formation, associated with the ancestral Jackson Lake fault, all demonstrate the salient features of sedimentary basins that develop in active tectonic settings. The depositional basins formed in high-relief zones similar to modern intermontane settings or rugged, high-relief coastlines, but the petrography and geochemistry shows that these deposits with a short residence time are quartz-rich. Weathering

**Table 4.** Examples of Phanerozoic and modern counterparts for the Keskarrah, Beaulieu Rapids and Jackson Lake formations.

| Reference | Location/age | Methods | Objectives | Results |
|---|---|---|---|---|
| Mack (1981) | Modern stream sand from North Georgia, USA | Point counting | Present compositions of modern stream sands forming under humid conditions and sourced by a mixed sedimentary/low-grade metamorphic terrain | Differences in composition are a function of the mixed source terrain (e.g. monocrystalline quartz derived from sedimentary terrain, polycrystalline from metamorphic terrain) |
| Franzinelli & Potter (1983) | Modern river sands, Amazon River System, South America | Point couting, K-feldspar staining | Determine the relationship between sand composition, source rock composition and climate | Rock fragments are easily destroyed under tropical conditions. Tropical weathering and low relief can result in first cycle quartz arenites. Quartz increases and rock fragments decrease downstream |
| Grantham & Velbel (1988) | Modern fluvial sands from the southern Blue Ridge Mountains, North Carolina, USA | Point counting, petrography | Examine the relationship between chemical weathering and rock fragment abundance | Total chemical weathering is a function of duration and intensity of chemical weathering. Intensity is greater under humid conditions and duration is increased on low slopes |
| Johnsson & Stallard (1989) | Modern sands from the Barro Colorado Island, Panama | Point counting, K-feldspar staining, X-ray diffraction | Determine the influence of lithology and physiography on sediment composition in a tropical weathering environment | On steep slopes, erosion rates are greater than rates of chemical weathering. Residence times on steep slopes are low, resulting in an increase in labile minerals |
| McBride *et al.* (1996) | Modern beach sand along South Texas Barrier Island, USA | Point counting, K-feldspar staining, roundness | Determine whether rock fragments or feldspar decrease as sand drifts north from where the Rio Grande meets the beach | Quartz increases northward, and feldspar and rock fragments decrease. This is the result of wave action (abrasion/breakage, selective sorting) |
| Nesbitt *et al.* (1996) | Recent sediments from the Mallacoota Basin, SE Australia | Mass balance comparisons, X-ray diffraction, X-ray fluorescence | Determine the influence of sorting and chemical weathering on the bulk composition of sediments | Sediments reflect the composition of the source rock weathering profiles. Sands are quartz-rich compared to source rocks. Subsequent sorting produced quartz-rich sediments |
| Nesbitt & Young (1996) | Recent sediments from the glacio-fluvial environment of the Guys Bight Basin, north-east Baffin Island | X-ray fluorescence, petrography | Determine the efects of selective sorting and abrasion on the bulk composition of sediments | Abrasion and sorting had minimal effect on composition in the absence of chemical weathering |
| Basu (1976) | Holocene fluvial sand from the Appalachian and Rocky Mountains, USA | Point counting, K-feldspar staining | Petrographically compare sands derived from plutonic rocks under humid and dry climatic conditions | Rock fragments are more abundant in arid Rockies, quartz more abundant in humid Appalachians. Framework mineral percentages are dependent on source rock composition and loss of plagioclase in humid climates |
| Suttner & Dutta (1986) | Permian Cutler Formation and Permian–Pennsylvanian Fountain Formation in Colorado, and Permian–Triassic Gondwana Supergroup in India | Point counting | Test the influence of palaeoclimate on sandstone composition | Sandstones that formed under arid to semiarid conditions are the most immature compositionally. Quartz percentages are higher in sandstones that formed in humid climates |
| Kairo *et al.* (1993) | Pennsylvanian Fountain and Minturn formations, Colorado, USA | Point counting | Elucidate the effects of depositional environment on sedimentary rock composition | Quartz is more abundant in foreshore sandstones than offshore/transition sandstones in a wave-dominated delta (result of shoreline reworking). Small feldspar grains were removed from the foreshore and transported offshore during storms. Composition modification occurred more in wave-dominated (Fountain) than river-dominated (Minturn) delta |

processes and sorting must have therefore played a far more prominent role than hitherto perceived.

*In situ* weathering of sources and of sediments during transport played a crucial role in the development of quartz-rich sandstones and quartz arenites. Ubiquitous unconformable contacts between the sedimentary basins and volcanic-dominated or volcano-plutonic hinterlands, clast compositions in the conglomerates and the mineralogy of sandstones and siltstone–sandstones support a plutonic-dominated source for the Keskarrah Formation, a mixed mafic volcanic–plutonic source for the Beaulieu Rapids Formation and a predominant mafic volcanic source for the Jackson Lake Formation. The development of quartz-rich sandstones and quartz arenites in the Keskarrah Formation is attributed to a combination of chemical weathering, sorting and a plutonic source, whereas a lower degree of sorting in the Beaulieu Rapids basin is reflected in the paucity of quartz arenites. Lower quartz percentages in the Jackson Lake Formation, in which feldspars and volcanic rock fragments predominate, are the results of a mafic-dominated source. Geochemical results corroborate the petrography, as indicated by high Th values and $La_n/Yb_n$ and $Sm_n/Yb_n$ ratios, and low Ti, Sc and Ni values for the Keskarrah Formation, low Th values and $La_n/Yb_n$ and $Sm_n/Yb_n$ ratios and higher Ti, Sc and Ni values for the Jackson Lake Formation, and intermediate proportions of all elements for the Beaulieu Rapids Formation. Samples from the Keskarrah, Beaulieu Rapids and Jackson Lake formations plot in the active continental margin and continental volcanic arc settings on tectonic discrimination diagrams, supporting active tectonism, as suggested by the lithofacies types and distribution in all three basins. A plutonic source, an aggressive Archaean atmosphere and high degrees of sorting are necessary to produce quartz-rich sandstones and quartz arenites in high relief, tectonically active settings. As pointed out by Mueller *et al.* (this volume, pp. 153–182), Archaean Earth–Moon dynamics would have affected the marine coastline considerably, even in high relief settings. Tide-influence is considered a major Archaean sorting process and, in addition to chemical weathering and source composition, can account for the extreme quartz enrichment in the Keskarrah Formation.

## ACKNOWLEDGEMENTS

Logistical support for the project was provided by the Geology Division of Indian and Northern Affairs (contribution #00-002), Canada, and NSERC and LITHOPROBE (contribution #1158) operating grants to W.U. Mueller. Special thanks to Grant Young, Ned Chown and Jarda Dostal for constructive reviews that significantly improved the manuscript. Great appreciation goes out to field assistants Sherley Belley, Fred Savoie, Joe Bucher, Sophie Turcotte, Jean-François Tremblay and Yves Boulianne. Clarence Pickett prepared the samples for geochemistry.

## REFERENCES

Basu, A. (1976) Petrology of Holocene fluvial sand derived from plutonic source rocks: implications to paleoclimatic interpretation. *J. sediment. Petrol.*, **46**, 694–709.

Bhatia, M.R. (1983) Plate tectonics and geochemical composition of sandstones. *J. Geol.*, **91**, 611–627.

Bhatia, M.R. & Crook, K.A.W. (1986) Trace element characteristics of graywackes and tectonic setting discrimination of sedimentary basins. *Contrib. Miner. Petrol.*, **92**, 181–193.

Corcoran, P.L. (2000) Recognizing distinct portions of seamounts using volcanic facies analysis: examples from the Archean Slave Province, Northwest Territories, Canada. *Precam. Res.*, **101**, 237–261.

Corcoran, P.L., Mueller, W.U. & Chown, E.H. (1998) Climatic and tectonic influences on fan deltas and wave- to tide-controlled shoreface deposits: evidence from the Archean Keskarrah Formation, Slave Province, Canada. *Sediment. Geol.*, **120**, 125–152.

Corcoran, P.L., Mueller, W.U. & Padgham, W. (1999) Influence of tectonism and climate on lithofacies distribution and sandstone and conglomerate composition in the Archean Beaulieu Rapids Formation, Northwest Territories, Canada. *Precam. Res.*, **94**, 175–204.

Cousens, B.L. (2000) Geochemistry of the Archean Kam Group, Yellowknife Greenstone Belt, Slave Province, Canada. *J. Geol.*, **108**, 181–197.

Des Marais, D.J. (1994) The Archean atmosphere: its composition and fate. In: *Archean Crustal Evolution* (Ed. Condie, K.C.), pp. 505–519. Developments in Precambrian Geology, 11. Elsevier, Amsterdam.

Dickinson, W.R., Beard, L.S., Brakenridge, G.R. *et al.* (1983) Provenance of North American Phanerozoic sandstones in relation to tectonic setting. *Geol. Soc. Am. Bull.*, **94**, 222–235.

Dickinson, W.R. & Suczek, C.A. (1979) Plate tectonics and sandstone compositions. *Am. Ass. Petrol. Geol. Bull.*, **63**, 2164–2182.

Easton, R.M. (1985) The nature and significance of pre-Yellowknife Supergroup rocks in the Point Lake area, Slave Structural Province, Canada. In: *Evolution of Archean Supracrustal Sequences* (Eds Ayres, L.D., Thurston, P.C., Card, K.D. & Weber, W.), Spec. Publ. geol. Ass. Can., St John's, **28**, 153–167.

Fedo, C.M., Nesbitt, W.H. & Young, G.M. (1995) Unraveling the effects of potassium metasomatism in sedimentary rocks and paleosols, with implications for paleoweathering conditions and provenance. *Geology*, **23**, 921–924.

FEDO, C.M., YOUNG, G.M., NESBITT, H.W. & HANCHAR, J.M. (1997a) Potassic and sodic metasomatism in the Southern Province of the Canadian Shield: evidence from the Paleoproterozoic Serpent Formation, Huronian Supergroup, Canada. *Precam. Res.*, **84**, 17–36.

FEDO, C.M., YOUNG, G.M. & NESBITT, H.W. (1997b) Paleoclimatic control on the composition of the Paleoproterozoic Serpent Formation, Huronian Supergroup, Canada: a greenhouse to ice-house transition. *Precam. Res.*, **86**, 201–223.

FENG, R. & KERRICH, R. (1990) Geochemistry of fine-grained clastic sediments in the Archean Abitibi greenstone belt, Canada: implications for provenance and tectonic setting. *Geochim. Cosmochim. Acta*, **54**, 1061–1081.

FOLK, R.L. (1974) *Petrology of Sedimentary Rocks*. Hemphill's, Austin, TX, 170 pp.

FRALICK, P.W. & KRONBERG, B.I. (1997) Geochemical discrimination of clastic sedimentary rock sources. *Sediment. Geol.*, **113**, 111–124.

FRANZINELLI, E. & POTTER, P.E. (1983) Petrology, chemistry, and texture of modern river sands, Amazon River system. *J. Geol.*, **91**, 23–39.

GIRTY, G.H., BARBER, R.W. & KNAACK, C. (1993) REE, Th, and Sc evidence for the depositional setting and source rock characteristics of the Quartz Hill Chert, Sierra Nevada, California. In: *Processes Controlling the Composition of Clastic Sediments* (Eds Johnsson, M.J. & Basu, A.), Spec. Publ. geol. Soc. Am., Boulder, **284**, 109–119.

GRANTHAM, J.H. & VELBEL, M.A. (1988) The influence of climate and topography on rock-fragment abundance in modern fluvial sands of the southern Blue Ridge Mountains, North Carolina. *J. sediment. Petrol.*, **58**, 219–227.

HEINS, W.A. (1995) The use of mineral interfaces in sand-sized rock fragments to infer ancient climate. *Geol. Soc. Am. Bull.*, **107**, 113–125.

HELMSTAEDT, H. & PADGHAM, W.A. (1986) A new look at the stratigraphy of the Yellowknife Supergroup at Yellowknife, NWT—implications for the age of gold-bearing shear zones and Archean basin evolution. *Can. J. Earth Sci.*, **23**, 454–475.

HEMPTON, M.R., DUNNE, L.A. & DEWEY, J.F. (1983) Sedimentation in an active strike-slip basin, southeastern Turkey. *J. Geol.*, **91**, 401–412.

HENDERSON, J.B. (1970) Stratigraphy of the Archean Yellowknife Supergroup, Yellowknife Bay–Prosperous Lake area, District of Mackenzie. *Paper 70-26*, Geological Survey of Canada, 12 pp.

HENDERSON, J.B. (1981) Archean basin evolution in the Slave Province, Canada. In: *Precambrian Plate Tectonics* (Ed. Kroner, A.). Elsevier, Amsterdam, pp. 213–236.

HENDERSON, J.B. (1985) *Geology of the Yellowknife–Hearne Lake area, District of Mackenzie: a Segment across an Archean Basin*. Mem. Geol. Surv. Can., Ottawa, **414**.

HENDERSON, J.B. (1988) Geology, Keskarrah Bay Area, District of Mackenzie, Northwest Territories. Map 1679A, scale 1:50 000. Geological Survey of Canada, Ottawa.

HENDERSON, J.B. (1998) Geology of the Keskarrah Bay Area, District of Mackenzie, Northwest Territories. *Geol. Surv. Can. Bull.*, **527**, 122 pp.

HENDERSON, J.B., LOVERIDGE, W.D. & SULLIVAN, R.W. (1982) A U–Pb study of zircon from granitic basement beneath the Yellowknife Supergroup, Point Lake, District of Mackenzie. In: *Current Research, Part C. Paper 82-1C*, pp. 173–177. Geological Survey of Canada, Ottawa.

HENDERSON, J.B., VAN BREEMEN, O. & LOVERIDGE, W.D. (1987) Some U–Pb zircon ages from Archean basement, supracrustal and intrusive rocks, Yellowknife–Hearne Lake area, District of Mackenzie. In: *Radiogenic Age and Isotopic Studies Report 1. Paper 87-2*, pp. 111–121. Geological Survey of Canada, Ottawa.

HOLAIL, H.M. & MOGHAZI, A.-K.M. (1998) Provenance, tectonic setting and geochemistry of greywackes and siltstones of the late Precambrian Hammamat Group, Egypt. *Sediment. Geol.*, **116**, 227–250.

HOLLAND, H.D. (1978) *The Chemistry of the Atmosphere and Oceans*. Wiley, New York, 351 pp.

ISACHSEN, C.E. & BOWRING, S.A. (1994) Evolution of the Slave Craton. *Geology*, **22**, 917–920.

ISACHSEN, C.E. & BOWRING, S.A. (1997) The Bell Lake group and Anton Complex: a basement–cover sequence beneath the Archean Yellowknife greenstone belt revealed and implicated in greenstone belt formation. *Can. J. Earth Sci.*, **34**, 169–189.

ISACHSEN, C.E., BOWRING, S.A. & PADGHAM, W.A. (1991) U–Pb zircon geochronology of the Yellowknife volcanic belt, NWT, Canada: new constraints on the timing and duration of greenstone belt magmatism. *J. Geol.*, **99**, 55–67.

JENNER, G.A., FRYER, B.J. & MCLENNAN, S.M. (1981) Geochemistry of the Archean Yellowknife Supergroup. *Geochim. Cosmochim. Acta*, **45**, 1111–1129.

JOHNSON, S.Y. (1985) Eocene strike-slip faulting and nonmarine basin formation in Washington. In: *Strike-slip Deformation, Basin Formation, and Sedimentation* (Eds Biddle, K.T. and Christie-Blick, N.), Spec. Publ. Soc. econ. Paleont. Miner., Tulsa, **37**, 283–302.

JOHNSSON, M.J. (1993) The system controlling the composition of clastic sediments. In: *Processes Controlling the Composition of Clastic Sediments* (Eds Johnsson, M.J. and Basu, A.), Spec. Publ. geol. Soc. Am., Boulder, **284**, 1–19.

JOHNSSON, M.J. & STALLARD, R.F. (1989) Physiographic controls on the composition of sediments derived from volcanic and sedimentary terrains on Barro Colorado Island, Panama. *J. sediment. Petrol.*, **59**, 768–781.

KAIRO, S., SUTTNER, L.J. & DUTTA, P.K. (1993) Variability in sandstone composition as a function of depositional environment in coarse-grained delta systems. In: *Processes Controlling the Composition of Clastic Sediments* (Eds Johnsson, M.J. & Basu, A.), Spec. Publ. geol. Soc. Am., Boulder, **284**, 263–283.

KASTING, J.F. (1993) Earth's early atmosphere. *Science*, **259**, 920–926.

KRAPEZ, B. & BARLEY, M.E. (1987) Archaean strike-slip faulting and related ensialic basins: evidence from the Pilbara Block, Australia. *Geol. Mag.*, **124**, 555–567.

KROGH, T.E. & GIBBINS, W. (1978) U–Pb isotopic ages of basement and supracrustal rocks in the Point Lake area of the Slave Province. *Geol. Ass. Can., Prog. Abstr.*, **3**, 38.

KUSKY, T.M. (1989) Accretion of the Archean Slave province. *Geology*, **17**, 63–67.

KUSKY, T.M. (1991) Structural development of an Archean orogen, western Point Lake, Northwest Territories. *Tectonics*, **10**, 820–841.

LAMBERT, M.B., ERNST, R.E. & DUDAS, F.O.L. (1992) Archean mafic dyke swarms near the Cameron River and

Beaulieu River volcanic belts and their implications for tectonic modeling of the Slave Province, Northwest Territories. *Can. J. Earth Sci.*, **29**, 2226–2248.

LONGERICH, H.P., JENNER, G.A., FRYER, B.J. & JACKSON, S.E. (1990) Inductively coupled plasma-mass spectrometric analysis of geological samples: a critical evaluation based on case studies. *Chem. Geol.*, **83**, 105–118.

MCBRIDE, E.F., ABEL-WAHAB, A. & MCGILVERY, T.A. (1996) Loss of sand-size feldspar and rock fragments along the South Texas Barrier Island, USA. *Sediment. Geol.*, **107**, 37–44.

MACK, G.H. (1981) Composition of modern stream sand in a humid climate derived from a low-grade metamorphic and sedimentary foreland fold–thrust belt of north Georgia. *J. sediment. Petrol.*, **51**, 1247–1258.

MACLACHLAN, K. & HELMSTAEDT, H. (1995) Geology and geochemistry of an Archean mafic dike complex in the Chan Formation: basis for a revised plate-tectonic model of the Yellowknife greenstone belt. *Can. J. Earth Sci.*, **32**, 614–630.

MCLENNAN, S.M., HEMMING, S., MCDANIEL, D.K. & HANSON, G.N. (1993) Geochemical approaches to sedimentation, provenance, and tectonics. In: *Processes Controlling the Composition of Clastic Sediments* (Eds Johnsson, M.J. & Basu, A.), Spec. Publ. Geol. Soc. Am., Boulder, **284**, 21–40.

MORTENSEN, J.K., HENDERSON, J.B., JACKSON, V.A. & PADGHAM, W.A. (1992) U–Pb geochronology of Yellowknife Supergroup felsic volcanic rocks in the Russell Lake and Clan Lake areas, southwestern Slave Province, NWT. In: *Radiogenic Age and Isotopic Studies 5, Paper 91-2*, pp. 1–7. Geological Survey of Canada, Ottawa.

MORTON, A.C. & HALLSWORTH, C.R. (1999) Processes controlling the composition of heavy mineral assemblages in sandstones. *Sediment. Geol.*, **124**, 3–29.

MUELLER, W.U., BOWRING, S.A., CORCORAN, P.L. & PICKETT, C. (1998) Unconformities, major faults and the evolution of volcano-sedimentary basins on the Slave craton. *Slave–Northern Cordillera Lithospheric Evolution Transect and Cordilleran Tectonics Workshop Meeting, 6–8 March, Simon Fraser University*, pp. 15–16.

MUELLER W.U. & CORCORAN, P.L. (1998) Late-orogenic basins in the Archean Superior Province, Canada: characteristics and inferences. *Sediment. Geol.*, **120**, 177–203.

NESBITT, H.W. & YOUNG, G.M. (1982) Early Proterozoic climates and plate motions inferred from major element chemistry of lutites. *Nature*, **299**, 715–717.

NESBITT, H.W. & YOUNG, G.M. (1984) Prediction of some weathering trends of plutonic and volcanic rocks based on thermodynamic and kinetic considerations. *Geochim. Cosmochim. Acta*, **48**, 1523–1534.

NESBITT, H.W. & YOUNG, G.M. (1996) Petrogenesis of sediments in the absence of chemical weathering: effects of abrasion and sorting on bulk composition and mineralogy. *Sedimentology*, **43**, 341–358.

NESBITT, H.W., YOUNG, G.M., MCLENNAN, S.M. & KEAYS, R.R. (1996) Effects of chemical weathering and sorting on the petrogenesis of siliclastic sediments, with implications for provenance studies. *J. Geol.*, **104**, 525–542.

NORTHRUP, C.J., ISACHSEN, C. & BOWRING, S.A. (1999) Field relations, U–Pb geochronology and Sm–Nd isotope geochemistry of the Point Lake greenstone belt and adjacent gneisses, central Slave craton, NWT, Canada. *Can. J. Earth Sci.*, **36**, 1043–1059.

PADGHAM, W.A. & FYSON, W.K. (1992) The Slave Province: a distinct Archean craton. *Can. J. Earth Sci.*, **29**, 2072–2086.

PETTIJOHN, F.J. (1975) *Sedimentary Rocks*. Harper & Row, New York, 628 pp.

PICARD, M.D. & MCBRIDE, E.F. (1993) Beach sands of Elba Island, Tuscany, Italy: roundness study and evidence of provenance. In: *Processes Controlling the Composition of Clastic Sediments* (Eds Johnsson, M.J. & Basu, A.), Spec. Publ. Geol. Soc. Am., Boulder, **284**, 235–245.

POTTER, P.E. (1986) South America and a few grains of sand. Part 1: beach sands. *J. Geol.*, **94**, 301–319.

RELF, C., CHOUINARD, A., SANDEMAN, H. & VILLENEUVE, M. (1994) Contact relationships between the Anialik River volcanic belt and the Kangguyak gneiss belt, northwestern Slave Province, Northwest Territories. In: *Current Research 1994-C*, pp. 49–59. Geological Survey of Canada, Ottawa.

ROSER, B.P. & KORSCH, R.J. (1988) Provenance signatures of sandstone–mudstone suites determined using discriminant function analysis of major-element data. *Chem. Geol.*, **67**, 119–139.

RUST, B.R. & KOSTER, E.H. (1984) Coarse clastic deposits. In: *Facies Models*, 2nd edn (Ed. Walker, R.G.). Geosci. Can. Reprint Ser. 1, 53–69.

SCHWAB, F.L. (1981) Evolution of the western continental margin, French–Italian Alps: sandstone mineralogy as an index of plate tectonic setting. *J. Geol.*, **89**, 349–368.

SUN, S.S. (1982) Chemical composition and origin of the Earth's primitive mantle. *Geochim. Cosmochim. Acta*, **46**, 179–192.

SUTTNER, L.J. & DUTTA, P.K. (1986) Alluvial sandstone composition and paleoclimate, I. Framework mineralogy. *J. sediment. Petrol.*, **56**, 329–345.

VAN BREEMEN, O., DAVIS, W.J. & KING, J.E. (1992) Temporal distribution of granitoid plutonic rocks in the Archean Slave Province, northwest Canadian Shield. *Can. J. Earth Sci.*, **20**, 2186–2199.

VILLENEUVE, M., HRABI, B., JACKSON, V. & RELF, C. (1993) Geochronology of the supracrustal sequences in the central and northern Slave Province. Exploration Overview. Northwest Territories, NWT Geology Division, Indian and Northern Affairs Canada, 53.

WILLIAMS, G.E. (1998) Precambrian tidal and glacial clastic deposits: implications for Precambrian Earth–Moon dynamics and paleoclimate. *Sediment. Geol.*, **120**, 55–74.

YOUNG, G.M. (1991) The geologic record of glaciation: relevance to the climatic history of the Earth. *Geosci. Can.*, **18**, 100–108.

*Spec. Publs int. Ass. Sediment.* (2002) **33**, 213–234

# Stratigraphic evolution of Archaean volcanic rock-dominated rift basins from the Whim Creek Belt, west Pilbara Craton, Western Australia

G. PIKE* *and* R. CAS

*School of Geosciences, PO Box 28E, Monash University, Victoria 3800, Australia*

## ABSTRACT

The *c*.3009–2945 Ma Whim Creek Belt, Pilbara Craton, Western Australia, contains volcano-sedimentary rocks deposited within Archaean extensional basins. The stratigraphy of the Whim Creek Belt is redefined from detailed facies analysis and is subdivided into two groups, the Whim Creek Group and the overlying Bookingarra Group. The stratigraphy presented here recognizes volcanic and tectonic events as the dominant controls on facies and stratigraphic development. The Whim Creek Group is subdivided into three lithostratigraphic formations. The *c*.200 m thick Warambie Basalt unconformably overlies Mesoarchaean amphibolite facies basement and is overlain by *c*.290 m of volcaniclastic rocks of the Red Hill Formation. These formations are intruded by 160 m of dacitic intrusions of the Mons Cupri Dacite. The Bookingarra Group overlies the Whim Creek Group with a low angle unconformity or disconformity. Three formations are recognized. The *c*.300 m Cistern Formation and *c*.300 m Rushall Slate comprise volcaniclastic and siliciclastic sedimentary rocks and are conformably overlain by the Negri Volcanics, a *c*.2000 m thick package of dominantly basaltic volcanic and volcaniclastic rock. The Whim Creek Belt preserves evidence for two Mesoarchaean rift basins. The older Whim Creek sub-basin shows some stratigraphic similarity to modern back-arc basins. The recognition of bimodal volcanism and intrusion (Warambie Basalt and Mons Cupri dacite), coupled with resedimented volcaniclastic debris (Red Hill volcanics), is similar to modern back arc basins within continental crust and is consistent with the recognition of amphibolite and granitoid basement. Uplift and erosion of the Whim Creek Group preceded deposition of polymictic sedimentary rocks of the Cistern and Rushall Slate formations (Bookingarra Group) within the younger Mallina Basin. The Cistern Formation comprises coarse, polymictic conglomerate and breccia associated with early rifting of the Bookingarra Basin. These coarse sedimentary rocks contain volcaniclastic material derived from the eroded Whim Creek Group and an upward increasing proportion of siliciclastic debris. The Cistern Formation passes conformably upwards into siliciclastic-dominated sandstone, siltstone and shale of the Rushall Slate. The Cistern Formation/Rushall Slate contact reflects the decreasing importance of epiclastic volcanic debris derived from the Whim Creek Group. The Mallina Basin underwent a second rift phase that allowed the emplacement of basaltic volcanic rocks and subvolcanic sills of the Negri Volcanics. Rare peperitic contacts with the underlying Cistern Formation indicate that this late rifting overlapped with sedimentation within the Mallina Basin.

## INTRODUCTION

The Archaean Whim Creek Belt, Pilbara Craton, Western Australia (Fig. 1), preserves a complex stratigraphy of basalt lava and intrusions, dacite intrusions and volcaniclastic and siliciclastic terrigenous sedimentary rocks. The Whim Creek Belt forms the western part of the central Pilbara Craton (Mallina Basin) and formed over *c*.65 Myr from *c*.3010 to 2945 Ma (e.g. Nelson, 1998; Huston *et al.*, 2000). The Whim Creek Belt has been interpreted as a remnant of a strike-slip basin (Barley, 1987) or a transtensional rift basin (Krapez & Eisenlohr, 1998), and has some similarity with modern back-arc rift basins.

The stratigraphy of the Whim Creek Belt has been relatively well documented (e.g. Ryan & Kriewaldt, 1964; Fitton *et al.*, 1975; Hickman, 1983; Barley, 1987; Krapez & Eisenlohr, 1998; Smithies, 1998). However,

* Corresponding author: G.pike@shell.com

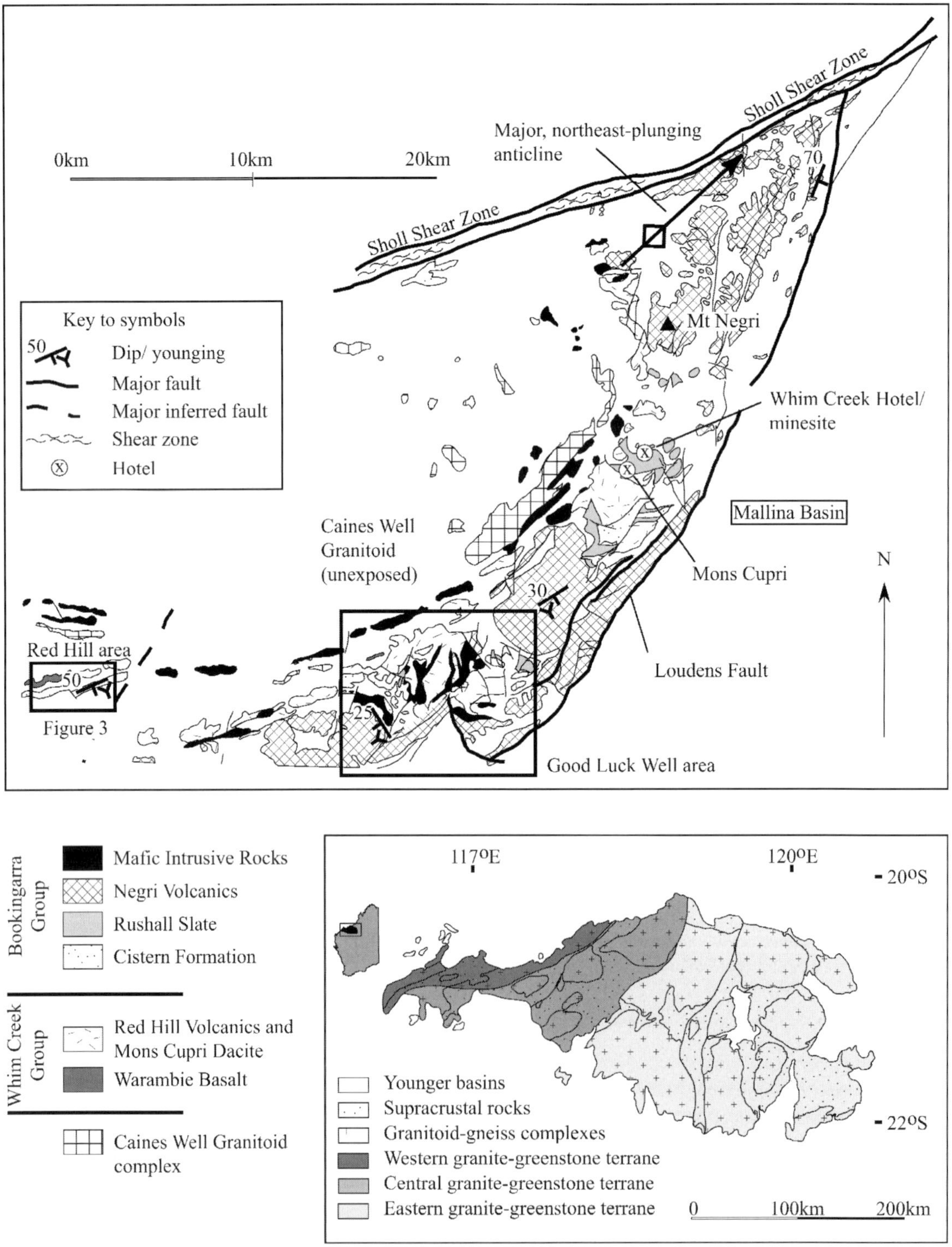

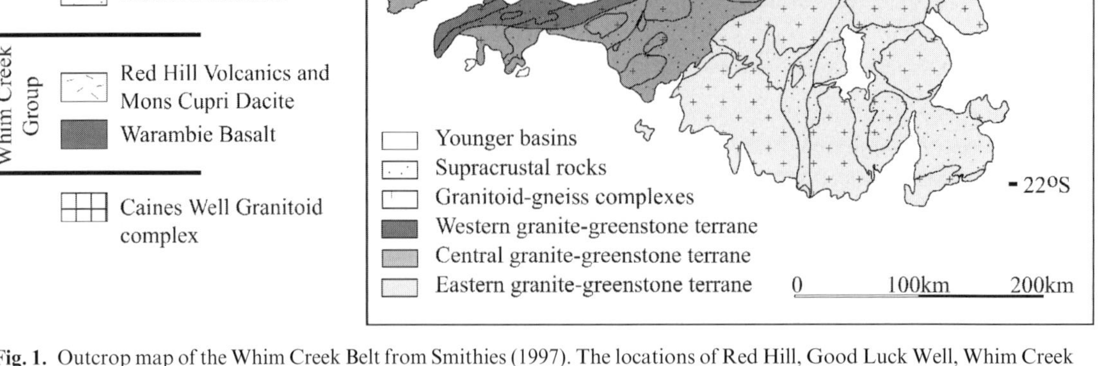


**Fig. 1.** Outcrop map of the Whim Creek Belt from Smithies (1997). The locations of Red Hill, Good Luck Well, Whim Creek and Mons Cupri are indicated; they are discussed in the text. The Bookingarra Group is redefined to include the Negri Volcanics. Lithostratigraphy and geology after Smithies (1998). Bottom: regional geology of the Pilbara Craton from Huston *et al.* (2000).

no previous interpretation fully defines both the spatial and temporal relationships between lithofacies units. The aims of this paper are to:

**1** Summarize previous lithostratigraphic work and present a revised interpretation based on detailed field mapping.

**2** Consider the causes of lithostratigraphic change in terms of eustatic, tectonic and volcanic influences.

**3** Discuss the validity of sequence stratigraphic models previously applied to the Whim Creek Belt.

**4** Assess whether Phanerozoic basin and tectonic models are applicable to the Whim Creek Belt.

## REGIONAL GEOLOGY

The Pilbara Craton exposes 60 000 km$^2$ of Mesoarchaean rocks in the centre-west of Western Australia (Fig. 1). The main outcrop of granitoid–greenstone terrane is restricted to 116° to 121°E, 20° to 22°30′S with inliers to the south. This Mesoarchaean block is unconformably overlain by Neoarchaean and Palaeoproterozoic volcanic and sedimentary rock successions. Approximately 1.1 billion years of crustal evolution from *c*.3650 Ma to post-2500 Ma are recorded. Nelson *et al.* (1999) define three stages of evolution: (i) formation of granitoid–greenstone crust during 3650–3100 Ma; (ii) craton-wide erosion from 3125 to 3000 Ma; and (iii) flood basalt volcanism during 2760–2680 Ma. The *c*.3009–2945 Ma Whim Creek Belt and Mallina Belt (Smithies *et al.*, 1999; Huston *et al.*, 2000), formed during the craton-wide erosion of Nelson *et al.* (1999). Hence, the evolution proposed by those authors should include a fourth stage that includes rocks of the Whim Creek Belt and time-equivalent successions (*c*.3009–2945 Ma). There is little evidence of post-2945 Ma sedimentation or intrusive activity within the Pilbara Craton until the 2760–2680 Ma event of Nelson *et al.* (1999). However, Smithies *et al.* (2001) suggest a post-2941 ± 9 Ma depositional cycle in the Mallina Basin.

The Whim Creek Belt crops out over *c*.500 km$^2$ of the west Pilbara Craton (Fig. 1). Rocks of the Whim Creek Belt unconformably overlie amphibolite facies metamorphic rocks that are intruded by the 3093 ± 4 Ma (Nelson, 1997) Caines Well Granite. Three regionally significant structural features form boundaries to the Whim Creek Belt (Fig. 1). The Sholl Shear Zone is an ENE–WSW trending, largely sinistral strike-slip fault with late (post-Whim Creek Belt formation) dextral displacement of *c*.30 km (Smith *et al.*, 1998; Smithies, 1998). The Loudens Fault is a NE–SW trending fault of uncertain character and forms the boundary between the present-day Whim Creek Belt and Mallina Basin to the east (Fig. 1). The Mallina Shear separates the same two units to the south of the Whim Creek Belt. Rocks of the Mallina Basin are sandstone (Constantine Sandstone) and shale (Mallina Formation) intruded by large granitic bodies. Possible correlation between the Whim Creek and Mallina basins is discussed by Fitton *et al.* (1975), Smithies *et al.* (1999), Huston *et al.* (2000) and Smithies *et al.* (2001), although Krapez & Eisenlohr (1998) regard the Mallina Basin as a significantly older system than the Whim Creek Belt.

Structural analysis of the Pilbara Craton suggests nine deformational events (Blewett, 1999; Blewett, personal communication, 2000). Greenschist to amphibolite facies metamorphism of the west Pilbara was largely confined to the D3b event at *c*.3015 Ma (Blewett, 1999; Blewett, personal communication, 2000) and did not affect the oldest rocks of the Whim Creek Belt. In the Whim Creek Belt, metamorphic grade is typically greenschist facies and does not mask primary igneous or sedimentary textures. Structural deformation varies in intensity and is dominated by gently NE-plunging folds, including the antiformal structure that defines the present shape of the Whim Creek Belt (Fig. 1; D4 of Blewett, 1999). Parasitic folds are kilometre-scale but few outcrop-scale folds are noted. The limbs of the main fold dip at low angles (<45°) away from the fold axis such that rocks typically young away from the Caines Well/Bookingarra Granites (Fig. 1). The dominant brittle features are NE–SW striking, subvertical, largely sinistral strike-slip faults (Fig. 1).

## THE GEOLOGY OF THE WHIM CREEK BELT

### Lithostratigraphy

The Whim Creek Group was described by Fitton *et al.* (1975) who identified four subdivisions: (i) Warambie Basalt, (ii) Mons Cupri Volcanics, (iii) Constantine Sandstone and (iv) Mallina Formation (Table 1). The latter units (Constantine Sandstone and Mallina Formation) outcrop in the Mallina Basin (Fig. 1) and are described by Smithies *et al.* (1999). Fitton *et al.* (1975) regard the Whim Creek Belt and Mallina Basin (Fig. 1) as a single depositional basin (see discussion in Krapez and Eisenlohr, 1998; Smithies *et al.*, 1999; Huston *et al.*, 2000).

**Table 1.** Stratigraphic interpretations of the Whim Creek Belt. Note that Fitton *et al.* (1975) originally correlated the Mallina Basin and Whim Creek Belt as a single unit. The base of the Warambie Basalt is a regional unconformity. Fitton *et al.* (1975) and Smithies *et al.* (1999) suggest an unconformity between the Rushall Slate and Negri Volcanics, but this unconformity is rejected based on the recognition of a conformable contact between these units (see text).

| Stratigraphy (Fitton *et al.*, 1975) | | Stratigraphy (Smithies, 1997) | | Stratigraphy (this study) | |
|---|---|---|---|---|---|
| Group | Formation | Group | Formation | Group | Formation |
| | Negri Volcanics | | Loudens and Negri Volcanics | Bookingarra Group | Negri Volcanics<br>Rushall Slate<br>Cistern Formation |
| Whim Creek Group | Mallina Formation<br>Constantine Sandstone<br>Mons Cupri Volcanics<br>Warambie Basalt | Whim Creek Group | Rushall Slate<br>Cistern Formation<br>Mons Cupri Volcanics | Whim Creek Group | Mons Cupri dacite<br>Red Hill volcanics<br>Warambie Basalt |

Smithies's (1998) Whim Creek Group (Table 1, Fig. 2) includes only the rocks to the west of the Loudens Fault (Fig. 1) and therefore separates the Whim Creek Group and Mallina Basin. Smithies (1998) recognized three formations within the Whim Creek Group (Table 1): (i) Mons Cupri Volcanics; (ii) Cistern Formation; and (iii) Rushall Slate. Fitton *et al.* (1975) include the upper Cistern Formation and Rushall Slate within the Constantine Sandstone and Mallina Formation, respectively (Table 1). Overlying the Rushall Slate is a thick, basaltic unit referred to as the Negri Volcanics in this work. Smithies (1998) subdivides the basaltic rocks into the Negri Volcanics and Loudens Volcanics based on the recognition of compositionally distinct lithostratigraphic units. However, Krapez & Eisenlohr (1998) place both units into a unified Negri Basin. Barley (1987) regards the Loudens and Negri Volcanics as compositional variations within a single eruptive series, such that stratigraphic subdivision is unnecessary.

New work records six formations within two groups (Fig. 2, Table 1). The lower group is the Whim Creek Group and comprises the Warambie Basalt, Red Hill volcanics and Mons Cupri dacite. The Whim Creek Group is dominated by volcanic and volcaniclastic rocks that Smithies (1998) assigns to the Mons Cupri Volcanics and parts of the Cistern Formation (Fig. 2). The Bookingarra Group overlies the Whim Creek Group and is dominated by sedimentary rocks of the Cistern Formation and Rushall Slate and basalt lava of the Negri Volcanics. The definition of the Negri Volcanics is in accordance with Barley (1987), as there is no obvious hiatus in formation and some overlap in eruptive style between the Loudens and Negri Volcanics of Smithies (1997). Previous work (e.g. Fitton *et al.*, 1975; Barley, 1987; Krapez & Eisenlohr, 1998; Smithies, 1998) describes, or agrees with, a significant hiatus between the Whim Creek Group and the Negri Volcanics. The boundary has been described as a low-angle unconformity (e.g. Krapez & Eisenlohr, 1998). This study recognizes a conformable and gradational or soft-sediment intrusive contact between the Cistern Formation/Rushall Slate and Negri Volcanics and includes the latter unit within the Bookingarra Group.

The stratigraphic thickness of individual units varies considerably throughout the Whim Creek Belt. The maximum thickness of each formation defines a total thickness of 3250 m. However, this thickness is not preserved in any one section due to lateral thickness changes within individual units. At Red Hill (Fig. 1) the Warambie Basalt (*c.*200 m), Mons Cupri dacite (160 m) and Red Hill volcanics (*c.*290 m) are 650 m in stratigraphic thickness and are overlain by 2000 m (Krapez & Eisenlohr, 1998) of Negri Volcanics. Hence, at Red Hill the Cistern Formation and Rushall Slate are not present. This 2650 m section is, perhaps, the thickest preserved section of the Whim Creek Belt. The Cistern Formation reaches a maximum thickness of 300 m south-east of Good Luck Well (Fig. 1). At Whim Creek (Fig. 1), the Rushall Slate is 200–300 m thick (Black, 1999).

### The Red Hill and Good Luck Well areas

Red Hill (Figs 1 & 3) ('Copper Bore' of Krapez & Eisenlohr, 1998) is selected as the type locality for the stratigraphic interpretation of the Whim Creek Group

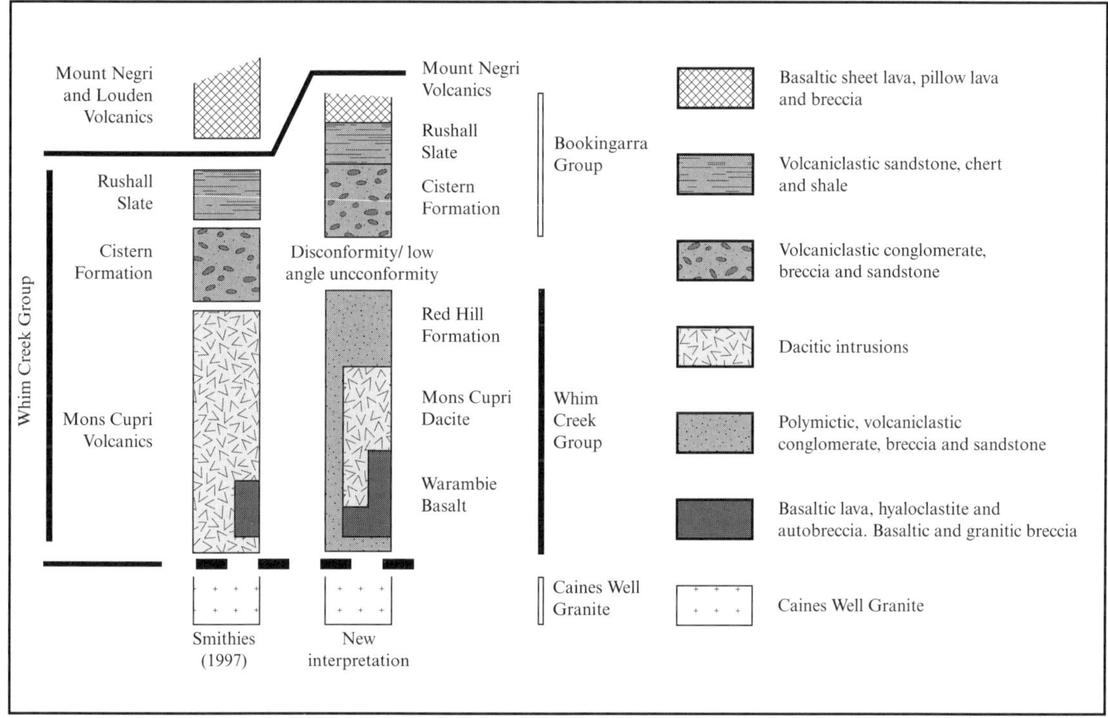


**Fig. 2.** Comparison of stratigraphic interpretations from Smithies (1998) and this study. The main differences are the recognition of the Red Hill volcanics as a distinct unit (separate from the Mons Cupri dacite and Cistern Formation), the subdivision of the 'Whim Creek Group' into the Whim Creek and Bookingarra groups and the recognition of a conformable Rushall Slate–Negri Volcanics contact.

(Warambie Basalt, Red Hill volcanics Formation, Mons Cupri Dacite: Table 1 & Fig. 2). The Red Hill area (Fig. 3) preserves an unconformity-bound package of volcanic and volcaniclastic rocks that represent all of the Whim Creek Group facies (Table 2). The Whim Creek Group is dominated by basaltic rocks (Warambie Basalt, ≤194 m) and pumiceous sedimentary rocks (Red Hill Volcanics, *c*.200 m) to the west of Red Hill (in the west of Fig. 3). In the centre and east of Fig. 3, dacite intrusions (Mons Cupri dacite) form up to 160 m thickness and immediately overlie <50 m of basalt lava. In this section, the Red Hill volcanics formation is *c*.290 m in thickness. Nelson (1998) gives a magmatic and depositional age of 3009 ± 4 Ma for rocks of the Red Hill volcanics formation in the Red Hill area. This interpretation is based on the recognition of a single zircon population within pumice breccia that must have been deposited syn- or immediately post-volcanism to preserve angular pumice and glass shards. The 3009 ± 4 Ma date gives an approximate age of the Whim Creek Group. Barley *et al.* (1994) give an age of 2991 ± 12 Ma for the Mons Cupri dacite from Red Hill. These dates do not overlap and suggest either a time gap of 2–34 Myr between the two formations or some error in either date. The area is well exposed and relatively undeformed (no significant cleavage) with a low dip of 15–30° to the SSE. Two sets of faults cross cut the area. Early NNE–SSW normal faults with hundreds of metres spacing are cross cut by NE–SW trending, subvertical, dominantly sinistral strike-slip faults (Fig. 3).

The Good Luck Well area (Fig. 1) preserves examples of the lower stratigraphy of the Bookingarra Group (Cistern Formation and Rushall Slate) as well as the underlying Whim Creek Group. Kilometre-scale, gently NE-plunging folds complicate facies interpretations, but Cistern Formation rocks are well preserved on fold limbs. These rocks pass upward gradationally into thin units of Rushall Slate and these have well preserved contacts with intrusive facies of the Negri Volcanics. Hence, the Good Luck Well area contains fragmented examples of the stratigraphic succession of the whole Bookingarra Group and much of the Whim Creek Group. As well as kilometre-scale

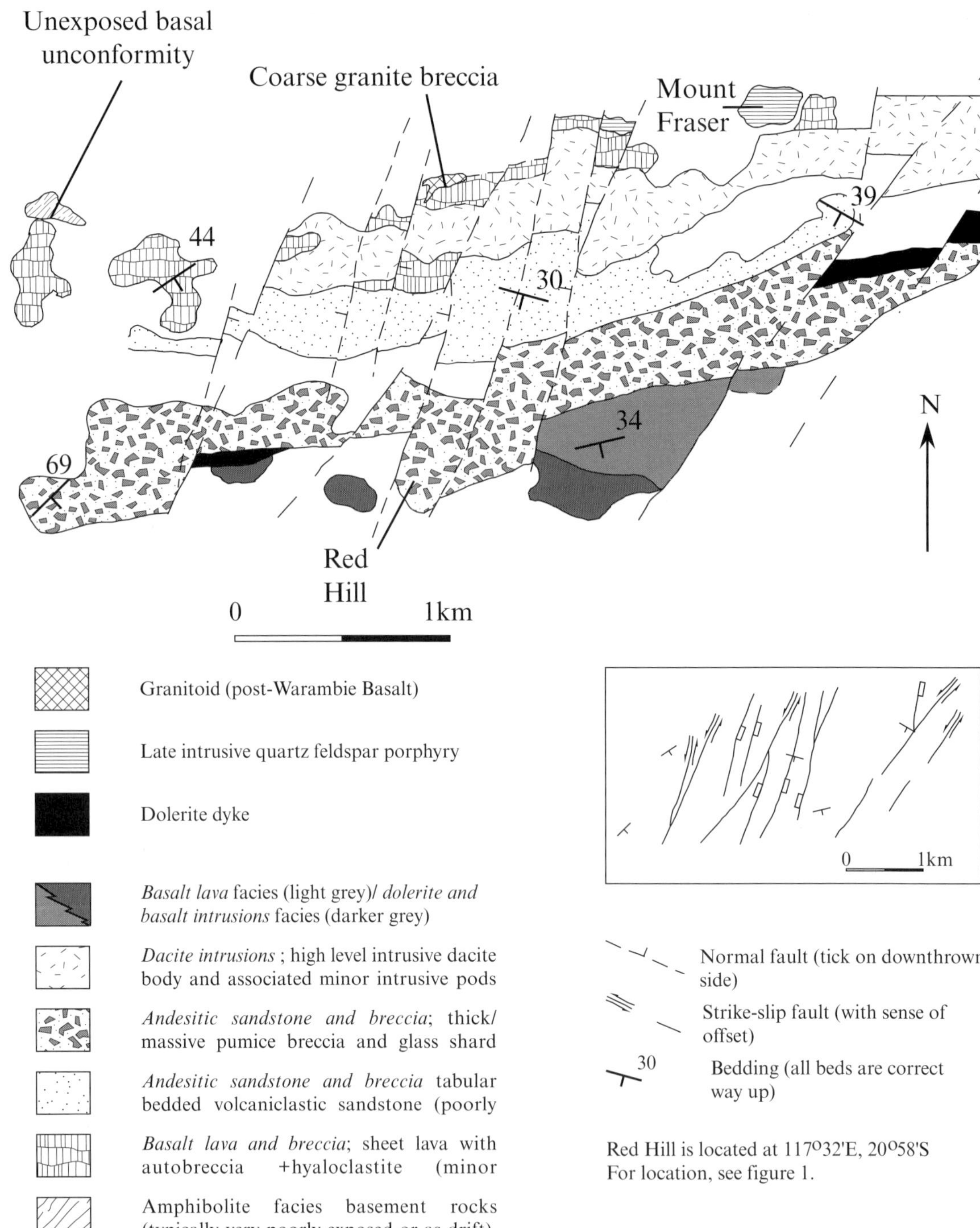


**Fig. 3.** Simplified geology of the Red Hill area showing ENE–WSW strike of facies and a general younging to the south (all units are correct way up). (Inset) Early SSW–NNE trending normal faults and late SW–NE, subvertical, sinistral, strike-slip faults dominate the brittle deformation (see Fig. 1 for location).

**Table 2.** Facies associated with lithostratigraphic units from the Whim Creek Belt (see Table 1 & Fig. 2 for detailed stratigraphy).

| Group | Formation | Facies |
|---|---|---|
| Bookingarra Group | Negri Volcanics | Basalt lava |
| | | Dolerite and basalt intrusions |
| | Rushall Slate | Siliciclastic-dominated sedimentary rock |
| | Cistern Formation | Volcaniclastic-dominated sedimentary rock |
| Whim Creek Group | Mons Cupri dacite | Dacite intrusions |
| | Red Hill volcanics | Andesitic sandstone and breccia |
| | | Polymictic conglomerate, breccia and sandstone |
| | Warambie Basalt | Grey dolerite and basalt |
| | | Arkosic sandstone and conglomerate |
| | | Basalt lava and breccia |
| | | Coarse granite breccia |

folds, the area is cross cut by E–W trending normal faults and the same NE–SW trending, sinistral, strike-slip fault set observed in the Red Hill area.

Barley (1987) describes rocks from between the Whim Creek and Mons Cupri minesites (Fig. 1). This area contains significant quartz–sericite–chlorite alteration but most facies are relatively well preserved. The Rushall Slate (*c*.200–300 m) surrounds and hosts the Whim Creek Mine (Fig. 1). The geology of the Whim Creek Mine is described by Black (1999), and this area represents the largest exposed section of Rushall Slate. Detailed facies description, logging and mapping of the Cistern Formation and Rushall Slate from these areas have been carried out during this study, and these observations and interpretations are summarized here.

The Negri Volcanics are discussed by Smithies (1997) and Krapez & Eisenlohr (1998). This paper does not focus on these rocks, as they require a fuller discussion in a separate manuscript. However, the lower contacts and internal facies characteristics are presented in order to define the stratigraphic position and relationship of the formation. Observations are taken from a variety of locations described in the text. The minimum age of the Negri Volcanics (and therefore the Bookingarra Group) is given by the Kialrah Rhyolite that either intrudes or overlies rocks of the Negri Volcanics west of Red Hill. Nelson (1998) gives a date of 2975 ± 4 Ma for this unit.

## Facies architecture of the Whim Creek Group

The Whim Creek Group is dominated by volcanic and volcaniclastic rocks. Seven facies are described, each of which contain a number of subfacies. Important subfacies are described where they have particular relevance to palaeo-environmental reconstruction. The facies described are: coarse granite breccia, basalt lava and breccia, arkosic sandstone and conglomerate, grey dolerite and basalt (Warambie Basalt), andesitic sandstone and breccia, polymictic conglomerate, breccia and sandstone (Red Hill volcanics) and dacite intrusions (Mons Cupri dacite). These facies are also listed in Table 2, where their position and relationship to stratigraphic units are depicted.

*Coarse granite breccia* at Red Hill (Fig. 3) contains boulder-size angular clasts of granite (60%) and basalt (40%), and constitutes the base of the Warambie Basalt. It is restricted to a single exposure of around 5 m thickness (Fig. 4A). The breccia is clast-supported, with a matrix of poorly sorted sand–pebble size granite and basalt lithic clasts and quartz–feldspar–lithic sandstone. Underlying granite shows strong hydrofracturing at the contact and has thin (cm-thick) pegmatitic apophyses into the coarse granite breccia, indicating that it is a later granite, and not basement to the Whim Creek Group. At the contact, basalt xenoliths or clasts from the breccia are incorporated into this granite. This unnamed granite is therefore younger than the Caines Well Granite (Fig. 1) from which the granite clasts may have been derived. The basaltic clasts are interpreted as products of contemporaneous basalt units, described below.

*Basalt lava and breccia* facies is restricted to the Red Hill area (Figs 1 & 3) and forms a stratigraphic interval of around 100 m thickness. The facies is characterized by coherent, basalt units with interbedded breccia and conglomerate. Coherent basalt units are up to 25 m in thickness. The thicker units are tabular and extend for >100 m along strike whereas thin units

M
BC
GC
A

BL
VH
VH
B

CRB
JFB
CC
C

SI
AR
AC
D
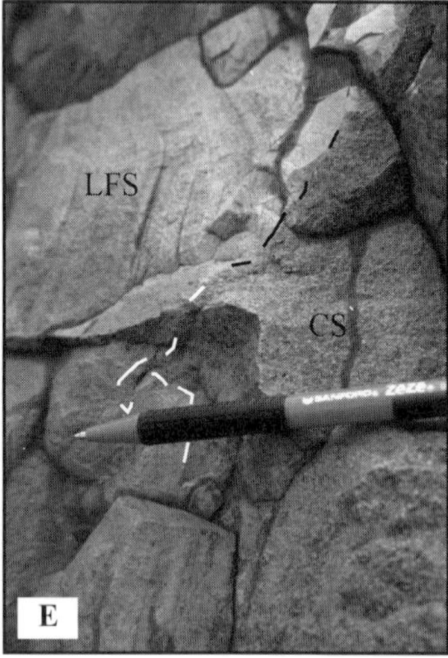
LFS
CS
E
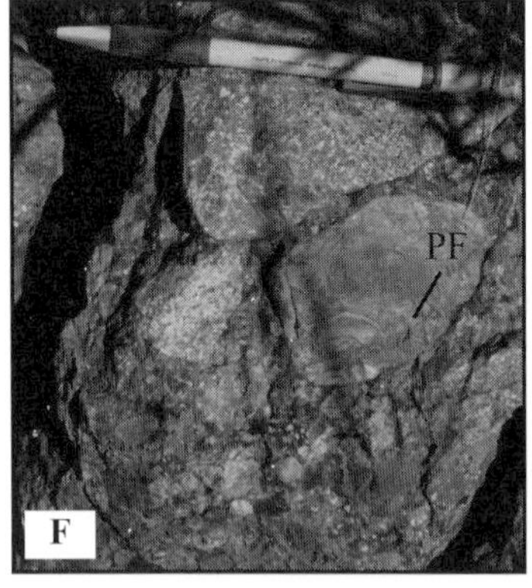
PF
F
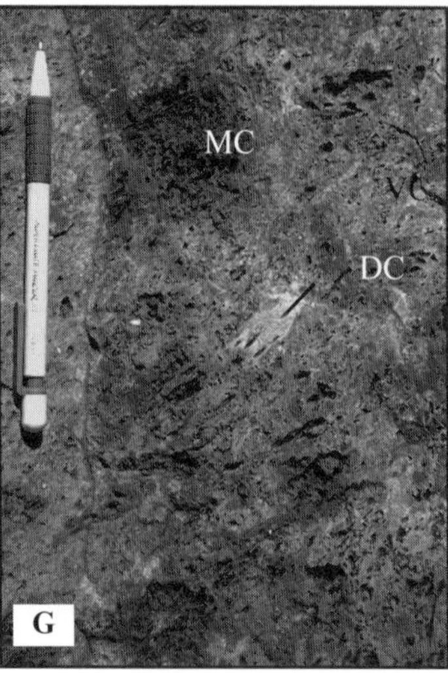
MC
DC
G

(*c*.0.5 m thick) are irregular in shape and cannot be correlated between logged sections (i.e. <100 m strike). The basalts range from aphyric to 30% plagioclase–phyric/glomeroporphyritic basalt. They are also variably vesicular (0–40%) and some outcrops contain bedding-parallel vesicle layers (Fig. 4B). Interbedded clastic rocks are composed of basalt clasts that show similar groundmass and phenocryst compositions to the coherent units. Thin (<1m) clast-supported, monomictic conglomerate is present in the lowest stratigraphic sections (<100 m above the base of the Warambie Basalt) and contains subrounded pebble clasts within a basalt-derived granule matrix. In contrast, breccia in the lower sections is composed of angular, clast-rotated basalt boulders in finer (granule) jigsaw-fit and clast-rotated breccia (Fig. 4C). The abundance of basalt breccia increases at >100 m above the base of the Warambie Basalt, with up to 60 m thickness of breccia-dominated basalt rocks at the top of the formation. The breccia is granule- to pebble-sized, with angular, elongate, jigsaw-fit and clast-rotated textures. It contains highly irregular, whispy, basalt inclusions of similar composition to the breccia. The margins of the inclusions show evidence of fragmentation and grade from fractured, coherent basalt into basalt breccia.

*Arkosic sandstone and conglomerate* is restricted to the Warambie Basalt in the Red Hill area and is volumetrically insignificant in comparison with the basalt lava and breccia facies. It locally forms a 6 m stratigraphic thickness but has little lateral extent (<50 m). Arkosic sandstone comprises plutonic quartz (50%), feldspar (35%) and amphibole grains (15%). Individual beds are decimetre-scale, massive and graded, and contain basal, monomictic, granitic microbreccia zones. They have no interbedded fine-grained layers. Packages are laterally discontinuous (lensoidal?) and are generally only a few metres in thickness. Conglomerate is matrix-supported with rare, subrounded basalt clasts of similar composition to the basalt lava and breccia facies and a matrix of the quartz–feldspar sandstone. Basalt clasts are randomly dispersed within the sandstone matrix and show no obvious layering or concentration.

*Grey dolerite and basalt* is restricted to the Good Luck Well area and lies at the base of the Warambie Basalt (Whim Creek Group). The facies comprises crystalline, dark charcoal grey dolerite dominated by feldspar and pyroxene. The dolerite is interlayered with basalt rocks of identical colour but finer grain size. Packages of dolerite and basalt are typically tens of metres in thickness and have a tabular geometry. Dark grey, basalt sandstone may overlie the coherent rocks but is poorly preserved and strongly chlorite altered. In a single exposure, basalt apophyses intrude the base of andesitic sandstone and breccia facies sandstone and display a transition from coherent basalt through to mixed, angular basalt breccia within a structureless sandstone matrix. This breccia is interpreted to be a peperite and is evidence of an intrusive phase of the Warambie Basalt.

*Polymictic conglomerate, breccia and sandstone* form the lowest stratigraphic interval within the Red Hill volcanics. Clasts are typically subrounded and coarser (pebble–cobble size) than clasts within the andesitic sandstone and breccia facies breccias described below. The facies is dominated (*c*.90%) by dacite clasts similar to the dacite intrusions facies (see below) and contains lesser amounts of volcaniclastic breccia similar to that of the andesitic sandstone and breccia. Individual beds are not seen, owing in part to poor exposure. Clast-supported subfacies grade into matrix-supported subfacies and sandstone, with no indication of breaks in sedimentation. Quartz is again rare and the sandstone or granulestone matrix is volcaniclastic and similar in appearance to that of the andesitic sandstone and breccia. The facies is locally thick (75 m) but is of limited lateral extent (*c*.2 km preserved strike).

A full description of the variety of sedimentary facies within the *andesitic sandstone and breccia* is not

**Fig. 4.** (*opposite*) Lithologies from the Whim Creek Group. (A) Coarse granite breccia facies (Fig. 3) with coarse, angular clasts of granite (GC) and subangular basalt clasts (BC) within a dark sandstone to granule matrix (M). (B) Basalt lava and breccia facies lava (BL) with laterally continuous vesicle horizons (VH) parallel to the subhorizontal layering of the rock. Vesicle horizons contain up to 30% vesicles. (C) Basalt lava and breccia facies coarse breccia containing clast-rotated breccia (CRB), jigsaw-fit breccia (JFB) and abundant cobble to boulder size angular clasts (CC). The breccia is interpreted as a mixed autobreccia (coarse clasts) and hyaloclastite (jigsaw-fit and clast-rotated breccia). (D) Andesitic sandstone and breccia facies angular breccia. This example shows highly angular clasts (AC), some with well developed alteration rims (AR). Sedimentary inclusions (SI) are indicative of volcanism–sediment interaction during formation of the breccia. (E) Andesitic sandstone and breccia facies sandstone with a small (cm-scale) sandstone dyke (indicating younging to the left). The dyke consists of coarse sandstone (CS) and intrudes related, laminated fine sandstone (LFS). (F) Andesitic sandstone and breccia coarse, angular breccia similar to D and containing breccia with circular perlite fractures (PF). (G) Andesitic sandstone and breccia facies pumiceous breccia. The breccia contains mafic clasts (MC) and dense (vesicle poor) clasts (DC) but is dominated by tube pumice and devitrified glass shards (seen in thin section).

warranted as (i) there are a large number of subfacies and (ii) most subfacies contain similar clast types and sedimentary structures. All rocks of the facies are sedimentary and are composed of variable amounts of volcaniclastic material. Within coarse examples, clasts are dominated by light grey, aphanitic volcanic lithics with lesser amounts of beige coloured subrounded aphanitic dacite lithics, flow-banded dacite lithics and rare pumice, scoriaceous basalt and quartz. The matrix material is similar in appearance to the light grey, aphanitic volcanic clasts. The bulk rock geochemistry is andesite to rhyodacite based on limited, unpublished geochemistry and is consistent with hand specimen and thin section petrography. Typically, quartz is a minor component and does not exceed 20% of the rock. The facies is dominated by coarse sand size debris with large volumes (*c*.25%) of granule to cobble breccia and minor amounts of mudstone. Volcaniclastic lithic, pumiceous and glass shard grains are a light grey colour and are aphanitic to weakly plagioclase-phyric. The majority of the clasts are dense material without vesicles but they may contain subspherical perlite fractures. In contrast, the pumiceous material is dominated by tube pumice with quartz-filled vesicles. Breccia and sandstone are interbedded, with beds being massive and decimetre to a few metres in scale. Breccia is typically highly angular and may contain sedimentary inclusions within clasts (Fig. 4D). Most breccia is matrix-supported with floating clasts and, commonly, diffuse planar stratification. Sandstone is typically structureless to diffuse planar stratified within decimetre-scale beds. Some sandstone is gradationally capped by thin (5–10 cm), laminated mudstone. The sandstone–mudstone units contain flame structures and sandstone dykes (Fig. 4E) indicative of rapid sedimentation and overpressuring of pore water during deposition. Evidence of aggradation from tractional currents is rare, although some units contain imbricated pebble clasts and low-angle cross-bedding. However, these units form less than 5% of the total facies exposed.

Overall, the andesitic sandstone and breccia facies is dominated by vesicle-poor, commonly perlite-fractured (Fig. 4F), angular breccia. However, pumiceous clasts (Fig. 4G) may dominate dense breccia and sandstone in some parts of the Red Hill volcanics, although these units typically form less than 20 m stratigraphic thickness and are of limited preserved strike. However, a subfacies in the Red Hill area (monomictic pumice breccia) forms a tabular stratigraphic unit of up to 145 m thickness and extends for at least 4.5 km along strike. The subfacies represents sedimentation of a large volume of vesicular debris within upward-fining graded beds up to tens of metres thick. Thin section analysis reveals a matrix of devitrified, angular shards.

*Dacite intrusions* dominate the Mons Cupri dacite. Field relationships, including bedding parallel margins and apophyses into overlying sedimentary rocks of both Warambie Basalt and Red Hill volcanics, show that the facies is dominated by sills. At Red Hill (Fig. 3), a dacite body lies stratigraphically within breccia of the basalt lava and breccia facies (Fig. 5). The 3.5 km × 160 m intrusion has an apparent aspect ratio of 0.046. This is similar to other low aspect ratio felsic intrusions and extrusions (e.g. Enebro Mountain Rhyolite, Arizona: Shroeder & Riggs, 1996). Dacite bodies are typically plagioclase-phyric (<5%) with a spherulitic and felsitic groundmass. Spherulites dominate the groundmass of the rock (10–80%) and are characteristically irregular, spherical and cauliflower-shaped. Amygdales are rare (0–5%) and are more abundant at the margins but are also present at the centre of the main body. Small dacite bodies, petrographically similar to the main body, typically extend 1–50 m along strike, are 0.2–10 m thick and have been interpreted as fault scarp rockfall megaclasts (Krapez & Eisenlohr, 1998). They have relatively rounded, weakly flow banded margins and are hosted within basalt breccia at a similar stratigraphic level to the main dacite unit.

## Interpretation of facies from the Whim Creek Group

The Whim Creek Group contains seven main facies, each with considerable internal variation. The facies are dominantly volcanic and volcaniclastic and may be grouped into *in situ* (basalt lava and breccia, grey dolerite and basalt and dacite intrusions) and reworked and resedimented (coarse granite breccia, arkosic sandstone and conglomerate, polymictic conglomerate, breccia and sandstone, and andesitic sandstone and breccia) types. Basalt and dacite rocks are *in situ* or form the coarsest clast fraction of sedimentary rocks indicating proximity to source. There is no *in situ* coherent andesite facies suggesting that the volcaniclastic debris is not autoclastic and was transported from a distal volcanic centre. Figure 5 illustrates the lower stratigraphy at Red Hill and is a reconstruction based on detailed logging.

The basal unconformity of the Whim Creek Group is described from Red Hill. Krapez & Eisenlohr (1998) suggest that this unconformity overlies granitic rocks (Fig. 3). Although the 3093 ± 4 Ma (Nelson, 1997)

**Fig. 5.** Palinspastic reconstruction of the facies architecture of the lower stratigraphy (Warambie Basalt and Mons Cupri dacite) of the Red Hill area (Figs 1 & 3). The basal unconformity over Caines Well Granite is illustrated, although it is an inferred feature. Basalt lavas and intrusions dominate the lower stratigraphy, with larger amounts of basalt breccia at the top.

Caines Well Granite is clearly much older than rocks of the Whim Creek Belt, the contact between granite and coarse granite breccia is not an unconformity. Pegmatite from the granite intrudes this breccia and overlying basalt lava and breccia and the granite contains rare xenoliths of the basalt. Hence, the granite at Red Hill must be younger than the Caines Well Granite and the Whim Creek Group. Strongly cleaved amphibolite surficial drift and subcrop is recorded in the west of the Red Hill area (Fig. 3) and the unexposed contact of this amphibolite with basalt lava and breccia represents the true unconformity (Fig. 5).

Basalt rocks typically form the stratigraphic base of the Whim Creek Group. The grey dolerite and basalt, and basalt lava and breccia are both associated with clastic material that, in the latter facies, is interpreted as an autoclastic facies of lavas and intrusions. Within the basalt lava and breccia, jigsaw-fit and clast-rotated breccia are hyaloclastite, with intrusive basalt lobes that show evidence of gradation from coherent to clastic textures. Highly vesicular basalt units with distinct vesicle layers (Fig. 4B) are characteristic of lava flows. Rising vapour bubbles are trapped under the surface of the lava against the base of a solid crust, which thickens through heat loss to the ambient environment (e.g. model of Self *et al.*, 1997, for continental flood basalt). In addition, coarse, angular breccia with no rounding of clast margins (Fig. 4C) appears not to have been transported more than a few metres and is interpreted as autobreccia. Some of the basaltic rocks are vesicle-poor, tabular bodies with no associated breccia and may represent subvolcanic sills. Thin, decimetre-scale horizons of poorly sorted, sub-rounded basalt pebble conglomerate indicate energetic reworking of breccia or coherent basalt lava and may explain the lack of abundant primary breccia near the base of the Warambie Basalt. These conglomerates are restricted to the base of the Warambie Basalt (≤50 m). Basalt pebble conglomerate is entirely monomictic and clasts are derived from the basalt lava. Hence, the only evidence of transportation of material into the basin during basaltic volcanism is the presence of the arkosic sandstone and conglomerate facies. This facies is of limited volume (*c*.20 $m^2$ exposure) and may have been derived from a granitic body (e.g. Caines Well Granite) from within the Whim Creek Basin. Hence, this facies is not evidence of significant allochthonous sedimentary material.

Andesitic sandstone and breccia are dominated by dense clast breccia and sandstone with lesser tube pumice breccia. The dense clasts suggest that explosive volcanism was not the dominant fragmentation mechanism. Highly angular, aphanitic clasts, some with perlite fractures (Fig. 4G), alteration rims or sediment inclusions (Fig. 4D), dominate the breccia. The most likely fragmentation mechanisms were quench fragmentation and autobrecciation. Matrix-supported, massive to diffuse planar-bedded breccia is interpreted as a deposit from debris flows with lesser fine sandstone and mudstone turbidite subfacies developed between these units. These facies were probably part of a dome margin volcaniclastic apron deposited in relatively deep water (below storm wave base), as evidenced by the preservation of the mudstone turbidite subfacies.

Monomictic tube pumice breccia of the andesitic sandstone and breccia forms a significant facies interval (*c*.145 m) at Red Hill. Such a large stratigraphic thickness over 4.5 km strike, coupled with the abundance of devitrified glass-shard sandstone matrix, is compatible only with contemporaneous resedimentation of non-welded ignimbrite or water modification of a pyroclastic flow. No evidence of clast rounding or modification is noted. Thick, upward-fining packages suggest that transport and deposition was by turbidity currents rather than debris flows. The large volume of these megaturbidite beds is consistent with pyroclastic flows entering the basin margins and contemporaneously transforming into large-scale turbidity currents (see Cas & Wright, 1991).

The dacite intrusions have been interpreted as lava flows (Fitton *et al.*, 1975) and as pyroclastic rocks (Barley, 1987). However, a number of features are inconsistent with these interpretations. Felsic lava typically exhibits a marginal clastic facies (e.g. Badlands Lava, Idaho: Manley, 1996) of autoclastic and talus origin (subaerial settings) or autoclastic/quench fragmentation origin (subaqueous settings). No such breccia is observed at Red Hill or in any other area. Although this may be attributed to erosion, the presence of thin (<10 m) outer zones of strong flow banding parallel to the upper and lower margins of the bodies suggests that the entire bodies are preserved and that the flow banded zones represent shear-induced laminar flow zones at the margins of an intrusive body. In addition, the main dacite body has associated apophyses and pods (tens of metres strike length) into overlying basalt breccia. The presence of spherulites and lithophysae indicates high-temperature crystallization or devitrification of glass (Lofgren, 1971; McArthur *et al.*, 1998). The formation of a glass groundmass up to 80 m from the cooling margin of the main intrusion is consistent only with emplacement at very high level as a lava flow or shallow intrusion.

At Red Hill, thin flow-banded margins and the absence of marginal breccia suggest an intrusive origin. Devitrification textures are consistent with a relatively high level, shallow intrusive origin, and the presence of petrographically similar, flow-banded, quenched dacite debris within the andesitic sandstone and breccia shows that dacite rocks were locally emergent. At Red Hill, only intrusive rocks are observed and are hosted within thick breccia and lava of the basalt lava and breccia facies (Fig. 5). However, the contacts are poorly exposed. Dacite 'pods' are not rockfall debris (e.g. Krapez & Eisenlohr, 1998) but smaller intrusive bodies related to the main dacite intrusion. The presence of resedimented dacite breccia suggests that dacite rocks were emergent in the basin (now preserved out of section?) or that shallow intrusion related peperite and hyaloclastite were uplifted and resedimented during basin formation.

## Facies architecture of the Bookingarra Group

The Bookingarra Group is subdivided into four component facies. These facies, as with those of the Whim Creek Group, show a number of associated subfacies. The facies (Table 2) are volcaniclastic-dominated sedimentary rock (Cistern Formation), siliciclastic-dominated sedimentary rock (Cistern Formation and Rushall Slate), dolerite and basalt intrusions and basalt lava (Negri Volcanics).

Much of the *volcaniclastic-dominated sedimentary rock* facies is exposed to the south of Whim Creek (Fig. 1). The facies is found at the base of the Cistern

Formation and is dominated by upward-fining conglomerate and lesser breccia that forms a total thickness of >50 m. Conglomerate is typically cobble- to boulder-sized and may be clast or matrix supported. The sandstone matrix is a mixture of quartz, volcaniclastic lithic grains and rare pumice. Clasts are dominated (*c*.90%) by dacite similar to the dacite intrusions facies, with lesser basalt and rare granitoid. Clasts are typically rounded with increasing volumes of breccia at stratigraphically higher levels (Fig. 6A). The facies is massive but fines upwards into breccia, granule beds and upward-fining sandstone units with a higher siliciclastic component. Although conglomerates are thought to form the stratigraphic base of the Bookingarra Group, the actual base is not exposed.

Granule beds and sandstone from south of Whim Creek (Fig. 6B) are dominated by a siliciclastic component and are the stratigraphically lowest rocks of the siliciclastic-dominated sedimentary rock facies. They represent a conformable transition from the volcaniclastic-dominated sedimentary rock facies conglomerate. These rocks form upward-fining beds (*c*.1 m thickness) with scoured bases and a transition from granule-rich base to fine sandstone top. The units show diffuse planar bedding and laminated, fine sandstone tops (Fig. 6B). In the Whim Creek area (Fig. 1), the coarse sandstone passes upwards into quartz arenite sandstone, pumiceous sandstone, chert and shale. The shale (Fig. 6c) is interlaminated with mafic and quartz grain-rich fine sandstone laminae and is mapped as the Rushall Slate. The total thickness of the facies is around 300 m in the Whim Creek area.

In the Good Luck Well area, the siliciclastic-dominated sedimentary rock facies forms metre scale beds of diffuse planar bedded coarse sandstone and breccia characterized by an upward increasing siliciclastic component (from 10 to 60%, rarely 85% quartz) within individual packages. Volcaniclastic material (typically light grey, siliceous) decreases upwards within individual packages and throughout the stratigraphy. Mudstone rip-up clasts are common components and may be weakly imbricated. Locally abundant pebble to cobble breccia contains clasts of black sedimentary chert (Fig. 6D), basalt, dacite and granite. The uppermost sedimentary rocks comprise quartz-rich (>90%) sandstone, shale and blue chert that are correlated with quartz arenite sandstone and shale in the Whim Creek area. The thickest exposed section is 70 m, although the recognition of similar, poorly exposed rocks suggests that the facies may be up to 300 m in thickness.

Coherent, bedding-parallel *dolerite and basalt intrusions* of the Negri Volcanics are preserved in the Good Luck Well area. They typically show planar contacts with surrounding rocks with no associated breccia and have been interpreted as intrusive rocks (e.g. Smithies, 1997). Bedding-parallel margins show that these rocks are dominantly sills. Dolerite sills may be >100 m in thickness and basalt sills are generally tens of metres thick. Basalt sills in the south of the Good Luck Well area intrude quartz-rich sandstone, shale and blue chert of the siliciclastic-dominated sedimentary rock facies and give constraints on the timing of mafic magmatism. At the contact between basalt and blue chert, vesicular basalt shows weak brecciation and a distinctive subfacies, at the upper surface of the basalt unit, comprising angular, irregular pebble clasts of scoriaceous basalt within a structureless blue chert matrix (Fig. 6E). Poorly exposed contacts show thin (centimetre-long) flame structures of chert into the basalt. Mafic sills in the Good Luck Well and Whim Creek areas intrude shale at a particular horizon around 10 m stratigraphically above the transition from sandstone to shale.

*Basalt lava* facies rocks (Negri Volcanics) dominate much of the outcrop of the Whim Creek Belt (Fig. 1). At Mount Negri (Fig. 1) the poorly preserved contact between shale of the siliciclastic-dominated sedimentary rock facies and basalt lava facies is a transitional unit composed of interbedded shale and basalt, matrix-supported conglomerate with shale matrix. Overlying rocks are basalt sheets with little associated breccia. At Mount Negri, pillow lava and breccia are rare but present within the lower 100 m of stratigraphy. An exposure of monomictic basalt breccia is illustrated in Fig. 6F and was first noted by Smithies (personal communication, 2000). The breccia comprises pebble- to cobble-sized material with subrounded, commonly elongate clasts, some of which are surrounded by raised rims. Pillow lava is very common to the north of Mount Negri (Fig. 1) in basalt rocks mapped as Loudens Volcanics by Smithies (1998). Pillows have very little associated breccia (Fig. 6G) and are interbedded with quartzite, black chert breccia and shale analogous to siliciclastic-dominated sedimentary rock facies sedimentary rocks from the Good Luck Well area. The total thickness of 2000 m (Krapez & Eisenlohr, 1998) of the Negri Volcanics is dominated by this facies.

### Interpretation of facies from the Bookingarra Group

The Bookingarra Group sedimentary rocks show an upward decrease in volcaniclastic material, some of which is clearly derived from the underlying Whim Creek Group. This indicates that significant uplift

VBC
DC
GM
A

BCC
QM
D

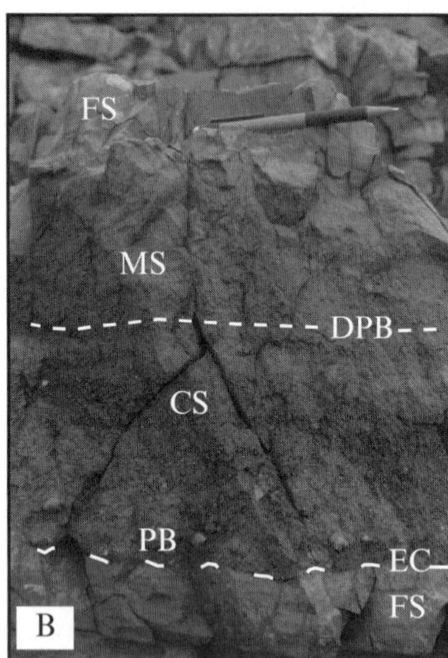
FS
MS
DPB
CS
PB
EC
FS
B

SBC
SCM
E

C

PB
BC
IR?
F

PM
PB
G

and erosion took place between the formation of the two units and is consistent with rounded volcanic clasts that suggest an epiclastic origin. The Whim Creek/Bookingarra Group contact is unexposed but the similarity of structural style between the two groups, and roughly parallel bedding suggests that the contact is a disconformity or low angle unconformity. Within the Cistern Formation and Rushall Slate, the transition from volcaniclastic-rich basal rocks to siliciclastic-rich upper rocks reflects the decreasing influence of the underlying, igneous-dominated Whim Creek Group as source rocks. Basal, volcaniclastic conglomerate south of Whim Creek represents a proximal deposit derived from exposed Whim Creek Group rocks that may have formed topographic highs. The facies contains massive, clast and matrix-supported conglomerate, with little internal structure. The conformably overlying graded sandstones are interpreted as turbidites (Fig. 6B) and indicate a subaqueous origin with conglomerates formed by debris flow. Upward fining of this stratigraphic succession is noted by Barley (1987) and Krapez & Eisenlohr (1998), and suggests progressive deepening of the basin. Laminated shale and chert suggest deposition under low energy conditions below storm wave base, although the presence of sandstone laminae suggests periodic energetic currents, most likely to be distal turbidity currents. These rocks represent low energy, probably deep-water, conditions, similar to a modern abyssal plain.

Sedimentary rocks from the Good Luck Well area commonly display compositional variation within a bed or package of beds. This suggests aggradation of the unit from a current with changing sediment composition with time. All of the sedimentary rocks contain abundant siliciclastic (typically quartz and clay minerals) debris, suggesting a relatively uniform input of this material over the depositional lifetime of the basin. Volcaniclastic material is typically concentrated in the base of a sedimentary package and decreases in abundance upwards. This may be explained by tectonic uplift of basement (Whim Creek Group) blocks that supplied some of the clastic material to that developing basin. Abundant mudstone rip-up clasts are evidence of the upstream erosional capability of moderate to high energy flows.

The sedimentary rocks (volcaniclastic- and siliciclastic-dominated sedimentary rock facies) are restricted to the central Whim Creek Belt, whereas the doleritic and basalt rocks (dolerite and basalt intrusions and basalt lava facies) are of much greater volume and wider distribution. The contacts between the siliciclastic-dominated sedimentary rock and dolerite and basalt intrusions facies are conformable or are syndepositional intrusions. In the Good Luck Well area, one such contact is an upper margin showing dispersed basalt clasts within structureless shale (Fig. 6E). The dispersed basaltic material shows that mixing took place as mafic magma was emplaced into unlithified, fine-grained sediment within a deep-water or low energy subaqueous environment. At Mount Negri, the siliciclastic-dominated sedimentary rock–basalt lava facies contact is also conformable and shows upward-increasing amounts of basalt conglomerate with a shale matrix. Krapez & Eisenlohr (1998) interpret the basalt rocks as subaqueous and subaerial lava flows from a shield volcano. This interpretation is consistent with the recognition of both subaerial or shallow water spatter breccia (Fig. 6F) and subaqueous (Fig. 6G) pillow lava.

**Fig. 6.** (*opposite*) Lithologies from the Bookingarra Group. (A) Volcaniclastic-dominated sedimentary rock facies coarse breccia containing relatively high proportions of vesicular basalt clasts (VBC) with dacite clasts (DC) and a granule matrix (GM). The dacite clasts show no evidence of quenching (cf. dacite breccia within the Red Hill volcanics) and have subangular to rounded margins, indicating significant reworking. (B) Siliciclastic-dominated sedimentary rock facies pebble-bed to fine sandstone. The facies contains upward-fining packages comprising granule and pebble beds (PB) overlying an erosional contact (EC) with fine sandstone (FS). The beds are diffuse planar bedded (DPB) and fine upwards through coarse and medium sandstone (CS and MS) into fine sandstone. These rocks are considered to be deposits from turbidity currents and conformably overlie conglomerate and breccia of the volcaniclastic-dominated sedimentary rock facies. (C) Siliciclastic-dominated sedimentary rock facies shale folded into an upright anticline. Well developed cleavage shows that the rock is fine-grained with abundant clay minerals. (D) Siliciclastic-dominated sedimentary rock facies quartzite (QM) with quartz-veined, sedimentary, black chert clasts (BCC). These rocks show no volcaniclastic component and are evidence for the exposure of continental crust adjacent to the Mallina Basin. (E) Mixed dolerite and basalt intrusions and siliciclastic-dominated sedimentary rock facies angular basalt breccia. Scoriaceous basalt clasts (SBC) are hosted within a structureless chert matrix (SCM) suggesting mixing between the basalt breccia and chert prior to lithification. (F) Basalt lava facies breccia. Basalt clasts (BC) may be hosted within depressions containing a raised rim, possibly an impact rim (IR). Other pebble breccia (PB) is elongate and subrounded without a significant raised rim. The breccia has been considered to represent semi-agglutinated spatter formed in a subaerial setting (Smithies, personal communication, 2000). (G) Basalt lava facies pillow lava. Pillow basalt (PB) shows little associated breccia at the pillow margin (PM).

**Table 3.** Depositional, erosional and volcanic events from the Whim Creek Belt.

| Event | Formation | Bulk composition | Style and evidence | Possible cause |
|---|---|---|---|---|
| **8** | Negri Volcanics | Basalt | Subaqueous + subaerial lava sheets | Basic volcanism. Regression caused by volcanic input leading to subsidence |
| **7** | Cistern Fm/ Rushall Slate | Siliciclastic | Upward-fining | Transgression, possibly induced by regional subsidence |
| **6** | Base Cistern Fm | Unknown | Granule pumice matrix | Transgression, tectonic half-graben subsidence |
| **5** | Disconformity | | Erosional, low angle bedding discordance, geochronology | Regression, tectonic uplift or eustatic sea-level fall |
| **4** | Mons Cupri Dacite | Dacite | High-level intrusions | Intrusion during or post subsidence |
| **3** | Red Hill Volcanics | Andesite/ rhyodacite | Resedimented volcaniclastic sandstone and breccia | Felsic volcanism. Transgression, basin subsidence or eustatic controls |
| **2** | Warambie Basalt | Basalt | Subaqueous lava with abundant hyaloclastite | Transgression, tectonic half-graben subsidence |
| **1** | Basal unconformity | | Flat, erosional surface over amphibolite basement | Shallow subaqueous or subaerial erosion |

## STRATIGRAPHY

### Event stratigraphy and basin evolution of the Whim Creek Basin

Individual volcanic episodes may be thought of as relatively instantaneous events (i.e. typically of shorter duration than U/Pb SHRIMP errors) that may produce large thicknesses of volcanic rock and volcaniclastic debris. These events separate erosional and depositional events that may be of longer duration. Where precise geochronology is available, the volcanic events may be used to constrain the timing and duration of depositional and erosional events.

Within the Whim Creek and Bookingarra groups, evidence of eight volcanic, depositional and erosional events is preserved (Table 3). Volcanic, intrusive and depositional events broadly correlate with lithostratigraphic formations because each is compositionally distinct, and therefore mapped as a separate unit. The first event is marked by the unconformity at the base of the Whim Creek Group. This surface records a *c.*75–90 Myr time gap between intrusion of the Caines Well Granite (3093 ± 4 Ma: Nelson, 1998) and deposition of the Whim Creek Group. The unconformity suggests that the amphibolite–granite basement was subaerial/shallow subaqueous immediately prior to formation of the Whim Creek sub-basin. Basalt rocks (basalt lava and breccia, and grey dolerite and basalt facies, respectively) form the base of the succession in the Red Hill and Good Luck Well areas and are grouped together as the Warambie Basalt. These rocks resulted from event 2 (Table 3) that was characterized by eruption of basalt lava with lesser high level intrusions. Abundant hyaloclastite indicates subaqueous conditions that, along with the thickness of basalt material (*c.*200 m), indicates that event 2 was associated with transgression. In the Red Hill area, rocks of event 2 are overlain by resedimented rocks composed of volcanic debris produced by event 3, the Red Hill volcanics. Intrusive contacts of basalt and dolerite into Red Hill volcanics sedimentary rocks in the Good Luck Well area suggest that Warambie Basalt volcanic activity was approximately coeval with deposition of Red Hill volcanics sediments. Consequently, events 2 and 3 were approximately time-equivalent but geographically separate. A U/Pb SHRIMP date of 3009 ± 4 Ma (Nelson, 1998) was obtained from a single zircon population extracted from pumice breccia (volcaniclastic dominated sedimentary rock) at Red Hill. The presence of pumice indicates that volcanism and deposition were probably contemporaneous and, therefore, 3009 ± 4 Ma dates both magmatism and deposition. In summary, events 2 and 3 were approximately coeval at *c.*3009 ± 4 Ma. The presence of large volumes of subaqueous volcaniclastic rocks indicates continued subsidence (transgression), consistent with the opening of a rift basin (e.g. Barley, 1987; Krapez & Eisenlohr, 1998).

Intrusion of the Mons Cupri dacite is the fourth event (Table 3) and the style of intrusion is constrained by a number of contact relationships. In the Red Hill area, dacite intrusions are hosted within Warambie Basalt hyaloclastite breccia or between this hyaloclastite and sandstone of the Red Hill volcanics. Upper surface contacts show distinct lobes into the breccia, suggesting intrusion prior to lithification. The intrusions were, therefore, emplaced before burial and lithification of the Warambie Basalt and Red Hill volcanics. Barley *et al.* (1994) give an age of 2991 ± 12 Ma for the Mons Cupri dacite, which supports field evidence that the magmatic products of event 4 intruded clastic rocks of events 2 and 3. Errors on these dates suggest that the gap between events 2 and 3 is in the range of 2–34 million years. However, event 4 is interpreted as a high level intrusive event into unlithified sediments and a lower value is preferred.

Evidence of a disconformity or low angle unconformity (event 5) between the Whim Creek and Bookingarra groups includes the recognition of epiclastic material (derived from the Whim Creek Group) within the Cistern Formation, a change in provenance and low angles of bedding discordance. The disconformable surface records an erosional event and suggests that uplift and regression separated the Whim Creek and Bookingarra basins (Table 3). The overlying volcaniclastic-dominated sedimentary rock facies (lower Cistern Formation) was deposited during event 6 and is dominated by conglomerate with pumiceous matrix. This well rounded epiclastic material is evidence of subaerial or shallow-water reworking of basement (Whim Creek Group) and is further evidence of the event 5 regression. The volcaniclastic-dominated sedimentary rock facies shows extreme lateral thickness changes and preserves boulder size clasts, with Whim Creek Group-derived clasts indicating possible proximal deposition adjacent to palaeotopographic highs. The sedimentary products of events 6 and 7 are upward-fining and suggest subsidence and basin deepening. This evidence, and the presence of pumiceous material, is consistent with the opening of a rift-basin with associated local or regional volcanism. Event 7 is recorded by the upper Cistern Formation and Rushall Slate that are characterized by siliciclastic sedimentary rocks with no associated juvenile volcanic component. The upward-fining stratigraphy associated with event 7 is consistent with basin deepening/widening during transgression. This transgression was probably initiated during basin opening (event 6, Table 3). Event 8 resulted in the emplacement of the Negri Volcanics lavas, intrusions and associated breccia. The event is separated stratigraphically from pumice of event 6 by up to 300 m of shale (Black, 1999). Consequently, the volcanism of event 8 is significantly younger than the volcanism of event 6, although there are no absolute data available to constrain the time gap. Nelson *et al.* (1999) suggest that the minimum age of the Negri Volcanics (event 8) is *c.*2970 Ma based on the recognition of a zircon population of 2975 ± 4 Ma (Nelson, 1998) within a rhyolite that overlies these rocks. The presence of both subaerial and subaqueous facies (e.g. Krapez & Eisenlohr, 1998) suggests that basalt volcanism filled the basin at a greater rate than subsidence, resulting in a regressive phase of basin evolution (Table 3).

## DISCUSSION

### Stratigraphic models applied to volcanic rock-dominated basins

Krapez & Eisenlohr (1998) have interpreted the geology of the Whim Creek Belt in terms of sequence stratigraphy. Krapez & Eisenlohr (1998) recognize four sequences (VL-$4^1$ to VL-$4^4$), which unconformably underlie the Negri Basin (their supercycle VU-1). In the sequence stratigraphic model of Krapez & Eisenlohr (1998), sequences VL-$4^1$ to VL-$4^4$ are separated from the overlying Negri Basin (Negri Volcanics) by a megacycle boundary. Sequence stratigraphic interpretations originate from, and are commonly applied to, Phanerozoic sedimentary rock-dominated successions (e.g. Sloss, 1988). In these interpretations, detailed palaeontological and seismic data allow the definition of unconformity- or conformity-bounded stratigraphic packages. Sequence stratigraphy is most applicable to relatively shallow water sedimentary basins in which the migration of the shoreline tracks relative sea-level changes (e.g. Vail *et al.*, 1977). In these systems, deposition is dominantly controlled by eustacy, tectonics and sediment supply. The recognition of deep-water facies, lack of subsurface data and poor resolution of dating suggest that a conventional sequence stratigraphic interpretation is at best difficult and perhaps even inappropriate to the geology of the Whim Creek Belt.

Within volcanically active basins, volcanic events exert a control on lithofacies-stratigraphic architecture (Cas & Wright, 1987). These events produce large volumes of volcanic and volcaniclastic material and may be considered instantaneous relative to the development of third-order sequences (e.g. sequences VL-$4^1$ to VL-$4^4$: Krapez & Eisenlohr, 1998). Consequently, these events may modify or overwhelm the effects of eustacy, tectonics and background sediment supply. The lithofacies and stratigraphic architecture may therefore reflect, and be the product of, diverse events. The event stratigraphy presented above and in Table 3 is considered a more appropriate stratigraphic interpretation of the geology of the Whim Creek Belt. The contact between Rushall Slate and Negri Volcanics illustrates the advantage of the event stratigraphic approach over sequence stratigraphy. Detailed facies analysis reveals that intrusive rocks of the dolerite and basalt intrusions facies (Negri Volcanics) intrude the siliciclastic-dominated sedimentary rock facies (Cistern Formation and Rushall Slate) with peperitic contacts. These contacts show that intrusive activity of event 5, which formed the Negri Volcanics, took place prior to lithification of the Rushall Slate sediments. Consequently, the Negri Volcanics and Rushall Slate were deposited in the same basin. A sequence stratigraphic consequence is that this contact cannot be a first- or second-order boundary as defined by Krapez (1996), as these record ocean opening/closure systems and development of depositional basins, respectively. Hence, the interpretation that the boundary represents a (first-order) megasequence boundary (megacycle VL to megacycle VU: Krapez & Eisenlohr, 1998) is refuted. The event stratigraphy approach emphasizes only the relative order of events where no precise geochronology is available. In this case, the products of event 5 (basalt volcanism) clearly overlie the products of event 4 but formed in the same depositional basin (Table 3). Consequently, if sequence stratigraphic interpretations cannot resolve first-order boundaries (i.e. stratigraphic units with a periodicity 182 Myr: Krapez, 1996) within volcanic-dominated basins, they are clearly of little use in the reconstruction of the internal stratigraphic architecture of these basin successions.

## Modern and Phanerozoic arc-related basins: analogues for the Whim Creek Group?

The extent to which plate tectonics drove crustal processes in the Archaean is of considerable debate. Excellent reviews of the evidence for and against plate tectonic models for Archaean successions are provided by de Wit (1998) and Hamilton (1998). The Whim Creek Group shows some similarity to modern back-arc basins and this model is tested against the facies architecture and stratigraphy discussed previously.

The Whim Creek Group was deposited within a basin with an ensialic crust of amphibolite facies metamorphic rocks and granite. The nature of this crust is poorly constrained but deposition took place over submerged continental crust, a common feature of Archaean successions (Arndt, 1999). The size of the original basin is unknown but the preserved stratigraphy outcrops over 40 km strike length. Smithies *et al.* (1999) suggest the Mallina Basin (Fig. 1) to the east of the Whim Creek Belt was formed and infilled between 3015 and 2950 Ma, and suggest possible correlation between components of the Whim Creek Belt and the Mallina Basin. These interpretations are discussed in Fitton *et al.* (1975), Krapez & Eisenlohr (1998), Smithies *et al.* (1999) and Huston *et al.* (2000). Hence, the original Whim Creek sub-basin system may have been significantly larger than the preserved 40 km.

The recognition of amphibolite and granite basement and the early development of subaqueous environments are consistent with the presence of relatively thin continental crust. This may have resulted from heating and extension of 'normal' crust or simply thinner continental crust in the Archaean. A comparison with a modern, ensimatic back-arc basin such as the Lau Basin, south-west Pacific (e.g. Clift *et al.*, 1995), shows some stratigraphic similarity despite the recognition of an ensialic basement. In the Lau back-arc basin, Clift *et al.* (1995) record basal volcaniclastic, dominantly dacite conglomerate and basalt hyaloclastite, abundant volcaniclastic turbidite sandstone and fine pelagic sediments (nanofossil ooze, chalk and volcanic ash) overlying a basement of oceanic crust. This stratigraphy shows crude upward-fining, volcaniclastic composition and similar fragmentation/transportation mechanisms to those of the Whim Creek Group. However, the Whim Creek Group formed on ensialic basement, is dominated by *in situ* lavas and intrusions in comparison with the dominantly resedimented volcaniclastic rocks of the Lau Basin and has a substantial felsic component and no pelagic sedimentary facies. The recognition of a variety of rock compositions within the Whim Creek Basin (basalt, andesite and dacite) is more consistent with a continental arc-basin setting (e.g. Gamble *et al.*, 1995). Hence, it is reasonable to conclude that the Whim Creek Basin has more affinity with modern, ensialic back-arc settings. However, there is very little or no allochthonous sediment within the Whim Creek Group, suggesting that the basin was a significant distance from, or separated from, emergent continental crust.

Busby *et al.* (1998) propose a three-stage model of arc-evolution based on studies of Baja California, Mexico. The model involves the development of a highly extensional intraoceanic arc followed by mildly extensional fringing arc and finally a compressional continental arc system. An extensional setting for the Whim Creek Basin is indicated by bimodal volcanism (e.g. Barley, 1987; Krapez & Eisenlohr, 1998). The Whim Creek Group lacks significant plutonic or metamorphic clasts indicative of the phase 3 compressional continental arc system (Busby *et al.*, 1998). There is no evidence of oceanic crust basement and, hence, an intraoceanic arc model is considered unlikely. However, the phase 2 extensional fringing-arc system that models the development of a back-arc basin over extended continental crust (Busby *et al.*, 1998) is consistent with the preserved Whim Creek Group. The model of arc development of Fackler-Adams & Busby (1998) involves intermediate to silicic explosive and effusive volcanism resulting in caldera-forming silicic ignimbrite eruption (stage I), followed by mafic, effusive hydroclastic rocks and dyke swarms (stage II). Voluminous, andesitic debris, including pumice breccia, may have formed in stage I of Fackler-Adams & Busby's (1998) model, although andesitic breccia is not restricted to such environments. Pumice breccia at Red Hill may be related to a caldera-forming eruption and has an arc-related analogue in the subaqueous, pumice rich deposits of the 1883 Krakatau eruption (Mandeville *et al.*, 1996). The authors describe subaqueously-deposited, mixed pumice and lithic fragments with sandy matrices that represent water-deposited products sourced from pyroclastic flows. Mandeville *et al.* (1996) calculate that 13.6 $km^3$ of debris is located within their mapped area, most of which lies less than 15 km from source. Carey *et al.* (1996), however, show that pyroclastic flows may travel up to 80 km across water before depositing sediment. The proximity to source of the pumice breccia is therefore unclear. Cas & Wright (1991) suggest that most pyroclastic flows explosively disrupt and/or transform into water-supported mass-flows on entering water. This suggestion is consistent with a turbidity current interpretation for the transport and deposition of pumice breccia at Red Hill.

The stratigraphy of the Whim Creek Group is consistent with an ensialic arc-related basin (e.g. back-arc basin: Horwitz, 1990). However, there are no preserved features that are definitive evidence of a subduction-related tectonic setting (e.g. accretionary wedge, obducted oceanic crust, high-P metamorphic belt; see Hamilton, 1998). In addition, the basin is relatively small (minimum *c.*40 km strike) and, although a larger section may be preserved below the adjacent Mallina Basin (Fig. 1), this would still be smaller than modern arc-related systems (*c.*$10^3$ km length). Hence, it is not possible at this stage to define accurately an analogue for the tectonic setting of the Whim Creek Basin despite a stratigraphic similarity to a modern back-arc sub-basin. An ensialic, volcanically active rift or transtensional basin is also consistent with the characteristics of the Whim Creek Group.

## Stratigraphy, age and tectonic setting of the Bookingarra Group

The Bookingarra Group was deposited in the Mallina Basin over a basement dominated by rocks of the Whim Creek Group. The lower stratigraphic unit of the Bookingarra Group is the Cistern Formation. The volcaniclastic-dominated sedimentary rock facies forms the base of the formation and is related to event 6. It is overlain by the siliciclastic-dominated sedimentary rock facies (Cistern Formation) and a thick shale horizon (Rushall Slate), which represent a lack of volcanic activity. The absolute minimum age of deposition of the Rushall Slate is *c.*2945 Ma from the Pb-model age of galena from the Whim Creek Deposit (Huston *et al.*, 2000). However, the true minimum age of the Bookingarra Group is probably 2975 ± 4 Ma from the age of a rhyolite that overlies the Negri Volcanics (Nelson, 1998; Nelson *et al.*, 1999). The Cistern Formation and Rushall Slate are a conformable, upward-fining package but relatively poor exposure of these rocks does not allow accurate reconstruction of lateral variations within the depositional basin.

The Cistern Formation and Rushall Slate were deposited at the base of the Mallina Basin. Upward-deepening (fining), early volcanic influence (event 6) and erosion of Whim Creek Group basement are most consistent with a subsiding, extending rift origin (Fig. 7A). The relation of these rocks to those of the Mallina Basin (Fig. 1) is discussed by Fitton *et al.* (1975), Krapez & Eisenlohr (1998), Smithies *et al.* (1999) and Huston *et al.* (2000). Deposition of the Cistern Formation and Rushall Slate was followed by voluminous effusive basalt volcanism of the Negri Volcanics that produced up to 2.0 km of basalt lava and associated volcaniclastic material (Fig. 7B). Subvolcanic intrusive rocks formed dykes into the subjacent Whim Creek Group (e.g. mapping of Smithies, 1997) and synsedimentary sills into the Cistern Formation and Rushall Slate. Hence, the Whim Creek Group was relatively lithified rock and the Cistern Formation and Rushall Slate were unlithified sediment. Much of the sedimentary material from the Cistern Formation (siliciclastic-dominated sedimentary

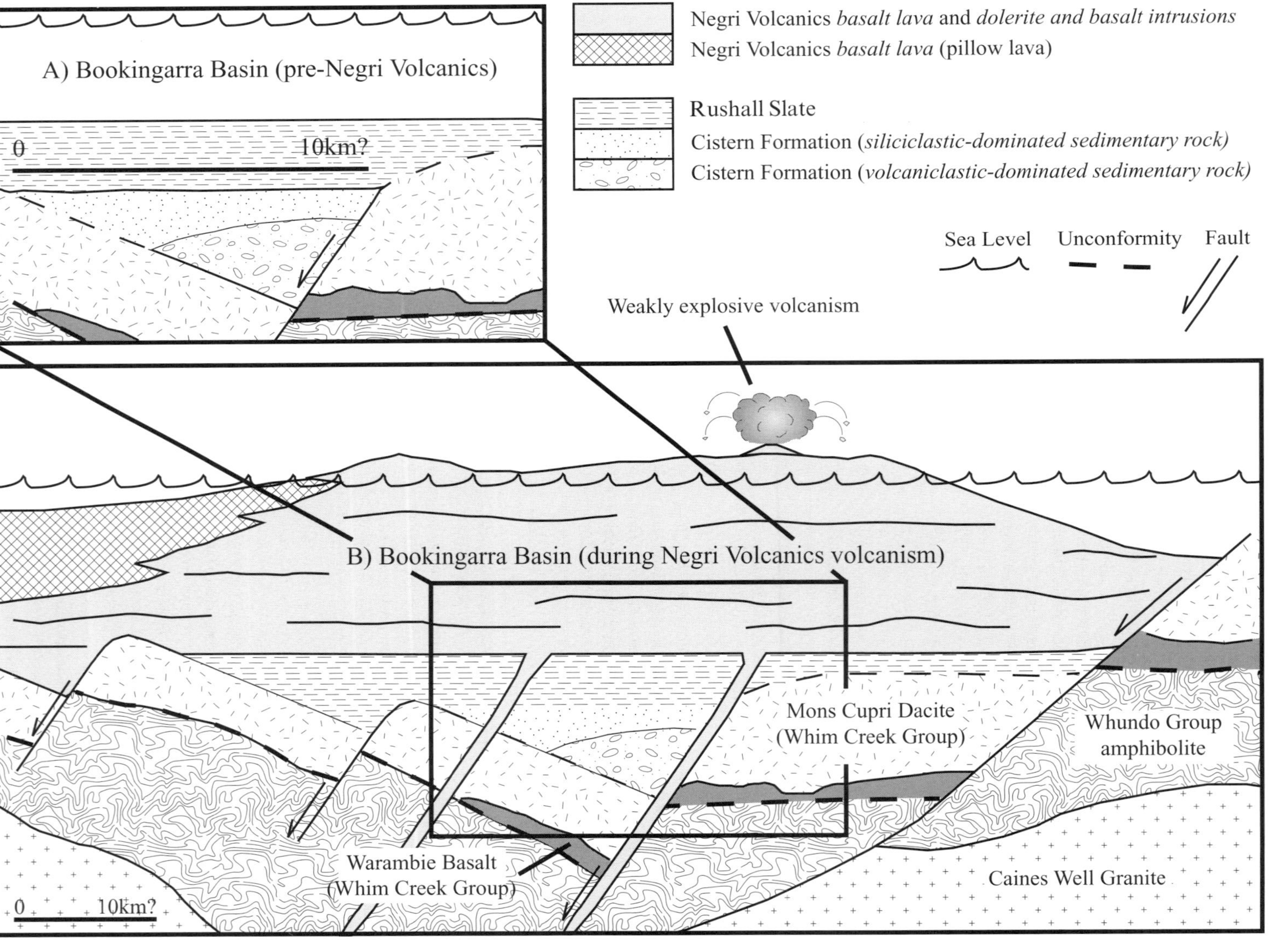


**Fig. 7.** Model for development of the Bookingarra Basin. (A) Opening of rift basins over continental crust (including the Whim Creek Group) and sedimentary infill. (B) Further rifting (crustal scale?) and basalt volcanism.

rock) and Rushall Slate was derived from a granitic source, presumably exposed continental crust. This suggests that the region had undergone extensive uplift or had been emplaced near to such crust since deposition of the Whim Creek Group. The restricted distribution of sedimentary rocks suggests that they were deposited within fault or topography-controlled sub-basins. This is consistent with the early stages of rifting of continental crust that included the Whim Creek Group. The thick (*c*.2 km) basalt lavas and intrusive rocks of the Negri Volcanics are more widely distributed and may be related to more extensive rifting. The basalt rocks are conformable with underlying sedimentary rocks and it is likely that the Cistern Formation and Rushall Slate were deposited within the initial sub-basins formed at the onset of rifting. Hence, the Mallina Basin shows an initial, localized rift stage, followed by crustal rifting and basalt volcanism within a continental setting.

## CONCLUSIONS

The lower stratigraphic succession exposed in the Whim Creek Belt is the Whim Creek Group which was deposited between ≥3009 ± 4 and 2991 ± 12 Ma. The upper stratigraphic succession is the Bookingarra Group that formed between 2991 ± 12 and *c*.2945 Ma. A low angle unconformity or disconformity separates the two groups.

Volcanic events related to the Warambie Basalt (event 2), Red Hill volcanics (event 3), Mons Cupri dacite (event 4) and Negri Volcanics (event 8) produced large volumes ($10^0$ to $10^7$ $km^3$) of volcanic and volcaniclastic material that locally overwhelmed tectonic and eustatic influences and fundamentally controlled the stratigraphic architecture of the Whim Creek Belt.

The Rushall Slate/Negri Volcanics boundary in the Bookingarra Group is a conformable and intrusive contact related to the onset of basalt volcanism of event 8 (now preserved as the Negri Volcanics).

Sequence stratigraphy is based on the recognition of facies and physical boundaries with temporal significance (Krapez, 1996). Within volcanic successions, eruption phases produce distinctive lithofacies-stratigraphic packages defined by unconformities and/or sudden changes in lithofacies aspect. Hence, these units are best described using a lithostratigraphic and event-stratigraphic framework.

Phanerozoic basin models can explain the preserved stratigraphy of both the Whim Creek and Bookingarra Groups. However, there is no definitive evidence that the Whim Creek Belt formed within a plate subduction, arc-related setting.

The stratigraphy of volcanic rock-dominated successions is best described using facies architecture and lithostratigraphy, followed by an event-stratigraphic model that accounts for the dominant controls on stratigraphic development. At outcrop to basin scale, lithostratigraphy correlates well with event-stratigraphic interpretations.

## ACKNOWLEDGEMENTS

The authors wish to acknowledge financial support from a Monash Graduate Scholarship and International Postgraduate Research Scholarship for G.P. and field assistance from Straits Resources Ltd. Thanks for endless field discussions go to Hugh Smithies (Geological Survey of Western Australia), Dave Huston (Australian Geological Survey Organisation), Bruce Hooper (Straits Resources Ltd) and Mark Barley (University of Western Australia). Drs W. Mueller and A. Thorne are thanked for fair and thorough reviews. Thanks also go to Steve Beresford and Mark Tait for reviews and assistance with the manuscript.

## REFERENCES

Arndt, N. (1999) Why was flood volcanism on submerged continental platforms so common in the Precambrian? *Precam. Res.*, **97**, 155–164

Barley, M.E. (1987) The Archaean Whim Creek Belt, an ensialic fault bounded basin in the Pilbara Block, Australia. *Precam. Res.*, **37**, 199–125.

Barley, M.E., McNaughton, N.J., Williams, I.S. & Compston, W. (1994) Age of Archaean volcanism and sulphide mineralization in the Whim Creek Belt, West Pilbara. *Aust. J. Earth Sci.*, **41**(2), 175–177.

Black, S.J. (1999) *Genesis of the Whim Creek sediment-hosted volcanogenic massive sulphide deposit, Western Australia*. Unpublished honours thesis, Curtin University of Technology.

Blewett, R. (1999) Characteristics of Pilbara Deformation. http://www.agso.gov.au/minerals/pilbara/table.html

Busby, C., Smith, D., Morris, W. & Fackler-Adams, B. (1998) Evolutionary model for convergent margins facing large ocean basins; Mesozoic Baja California, Mexico. *Geology*, **26**(3), 227–230.

Carey, S., Sigurdsson, H., Mandeville, C. & Bronto, S. (1996) Pyroclastic flows and surges over water; an example from the 1883 Krakatau eruption. *Bull. Volcanol.*, **57**(7), 493–511.

Cas, R.A.F. & Wright, J.V. (1987) *Volcanic Successions*. London, Chapman and Hall, 528 pp.

CAS, R.A.F. & WRIGHT, J.V. (1991) Subaqueous pyroclastic flows and ignimbrites; an assessment. *Bull. Volcanol.*, **53**(5), 357–380.

CLIFT, P.D. & LEG 135 SCIENTIFIC PARTY (1995) Volcanism and sedimentation in a rifting island arc terrain; an example from Tonga, SW Pacific. In: *Volcanism Associated with Extension at Consuming Plate Margins* (Ed. Smellie, J.L.), pp. 29–51. Geological Society, London.

DE WIT, M.J. (1998) On Archean granites, greenstones, cratons and tectonics: does the evidence demand a verdict? *Precam. Res.*, **91**, 181–226.

FACKLER-ADAMS, B. & BUSBY, C.J. (1998) Structural and stratigraphic evolution of extensional oceanic arcs. *Geology*, **26**(8), 735–738.

FITTON, M.J., HORWITZ, R.C. & SYLVESTER, G.C. (1975) Stratigraphy of the early Precambrian in the west Pilbara, Western Australia. *CSIRO Australia Mineral Research Laboratory Report FP11*, 41 pp.

GAMBLE, J.A., WRIGHT, I.C., WOODHEAD, J.D. & MCCULLOCH, M.T. (1995) Arc and back-arc geochemistry in the southern Kermadec arc–Ngatoro Basin and offshore Taupo Volcanic Zone, SW Pacific. In: *Volcanism Associated with Extension at Consuming Plate Margins* (Ed. Smellie, J.L.), pp. 193–212. Geological Society, London.

HAMILTON, W.B. (1998) Archean magmatism and deformation were not products of plate tectonics. *Precam. Res.*, **91**, 143–179.

HICKMAN, A.H. (1983) Geology of the Pilbara Block and its environs. *Geol. Surv. W. Aust. Bull.*, **127**, 268 pp.

HORWITZ, R.C. (1990) Paleogeographic and tectonic evolution of the Pilbara Craton, Northwestern Australia. *Precam. Res.*, **48**, 327–340.

HUSTON, D.L., SMITHIES, R.H. & SUN, S.-S. (2000) Correlation of the Archaean Mallina–Whim Creek Basin: implications for base-metal potential of the central part of the Pilbara granite-greenstone terrane. *Aust. J. Earth Sci.*, **47**, 217–230.

KRAPEZ, B. (1996) Sequence stratigraphic concepts applied to the identification of basin-filling rhythms in Precambrian successions. *Aust. J. Earth Sci.*, **43**, 355–380.

KRAPEZ, B. & EISENLOHR, B. (1998) Tectonic settings of Archaean (3325–2775 Ma) crustal–supracrustal belts in the West Pilbara Block. *Precam. Res.*, **88**, 173–205.

LOFGREN, G. (1971) Spherulitic textures in glassy and crystalline rocks. *J. geophys. Res.*, **76**, 5635–5648.

MCARTHUR, A.W., CAS, R.A.F. & ORTON, G.J. (1998) Distribution and significance of crystalline, perlitic and vesicular textures in the Ordovician Garth Tuff (Wales). *Bull. Volcanol.*, **60**(4), 260–285.

MANDEVILLE, C.W., CAREY, S. & SIGURDSSON, H. (1996) Sedimentology of the Krakatau 1883 submarine pyroclastic deposits. *Bull. Volcanol.*, **57**(7), 512–529.

MANLEY, C.R. (1996) Physical volcanology of a voluminous rhyolite lava flow; the Badlands lava, Owyhee Plateau, southwestern Idaho. *J. Volcanol. geothermal. Res.*, **71**(2–4), 129–153.

NELSON, D.R. (1997) Compilation of chronology data, 1996. *Western Australia Geological Survey Record 1997/2.*

NELSON, D.R. (1998) Compilation of chronology data, 1997. *Western Australia Geological Survey Record 1998/2.*

NELSON, D.R. (2000) Compilation of SHRIMP U–Pb zircon geochronology data, 1999. *Western Australia Geological Survey Record 2000/2.*

NELSON, D.R., TRENDALL, A.F. & ALTERMANN, W. (1999) Chronological correlations between the Pilbara and Kaapvaal cratons. *Precam. Res.*, **97**, 165–189.

RYAN, G.R. & KRIEWALDT, M. (1964) Facies changes in Archaean of the West Pilbara Goldfield. *Western Australia Geological Survey Annual Report 1963*, pp. 28–30.

SCHROEDER, T.J. & RIGGS, N.R. (1996) Morphology and structure of a small-volume rhyolitic dome field in eastern Arizona. *Abstr. Prog. geol. Soc. Am.*, **28**, 110.

SELF, S., THORDARSON, T. & KESZTHELYI, L. (1997) Emplacement of continental flood basalt lava flows. *Geophys. Mono.*, **100**, 381–410.

SLOSS, L.L. (1988) Forty years of sequence stratigraphy. *Geol. Soc. Am. Bull.*, **100**(11), 1661–1665.

SMITH, J.B., BARLEY, M.E., GROVES, D.I. *et al.* (1998) The Sholl Shear Zone, West Pilbara: evidence for a domain boundary structure from integrated tectonostratigraphic analyses, SHRIMP U–Pb dating and isotopic and geochemical data of granitoids. *Precam. Res.*, **88**, 143–171.

SMITHIES, R.H. (1998) Geology of the Sherlock 1 : 100 000 sheet. *Western Australia Geological Survey, 1 : 100 000 Series Explanatory Notes*, 29 pp.

SMITHIES, R.H., HICKMAN, A.H. & NELSON, D.R. (1999) New constraints on the evolution of the Mallina Basin, and their bearing on relationships between the contrasting eastern and western granite–greenstone terranes of the Archaean Pilbara craton, Western Australia. *Precam. Res.*, **94**, 11–28.

SMITHIES, R.H., NELSON, D.R. & PIKE, G. (2001) Development of the Archaean Mallina Basin, Pilbara Craton, northwestern Australia: a study of detrital and inherited zircon ages. *Sediment. Geol.*, **141–142**, 79–94.

TAYLOR, R.N. & NESBITT, R.W. (1995) Arc volcanism in an extensional regime at the initiation of subduction: a geochemical study of Hahajima, Bonin Islands, Japan. In: *Volcanism Associated with Extension at Consuming Plate Margins* (Ed. Smellie, J.L.), pp. 115–134. Geological Society, London.

VAIL, P.R., MITCHUM, R.M. JR & THOMPSON, S. (1977) Seismic stratigraphy and global changes of sea level. Part 3: Relative changes of sea level from coastal onlap. In: *Seismic Stratigraphy; Applications to Hydrocarbon Exploration* (Ed. Payton, C.E.), Mem. Am. Ass. petrol. Geol., Tulsa, **26**, 63–81.

*Spec. Publs int. Ass. Sediment.* (2002) **33**, 235–258

# Syn- and post-eruptive volcaniclastic sedimentation in Late Archaean subaqueous depositional systems of the Black Flag Group, Eight Mile Dam, Kalgoorlie, Western Australia

J. L. HAND*, R. A. F. CAS*, L. ONG*, S. J. A. BROWN†, B. KRAPEZ† *and* M. E. BARLEY†

**Department of Earth Sciences, Monash University, Clayton, Victoria 3168, Australia; and*
*†Centre for Teaching and Research in Strategic Mineral Deposits, Department of Geology and Geophysics, University of Western Australia, Nedlands, WA 6907, Australia*

## ABSTRACT

The Late Archaean Black Flag Group is widespread but poorly exposed throughout the Kalgoorlie Terrane of the Eastern Goldfields Province in the Yilgarn Craton. From drill core analysis, five distinct facies can be defined within the succession: massive to laminated mudstone, diffusely bedded sandstone, polymictic conglomerate, massive sandstone and rhyolitic breccia. Each of the coarser-grained facies is associated with the mudstone facies, indicating a consistent, below wave base depositional setting in which mass flow processes were responsible for depositing the majority of detritus. Three distinct depositional systems are identified within the Eight Mile Dam succession, each of which is separated by periods of tranquil water sedimentation. Depositional system 1 is defined by an upward-coarsening trend, consisting of vertically stacked mudstone and sandstone, overlain by polymictic conglomerate. This facies arrangement is characteristic of vertically stacked outer lobe, proximal lobe and channel deposits, defining a progradational submarine fan system. This progradational submarine fan consists of polymictic debris and has a maximum depositional age of 2666 ± 6 Ma. Depositional system 2 consists of amalgamated massive sandstone beds with erosive basal contacts, indicative of a proximal channelized submarine fan setting. The debris is juvenile and dacitic in origin and appears to have been derived from contemporaneous pyroclastic volcanism from a single dacitic centre. Depositional system 3 consists of unchannelized rhyolitic breccias that have characteristics indicative of deposition on a slope-apron environment, likely to be closely associated with a subaqueous rhyolitic centre. The rhyolitic slope apron system has a maximum depositional age, and likely volcanic eruption age of 2669 ± 8 Ma. The SHRIMP U/Pb detrital zircon geochronology constrains the age of the Eight Mile Dam Black Flag Group to 2677–2660 Ma.

## INTRODUCTION

The Eastern Goldfields Province of the Yilgarn Craton, Western Australia, is characterized by extensive granitoids and narrow, elongate belts of Late Archaean greenstones (Fig. 1). Within the Kalgoorlie region, the greenstones consist of a lower ultramafic and mafic succession, overlain by the Black Flag Group (BFG), a felsic volcaniclastic and sedimentary succession, but also including felsic and mafic intrusions. Mineral exploration has focused on the ultramafic and mafic units, whereas the felsic rocks have been largely ignored. Combined with limited surface exposure (<5%) and deep weathering, the stratigraphic architecture and palaeoenvironmental setting of the BFG is poorly understood. Prior to the study of Hand (1998) little was known about the regional stratigraphic architecture of the BFG. Previous research on the BFG was limited to localized studies of the sedimentology and volcanology (Brauns, 1991; Ong, 1994) and locality descriptions were reported as part of the Geological Survey of Western Australia (GSWA) mapping projects (Keats, 1987; Hunter, 1993). Recently, Morris (1998) presented detailed accounts of volcanological and sedimentological aspects of many BFG rocks as

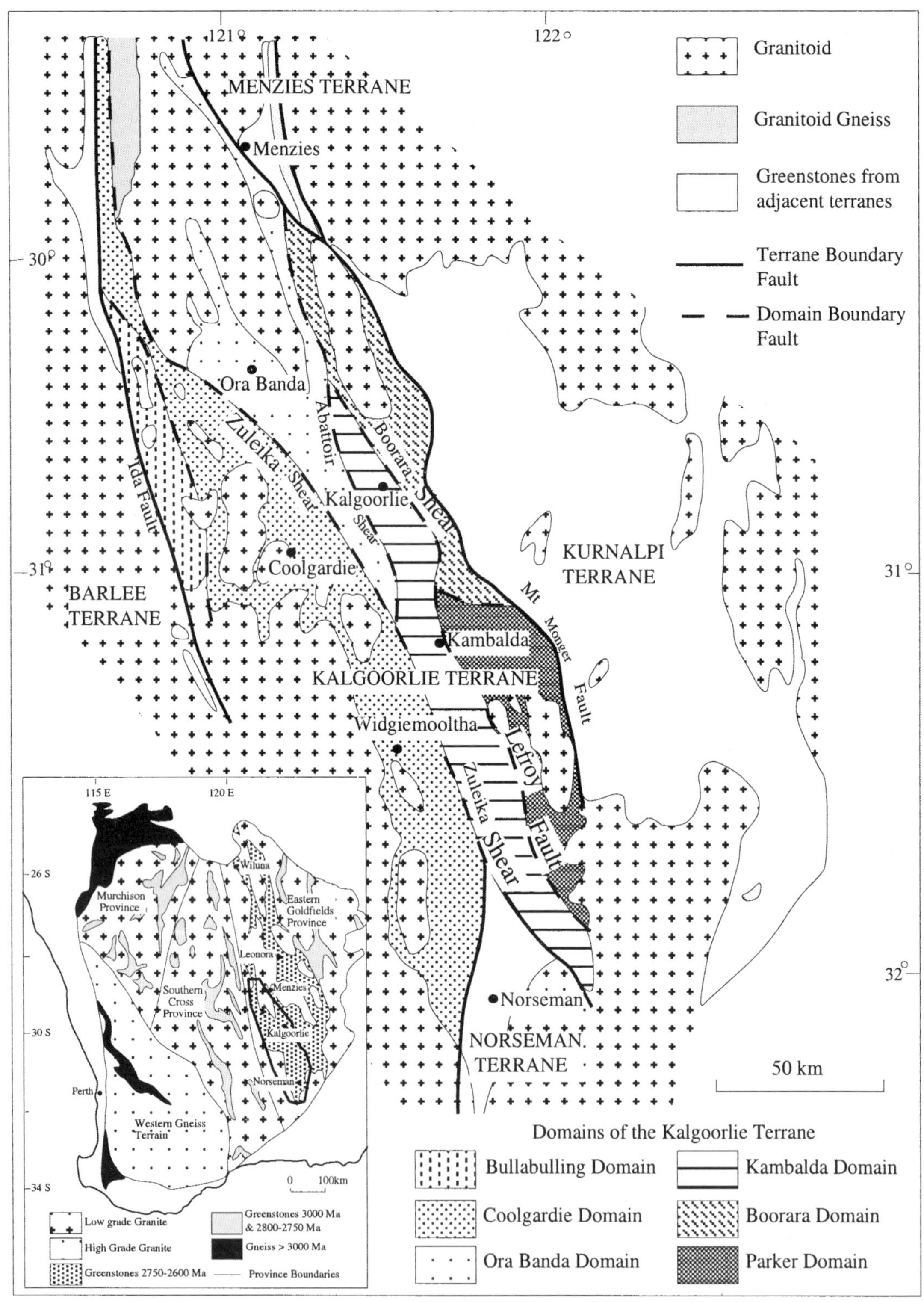


**Fig. 1.** Tectonostratigraphic domains of the Kalgoorlie Terrane (after Swager *et al.*, 1990) and the Yilgarn Craton divisions of Gee *et al.* (1981).

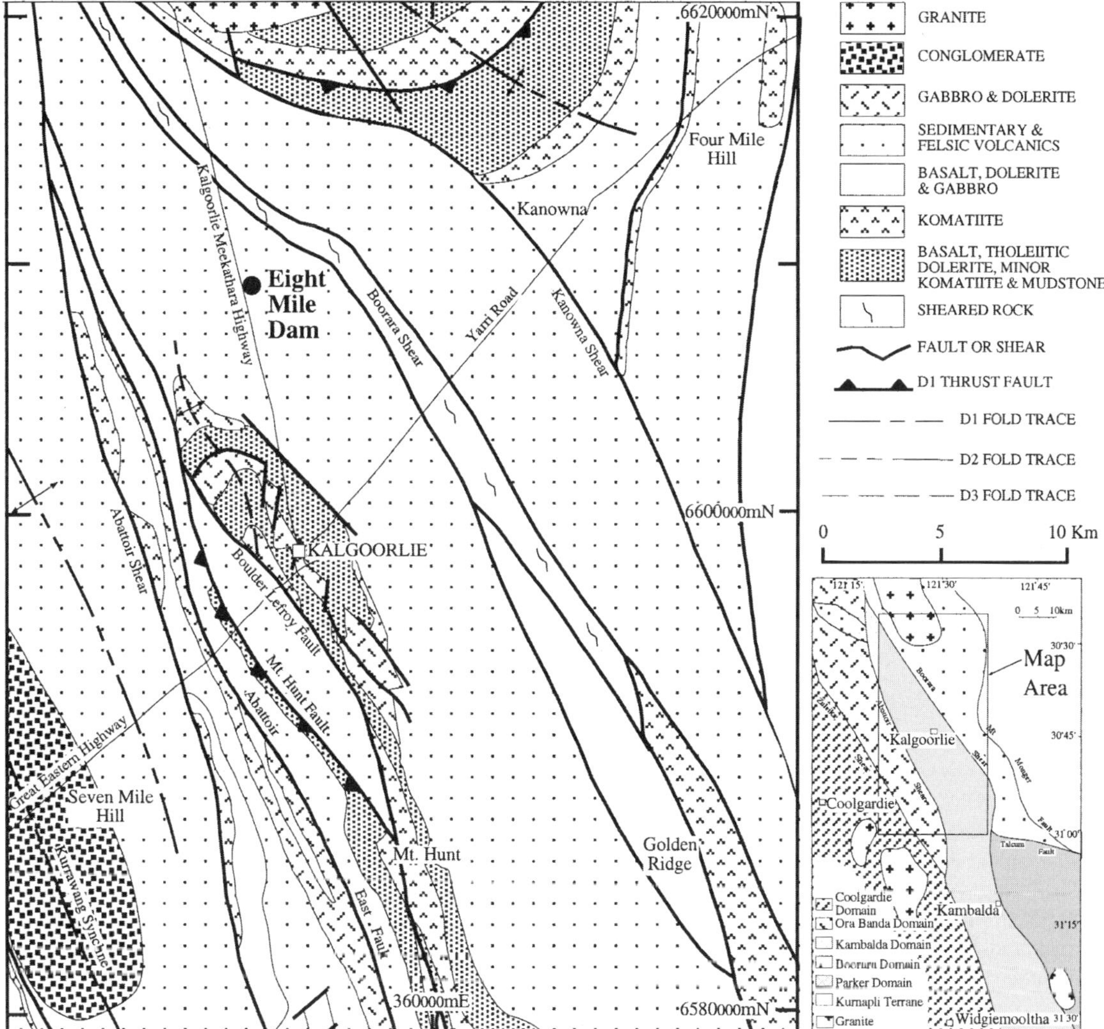


**Fig. 2.** Location of Eight Mile Dam and the main structural features of the Kalgoorlie region (after Swager *et al.*, 1990).

part of his GSWA report on Archaean felsic volcanism throughout the Eastern Goldfields of Western Australia.

At Eight Mile Dam (EMD), 8 km north of Kalgoorlie (Fig. 2), ten diamond drill holes through the upper stratigraphic levels of the BFG were studied, resulting in a well constrained stratigraphy over 400 m in thickness (Figs 3 & 4). The EMD succession is interpreted to be a coherent stratigraphic package, as no evidence of major faults or shear zones was found in the drill core. The succession dips steeply to the west and is interpreted to be slightly overturned. Lateral variations in the thickness of facies may reflect original changes in bedding thickness. However, the drill core was not orientated, preventing calculation of the true thickness of individual facies.

This paper studies the volcanology and sedimentology of the BFG at EMD, based entirely on analysis of drill core data, and shows that use of such data need not be limited to descriptive purposes, but can be used to interpret large-scale depositional systems and palaeoenvironmental settings. The description and interpretation of facies is followed by an interpretation on the depositional systems and palaeoenvironmental setting. New U/Pb zircon dates and provenance geochemical data on the BFG are presented.

## GEOLOGICAL SETTING

The greenstones in the Kalgoorlie region of the Eastern Goldfields Province have been divided into a number of tectonostratigraphic terranes bounded by major faults or granitoid intrusions (Swager *et al.*, 1990) (Fig. 1). The Kalgoorlie Terrane is well defined, consisting of four major domains, the Coolgardie, Ora Banda, Kambalda and Boorara domains, and two minor domains (Bullabulling and Parker) (Fig. 1), which are thought to have originally been part of a single volcano-sedimentary basin (Swager *et al.*, 1990). The base of the greenstone sequence is not known as it is truncated by a subhorizontal detachment fault 4–7 km below surface (Goleby *et al.*, 1993; Swager *et al.*, 1997). A relatively simple greenstone stratigraphy has been defined for the four major domains of the Kalgoorlie Terrane (Table 1). SHRIMP U/Pb zircon dating of interbedded felsic units within the regional marker komatiite unit indicate an age between 2700 and 2710 Ma (Nelson, 1997a). This is consistent with an age of 2706 ± 36 Ma, obtained by Re–Os isotopes from komatiites at Kambalda (Foster *et al.*, 1996). The minimum age of komatiitic volcanism is 2692 ± 4 Ma, based on SHRIMP U/Pb zircon dating from the overlying Kapai Slate (Claoué-Long *et al.*, 1988). The final episode of basaltic magmatism was followed by felsic volcanism and sedimentation (BFG) between 2685 and 2670 Ma, based on SHRIMP U/Pb zircon dating (Claoué-Long, 1990; Nelson, 1995, 1997a,b).

Regional deformation includes: D1, recumbent folding and thrusting resulting in large-scale stratigraphic repetition; D2, ENE–WSW regional shortening producing upright F2 folds; D3, transcurrent faults; and D4, oblique faults (Swager, 1989). The minimum age of D1 is 2674 ± 6 Ma (Kent & McDougall, 1995), overlapping with felsic volcanism. The metamorphic grade in the Kalgoorlie region, including the EMD area, is low–mid greenschist facies (Binns *et al.*, 1976). Peak metamorphism occurred synchronously with D2 and widespread granite emplacement at *c.*2660 Ma (Swager, 1997). Regional (D2–D3) compression appears to have been long lived, with major north–south shear zones active until *c.*2635 Ma (Nelson, 1997a). Granite continued to be emplaced until *c.*2600 Ma (Hill *et al.*, 1992) and gold mineralization was late, occurring between 2650 and 2600 Ma (Barley, 1998a).

The Late Archaean tectonic setting of the Eastern Goldfields was extensional and floored by continental crust. Tectonic models are divided between extension in either an intracontinental rift (Archibald *et al.*, 1978, 1981; Gee *et al.*, 1981; Groves & Batt, 1984; Hallberg, 1986; Hammond & Nisbet, 1993; Swager, 1997; Smithies & Witt, 1997) or a marginal basin (Barley & McNaughton, 1988; Barley *et al.*, 1989, 1998a,b; Barley & Groves, 1990; Perring *et al.*, 1990; Cassidy *et al.*, 1991; Morris, 1993; Swager *et al.*, 1995; Witt, 1995; Myers, 1995; Champion & Sheraton, 1997; Krapez, 1997; Morris & Witt, 1997; Nelson, 1997a; Barley *et al.*, 1998a,b).

On a regional scale the BFG is a dominantly felsic volcaniclastic succession composed of multiple submarine fan and slope apron depositional systems, which represent distinct depositional stages within different parts of a large basin that developed between 2690 and 2665 Ma (Hand, 1998; Barley *et al.*, 1998b). Composite stratigraphic sections indicate the BFG is up to 2.5 km thick in parts of the Kalgoorlie Terrane. The provenance characteristics indicate that multiple isolated sources contributed to the succession, including older exposed terranes and contemporaneous volcanic centres of dominantly andesitic to dacitic composition (Hand, 1998). The BFG is preserved within all four major domains of the Kalgoorlie Terrane, and has a distinct geochemistry when compared with other felsic successions of the Eastern Goldfields (Hand, 1998). Although the BFG is known as a 'group', based on the division of felsic rocks north of Coolgardie into the Spargoville and White Flag Formations (Hunter, 1993), Hand (1998) found it difficult to establish a regionally extensive, coherent internal stratigraphy that could be used to subdivide the group into formations. Therefore, it is suggested that the BFG is more appropriately named the Black Flag Formation.

## STRATIGRAPHY OF THE BFG AT EIGHT MILE DAM

Ten EMD holes (Fig. 3) were drilled from west to east from collars located on a north–south line along strike, extending over a distance of about 1 km. The holes intersected slightly different levels of the stratigraphy but can be correlated (Fig. 3) to construct a composite stratigraphy (Fig. 4). Deformational overprinting of primary textures and structures is absent and primary mineralogy is largely intact, with secondary carbonate alteration and metamorphic minerals minimal.

The total stratigraphic thickness intersected is about 420 m and comprises seven main units composed of

**Table 1.** Stratigraphy of the main tectonostratigraphic domains in the Kalgoorlie Terrane. Units in italics are intrusive in origin (after Swager *et al.*, 1990).

| Stratigraphic succession | Age (Shrimp U–Pb zircon) | Kambalda Domain | Ora Banda Domain | Coolgardie Domain | Boorara Domain |
|---|---|---|---|---|---|
| Polymictic conglomerate unit | No age data at present | Merougil conglomerate | Kurrawang Formation | Absent | Absent |
| Felsic volcaniclastic and sedimentary unit | 2681 ± 5 Ma (Nelson, 1997a)<br>2676 ± 4 Ma (Claoue-Long, 1990) | BLACK FLAG BEDS:<br>*Junction dolerite*<br>*Condenser dolerite*<br>*Golden Mile dolerite*<br>Triumph gabbro | BLACK FLAG BEDS:<br>Pipeline andesite<br>*Orinda sill*<br>*Ora Banda sill* | BLACK FLAG BEDS:<br>White Flag Formation<br>*Powder sill*<br>Spargoville Formation | Felsic volcaniclastic and sedimentary rocks<br>(Undivided) |
| Upper basalt unit | 2691 ± 6 Ma (Claoue-Long *et al.*, 1988)<br>2692 ± 4 Ma (Claoue-Long *et al.*, 1988) | KALGOORLIE GROUP:<br>Paringa basalt<br>*Defiance dolerite*<br>*Williamstown dolerite*<br>Kapai slate | GRANTS PATCH GROUP:<br>Victorious basalt<br>Bent Tree basalt<br>*Mt. Pleasant sill*<br>*Mt Ellis sill* | COOLGARDIE GROUP:<br>Absent or not prominent | Absent or not prominent |
| Komatiite unit | 2708 ± 7 Ma (Nelson, 1997a) | KALGOORLIE GROUP:<br>Devon Consols basalt<br>Kambalda komatiite | LINGER AND DIE GROUP:<br>Big Dick basalt<br>Siberia komatiite<br>Walter Williams Formation | COOLGARDIE GROUP:<br>Hampton Formation | Highway ultramafics<br>Big Blow chert |
| Lower basalt unit | No age data at present | KALGOORLIE GROUP:<br>Lunnon basalt | POLE GROUP:<br>Missouri basalt<br>Wongi basalt | COOLGARDIE GROUP:<br>*Golden Bar sill*<br>Burbanks Formation<br>*Three Mile sill* | Scotia basalt |

five distinct facies (Fig. 4). The oldest unit is diffusely stratified, polymictic sandstone at least 100 m thick, the base of which was not intersected. This is overlain by 70 m of diffusely bedded sandstone with pebble horizons, and then 110 m of diffusely bedded, polymictic volcanic conglomerate, in the top of which are thin beds of mudstone. A mudstone unit about 20 m thick separates the polymictic conglomerate from a 50 m thick unit of monomictic, massive volcanic sandstone. Above is a 6 m thick mudstone unit, overlain by a 50 m thick, massive to diffusely bedded rhyolite breccia unit.

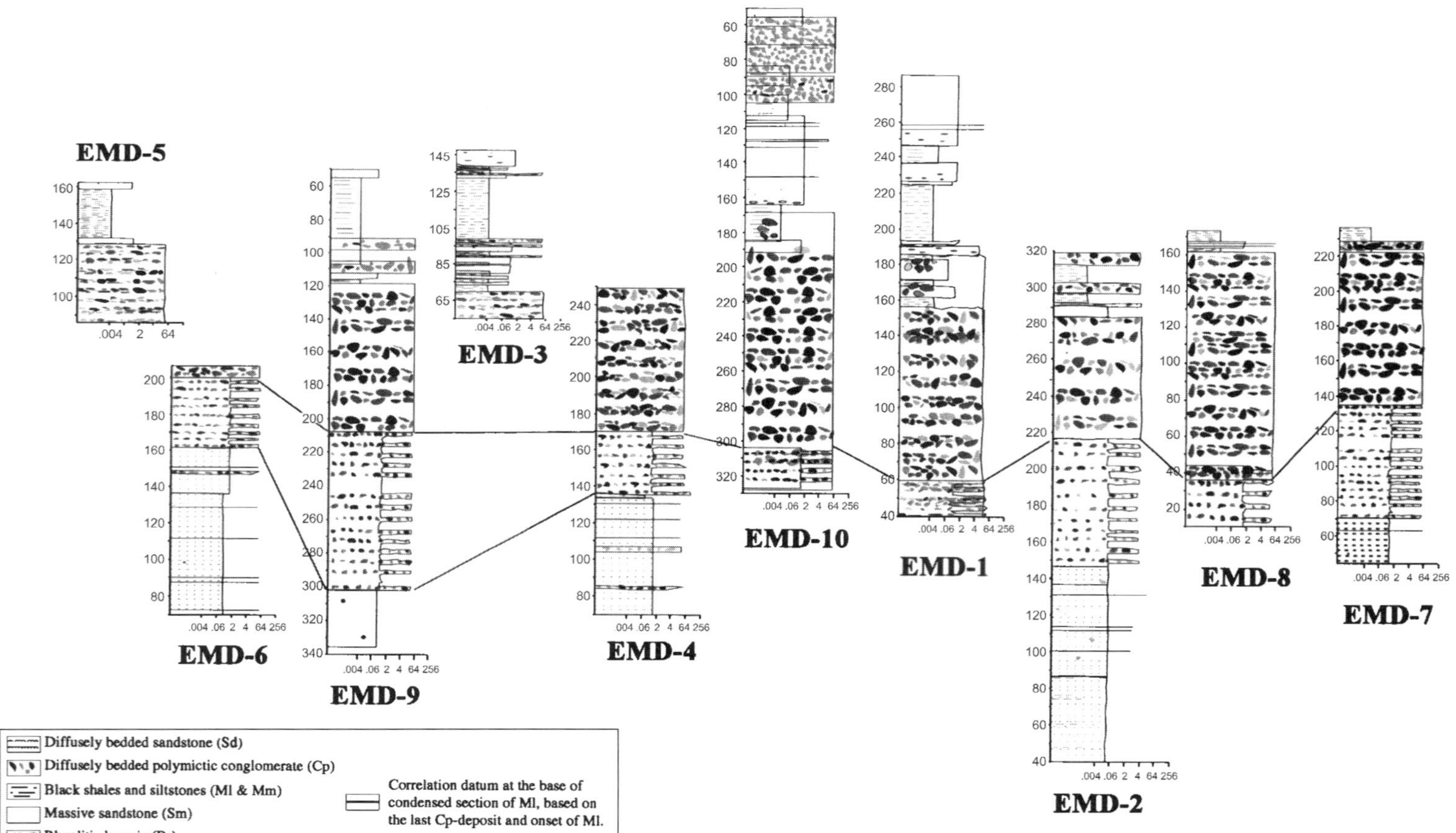


**Fig. 3.** Correlation profile of the 10 drill holes that comprise the BFG at EMD. The holes show consistent stratigraphy, indicating that they intersect a relatively continuous section. Original depths of the drill core have been retained for future reference.

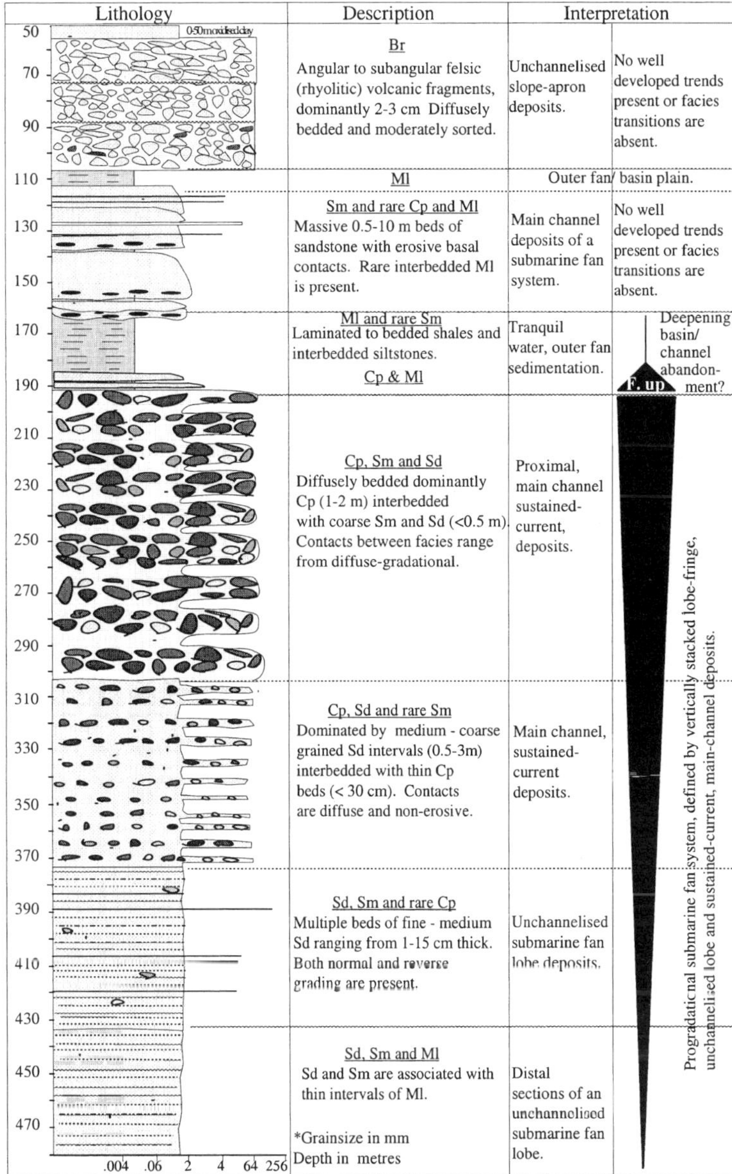


**Fig. 4.** Simplified lithofacies succession and facies interpretation of the BFG at EMD. Note: facies thicknesses are apparent as the drill core was not structurally orientated to obtain true thickness measurements.

## FACIES DESCRIPTIONS AND DEPOSITIONAL PROCESSES

This section describes the sedimentary facies preserved at EMD and interprets the processes of deposition. The five facies (Fig. 5) in the EMD succession are defined by grain size, sedimentary structures and composition, and may occur at various intervals throughout the succession. Each facies has been given a code that consists of two parts: the first letter, in upper case, designates the textural class and the second letter, in lower case, designates the dominant clast type in the breccia and conglomerate facies, and the structure in the sandstone and mudstone facies.

Selected samples were analysed using X-ray fluorescence spectrometry (XRF) in order to assess further the provenance of the different lithofacies within the EMD succession. The results of these analyses have been recalculated to anhydrous values and are plotted on the major element, total alkalis versus silica (TAS)

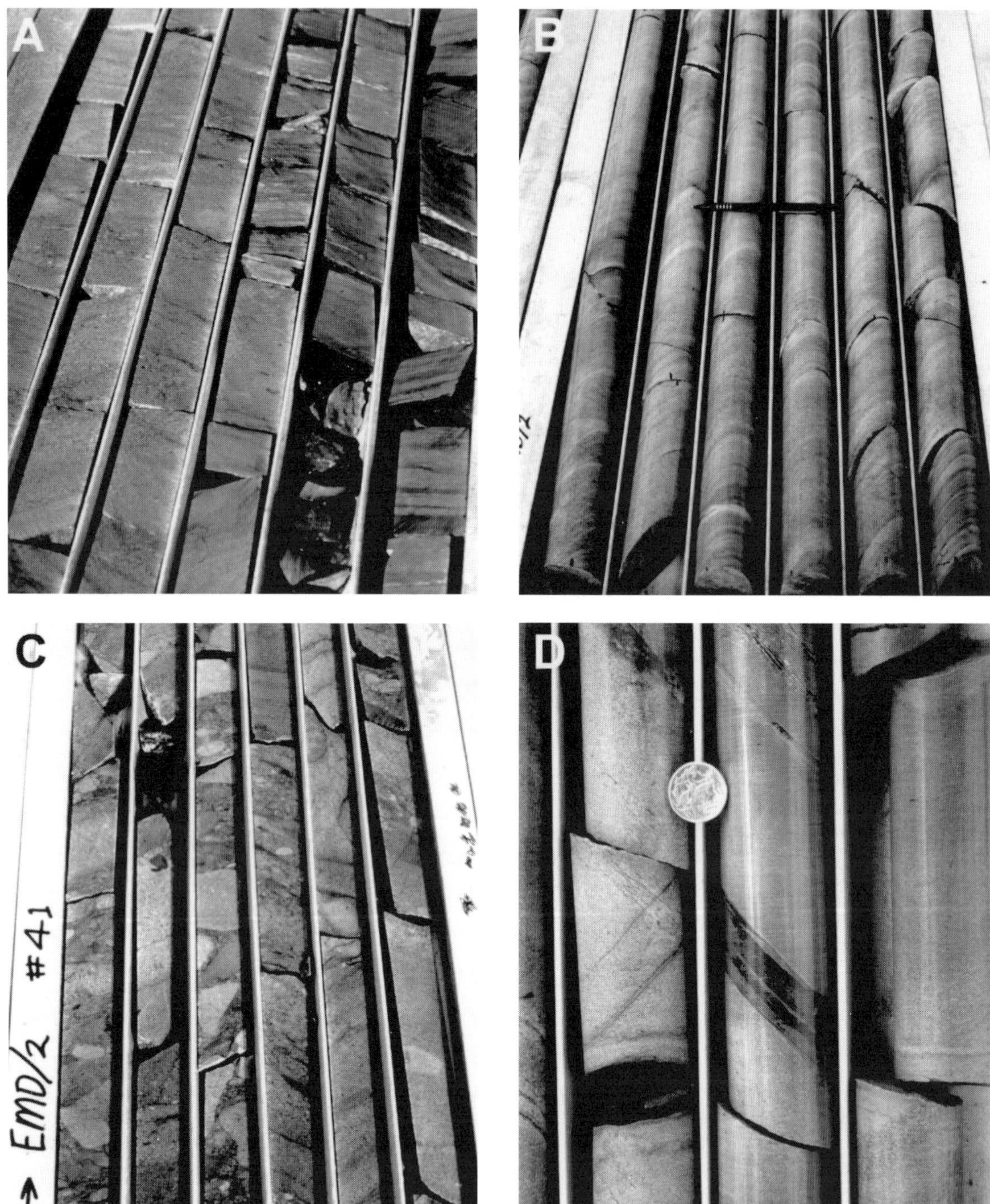


**Fig. 5.** Common lithofacies of the BFG at EMD. (A) Facies 1 Ml, thinly laminated black shale within EMD-8, 164–170 m. (B) Facies 2 Sd, diffusely bedded, fine to medium sandstone from EMD-2, 81.3–87.0 m. (C) Facies 3 Cp, massive polymictic conglomerate within EMD-8, 271.6–277.3 m. (D) Facies 4 Sm, massive, well sorted medium sandstone within EMD-1, 257 m. (*continued*)

**Fig. 5.** (*cont'd*) (E) Facies 5 Br, diffusely bedded rhyolite breccia in EMD-10, 66 m.

plot (Fig. 6A & B) of Le Maitre *et al.* (1989). The TAS classification scheme was found to be the most appropriate for classifying these rocks, based on the low levels of metamorphism and alteration in this region. Trace elements Ti, Zr, Y and Nb, considered to be immobile (Cann, 1970; Field & Elliot, 1974; Hallberg, 1985; Morris & Witt, 1997), were used to attempt to classify these rocks, using the $Zr/TiO_2$ versus Nb/Y (Winchester & Floyd, 1977) discrimination diagram. This scheme did not successfully discriminate the composition of the rocks, as the samples consistently plotted in the intermediate field, due to the presence of significant Ti anomalies. This classification was not consistent with the quartz, feldspar-rich mineralogy of these rocks (Table 2).

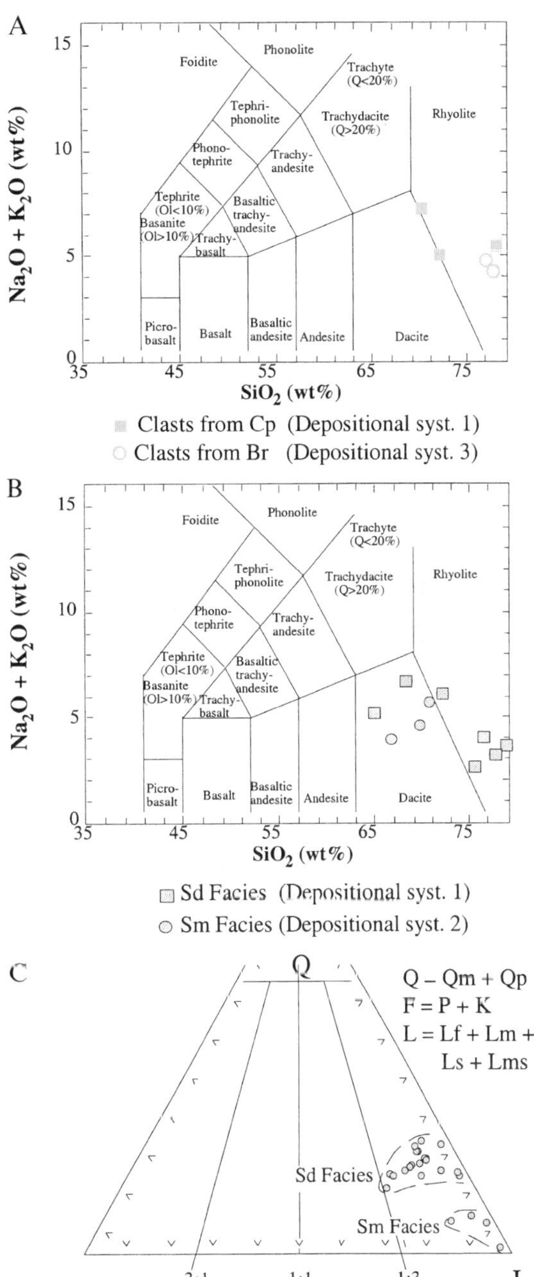


**Fig. 6.** Provenance data from EMD. (A) Total alkalis versus silica plot of clasts after Le Maitre *et al.* (1989). (B) Total alkalis versus silica plot of sandstones after Le Maitre *et al.* (1989). (C) QFL plot of sandstones after Folk (1974).

## Facies 1: massive to laminated mudstone (Ml)

### *Description*

This facies includes massive to planar laminated black shale and siltstone (Fig. 5A), to interlaminated siltstone and black shale. Laminae range from <1 to 4 mm

**Table 2.** Petrographic summary of diffusely bedded sandstone (Sd) and massive sandstone facies (Sm). A, angular; Sa, subangular; Sr, subrounded; R, rounded; Ind, indistinguishable.

| | | Framework grains | | | | | | | | | | | | | |
|---|---|---|---|---|---|---|---|---|---|---|---|---|---|---|---|
| | | Quartz | | | Plagioclase | | | | Felsic lithics | | | Other lithics | | | |
| Sample number | Location (AMG) | Vol. (%) | Size (mm) | Shape | Vol. (%) | Size (mm) | Shape | An. (%) | Vol. (%) | Size (mm) | Shape | Vol. (%) | Size (mm) | Shape | Matrix |
| EMD-1 71.0 m<br>Sd-Facies 2 | 352 496.12 mE<br>6 608 457.90 mN | 26 | 0.5–1.5 | A | 5 | 0.5 | A | Ind | Microg. qtz & feld.<br>48 | <br>1 | <br>Sa–Sr | Metasedimentary<br>5<br>Mafic<br>5 | <br>1<br><br>1 | <br>Sr–R<br><br>Sr | 11% altered to sericite |
| EMD-2 74.0 m<br>Sd-Facies 2 | 352 436.35 mE<br>6 608 409.0 mN | 25 | 0.5 | A–Sa | 9 | <0.3 | Sa–Sr | Ind | Microg. qtz & feld.<br>50 | <br>1 | <br>Sa–Sr | Metasedimentary<br>5<br>Mafic<br>2 | <br>1<br><br>0.5 | <br>Sr<br><br>Sa | 9% carbonate and sericite |
| EMD-2 84.2 m<br>Sd-Facies 2 | 352 436.35 mE<br>6 608 409.0 mN | 23 | 0.5–1 | Sa–Sr | 15 | 1 | Sr–R | Ind | Microg. qtz & feld.<br>49 | <br>1 | <br>Sa–Sr | Metasedimentary<br>3<br>Mafic<br>2 | <br>1<br><br>0.5 | <br>Sr<br><br>Sa | 8% pervasively carbonate and sericite altered + silicified |
| EMD-2 125.2 m<br>Sd-Facies 2 | 352 436.35 mE<br>6 608 409.0 mN | 18 | 0.5 | Sa–R | 16 | 0.5 | Sa | Ind | Microg. qtz & feld.<br>37<br>Plagioclase-phyric<br>7 | <br>1<br><br>1 | <br>Sa–Sr<br><br>Sa | Metasedimentary<br>4<br>Mafic<br>3 | <br>0.5<br><br><0.5 | <br>Sr–R<br><br>Sr | 15% sericite and microgranular quartz |
| EMD-2 140.0 m<br>Sd-Facies 2 | 352 436.35 mE<br>6 608 409.0 mN | 25 | 0.5–1 | A–Sa | 10 | 0.5 | A–Sa | Ind | Microg. qtz & feld.<br>50<br>Plagioclase-phyric<br>5 | <br>0.5–1<br><br>1 | <br>Sa<br><br>Sa | Metasedimentary<br>2<br>Mafic<br>1 | <br>0.5<br><br>0.5 | <br>Sr<br><br>Sa | 7% pervasively carbonate and sericite altered |
| EMD-2 216.0 m<br>Sd-Facies 2 | 352 436.35 mE<br>6 608 409.0 mN | 29 | 1 | A–Sa | 6 | 1 | A | Ind | Microg. qtz & feld.<br>37<br>Plag-phyric<br>9 | <br>1–2<br><br>1–2 | <br>Sa–Sr<br><br>Sr | Metasedimentary<br>5<br>Mafic<br>4 | <br>1<br><br>0.5–1 | <br>Sr<br><br>Sa | 10% carbonate and sericite altered |
| EMD-4 90.8 m<br>Sd-Facies 2 | 352 617.12 mE<br>6 608 302.86 mN | 27 | 0.5–1 | Sa | 6 | 0.5 | A–Sa | Ind | Microg. qtz & feld.<br>45<br>Plagioclase-phyric<br>5 | <br>1<br><br>1 | <br>Sa–Sr<br><br>Sr | Metasedimentary<br>3<br>Mafic<br>2 | <br>0.5<br><br>0.5 | <br>Sr<br><br>Sr | 12% carbonate and sericite |
| EMD-4 125.9 m<br>Sd-Facies 2 | 352 617.12 mE<br>6 608 302.86 mN | 25 | 0.5 | Sa–Sr | 5 | 0.5 | A | Ind | Microg. qtz & feld.<br>51 | <br>0.5–1 | <br>Sr | Metasedimentary<br>3<br>Mafic<br>2 | <br>0.5<br><br>0.5 | <br>Sr<br><br>Sr | 14% carbonate and sericite altered |
| EMD-4 154.6 m<br>Sd-Facies 2 | 352 617.12 mE<br>6 608 302.86 mN | 29 | 1–2 | A | | | | | Microg. qtz & feld.<br>42<br>Qtz+ plag–phyric<br>10 | <br>1–2<br><br>2 | <br>Sa–Sr<br><br>Sa–Sr | Metasedimentary<br>4<br>Mafic<br>3 | <br>1<br><br>0.5 | <br>Sr<br><br>Sr | 12% sericite altered |

| | | | | | | | | | | |
|---|---|---|---|---|---|---|---|---|---|---|
| EMD-5 100.0 m<br>Sd-Facies 2 | 352 312.18 mE<br>6 608 837.84 mN | 30 | 1.5 | Sa–Sr | 4 | 0.5 | A | Ind | Microg. qtz & feld.<br>50 1–2 Sa–Sr | Metasedimentary<br>5 2 Sr | 11% calcite and sericite |
| EMD-6 97.7 m<br>Sd-Facies 2 | 352 238.71 mE<br>6 608 776.48 mN | 20 | 0.5–1 | A–Sa | 10 | 0.5 | A | Ind | Microg. qtz & feld.<br>45 0.5–1 Sa–Sr | Sedimentary<br>2 1 Elongate<br>Metasedimentary<br>2 0.5 Sr<br>Mafic<br>1 0.5 Sa | 10% sericite and carbonate altered |
| EMD-6 168.6 m<br>Sd-Facies 2 | 352 238.71 mE<br>6 608 776.48 mN | 22 | 0.5 | A–Sa | 14 | 0.5 | A–R | Ind | Microg. qtz & feld.<br>52 0.5–1 Sa | Metasedimentary<br>2 0.5 Sr | 10% sericite altered |
| EMD-6 172.7 m<br>Sd-Facies 2 | 352 238.71 mE<br>6 608 776.48 mN | 23 | 0.5 | A–R | 11 | 0.5 | Sa–Sr | Ind | Microg. qtz & feld.<br>45 0.5–1 Sa–Sr | Metasedimentary<br>5 0.5 Sr<br>Mafic<br>5 0.5 Sa | 11% sericite and carbonate altered |
| EMD-6 207.7 m<br>Sd-Facies 2 | 352 238.71 mE<br>6 608 776.48 mN | 24 | 0.5–1 | Sa | 7 | 0.5 | Sa | Ind | Microg. qtz & feld.<br>40 1 Sa–Sr<br>Plag + Qtz-phyric<br>10 1 Sr | Metasedimentary<br>3 0.5–1 Sr<br>Mafic<br>3 0.5 Sa | 13% calcite and sericite |
| EMD-7 84.4 m<br>Sd-Facies 2 | 352 436.35 mE<br>6 608 409.03 mN | 27 | 0.5 | A–Sr | 6 | 1 | Sa | $An_3$ | Microg. qtz & feld.<br>44 1–2 Sa–Sr<br>Plag-phyric<br>5 1 Sa | Metasedimentary<br>2 0.5–1 Sr<br>Mafic<br>3 1 Sa | 13% sericite and microgranular quartz |
| EMD-7 102.0 m<br>Sd-Facies 2 | 352 436.35 mE<br>6 608 409.03 mN | 21 | 0.5–1 | A–Sa | 3 | 0.5 | Sa | Ind | Microg. qtz & feld.<br>47 1–2 Sa<br>Qtz + plag-phyric<br>7 2 Sr | Mafic<br>3 1 Sa | 19% sericite and microgranular quartz |
| EMD-7 112.9 m<br>Sd-Facies 2 | 352 436.35 mE<br>6 608 409.03 mN | 25 | 1 | A–Sa | 5 | 1 | Sa | Ind | Microg. qtz & feld.<br>30 1 Sa–Sr<br>Qtz + plag-phyric<br>15 0.5–1.5 Sa | Metasedimentary<br>7 1 Sr<br>Mafic<br>5 1 Sa | 13% carbonate and sericite altered |
| EMD-1 327.5 m<br>Sm-Facies 4 | 352 496.12 mE<br>6 608 457.90 mN | 11 | 0.1–1 | A–Sr | 7 | 1 | Sa | Ind | Microg. qtz & feld.<br>50 1 Sa<br>Plagioclase-phyric<br>10 1 Sr | Sedimentary<br>10 2 Elongate | 12% microgranular quartz and carbonate |
| EMD-3 170.6 m<br>Sm-Facies 4 | 352 443.05 mE<br>6 608 685.13 mN | 12 | 1–2 | Em–Sa–Sr | 2 | 0.5 | Sa | Ind | Microg. qtz & feld.<br>66 1–2 Sa | Sedimentary<br>5 0.5–1 Elongate | 15% silicified and carbonate altered |
| EMD-5 171.4 m<br>Sm-Facies 4 | 352 312.18 mE<br>6 608 837.84 mN | 5 | 0.2–0.5 | Sr | | | | | Microg. qtz & feld.<br>80 1–2 Sa | | 10% silicified and carbonate altered |
| EMD-10 124.2 m<br>Sm-Facies 4 | 352 503.05 mE<br>6 608 636.38 mN | 11 | 1–2 | A–R | | | | | Microg. qtz & feld.<br>77 1 Sa<br>Plagioclase-phyric<br>2 1 Sa | | 10% silicified and carbonate altered |

thick and are commonly graded, and massive units are up to 10 cm thick. Basal laminae contacts are generally sharp and planar, whereas upper contacts with massive units are commonly scoured and irregular, and flame structures may be present. Soft-sediment deformation structures, including folds, slumps and fractures, are common throughout the mudstone intervals.

*Interpretation*

This facies represents deposition by low-energy, tranquil water sedimentation processes such as low-concentration turbidity currents and hemipelagic settling from suspension. Bouma Sequence divisions $T_d$ and $T_e$ (Bouma, 1962) are common. The thick intervals of Ml within the EMD succession (e.g. 162–185 m, Fig. 4) represent deposition below wave base, in relatively deep water (as indicated by black shale beds representing anoxic conditions). Ml occurs at several stratigraphic levels (Figs 3 & 4), suggesting that the environment of deposition consistently remained below wave base. Ml is closely associated with all the other lithofacies at EMD (Figs 3 & 4), providing important constraints on the depositional processes and setting.

## Facies 2: diffusely bedded sandstone (Sd)

*Description*

This facies consists of up to 12 m thick intervals of diffusely stratified planar beds (Fig. 5B), 0.5–15 cm thick, of medium- to coarse-grained sandstone. Non-graded, reversely graded and normally graded beds are present. Sd is commonly associated with thin beds of diffusely stratified conglomerate and thin beds of Ml. Lower contacts with conglomerate are diffuse to gradational, and upper contacts range from diffuse to sharp and erosional to non-erosional. Upper and lower contacts with Ml facies are commonly sharp.

Sd on average consists of volcanic quartz (30%), plagioclase (8%), K-feldspar (1%), felsic porphyritic and aphyric volcanic lithic grains (52%), sedimentary lithic grains (2%) and mafic lithic grains (2%) in a quartz–sericite matrix (5%) (Fig. 7A & Table 2). The majority of primary framework grains are angular to subangular, although the quartz and lithic grains may be rounded. Porphyritic felsic volcanic grains have a range of compositions and phenocryst assemblages, including K-feldspar + quartz + plagioclase, plagioclase + quartz, and plagioclase. In accordance with the IUGSS recommendations (Le Maitre *et al.*, 1989) the mineralogy of the felsic lithic grains reflects derivation from a mixture of rhyolitic to dacitic volcanic sources.

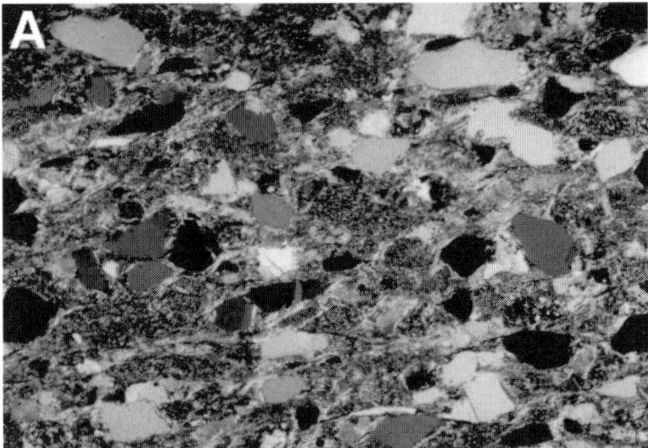


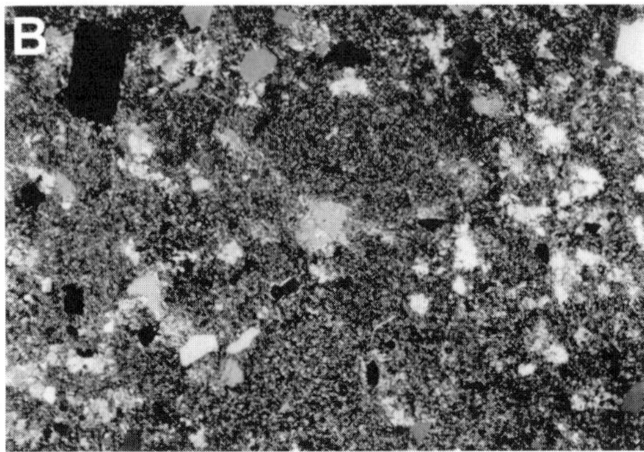


**Fig. 7.** (A) Photomicrograph of Sd (EMD-8, 87.1 m). Note the quartz and felsic lithic-rich composition, with most of the grains relatively angular, XPL, field of view 5 mm. (B) Photomicrograph of dacite derived Sm horizons (EMD-10, 124.2 m). In comparison to the sandstones of the basal progradational suite (Fig. 7A), these sandstones are quartz-poor and are dominated by felsic lithic grains. Partial silicification makes identification of lithic boundaries difficult. XPL, field of view 5 mm.

The whole rock composition of Sd is variable, as indicated by the scattered data on the TAS diagram (Fig. 6B). Bulk compositions are dacite–rhyolite (Fig. 6B), although the petrographical data (Fig. 7A & Table 2) indicate derivation from multiple sources. Sd forms a distinct population on the QFL diagram (Fig. 6C), plotting as litharenites in the Folk (1984) classification scheme. These sandstones are a mixture of older rounded and reworked epiclastic debris and angular juvenile volcanic debris.

*Interpretation*

Sd is indicative of deposition by low- to high-concentration turbidity currents. Sd may grade downwards into massive sandstone (Sm), reflecting deposition from a high-concentration turbidity current that was waning in velocity. Although the beds commonly have sharp basal contacts, evidence of scouring and irregular contacts representing erosional channels is absent. Intervals of Sd beds with planar basal contacts, in places associated with thin black shale beds (Fig. 4, 460 m), represent deposits of unchannelized high-concentration turbidity currents, which are laterally continuous and possibly have sheet-like geometries.

### Facies 3: polymictic conglomerate (Cp)

*Description*

Cp beds (Fig. 5C) are dominantly 1–2 m thick, commonly crudely interbedded with sandstone beds (5–30 cm thick), and inverse to normal grading is common. Cp beds may also be interbedded with mudstone facies (Fig. 4, 183–191 m). Cp beds are generally poorly sorted and diffusely stratified to massive, and range from clast-supported to supported by a coarse-grained sandstone matrix. The clasts range from 0.5 to 15 cm in diameter and from ultramafic to rhyolitic in composition. Clast abundances are: dacitic–rhyolitic (Fig. 6A) porphyritic and aphyric (75%), komatiite (5%), basalt (5%), metasedimentary (5%), vein quartz (5%) and intraformational mudstone (5%). The felsic, metasedimentary and vein quartz clasts are subrounded to well rounded, whereas the ultramafic and mafic clasts are angular to subangular. Angular to rounded volcanic quartz, felsic volcanic lithic clasts and plagioclase grains dominate the sandstone matrix.

*Interpretation*

The common association of Cp horizons with mudstone facies indicates the site of deposition was subaqueous. The coarse sandstone matrix, the often high concentration of clasts and the common diffuse stratification are most characteristic of deposition by high-concentration turbidity currents or density-modified grain flows. High-concentration turbidity currents are capable of transporting cobbles and produce poorly graded, normal to reverse graded, diffusely bedded, sand-supported deposits (e.g. Lowe, 1982, 1988; Sohn, 1997). The interbedded nature of Cp and sandstone is a common feature of deposits produced by high-concentration turbidity currents (Sohn, 1997).

In a high-concentration turbidity current, the coarse gravel component is largely transported at the base of the current within a highly concentrated bedload layer, commonly known as a traction carpet (Walker, 1975, 1977; Aalto, 1976; Hiscott & Middleton, 1979, 1980; Lowe, 1982). Aggradation from the bottom upward can explain thick traction carpet deposits, as the thickness is dependent on the duration and deposition rate, not the thickness of the traction carpet (Hiscott, 1994; Sohn, 1997). The common stratification in the EMD succession, between m thick Cp beds and cm thick sandstone beds, which lack distinct layer boundaries, is probably a result of the variation in the grain size supplied under sustained traction carpet sedimentation. Where successive Cp beds occur, deposition was probably by collapse and regeneration of traction carpets through flow pulsation or surging, associated with rapid suspension fallout (e.g. Sohn, 1997). Inversely graded Cp beds are possibly traction carpet deposits produced by an upwards reduction in dispersive pressure (Lowe, 1982).

Although high-concentration turbidity currents provide an explanation for the conglomerate facies, thick beds of poorly sorted pebbles to boulders can result from density-modified grain flows (e.g. Lowe, 1976; Lash, 1984). Density-modified grain flow deposits described in the literature (e.g. Mullins & Van Buren, 1979; Lash, 1984) often contain sand-dominated beds overlying conglomerate beds. These result from turbidity currents developing on top of the main sediment gravity flow (Krause & Oldershaw, 1979). Therefore, the Cp beds overlain by sandstone beds may represent the deposits of a single event, involving two sedimentary process end members, in a process spectrum. A basal, denser, faster moving density-modified grain flow deposit, with grains supported by dispersive pressure, is thus overlain by a high-concentration turbidity current deposit. It is likely that a continuum existed between high-concentration turbidity currents and density-modified grain flows for deposition of Cp beds.

### Facies 4: massive sandstone (Sm)

*Description*

Sm intervals are composed of multiple, medium- to coarse-grained sandstone beds that range in apparent thickness from 0.5 to 20 m, with most beds being 2–5 m thick. Sm beds are commonly well sorted (Fig. 5D), and may contain numerous intraformational clasts of mudstone. Basal contacts with associated conglomerate and mudstone facies range from sharp, erosional to non-erosional, to gradational. Single Sm beds are commonly normally graded, and they may also grade up into Sd.

Sm beds are composed of, in decreasing abundance, microgranular felsic volcanic and plagioclase-phyric lithic grains, volcanic quartz and feldspar, within a microgranular quartz/feldspar matrix (Fig. 7B & Table 2). The framework grains of Sm are angular, and indicate derivation from a single felsic source (Table 2). The whole rock composition of the upper Sm is dacitic (Fig. 6B). The Sm beds plot in a distinct field from Sd on the Folk (1984) QFL diagram (Fig. 6C).

*Interpretation*

Sm beds are closely associated with mudstones, as indicated by intraformational mudstone clasts and common interbedded mudstone intervals (Fig. 4), implying deposition in a subaqueous, below wave base setting. The amalgamated, normally graded Sm beds, with erosional basal contacts (Figs 3 & 4), represent deposition from pulsating, single-surge, high-concentration turbidity currents (Middleton, 1967; Middleton & Hampton, 1973, 1976). Periods of sustained current activity are represented by thicker, massive Sm beds (Fig. 4, 137–159 m), which were deposited through sediment aggradation (Kneller & Branney, 1995). The intervals of bedded Sm display erosional contacts and contain numerous intraformational mudstone clasts, reflecting a channelized environment, possibly within a relatively proximal setting, relative to a feeder channel, of a subaqueous depositional system.

### Facies 5: rhyolitic breccia (Br)

*Description*

Intervals of Br consist of diffuse beds tens of cm thick that may have poorly developed normal grading, and have non-erosional basal contacts with either interbedded sandstone or successive Br beds. Beds of Br range from being clast-supported to matrix-supported, and are moderately sorted (Fig. 5E). The clasts are leucocratic, microporphyritic, fine-grained and dominantly 1–3 cm long, and may be flow-banded. Minor volumes (<5%) of intraformational Ml clasts are present, which become less abundant up-section. The Br facies occurs at the top of the section (Fig. 3, EMD 10), where it is 50 m thick.

In thin section, abundant pseudomorphed potassium-feldspar and plagioclase phenocrysts and minor quartz phenocrysts are present in a partially silicified groundmass. Based on the geochemical and petrographic data, the clasts are interpreted to have been rhyolitic in composition (Fig. 6A). The Br sandstone matrix and thin interbedded sandstone beds are dominated by felsic lithic grains and angular fragments of volcanic quartz, possibly reflecting derivation from the same source as the clasts.

*Interpretation*

The angularity and monomictic nature of the clasts indicates minimal if any reworking. The diffusely bedded, non-erosional, amalgamated nature of the Br beds possibly represents deposition through a continuum between high-concentration turbidity currents and density-modified grain flows, as outlined above in the interpretation of Cp. The depositional setting is likely to be relatively proximal to the rhyolitic volcanic source, considering the coarse grain size and lack of interbedded Ml.

## SHRIMP U/Pb ZIRCON AGE DATA FROM THE BFG

Analyses were performed using a SHRIMP II instrument at the Curtin University of Technology in Perth, Western Australia, under the general operating conditions described by Smith *et al.* (1998). For sample EMD-10, six data collection cycles (scans) were performed per analysis, and count times (per scan) were 10 s for the $^{204}$Pb, $^{206}$Pb and $^{208}$Pb mass peaks and background, and 20 s for the $^{207}$Pb mass peak. Sample EMD-2 was analysed with five data collection cycles, and 30 s counting on the $^{207}$Pb mass peak. Analyses were referenced to multiple analyses of the CZ3 Sri Lankan zircon standard ($^{206}$Pb/$^{238}$U = 0.0914, corresponding to an age of 564 Ma) for U/Pb calibration. Analyses which showed a high common Pb component (>2%, expressed as % f206 in Table 4), more than 5% discordance or high U contents (>350 ppm) were excluded from mean age calculations. Nelson (1997b) found that zircon grains with high U contents often yield spuriously young radiogenic ages suggesting Pb loss from these grains. Data corrected for common Pb (from $^{204}$Pb counts) are listed in Table 4. Errors reported for individual analyses are 1$\sigma$, while errors for mean ages are $t\sigma$ at 95% confidence.

A sample of Sd (drillhole EMD-2 at 85 m, Fig. 3) yielded a wide range in zircon grain morphologies. Many are large (100–150 μm) equant to slightly elongate euhedral magmatic grains with oscillatory zoning and common metamictization of high U zones. Other

**Table 3.** XRF whole rock, major and trace element data of mudstones, sandstones and clasts from EMD. Analyses were conducted at the Department of Earth Sciences, Melbourne University, and all samples have been recalculated to anhydrous values.

| | EMD2 84.2 m Sd | EMD6 97.7 m Sd | EMD8 87.1 m Sd | EMD10 272.2 m Sd | EMD2 59 m Mudstone | EMD8 197.3 m Mudstone | EMD9 73.8 m Mudstone | EMD8 156 m Cp-Felsic clast | EMD8 156 m Cp-Felsic clast | EMD9 148.7 m Cp-Felsic clast | EMD3 170.6 m Sm | EMD10 124.2 m Sm | EMD10 143.3 m Sm | EMD10 <80 m Br-Felsic clast | EMD10 <80 m Br-Felsic clast | EMD10 <80 m Br-Felsic clast |
|---|---|---|---|---|---|---|---|---|---|---|---|---|---|---|---|---|
| wt % | | | | | | | | | | | | | | | | |
| $SiO_2$ | 71.40 | 68.33 | 77.99 | 79.24 | 65.03 | 75.34 | 76.71 | 70.30 | 78.00 | 72.02 | 69.90 | 69.45 | 66.87 | 76.90 | 76.74 | 77.45 |
| $TiO_2$ | 0.34 | 0.34 | 0.25 | 0.31 | 0.51 | 0.39 | 0.33 | 0.51 | 0.09 | 0.34 | 0.44 | 0.47 | 0.45 | 0.37 | 0.35 | 0.35 |
| $Al_2O_3$ | 14.08 | 14.28 | 13.56 | 10.77 | 15.13 | 12.77 | 13.12 | 18.44 | 11.49 | 16.14 | 15.21 | 16.61 | 15.21 | 13.93 | 12.73 | 12.78 |
| $Fe_2O_3$ | 3.09 | 3.02 | 1.69 | 2.32 | 5.01 | 4.00 | 3.61 | 1.80 | 2.15 | 3.10 | 3.02 | 2.90 | 4.50 | 1.04 | 1.32 | 1.12 |
| MnO | 0.05 | 0.05 | 0.04 | 0.08 | 0.09 | 0.03 | 0.10 | 0.04 | 0.06 | 0.07 | 0.04 | 0.05 | 0.11 | 0.01 | 0.01 | 0.01 |
| MgO | 1.33 | 1.92 | 1.77 | 0.69 | 2.97 | 2.18 | 1.09 | 0.35 | 0.58 | 0.86 | 1.63 | 1.59 | 3.47 | 0.99 | 1.28 | 1.18 |
| CaO | 2.69 | 4.54 | 2.81 | 2.19 | 4.44 | 1.80 | 0.25 | 0.78 | 1.49 | 0.94 | 2.34 | 2.31 | 3.89 | 1.25 | 1.98 | 1.80 |
| $Na_2O$ | 3.75 | 3.74 | 1.46 | 1.31 | 2.43 | 0.92 | 0.22 | 3.39 | 4.43 | 0.69 | 2.47 | 0.52 | 0.36 | 2.18 | 2.73 | 1.81 |
| $K_2O$ | 2.28 | 2.55 | 1.67 | 2.19 | 2.70 | 1.76 | 3.76 | 3.66 | 1.00 | 4.51 | 3.24 | 4.08 | 3.51 | 2.55 | 2.04 | 2.44 |
| $P_2O_5$ | 0.09 | 0.09 | 0.08 | 0.07 | 0.13 | 0.12 | 0.15 | 0.25 | 0.02 | 0.14 | 0.21 | 0.24 | 0.24 | 0.15 | 0.15 | 0.16 |
| $SO_3$ | 4081.40 | 3706.26 | 677.78 | 3653.40 | 6071.41 | 5243.16 | 2712.35 | 891.85 | 2254.37 | 4543.84 | 7622.66 | 12 996.66 | 10 790.76 | 496.09 | 755.21 | 355.57 |
| LOI | 5.02 | 6.85 | 5.66 | 3.86 | 8.15 | 3.99 | 4.06 | 2.67 | 2.89 | 3.87 | 4.08 | 3.99 | 7.37 | 3.03 | 3.85 | 4.07 |
| Total | 99.75 | 99.46 | 101.54 | 99.73 | 99.43 | 100.02 | 99.79 | 99.83 | 99.66 | 99.44 | 99.47 | 99.79 | 99.94 | 99.59 | 99.54 | 99.33 |
| ppm | | | | | | | | | | | | | | | | |
| Ba | 911.89 | 962.27 | 647.07 | 952.88 | 1230.84 | 815.53 | 801.61 | 1277.16 | 366.63 | 1096.68 | 894.70 | 1422.89 | 1461.80 | 934.43 | 640.78 | 893.62 |
| Ce | 62.13 | 48.33 | 72.01 | 43.69 | 72.98 | 49.99 | 55.25 | 89.39 | 25.75 | 23.93 | 119.92 | 121.87 | 124.16 | 34.04 | 27.05 | 27.11 |
| Cr | 134.78 | 139.62 | 109.08 | 139.40 | 243.99 | 141.65 | 46.91 | 91.45 | 9.27 | 31.21 | 60.48 | 64.58 | 59.38 | 18.56 | 16.64 | 17.73 |
| Nb | 3.16 | 3.22 | 5.30 | 2.08 | 4.36 | 3.12 | 3.13 | 3.08 | 3.09 | 0.00 | 8.34 | 7.29 | 4.32 | 2.06 | 2.08 | 2.09 |
| Ni | 111.62 | 53.70 | 40.24 | 45.77 | 105.66 | 117.70 | 34.40 | 52.40 | 42.22 | 44.74 | 31.28 | 30.21 | 34.55 | 12.38 | 100.90 | 6.26 |
| Pb | 6.32 | 21.48 | 8.47 | 8.32 | 17.43 | 11.46 | 15.64 | 4.11 | 62.82 | 6.24 | 11.47 | 25.00 | 20.51 | 14.44 | 8.32 | 12.51 |
| Rb | 68.44 | 76.25 | 48.72 | 69.70 | 88.23 | 52.08 | 112.58 | 108.91 | 30.90 | 137.35 | 88.64 | 113.54 | 97.17 | 77.35 | 61.37 | 72.99 |
| Sr | 743.41 | 815.14 | 475.51 | 399.46 | 776.63 | 294.76 | 172.00 | 286.67 | 330.59 | 307.99 | 423.37 | 366.66 | 291.50 | 338.29 | 363.04 | 374.34 |
| Th | 3.16 | 7.52 | 7.41 | 6.24 | 13.07 | 12.50 | 14.59 | 6.16 | 14.42 | 9.36 | 16.68 | 16.67 | 15.11 | 9.28 | 7.28 | 6.26 |
| U | 0.00 | 1.07 | 0.00 | 2.08 | 4.36 | 5.21 | 4.17 | 1.03 | 3.09 | 2.08 | 6.26 | 8.33 | 6.48 | 0.00 | 4.16 | 0.00 |
| V | 48.44 | 47.25 | 34.95 | 43.69 | 88.23 | 62.49 | 44.82 | 86.31 | 30.90 | 53.07 | 56.31 | 59.37 | 65.86 | 26.82 | 28.09 | 23.98 |
| Y | 9.48 | 13.96 | 3.18 | 8.32 | 9.80 | 9.37 | 10.42 | 6.16 | 5.15 | 10.40 | 15.64 | 10.42 | 15.11 | 8.25 | 9.36 | 5.21 |
| Zn | 72.66 | 74.10 | 12.71 | 29.13 | 98.03 | 54.16 | 53.16 | 26.71 | 32.96 | 22.89 | 60.48 | 63.54 | 100.40 | 12.38 | 9.36 | 9.38 |
| Zr | 104.25 | 108.47 | 83.66 | 85.30 | 150.31 | 111.45 | 133.43 | 115.08 | 73.12 | 96.77 | 168.93 | 167.71 | 159.78 | 151.61 | 141.47 | 139.73 |
| Cl | 71.60 | 53.70 | 65.66 | 43.69 | 135.07 | 53.12 | 116.75 | 70.90 | 89.60 | 55.15 | 95.94 | 122.91 | 62.62 | 111.39 | 93.62 | 77.16 |
| Sc | 5.26 | 4.30 | 3.18 | 6.24 | 11.98 | 11.46 | 13.55 | 10.27 | 12.36 | 11.45 | 7.30 | 7.29 | 12.96 | 3.09 | 1.04 | 4.17 |
| Nd | 27.38 | 22.55 | 21.18 | 14.56 | 27.23 | 18.75 | 28.15 | 39.04 | 5.15 | 10.40 | 52.14 | 51.04 | 47.50 | 26.82 | 16.64 | 15.64 |
| Co | 51.60 | 68.73 | 31.77 | 59.29 | 46.84 | 47.91 | 43.78 | 35.96 | 31.93 | 45.78 | 38.58 | 30.21 | 35.63 | 40.22 | 105.06 | 111.57 |
| Cu | 31.59 | 31.15 | 14.83 | 15.60 | 47.93 | 45.83 | 19.81 | 11.30 | 63.85 | 16.65 | 20.86 | 22.92 | 26.99 | 6.19 | 10.40 | 6.26 |
| Ga | 14.74 | 19.33 | 15.89 | 9.36 | 18.52 | 15.62 | 16.68 | 20.55 | 11.33 | 16.65 | 16.68 | 19.79 | 14.04 | 19.60 | 16.64 | 16.68 |
| As | 91.61 | 41.88 | 48.72 | 45.77 | 75.16 | 88.53 | 114.67 | 67.81 | 36.05 | 62.43 | 35.45 | 45.83 | 97.17 | 19.60 | 6.24 | 3.13 |
| Mo | 0.00 | 0.00 | 0.00 | 0.00 | 2.18 | 0.00 | 0.00 | 0.00 | 2.06 | 0.00 | 0.00 | 2.08 | 0.00 | 0.00 | 0.00 | 0.00 |

**Table 4.** SHRIMP isotopic and age data from EMD-2 (85 m depth) and EMD-10 (75 m depth). Uncertainties in 207/206 ages are 1σ. Analyses that were excluded from mean age calculations owing to excessive discordance, high U or high common Pb are marked with an asterisk and italicized. Standard analyses performed throughout each analysis session gave errors on Pb*/U of 1.14% (EMD-2; $n = 11$) and 1.25% (EMD-10; $n = 11$).

| Analysis | U (ppm) | Th (ppm) | Th/U | f206% | 207/206 | ± | 208/206 | ± | 206/238 | ± | 207/235 | ± | %conc | Age (207/206) | ± |
|---|---|---|---|---|---|---|---|---|---|---|---|---|---|---|---|
| *EMD-2, sandstone, Eight Mile Dam, Hole EMD2* | | | | | | | | | | | | | | | |
| *4-1** | *187* | *129* | *0.69* | *2.46* | *0.1811* | *0.0018* | *0.1704* | *0.0040* | *0.3142* | *0.0038* | *7.8467* | *0.1308* | *66* | *2663* | *17* |
| *1-1** | *154* | *113* | *0.73* | *2.93* | *0.1806* | *0.0019* | *0.2020* | *0.0042* | *0.4674* | *0.0058* | *11.6355* | *0.2006* | *93* | *2658* | *17* |
| 5-1 | 11 | 4 | 0.36 | 0.00 | 0.1886 | 0.0033 | 0.1081 | 0.0043 | 0.5010 | 0.0127 | 13.0280 | 0.4222 | 96 | 2730 | 28 |
| *2-1** | *180* | *160* | *0.89* | *4.83* | *0.1781* | *0.0024* | *0.2076* | *0.0053* | *0.3888* | *0.0048* | *9.5460* | *0.1836* | *80* | *2635* | *22* |
| 6-1 | 142 | 107 | 0.75 | 0.21 | 0.1808 | 0.0010 | 0.2018 | 0.0021 | 0.5043 | 0.0063 | 12.5728 | 0.1796 | 99 | 2660 | 9 |
| 7-1 | 65 | 80 | 1.23 | 0.00 | 0.1891 | 0.0013 | 0.3301 | 0.0031 | 0.5306 | 0.0076 | 13.8375 | 0.2296 | 100 | 2735 | 11 |
| *9-1** | *151* | *80* | *0.53* | *0.81* | *0.1838* | *0.0013* | *0.1465* | *0.0027* | *0.4461* | *0.0055* | *11.3080* | *0.1714* | *88* | *2688* | *12* |
| *10-1** | *195* | *93* | *0.47* | *3.38* | *0.1855* | *0.0020* | *0.1378* | *0.0044* | *0.3655* | *0.0045* | *9.3475* | *0.1625* | *74* | *2702* | *18* |
| 11-1 | 65 | 28 | 0.43 | 0.05 | 0.1858 | 0.0015 | 0.1142 | 0.0025 | 0.5125 | 0.0073 | 13.1281 | 0.2258 | 99 | 2705 | 13 |
| *14-1** | *157* | *68* | *0.43* | *0.67* | *0.2810* | *0.0012* | *0.1160* | *0.0019* | *0.6250* | *0.0077* | *24.2149* | *0.3269* | *93* | *3368* | *7* |
| 13-1 | 80 | 38 | 0.47 | 0.17 | 0.1793 | 0.0014 | 0.1288 | 0.0026 | 0.5125 | 0.0070 | 12.6724 | 0.2114 | 101 | 2647 | 13 |
| 13-2 | 42 | 23 | 0.56 | 0.39 | 0.1804 | 0.0024 | 0.1370 | 0.0048 | 0.5134 | 0.0082 | 12.7722 | 0.2812 | 101 | 2657 | 22 |
| 15-1 | 26 | 8 | 0.29 | 0.07 | 0.1889 | 0.0033 | 0.0812 | 0.0062 | 0.5265 | 0.0098 | 13.7153 | 0.3697 | 100 | 2733 | 29 |
| 16-1 | 140 | 116 | 0.83 | 0.15 | 0.1804 | 0.0010 | 0.2232 | 0.0020 | 0.5051 | 0.0063 | 12.5639 | 0.1787 | 99 | 2656 | 9 |
| 18-1 | 145 | 83 | 0.57 | 0.11 | 0.1806 | 0.0010 | 0.1517 | 0.0018 | 0.4983 | 0.0062 | 12.4088 | 0.1761 | 98 | 2659 | 9 |
| 19-1 | 82 | 28 | 0.34 | 0.16 | 0.1803 | 0.0014 | 0.0899 | 0.0025 | 0.5082 | 0.0070 | 12.6342 | 0.2126 | 100 | 2656 | 13 |
| 20-1 | 86 | 89 | 1.04 | 0.12 | 0.1826 | 0.0013 | 0.2847 | 0.0030 | 0.5070 | 0.0069 | 12.7668 | 0.2051 | 99 | 2677 | 12 |
| 22-1 | 116 | 121 | 1.05 | 0.08 | 0.1824 | 0.0011 | 0.2825 | 0.0024 | 0.5261 | 0.0068 | 13.2322 | 0.1950 | 102 | 2675 | 10 |
| 23-1 | 91 | 101 | 1.11 | 0.28 | 0.1889 | 0.0014 | 0.2969 | 0.0031 | 0.5396 | 0.0072 | 14.0558 | 0.2231 | 102 | 2733 | 12 |
| 25-1 | 36 | 15 | 0.42 | 0.24 | 0.1784 | 0.0022 | 0.1085 | 0.0040 | 0.5241 | 0.0088 | 12.8940 | 0.2833 | 103 | 2638 | 20 |
| 27-1 | 119 | 53 | 0.44 | 0.08 | 0.1872 | 0.0010 | 0.1220 | 0.0016 | 0.5449 | 0.0069 | 14.0613 | 0.2030 | 103 | 2717 | 9 |
| 28-1 | 168 | 145 | 0.86 | 0.78 | 0.1839 | 0.0012 | 0.2357 | 0.0025 | 0.5173 | 0.0063 | 13.1180 | 0.1887 | 100 | 2689 | 10 |
| *29-1** | *106* | *128* | *1.20* | *2.82* | *0.1875* | *0.0022* | *0.3229* | *0.0051* | *0.5229* | *0.0068* | *13.5217* | *0.2502* | *100* | *2721* | *19* |
| 31-1 | 82 | 69 | 0.85 | 0.10 | 0.1907 | 0.0014 | 0.2347 | 0.0029 | 0.5245 | 0.0072 | 13.7934 | 0.2248 | 99 | 2748 | 12 |
| 34-1 | 98 | 62 | 0.63 | 0.40 | 0.1794 | 0.0013 | 0.1662 | 0.0026 | 0.5190 | 0.0068 | 12.8401 | 0.2027 | 102 | 2648 | 12 |
| 36-1 | 98 | 107 | 1.09 | 0.14 | 0.1820 | 0.0012 | 0.2964 | 0.0027 | 0.5139 | 0.0067 | 12.8945 | 0.1971 | 100 | 2671 | 11 |
| 37-1 | 67 | 42 | 0.63 | 0.09 | 0.1879 | 0.0014 | 0.1690 | 0.0026 | 0.5249 | 0.0075 | 13.5976 | 0.2298 | 100 | 2724 | 12 |
| 38-1 | 116 | 42 | 0.37 | 0.00 | 0.1816 | 0.0009 | 0.1022 | 0.0012 | 0.5068 | 0.0064 | 12.6899 | 0.1820 | 99 | 2668 | 9 |
| 39-1 | 68 | 58 | 0.85 | 0.06 | 0.1845 | 0.0015 | 0.2334 | 0.0033 | 0.5099 | 0.0072 | 12.9744 | 0.2254 | 99 | 2694 | 14 |
| 40-1 | 94 | 84 | 0.89 | 0.06 | 0.1836 | 0.0013 | 0.2451 | 0.0028 | 0.5159 | 0.0068 | 13.0578 | 0.2060 | 100 | 2685 | 12 |
| 41-1 | 145 | 107 | 0.74 | 0.36 | 0.1824 | 0.0011 | 0.2034 | 0.0023 | 0.5116 | 0.0063 | 12.8680 | 0.1854 | 100 | 2675 | 10 |
| 41-2 | 29 | 35 | 1.21 | 0.00 | 0.1813 | 0.0019 | 0.3369 | 0.0048 | 0.5138 | 0.0092 | 12.8443 | 0.2804 | 100 | 2665 | 17 |
| 42-1 | 81 | 64 | 0.79 | 0.35 | 0.1824 | 0.0015 | 0.1866 | 0.0031 | 0.4941 | 0.0068 | 12.4279 | 0.2106 | 97 | 2675 | 14 |
| 43-1 | 70 | 34 | 0.49 | 1.27 | 0.1807 | 0.0021 | 0.1372 | 0.0043 | 0.4857 | 0.0069 | 12.1055 | 0.2368 | 96 | 2660 | 19 |

| | | | | | | | | | | | | | | | |
|---|---|---|---|---|---|---|---|---|---|---|---|---|---|---|---|
| 44-1 | 67 | 32 | 0.48 | 0.12 | 0.1800 | 0.0015 | 0.1287 | 0.0026 | 0.5193 | 0.0074 | 12.8886 | 0.2231 | 102 | 2653 | 14 |
| 46-1 | 66 | 24 | 0.36 | 0.75 | 0.1840 | 0.0021 | 0.0943 | 0.0040 | 0.5111 | 0.0074 | 12.9673 | 0.2514 | 99 | 2689 | 19 |
| *47-1** | *126* | *65* | *0.51* | *1.68* | *0.1793* | *0.0018* | *0.1317* | *0.0038* | *0.4433* | *0.0057* | *10.9574* | *0.1888* | *89* | *2646* | *17* |
| 48-1 | 98 | 78 | 0.80 | 0.03 | 0.1897 | 0.0012 | 0.2175 | 0.0024 | 0.5330 | 0.0070 | 13.9418 | 0.2131 | 101 | 2740 | 10 |
| 49-1 | 126 | 59 | 0.47 | 0.68 | 0.1822 | 0.0014 | 0.1299 | 0.0027 | 0.4830 | 0.0061 | 12.1343 | 0.1885 | 95 | 2673 | 13 |
| 50-1 | 70 | 51 | 0.72 | 0.20 | 0.1810 | 0.0016 | 0.1929 | 0.0031 | 0.4983 | 0.0071 | 12.4340 | 0.2180 | 98 | 2662 | 14 |
| *51-1** | *8* | *2* | *0.22* | *5.59* | *0.1712* | *0.0150* | *0.0139* | *0.0328* | *0.3866* | *0.0129* | *9.1248* | *0.8925* | *82* | *2569* | *147* |
| 52-1 | 40 | 27 | 0.69 | 0.50 | 0.1877 | 0.0025 | 0.1880 | 0.0053 | 0.5244 | 0.0086 | 13.5695 | 0.3052 | 100 | 2722 | 22 |
| 53-1 | 101 | 89 | 0.87 | 0.25 | 0.1828 | 0.0014 | 0.2570 | 0.0030 | 0.4872 | 0.0064 | 12.2782 | 0.1951 | 96 | 2678 | 12 |
| 54-1 | 61 | 37 | 0.60 | 1.43 | 0.1776 | 0.0025 | 0.1713 | 0.0053 | 0.5032 | 0.0074 | 12.3245 | 0.2646 | 100 | 2631 | 23 |
| 55-1 | 56 | 65 | 1.15 | 0.08 | 0.1779 | 0.0018 | 0.3086 | 0.0042 | 0.5221 | 0.0078 | 12.8033 | 0.2433 | 103 | 2633 | 17 |
| *EMD-10, dacite breccia, Eight Mile Dam, Hole EMD2* | | | | | | | | | | | | | | | |
| *1-1** | *143* | *77* | *0.54* | *0.85* | *0.1823* | *0.0014* | *0.1561* | *0.0028* | *0.4672* | *0.0062* | *11.7474* | *0.1920* | *92* | *2674* | *13* |
| *2-1** | *176* | *106* | *0.60* | *1.38* | *0.1806* | *0.0016* | *0.1657* | *0.0032* | *0.4268* | *0.0055* | *10.6251* | *0.1749* | *86* | *2658* | *14* |
| 3-1 | 169 | 97 | 0.57 | 0.17 | 0.1812 | 0.0010 | 0.1616 | 0.0016 | 0.4984 | 0.0065 | 12.4545 | 0.1842 | 98 | 2664 | 9 |
| 4-1 | 92 | 132 | 1.44 | 0.03 | 0.1821 | 0.0012 | 0.3974 | 0.0028 | 0.5071 | 0.0071 | 12.7302 | 0.2053 | 99 | 2672 | 11 |
| *5-1** | *114* | *95* | *0.83* | *0.49* | *0.1821* | *0.0014* | *0.2368* | *0.0028* | *0.4781* | *0.0066* | *12.0042* | *0.1979* | *94* | *2672* | *13* |
| 6-1 | 134 | 194 | 1.44 | 0.01 | 0.1802 | 0.0010 | 0.3898 | 0.0024 | 0.5162 | 0.0070 | 12.8256 | 0.1955 | 101 | 2655 | 9 |
| 7-1 | 84 | 79 | 0.94 | 0.61 | 0.1801 | 0.0017 | 0.2539 | 0.0035 | 0.4975 | 0.0073 | 12.3569 | 0.2251 | 98 | 2654 | 15 |
| *8-1** | *108* | *88* | *0.82* | *16.44* | *0.1793* | *0.0061* | *0.1927* | *0.0137* | *0.4701* | *0.0071* | *11.6231* | *0.4518* | *94* | *2647* | *56* |
| 9-1 | 54 | 53 | 0.99 | 0.03 | 0.1824 | 0.0018 | 0.2687 | 0.0037 | 0.5037 | 0.0080 | 12.6672 | 0.2506 | 98 | 2675 | 16 |
| 10-1 | 51 | 54 | 1.06 | 0.05 | 0.1815 | 0.0017 | 0.3003 | 0.0036 | 0.5329 | 0.0084 | 13.3355 | 0.2587 | 103 | 2667 | 16 |
| *11-1** | *10 226* | *4775* | *0.47* | *3.13* | *0.1288* | *0.0003* | *0.1611* | *0.0006* | *0.4496* | *0.0052* | *7.9874* | *0.0960* | *115* | *2082* | *4* |
| *12-1** | *80* | *48* | *0.59* | *2.51* | *0.1812* | *0.0029* | *0.1574* | *0.0063* | *0.4687* | *0.0070* | *11.7089* | *0.2735* | *93* | *2664* | *27* |
| *13-1** | *413* | *86* | *0.21* | *54.39* | *0.1454* | *0.0161* | *−0.0343* | *0.0366* | *0.0994* | *0.0020* | *1.9932* | *0.2299* | *27* | *2292* | *192* |
| 14-1 | 68 | 83 | 1.21 | 0.04 | 0.1817 | 0.0014 | 0.3301 | 0.0031 | 0.5091 | 0.0078 | 12.7537 | 0.2299 | 99 | 2668 | 13 |
| 15-1 | 49 | 13 | 0.27 | 0.08 | 0.1859 | 0.0018 | 0.0718 | 0.0028 | 0.5147 | 0.0087 | 13.1948 | 0.2715 | 99 | 2706 | 16 |
| *16-1** | *124* | *150* | *1.21* | *16.52* | *0.1751* | *0.0066* | *0.3795* | *0.0152* | *0.3501* | *0.0053* | *8.4505* | *0.3576* | *74* | *2607* | *63* |
| 17-1 | 82 | 87 | 1.07 | 0.03 | 0.1816 | 0.0013 | 0.2889 | 0.0027 | 0.5124 | 0.0076 | 12.8290 | 0.2215 | 100 | 2667 | 12 |
| *18-1** | *225* | *175* | *0.78* | *15.27* | *0.1744* | *0.0057* | *0.2851* | *0.0130* | *0.2706* | *0.0038* | *6.5058* | *0.2421* | *59* | *2600* | *54* |
| *19-1** | *74* | *28* | *0.38* | *2.99* | *0.1842* | *0.0036* | *0.1291* | *0.0076* | *0.3861* | *0.0060* | *9.8046* | *0.2576* | *78* | *2691* | *32* |
| *20-1** | *434* | *144* | *0.33* | *29.38* | *0.1726* | *0.0069* | *0.1329* | *0.0156* | *0.1684* | *0.0024* | *4.0092* | *0.1773* | *39* | *2583* | *67* |
| *21-1** | *179* | *144* | *0.80* | *2.63* | *0.1712* | *0.0028* | *0.2642* | *0.0062* | *0.2922* | *0.0041* | *6.8973* | *0.1560* | *64* | *2569* | *27* |
| 22-1 | 60 | 57 | 0.94 | 0.02 | 0.1811 | 0.0015 | 0.2552 | 0.0030 | 0.5061 | 0.0079 | 12.6386 | 0.2354 | 99 | 2663 | 14 |
| *23-1** | *159* | *78* | *0.49* | *13.21* | *0.1767* | *0.0051* | *0.1480* | *0.0113* | *0.3610* | *0.0052* | *8.7966* | *0.2958* | *76* | *2623* | *48* |
| 24-1 | 141 | 84 | 0.59 | 0.88 | 0.1816 | 0.0015 | 0.1678 | 0.0031 | 0.4830 | 0.0066 | 12.0910 | 0.2044 | 95 | 2667 | 14 |
| 15-2 | 105 | 58 | 0.55 | 0.26 | 0.1843 | 0.0013 | 0.1540 | 0.0024 | 0.4923 | 0.0069 | 12.5084 | 0.2081 | 96 | 2692 | 12 |
| *20-2** | *314* | *84* | *0.27* | *20.40* | *0.1792* | *0.0053* | *0.1010* | *0.0120* | *0.3118* | *0.0043* | *7.7062* | *0.2647* | *66* | *2646* | *50* |

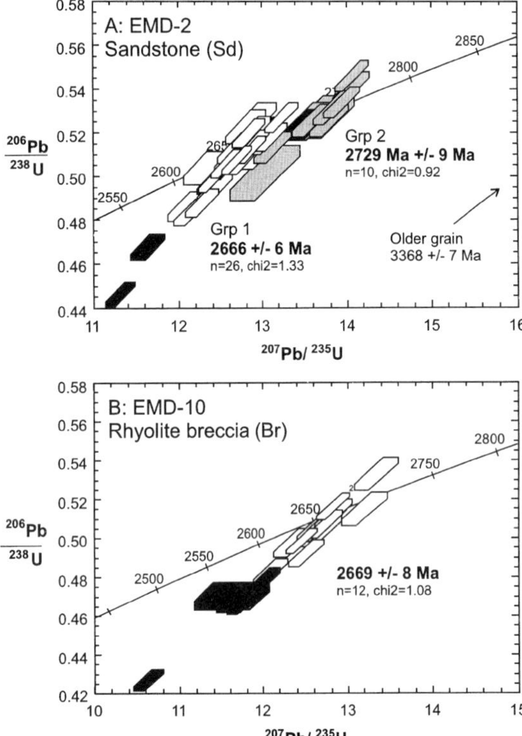


**Fig. 8.** (A) Concordia plot of SHRIMP U/Pb in zircon data from Sd (EMD-2, 85 m, AMG: 352 436 mE-6 608 409 mN). (B) Concordia plot showing SHRIMP U/Pb in zircon data from Br beds (EMD-10, 75 m, AMG: 352 503 mE-6 608 636 mN). Analyses plotted in black are those excluded from mean age calculations owing to excessive discordance, high U or high common Pb contents. Many highly discordant analyses occur outside of the plotted isotopic range. Error boxes represent 1σ.

grains are anhedral or subhedral, with broken or resorbed margins and extensively recrystallized interiors. Forty-five analyses were obtained from 43 grains, of which nine were excluded from mean age calculations owing to discordance, high U or high common Pb (Table 4). Data fall into two major populations, with a much older discordant (93%) outlier at 3368 ± 7 Ma (Fig. 8A). The largest population (26 analyses) has a mean age of 2666 ± 6 Ma ($\chi^2 = 1.33$). The possibility of more than one population being present, or of scatter having been produced by Pb-loss in the younger grains, is indicated by the high $\chi^2$ value. The second population (10 analyses) has a mean age of 2729 ± 9 Ma ($\chi^2 = 0.92$).

Zircons from a sample of Br from drillhole EMD-10 (75 m depth) are generally small (<100 μm) euhedral grains or anhedral broken crystal fragments. Many were fractured or metamict. Most grains contain less than 200 ppm U and Th, but several grains have higher U contents (up to 434 ppm), and these grains tended to yield lower ages. From this sample, 26 analyses from 24 grains were obtained (Table 4), of which 14 were excluded from the mean calculation owing to excessive discordance, high U or high common Pb. Concordant data forms a single population (Fig. 8B) with a mean age of 2669 ± 8 Ma ($n = 12, \chi^2 = 1.08$).

## THE BFG DEPOSITIONAL SYSTEMS AT EIGHT MILE DAM

This section interprets the depositional systems preserved at EMD. Three distinct periods of coarse-grained sedimentation (Fig. 4), separated by periods of tranquil water sedimentation, represent three distinct depositional systems. The resedimented nature of the dominantly volcaniclastic debris, and the common association with mudstone facies indicate that the sediments were deposited in an environment below wave base, dominantly through turbidity current processes. Turbidity currents operate in lakes, delta fronts, continental shelves, bases of slopes and deep ocean basin settings. However, thick sequences of turbidites will only be preserved in deep-water environments (i.e. large lakes or deep-water basins) where reworking of the sediment is rare (Walker, 1979).

The range of sedimentary facies preserved within the EMD succession is representative of proximal to distal settings relative to a feeder channel. Recognition of proximal to distal settings is based on grain size, sedimentary structures, facies transitions, the thickness of beds and the ratio of sand to mud. Proximal environments are characterized by channelized Cp and Sm deposits. In more distal environments, evidence of channels is absent, and laterally continuous Sd and Ml reflect unconfined deposits common in subaqueous lobe and basin plain settings.

The proximal Cp to distal Ml are interpreted to represent deposition in subenvironments of submarine fan depositional systems. Submarine fans are commonly characterized by channels in proximal settings and lobes in distal settings (Nelson *et al.*, 1991). Features of the EMD succession consistent with deposition on submarine fans include the abundance of turbidites, which are volumetrically most abundant in submarine fans within deep ocean basins (Walker, 1979), and the recognition of channels and lobe deposits, which are integral to defining a submarine fan system (Shanmugam & Moiola, 1988).

In this study stratigraphical intervals containing facies characteristic of lobes, channels and tranquil water elements of a submarine fan were identified and then put into context (i.e. proximal versus distal settings) using the 'classic' fan model, without inferring the geometry or size of the fan. The vertical stacking of the elements can then be used to interpret the evolution of the submarine fan system through time. In the BFG at EMD, a fan model is required to put the elements or facies assemblages into context, because no time-equivalent elements (see Mutti & Normark, 1987, 1991) can be recognized.

### Depositional system 1: progradational epiclastic submarine fan

This system is represented by Sd, Sm, Cp and rare Ml and is intersected at EMD from 480 to 185 m (Fig. 4). This interval is interpreted to represent the deposits of a single depositional system, as a gradual upward-coarsening and upward-thickening trend composed of the same debris is preserved. The coarsening upward succession of depositional system 1 is polymictic. The base of this depositional system consists of Sd and Sm associated with Ml (Fig. 4). In comparison to the morphological elements defined in the 'classic' submarine fan models, the presence of amalgamated, unchannelized Sd and Sm beds and rare Ml is common in mid-distal sections of submarine fan lobes. Thick intervals of Sd beds (Fig. 4, 480–430 m) are interpreted to represent deposits on a submarine fan lobe, defined as a lobe-shaped, low relief, depositional mound, which is generally located at the end of a submarine channel (Stow *et al.*, 1996). Overlying the unchannelized lobe deposits, from 434 to 372 m (Fig. 4), Sd beds are thicker and coarser-grained than the underlying Sd beds associated with Ml. The coarser grain and thicker bedding suggest that the environment was more proximal than that of the underlying Sd–Ml association. The Cp and Sd interval, from *c.*372 to 190 m (Fig. 4), becomes progressively coarser-grained and dominated by Cp upsection. The diffuse, non-erosional contacts between Cp and Sd, and the coarse grain size of this interval are indicative of deposition from sustained currents within a main channel. The thick, diffusely stratified pebble–cobble Cp beds and associated sandstone beds (Fig. 4, 192–302 m) are characteristic of deposits within a fixed channel, where long-term, high-volume currents commonly operate (Shanmugam *et al.*, 1988). To form and sustain leveed, fixed channels, the turbidity currents must be relatively steady, long term and of high volume (Shanmugam *et al.*, 1988), rather than short intermittent flows traditionally associated with turbidity current events.

The initial upward-coarsening trend of depositional system 1, characterized by vertically stacked submarine fan lobe and channel deposits, represents a progradational submarine fan system. Overlying the progradational submarine fan deposits, a thin retrogradational stage is represented by vertically stacked, thinner-bedded and finer-grained Sm and Cp, interbedded with Ml, fining into *c.*25 m of Ml (Fig. 4), which is likely to represent outer submarine fan or basin plain setting deposits. The abandonment of the polymictic submarine fan system may have occurred due to changes in sediment supply rate or transport direction, changing rates of basin floor subsidence and source region uplift, climate, eustasy or a combination of these factors.

From the SHRIMP U/Pb zircon dating of Sd from this system, the younger population (2666 ± 6 Ma) is interpreted to represent the youngest volcanic component in the sandstones and therefore represents the maximum depositional age of the progradational epiclastic submarine fan system. This interpretation is based on the euhedral magmatic grains being associated with the younger ages. This age is younger than the ages of 2681 ± 5 Ma (Nelson, 1997a) and 2676 ± 4 Ma (Claoué-Long, 1990), previously determined for the BFG. The older age (2729 ± 9 Ma) is represented by subhedral zircons interpreted to have been eroded from a significantly older source terrain. This large range in detrital zircon ages is consistent with the petrographical evidence, which indicates the presence of reworked and juvenile volcanic debris, implying that both old and young sources occurred in the source volcanic edifice. The provenance possibly reflects erosion of older terrains represented by rounded grains and clasts, as well as contemporaneous volcanic sediment influx represented by very angular crystal fragments. The rounded clasts indicate reworking in fluvial or shoreline environments before resedimentation by mass-flow processes into deeper water. The dominance of dense felsic porphyritic clasts in Cp suggests a felsic volcanic source terrane with a lava component, perhaps a lava-dome complex.

### Depositional system 2: dacitic submarine fan contemporaneous with explosive volcanism

Overlying the outer submarine fan or basin plain deposits of Ml, from *c.*162 m, is the second coarse-grained depositional system preserved in the EMD succession (Fig. 4). It is composed of a 50 m thick interval of amalgamated, medium- to coarse-grained Sm beds (Fig. 4), interpreted to represent deposits of single surge to sustained high-concentration turbidity currents within a proximal main channel of a submarine fan. The currents at times were intermittent, allowing finer-grained overbank deposits of Ml, derived from adjacent channels, to be deposited between Sm horizons.

The juvenile uncontaminated nature and the grain size of the Sm beds suggests derivation from contemporaneous explosive volcanism, and the sandstone composition directly reflects the composition of the source. The angular nature of the grains indicates little, if any, reworking. The beds are considerably thicker than most terrigenous turbidites, which are rarely more than a few metres thick. However, volcaniclastic megaturbidites are not uncommon in marine basins adjacent to subaerial to shallow marine volcanic centres experiencing large volume explosive eruptions (Cas & Wright, 1987). Subaerial pyroclastic flows can transform into water-supported megaturbidity currents, or form temporary deposits in near shore areas that are later resedimented into deeper water. Such processes may produce thick fans or sheets of syneruptive volcaniclastic megaturbidite deposits (Cas, 1979; Cas *et al.*, 1981; White & McPhie, 1996).

Deposition in the channels suddenly ceased (Fig. 4, *c.* 112 m) and the system was capped by deposits of Ml, signifying a return to tranquil water processes. The channel abandonment may be directly related to volcanic flux, if the depositional system developed synchronous with volcanism at the source.

### Depositional system 3: slope apron contemporaneous with rhyolite lava dome growth

The termination of the second period of tranquil water sedimentation is marked by an influx of rhyolitic breccia (Fig. 4, 108 m), representing the development of the third coarse-grained depositional system. Amalgamated horizons of Br, lacking erosional contacts, may have been deposited in proximal channels of a submarine fan system, or alternatively in a subaqueous slope apron system. Slope aprons develop in the area between the shelf and the basin floor (Stow *et al.*, 1996), and may be associated with active syndepositional faulting (e.g. Nelson & Maldonado, 1988; Inenson, 1989; Tanaka & Maejima, 1995), or intrabasinal volcanic centres (e.g. Cas *et al.*, 1981). Slope apron deposits commonly consist of sheet-like breccia horizons, and characteristically lack channels (Nelson *et al.*, 1991). The coarse-grained nature, lack of erosional contacts and monomictic composition of the Br are more consistent with deposits of unconfined density-modified grain flows operating on a slope, rather than channelized deposits of high-concentration turbidity currents. A slope setting closely associated with lava dome growth is favoured as the depositional site of the Br horizons.

Due to the evidence for subaqueous environments, combined with the slightly reworked nature of the clasts, the probable fragmentation mechanism was quenching associated with contemporaneous basin margin, or intrabasinal, rhyolitic volcanism. However, storage and subsequent resedimentation are possible, as the overall facies character suggests final deposition by density-modified grain flow or high-concentration turbidity current activity. This resulted in deposition at the foot of an apron, marginal to the dome complex, at the level of basin floor hemipelagic sedimentation. The 2669 ± 8 Ma age of Br is interpreted to reflect the crystallization age of the rhyolite magma, and based on the monomictic and juvenile nature of the breccia horizons, possibly closely approximates the depositional age.

## CONCLUSIONS

Depositional system 1 (Fig. 4) can be interpreted as a progradational epiclastic submarine fan system, defined by unchannelized fan lobe (facies Sd, Sm and Ml) and overlying sustained-current, main channel deposits (facies Cp, Sm and Sd). The lobes lack braided channels and are characteristic of depositional lobes which form at the mouths of sustained-current channels. The textural characteristics of the sandstones indicate a mixture of mature and juvenile volcanic detritus, and the zircon dates support this, with older (2729 ± 9 Ma) and younger (2666 ± 6 Ma) felsic sources having contributed to the sandstones. The older provenance of depositional system 1 (2729 ± 9 Ma) clearly indicates derivation of an older erosionally degraded felsic volcanic complex. The younger provenance age of depositional system 1 (2666 ± 6 Ma) indicates new volcanic activity, and is unrelated to the older centre in terms of magma chamber genesis, given the significant age difference. Post-eruptive erosion and mixing of older and newer volcanic debris, and deposition into a proximal volcaniclastic fan setting marginal to the edifice, are indicated. Degradation of the edifice to below sea-level, sea-level rise or tectonic subsidence appear to have briefly terminated volcaniclastic sedimentation, leading to a sustained period of mud deposition. The progradational submarine fan system has a maximum depositional age of 2666 ± 6 Ma.

Depositional system 2 is defined by voluminous deposits of Sm, composed of juvenile dacitic debris, likely to have been derived from contemporaneous explosive volcanism. These deposits are channelized, and are characteristic of proximal channel deposits of a submarine fan system. Depositional system 2 deposits may represent renewed volcanism from the 2666 ± 6 Ma dome complex or the development of an entirely separate edifice. Given the thick nature of the volcaniclastic turbidites, the relatively monomictic dacitic composition and the texturally juvenile nature of the debris, this appears to reflect major explosive eruptions of dacitic magma. Contemporaneous resedimentation produced a channelized volcaniclastic fan, superimposed on the older fan system. Another hiatus in eruptive activity is represented by another mudstone interval.

Depositional system 3 consists of resedimented, variably reworked juvenile rhyolitic debris, organized into sheet-like breccia beds, characteristic of deposition on a slope apron. This appears to represent growth of a new dome complex to above sea-level. Although much of the debris is texturally consistent with a quench fragmented origin, some of it appears to have been rounded by shoreline processes. Resedimentation of variably reworked rhyolitic debris into a slope apron at the base of the edifice seems the most likely setting. This slope apron was superimposed on and prograded over the pre-existing volcaniclastic fan. The age of rhyolitic volcanism and associated deposition on the slope setting is 2669 ± 8 Ma.

Each depositional system is separated by a period of tranquil water sedimentation, has distinct facies and facies relationships and is characterized by distinct provenance affinities. The two maximum depositional ages of the progradational epiclastic submarine fan and the rhyolitic slope apron systems are within error of each other, and constrain the depositional age of the EMD succession to an age range of 2677–2660 Ma, but it could have been narrower. The maximum depositional ages obtained by SHRIMP U/Pb zircon dating at EMD are significantly younger than the previous youngest (2676 ± 4 Ma; Cloué-Long, 1990) age obtained from BFG rocks.

This study illustrates that palaeoenvironmental reconstructions of poorly exposed Archaean depositional systems based entirely on drill core analysis can be achieved. Although the lateral constraints of the palaeoenvironment and the depositional systems are limited due to the spacing of the drill holes, this should not be seen as a major constraint to interpreting the systems. The drill holes intersect a section of each system, which can still be identified as being proximal or distal within a much more laterally extensive submarine depositional system. This study of limited lateral extent should be viewed as a building block, and as further information becomes available (i.e. continued exploration and drilling) it can be expanded to aid understanding of the true scale of the Archaean depositional systems. Only by utilizing pockets of information such as the drill holes at EMD can we start to understand the scale, architecture and distribution of these poorly exposed Archaean depositional systems.

## ACKNOWLEDGEMENTS

This research is from the 1998 PhD thesis of J.L.H. under the supervision of R.A.F.C. at Monash University. The research was supported by AMIRA grant P437, an ARC collaborative research grant to M.E.B. and R.A.F.C. We would like to thank Kalgoorlie Consolidated Gold Mines and Dr Paul Sauter for access to the Eight Mile Dam drill core. Critical comments by reviewers (Drs J. McPhie and D.R. Nelson) greatly improved the earlier version of the manuscript. This paper was written while J.L.H. was a recipient of a Monash University Postgraduate Writing Up Award.

## REFERENCES

Aalto, K.R. (1976) Sedimentology of a mélange: Franciscan of Trinidad, California. *J. sediment. Petrol.*, **46**, 913–929.

Archibald, N.J., Bettenay, L.F., Binns, R.A., Groves, D.I. & Gunthorpe, R.J. (1978) The evolution of Archaean greenstone terrains, Eastern Goldfields Province, Western Australia. *Precam. Res.*, **6**, 103–131.

Archibald, N.J., Bettenay, L.F., Bickle, M.J. & Groves, D.I. (1981) *Evolution of Archaean Crust in the Eastern Goldfields Province of the Yilgarn Block*. Spec. Publs Geol. Soc. Aust., **7**, 491–504.

Barley, M.E., Eisenlohr, B.N., Groves, D.I., Perring, C.S. & Vearncombe, J.R. (1989) Late Archaean convergent margin tectonics and gold mineralisation: evidence from the Norseman–Wiluna Belt. *Geology*, **17**, 826–829.

Barley, M.E. & Groves, D.I. (1990) Deciphering the tectonic evolution of Archaean greenstone belts: the importance of contrasting histories to the distribution of mineralization in the Yilgarn Craton, Western Australia. *Precam. Res.*, **46**, 3–20.

Barley, M.E., Krapez, B., Brown, S.J.A., Hand, J.L. & Cas, R.A.F. (1998b) *Mineralised Volcanic and Sedimentary*

*Successions in the Eastern Goldfields Province, Western Australia.* AMIRA Project P437 Final Report, 210 pp.

Barley, M.E., Krapez, B., Groves, D.I. & Kerrich, R. (1998a) The Late Archaean bonanza: metallogenic and environmental consequences of the interaction between mantle plumes, lithospheric tectonics and global cyclicity. *Precam. Res.*, **91**, 65–90.

Barley, M.E. & McNaughton, N.J. (1988) The tectonic evolution of greenstone belts and setting of Archaean gold mineralisation in Western Australia: geochronological constraints on conceptual models. In: *Advances in Understanding Precambrian Gold Deposits*, Vol. 2 (Eds Ho, S.E. & Groves, D.I.). University of Western Australia, Geology Dept and Uni. Extension, Publication **12**, 23–40.

Binns, R.A., Gunthorpe, R.J. & Groves, D.I. (1976) Metamorphic patterns and development of greenstone belts in the Eastern Yilgarn Block. In: *The Early History of the Earth* (Ed. Windley, B.F.), pp. 303–316. John Wiley and Sons, London.

Bouma, A.H. (1962) *Sedimentology of Some Flysch Deposits: a Graphic Approach to Facies Interpretation.* Elsevier, Amsterdam, 168 pp.

Brauns, K. (1991) *Paleovolcanological and environmental setting of the Archaean Black Flag Beds, Kambalda, Western Australia.* BSc honours thesis, Monash University, Clayton.

Cann, J.R. (1970) Rb, Sr, Y, Zr, Nb in some ocean floor basaltic rocks. *Earth planet. Sci. Lett.*, **10**, 7–11.

Cas, R.A.F. (1979) Mass-flow arenites from a Paleozoic interarc basin, New South Wales, Australia: Mode and environment of emplacement. *J. sediment. Petrol.*, **49**, 29–44.

Cas, R.A., Powell, C.McA., Fergusson, C.L., Jones, J.G., Roots, W.D. & Fergusson, J. (1981) The Lower Devonian Kowmung Volcaniclastics: a deep-water succession of mass-flow origin, northeastern Lachlan Fold Belt, NSW. *J. geol. Soc. Aust.*, **28**, 271–288.

Cas, R.A.F. & Wright, J.V. (1987) *Volcanic Successions: Modern and Ancient.* Chapman & Hall, London, 528 pp.

Cassidy, K.F., Barley, M.E., Groves, D.I., Perring, C.S. & Hallberg, J.A. (1991) An overview of the nature, distribution and inferred tectonic setting of granitoids in the Late-Archaean Norseman–Wiluna. *Precam. Res.*, **51**, 51–83.

Champion, D.C. & Sheraton, J.W. (1997) Geochemistry and Nd isotope systematics of Archaean granites of the eastern goldfields, Yilgarn Craton, Western Australia: implications for crustal growth processes. *Precam. Res.*, **83**, 109–132.

Claoué-Long, J.C. (1990) High resolution timing constraints on the evolution of the Kalgoorlie-Kambalda mineral belt. *Third International Archaean Symposium Perth 1990, Extended Abstracts Volume, Geoconferences (W.A.) Inc. Perth*, 355–356.

Claoué-Long, J.C., Compston, W. & Cowden, A. (1988) The age of the Kambalda greenstones resolved by ion-microprobe: implications for Archaean dating methods. *Earth planet. Sci. Lett.*, **89**, 239–259.

Field, D. & Elliot, R.B. (1974) The chemistry of gabbro/amphibolite transitions in south Norway. *Contrib. Mineral. Petrol.*, **47**, 63–76.

Folk, R.L. (1984) *Petrology of Sedimentary Rocks.* Hemphills, Austin, TX, 178 pp.

Foster, J.G., Lambert, D.D., Frick, L.R. & Maas, R. (1996) Re–Os isotopic evidence for genesis of Archaean nickel ores from uncontaminated komatiites. *Nature*, **382**, 703–706.

Gee, R.D., Baxter, J.L., Wilde, S.A. & Williams, I.R. (1981) *Crustal Development in the Archaean Yilgarn Block, Western Australia.* Spec. Publ. geol. Soc. Aust., **7**, 43–56.

Goleby, B.R., Rattenbury, M.S., Swager, C.P. *et al.* (1993) Archaean crustal structure from seismic reflection profiling, Eastern Goldfields, Western Australia. *Aust. geol. Surv. Org. Rec.*, **15**, 54 pp.

Groves, D.I. & Batt, W.D. (1984) Spatial and temporal variations of Archaean metallogenic associations in terms of evolution of granitoid–greenstone terrains with particular emphasis on the Western Australian Shield. In: *Archaean Geochemistry* (Ed. Kröner, A.), pp. 73–98. Springer-Verlag, Berlin.

Hallberg, J.A. (1985) *Geology and Mineral Deposits of the Leonora–Laverton Arc, Northeastern Yilgarn Block, Western Australia.* Hesperian Press, Perth, 140 pp.

Hallberg, J.A. (1986) Archaean basin development and crustal extension in the northeastern Yilgarn Block, Western Australia. *Precam. Res.*, **31**, 133–156.

Hammond, R.L. & Nisbet, B.W. (1993) Archaean crustal processes as outlined by structural geology, Eastern Goldfields Province of Western Australia. In: *Kalgoorlie '93—An International Conference on Crustal Evolution, Metallogeny, and Exploration of the Eastern Goldfields*, 39–46.

Hand, J.L. (1998) *The sedimentological and stratigraphic evolution of the Archaean Black Flag Beds, Kalgoorlie, Western Australia: implications for regional stratigraphy and basin setting of the Kalgoorlie Terrane.* PhD thesis, Monash University, Clayton.

Hill, R.I., Chappell, B.W. & Campbell, I.H. (1992) Late Archaean granites of the southeastern Yilgarn Block, Western Australia: age, geochemistry and origin. *Trans. R. Soc. Edinburgh, Earth Sci.*, **83**, 211–226.

Hiscott, R.N. (1994) Traction-carpet stratification in turbidites—fact or fiction? *J. sediment. Res.*, **A64**, 204–208.

Hiscott, R.N. & Middleton, G.V. (1979) Depositional mechanics of thick-bedded sandstones at the base of a submarine slope, Tourelle Formation (Lower Ordovician), Quebec, Canada. In: *Geology of Continental Slopes* (Eds Doyle, L.J. & Pilkey, O.H.). Spec. Publs Soc. econ. Paleont. Miner., Tulsa, **27**, 307–326.

Hiscott, R.N. & Middleton, G.V. (1980) Fabric of coarse deep-water sandstones, Tourelle Formation, Quebec, Canada. *J. sediment. Petrol.*, **50**, 703–722.

Hunter, W.M. (1993) Geology of the granite-greenstone terrane of the Kalgoorlie and Yilmia 1 : 100 000 sheets, Western Australia. *Geol. Surv. W. Aust. Rep.*, **35**, 80 pp.

Inenson, J.R. (1989) Coarse-grained submarine fan and slope apron deposits in a Cretaceous back-arc basin, Antarctica. *Sedimentology*, **36**, 793–819.

Keats, W. (1987) Regional geology of the Kalgoorlie-Boulder gold mining district. *Geol. Surv. W. Aust. Rep.*, **21**, 44 pp.

Kent, A.J.R. & McDougall, I. (1995) $^{40}Ar$–$^{39}Ar$ and U–Pb age constraints on the timing of gold mineralization in the Kalgoorlie Gold Field, Western Australia. *Econ. Geol.*, **90**, 845–859.

Kneller, B.C. & Branney, M.J. (1995) Sustained high-density turbidity currents and the deposition of massive sands. *Sedimentology*, **42**, 607–616.

KRAPEZ, B. (1997) Sequence-stratigraphic concepts applied to the identification of basin filling rhythms in Precambrian successions. *Aust. J. Earth Sci.*, **43**, 355–380.

KRAUSE, F.F. & OLDERSHAW, A.E. (1979) Submarine carbonate breccia beds—a depositional model for two-layer, sediment gravity flows from the Sekwi Formation (Lower Cambrian), Mackenzie Mountains, Northwest Territories, Canada. *Can. J. Earth Sci.*, **16**, 189–199.

LASH, G.G. (1984) Density-modified grain-flow deposits from an early Paleozoic passive margin. *J. sediment. Petrol.*, **54**, 557–562.

LE MAITRE, R.W., BATEMAN, P., DUDEK, A. *et al.* (1989) *A Classification of Igneous Rocks and Glossary of Terms.* Blackwell Scientific Publications, Oxford, 193 pp.

LOWE, D.R. (1976) Grain flow and grain flow deposits. *J. sediment. Petrol.*, **46**, 188–199.

LOWE, D.R. (1982) Sediment Gravity Flows: II. Depositional models with special reference to the deposits of high-density turbidity currents. *J. sediment. Petrol.*, **52**, 279–297.

LOWE, D.R. (1988) Suspended-load fallout rate as an independent variable in the analysis of current structures. *Sedimentology*, **35**, 765–776.

MIDDLETON, G.V. (1967) Experiments on density and turbidity currents: III. Deposition of Sediment. *Can. J. Earth Sci.*, **4**, 475–505.

MIDDLETON, G.V. & HAMPTON, M.A. (1973) Sediment gravity flows: mechanics of flow and deposition. In: *Turbidites and Deep-water Sedimentation* (Eds Middleton, G.V. & Bouma, A.H.), pp. 1–38. Pacific Section, Society of Economic Paleontologists and Goelogists, Los Angeles.

MIDDLETON, G.V. & HAMPTON, M.A. (1976) Sub-aqueous sediment transport and deposition by sediment gravity flows. In: *Marine Sediment Transport and Environmental Management* (Eds Stanley, D.J. & Swift, D.J.), pp. 197–218. John Wiley & Sons, New York.

MORRIS, P.A. (1993) Physical volcanology and geochemistry of Archaean mafic and ultramafic volcanic rocks between Menzies and Norseman, Eastern Yilgarn Craton, Western Australia. *W. Aust. geol. Surv. Rep.*, **36**, 107 pp.

MORRIS, P.A. (1998) Archaean felsic volcanism in parts of the Eastern Goldfields region Western Australia. *W. Aust. geol. Surv. Rep.*, **55**, 80 pp.

MORRIS, P.A. & WITT, W.K. (1997) Geochemistry and tectonic setting of two contrasting Archaean felsic volcanic associations in the Eastern Goldfields, Western Australia. *Precam. Res.*, **83**, 83–107.

MULLINS, H.T. & VAN BUREN, H.M. (1979) Modern modified carbonate grain-flow deposit. *J. sediment. Petrol.*, **49**, 747–752.

MUTTI, E. & NORMARK, W.R. (1987) Comparing examples of modern and ancient turbidite systems: problems and concepts. In: *Marine Clastic Sedimentology* (Eds Leggett, J.K. & Zuffa, G.G.), pp. 1–38. Graham & Trotman, London.

MUTTI, E. & NORMARK, W.R. (1991) An integrated approach to the study of turbidite systems. In: *Seismic Facies and Sedimentary Processes of Submarine Fans and Turbidite Systems* (Eds Weimer, P. & Link, H.), pp. 75–106. Springer-Verlag, Berlin.

MYERS, J.S. (1995) The generation and assembly of an Archaean supercontinent: evidence from the Yilgarn craton, Western Australia. In: *Early Precambrian Processes* (Eds Coward, M.P. and Ries, A.C.), Spec. Publs geol. Soc. London, No. 95, pp. 143–154. Geol. Soc. London, Bath.

NELSON, C.H. & MALDONADO, A. (1988) Factors controlling depositional patterns of Ebro turbidite systems, Mediterranean Sea. *AAPG Bull.*, **72**, 698–716.

NELSON, C.H., MALDONADO, A., BARBER, J.H. JR & ALONSO, B. (1991) Modern sand-rich and mud-rich siliclastic aprons: alternative base-of-slope turbidite systems to submarine sans. In: *Seismic Facies and Sedimentary Processes of Submarine Fans and Turbidite Systems* (Eds Weimer, P. & Link, H.), pp. 171–190. Springer-Verlag, Berlin.

NELSON, D.R. (1995) Compilation of SHRIMP U–Pb zircon geochronology data, 1994. *W. Aust. geol. Surv. Rec.*, **3**, 244 pp.

NELSON, D.R. (1997a) Evolution of the Archaean granite-greenstone terranes of the Eastern Goldfields, Western Australia: SHRIMP U–Pb zircon constraints. *Precam. Res.*, **83**, 57–81.

NELSON, D.R. (1997b) Compilation of SHRIMP U–Pb zircon geochronology data, 1996. *W. Aust. geol. Surv. Rec.*, **2**.

ONG, L. (1994) *A study of the Archaean Black Flag Beds in the Kalgoorlie area, Western Australia.* BSc honours thesis, Monash University, Clayton.

PERRING, C.S., BARLEY, M.E., CASSIDY, K.F. *et al.* (1990) The association of linear orogenic belts, mantle-crustal magmatism, and Archaean gold mineralisation in the Eastern Yilgarn Block of Western Australia. In: *The Geology of Gold Deposits: the Perspective in 1988* (Eds Keays, R., Ramsay, W.R.H. & Groves, G.D.I.). Economic Geology Monograph, No. 6, pp. 571–584.

SHANMUGAM, G. & MOIOLA, R.J. (1988) Submarine fans: characteristics, models, classification and reservoir potential. *Earth Sci. Rev.*, **24**, 383–428.

SHANMUGAM, G., MOIOLA, R.J., MCPHERSON, J.G. & O'CONNELL, S. (1988) Comparison of turbidite facies associations in modern, passive-margin Mississippi fan with ancient active margin fans. *Sediment. Geol.*, **58**, 63–77.

SMITH, J.B., BARLEY, M.E., GROVES, D.I. *et al.* (1998) The Sholl Shear Zone, West Pilbara: evidence for a domain boundary structure from integrated tectonostratigraphic analyses, SHRIMP U/Pb dating and isotopic and geochemical data of granitoids. *Precam. Res.*, **88**, 143–171.

SMITHIES, R.H. & WITT, W.K. (1997) Distinct basement terranes identified from granite geochemistry in Late Archaean granite–greenstones, Yilgarn Craton, Western Australia. *Precam. Res.*, **83**, 185–201.

SOHN, Y.K. (1997) On traction-carpet sedimentation. *J. sediment. Res.*, **67**, 502–509.

STOW, D.A.V., READING, H.G. & COLLINSON, J.D. (1996) Deep seas. In: *Sedimentary Environments: Processes, Facies and Stratigraphy*, 3rd edn (Ed. Reading, H.G.), pp. 395–453. Blackwell Science, Oxford.

SWAGER, C.P. (1989) Structure of the Kalgoorlie greenstones —regional deformation history and implications for the structural setting of gold deposits within the Golden Mile. *W. Aust. geol. Surv. Rep.*, **25**, 59–84.

SWAGER, C.P. (1997) Tectono-stratigraphy of Late Archaean greenstone terranes in the southern Eastern Goldfields, Western Australia. *Precam. Res.*, **83**, 11–42.

SWAGER, C.P., GOLEBY, B.R., DRUMMOND, B.J., RATTENBURY, M.S. & WILLIAMS, P.R. (1997) Crustal structure of granite–greenstone terranes in the Eastern Goldfields, Yilgarn Craton, as revealed by seismic reflection profiling. *Precam. Res.*, **83**, 43–56.

Swager, C., Griffin, T.J., Witt, W.K. *et al.* (1990) Geology of the Archaean Kalgoorlie Terrane—an explanatory note. *W. Aust. Geol. Surv. Rec.*, **12**, 55 pp.

Swager, C., Griffin, T.J., Witt, W.K. *et al.* (1995) Geology of the Archaean Kalgoorlie Terrane—an explanatory note. *Geol. Surv. West. Aust. Rep.*, **48**, 26 pp.

Tanaka, J. & Maejima, W. (1995) Fan-delta sedimentation on the basin margin slope of the Cretaceous, strike-slip Izumi Basin, southwestern Japan. *Sediment. Geol.*, **98**, 205–213.

Walker, R.G. (1975) Generalized facies models for resedimented conglomerates of turbidite association. *Geol. Soc. Am. Bull.*, **86**, 737–748.

Walker, R.G. (1977) Deposition of upper Mesozoic resedimented conglomerates and associated turbidites in southwestern Oregon. *Geol. Soc. Am. Bull.*, **88**, 273–285.

Walker, R.G. (1979) Turbidites and associated coarse clastic deposits. In: *Facies Models* (Ed. Walker, R.G.). Geosci. Can. Reprint Ser. **1**, 91–103.

White, M. & McPhie, J. (1996) Stratigraphy and palaeovolcanology of the Cambrian Tyndall Group, Mt Read Volcanics, western Tasmania. *Aust. J. Earth Sci.*, **43**, 147–159.

Winchester, J.A. & Floyd, P.A. (1977) Geochemical discrimination of different magma series and their differentiation products using immobile elements. *Chem. Geol.*, **20**, 325–343.

Witt, W.K. (1995) Tholeiite and high-Mg mafic/ultramafic sills in the Eastern Goldfields Province, Western Australia: implications for tectonic settings. *Aust. J. Earth Sci.*, **42**, 407–422.

*Spec. Publs int. Ass. Sediment.* (2002) **33**, 259–273

# Sedimentary environment of the amphibolite-grade Early Proterozoic Keiva and Kukas basins (north-east Baltic Shield): normative mineral MINLITH analysis

O. M. ROSEN*, V. T. SAFRONOV* *and* A. A. ABBYASOV†

**Geological Institute, Russian Academy of Sciences, Pyzhevsky per., 7, Moscow 109017, Russia; and*
*†Institute of the Lithosphere of the Marginal Seas, Russian Academy of Sciences, Staromonetny per., 22, Moscow 109180, Russia*

## ABSTRACT

The sedimentary environments of two Early Proterozoic basins within the north-east Baltic Shield have been studied using the MINLITH program for normative mineral calculations. The Keiva Basin forms a recumbent syncline on the Archaean footwall of the Kolmozero–Voronya Suture. This amphibolite-grade sequence contains a high-Al–kyanite schist member, covered by meta-carbonates. The calculated primary mineral composition of the schist member shows that it was originally composed mainly of kaolinite and quartz fragments with an iron oxide admixture, whereas dolomite and feldspar fragments predominated in the meta-carbonate member. The shallow open shelf clastic sediments were derived from humid weathering of continental crust and were followed by sedimentation of dolomite with some feldspar clasts in a closed basin, under arid conditions. The Kukas Basin now forms a recumbent syncline within a suture between the Archaean Karelian and Belomorian palaeocontinents. It is composed of amphibolite-grade biotite schists and meta-carbonates, both enriched with carbonaceous (graphite) matter. These meta-sediments are associated with meta-basalt and para-amphibolite rocks. Calculated primary mineral composition of the schist hints towards a tuffaceous sediment composed mainly of quartz, feldspar and chlorite. The meta-carbonates represent former dolomitic limestones. Both protoliths were probably formed in a partly closed, euxinic shelf basin, situated close to volcanic centres.

## INTRODUCTION

The Precambrian represents more than 85% of the Earth's evolution. Many Precambrian sedimentary basins underwent various grades of metamorphism and thus the establishment of depositional settings is a fundamental problem. This paper introduces new data on two Early Precambrian amphibolite-grade sedimentary basins on the north-east Baltic Shield. The data are derived from chemical and microscopic analyses and field observations, and aim at facilitating environmental interpretation of pre-metamorphic depositional conditions. The interpretation is based on calculation of mineral compositions in the meta-sedimentary rocks from bulk chemical analyses using the MINLITH program. In the Keiva basin, the depositional setting appears to have involved humid weathering products and mainly kaolinitic sedimentation on a shallow shelf of a passive palaeocontinental margin. The Kukas Basin meta-sediments represent greywacke sedimentation under euxinic shelf conditions of an active continental margin.

### Comments on the MINLITH method

Interactive computer programs to calculate mineral composition from chemical and X-ray diffraction data of all varieties of sedimentary rocks were proposed in several publications (Cohen & Ward, 1991; Currie, 1991; de Caritat *et al.*, 1994; Pactung, 1998). Several high-precision methods for calculating mineral composition from chemical composition of rocks were proposed for pelites (Pearson, 1978; Meridio *et al.*, 1992; Kolka *et al.*, 1994; Laird & Dowdy, 1994), but these required X-ray diffraction data for estimation of mineral contents. All these standardless calculation programs are based on X-ray data of minerals present in the rock (Starks *et al.*, 1984; Braun, 1986; Zangalis,

1998). It is thus impossible to use these programs for Precambrian metamorphic rocks.

The MINLITH program was proposed to calculate mineral compositions of sedimentary rocks from bulk chemical analyses in a Microsoft *Excel* for Windows user environment (Rosen *et al.*, 2000). The program calculates mineral weight per cent (wt%) using molecular (weight per cent divided by the molar weight, molar ratio) amounts of oxides and offers a sequential solution for the following separate tasks:

**1** Contents of accessory minerals and minerals of secondary importance are evaluated according to their stoichiometric compositions: apatite (on $P_2O_5$), named below as Ap; pyrite (on $S_{py}$·), Py; $TiO_2$ (rutile), Rt; and $C_{org}$, carbon and others.

**2** Amounts of aluminosilicate minerals according to the accepted formulas: plagioclase (Pl), $Na_2O{\cdot}Al_2O_3{\cdot}6SiO_2$ + $0.2CaO{\cdot}Al_2O_3{\cdot}2SiO_2$; orthoclase (Or), $K_2O{\cdot}Al_2O_3{\cdot}6SiO_2$; illite (Ill), $2K_2O{\cdot}MgO{\cdot}FeO{\cdot}6.5Al_2O_3{\cdot}16SiO_2{\cdot}5H_2O$; montmorillonite (Mm), $Na_2O{\cdot}2MgO{\cdot}5Al_2O_3{\cdot}24SiO_2{\cdot}6H_2O$; chlorite (Chl), $k1MgOk2FeO{\cdot}k3{\cdot}Al_2O_3k4{\cdot}SiO_2{\cdot}4H_2O$; kaolinite (Kn), $Al_2O_3{\cdot}2SiO_2{\cdot}2H_2O$ and others. In plagioclase, the anorthite content is accepted as 20%, which corresponds to the observed content in the overwhelming majority of sedimentary rocks; compositions of the illite and montmorillonite are accepted from published formulas, especially for authigenic varieties of these phases (Rosen & Nistratov, 1984), and the coefficients in the chlorite formula are accepted as a function of the magnitudes of the relative iron content in the bulk composition of the rocks (Rosen, 1976): k1, k2, k3, k4 = f·(Fe/[Fe + Mg]).

**3** Contents of either silicates and oxides or carbonates are evaluated according to lack or excess of calcium. Silicates and the oxides include quartz (Q), goetithe (Ht), pyrolusite (Prl) and serpentine (Srp). Carbonates are calcite (Cc), dolomite (Dl), ankerite (Ank), siderite (Sd), magnesite (Mst) and rhodochrosite (Rdh). The $H_2O$ (in any form) and $CO_2$ contents, when present at analysis, are not used in calculations and this is the reason why the MINLITH program for metamorphic rocks, which mainly lose volatile components, is useful (Rosen *et al.*, 2000).

The authentic mineral composition of sedimentary rocks is more complicated and diverse. Normative illite corresponds to the real composition of authigenic illite, and also hydro-mica and muscovite; the normative montmorillonite may include montmorillonite, smectite and a mixed-layered phase. The last, in turn, will go into particulate normative chlorite. The set, accepted in calculation, and composition of normative minerals correspond sufficiently well to mature deposits, elapsed long-term transport, recycling and deep diagenesis. These processes resulted in coloured minerals (pyroxene, olivine etc.) and clasts of volcanics, magmatic rocks etc. being completely replaced by minerals like illites, chlorites, feldspars and quartz, typical of mature sedimentary rocks.

The accuracy of the calculations determined as a relative error, compared to 115 published standard rock analyses, appears to be ±5–15 rel.% for most minerals (Rosen *et al.*, 2000). This value is of the same order of magnitude as provided by instrumental estimations of the high precision X-ray diffraction method (Hill *et al.*, 1993). In some cases, the error value amounts to 40 rel.% for normative feldspars, comparable to that for some instrumental phase estimations (Creelman & Ward, 1996; Ward & Taylor, 1996). When normative mineral contents are less than 5 wt%, the relative error is up to 60–70 rel.%. This decreased accuracy for minor normative components is comparable to the instrumental measurements shown in Hill *et al.* (1993). The data are used as semi-quantitative assessments.

## KEIVA BASIN

### Geological setting

The Keiva synform, located on the Archaean basement of the Keiva–Lebyazhya block, can be traced from west to east for a distance of about 200 km in the central part of the Kola Peninsula (Figs 1–3). The Keiva synform is

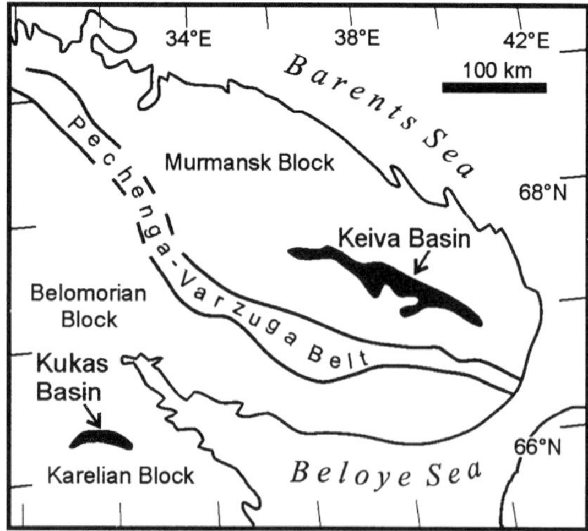


**Fig. 1.** Distribution of the Keiva and Kukas basins among the main tectonic elements of the north-eastern Baltic Shield.

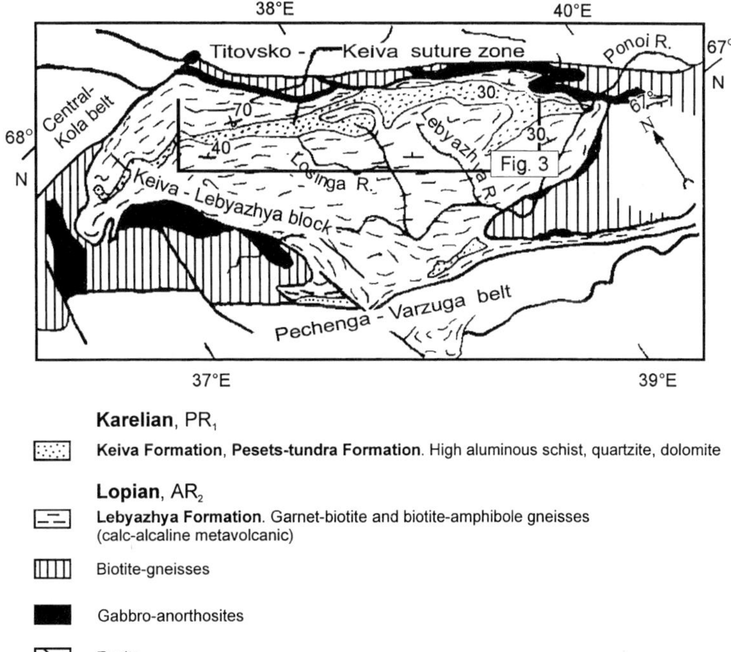


**Fig. 2.** Geological sketch map of the Keiva basin forming a cover complex on the Keiva–Lebyazhya Block. Inset corresponds to Fig. 3.

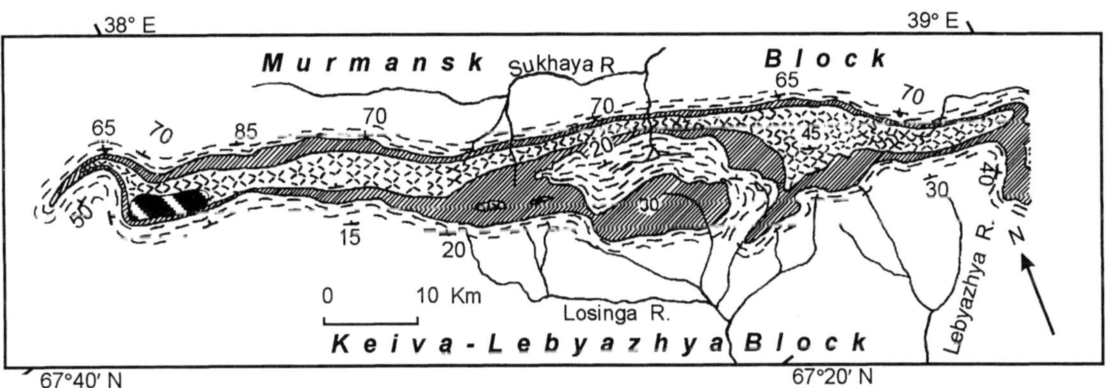


**Karelian**, $PR_1$

**Pesets-tundra Formation**. Unit F. Quartz-feldspar (sandy) dolomite covered by two mica schist with garnet and staurolite, muscovite-quartz schist (Unit F).

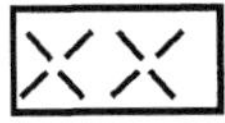

**Pesets-tundra Formation**. Unit D and **Keiva Formation**. Unit D. Porphyroblastic kyanite-staurolite schist, muscovite schist with kyanite and staurolite.

**Keiva Formation**. Unit C-muscovite quartzite; Unit B-kyanite-staurolite schist; Unit A-garnet-muscovite schist, mica schist.

**Lopian**, $AR_2$

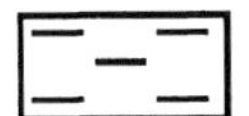

**Lebyazhya Formation**. Garnet-biotite- and biotite-amphibole gneiss.

**Fig. 3.** Distribution of the metamorphosed sedimentary rocks in the Keiva basin.

sandwiched between the Murmansk granite–migmatite block, with the Titovsko–Keiva Suture, in the north and the Belomorian gneiss belt, with the Pechenga–Varzuga suture, in the south. In the west, that structure abuts against the Central–Kola granulite-gneiss belt, and in the east, it disappears beneath the Beloye Sea.

The general strike of the Keiva synform is northwest 290–300° (Fig. 3). The Keiva synform is formed by a system of major conjugated folds trending parallel to the general structure. Folds on the northern flank are overturned southward (dip angles 60–80°). The intensely compressed isoclinal folds are developed over a large distance along the northern flank. The southern flank exhibits the stratigraphic base and overlies the crystalline basement at an angle of 15–45° (Fig. 3). There, the open brachysynclinal folds show a low-angle dip of 0–30°.

The Keiva Formation is subdivided into the following units (Belkov, 1963), from the base upwards. Unit A, 5–30 m thick, and composed of garnet–muscovite and staurolite–garnet schists, contains commonly carbonaceous matter ($C_{org}$ up to 1%). Unit B, 25–325 m thick, consists of kyanite, staurolite–kyanite schists ($C_{org}$ up to 2.5–3%) and kyanite (up to 40%). Unit C, 0–60 m thick, is represented by muscovite quartzites. Unit D, 0–250 m thick, is formed by plagioclase–staurolite and plagioclase–kyanite–staurolite porphyroblastic schists. The overlying rocks of the Pesets–Tundra Formation (Units E and F) differ from the schists of the Keiva Formation in petrographic and chemical composition. Unit E contains muscovite and biotite–muscovite schists, sometimes with garnet and staurolite and also lenses of conglomerates in the eastern margin (thickness up to 200–250 m). Unit F is mainly composed of meta-carbonates (40–100 m) locally preserving stromatolitic laminae.

The Keiva Formation is dated as Early Proterozoic (>2300 Ma; Sumian and Sariolian Horizons, Table 1).

**Table 1.** Stratigraphic position of the sequences studied.

| Eon | Era | Period | Subperiod | Horizon | Age boundaries (Ma) | Keiva basin: Formation | Keiva basin: Unit | Kukas basin: Formation | Kukas basin: Unit |
|---|---|---|---|---|---|---|---|---|---|
| Proterozoic-PR | Lower $PR_1$ | Karelian | Upper | Vepsian | 1650 | | | | |
| | | | | | 1800 | | | | |
| | | | | Kalevian | | | | | |
| | | | | | 1950 | | | | |
| | | | Middle | Ludikovian | | | | Khirvinavolok | V |
| | | | | | | | | | IV |
| | | | | | | | | | III |
| | | | | | | | | | II |
| | | | | | | | | | I |
| | | | | | 2100 | | | | |
| | | | | Yatulian | | Pesets-tundra | F | Kukas | Upper |
| | | | | | | | E | | Lower |
| | | | | | 2300 | | | | |
| | | | Lower | Sariolian | | Keiva | D | Amphibolitic | Upper |
| | | | | | | | C | | |
| | | | | | 2450 | | | | |
| | | | | Sumian | | | B | | Lower |
| | | | | | | | A | | |
| | | | | | 2600 | | | | |
| Archaean $AR_2$ | | Lopian | Upper | | | Lebyazhya | Losinga | Irinogor | |
| | | | | | | | Patcherv-tundra | | |
| | | | | | | | Kolovay | | |
| | | | | | 2800 | | | | |
| | | | Lower | | | | | | |
| | | | | | 3200 | | | | |
| Archaean $AR_1$ | | Saamian | | | | | | Belomorian Group | |

The Keiva–Lebyazhya block sequence (Figs 2 & 3) starts with amphibole–biotite gneiss, biotite gneiss and mica–quartz schists of the Kolovay Unit (~1000 m) (Table 1). This unit is conformably overlain by amphibolites, biotite–amphibole plagioclase schists and two-mica gneisses of the Patcherv–Tundra Unit (1000–1100 m). The hanging Losinga Unit rocks (1400 m) contain garnet–biotite gneisses, representing metamorphosed felsic lavas (Mints *et al.*, 1996), and are dated at 2.6–2.8 Ga (Pushkarev, 1990), corresponding to the Late Archaean (Upper Lopian Horizon; Negrutsa, 1999). The Keiva–Lebyazhya block is framed by variable (in thickness and form) Late Archaean (2.61 Ma) gabbro-anorthosites (Mitrofanov *et al.*, 1993).

The Keiva Formation unconformably overlies the Lebyazhya gneisses which have a weathered crust at the contact (Golovenok, 1977; Negrutsa, 1984). The age of Lebyazhya gneisses near the contact with the Keiva Formation is estimated to be 2780 ± 100 Ma (Pb–Pb isochron whole-rock dating; Pushkarev, 1990), which is confirmed by Pb–Pb data from clastic zircons of the Keiva meta-sediments (Bridgwater *et al.*, 1999). This age is interpreted to correspond to the age of source rocks for the Keiva Formation. An accepted age of these meta-sediments lies within the interval of 2600–2100 Ma (Negrutsa, 1999; Table 1). Recent data appear to be younger. Zircons from the Keiva kyanite schists dated at 1790 ± 50 Ma (Pb–Pb method), and a U–Pb age of 2100 ± 20 Ma (zircon upper discordia intercept; Mints *et al.*, 1996), are probably related to metamorphic overprinting.

Rocks of the Keiva Formation were metamorphosed to amphibolite grade at $T = 510–600°C$ and $P = 4.1–5.4$ kbar (Petrov *et al.*, 1986). On a regional stratigraphic scale, the Keiva Formation corresponds to the base of the Lower Proterozoic—the Sumian and Sariolian Horizons of 2300–2600 Ma—and the Pesets–Tundra Formation to the Yatulian Horizon of 2100–2300 Ma (Table 1).

## Depositional model

A pre-metamorphic mineral composition calculated using chemical analysis data on the Keiva meta-sediments with the MINLITH program is shown in Table 2. The primary composition of Unit A schists corresponds to quartz sandstones with a noticeable admixture of illite (garnet–muscovite schists) and quartz-bearing pelite (staurolite–garnet schists), whose pelitic component is represented by a mixture of kaolinite, montmorillonite and high-Fe chlorite. The uppermost part of the essentially kaolinitic zone of weathering crust (cuirass) served as a source for these rocks. High-alumina schists of Unit B were initially quartz-bearing kaolinite pelites (Table 2 & Fig. 4), locally with small fractions of montmorillonite and illite. Muscovite quartzites of Unit C represent metamorphosed quartz sandstone with an illite matrix. Unit D sedimentary protoliths contain significant amounts of feldspars (mainly Na-plagioclase) and montmorillonite, apparently resulting from the input of a less weathered substance. Unit E was probably originally formed by pelite rich siltstones or sandstones, now metamorphosed to muscovite schists, and by pelites composed of kaolinite, montmorillonite and illite, now metamorphosed to the two-mica schists with garnet and staurolite (Table 2 & Fig. 4). Sedimentation in the Keiva basin terminated with deposition of carbonate sediments (Unit F), mostly dolomite, with a siliciclastic admixture of quartz and feldspars.

It is assumed that the Keiva basin was slowly filled with material derived from the well developed weathering crust of a flat, deeply denudated palaeocontinent. The terrigenous material was well differentiated (kaolinite pelite–quartz sand) and the palaeoclimate of that epoch is assumed to have been hot and humid. A decrease in quartz and kaolinite content in the upper part of the formation, an increase in montmorillonite and feldspar and the appearance of carbonates containing stromatolites (Unit F) suggest that less weathered rocks were involved in the sedimentation process and indicate a change from a humid to an arid palaeoclimate. At that time, the open shelf environment may have altered to a closed basin (lagoon?).

Based on the observed longitudinal profile of the Keiva metamorphosed sequence (Fig. 5A), the palaeobasin profile was reconstructed based on primary mineral composition calculated from chemical analyses of metamorphic rocks discussed above (Fig. 5B). The sedimentary rhythms can be clearly seen: the base is represented by sands and the top by pelites (fining upward). The first rhythm corresponds to Units A and B, the second to Units C and D and the third to Unit E, which is locally overlain by the carbonates of Unit F.

The sedimentary basin was stretched in latitudinal direction (in present coordinates) and occupied a larger area when compared with the distribution of present outcrops. This is evident from the distribution of similar rocks southward and north-westward of the area characterized by metamorphic rocks of the Keiva Formation. This basin had a trough in its central part (Figs 2 & 5), where the thick succession of essentially

**Table 2.** Bulk chemical composition and the normative pre-metamorphic mineral content of the Keiva basin metasediments (in wt %).

| | Stratigraphic unit and metamorphic rock name | | | | | |
|---|---|---|---|---|---|---|
| | Unit A Garnet–muscovite schist ($n = 5$) | Unit B Kyanite–staurolite schist ($n = 7$) | Unit C Muscovite–quartzite ($n = 6$) | Unit D Kyanite–staurolite schist ($n = 10$) | Unit E Two-mica schist ($n = 2$) | Unit F Sandy dolomite ($n = 2$) |
| *Chemical composition* | | | | | | |
| $SiO_2$ | 77.60 | 62.84 | 93.93 | 62.97 | 67.51 | 25.37 |
| $TiO_2$ | 0.38 | 0.72 | 0.07 | 1.45 | 1.08 | 0.06 |
| $Al_2O_3$ | 10.47 | 33.53 | 4.09 | 24.56 | 18.69 | 1.55 |
| $Fe_2O_3$ | 2.12 | 0.87 | 0.41 | 2.46 | 3.89 | 0.87 |
| FeO | 4.63 | 0.56 | 0.71 | 1.39 | 1.73 | 0.35 |
| MnO | 0.05 | 0.01 | 0.03 | 0.02 | 0.07 | 0.04 |
| MgO | 0.63 | 0.12 | 0.21 | 0.81 | 1.1 | 18.62 |
| CaO | 0.88 | 0.17 | 0.38 | 0.88 | 1.01 | 26.03 |
| $Na_2O$ | 0.32 | 0.9 | 0.23 | 1.84 | 1.74 | 0.69 |
| $K_2O$ | 2.06 | 0.34 | 0.85 | 1.65 | 1.79 | 0.67 |
| Total | 99.14 | 99.25 | 100.91 | 98.03 | 98.61 | 96.72 |
| *Normative components* | | | | | | |
| Q | 58.77 | 20.05 | 85.39 | 20.81 | 32.20 | 16.52 |
| Pl | 1.58 | 0.45 | 1.91 | 13.24 | 11.31 | 4.78 |
| Or | – | – | – | – | – | 3.24 |
| Total clastics | 60.36 | 20.50 | 87.30 | 34.05 | 43.51 | 24.54 |
| Kn | 4.13 | 72.74 | 1.27 | 31.11 | 14.22 | – |
| Ill | 21.36 | 3.27 | 8.90 | 16.77 | 14.47 | – |
| Mm | 5.01 | 1.16 | 0.75 | 11.86 | 19.52 | – |
| Total pelites | 30.50 | 77.17 | 10.92 | 59.74 | 48.21 | – |
| Ht | 5.59 | 1.12 | 0.32 | 2.58 | 4.05 | 0.98 |
| Ank | 3.17 | 0.56 | 1.29 | 2.23 | 3.09 | 0.38 |
| Dl | – | – | – | – | – | 74.04 |
| Total carbonates | 3.17 | 0.56 | 1.29 | 2.23 | 3.09 | 74.42 |
| Rt, Ap | 0.37 | 0.65 | 0.11 | 1.40 | 1.13 | 0.06 |
| Pre-metamorphic sediments assumed | Pelitic quartz sandstone | Quartz–kaolinite pelite | Kaolinite–quartz sandstone | Sandy pelite | Sandy pelite | Sandy dolomite |

*n*, number of samples analysed.

quartz–kaolinite composition accumulated. The eastern part of the basin was characterized by some peculiar features, including the presence of plagioclase schists in Unit B (originally sandy pelite with a significant fraction of feldspars, illite and montmorillonite), the presence of beds with pebbles in Unit E and the lack of Units C and D in the sequence. Characteristics of this basin also include the uniform composition of rocks represented by a combination of kaolinite pelite with quartz sand in the sequence that is traceable over a large distance. These data are suggestive of a hydrodynamically quiet shallow shelf.

Most of the Keiva Formation is enriched in organic matter, highest contents being most characteristic of schists of Unit F. This suggests that the palaeobasin was probably densely inhabited by cyanophytes, an assumption supported by the presence of stromatolite remnants (organo-sedimentary traces of the protozoa activity) in dolomites of Unit F.

## KUKAS BASIN

### Geological setting

The Kukasozero synform is the most representative area of the North Karelian structural zone of the Lower Proterozoic sedimentary cover (Fig. 1). It is located in the suture zone between the Belomorian and Karelian blocks (Gorbatschev & Bogdanova, 1993).

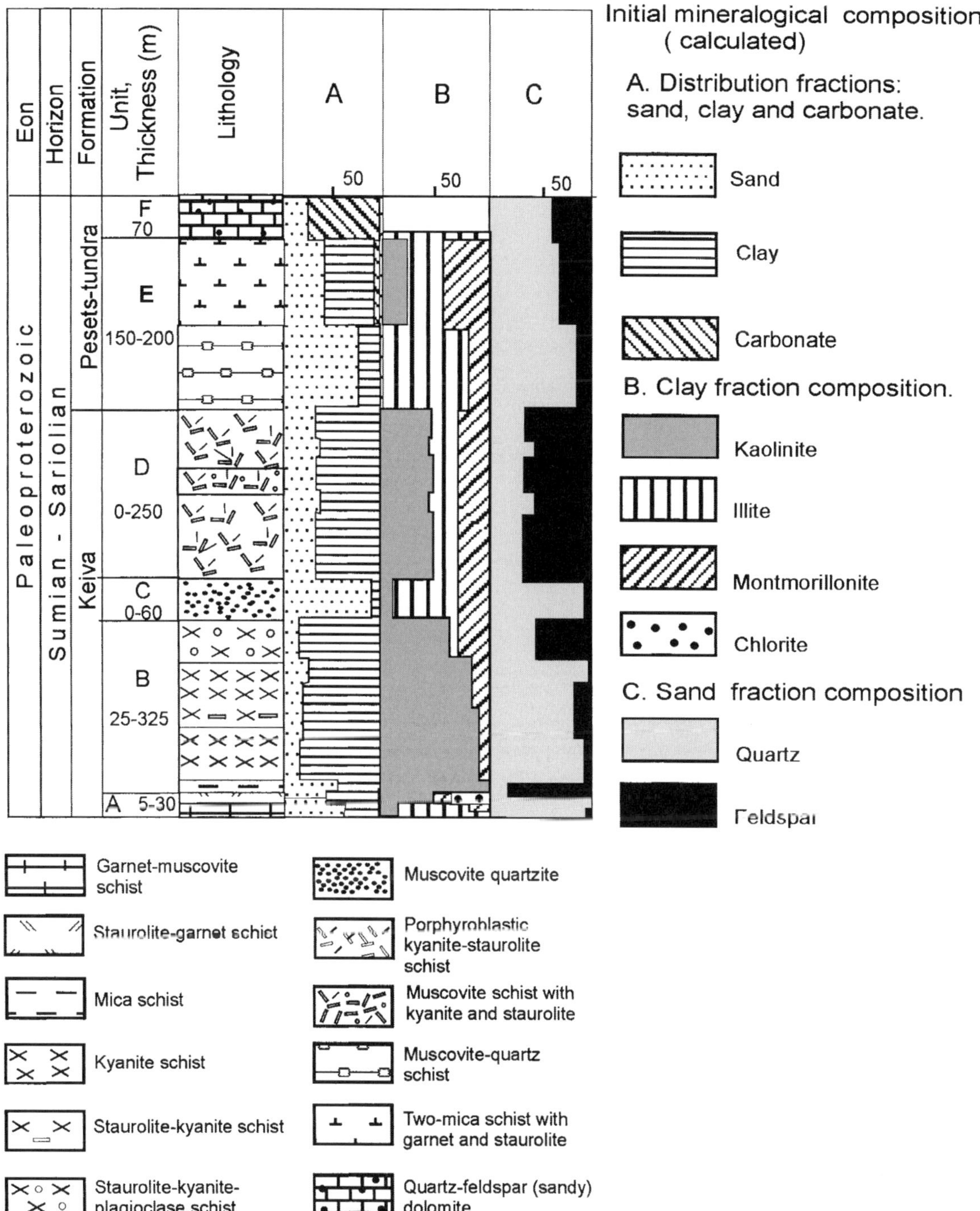


**Fig. 4.** Metamorphic rocks and their calculated primary mineral compositions of sediments (protoliths) in the stratigraphic column of the Keiva basin.

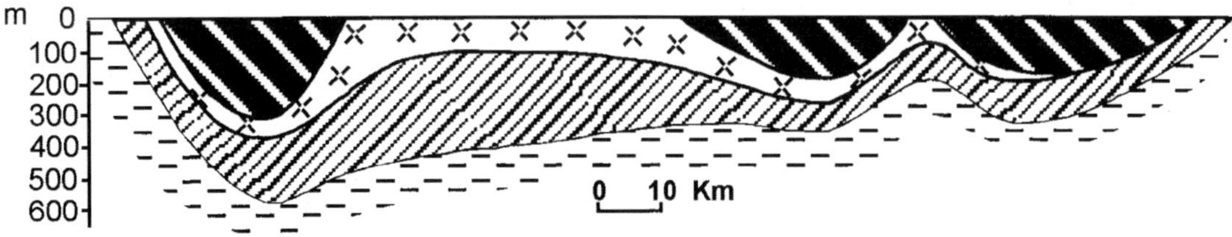


**Karelian**, $PR_1$

**Pesets-tundra Formation**. Unit F. Quartz-feldspar (sandy) dolomite covered by two mica schist with garnet and staurolite, muscovite-quartz schist (Unit F).

**Pesets-tundra Formation**. Unit D and **Keiva Formation**. Unit D. Porphyroblastic kyanite-staurolite schist, muscovite schist with kyanite and staurolite.

**Keiva Formation**. Unit C-muscovite quartzite; Unit B-kyanite-staurolite schist; Unit A-garnet-muscovite schist, mica schist.

**Lopian**, $AR_2$

**Lebyazhya Formation**. Garnet-biotite- and biotite-amphibole gneiss.

B

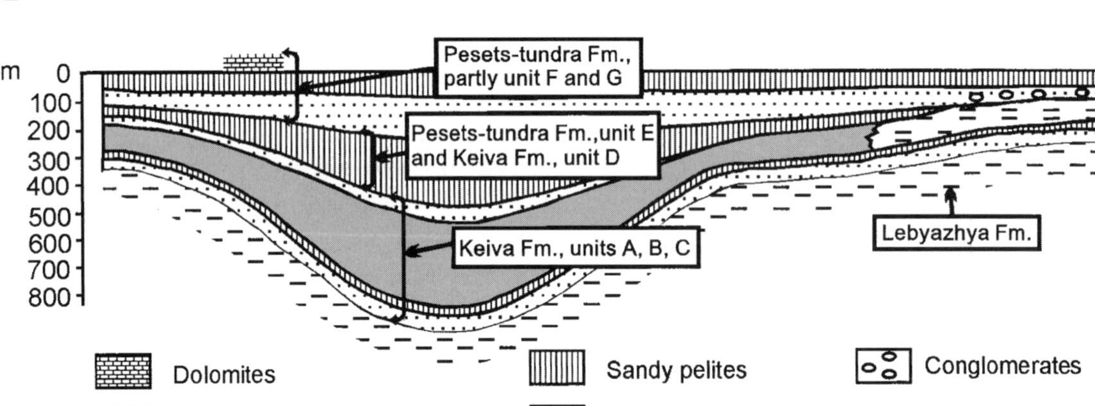


**Fig. 5.** Observed metamorphic (A) and calculated, pre-metamorphic reconstruction (B) cross-sections of lateral facies distribution in the Keiva Basin.

The Kukasozero synform represents a narrow trough, up to 4–6 km wide and 70 km long. Metamorphosed volcanogenic mafic rocks are most abundant in this structure, and meta-sediments are widespread in its axial part (the Khirvinavolok Formation). Metamorphic rocks are intensely dismembered by tectonism. Most common are isoclinal folds with limb dips of 35–80°; low-angle dips were occasionally observed. Folds, 0.5–1.0 km wide and less, are well expressed. Their cores are composed of the Kukas and Khirvinavolok formations with intense small-scale folds. The folds are mainly overturned to the south, with axial planes dipping steeply northward. In the axial part of the Kukasozero synform, the axes of small folds and linear structural elements are characterized by sustained westward dip at an angle of 20–30°. On the whole, the body

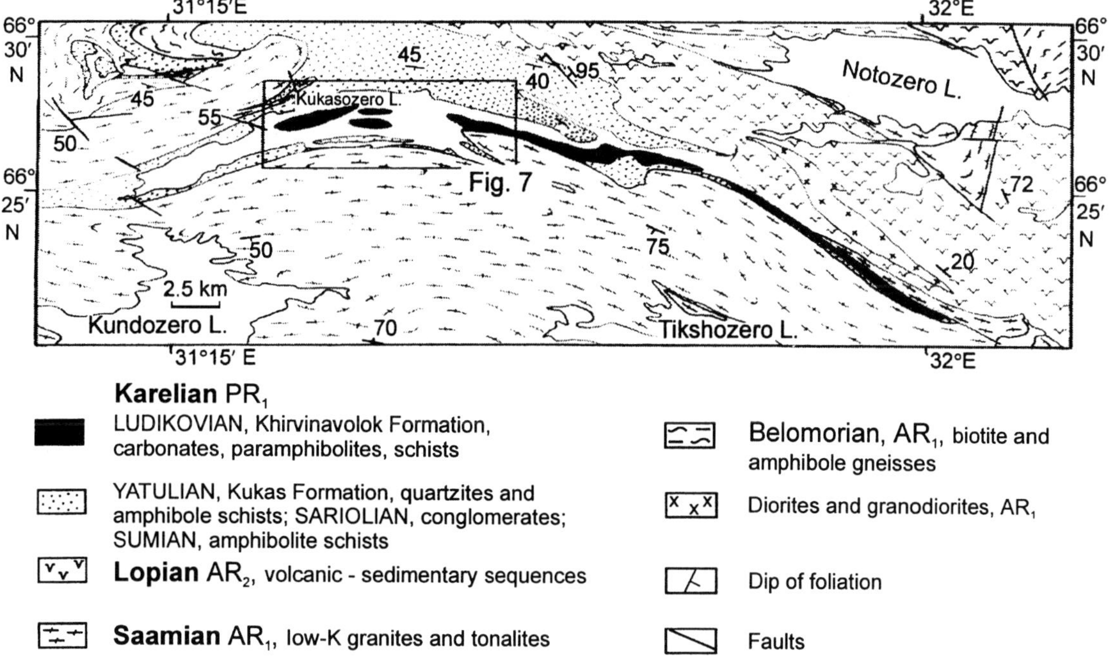


**Fig. 6.** Distribution of the Kukas basin metasediments among the Archaean complexes ($AR_1$ and $AR_2$).

of the synform is inclined to the west. Abundant small faults distort and complicate the synform structure.

In the Kukasozero synform fringing area, Archaean migmatite plagioclase gneisses and granite–gneisses are widespread (Fig. 6). Biotite and amphibole gneisses and granite gneisses (Belomorian Group) are predominant along the northern margin, forming a part of the Belomorian block. Biotite gneiss granites (tonalite and plagiogranite) developed along the southern margin are referred to as the Karelian block (Systra, 1991). Metamorphism of both blocks was traditionally considered as being most ancient, corresponding to the Lower Archaean (Saamian) highly metamorphosed sedimentary–volcanic rocks (Table 2). Recently, additional data on the Belomorian Group appeared to suggest a younger, Late Archaean age of 2850–3000 Ma (Bibikova *et al.*, 1996, 1999). In the north-eastern part of the Kukasozero synform, the Upper Archaean (Lopian) sedimentary–volcanic deposits of the Irinogor Formation are widespread. They are now represented by biotite–amphibole–plagioclase gneisses, garnet, feldspar amphibolites and kyanite–garnet–biotite gneisses (Table 1 & Fig. 6) associated with small lenses of plagio-granites, granodiorites and diorites ranging in age from 2830 to 2900 Ma (Systra, 1991).

The Archaean complexes are overlain by Lower Proterozoic volcano-sedimentary rocks (the Karelian complex) that occupy the central part of the Kukasozero synform. According to the regional stratigraphic scale (Table 2), the Karelian period corresponds to the time interval of 2600–1650 Ma and is subdivided into several subperiods. In the Karelian block basement, the Lower Sumian deposits (conglomerates, sandstones and others) are intruded by the Nuronen granite (2500 Ma) and layered mafic–ultramafic intrusions of the Olang Group (2453–2439 Ma), and overlain by basaltic andesite lava flows (2436 Ma) (Systra, 1991; Glebovitsky & Shemyakin, 1995).

The Sumian Amphibolitic Formation overlies the Archaean complex with a weathering crust at the base marked by meta-arkose, quartzite, local muscovite–quartz schist. It is composed of basaltic andesite altered to biotite–amphibole, amphibole schists, sometimes with garnet, and amphibolites (Table 1 & Fig. 6). Relics of pillow and vesicular structures are visible, together with tuff breccia. The maximum thickness (up to 2000 m) of the Amphibolitic Formation is characteristic of the northern flank of the Kukasozero synform, whereas in its southern flank it is much thinner (100–200 m). Westward, where the formation pinches out, Yatulian rocks (the Kukas Formation) overlie the Archaean basement. Polymictic conglomerates with quartzite, plagio-granite and meta-basalt pebbles forming lenses up to 200 m thick appear in the upper part of the formation, in the northern flank of the Kukasozero synform.

Subdivision of the overlying horizons of Yatulian and Ludikovian age (Middle Proterozoic, Table 1) is based on the stratigraphic scheme proposed earlier (Demidov & Kratz, 1974). Rocks of the Kukas and Khirvinavolok formations are assumed to occupy a higher stratigraphic level than the Amphibolitic Formation, are assigned to the Yatulian and Ludikovian horizons, respectively, and are overlain by polymictic conglomerates of presumably Kalevian age (Tables 1 & 3).

**Table 3.** Bulk chemical composition and the normative pre-metamorphic mineral content of the Kukas basin metasediments (in wt %).

| | Stratigraphic unit and metamorphic rock name | | | |
|---|---|---|---|---|
| | Unit I Feldspathic paramphibolite ($n = 1$) | Unit III Biotite-amphibole schist ($n = 6$) | Unit IV Marble ($n = 4$) | Unit V Graphitic rock ($n = 7$) |
| *Chemical composition* | | | | |
| $SiO_2$ | 47.85 | 57.70 | 9.03 | 43.32 |
| $TiO_2$ | 1.07 | 1.35 | 0.16 | 0.68 |
| $Al_2O_3$ | 13.60 | 12.49 | 0.49 | 9.14 |
| $Fe_2O_3$ | 1.58 | 2.71 | 0.53 | 3.67 |
| FeO | 8.04 | 6.85 | 1.42 | 0.71 |
| MnO | 0.11 | 0.08 | 0.23 | 0.02 |
| MgO | 12.09 | 4.73 | 6.82 | 2.28 |
| CaO | 9.74 | 4.28 | 40.15 | 3.46 |
| $Na_2O$ | 1.13 | 2.18 | 0.66 | 1.49 |
| $K_2O$ | 0.50 | 1.32 | 0.10 | 1.71 |
| $P_2O_5$ | 0.20 | 0.09 | 0.09 | 0.02 |
| $H_2O^+$ | – | 1.10 | 0.24 | 1.11 |
| $H_2O^-$ | 0.02 | 0.33 | 0.16 | 0.33 |
| $CO_2$ | 0.96 | 0.21 | 36.70 | 0.37 |
| $C_{org}$ | – | 4.22 | 2.92 | 30.20 |
| $S_{tot}$ | – | – | – | 0.75 |
| LOI | 3.11 | – | – | 1.29 |
| $SO_3$ | 0.37 | – | – | – |
| Total | 100.37 | 99.64 | 99.70 | 100.55 |
| *Normative components* | | | | |
| Q | 24.44 | 27.48 | 7.13 | 23.26 |
| Pl | 9.32 | 17.47 | 3.00 | 12.87 |
| Or | – | 1.85 | 0.14 | 2.19 |
| Total clastics | 33.76 | 46.80 | 10.27 | 38.32 |
| Kn | – | – | 0.31 | 0.17 |
| Ill | 3.36 | 8.18 | 3.06 | 13.74 |
| Mm | – | 3.99 | – | 1.87 |
| Chl | 35.80 | 21.94 | – | 3.52 |
| Srp | – | 0.61 | 1.34 | 0.26 |
| Total pelites | 39.16 | 34.72 | 4.71 | 19.55 |
| Ht | – | 2.35 | 0.45 | 1.56 |
| Cc | 3.40 | 1.33 | 54.65 | 1.89 |
| Dl | 15.63 | 5.58 | 23.58 | 6.09 |
| Ank | 6.64 | 3.78 | 3.84 | 0.69 |
| Total carbonates | 25.67 | 10.69 | 82.07 | 8.67 |
| Carbon | – | 4.03 | 1.87 | 29.37 |
| Rt, Ap | 1.41 | 1.41 | 0.63 | 2.53 |
| Pre-metamorphic sediments assumed | Dolomite–sandy pelite | Dolomite–pelitic sandstone | Sandy dolomitic limestone | Carbonaceous pelitic sandstone |

*n*, number of samples analysed.

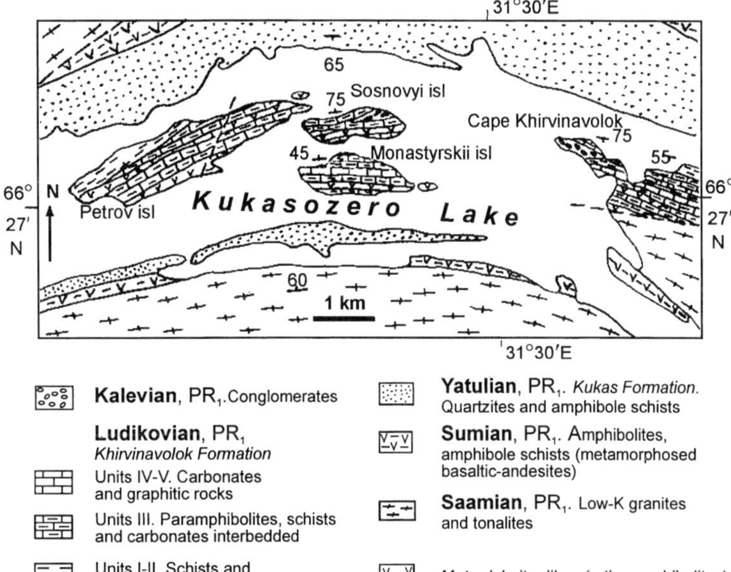


**Fig. 7.** Distribution of the metamorphosed sedimentary rocks within the central part of the Kukas basin.

The Kukas Formation overlies all older complexes with a structural unconformity. Thus, the lowermost Kukas carbonate rocks (up to 2–3 m thick) and conglomerates (up to 10–15 m) overlie both the Archaean and Sumian–Sariolian strata. The lower volcanogenic part of this formation, up to 1600 m thick, is composed of amphibole- and biotite–amphibole schists, which represent metamorphosed tuffs, tuff breccia and basalts with vesicular structures, intercalated with meta-carbonate and quartzite beds (Demidov & Kratz, 1974). The upper part, up to 400 m thick, is composed of massive feldspar quartzite, gradually changing to micaceous quartzite, sericite, biotite schists and carbonates.

The Khirvinavolok Formation is found in the central part of the Kukasozero synform and exposed on islands in Kukasozero Lake and Khirvinavolok Peninsula continuing south-eastward of the lake (Fig. 7). Five units can be distinguished in the section from the base upward (Demidov & Kratz, 1974). Unit I, 200–250 m thick, is composed of par-amphibolites represented by dark grey, massive, coarse-grained, porphyroblastic garnet, feldspar and monomineralic amphibolites, commonly interbedded with fine-grained schistose amphibolites. At the top, there are beds of biotite- and amphibole–biotite schists. Unit II corresponds to interbedded par-amphibolites, various schists (biotite, quartz–biotite, biotite–amphibole) and local quartzites. The unit is 100–150 m thick. Unit III, 50–200 m thick, contains interbedded schists, carbonate rocks (crystalline dolomites, limestones) and coarse-grained amphibolites. Bed thicknesses range from 5 to 20 cm, and are occasionally as much as 1–1.5 m. The characteristic feature of this unit is its rhythmic, often bimodal, stratification, with carbonate rocks interlayered with schists or amphibolites. In the upper part of the unit, carbonates predominate over schists and amphibolites. Unit IV, 50–100 m thick, is composed of massive, mostly medium-grained crystalline dolomites, partly limestones, commonly containing an admixture of carbonaceous (graphitic) matter (Figs 7 & 8). The uppermost Unit V is up to 10 m thick, conformably overlies black marbles of Unit IV and contains chiefly graphitic rocks represented by black massive mica–feldspar schists with graphitic matter constituting up to 40%. The total thickness of the Khirvinavolok Formation is estimated to be 500–700 m. The presence of organic (graphitic) matter is a characteristic feature of the entire formation.

The Khirvinavolok Formation exhibits strong metamorphic alteration under high temperature and pressure. The thermodynamic conditions of metamorphism correspond to the middle-temperature interval of the almandine–amphibolite facies ($T = 530$–$600°C$ and $P = 7$–$8$ kbar) (Moskovchenko & Turchenko, 1975). On a regional stratigraphic scale, the Khirvinavolok Formation corresponds to the

| Eon | Horizon | Formation | Unit, Thickness (m) | Lithology | Rock description |
|---|---|---|---|---|---|
| Paleoproterozoic | Ludikovian | Khirvinavolok | V 10 | | Graphitic rocks |
| | | | IV 50-100 | | Massive crystalline dolomites, subordinate limestones |
| | | | III 50-200 | | Carbonates, schists and amphibolites interlayered, thickness are of 1-2 cm up to 1-1.5 m. Carbonates dominate in the upper part of the unit |
| | | | II 100-150 | | Interlayered amphibolites and quartz-biotite-, biotite-amphibole-, garnet-biotite-amphibole schists. Quartzites appear occasionally |
| | | | I 200-250 | | Monomineralic, garnet and feldspatic paramphibolites. Biotite schists present in the upper part of the unit |

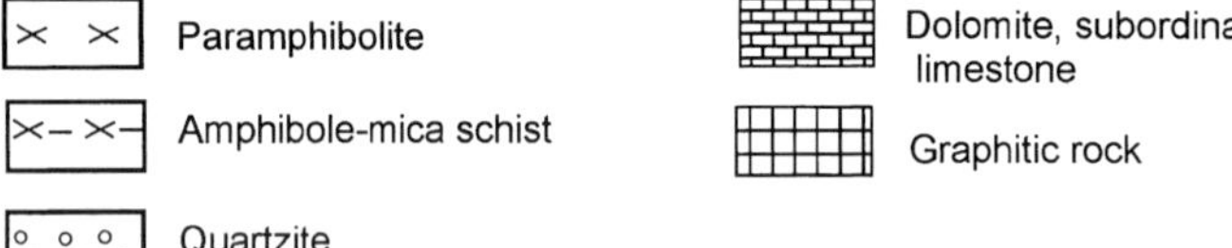


**Fig. 8.** Stratigraphic column and description of the metasedimentary rocks of the Khirvinavolok Formation in the Kukas basin.

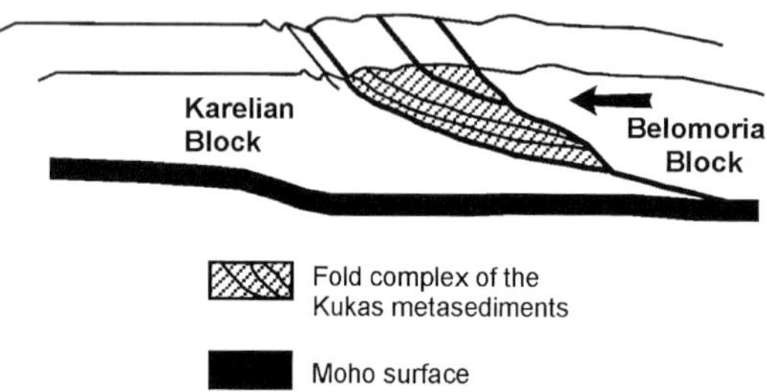


**Fig. 9.** Schematic palaeotectonic profile across the collisional boundary of the Karelian and Belomorian blocks during the deformation and metamorphism of the Kukas basin sediments (after Babarina, 1998).

middle part of the Lower Proterozoic, the Ludikovian Horizon (1950–2400 Ma; Table 2). The metamorphic event was caused by thrusting of the Belomorian block on to the Karelian block. At that time, the Kukas sediments were metamorphosed and deformed to recumbent isoclinal folds, thrust upon the continental margin of the Karelian block (Babarina, 1998) (Fig. 9).

## Depositional model

Lithochemical recalculation results obtained by application of the MINLITH program (Rosen *et al.*, 2000) are shown in Table 3. The calculated primary mineral composition of the Khirvinavolok Formation metamorphic rocks is as follows.

The paramphibolite–schist rock associations of Units I–III (paramphibolites, biotite, amphibole–biotite and other schists) were primarily dolomitic–sandy pelite and dolomitic–pelitic sandstones (Table 3). Locally, they contain significant amounts of feldspars and ferro-magnesian chlorites. This feature suggests the presence of volcanogenic material in the rocks. The quantitative primary mineral composition of these sediments corresponds to greywacke and subgreywacke. Metacarbonate rocks of Units III–IV were originally pure carbonate sediments ($CO_2$ > 40%) and locally contained a noticeable mixture of clastic material (quartz, feldspar, clay minerals). Graphitic rocks of Unit V originally contained, in addition to organic matter, a significant amount of illite, plagioclase and quartz. They are assumed to represent greywackes (Table 3).

The lateral facies cross-section of the Kukas basin for the Ludikovian and Yatulian time periods (Khirvinavolok and Kukas Formations, respectively) shown in Fig. 10 reveals regularities in spatial distribution of the primary rocks. From the base upward, sandy sediments comprising the upper unit of the Kukasozero Formation are replaced by pelitic–sandy sediments (Units I and II, Khirvinavolok Formation), and then successively by sandy-clayey-carbonate rocks (Unit III), carbonates (Unit IV) and carbonate–clastic deposits enriched in organic matter (Unit V). The calculated primary mineral compositions suggest, in general, a transgressive cycle of sedimentation in the Early Proterozoic. The cycle terminated with stagnant conditions of deposition in a reduced environment accompanied by $H_2S$ contamination.

Mineral (normative) composition indicates low maturity of the primary sediments that presumably contained feldspars, illites and chlorites. General features of this Early Proterozoic environment appear to be a differentiated provenance, low chemical weathering, intense denudation of the source area and transportation and burial of the eroded material. The last came from Archaean metamorphic rocks as well as from coeval volcanics.

The sedimentary basin was possibly of an elongated form controlled by patterns of the Karelian continental margin. Sedimentation occurred in a relatively small basin. During the period when organic-rich sediments accumulated, the basin decreased in size and shifted slightly westward (Fig. 10). The basin in

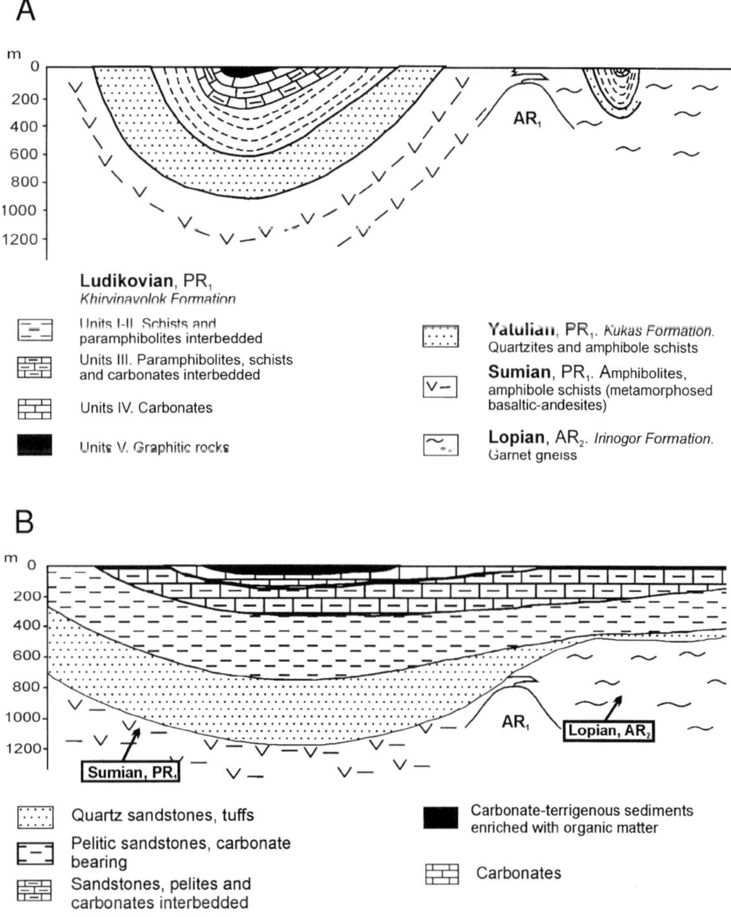


**Fig.10.** Cross-sections of (A) observed metamorphic and (B) calculated, pre-metamorphic reconstruction of lateral facies distribution in the Kukas basin.

general was probably not deep, as confirmed by hints of cyanobacterial activity (stromatolitic lamination) in carbonate rocks.

## CONCLUSIONS

The Keiva basin corresponds to a recumbant syncline on the Archaean footwall of the Kolmozero–Voronya suture. The amphibolite-grade sequence contains high-Al kyanite schists overlain by meta-carbonate. The calculated primary mineral composition of the schists indicates that they were derived from clastic protolith rocks, composed mainly of kaolinite and quartz fragments with some iron oxide admixture, whereas dolomite and feldspar fragments predominated in the present meta-carbonates. The shallow open shelf clastic sediments were derived from continental crust weathered under humid climatic conditions, later followed by sedimentation of dolomite with feldspar-rich clasts in a closed basin under arid conditions. The basin was located on the passive continental margin of the Keiva–Lebyazhya microcontinent.

The Kukas basin now forms a recumbent syncline in the suture zone between the Archaean Karelian and Belomorian blocks (palaeocontinents). It comprises amphibolite-grade biotite schist and meta-carbonates, both enriched in carbonaceous (graphite) matter associated with meta-basalt and paramphibolite. The calculated primary mineral composition of the schist corresponds to tuffaceous sediment composed mainly of quartz, feldspar and chlorite. The meta-carbonate represents dolomitic limestone. Both these rock types probably formed in a partly closed, anoxic shelf basin located near volcanic centres, on the active continental margin of the Karelian continent.

Lithochemical recalculation of analytical data from metamorphic rocks makes it possible to establish the mineral compositions of the source rocks during several stages of sedimentation in the Early Proterozoic Keiva and Kukas basins of the north-eastern Baltic Shield, and to shed light on broad depositional conditions.

## ACKNOWLEDGEMENTS

Special thanks are due to A.A. Migdisov, N.V. Bredanova, V.L. Zlobin and P.A. Chekhovich for useful discussions on mineral compositions of sedimentary rocks and general regularities in the structure of large sedimentary basins of the Russian platform and its foredeeps and of the Siberian platform. This study was supported by Russian Basic Research Foundation Grants 99-05-65154 and 98-05-65081.

## REFERENCES

Babarina, I.I. (1998) Structural evolution Kukasozero segment of the North-Karelian collision zone. *Geotektonika*, **3**, 80–96 (in Russian).

Belkov, I.V. (1963) *Kyanite Schists of Keiva Formation*. AN USSR Press, Moscow, 322 pp. (in Russian).

Bibikova, E.V., Skiold, T. & Bogdanova, S.V. (1996) Age and geodynamic aspect of the oldest rocks in the Precambrian Belomorian belt, Baltic Shield. In: *Precambrian Crustal Evolution in the North Atlantic Region* (Ed. Brewer, T.), Spec. Publs geol. Soc. London, No. 112, pp. 58–96. Blackwell Science, Oxford.

Bibikova, E.V., Slabunov, A.I., Bogdanova, S.V., Skiold, T., Stepanov, V.S. & Borisova, E.Yu. (1999) Early magmatism of Belomorian mobile belt, Baltic Shield: lateral zonation and isotope age. *Petrologiya*, **7**, 115–140 (in Russian).

Braun, G.M. (1986) Quantitative analysis of mineral mixtures using linear programming. *Clay and Clay Minerals*, **34**, 330–337.

Bridgwater, D., Scott, D., Balaganskii, V.V. *et al.* (1999) Provenance of Early Precambrian metasediments in the Lapland–Kola Belt as shown by $^{207}Pb/^{206}Pb$ dating of single grains of zircon and whole rock Sm/Nd isotope studies. *Doklady Akad. Nauk*, **366**(5), 664–668 (in Russian).

Cohen, D. & Ward, C.R. (1991) SEDNORM—a program to calculate a normative mineralogy for sedimentary rocks based on chemical analyses. *Comp. Geosci.*, **17**, 1235–1253.

Creelman, R.A. & Ward, C.R. (1996) A scanning electron microscope method for automated, quantitative analysis of mineral matter in coal. *Int. J. Coal Geol.*, **30**, 249–269.

Currie, K.L. (1991) GENORM: a generalized norm calculation. *Comp. Geosci.*, **17**, 77–89.

de Caritat, P., Bloch, J. & Hutcheon, I. (1994) LPNORM: a linear programming normative analysis code. *Comp. Geosci.*, **20**, 313–347.

Demidov, N.F. & Kratz, K.O. (1974) Stratigraphy and tectonics of the Kukasozero–Tikshozero zone of North Karelia. In: *Geological Problems Lower Proterozoic of Karelia* (Ed. Robonen, V.I.), pp. 95–116. Karelia Press, Petrozavodsk (in Russian).

Glebovitsky, V.A. & Shemyakin, V.M. (1995) *Main Frontiers in the Early Precambrian Earth's Evolution*. Geoinformark, Moscow, 47 pp. (in Russian).

Golovenok, V.K. (1977) *High-aluminous Precambrian Formations*. Nedra, Leningrad, 268 pp. (in Russian).

Gorbatschev, R. & Bogdanova, S.V. (1993) Frontiers in the Baltic Shield. *Precam. Res.*, **64**, 3–21.

Hill, R.J., Tsambourakis, G. & Madsen, I.C. (1993) Improved petrological modal analyses from X-ray powder diffraction data by use of the Rietveld method. Part I. Selected igneous, volcanic and metamorphic rocks. *J. Petrol.*, **34**, 867–900.

Kolka, R.K., Laird, D.A. & Nater, E.A. (1994) Comparison of four elemental mass methods for clay mineral quantification. *Clay and Clay Minerals*, **42**, 437–443.

Laird, D.A. & Dowdy, R.N. (1994) Simultaneous mineralogical quantification and chemical characterization of soil clays. *Clay and Clay Minerals*, **42**, 745–754.

Merodio, J.C., Spalletti, L.A. & Bertone, L.M. (1992) A FORTRAN program for the calculation of normative composition of clay minerals and pelitic rocks. *Comp. Geosci.*, **18**, 47–61.

Mints, M.V., Glaznev, V.N., Konilov, A.N. *et al.* (1996) *Early Precambrian of Northeast of a Baltic Shield: Paleodynamic, Structure and Evolution of the Continental Crust*. Nauchny mir, Moscow, 287 pp. (in Russian).

Mitrofanov, F.P., Balaganskii, V.V., Balashov, Yu.A. *et al.* (1993) U–Pb age of gabbro–anorthosite of the Kola Peninsula. *Doklady Acad. Nauk*, **331**(1), 98 (in Russian).

Moskovchenko, N.I. & Turchenko, S.I. (1975) *Kyanite–Sillimanite Metamorphism and Sulphide Ore Deposits (North Karelia)*. Nauka, Leningrad, 140 pp. (in Russian).

Negrutsa, V.Z. (1984) *Early Proterozoic Stages of Evolution of the East Part of the Baltic Shield*. Nedra, Leningrad, 270 pp. (in Russian).

Negrutsa, V.Z. (1999) *The Early Precambrian Stratigraphic Scale Recommended for the North-western Baltic Shield*. Matter of Lower Precambrian Subcomission Chairman, the Stratigraphic Committee of the Russian Academy of Sciences. The Kola Scientific Center, Apatity, 5 pp. (in Russian).

Pactung, A.D. (1998) MODAN: an interactive computer program for estimating mineral quantities based on bulk composition. *Comp. Geosci.*, **24**, 425–431.

Pearson, M.J. (1978) Quantitative clay mineralogical analyses from the bulk chemistry of sedimentary rocks. *Clay and Clay Minerals*, **26**, 423–433.

Petrov, V.P., Belyaev, O.A. & Voloshina, Z.M. (1986) *Metamorphism of the Supracrustal of Complexes of the Early Precambrian.* Nauka, Leningrad, 272 pp. (in Russian).

Pushkarev, Yu.D. (1990) *Megacycle Crust–Mantle System.* Nauka, Leningrad, 126 pp. (in Russian).

Rosen, O.M. (1976) Chlorites in sedimentary rocks: compositional typification depending on the bulk chemistry. *Doklady Akad. Nauk*, **228**(3), 689–692 (in Russian).

Rosen, O.M., Migdisov, A.A., Abbyasov, A.A. & Yaroshevskii, A.A. (2000) MINLITH—a program to calculate the normative mineralogy of sedimentary rocks: the reliability of result obtained for deposits of old platforms. *Geochem. Int.*, **38**, 388–400.

Rosen, O.M. & Nistratov, Yu.A. (1984) Determination of the mineral composition of sedimentary rocks based on their chemical analyses. *Sov. geol.*, **3**, 76–83 (in Russian).

Starks, T.H., Fang, J.H. & Zevin, L.S. (1984) A standardless method of quantitative X-ray diffractometry using target-transformation factor analysis. *Math. Geol.*, **16**, 351–367.

Systra, Yu.J. (1991) *Tectonics of the Karelian Region.* Nauka, Lenigrad, 176 pp. (in Russian).

Ward, C.R. & Taylor, J.C. (1996) Quantitative mineralogical analysis of coals from the Callide Basin, Queensland, Australia using X-ray diffractometry and normative interpretation. *Int. J. Coal Geol.*, **30**, 211–229.

Zangalis, K.P. (1998) Standardless quantitative mineralogical analysis of rocks. *Powder Diffraction*, **13**, 74–84.

*Spec. Publs int. Ass. Sediment.* (2002) **33**, 275–291

# Two meta-sedimentary basins in the Early Precambrian granulites of the Anabar Shield (polar Siberia): normative mineral compositions calculated by the MINLITH program and basin facies interpretations

V. L. ZLOBIN, O. M. ROSEN* *and* A. A. ABBYASOV

*Institute of the Lithosphere of Marginal Seas, Staromonetny per. 22, Moscow 109180, Russia*

## ABSTRACT

The Anabar Shield represents the exposed basement of the north-eastern Siberian craton. It is subdivided into several Precambrian granulite and granite–greenstone terranes (primary microcontinents), which are separated by amphibolite grade shear zones, probably representing preserved collisional sutures. These structures are complicated by SW-trending folds. The Early Proterozoic Vyurbyur and Hapschan sedimentary basins are located within the granulite Magan and Birekte terranes, respectively. The Vyurbyur basin formed at 2.4 Ga, due to erosion of the Magan terrane basement, of T(DM)Nd = 2.9 Ga. Its fragments are identified within a band of meta-sedimentary and meta-volcanogenic rocks of the Vyurbyur Group. This group represents an intercalation of predominantly garnet schists (volcaniclastic, feldspathic greywackes according to MINLITH) with clinopyroxene–plagioclase meta-carbonate rocks (limestones, possibly with volcanic impurity, as indicated by the considerable content of normative chlorite) and meta-basalts and meta-andesites. It was formed during the Early Proterozoic under shelf conditions on the active continental margin of the Archaean Magan microcontinent. The Hapschan basin formed at 2.1 Ga, on the basement of the Birekte terrane, dated at T(DM)Nd = 2.4 Ga, and consists of meta-sedimentary rocks of the Hapschan Group. This group is made up of garnet–biotite gneisses (feldspar–quartz greywackes with somewhat subordinate amounts of volcaniclastic greywackes), which are associated with clinopyroxene–plagioclase and thick clinopyroxene–forsterite meta-carbonates in the upper part of the sequence (initially limestones and dolomitic limestones). This Lower Proterozoic sequence presumably formed under shelf conditions on the passive continental margin of the Birekte microcontinent. The carbonate rocks and graywackes form individual petrochemical series. Their alternation in stratigraphic sections is explained by the supply of material from different sources, which was controlled by the tectonic activity in the region. Carbonate accumulation occurred during tectonically stable phases and was accompanied by supply of fine terrigenous admixture from distal sources. This regime changed to one characterized by the abundant supply of immature terrigenous material from proximal sources during tectonic reactivation.

## INTRODUCTION

The MINLITH program was developed, based on a LITHOCHEM program, which was proposed by O.M. Rosen for determining the nature of meta-sedimentary rocks by calculation of mineral composition from chemical analysis (Rosen, 1970, 1975). This method makes it possible to compare meta-sedimentary rocks with their unmetamorphosed analogues. Such a comparison in terms of mineral composition serves as a basis for a fairly well developed classification of sedimentary rocks. The results of the application of the method are reported in several publications (Rosen & Dimroth, 1982; Rosen & Zlobin, 1990; Zlobin, 1990). The use of these calculations is constrained by the considerable variations in the initial chemical composition of meta-sedimentary rocks. This technique makes it possible to obtain additional data on palaeogeographic and palaeogeodynamic settings of sedimentary rocks, involved in the ancient metamorphic complexes. The MINLITH program was developed in

* Present address: Geological Institute, Russian Academy of Sciences, Pyzhevsky per., 7, Moscow, 109017, Russia.

*Excel* (Rosen *et al.*, 2000), and is described in the Appendix below.

## METHODS

The stratigraphic succession was determined by conventional field investigations with subsequent correlation of carbonate-bearing sections. Post-metamorphic processes were studied in thin sections. Altered (silicified, migmatized, skarnized, weathered etc.) rocks were excluded from the study, in particular the scarnized calc-silicate rocks of the Vyurbyur Group. For calculation of initial normative composition with a MINLITH program, 70 samples were selected from a total of more than 200 studied samples from the Early Proterozoic meta-sedimentary rocks of the Vyurbyur and Hapschan groups of the Anabar Shield.

Major chemical components were determined with the X-ray fluorescence technique using standard procedures. The exceptions were $Na_2O$ and $K_2O$, which were analysed with flame photometry. These analyses were performed at the Complex Expedition of the Geological Ministry, Narofominsk City, Russia.

According to the systematics of granulite grade metamorphism (Dobretsov *et al.*, 1992), the studied meta-carbonate rocks were subdivided into marbles (100–85% carbonates, calcite and/or dolomite, with olivine, clinopyroxene and spinel), impure marbles (calciphyres, 85–15% carbonates with clinopyroxene and plagioclase) and calc-silicate rocks (less than 15% carbonates with clinopyroxene and plagioclase). The rocks studied include 17 samples from the Vyurbyur Group (10 garnet gneisses, seven impure marbles) and 53 samples from the Hapschan Group (17 garnet gneisses, 15 calc-silicate rocks, 12 impure marbles, nine marbles).

## GEOLOGICAL POSITION AND AGE

The north-eastern Siberian craton can be subdivided (from west to east) into the granulite–gneiss Magan and Daldyn terranes and the granite–greenstone Birekte terrane. These terranes (originally microcontinents) were united by the collisional Kotuykan and Billyakh zones (suture), respectively (Rosen *et al.*, 1994). The Sayan–Taimyr zone separates this region from the western portion of the Siberian craton, while the Early Proterozoic Akitkan fold belt limits its south-eastern edge (Fig. 1). The rocks of these terranes crop out within the Anabar Shield.

The Magan terrane consists mainly of plagio-gneisses, enderbites and charnockites of the Upper Anabar Group, and of meta-carbonates, meta-mafics and plagio-gneisses of the Vyurbyur Group, developed in the easternmost portion of the terrane.

The Daldyn terrane consists of meta-mafics and plagio-gneisses of the Daldyn Group. The bedded structure of the sequences and geochemical properties indicate that the meta-mafic rock association of both terranes (two pyroxene–plagioclase schist) and plagioclase gneiss (quartz–plagioclase–orthopyroxene gneiss) were transformed during granulite metamorphism of former (original) basalts, andesites and dacites (Markov, 1988). The individual interbeds also contain sedimentary rocks: meta-greywackes–garnet gneisses, meta-carbonates to impure marbles, calc-silicate rocks and orthopyroxene quartzites. The rocks are deformed in isoclinal folds, which have limb amplitudes of a few kilometres and an ENE vergence, owing to collision-related overthrusting (Rosen, 1995). The Daldyn terrane is overthrust on to the Magan terrane along the Kotuykan fault zone.

The Birekte granite–greenstone terrane is situated in the basins of the Popigai, Anabar, Birekte and Olenek rivers and is mainly overlain by Riphean–Palaeozoic platformal cover. The Birekte terrane is overthrust on to the Daldyn terrane along the Billyakh collisional fault zone. Gravi-magnetic data indicate that it contains large, buried granite massifs and mafic (ultramafic) plutons and/or greenstone belts (Khoreva, 1987). All these structures are covered by Early Proterozoic fold belts. The frontal part of the Birekte terrane within the Anabar Shield is represented by the Hapschan belt, which is composed of granulite meta-greywackes (garnet gneisses) and carbonates (calc-silicate rocks, impure marbles and marbles) of the Hapschan Group. Meta-volcanics account for no more than 10% of this group (Markov, 1988). The eastern, rear part of the Birekte terrane contains the Aekit belt restricted to the Olenek Rise. This belt consists of acid volcanic, terrigenous and carbonaceous rocks, cut by granitoids (Fig. 1). They were metamorphosed under greenschist facies (Rosen *et al.*, 1994).

The oldest age values were obtained from rocks of the Daldyn terrane by the U–Pb zircon upper discordia intercept on plagioclase gneisses (3.0 ± 0.02 Ga; Markov, 1988) and the Sm–Nd whole rock isochron method (32 samples) on plagio-gneisses and meta-mafics (3.1 ± 0.08 Ga at $\varepsilon Nd(T) = +3.1 \pm 1.5$; Spiridonov *et al.*, 1993). The model T(DM)Nd age of meta-volcanics is 3.16 Ga, whereas that of associated sedimentary rocks (quartzites, meta-greywackes and impure marbles) is

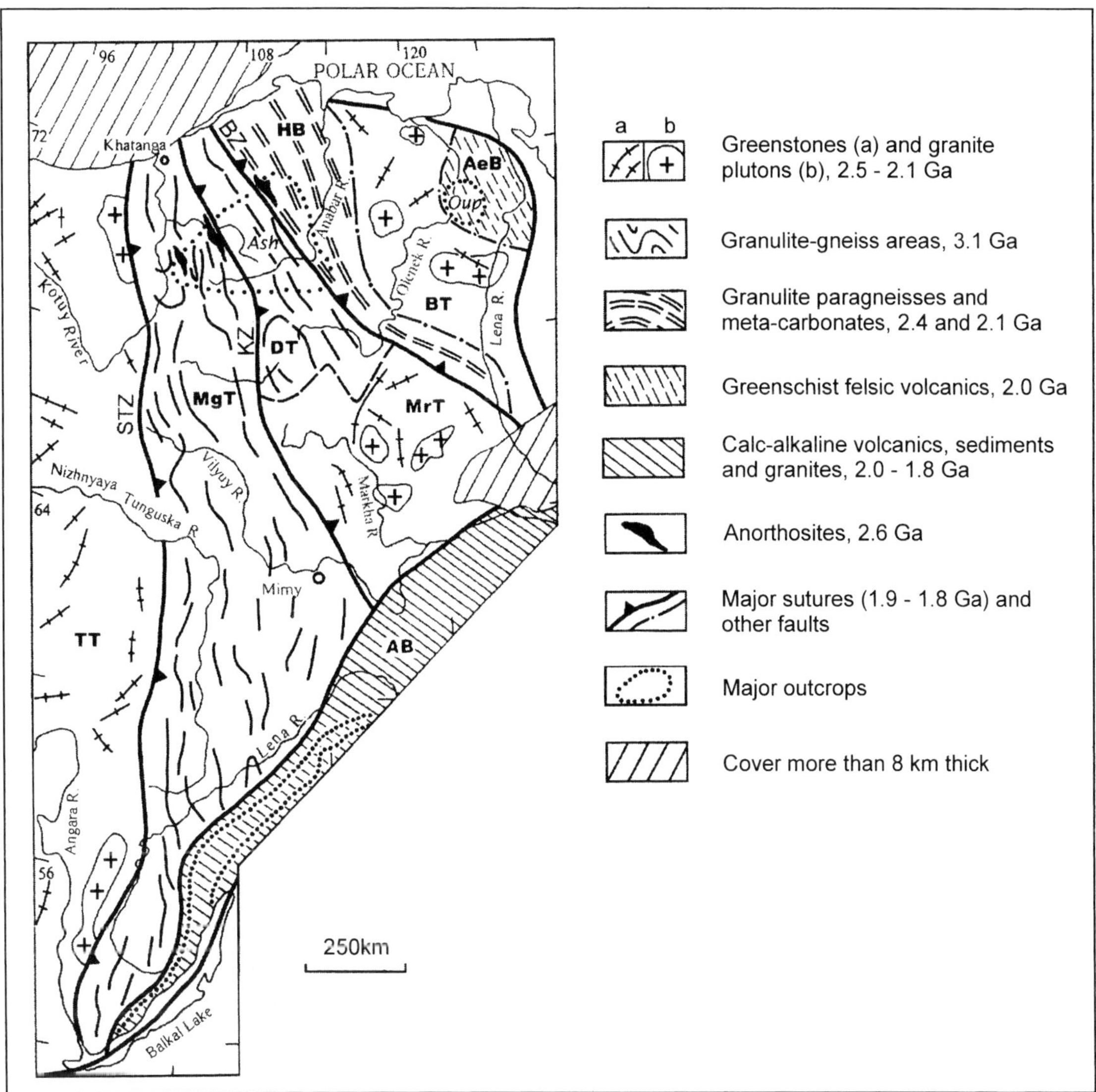


**Fig. 1.** Precambrian terrane map of the north-eastern part of the Siberian craton (modified after Rosen *et al.*, 1994). Terranes: BT, Birekte; DT, Daldyn; MrT, Markha; MgT, Magan; TT, Tungus. Belts: HB, Hapschan orogenic; AeB, Aekit orogenic; AB, Akitkan fold belt. Fault zones: BZ, Billakh; Kz, Kotuykan; STZ, Sayan–Taymyr. Outcropping basement: Oup, Olenek uplift; Ash, Anabar Shield.

3.00–3.19 Ga. This indicates a coeval separation of magmatic rocks from mantle to crust and accumulation of sedimentary rocks. The volcano-sedimentary protoliths of the Daldyn terrane are dated at 3.0–3.1 Ga.

The meta-andesite from the Magan terrane has a U–Pb zircon age of 2.42 ± 0.02 Ga. This rock is a quartz-free two-pyroxene plagio-gneiss, which is petrographically and mineralogically similar to meta-andesites of the Daldyn terrane (Markov, 1988). In addition, we studied the Sm–Nd systematics of the meta-volcanics (plagio-gneisses and crystalline schists) and metasedimentary rocks (garnet gneisses and impure marbles) of the Vyurbyur Group. The model T(DM)Nd ages of these rocks are from 2.8 to 3.09 Ga (Rosen *et al.*, 1999; Zlobin *et al.*, 1999). These data suggest that the Lower Proterozoic mantle melt, separated at 2.42 Ga, was considerably contaminated by Archaean lower crust. The T(DM)Nd values for meta-sedimentary rocks presumably record the average age of the source area—the old basement. The εNd values, ranging from −0.9 to −4.2 (T = 2.4 Ga) for the studied rocks, indicate a significant isotope evolution of basement by the time of erosion and accumulation of the volcano-sedimentary substrate of the Magan terrane.

The lower age limit (T(DM)Nd = 2.32–2.44) of the granite–greenstone basement of the Birekte terrane was constrained by a model age for its erosion products in the terrigenous–carbonate complex of the Hapschan terrane (Zhuravlev & Rosen, 1991). At this time, the basement was represented by mature continental crust with $^{147}Sm/^{144}Nd = 0.11$ according to Taylor & McLennan (1985). Near the Billyakh shear zone, these rocks experienced 1.97 Ga low-temperature metamorphism (Markov, 1988), which was followed by regional granulite metamorphism with a peak at 1.91–1.92 Ga (Rosen *et al.*, 1999). The rocks of the Aekit fold belt were dated only by the K–Ar method. The age values are 1.98 Ga for metamorphic micas from greenschist volcanics (Krylov *et al.*, 1963) and 1.85–2.08 Ga for the cross-cutting granites (Mokshantsev, 1979). The data suggest that meta-sedimentary rocks of the Hapschan and Aekit folded belts were formed at 2.1 Ga (Rosen *et al.*, 1999).

## STRUCTURE AND COMPOSITION OF THE LOWER PROTEROZOIC CARBONATE-BEARING SEQUENCES

The meta-sedimentary rocks of the Anabar Shield are mainly confined to the Magan and Birekte terranes. Within the former, these rocks mainly belong to the Vyurbyur Group, which is traced as a narrow NNW-trending band at the eastern margin of the terrane (Fig. 2). This sequence consists of two members. The lower member is represented by alternations of meta-terrigenous (garnet gneisses), meta-carbonate (marbles, impure marbles and calc-silicate–carbonate-terrigenous rocks) and meta-volcanics (two-pyroxene plagio-gneisses and mafic schists). The upper member is mainly represented by garnet gneisses with occasional individual impure marble interbeds or meta-volcanic beds (Grozdilov, 1983). The composition of the Vyurbyur section is as follows: garnet gneisses (30–45%), calc-silicate rocks (30–40%), impure marbles (3–7%), plagio-gneisses and crystalline schists (20–30%). The Vyurbyur Group normally has tectonic contacts with rocks of the Upper Anabar Group (Grozdilov, 1983; Markov, 1988). It shows high variations in strike. In particular, the amount of meta-volcanics considerably increases to the north and south (sections 1 and 3, Fig. 3). Mainly calcitic impure marbles occur as lenses from a few metres to a few tens of metres thick and from a few hundred metres to a few kilometres in length. Most of these lenses are confined to the lower part of the Vyurbyur Group. This sequence shows large compositional variations by dip (Markov, 1988). North of section 1, the amount of mafic meta-volcanics (crystalline schists) considerably increases. To the south, the sequence is mainly represented by meta-sedimentary rocks, dominated by calc-silicate (carbonate–terrigenous) rocks. To the south of section 2 (Fig. 3), there is an increase in meta-volcanics, which are dominated by andesites and dacites (plagio-gneisses), and meta-terrigenous rocks (garnet gneisses). The apparent thickness of the Vyurbyur Group is more than 3000 m.

Meta-sedimentary rocks of the Hapschan belt are regionally termed the 'Hapschan Group'. This group consists mainly of meta-sedimentary rocks: garnet gneisses (45–55%), calc-silicate rocks (30–40%), marbles and impure marbles (5–15%). The meta-volcanics (palgio-gneisses and crystalline schists) account for no more than 10%. The schists occur either as tectonic lenses among the meta-sedimentary rocks or at the base of the meta-sedimentary sequence. These rocks were previously classified as the Upper Anabar Group (Vishnevskii, 1978; Grozdilov, 1983; Markov, 1988). However, according to Sm–Nd data (Rosen *et al.*, 1991, 1994, 1999; Rosen, 1995), they are now recognized as individual regional stratigraphic units. The marbles and impure marbles form extended horizons, from a few metres thick in the lower portion of the Hapschan Group to a few hundreds of metres thick in its middle part. Based on sequence correlation, the Hapschan Group contains five meta-carbonate members and can be divided into several lithologic units (Fig. 4). Four former members compose the lower part of the Hapschan Group, whereas the fifth member, consisting of uniform garnet gneisses with scarce meta-carbonate interbeds, represents the upper part of the group. The total apparent thickness of the sequence is more than 5500 m (Markov, 1988).

The mineral assemblages of the meta-sedimentary rocks in the Vyurbyur and Hapschan groups correspond to granulite-facies conditions. The meta-terrigenous rocks of the Vyurbyur Group (garnet gneisses) are characterized by garnet–biotite–quartz–plagioclase–K-feldspar mineral assemblages. In

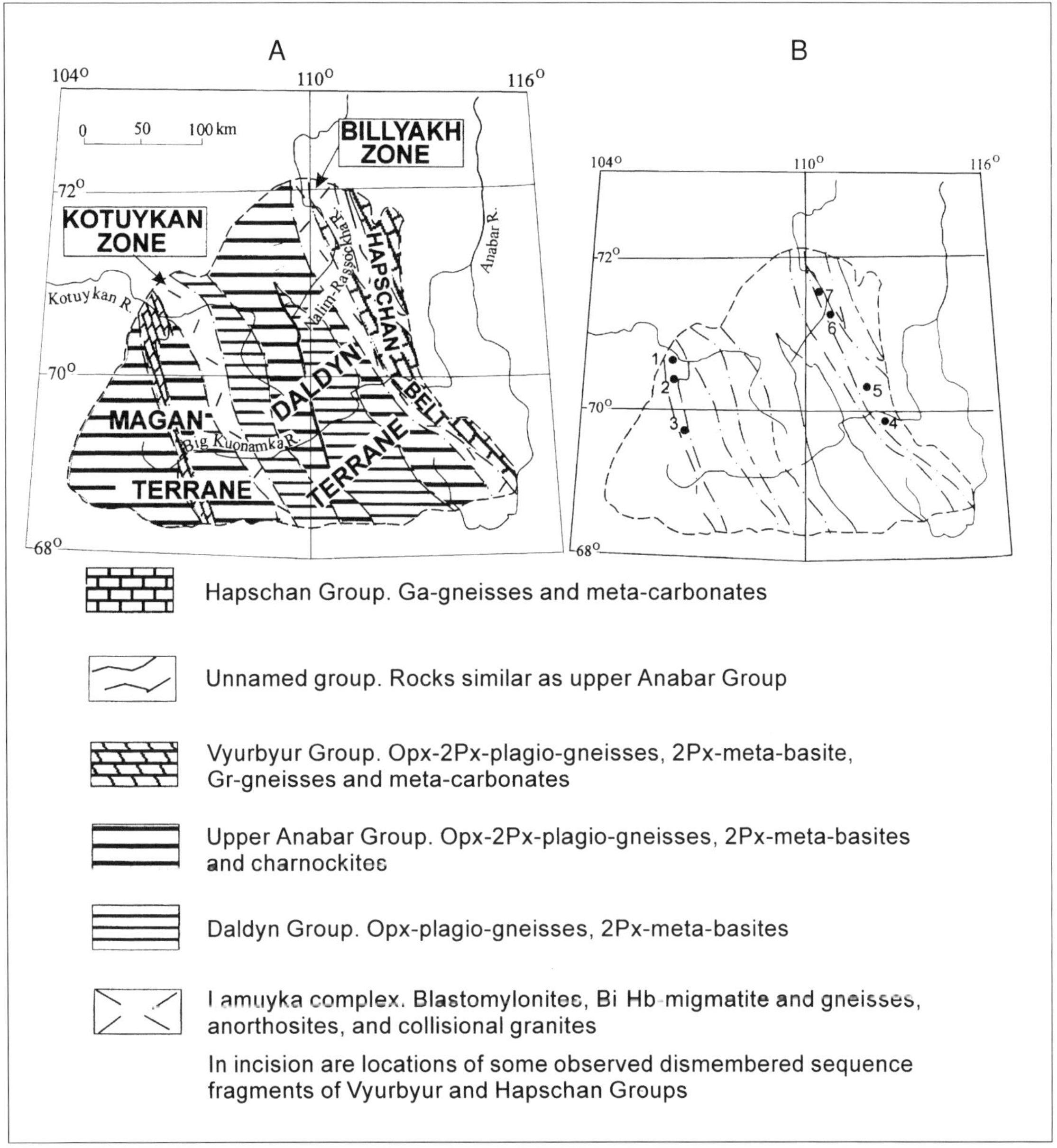


**Fig. 2.** Generalized geological map of Precambrian rocks in the Anabar Shield (modified after Rosen *et al.*, 1999; Zlobin *et al.*, 1999). The geological sketch map is shown in A. In B numbers of the sequences in Fig. 3 (Nos 1–3) and in Fig. 4 (Nos 4–7) are given.

addition to the mafic minerals, the similar rocks of the Hapschan Group contain orthopyroxene (hypersthene). Rocks with garnet as the only mafic mineral also occur. In most rocks, plagioclase dominates over K-feldspar. In the meta-carbonate rocks of the Vyurbyur Group, the diopside–titanite assemblage coexists with plagioclase or scapolite in the schists and with feldspar (mainly plagioclase) and quartz in gneisses (calc-silicate rocks). In the impure marbles of this group, the diopside–titanite assemblage coexists with calcite, scapolite and/or plagioclase. The mineral assemblages of calc-silicate rocks of the Hapschan and

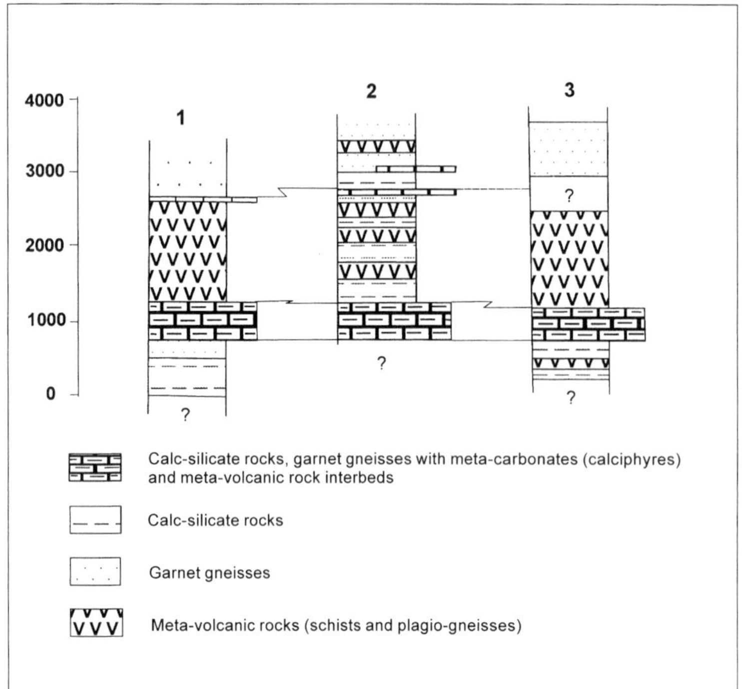


**Fig. 3.** Observed sections of the Vyurbyur Group and their possible correlation.

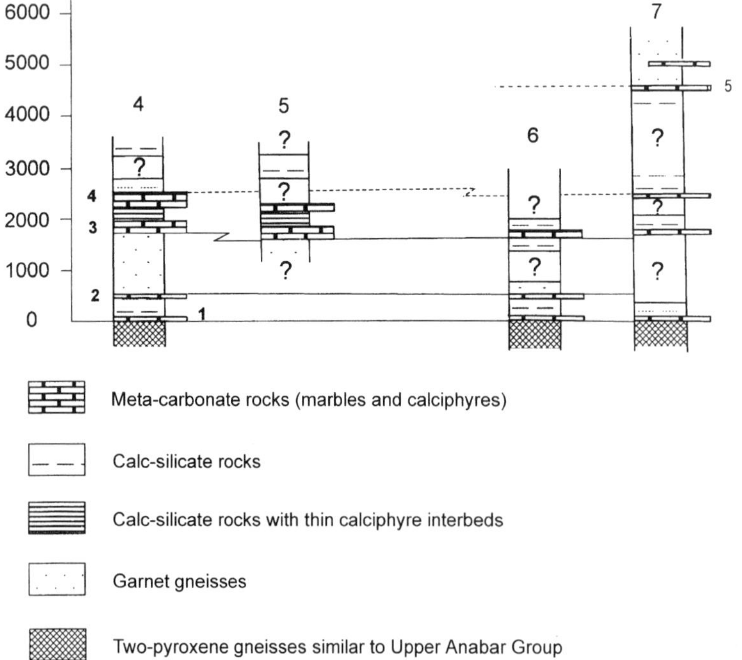


**Fig. 4.** Observed sections of the Hapschan Group and their possible correlation.

the Vyurbyur groups are similar. They correspond to a nearly continuous limestone–dolomite series dominated (to 80%) by low-Mg rocks typical of the Vyurbyur Group (Markov, 1988; Rosen & Zlobin, 1990). Calcite and dolomite are associated with clinopyroxene (diopside), phlogopite and olivine (forsterite), and locally with forsterite and spinel.

## RESULTS

The major-element composition and possible pre-metamorphic normative composition of the 41 representative samples of metasedimentary rocks from the Vyurbyur and Hapschan groups are listed in Tables 1–5 and interpreted in Figs 5–7.

**Table 1.** The representative analyses (%wt) and normative mineral composition of Early Proterozoic granulite-facies meta-sedimentary rocks of the Anabar Shield, Vuyrbuyr Group.

| | Meta-greywackes | | | | | Meta-carbonate rocks | | | |
|---|---|---|---|---|---|---|---|---|---|
| | 1 | 2 | 3 | 4 | 5 | 6 | 7 | 8 | 9 |
| $SiO_2$ | 50.1 | 56.66 | 65.89 | 66.84 | 68.95 | 19.73 | 29.49 | 35.44 | 40.72 |
| $TiO_2$ | 1.1 | 0.76 | 0.63 | 0.43 | 0.38 | 0.18 | 0.21 | 0.29 | 0.23 |
| $Al_2O_3$ | 23.36 | 18.5 | 15.08 | 15.35 | 15.03 | 4.31 | 4.6 | 7.24 | 6.7 |
| $Fe_2O_3$ | 8.06 | 8.75 | 7.63 | 3.38 | 3.81 | 2.17 | 2.76 | 2.97 | 2.69 |
| FeO | nd | nd | nd | nd | nd | nd | nd | nd | nd |
| MnO | 0.1 | 0.13 | 0.1 | 0.03 | 0.05 | 0.05 | 0.04 | 0.13 | 0.04 |
| MgO | 3.72 | 3.77 | 2.59 | 1.22 | 0.78 | 2.32 | 1.85 | 4.74 | 1.77 |
| CaO | 4.29 | 3.55 | 2.59 | 4.08 | 2.85 | 40.47 | 35.67 | 26.54 | 27.19 |
| $Na_2O$ | 4.22 | 3.22 | 1.7 | 3.92 | 4 | 0.57 | 0.42 | 0.6 | 1.29 |
| $K_2O$ | 3.25 | 3.11 | 2.54 | 3.06 | 2.85 | 0.36 | 0.63 | 0.52 | 1.2 |
| $P_2O_5$ | 0.06 | 0.06 | 0.04 | 0.13 | 0.31 | 0.04 | 0.04 | 0.06 | 0.04 |
| LOI | 1.31 | 0.95 | 0.6 | 1.06 | 0.59 | 29 | 23.82 | 20.97 | 17.73 |
| Total | 99.57 | 99.46 | 99.39 | 99.5 | 99.6 | 99.2 | 99.53 | 99.5 | 99.6 |
| $CO_2$ | nd | nd | nd | nd | nd | 24.51 | 20.21 | 19 | 15.5 |
| *Normative mineral composition of primary rocks* | | | | | | | | | |
| Ab | 34.44 | 26.42 | 13.97 | 31.94 | 33.04 | 4.68 | 3.4 | 4.95 | 10.28 |
| An | 3.65 | 2.8 | 1.48 | 3.39 | 3.51 | 0.5 | 0.36 | 0.53 | 1.09 |
| Or | 0.57 | 7.93 | 1.83 | 7.38 | 6.86 | 1.55 | 0.94 | 0 | 0.16 |
| Q | 2.07 | 15.94 | 37.33 | 27.5 | 30.35 | 11.37 | 20.41 | 23.93 | 25.08 |
| **Clastics** | **40.73** | **53.09** | **54.61** | **70.21** | **73.76** | **18.1** | **25.11** | **29.41** | **36.61** |
| Mm | 0 | 0 | 0 | 0 | 0 | 0 | 0 | 0 | 0 |
| Ill | 32.49 | 17.9 | 23.08 | 18.16 | 17.34 | 0.93 | 4.73 | 5.43 | 11.8 |
| Ch | 19.31 | 22.62 | 16.5 | 0 | 0 | 10.48 | 8.95 | 15.73 | 0 |
| Kn | 0 | 0 | 0 | 0 | 0 | 0 | 0 | 0 | 0 |
| Srp | 0 | 0 | 0 | 0 | 0 | 0 | 0 | 0 | 0 |
| **Pelites** | **51.8** | **40.52** | **39.58** | **18.16** | **17.34** | **11.41** | **13.68** | **21.16** | **11.8** |
| Cc | 5.78 | 4.73 | 2.94 | 0.32 | 0 | 69.58 | 60.38 | 42.72 | 39.08 |
| Dl | 0 | 0 | 0 | 3.72 | 1.89 | 0.56 | 0.47 | 4.73 | 6.55 |
| Ank | 0.33 | 0.59 | 2.01 | 6.84 | 4.24 | 0 | 0 | 1.35 | 5.59 |
| Rch | 0.16 | 0.2 | 0.16 | 0.05 | 0 | 0.08 | 0.06 | 0.21 | 0.06 |
| Sd | 0 | 0 | 0 | 0 | 0 | 0 | 0 | 0 | 0 |
| **Carbonates** | **6.27** | **5.52** | **5.11** | **10.93** | **6.13** | **70.22** | **60.91** | **49.01** | **51.28** |
| Ap | 0.14 | 0 | 0.09 | 0.3 | 0 | 0.09 | 0.09 | 0.14 | 0.09 |
| Rt | 1.06 | 0.74 | 0.61 | 0.41 | 0.37 | 0.17 | 0.2 | 0.28 | 0.2 |
| Prl | 0 | 0 | 0 | 0 | 0.06 | 0 | 0 | 0 | 0 |
| Ht | 0 | 0 | 0 | 0 | 1.63 | 0 | 0 | 0 | 0 |
| **Others** | **1.2** | **0.74** | **0.7** | **0.71** | **2.06** | **0.26** | **0.29** | **0.42** | **0.29** |
| Sample | 3318.1 | 3P14 | 3261.1 | 3262.4 | 3P16 | 3344.1 | 3320.3 | 3P14.3 | 3329.1 |
| Rock | Bi-Gr schist | Gr-Bi gneiss | Gr-Bi gneiss | Hp-Gr gneiss | Bi-Gr gneiss | Di-Sc calciphyre | Di-Sc calciphyre | Di-Sc calciphyre | Di-Pl calciphyre |

nd, not determined.

Mineral composition metamorphic rocks: Bi, Di, Fr, Gr, Pl, Sc, Sp, biotite, diopside, forsterite, garnet, plagioclase, scapolite, respectively.

Normative mineral composition of primary rocks in reference order: Al, An, Or, Q, Mm, Ill, Chl, Gb, Kn, Cc, Dl, Ank, Rch, Sd, Ap, Rt, Prl, Ht, albite, anorthite, orthoclase, quartz, montmorillonite, illite, chlorite, gibbsite, kaolinite, calcite, dolomite, ankerite, rhodochrosite, siderite, apatite, rutile, pyrolusite, goethite, respectively.

**Table 2.** The representative analyses (%wt) and normative mineral composition of Early Proterozoic granulite-facies meta-sedimentary rocks of the Anabar Shield, Hapschan Group, meta-greywackes.

| | Type I | | | | | Type II | | | |
|---|---|---|---|---|---|---|---|---|---|
| | 10 | 11 | 12 | 13 | 14 | 15 | 16 | 17 | 18 |
| $SiO_2$ | 56.63 | 64 | 67.39 | 69.75 | 70.84 | 60.02 | 62.79 | 64.14 | 68.15 |
| $TiO_2$ | 0.77 | 0.76 | 0.75 | 0.34 | 0.59 | 0.69 | 0.78 | 0.36 | 0.45 |
| $Al_2O_3$ | 20.84 | 15.6 | 14.17 | 14.74 | 13.13 | 15 | 16.28 | 14.19 | 15.94 |
| $Fe_2O_3$ | 7.21 | 7.74 | 5.95 | 4.1 | 5.42 | 10.44 | 7.88 | 12.44 | 5.55 |
| MnO | 0.04 | 0.09 | 0.08 | 0.12 | 0.08 | 0.11 | 0.11 | 0.09 | 0.07 |
| MgO | 2.27 | 3.95 | 1.93 | 1.4 | 2.1 | 5.91 | 3.93 | 4.29 | 2.78 |
| CaO | 1.69 | 1.01 | 1.7 | 2.06 | 3.62 | 3.45 | 3.46 | 1.41 | 2.48 |
| $Na_2O$ | 2.9 | 2.23 | 3 | 2.8 | 2.18 | 2.65 | 3.28 | 1.89 | 2.55 |
| $K_2O$ | 7.14 | 4.16 | 2.97 | 3.42 | 1.45 | 1.57 | 2.45 | 1.14 | 1.23 |
| $P_2O_5$ | 0.15 | 0.13 | 0.09 | 0.04 | 0.18 | 0.08 | 0.09 | 0.09 | 0.11 |
| LOI | 0.38 | 0.72 | 1.7 | 0.5 | 0.5 | 0.24 | 0.5 | 0.56 | 0.5 |
| Total | 100.02 | 100.39 | 99.73 | 99.27 | 100.09 | 100.16 | 101.55 | 100.6 | 99.81 |
| $CO_2$ | nd | nd | nd | nd | nd | nd | nd | nd | nd |
| *Normative mineral composition of primary rocks* | | | | | | | | | |
| Ab | 23.81 | 18.5 | 25.11 | 23.13 | 15.68 | 21.37 | 26.37 | 15.44 | 14.53 |
| An | 2.53 | 1.96 | 2.66 | 2.45 | 1.66 | 2.27 | 2.8 | 1.64 | 1.54 |
| Or | 23.55 | 8.18 | 5.27 | 6.67 | 0 | 5.62 | 7.26 | 0.72 | 0 |
| Q | 6.32 | 27.17 | 33.67 | 35.53 | 41.12 | 26.5 | 24.22 | 37.12 | 29.28 |
| **Clastics** | **56.21** | **55.81** | **66.71** | **67.78** | **58.46** | **55.76** | **60.65** | **54.92** | **45.35** |
| Mm | 0 | 0 | 0 | 0 | 8.45 | 0 | 0 | 0 | 25.91 |
| Ill | 31.48 | 28.82 | 21.89 | 23.64 | 14.87 | 5.83 | 11.75 | 10.46 | 12.62 |
| Chl | 0 | 0 | 0 | 0 | 10.34 | 32.15 | 21.74 | 31.24 | 9.2 |
| Kn | 0 | 0 | 0 | 0 | 0 | 0 | 0 | 0 | 0 |
| Srp | 0 | 5.61 | 1.38 | 0 | 0.57 | 0 | 0 | 0 | 0 |
| **Pelites** | **31.48** | **34.43** | **23.27** | **23.64** | **34.23** | **37.98** | **33.49** | **41.7** | **47.73** |
| Cc | 0 | 0 | 0 | 0 | 3.84 | 4.41 | 4.54 | 0.71 | 1.11 |
| Dl | 3.09 | 0 | 3.38 | 4.08 | 0 | 0.84 | 0.22 | 0 | 0 |
| Ank | 0 | 1.41 | 0 | 0.86 | 2.94 | 0 | 0 | 1.98 | 5.03 |
| Rch | 0 | 0 | 0 | 0 | 0.12 | 0.17 | 0.17 | 0.14 | 0.11 |
| Sd | 0 | 0 | 0 | 0 | 0 | 0 | 0 | 0 | 0 |
| **Carbonates** | **3.09** | **1.41** | **3.38** | **4.94** | **6.9** | **5.42** | **4.93** | **2.83** | **6.25** |
| Ap | 0.35 | 0 | 0.21 | 0.09 | 0.41 | 0.18 | 0.2 | 0.21 | 0.25 |
| Rt | 0.75 | 0.75 | 0.74 | 0.33 | 0.57 | 0.66 | 0.74 | 0.35 | 0.43 |
| Prl | 0.05 | 0.11 | 0.1 | 0.15 | 0 | 0 | 0 | 0 | 0 |
| Ht | 6.4 | 7.18 | 5.58 | 3.06 | 0 | 0 | 0 | 0 | 0 |
| **Others** | **7.55** | **8.04** | **6.63** | **3.63** | **0.98** | **0.84** | **0.94** | **0.56** | **0.68** |
| Sample | 80 330.5 | 1P1.2 | 9PA61 | 8355.5 | 81 355.3 | H74.1.81 | 3715 | 80 324.1 | 80 329 |
| Rock | Gr-Bi gneiss | Hp-Gr gneiss | Gr gneiss | Bi-Gr gneiss | Hp-Gr gneiss | Bi-Gr gneiss | Bi-Gr gneiss | Bi-Gr gneiss | Gr gneiss |

For abbreviations see Table 1.

**Table 3.** The representative analysis (%wt) and normative mineral composition of Early Proterozoic granulite-facies meta-sedimentary rocks of the Anabar Shield, Hapschan Group, meta-carbonate calc-silicate rocks.

| | 19 | 20 | 21 | 22 | 23 | 24 | 25 | 26 | 27 |
|---|---|---|---|---|---|---|---|---|---|
| $SiO_2$ | 69.39 | 44.84 | 49.6 | 51.47 | 52.45 | 52.98 | 55.8 | 57.26 | 58.8 |
| $TiO_2$ | 0.51 | 0.45 | 0.33 | 0.98 | 0.28 | 0.53 | 0.65 | 0.23 | 0.58 |
| $Al_2O_3$ | 16 | 13.38 | 11.41 | 15.25 | 10.9 | 14.39 | 14.27 | 11.08 | 14.8 |
| $Fe_2O_3$ | 4.35 | 0.2 | 0.28 | 0.93 | 0.76 | 0.74 | 0.1 | 0.12 | 0.42 |
| FeO | nd | 5.78 | 1.98 | 6.1 | 3.59 | 4.42 | 5 | 3.17 | 6.64 |
| MnO | 0.11 | 0.04 | 0.07 | 0.11 | 0.13 | 0.06 | 0.08 | 0.08 | 0.13 |
| MgO | 1.2 | 3.1 | 0.73 | 2.87 | 7.78 | 1.89 | 2.04 | 3.71 | 3.34 |
| CaO | 3.73 | 21.04 | 29.48 | 16.7 | 18.01 | 16.99 | 15.99 | 15.16 | 6.69 |
| $Na_2O$ | 3.4 | 2.05 | 1.4 | 1.8 | 1.73 | 2 | 1.62 | 1.9 | 3.67 |
| $K_2O$ | 1.37 | 1.37 | 0.55 | 0.82 | 1.88 | 3 | 0.67 | 3.28 | 2.65 |
| $P_2O_5$ | 0.09 | 0.11 | 0.04 | 0.1 | 0.05 | 0.04 | 0.12 | 0.1 | 0.24 |
| LOI | 0.36 | 7.24 | 3.8 | 2.5 | 1.56 | 2.5 | 3 | 2.9 | 1.5 |
| Total | 100.51 | 99.6 | 100.67 | 99.63 | 99.12 | 99.54 | 99.34 | 98.99 | 99.46 |
| $CO_2$ | nd | 6.62 | 2.2 | 1.13 | 0.59 | 2.0 | 2.1 | 1.92 | 0.62 |
| *Normative mineral composition of primary rocks* | | | | | | | | | |
| Ab | 21.28 | 13.88 | 6.42 | 5.14 | 12.01 | 14.69 | 1.9 | 14.14 | 28.11 |
| An | 2.26 | 1.47 | 0.68 | 0.55 | 1.28 | 1.56 | 0.2 | 1.5 | 2.98 |
| Or | 0 | 0 | 0 | 0 | 0 | 0.25 | 0 | 8.89 | 3.87 |
| Q | 27.27 | 15.25 | 20.92 | 12.9 | 25.6 | 21.98 | 16.45 | 27.2 | 21.21 |
| **Clastics** | **50.81** | **30.6** | **28.02** | **18.59** | **38.89** | **38.48** | **18.55** | **51.82** | **56.17** |
| Mm | 24.31 | 7.05 | 13.65 | 33.81 | 0 | 0 | 42.38 | 0 | 0 |
| Ill | 13.78 | 13.14 | 4.8 | 7.59 | 16.5 | 27.4 | 6.27 | 14.6 | 18.66 |
| Chl | 0.33 | 14.71 | 0 | 8.25 | 2.71 | 0 | 0.83 | 0 | 0 |
| Kn | 0 | 0 | 7.79 | 0 | 0 | 0 | 0 | 0 | 0 |
| Srp | 0 | 0 | 0 | 0 | 0 | 0 | 0 | 0 | 0 |
| **Pelites** | **38.42** | **34.9** | **26.24** | **49.65** | **19.21** | **27.4** | **49.48** | **14.6** | **18.66** |
| Cc | 0.66 | 32.18 | 40.33 | 20.98 | 8.68 | 18.18 | 19.11 | 12.38 | 0 |
| Dl | 0 | 0 | 0 | 0 | 24.94 | 5 | 0 | 13.58 | 12.11 |
| Ank | 9.92 | 1.61 | 4.97 | 9.57 | 7.78 | 10.33 | 11.94 | 7.1 | 5.73 |
| Rch | 0.17 | 0.06 | 0.09 | 0.15 | 0.17 | 0.08 | 0.11 | 0.11 | 0.19 |
| Sd | 0 | 0 | 0 | 0 | 0 | 0 | 0 | 0 | 6.1 |
| **Carbonates** | **10.75** | **34.9** | **26.24** | **49.65** | **19.21** | **27.4** | **49.48** | **14.6** | **18.66** |
| Ap | 0.2 | 0.23 | 0.08 | 0.21 | 0.1 | 0.08 | 0.25 | 0.21 | 0.52 |
| Rt | 0.48 | 0.4 | 0.27 | 0.85 | 0.23 | 0.46 | 0.57 | 0.2 | 0.53 |
| Prl | 0 | 0 | 0 | 0 | 0 | 0 | 0 | 0 | 0 |
| Ht | 0 | 0 | 0 | 0 | 0 | 0 | 0 | 0 | 0 |
| **Others** | **0.68** | **0.63** | **0.35** | **1.06** | **0.33** | **0.54** | **0.82** | **0.41** | **1.05** |
| Sample | 81 372.2 | 80 328.1 | 81 360.1 | 81 361.2b | 8036.1 | 362.5 | 81 311.5 | 81 311.6 | 81 334.12 |
| Rock | Bi-Gr gneiss | Di-Sc schist | Di-Sc schist | Di-Sc schist | Di-Sc schist | Di-Pl schist | Di-Sc schist | Di gneiss | Di plagio-gneiss |

For abbreviations see Table 1.

**Table 4.** The representative analysis (%wt) and normative mineral composition of Early Proterozoic granulite-facies meta-sedimentary rocks of the Anabar Shield, Hapschan Group, meta-carbonate calciphyre.

| | 28 | 29 | 30 | 31 | 32 | 33 | 34 | 35 | 36 |
|---|---|---|---|---|---|---|---|---|---|
| $SiO_2$ | 63.68 | 11.16 | 14.23 | 14.58 | 20.76 | 22.33 | 25.12 | 33.82 | 1.74 |
| $TiO_2$ | 0.55 | 0.14 | 0.09 | 0.17 | 0.54 | 0.33 | 0.46 | 0.31 | 0.04 |
| $Al_2O_3$ | 14.07 | 2.58 | 3.88 | 3.58 | 7.04 | 5.62 | 7.79 | 7.91 | 1.01 |
| $Fe_2O_3$ | 0.05 | 0.61 | 1.76 | 0.49 | 0.7 | 1.23 | 0.83 | 1.03 | 0.2 |
| FeO | 3.47 | 1.14 | 0 | 1.14 | 2.49 | 1.39 | 2.75 | 2.22 | 0 |
| MnO | 0.05 | 0.13 | 0.02 | 0.04 | 0.04 | 0.01 | 0.03 | 0.05 | 0.02 |
| MgO | 1.55 | 2.5 | 8.61 | 11.81 | 0.99 | 1.68 | 1.01 | 1.27 | 18.63 |
| CaO | 10.53 | 44.69 | 38.83 | 35.8 | 36.33 | 37.09 | 38.18 | 31.8 | 34.39 |
| $Na_2O$ | 2.3 | 0.33 | 0.27 | 0.5 | 1.4 | 1.15 | 1.4 | 1.53 | 0.1 |
| $K_2O$ | 1.43 | 0.22 | 0.38 | 0.67 | 1.01 | 1.25 | 1.49 | 1.86 | 0.04 |
| $P_2O_5$ | 0.16 | 0.02 | 0.11 | 0.09 | 0.02 | 0.02 | 0.04 | 0.04 | 0.06 |
| LOI | 1.78 | 35.32 | 32.43 | 30.38 | 28.16 | 27.86 | 20.86 | 17.96 | 43.68 |
| Total | 99.62 | 98.84 | 100.61 | 99.25 | 99.48 | 99.96 | 99.96 | 99.8 | 99.91 |
| $CO_2$ | nd | 33.8 | 32 | 29.5 | 26.06 | 26 | 20 | 17.9 | nd |
| *Normative mineral composition of primary rocks* | | | | | | | | | |
| Ab | 19.92 | 2.76 | 2.13 | 3.83 | 11.75 | 9.33 | 10.58 | 11.79 | 0.82 |
| An | 2.11 | 0.29 | 0.23 | 0.41 | 1.25 | 0.99 | 1.12 | 1.25 | 0.09 |
| Or | 6.67 | 0 | 0 | 0 | 2.42 | 2.44 | 0.47 | 3.66 | 0 |
| Q | 22.31 | 6.62 | 8.09 | 7.09 | 5.43 | 8.98 | 7.98 | 14.3 | 0.3 |
| **Clastics** | **51.01** | **9.67** | **10.45** | **11.33** | **20.85** | **21.74** | **20.15** | **31** | **1.21** |
| Mm | 0 | 0 | 0 | 0 | 0 | 0 | 0 | 0 | 0 |
| Ill | 26.23 | 2.32 | 3.79 | 6.49 | 6.34 | 8.4 | 13.38 | 11.51 | 0.42 |
| Chl | 0 | 4.52 | 6.66 | 0.75 | 7.8 | 0 | 0 | 0 | 2.23 |
| Kn | 0 | 0 | 0 | 0 | 0 | 0 | 0 | 0 | 0 |
| Srp | 0 | 0 | 0 | 0 | 0 | 0 | 0 | 0 | 0 |
| **Pelites** | **26.23** | **6.84** | **10.45** | **7.24** | **14.14** | **8.4** | **13.38** | **11.51** | **2.65** |
| Cc | 0 | 73.67 | 47.85 | 30.2 | 63.33 | 56.55 | 55.08 | 45.44 | 16.09 |
| Dl | 8.41 | 6.67 | 27.83 | 47.31 | 0 | 6.6 | 2.89 | 4.24 | 79.38 |
| Ank | 13.24 | 2.75 | 3.06 | 3.52 | 1.05 | 6.32 | 7.96 | 7.38 | 0.46 |
| Rch | 0.12 | 0.21 | 0.03 | 0.06 | 0.06 | 0.02 | 0.04 | 0.07 | 0.03 |
| Sd | 0.02 | 0 | 0 | 0 | 0 | 0 | 0 | 0 | 0 |
| **Carbonates** | **21.79** | **83.3** | **78.77** | **81.09** | **64.44** | **69.49** | **65.97** | **57.13** | **95.96** |
| Ap | 0.34 | 0.05 | 0.24 | 0.19 | 0.05 | 0.05 | 0.09 | 0.09 | 0.14 |
| Rt | 0.62 | 0.14 | 0.08 | 0.15 | 0.54 | 0.32 | 0.41 | 0.28 | 0.04 |
| Prl | 0 | 0 | 0 | 0 | 0 | 0 | 0 | 0 | 0 |
| Ht | 0 | 0 | 0 | 0 | 0 | 0 | 0 | 0 | 0 |
| **Others** | **0.96** | **0.19** | **0.32** | **0.34** | **0.59** | **0.37** | **0.5** | **0.37** | **0.18** |
| Sample | 80 314.1b | 79 321.5 | 81 369.1 | 9P29.3 | 81 371.1 | 81 377 | 81 372.1 | H7.6.81 | 81 382.2 |
| Rock | Di plagiogneiss | Di-Pl calciphyre | Di-Fr calciphyre | Di-Fr calciphyre | Di-Sc calciphyre | Di-Sc calciphyre | Di-Sc calciphyre | Di-Sc calciphyre | Sp-Fr-marble |

For abbreviations see Table 1.

**Table 5.** The representative analysis (%wt) and normative mineral composition of Early Proterozoic granulite-facies meta-sedimentary rocks of the Anabar Shield, Hapschan Group, meta-carbonate marble.

| | 37 | 38 | 39 | 40 | 41 |
|---|---|---|---|---|---|
| $SiO_2$ | 4.06 | 5.03 | 6.52 | 8.35 | 10.45 |
| $TiO_2$ | 0.1 | 0.18 | 0.12 | 0.14 | 0.1 |
| $Al_2O_3$ | 0.68 | 1.65 | 1.37 | 1.03 | 1.4 |
| $Fe_2O_3$ | 0.25 | 0.7 | 0.38 | 0.52 | 0.46 |
| FeO | 0.52 | 0.31 | 0.61 | 0.15 | 3.67 |
| MnO | 0.02 | 0.01 | 0.08 | 0.12 | 0.04 |
| MgO | 0.25 | 7.51 | 3.43 | 3.03 | 10.76 |
| CaO | 54.06 | 40.84 | 46.92 | 45.45 | 36.12 |
| $Na_2O$ | 0.06 | 0.12 | 0.26 | 0.2 | 0.13 |
| $K_2O$ | 0.1 | 0.09 | 0.51 | 0.29 | 0.07 |
| $P_2O_5$ | 0.02 | 0.01 | 0.02 | 0.02 | 0.1 |
| LOI | 40.14 | 42.5 | 38.5 | 40 | 36.22 |
| Total | 100.26 | 98.95 | 98.72 | 99.3 | 99.52 |
| $CO_2$ | 38.57 | 41.93 | 37.74 | 38.27 | 33.02 |
| *Normative mineral composition of primary rocks* | | | | | |
| Ab | 0.49 | 1.05 | 2.17 | 1.72 | 1.05 |
| An | 0.05 | 0.11 | 0.23 | 0.18 | 0.11 |
| Or | 0 | 0 | 2.25 | 0.95 | 0 |
| Q | 2.85 | 2.89 | 2.76 | 5.91 | 7.97 |
| **Clastics** | **3.39** | **4.05** | **7.41** | **8.76** | **9.13** |
| Mm | 0 | 0 | 0 | 0 | 0 |
| Ill | 1.04 | 1 | 1.31 | 1.43 | 0.71 |
| Chl | 0.95 | 4 | 0 | 0 | 3.1 |
| Kn | 0 | 0 | 0 | 0 | 0 |
| Srp | 0 | 0 | 0 | 0 | 0 |
| **Pelites** | **8.77** | **5** | **1.31** | **1.43** | **3.81** |
| Cc | 92.78 | 69.14 | 72.95 | 73.78 | 32.62 |
| Dl | 0.52 | 24.97 | 15.37 | 13.91 | 43.29 |
| Ank | 1.15 | 0.3 | 2.66 | 1.73 | 10.77 |
| Rch | 0.03 | 0.03 | 0.13 | 0.2 | 0.06 |
| Sd | 0 | 0 | 0 | 0 | 0 |
| **Carbonates** | **94.48** | **94.44** | **91.11** | **89.62** | **86.74** |
| Ap | 0.05 | 0.02 | 0.05 | 0.05 | 0.23 |
| Rt | 0.1 | 0.05 | 0.12 | 0.14 | 0.1 |
| Prl | 0 | 0 | 0 | 0 | 0 |
| Ht | 0 | 0 | 0 | 0 | 0 |
| **Others** | **0.15** | **0.07** | **0.17** | **0.19** | **0.33** |
| Sample | 386.1 | 81 369.2 | 81 366 | 81 334.9a | 81 334.3 |
| Rock | Di marble | Di-Fr marble | Di-Sc marble | Di-Sc marble | Fr marble |

For abbreviations see Table 1.

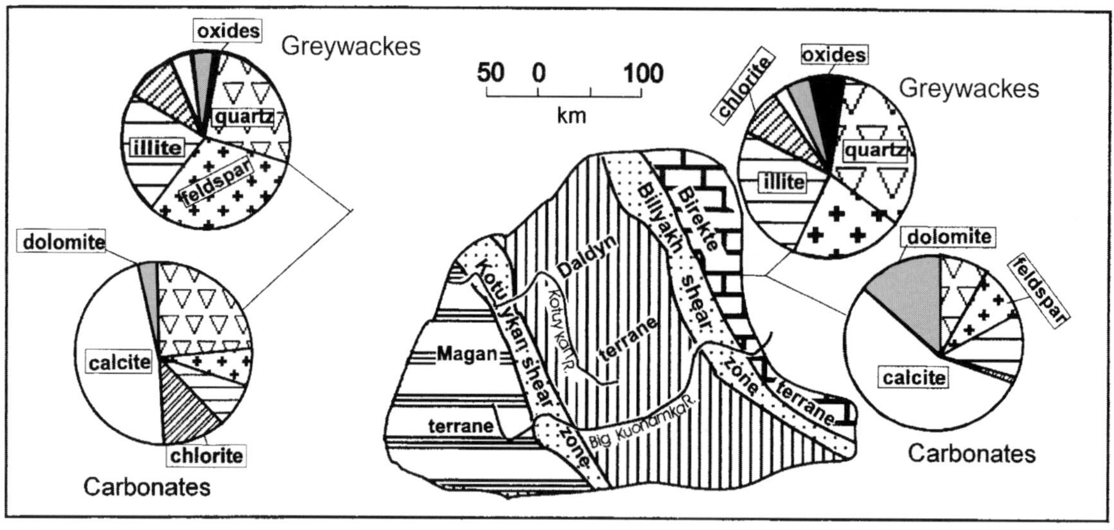


**Fig. 5.** Representative normative mineral compositions for Early Proterozoic meta-terrigenous and meta-carbonate rocks of the Magan and Birekte terranes, calculated with the MINLITH method.

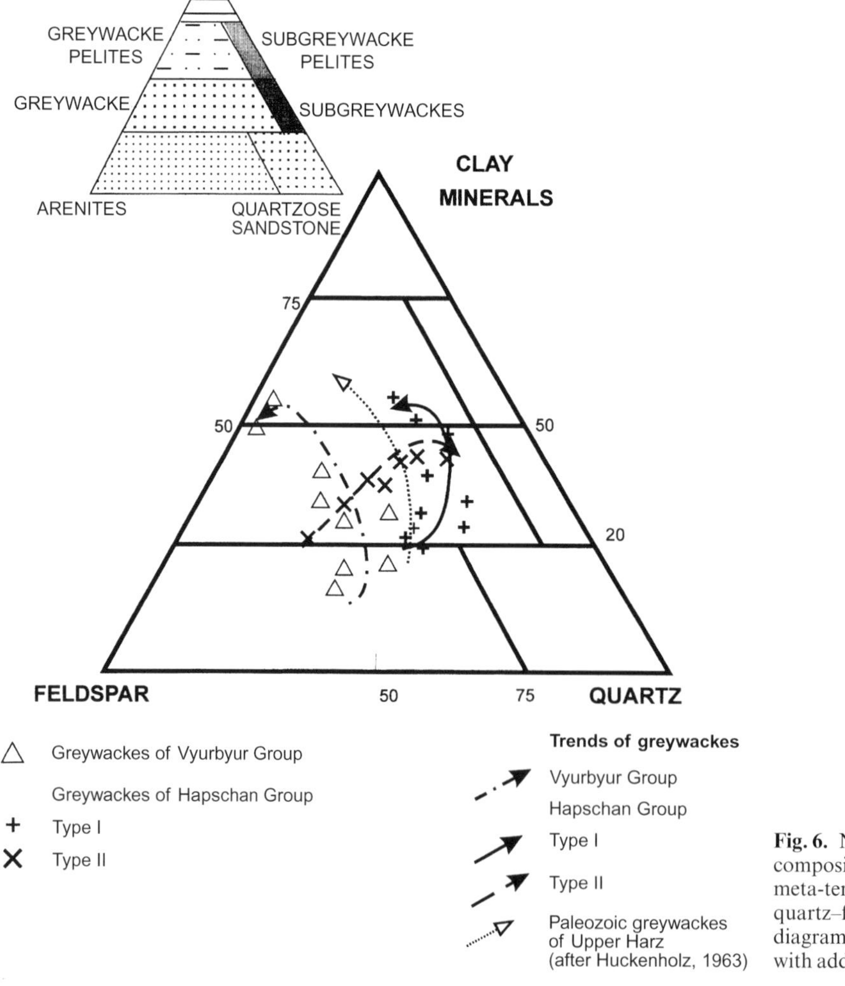


**Fig. 6.** Normative mineral composition of the Early Proterozoic meta-terrigenous rocks in the quartz–feldspar–clay minerals diagram (after Pettijhon *et al.*, 1972, with additions).

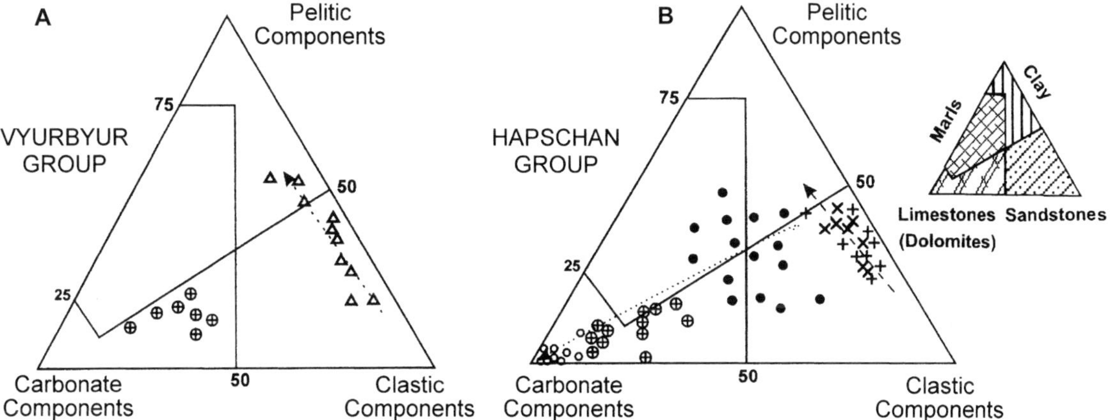


**Fig. 7.** Diagram of carbonate–pelitic–clastic components (modified after Rosen & Zlobin, 1990) for normative mineral composition of the Early Proterozoic metasedimentary rocks of the Anabar shield.

The meta-terrigenous rocks of the Vyurbyur Group are assumed to represent volcaniclastic greywackes because of their extreme enrichment in feldspar relative to the other studied clastic rocks (Fig. 6). From low- to high-silicate rocks, they show an increase in normative clastic components (feldspars and quartz) and a decrease in pelitic components (montmorillonite, illite, chlorite, kaolin, serpentine). The clastic component is characterized by an increase in its quartz constituent at a stable content of mainly plagioclase feldspar. The pelitic component has a high chlorite content.

Normative carbonate minerals (predominantly calcite, dolomite, ankerite, rhodochrosite) which are mainly found in impure marbles, range from 50 to 70% for high- to low-silicate rocks. The normative silicate minerals mainly consist of a quartz-dominated clastic constituent. Chlorite is the most common pelitic mineral.

The Hapschan Group predominantly contains quartz-feldspathic greywackes (Type I, Table 2) with a low volcanic component. The terrigenous and pelitic constituents in Hapschan Group rocks are similar to those in meta-greywackes of the Vyurbyur Group, but the increased $SiO_2$ content results in normative quartz increase and feldspar decrease. Illite is the most common pelitic mineral.

The other clastic rocks of the Hapschan Group show a cross-cutting trend in the greywacke mineral diagram (Fig. 6), beginning from the highest feldspar content field (Type II, Table 2). These rare greywackes show reversed proportions of coarse clastic and pelitic constituents and probably represent volcaniclastic sediments. The ratio of normative minerals in the terrigenous component is typical of greywackes, whereas the composition of the pelitic component is similar to that of volcaniclastic greywackes of the Vyurbyur Group.

The studied meta-carbonate rocks of the Hapschan Group are represented by marbles–impure marbles–calc–silicate rock series. In these series, the carbonate component decreases from 95% in marbles to 15% in calc-silicate rocks. The predominant normative mineral is calcite. However, each member of these series contains samples with dolomite as the predominant mineral. All studied rocks contain ankerite, whose content usually increases from marbles to calc-silicate rocks. The content of normative clastic and pelitic components increases from marbles to calc-silicate rocks. These rocks mainly consist of clastic components, where the quartz content somewhat exceeds the normative feldspar content. The pelitic component of these series is mainly represented by illite, but some rocks contain montmorillonite as the dominant pelitic mineral.

A considerable difference in the average compositions of the meta-terrigenous and meta-carbonate rocks of the Vyurbyur and Hapschan groups (Fig. 5) indicates different provenance and facies settings during the accumulation of the primary sedimentary rocks.

## DISCUSSION

As seen in the quartz–feldspar–pelitic minerals diagram (Fig. 6), most data points are plotted in the greywacke field. All distinguished greywacke groups have individual trends. In particular, the trends of greywackes from the Hapschan Group and volcaniclastic greywackes from the Vyurbyur Group are subparallel to the trend of Palaeozoic greywackes from the Upper Harz Mountain (genotype greywacke area). They show a decrease in pelitic component and an increase in clastic component, where normative quartz content increases relative to feldspar from low-Si, fine-grained to high-Si, coarse-grained rocks (Huckenholz, 1963). The Hapschan greywacke trend is displaced towards the quartz corner relative to the Harz greywackes, whereas the Vyurbyur Group trend of volcaniclastic greywackes is shifted towards the feldspar field. The Hapschan Group trend of volcaniclastic greywackes is discordant to Vyurbyur Group rocks and shows an increase in pelitic and a decrease in clastic (quartz, feldspar) components, mainly at the expense of feldspar from low-silicate to high-silicate rocks.

In the clastic (rock fragments)–carbonate–pelitic components (Fig. 7) the meta-terrigenous and meta-carbonate rocks of both the Vyurbyur and Hapschan groups are clustered as separate fields, allowing us to distinguish the greywacke and carbonate series. These features may be explained as follows. The calcium content of the Early Proterozoic Vyurbyur and Hapschan sedimentary basins was high enough to maintain the carbonate accumulation. The carbonate accumulation rate was very low (Garrels & McKenzie, 1971; Pettijohn, 1975), but because of a stable tectonic setting in the provenance area, the amount of precipitated carbonates exceeded the terrigenous input supplied from distal sources. The repeated tectonic reactivation of the provenance area resulted in an avalanche of immature terrigenous material (greywackes). This completely suppressed the carbonate accumulation. The low content of low-silicate rocks corresponding to fine-grained greywackes, which are normally represented by intrabed mudstones and siltstones, indicates an input of immature terrigenous material from proximal sources.

In the Vyurbyur basin, sedimentation was accompanied by intense mafic and intermediate to acid volcanism. This is confirmed by the displacement of the normative mineral trend of terrigenous rocks towards the feldspar field in the quartz–feldspar–pelitic minerals diagram (Fig. 6), as well as by the high chlorite content in the pelitic component of greywackes and carbonate rocks (Table 1). The considerable amount of normative quartz in the terrigenous constituent is presumably related to its deposition from colloidal solutions supplied by submarine volcanism.

Judging from the numerous thin lenses occurring in the lower portions of the Vyurbyur sequence, the carbonates were presumably accumulated in semi-enclosed basins with a limited supply of terrigenous and volcanic material. The relative isolation of these basins facilitated evaporite formation, which is deduced from the anomalously high carbon isotope composition ($\delta^{13}C = 7–10‰$; Rosen *et al.*, 1990).

As indicated by the Sm–Nd model ages in the meta-sedimentary rocks of the Daldyn and Magan terranes, the sources of the Vyurbyur basin sediments are represented by continental rocks similar to the exposed Archaean crust of the Daldyn terrane. Basaltic to andesitic volcanism with intercalated volcaniclastic sediments suggest that the Vyurbyur basin probably formed in an active continental margin setting.

The uniform composition of the Hapschan Group, with extensive carbonate horizons, indicates that sedimentation in the Hapschan basin occurred in a shallow shelf setting. This is implied by the similar isotope carbon composition ($\delta^{13}C$ ranges from −3 to −0.5‰) of these rocks and Phanerozoic marine carbonates (Rosen *et al.*, 1990). The presence of iron oxides in the normative greywacke composition indicates the higher maturity of the terrigenous component supplied to the sedimentary basin. The shift in the trend of these greywackes towards the quartz field in the quartz–feldspar–pelitic diagram (Fig. 6) supports this assumption.

The rare volcaniclastic greywackes were derived from proximal sources during eruptions or erosion of andesite, or andesite–dacite volcanic build-ups. This is shown by data plots of the low-silicate varieties near the feldspar field. The termination of these processes resulted in a change from a high feldspar content to a higher percentage of pelitic minerals (mainly montmorillonite and chlorite: Table 2) and quartz in the supplied material. This is confirmed by data plots of the high-silicate varieties (Fig. 6).

In the Hapschan basin, sedimentation was accompanied by an input of more mature terrigenous material than the admixture supplied to the greywacke series from the distal sources, as indicated by the trend of carbonate rocks in the clastic–pelitic–carbonate components diagram (Fig. 7B) and by the illite predominance in the pelitic component. A significant amount of terrigenous carbonate rock (calc-silicate rocks) in the Hapschan sequences indicates the existence of fairly long-term lapses related to relatively uplifted provenances, when input of terrigenous material was comparable to the amount of carbonate precipitation. The subsequent subsidence of the provenance led to a sharp decrease in the terrigenous component and to prevailing carbonate accumulation. Distal provenances were represented by a mature continental crust that is not observed at the present-day level of exposure. The types of protolith sediments and their intercalation with each other are in agreement with the interpretation that the Hapschan terrane was formed in a passive continental margin setting (Condie *et al.*, 1991).

## CONCLUSIONS

The spatially separated Vyurbyur and Hapschan sedimentary basins existed during the formation of the Early Proterozoic crust of the north-eastern Siberian Platform.

Sedimentation in the Vyurbyur basin occurred in an active continental margin setting and was accompanied by volcanism. The source areas of the accumulated terrigenous sediments were represented by a continental crust resembling the Archaean crust of the Daldyn terrane. Carbonates accumulated in half-closed, partly evaporitic basins.

Sedimentation in the Hapschan basin occurred in a shelf-related passive continental margin setting. Source areas were composed of mature continental crust, which is not observed at the present-day erosion level. Greywacke series accumulated during tectonic reactivation, whereas carbonates accumulated under stable tectonic conditions.

## ACKNOWLEDGEMENTS

This study was supported by Russian Foundation for Basic Research Projects 99-0565154, 98-05-65081 and 00-05-64130. We thank our colleague Maria M. Bogina for initial translation of the manuscript into English. We also thank Wladyslaw Altermann for help in the preparation of this manuscript.

## APPENDIX: DOMAIN PRINCIPLES OF THE MINLITH PROGRAM

The MINLITH is proposed as a universal (general) method to calculate mineral compositions of all the main types of sedimentary rocks (siliciclastics, carbonates etc.) from their chemical analyses. The procedure involves three main stages and several minor stages, which were combined in the program. Results are expressed in wt%. The algorithm (700 operations) includes the immediate calculation of each mineral from major chemical elements (excluding volatiles), and subsequent control with other mineral-forming elements. The deficiency of any element returns the program to the previous stage, and the distribution of elements between other minerals is recalculated according to empirical regularities of the distribution of these elements (in standard 600 reference samples). The obtained quantitative mineral composition represents a unique (single) result of arithmetic calculations (Rosen *et al.*, 2000, and references therein; Rosen *et al.*, 2002). The main stages in MINLITH are as follows.

First is calculation of contents of accessory and minor components using the stoichiometric compositions of apatite (from P content), pyrite ($S_{pyr}$), gypsum ($SO_3$) and fluorite (F), while $C_{org}$ carbonaceous matter and $TiO_2$ (rutile) are included directly in the final results.

Second is calculation of the amount of aluminosilicate and gibbsite on the basis of aluminium distribution between relevant minerals according to the observed contents of Na, K, Ca, Mg, Fe and Si. The accepted formulas are calculated in oxides as follows:

- Plagioclase: $0.8 \times Na_2O\text{-}Al_2O_3\text{-}SiO_2 + 0.2 \times CaO\text{-}Al_2O_3\text{-}2SiO_2$
- Orthoclase: $K_2O\text{-}Al_2O_3\text{-}6SiO_2$
- Illite: $2K_2O\text{-}MgO\text{-}FeO\text{-}6.5Al_2O_3\text{-}16SiO_2\text{-}5H_2O$
- Montmorillonite: $Na_2O\text{-}2MgO\text{-}5Al_2O_3\text{-}24SiO_2\text{-}6H_2O$
- Chlorite: k1 MgO-k2 FeO-k3 $Al_2O_3$-k4 $SiO_2$-$4H_2O$
- Kaolinite: $Al_2O_3\text{-}2SiO_2\text{-}2H_2O$
- Gibbsite: $Al_2O_3\text{-}3H_2O$.

Third, the remaining chemical components (after stages 1 and 2) are assigned to silicates, oxides and carbonates according to the excess or deficiency of Ca and amounts of Si, Fe, Mg and Mn. Silicates and oxides: quartz $SiO_2$, goethite $Fe_2O_3$-$H_2O$, pyrolusite MnO and serpentine $3MgO$-$2SiO_2$-$2H_2O$. Carbonates: calcite CaO-$CO_2$, dolomite CaO-MgO-$2CO_2$, ankerite CaOFeO-$2CO_2$, siderite FeO-$CO_2$, magnesite MgO-$CO_2$ and rhodochrosite MnO-$CO_2$.

Normative plagioclase composition is specified as statistically widespread plagioclase in most clastic rocks, compositions of illite and montmorillonite are accepted according to exactly authigenic examples. In the chlorite formula, coefficients k1, k2, k3 and k4 are functions of Fe/(Fe + Mg) and were obtained from compositional data on chlorites from sedimentary rocks, where iron contents and Fe/(Fe + Mg) ratio in bulk composition correlate with these contents in chlorite (Rosen *et al.*, 2000, and references therein). MINLITH supplies a facility to calculate a correct full formula of normative chlorite.

When compared, the data derived from X-ray powder diffraction on modal minerals, the clay phase identified as normative illite, denote a group of unexpanding clay minerals including authigenic illite, detritic (relic) di-octahedral K-mica of IMd type (where K deficiency is balanced by $H_2O$) and small amounts of mixed-layer phases. Normative montmorillonite includes montmorillonite proper, other smectites and mixed-layer phases, which consist of alternating illite–montmorillonite and partially ordered modifications with 10–20% expanding layers and disordered modifications with more than 40% expanding layers, which may be partially incorporated into normative chlorite as well.

However, the number of elements used for the calculation of mineral balance (Na, K, Ca, Mg, Fe and Si), i.e. the number of equations, is insufficient for an unambiguous solution (the number of minerals or the number of unknowns is higher than the number of equations). Therefore, we used additional empirical equations (based on standard 600 reference samples) relating the $Al_2O_3$ content of the rock to the association of $Al_2O_3$-bearing minerals. Kaolinite and feldspars are excluded from calculations at low and high $Al_2O_3$ contents, respectively.

$H_2O$ (in any form) and $CO_2$, if present in analyses, were ignored in the calculations. They may be counted up from rock normative composition according to formulas above. When compared with contents in chemical analyses of unmetamorphosed rocks these calculated volatiles reveal a good agreement. Calculations by MINLITH open a way to reconstruction of the initial mineral (sediment) associations and to palaeoenvironmental interpretations of metamorphosed sedimentary rock assemblages, which have lost volatiles under high temperature and pressure conditions.

The main point in the development of a universal (general) method is the concordance between the model (calculated or normative) and the observed (modal) mineral compositions of sedimentary rocks. Any method of quantitative investigation is characterized by two major parameters: (i) reliability (accuracy) and (ii) precision (reproducibility). Both chemical analytical methods and methods of direct determination of mineral composition (e.g. petrographic, X-ray and others) have their own errors in reliability (accuracy), affecting the results of mineral composition calculations. For a statistically straight assessment of reliability/accuracy (irrespective of bulk contents), the magnitude of a relative mean-square error, *Vd* (coefficient of variation), was evaluated, in rel.%, according to the equation:

$$Vd = \sqrt{\frac{\sum_{i=1}^{n}\left(\frac{d_i}{\bar{X}_i}\cdot 100\right)^2}{2n}}$$

where $n$ is the amount of parallel normative and modal values (weight contents, in wt%), $d_i = |x_i1 - x_i2|$ (absolute difference between normative and modal values for the pair) and $\bar{X}_i = (x_i1 + x_i2)/2$ is the arithmetic mean of these values.

In a series of pair parallel definitions of the recent 120 high-quality reference analyses, the reliability (accuracy) of calculations determined as relative error appears to be $Vd = \pm$ 5–15 rel.% for most minerals. It is of the same order of magnitude when compared with instrumental estimations of the highest precision (reference in Rosen *et al.*, 2002). In some cases that error rises to 40 rel.% for normative feldspars; comparable with that of some instrumental phase estimations. When normative mineral contents are less than 5 wt% the relative error is up to 60–70 rel.%. That decrease of accuracy for minor normative components is in good accordance with the instrumental measurements. Such data may be used as semi-quantitative assessments.

## REFERENCES

Condie, K.C., Wilks, M., Rosen, O.M. & Zlobin, V.L. (1991) Geochemistry of metasediments from the Precambrian Hapschan series, Earthten Anabar Shield, Siberia. *Precam. Res.*, **50**, 37–47.

Dobretsov, N.L., Bogatikov, O.A. & Rosen, O.M. (Eds) (1992) *Classification and Nomenclature of the Metamorphic Rocks, Handbook*. OIGGM Press, Novosibirsk, 205 pp. (in Russian).

Garrels, R.M. & Mc Kenzie, F.T. (1971) *Evolution of Sedimentary Rocks*. Norton, New York, 397 pp.

Grozdilov, L.A. (1983) Geological map of USSR, scale 1 : 1 000 000 (new series). Explanatory notes (sheet R 48–50). VSEGEI, Leningrad, 210 pp. (in Russian).

Huckenholz, H.G. (1963) Mineral composition and texture in graywackes from Harz Mountains (Germany) and in arkoses from Auvergne (France). *J. sediment. Petrol.*, **33**, 914–918.

Khoreva, B.Ya. (Ed.) (1987) Map of metamorphic complexes and granitoids of USSR, scale 1 : 10 000 000. Map Press, Ministry of Geology, Moscow (in Russian).

Krylov, A.Ya., Vishnevskii, A.N. & Silin B.L. (1963) Absolute age of the rocks of the Anabar Shield. *Geokhimiya*, **12**, 1140–1144 (in Russian).

Markov, M.S. (Ed.) (1988) *Archean Structures of the Anabar Shield and Problems of the Early Evolution of the Earth*. Nauka Press, Moscow, 255 pp. (in Russian).

Mokshantsev, K.B. (1979) *Proterozoic Structures of the North-eastern Siberian Platform*. Nauka Press, Novosibirsk, 127 pp. (in Russian).

Pettijohn, F.J. (1975) *Sedimentary Rocks*. Harper & Row, New York, 750 pp.

Pettijohn, F.J., Potter, P.E. & Siever, R. (1972) *Sands and Sandstone*. Springer-Verlag, New York, 618 pp.

Rosen, O.M. (1970) Recalculation of chemical composition of meta-sediment schists to components of the sedimentary rocks. *Sov. Geol.*, **7**, 31–41 (in Russian).

Rosen, O.M. (1975) Sedimentological interpretation by chemical composition of Precambrian meta-sediment schists. In: *Problems of Sedimentary Geology of the Precambrian* (Ed. Sidorenko, A.V.), pp. 60–75. Nedra, vyp.4, Moscow (in Russian).

Rosen, O.M. (1994) Graywackes from the metamorphic complexes of Precambrian: relation of composition and geodynamic formation settings. *Izv. Vyssh. Uchebn. Zaved., Geol. Razved*, **1**, 36–49 (in Russian).

Rosen, O.M. (1995) Metamorphic effects of tectonic movements at the lower crust level: Proterozoic collision zones and terranes of the Anabar shield. *Geotectonics*, **29**, 91–101 (in Russian).

Rosen, O.M., Abbyasov, A.A., Migdisov, A.A. & Yaroshevsky, A.A. (2000) MINLITH–a program to calculate the normative mineralogy of sedimentary rocks: the reliability of results obtained for deposits of old platforms. *Geochem. Int.*, **38**(4), 388–400.

Rosen, O.M., Abbyasov, A.A. & Tipper, J. (2002) MINLITH–general program to calculate mineralogical composition of sedimentary rocks from chemical analyses. *Comp. Geosci.* (in the press).

Rosen, O.M., Bibikova, E.V. & Zhuravlev, D.Z. (1991) Early crust of the Anabar Shield, age and formation models. In: *Early Crust: Its Composition and Age* (Ed. Shuklov, U.A.), Nauka Press, Moscow, 199–224 (in Russian).

Rosen, O.M., Condie, K.C., Natapov, L.M. & Nozshkin, A.D. (1994) Archean and early Proterozoic evolution of the Siberian craton: a preliminary assessment. In: *Archean Crustal Evolution* (Ed. Condie, K.C.), pp. 411–459. Elsevier, Amsterdam.

Rosen, O.M. & Dimroth, E. (1982) Old metamorphosed graywackes in the continental crust basement: the study of primary mineral composition (Canada, USSR). In: *Sedimentary Geology of the Highly Metamorphosed Precambrian Complexes* (Ed. Sidorenko, A.V.), pp. 155–178. Nauka Press, Moscow (in Russian).

Rosen, O.M., Zhuravlev, D.Z., Bibikova, E.V., Sukchanov, M.K. & Zlobin, V.L. (1999) Early Proterozoic amalgamation of continents: terranes, collisional zones and associated anorthosites in the North-Eastern Siberian craton (view from isotope cheochemistry and geochronology. In: *Early Precambrian: Genesis and Evolution of the Continental Crust (Geodynamics, Petrology, Geochronology, Regional Geology)*. Abstr. Vol., Geos M., 143–144.

Rosen, O.M. & Zlobin, V.L. (1990) Carbonate sediments of Early Archean granulite and greenstones belts. *Int. geol. Rev.*, **32**, 539–550.

Rosen, O.M., Zlobin, V.L. & Syngaevskii, E.D. (1990) Metamorphosed carbonate rocks from the granulite complex of the Anabar Shield: features of primary composition and sedimentation. *Litol. Polezn. Iskop.*, **5**, 72–81 (in Russian).

Spiridonov, V.G., Karpenko, S.F. & Lyalikov, A.V. (1993) Sm–Nd age and geochemistry of granulites from the central part of the Anabar Shield. *Geokhimiya*, **19**, 1412–1427 (in Russian).

Taylor, S.R. & McLennan, S.M. (1985) *The Continental Crust: Its Composition and Evolution*. Blackwell Scientific Publications, Oxford, 312 pp.

Vishnevskii, A.N. (1978) *Metamorphic Complexes of the Anabar Shield*. Nedra, Leningrad, 156 pp. (in Russian).

Zhuravlev, D.Z. & Rosen, O.M. (1991) Sm–Nd age of the metasediments of the granulite complex of the Anabar Shield. *Doklady AN*, **317**(1), 189–193 (in Russian).

Zlobin, V.L. (1990) Lithological reconstructions and correlation of sections of the carbonate-gneiss association of the Anabar Shield. In: *Lithological Investigations During Detailed Subdivision and Correlation of the Sedimentary Sequences* (Ed. Kazanskiy, U.P.), Nauka Press, Novosibirsk, pp. 6776 (in Russian).

Zlobin, V.L., Zhuravlev, D.Z. & Rosen, O.M. (1999) Sm–Nd model age of the meta-carbonate–gneiss association of the Anabar granulite complex of the western part of the Anabar Shield, Polar Siberia. *Dokl. Ross. Akad.*, **368**, 95–98 (in Russian).

*Spec. Publs int. Ass. Sediment.* (2002) **33**, 293–321

# Mixed siliciclastic–carbonate storm-dominated ramp in a rejuvenated Palaeoproterozoic intracratonic basin: upper Hurwitz Group, Nunavut, Canada

L. B. ASPLER* *and* J. R. CHIARENZELLI†

**23 Newton Street, Ottawa, Ontario, K1S 2S6, Canada; and*
*†Department of Geology, State University of New York at Potsdam, Potsdam, New York 13676, USA*

## ABSTRACT

The Hurwitz Group was deposited in an intracratonic basin that covered at least 140 000 km$^2$ of the Hearne domain (northern Canada) during the early Palaeoproterozoic. Continental siliciclastic rocks in the lower Hurwitz Group (sequences 1 and 2) record regional sagging due to <2.45–2.11 Ga lithospheric stretching during protracted break-up of a speculative Neoarchaean supercontinent (Kenorland). Upper Hurwitz Group strata (sequences 3 and 4) reflect basin rejuvenation after a long period of apparent tectonic quiescence (*c*.2.11–1.91 Ga). Post *c*.1.91 Ga, basin-margin arching, basin-centred subsidence and marine flooding, in part overlapping with assembly of Laurentia, led to growth of an extensive, gently inclined, storm-dominated, microtidal, mixed siliciclastic–carbonate ramp (Watterson Formation) that was paired to local deltas (Ducker Formation) and a low-relief sandy alluvial plain (lower Tavani Formation). Sequence 4 (upper Tavani Formation) reflects re-establishment of the ramp following marine flooding. Outer- to mid-ramp siliciclastic pelites and detrital carbonates represent background suspension sedimentation that was interrupted by siliciclastic and carbonate mass flows. Distal inner-ramp environments (stromatolitic bioherms isolated within fields of microbial mats) reflect fairweather calcilutite sedimentation and/or *in situ* precipitation punctuated by siliciclastic-contaminated tempestites. Proximal inner-ramp environments (fields of subtidal stratiform stromatolite interspersed with pools receiving silicilastic mud and sand derived from subaerial sheet floods) formed a broad transition linking marine and continental environments. The ramp geometry and environmental zonation of the upper Hurwitz Group are comparable to post-Precambrian mixed siliciclastic–carbonate systems, despite contrasts in the relative importance of specific depositional processes arising from differences in life forms, both on land and in the sea.

## INTRODUCTION

The diversity of carbonate platforms recognized in Precambrian basins closely matches the spectrum defined for Phanerozoic and modern successions, and the fundamental controls on platform evolution appear to have been established since the beginning of the Palaeoproterozoic (Grotzinger, 1989). Despite the lack of Precambrian carbonate-secreting skeletal metazoa, and despite possible differences in seawater chemistry, the depositional environments, overall extent and architecture of Palaeoproterozoic carbonate platforms appear remarkably similar to post-Precambrian examples (Grotzinger, 1989). In addition, rates of carbonate sedimentation may have been similar in the Precambrian, as indicated by *c*.2.55 Ga deposits on the Kaapvaal craton, where rates calculated for a wide range of depositional environments compare well to those in analogous Phanerozoic and modern settings (Altermann & Nelson, 1998).

During post *c*.1.91 Ga rejuvenation of Hurwitz Basin in the Western Churchill Province of northern Canada (Fig. 1), crustal flexure led to growth of an extensive mixed siliciclastic–carbonate ramp that was paired to a low-relief sandy alluvial plain adjacent to a basement-cored arch. Although Precambrian alluvial environments have been intensively investigated, including numerous treatments regarding conditions that existed before the appearance of vascular plants on land (see Eriksson *et al.*, 1998; Altermann, this

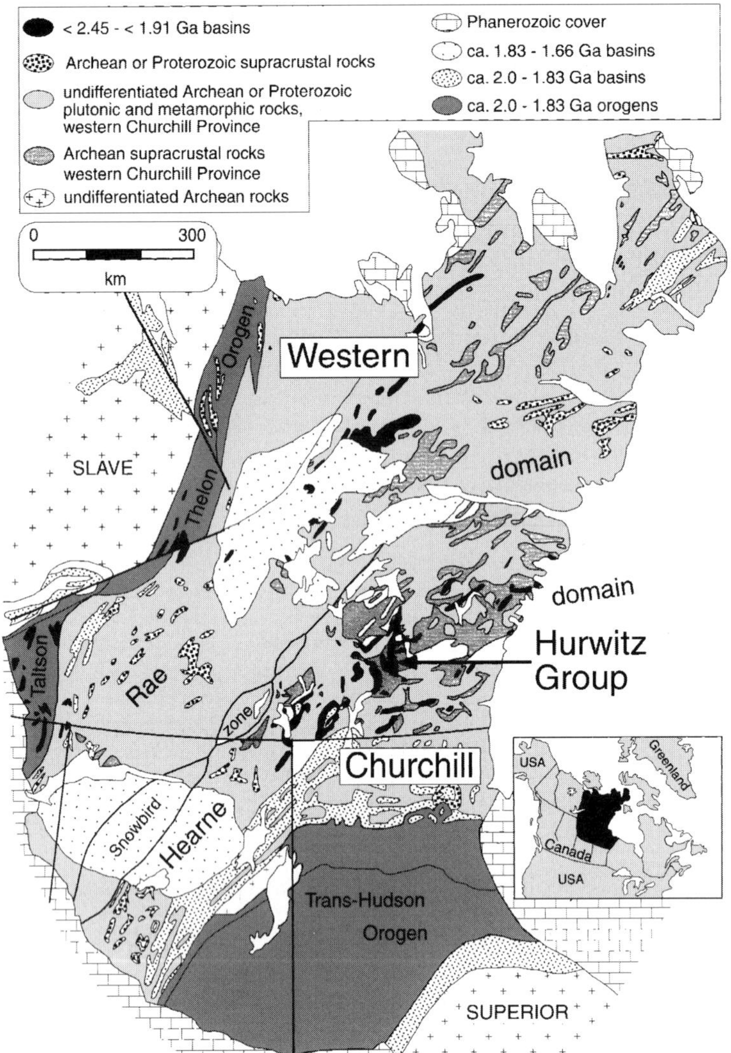


**Fig. 1.** Regional setting of the Hurwitz Group, Western Churchill Province.

volume, pp. 15–32, and references therein), Precambrian mixed siliciclastic–carbonate sequences have received relatively little attention. With notable exceptions (Grotzinger, 1986; Beukes, 1987; Ross & Donaldson, 1989; McCormick & Grotzinger, 1993; Tirsgaard, 1996; le Roux, 1997), few examples of continental to marine transitions marked by hybrid siliciclastic–carbonate depositional environments have been described from the Precambrian, despite increased awareness from work in Phanerozoic and modern basins (e.g. Mount, 1984; Doyle & Roberts, 1988; Testa & Bosence, 1998) that siliciclastic and carbonate sedimentation processes commonly occur together.

The main purpose of this paper is to document the stratigraphy, sedimentology and depositional environments that resulted from rejuvenation of Hurwitz Basin, with particular emphasis on mixing of siliciclastic and carbonate components on a storm-dominated ramp. Herein we stress the inherent similarities between ramp geometry and depositional environments inferred for sequences in the Hurwitz Group and those in post-Precambrian mixed siliciclastic–carbonate systems, while acknowledging differences in the relative importance of depositional processes resulting from changes in life forms, both on land and in the sea.

## HURWITZ GROUP: AGE AND SETTING

The Hurwitz Group is a succession of Palaeoproterozoic continental and marine deposits distributed as a series of outliers across an area of 200 by 700 km in the Hearne domain of the Western Churchill Province (Figs 1 & 2). It is infolded with, but lies unconformably above, an Archaean basement that includes: *c*.3.3–3.0 Ga gneisses; *c*.2.79–2.66 Ga volcanic, siliciclastic and chemogenic rocks (Ennadai–Rankin greenstone belt); and *c*.2.74–2.55 Ga plutonic and gneissic rocks (Aspler & Chiarenzelli, 1996a; Hanmer *et al.*, 2000). Separated from Neoarchaean supracrustal rocks and the Hurwitz Group by angular unconformities, the Montgomery Group forms a local siliciclastic wedge of uncertain age and tectonic significance (Fig. 2; Aspler & Chiarenzelli, 1996b; Aspler *et al.*, 2000). The Hurwitz Group is unconformably overlain by the Kiyuk Group, a succession of conglomerate, arkose and arkose–clast breccia that was deposited in a partitioned non-marine basin (Aspler *et al.*, 1989).

A maximum age for Hurwitz Basin is defined by *c*.2.45 Ga baddeleyite (Heaman, 1994) from the Kaminak dykes that cut basement rocks but not the Hurwitz Group (Davidson, 1970). Lithostratigraphic units in the Hurwitz Group define four major sequences (Fig. 3). Until recently, the entire Hurwitz Group was considered older than 2111 ± 1 Ma, the age of gabbro sills (Patterson & Heaman, 1991; Heaman & LeCheminant, 1993) emplaced after lithification of sequences 1 and 2 (Noomut, Padlei, Kinga and Ameto formations), and deformed with the Hurwitz Group (e.g. Aspler & Chiarenzelli, 1997a). However, new Sm–Nd and Pb–Pb isotopic data (Aspler *et al.*, 2001) and detrital zircon geochronologic data (Davis *et al.*,

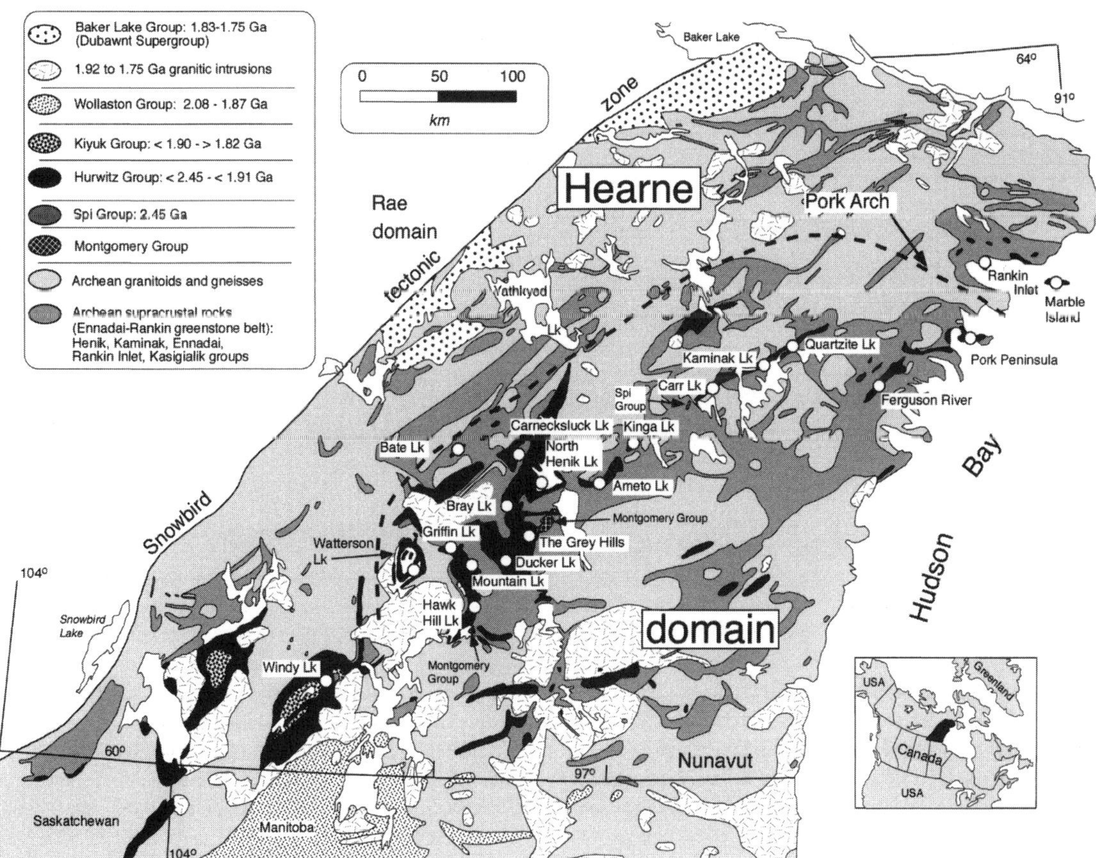


**Fig. 2.** Distribution of the Hurwitz Group and simplified geology of the Hearne domain, Western Churchill Province.

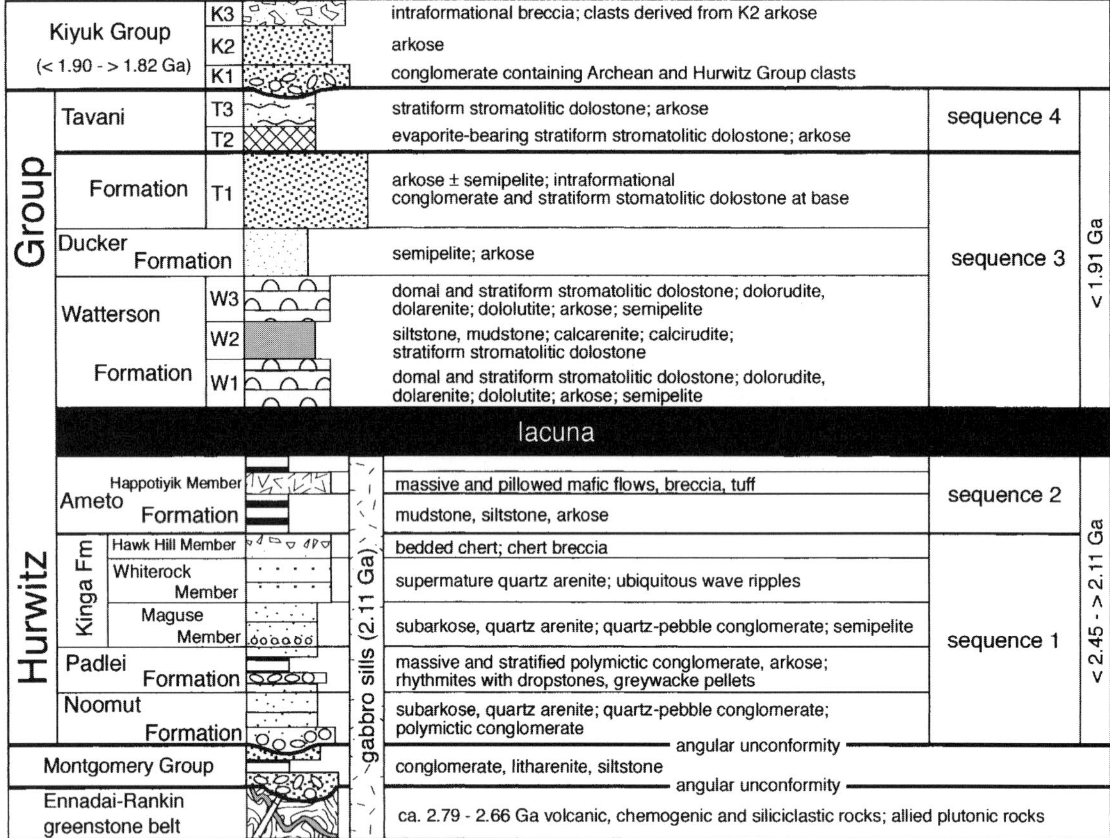


**Fig. 3.** Lithostratigraphy, sequence stratigraphy and regional context of the Hurwitz Group.

2000) document a hitherto unsuspected time break (at least 200 Myr) within the Hurwitz Group, and show that sequences 3 and 4 (Watterson, Ducker and Tavani formations) were deposited after 1907 ± 27 Ma (age of youngest detrital zircon). We maintain that sequences 1 and 2 were deposited in an intracratonic basin that occupied the interior of 'Kenorland' (Williams *et al.*, 1991), a possible Neoarchaean supercontinent interpreted to have undergone protracted break-up between *c.*2.4 and 2.1 Ga and whose daughter fragments ultimately dispersed at *c.*2.1–2.0 Ga (Aspler & Chiarenzelli, 1998). However, it now appears that deposition of sequences 3 and 4 overlapped with assembly of Laurentia on the western flank of the Rae domain (Fig. 1) and either late dispersion of Kenorland crustal blocks or assembly of Laurentia on the southern margin of Hearne domain (Aspler *et al.*, 2001).

## STRATIGRAPHIC AND SEDIMENTOLOGIC CONTEXT

### Sequences 1 and 2: Noomut, Padlei, Kinga and Ameto formations and emplacement of gabbro sills

Predominantly continental deposits that comprise sequence 1 (Noomut, Padlei and Kinga formations, Fig. 3) define a conformable pattern in which successive units onlap a low-relief basement. They record initiation of Hurwitz Basin by regional intracratonic sagging and radially expanding subsidence (Aspler & Chiarenzelli, 1997a). Sagging culminated with sheet deposition of supermature quartz arenites above a low-relief basement in a large, shallow, freshwater lake or series of lakes (Whiterock Member, Kinga Formation), ultimately blanketing an area of at least 100 000 km$^2$ (Aspler *et al.*, 1994a). Development of a

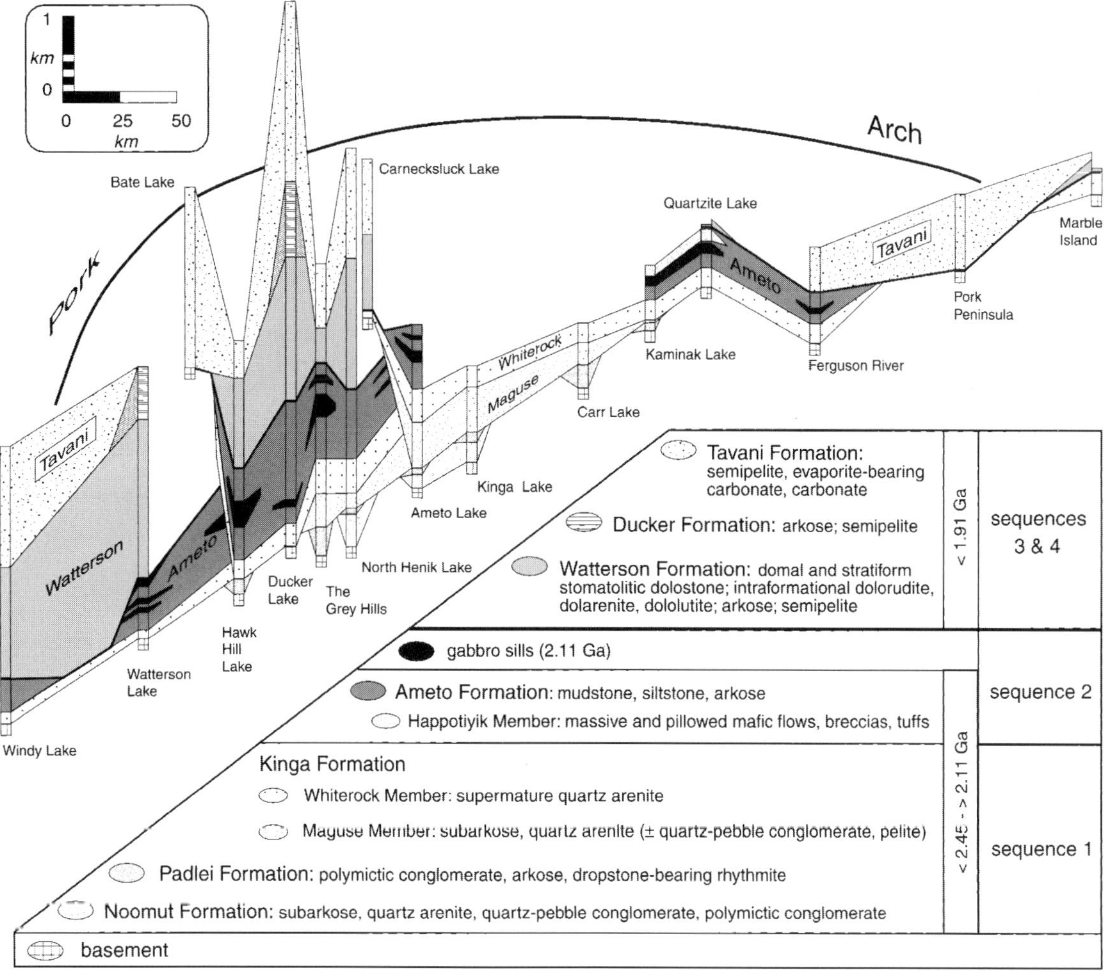


**Fig. 4.** Regional stratigraphic relationships (see Fig. 2 for locations). Thicknesses are estimates based on map data. Sources: Windy Lake, Aspler *et al.* (1989); Watterson Lake, Aspler *et al.* (1993b); Hawk Hill Lake, Aspler & Bursey (1990); Bate Lake, Eade (1974); Ducker Lake, Aspler *et al.* (1994b), The Grey Hills, Aspler *et al.* (1992); North Henik Lake, Aspler & Chiarenzelli (1997b); Ameto and Kinga lakes, Aspler *et al.* (1993a); Carr, Kaminak and Quartzite lakes, Bell (1968), Hofmann & Davidson (1998); Ferguson River and Pork Peninsula, Bell (1968), Heywood (1973); Marble Island, Tella *et al.* (1986).

broad shallow continental depression during sequence 1 is attributed to post *c.*2.45 Ga lithospheric stretching (Aspler *et al.*, 2001).

Sedimentation of continental quartz arenites was abruptly terminated and replaced by texturally and compositionally immature deep-water siliciclastic rocks of the Ameto Formation (sequence 2, Fig. 3). The abrupt change in depositional regime reflects sudden drowning in the central part of the basin, and the appearance of angular and labile grains represents concurrent arching and erosional stripping of basement near the margins of the basin. The Ameto Formation is inferred to have formed in response to a second episode of stretching during which flexural isostatic rebound led to basin-centred deepening and basin margin-arching (Aspler *et al.*, 2001).

Gabbro sills extend as discontinuous pods within the Hurwitz Group across an area of at least 30 000 km$^2$. The sills typically form multiple intrusions within weak Ameto pelites (Fig. 4) but also cut underlying units; feeder dykes are locally exposed (Aspler *et al.*, 2000). The gabbros are considered to have formed due to melting of enriched mantle during continued extension and thinning leading to dispersion of

Kenorland's daughter fragments shortly after *c*.2.1 Ga (Aspler *et al.*, 2001).

### Sub sequence 3 relationships

Recent geochronologic and isotopic data (Davis *et al.*, 2000; Aspler *et al.*, 2001) demonstrate that a time gap, at least 200 Myr long, separated gabbro sill and dyke emplacement from deposition of sequence 3 (*c*.2.1–1.91 Ga). Sub sequence 3 stratigraphic relationships define the locus of a basement palaeohigh ('Pork arch', Figs 2 & 4) near the western and northern margins of the basin, and sequence 3 strata (Fig. 3; Watterson and Ducker formations, lower Tavani Formation) define an upward-shallowing profile that records progradation of continental facies away from the Pork arch towards shallow- and deep-marine environments.

A record of cover-to-basement stripping is preserved in basal Tavani Formation strata near the Pork arch. On the north-west near Bate Lake (Eade, 1974; Eade & Chandler, 1975), and on the north-east at Pork Peninsula (Bell, 1970; Heywood, 1973), the Tavani Formation lies directly on Archaean basement (Figs 4 & 5), and beds of polymictic conglomerate with basement-derived clasts are found above the unconformity. In congruence with observations that the sub-Tavani unconformity cuts into progressively deeper stratigraphic levels, polymictic conglomerate beds at Pork Peninsula display an inverse sequence of clast compositions (Bell, 1970). Based on thicknesses at nearby sections, at least 200–300 m of Whiterock Member were lost beneath the sub-Tavani unconformity near Pork Peninsula, in addition to an unknown thickness of basement.

Direct field evidence of the sub sequence 3 unconformity is lacking on the flanks of the Pork arch and in the interior of the basin. Close to the flanks, the Ameto Formation pinches out and the Watterson Formation directly overlies the Kinga Formation (from west to east): between Windy and Watterson lakes; Hawk Hill and Griffin lakes; North Henik and Carnecksluck lakes; and at Marble Island (Figs 4 & 5). At the one locality where the Watterson–Kinga contact has been observed (north of Griffin Lake, Fig. 2), dm-scale interbeds of stratiform stromatolite and arkose concordantly drape rippled beds in underlying Whiterock Member quartz arenites, and signs of pre-Watterson tilting, weathering or erosion are lacking (Fig. 6A). In the interior of the basin, the Watterson Formation lies above rocks of sequence 2, and evidence of sub sequence 3 stripping is recorded only by local massive and stratified conglomerate lenses in basal Watterson Formation beds (Fig. 6B). These conglomerates contain angular to well rounded granules to cobbles derived from the Whiterock and Hawk Hill members (Kinga Formation) and from basement granites, as well as intraformational dolostone clasts.

In summary, mechanisms that could produce permanent accommodation space for the deposition and preservation of rocks in Hurwitz Basin were not operative between 2.11 and 1.91 Ga. The disconformable Watterson Formation–Whiterock Member contact suggests that the basin was relatively stable throughout this time despite active tectonics along the western margin of the Rae domain starting at *c*.2.0 Ga (Aspler *et al.*, 2001). Post 1.91 Ga basin rejuvenation and deposition of sequences 3 and 4 is considered to have been in response to flexural uplift along the Pork arch, concurrent with basin-centred subsidence and marine flooding.

### Sequences 3 and 4: Watterson, Ducker and Tavani formations

In contrast to other carbonate platform types (e.g. rimmed shelves, isolated platforms), carbonate ramps are characterized by low-gradient (<1°) depositional surfaces that pass from nearshore to basinal environments without a marked break in slope (e.g. Read, 1985; Burchette & Wright, 1992; Ahr, 1998). Similar to siliciclastic shelves (e.g. Johnson & Baldwin, 1996), the hydrodynamic regime of carbonate ramps may be viewed in terms of storm, wave and tidal end members. The palaeogeographic model for sequence 3 presented herein envisages deposition of the Watterson Formation on a storm-dominated, microtidal, mixed siliciclastic–carbonate ramp that passed landward into prograding coastal and continental deposits of the Ducker and Tavani formations (Fig. 7).

The Watterson Formation was deposited under strong storm influence, with frequent interruption of normal low wave and tidal energy conditions by high energy events. Two principal ramp environments are identified (Fig. 7): (i) outer ramp to mid-ramp (below wave base) siliciclastic siltstones, very fine-grained sandstones, mudstones, detrital carbonates and stratiform stromatolites; and (ii) inner ramp, with offshore domal and stratiform stromatolites, and siliciclastic sand-bearing dolorudites, dolarenites and nearshore stratiform stromatolites, siliciclastic sandstones, siltstones, mudstones and intraclastic dolostones. The Ducker Formation comprises deltaic arkoses and pelites, and the lower part of the Tavani Formation contains coastal to fluvial sand-plain arkoses, mudstones and intraformational conglomerates, and lacustrine semi-pelites. Sequence 4 (upper Tavani Formation) is preserved in local structural depressions and, consisting of shallow-marine evaporitic and carbonate rocks, is considered to represent re-establishment of the ramp following marine flooding of lower Tavani Formation continental deposits.

Siliciclastic–carbonate cycles are commonly interpreted in terms of reciprocal sedimentation (Wilson, 1967). According to this concept, predominantly siliciclastic deposition occurs during sea-level lowstands

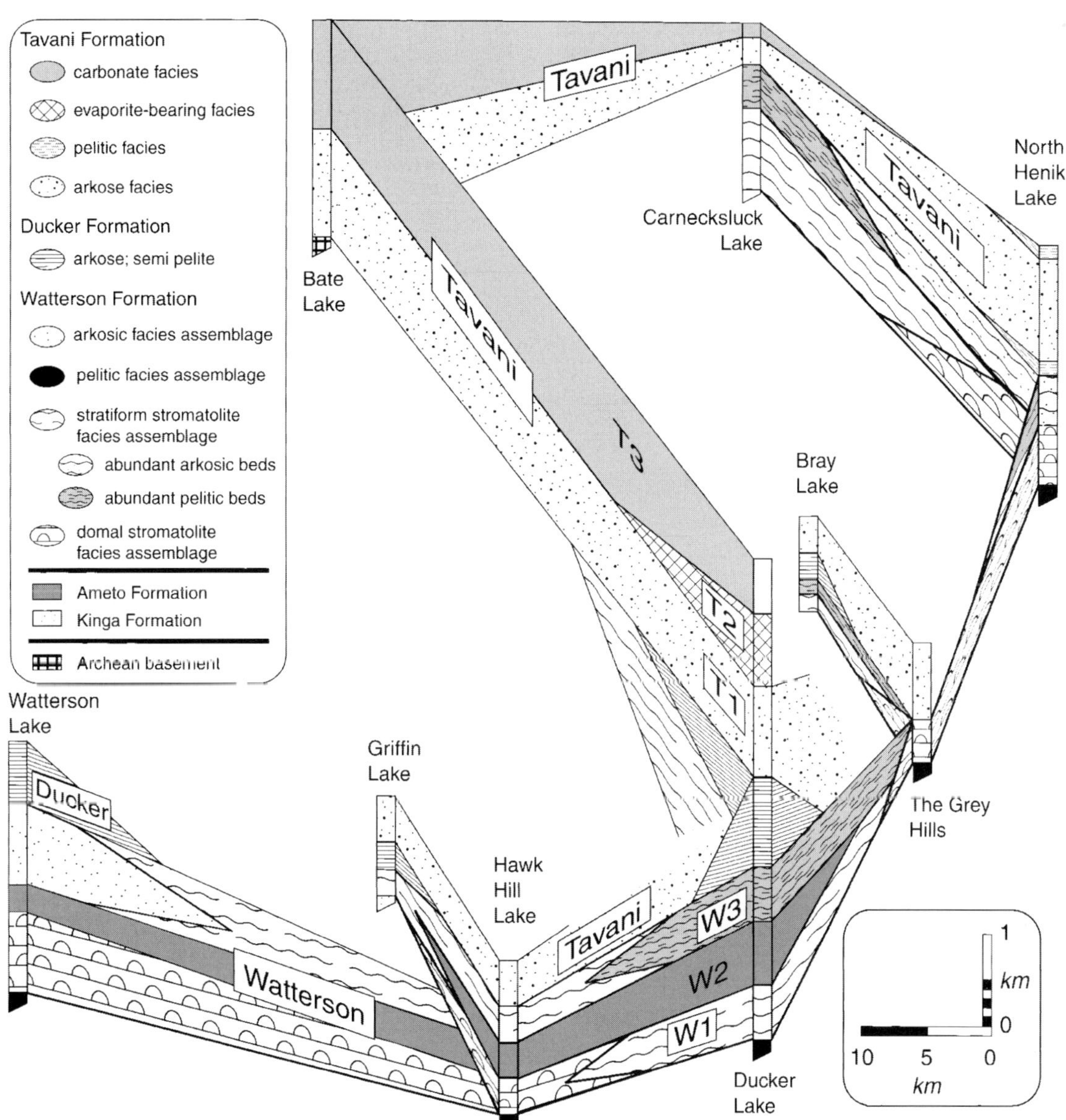


**Fig. 5.** Lithostratigraphic and facies relationships, Watterson, Ducker and Tavani formations (sequences 3 and 4). Sources: Watterson Lake, Aspler *et al.* (1993b); Griffin and Hawk Hill lakes, Aspler & Bursey (1990); Bate and Carnecksluck lakes, Eade (1974); Ducker Lake, Aspler *et al.* (1994b); The Grey Hills, Aspler *et al.* (1992); North Henik Lake, Aspler & Chiarenzelli (1997b).

**Fig. 6.** Sub sequence 3 relationships in outcrop. (A) Watterson Formation stratiform stromatolite and arkose layers concordantly drape Whiterock Member beds. (B) Local bed of massive conglomerate near base of the Watterson Formation containing clasts of Hawk Hill Member chert (ct), and intraformational dololutite (dol) set in a medium- to coarse-grained arkose matrix and cemented by dolomite.

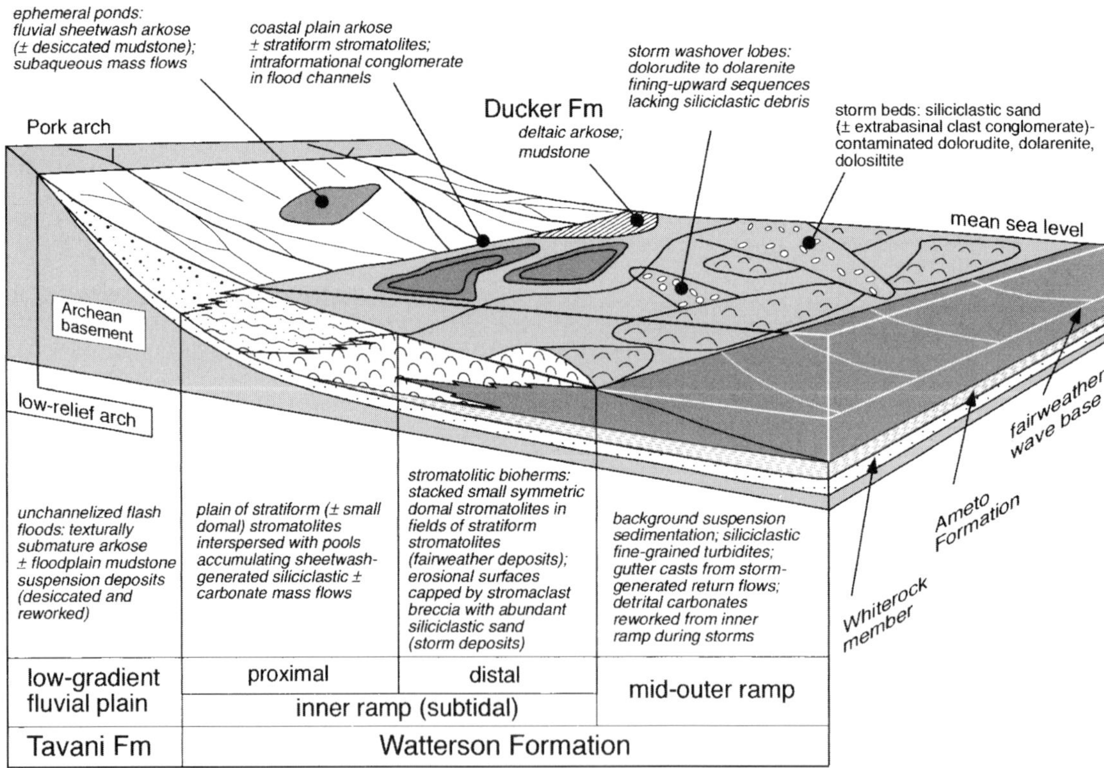


**Fig. 7.** Palaeogeographic model, sequence 3, upper Hurwitz Group: low-relief basement arch–low-gradient fluvial plain–gently inclined mixed siliciclastic–carbonate ramp.

(when base-level is lowered, source areas are exposed, carbonate production is shut down and siliciclastic sediments are transported across subaerially exposed shelves) and is replaced by predominantly carbonate deposition during highstands (when base-level is raised, siliciclastic sources are flooded, and carbonate production flourishes). However, in hybrid siliciclastic–carbonate systems, relating short-term cycles to relative sea-level fluctuations may not be straightforward. With a backdrop of continuous siliciclastic influx and carbonate production, changes in rates and sites of sedimentation induced by tectonic and climatic factors, or by shifts in patterns of continental drainage or oceanographic circulation, may result in small-scale cycles that are independent of relative sea-level history (Dolan, 1989; Bechstädt & Schweizer, 1991; Holmes & Christie-Blick, 1993; Osleger & Montañez, 1996; de Wet, 1998). The present study is based on outcrop and petrographic examinations derived from 1 : 50 000-scale mapping in a region of limited and discontinuous outcrop. Below we focus on establishing the environmental framework of major sequences in the upper Hurwitz Group, within which future sequence stratigraphic assessments of short-term cycles may be undertaken.

## WATTERSON FORMATION

The Watterson Formation (Eade & Chandler, 1975) is a regionally extensive succession of dolostones and siliciclastic rocks that is exposed across an area of *c.*80 000 km$^2$. It attains a maximum thickness of *c.*2000 m in the vicinity of Ducker Lake and Watterson Lake, and is subdivided into three informal lithostratigraphic units (‘W1’, ‘W2’, ‘W3’; Figs 3 & 5), based on the local occurrence of a middle section consisting predominantly of siliciclastic pelites (W2). Where this section is lacking, units W1 and W3 merge as a single lithostratigraphic unit (Fig. 5).

### Watterson Formation: facies description

Four principal facies assemblages are recognized in the Watterson Formation. The domal stromatolite, stratiform stromatolite and arkosic facies assemblages constitute inner-ramp environments, and the pelitic facies assemblage represents the mid-ramp to outer-ramp setting. Virtually all of the carbonate rocks have been altered to dolomicrite (<0.004 mm) and xenotopic to hypidiotopic dolomicrospar (0.004–0.05 mm).

#### *Domal stromatolite facies assemblage*

Isolated build-ups mainly consisting of domal stromatolite are prominent at Watterson, Hawk Hill and North Henik lakes and The Grey Hills (Fig. 5). These build-ups also contain abundant stratiform stromatolite, and local siliciclastic sand-bearing dolorudite, oolitic-intraclastic dolarenite and dololutite.

*Domal stromatolites* define dm-scale (typically <50 cm, rarely to 2 m), concentrically laminated, isolated to contiguous ovoids on bedding surfaces (Fig. 8A & B). In vertical cross-section, the stromatolites display a nodular to bulbous first-order macrostructure in which internal laminae have a high degree of inheritance (Fig. 8C). The internal laminae are generally symmetric, but some stromatolites display a weak asymmetry owing to slightly thicker laminations on one side. Individual stromatolites are laterally linked, with spacings that range from isolated (Fig. 8D) to contiguous, and interdome calcirudites or calcarenites are lacking. Relief is typically dm-scale (rarely to 1 m) and synoptic profiles are generally less than 15 cm. Hofmann & Davidson (1998) have illustrated second-order microstructures from stromatolites in the Quartzite Lake area (Fig. 2), in which pseudo-columnar to columnar ministromatolites extend radially outward within first-order convexities. Individual domal stromatolite beds are commonly superimposed and separated by erosional surfaces (Fig. 8E). Together with intimately interleaved dm- to m-scale beds of stratiform stromatolites (Fig. 8D & F), they form isolated compound structures up to several metres thick.

The stromatolites contain a wrinkly to wavy internal stratification on scales from 0.01 to 5 mm (Fig. 8G & H). This layering results mainly from: (i) alternations of dolomicrite and dolomicrospar; (ii) variation in dolomicrospar crystal size and inclusion content (alternating bands of dark, inclusion-rich, 0.01–0.03 mm-sized crystals and light, inclusion-poor, 0.03–0.05 mm-sized crystals); (iii) discontinuous concordant lamellae of diagenetic microquartz; (iv) diffuse carbonaceous seams *c.*0.03 mm thick; and (v) rare siliciclastic mudstone layers *c.*0.3 mm thick. Up to 3% angular, coarse silt to very fine sand quartz and feldspar grains are locally dispersed, independent of layering.

*Stratiform stromatolites* are interleaved with and envelop the domal stromatolites (Fig. 8D & F). They consist of dm- to m-scale beds of laterally continuous, crinkly to wavy dolostone that contain the same internal stratification as the domal stromatolites but,

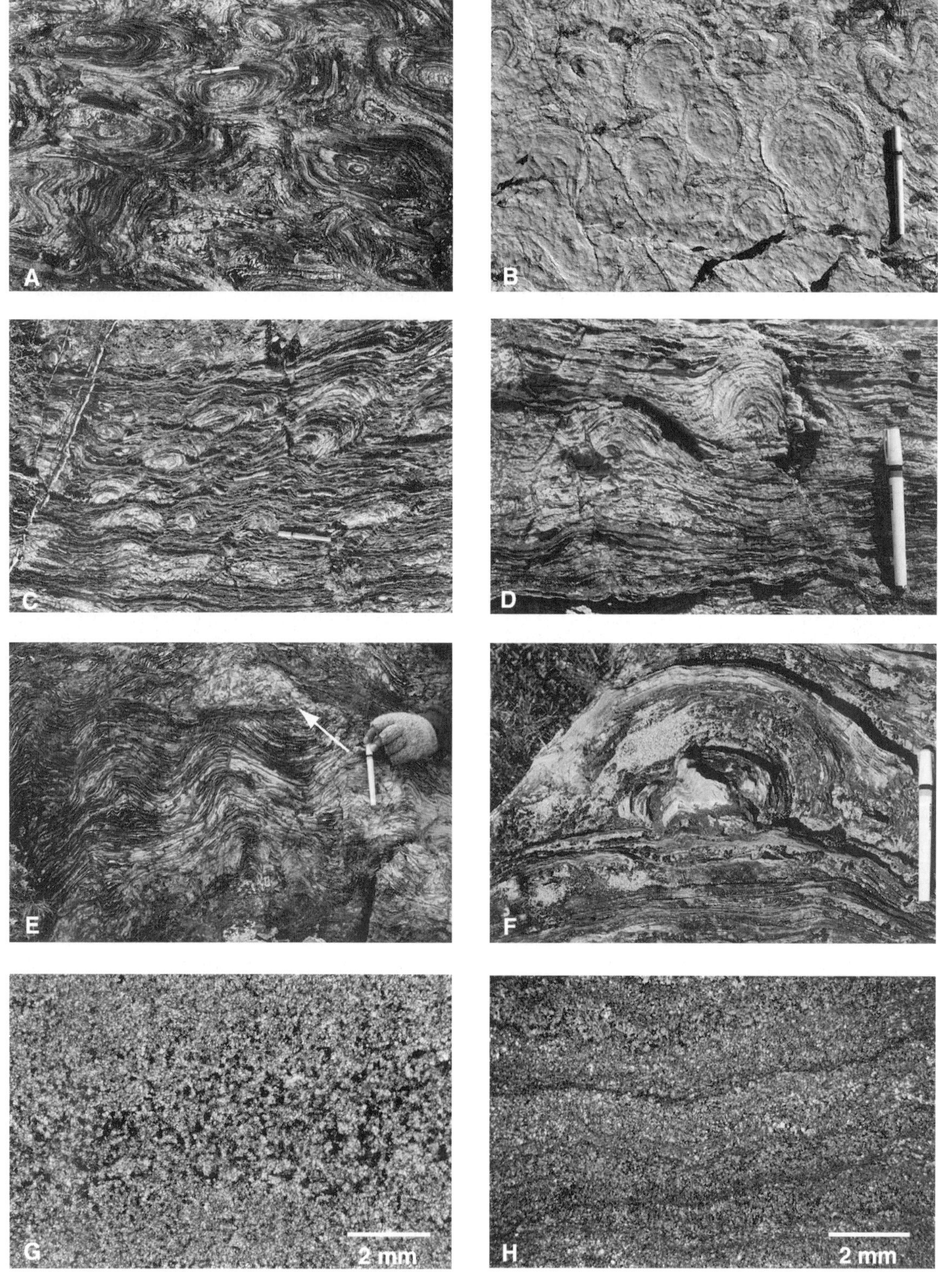
A
B
C
D
E
F
G
2 mm
H
2 mm

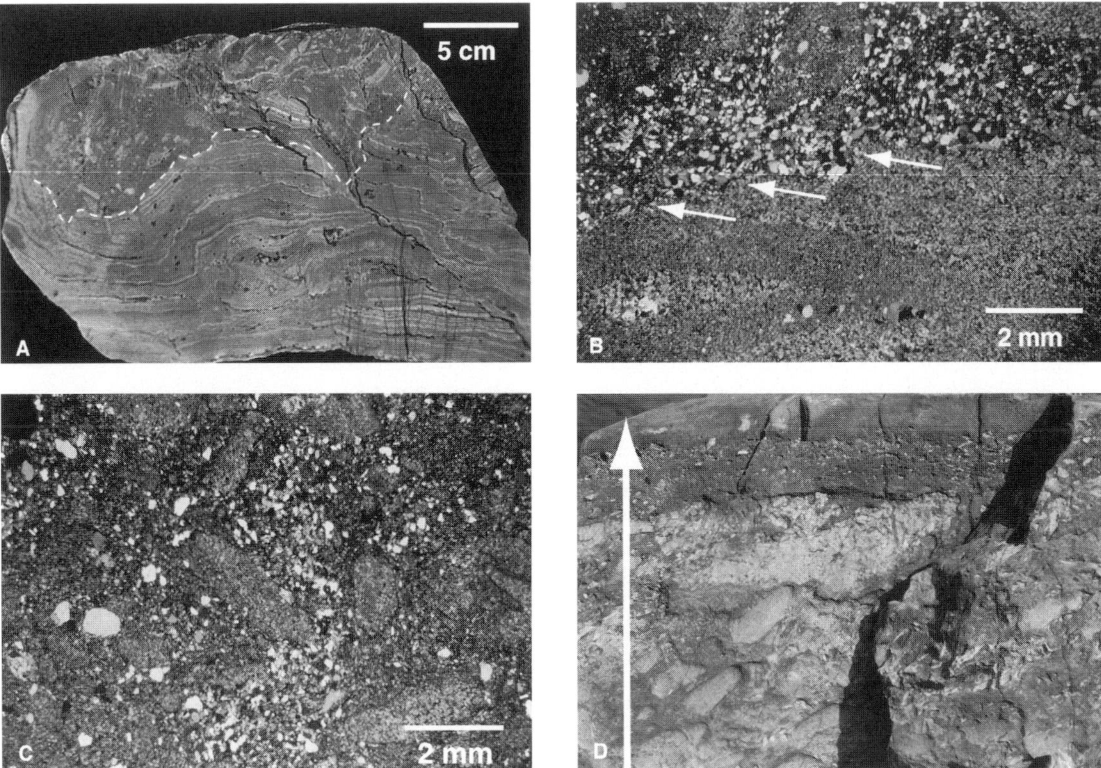


**Fig. 9.** Siliciclastic sand-bearing calcirudites of the domal stromatolite facies assemblage, Watterson Formation. (A) Erosional surface (dashed line) cut into small domal stromatolite overlain by stromaclast breccia containing abundant siliciclastic sand. (B) Large dolomicrospar stromaclast (lower part of photo) and siliciclastic sand matrix. Note distinct steps and risers (arrows) along clast margin that can be related to an internal layering defined by variation in crystal size. (C) Well rounded dolomicrospar clasts in siliciclastic sand-rich matrix. (D) Intraformational dolostone–clast conglomerate fining upward to granulestone and arenite.

**Fig. 8.** (*opposite*) Watterson Formation domal stromatolite facies assemblage. (A) Bedding-plane view of close to contiguous concentric elliptical laminae (long axes concordant with tectonic fabric). Internal lamination highlighted by pure dolostone layers (light-toned, recessive) and partially silicified dolostone (dark-toned, resistant). (B) Bedding-plane view of small, contiguous, ovoid stromatolites. (C) Bedding-oblique view (beds overturned) of laterally linked, spaced, nodular to bulbous stromatolites intercalated with stratiform stromatolites. Stromatolite asymmetry is concordant with tectonic fabric. (D) Bedding-normal view of isolated nodular stromatolites passing laterally and vertically to stratiform stromatolites. (E) Laterally linked, closely spaced, nodular stromatolites cut by a localized erosional surface (arrow) and recolonized. (F) Vertical profile of stratiform stromatolite passing upward to small bulbous stromatolite which provides the substrate for a nodular stromatolite. (G) Layering defined by variation in dolomicrospar crystal size. (H) Layering defined by variation in dolomicrospar crystal size, diagenetic microquartz lamellae and carbonaceous seams.

with the exception of rare cm-scale domes, lack synoptic relief.

*Siliciclastic sand-bearing dolorudites* form local dm- to m-scale lenses of breccia and stratified conglomerate within domal and stratiform stromatolite beds (Figs 8E & 9). The breccias consist of cm-scale (locally to 30 cm), angular to subrounded, tabular fragments that are typically self-supporting in siliciclastic sand (± dolarenite) matrix cemented by dolomicrospar. Most are massive, but in some beds a crude layering is defined by alignment of tabular fragments. Figure 9A illustrates a massive breccia deposited above an erosional surface cut into a small domal stromatolite. The breccia consists of fragments derived from the immediately underlying stromatolite, including massive dolomicrospar, layered dolomicrospar and interlayered dolomicrite and dolomicrospar ('stromaclasts' of Sami & James, 1994). Many of the clasts display a jigsaw fit, suggesting autobrecciation during transport, and some display distorted internal layering or have margins that step 90° along internal bedding surfaces (Fig. 9B), indicating varying degrees of lithification before reworking. The composition and shape of the clasts are significant, and imply penecontemporaneous neomorphism, cementation and dolomitization. Upsection the clasts become more spherical and better rounded (Fig. 9C).

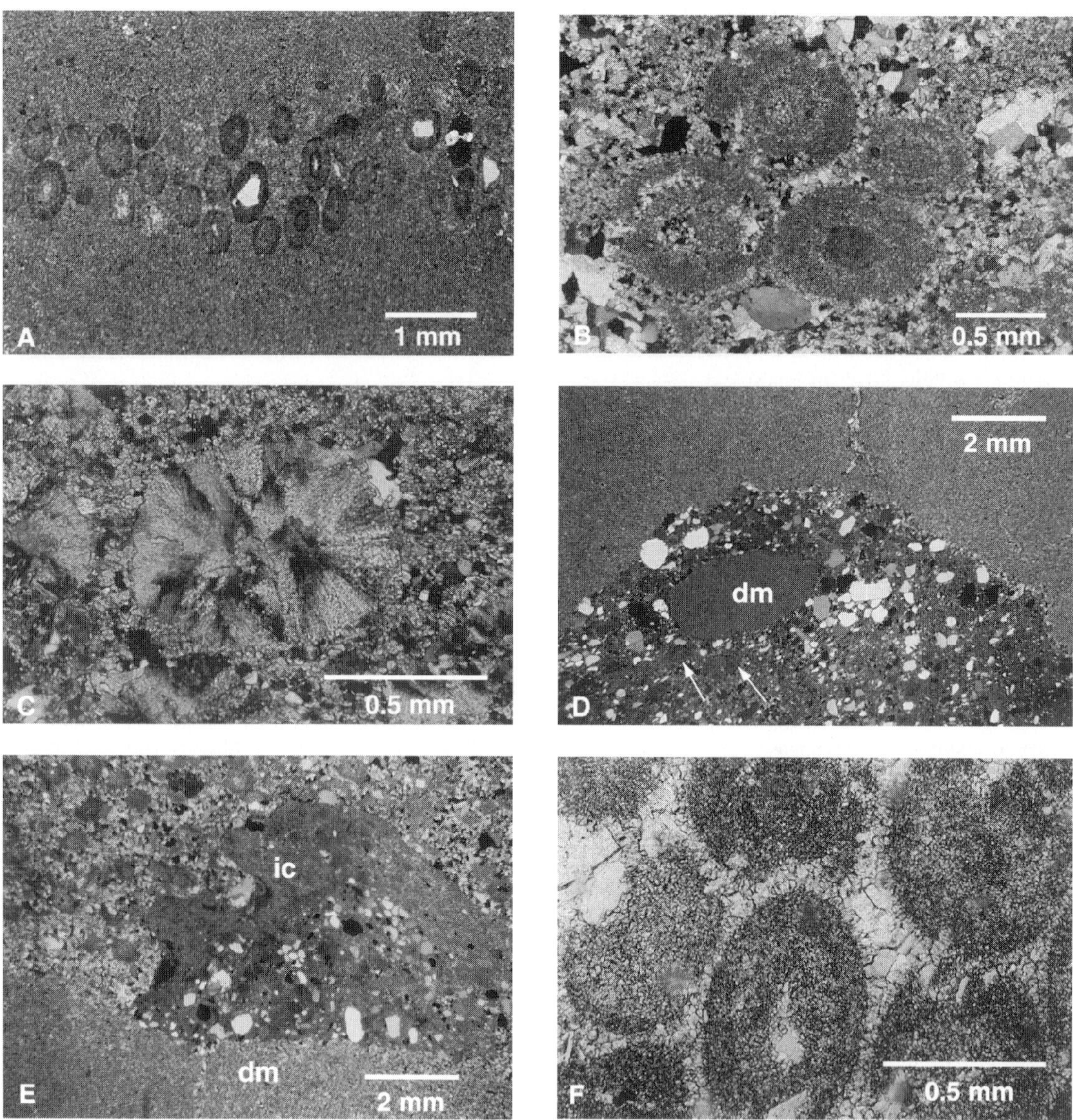


**Fig. 10.** Siliciclastic sand-bearing oolitic–intraclastic dolarenite of the domal stromatolite facies assemblage, Watterson Formation. (A) Thin oolitic packstone layer grading to dolomicrite with floating ooids (ooids are elongate along tectonic fabric). (B) Ooids with variably preserved concentric coats in grainstone with dolospar cement. (C) Ooid with coarse radiaxial fabric, possibly mimicking primary aragonite. (D) Rounded intraclast floating in dolomicrite; intraclast contains dolomicrite clast (dm), poorly preserved ooids (arrows) and siliciclastic sand grains in dolomicrite matrix. (E) Siliciclastic sand-bearing oolitic dolomicrite/wackestone intraclast (ic) in oolitic grainstone lying sharply above dolomicrite layer (dm). Irregular embayed-lobate margin and distorted internal layering (defined by variation in concentration of sand grains) suggest that the clast was incompletely lithified before reworking. (F) Oolitic grainstone with early pore-filling rim cement and blocky dolomite cement.

Dolostone granule and pebble conglomerates are framework-intact and commonly display a crude parallel stratification manifested by variation in clast size and concentration. They locally define upward-fining sequences of pebble conglomerate, granulestone and arenite (Fig. 9D). Some of the dolarenites display wave-rippled upper surfaces.

*Siliciclastic sand-bearing, oolitic–intraclastic dolarenite* intervals within the domal stromatolite facies assemblage are up to 2 m thick, and consist of mm to cm interlayers of grainstone, packstone, wackestone and dolomicrite (Fig. 10). Grainstone and packstone layers are sharp-based and locally grade upward to wackestone or dolomicrite containing sparse floating ooids. The tops of some grainstone beds display interference ripple marks and symmetric wave ripple marks (wavelengths 5–8 cm; ripple heights 1–3 cm).

The ooids are 0.5–2.0 mm in diameter and display wide variation in preservation of primary textures (Fig. 10A & B). Some exhibit well defined concentric coatings of dolomicrite, whereas others are almost completely replaced by dolomicrospar (± microquartz and megaquartz) and appear as dark-rimmed 'ghosts'. Rarely, a radiaxial array of dolosparite crystals, possibly inherited from a primary aragonitic fabric, is preserved (Fig. 10C). Siliciclastic coarse silt to coarse sand grains, including monocrystalline quartz, polycrystalline quartz with undulose extinction and twinned feldspar, constitute up to 15% of the grainstone and packstone layers, but also occur floating within the wackestones and dolomicrites. In addition, sparse intraclasts of wackestone to dolomicrite occur in both the grainstones and the dolomicrites. Some intraclasts are well rounded (Fig. 10D), but others have irregular margins, with rounded embayments and protuberances; many display a distorted internal layering defined by varying concentrations of ooids, siliciclastic sand grains and dolomicrite (Fig. 10E). The grainstones display primary pore-filling cement textures, in which an isopachous radiaxial rim is followed by idiotopic dolospar (Fig. 10F), whereas the packestones are cemented by a patchy mix of dolomicrospar and dolomicrite. The intraclasts indicate penecontemporaneous development of these intergranular cements.

*Siliciclastic sand-bearing dololutites* form local beds 1–2 m thick. Some define cm-scale upward-fining sequences in which parallel-stratified layers with 60–80% fine–medium grained siliciclastic sand grade to dololmicrite with dispersed (up to 20%) very fine siliciclastic sand and then to pure dolomicrite. The pure dolomicrite layers locally drape wave (± interference) ripple marks in underlying sandy layers. In addition, some dololutite intervals define cm-scale alternations of: (i) mm-scale laminae of coarse siltstone to very fine sandstone and dolomicrospar (with parallel stratification and ripple-drift cross-stratification); and (ii) siliciclastic-contaminated dololutite and pure dolomicrospar.

### *Stratiform stromatolite facies assemblage*

The most areally extensive facies assemblage in the Watterson Formation consists of stacked dm- to m-scale stratiform stromatolites, containing local beds of siliciclastic-free intraformational rudite, arenite and lutite (Fig. 5). In addition, arkose-rich and pelite-rich intervals up to several metres thick form relatively continuous (up to several km) tongues.

*Intraformational rudites, arenites and lutites* define local upward-fining sequences that follow the order: scoured base; granule–pebble conglomerate containing dolomicrospar clasts; parallel and ripple cross-stratified dolarenite; graded dolarenite to dolomicrospar (Fig. 11A). In contrast to the intraformational detrital dolostone beds in the domal stromatolite facies assemblage, these upward-fining sequences contain little to no siliciclastic sand.

*Arkose-rich intervals* contain laterally continuous, dm- to m-scale beds of very fine to fine grained arkose and siltstone that alternate with siliciclastic-bearing dololutite and siliciclastic-free stratiform stromatolites (Fig. 11B). The arkose beds have sharp lower contacts and commonly display parallel stratification or small-scale cross-stratification; some contain floating oncoids (Fig. 11C). Overlying dololutites bear local cm-scale sand lenses with ripple cross-stratification or graded dololutite chip breccias (Fig. 11D).

*Pelite-rich intervals*, such as at Ducker and Bray lakes (Fig. 5), contain up to 30% cm- to dm-scale, sharp-based, graded siltstone to mudstone upward-fining sequences and mm- to cm-scale rhythmites that separate beds of siliciclastic-free stratiform stromatolite.

### *Arkosic facies assemblage*

The arkosic facies assemblage is developed at the top of the Watterson Formation near the eastern flank of the Pork arch in the Watterson Lake area (Figs 4 & 5). It is characterized by m-scale arkose beds that are interlayered with intensely silicified stratiform stromatolite, and channel-filling intraformational breccia and conglomerate. The arkoses are medium to coarse grained (Fig. 12A) and locally display parallel and low-angle cross-stratification defined by heavy mineral concentrations. Stratiform stromatolite beds contain

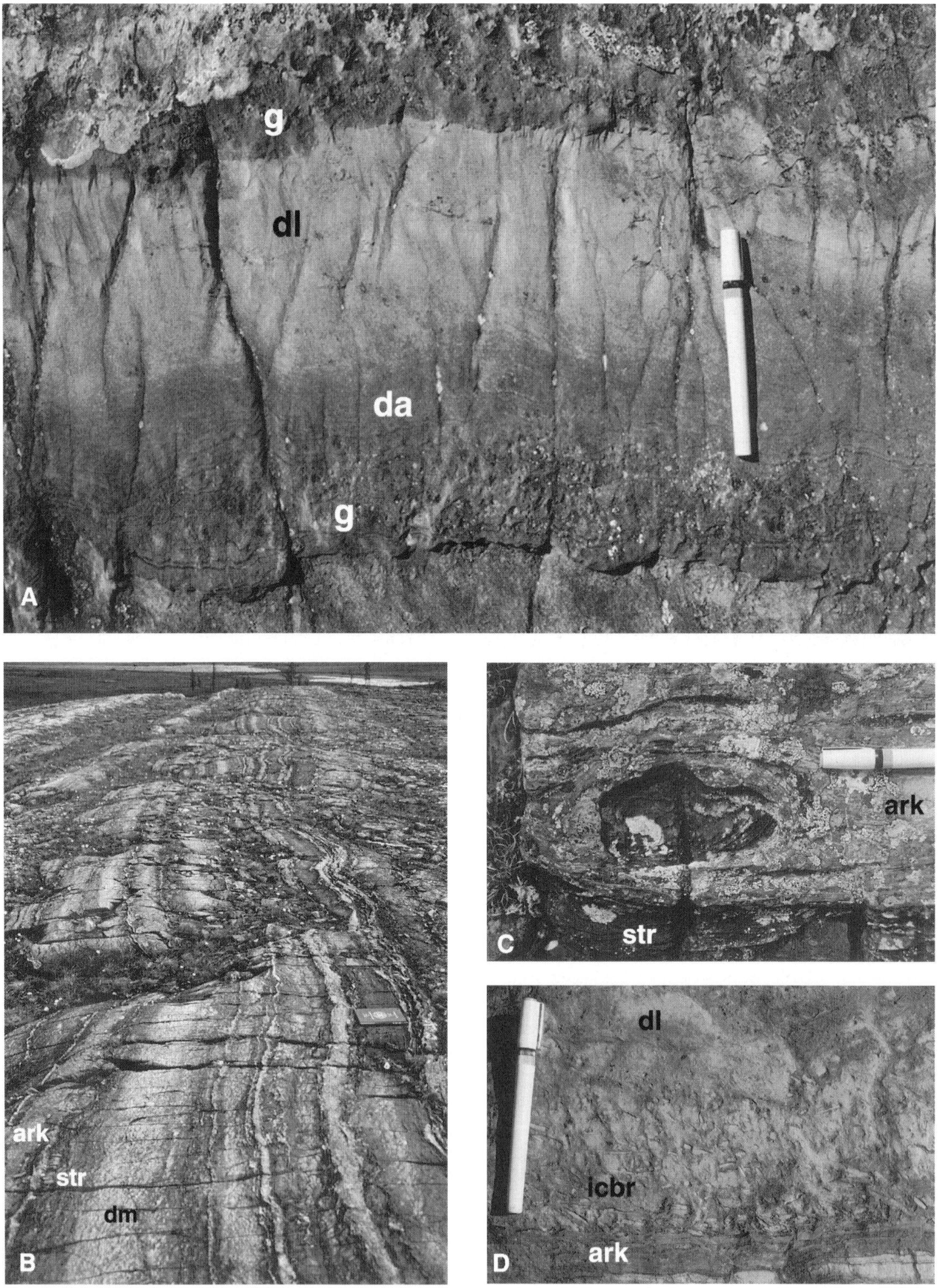

ark
str
dm
B

ark
C
str

dl
icbr
D
ark

**Fig. 12.** Arkosic facies assemblage, Watterson Formation. (A) Texturally submature medium- to coarse-grained arkose. (B) Semi concordant chert layers (light-toned, resistant) in unit of stratiform stromatolites. (C) Intraformational conglomerate containing variably silicified dolostone clasts. (D) Stratified beds of conglomerate (cg) containing arkose and variably silicified dolostone clasts, interlayered with stratiform stromatolite-bearing dolostone (str) and parallel-stratified arkosic sandstone (ark).

up to 20% white chert as continuous to discontinuous cm- to dm-scale layers that are broadly parallel to, but transect, bedding (Fig. 12B). Lying above channel surfaces cut into stratiform stromatolite beds are m-scale beds of massive breccia or stratified conglomerate and parallel-stratified arkose (Fig. 12C & D). Clasts comprise variably silicified dolostone, some with internal zones truncated at clast boundaries, and arkose.

**Fig. 11.** (*opposite*) Stratiform stromatolite facies assemblage, Watterson Formation. (A) Detrital dolostone upward-fining sequence, lacking siliciclastic sand, consisting of sharp-based granulestone (g) containing dololutite intraclasts, overlain by parallel-stratified dolarenite (da), which grades to dololutite (dl); in turn sharply overlain by a graded granulestone layer. (B) Laterally continuous interbeds of fine-grained arkose (ark), stratiform stromatolite with crenulate silicified layers (str) and pure massive dolomicrospar (dm). (C) Oncoid in arkosic bed (ark) overlying stratiform stromatolitic dolostone (str). Note similarity of concentric internal lamination in the oncoid to lamination in underlying stratiform stromatolite. (D) Parallel-stratified arkose (ark) overlain by graded intraclastic dololutite breccia bed (icbr) and dololutite (dl).

### *Pelitic facies assemblage*

The pelitic facies assemblage consists predominantly of fine-grained sandstone, siltstone and mudstone in sharp-based cm- to dm-scale upward-fining sequences, and mm- to cm-scale rhythmites. Locally, rhythmites contain siliciclastic siltstone laminae that grade to dolosiltite (Fig. 13A). Basal portions of the upward-fining sequences commonly display graded bedding or parallel stratification and ripple-drift cross-stratification (Fig. 13B). Some of the siltstones contain granules of intraclastic dolomicrospar and oolitic dolarenite. The pelitic beds contain erosional horizons defined by isolated dm-scale gutter casts (Fig. 13C). These display steep (partly stepped) sides cut into underlying

**Fig. 13.** Watterson Formation, pelitic facies assemblage. (A) Siliciclastic siltstone and dolosiltite rhythmites. (B) Parallel and cross-stratified sandstone–mudstone upward-fining sequences; sandstone ball and pillow structures in mudstone (arrows). (C) Isolated gutter casts (arrows) cut into fine-grained siliciclastic upward-fining sequences; basal dolorudite fill is draped by fine-grained siliciclastic layers.

rhythmites. Basal parts are filled with massive or stratified dolarenite and dolorudite; upper parts are draped by fine grained siliciclastic upward-fining sequences. Stratiform stromatolites persist as dm- to m-scale layers in the pelitic facies assemblage, separating pelitic intervals ranging in thickness from a few decimetres to several tens of metres. In the Hawk Hill Lake and Ducker Lake areas (Fig. 2), the assemblage contains rare discontinuous beds with cm-scale layers of quartz (± jasper)–magnetite–haematite, magnetite-bearing siltstone and dolostone, and ironstone-clast breccia (Eade, 1974; Miller & Reading, 1993).

## Watterson Formation: interpretation of depositional environments

The Watterson Formation is inferred to have been deposited on a storm-dominated, microtidal mixed siliciclastic–carbonate ramp (Fig. 7). As for other Precambrian carbonate successions, a marine setting cannot be unequivocally asserted in the absence of diagnostic palaeontologic data. However, although stromatolite-bearing ramp-like platforms may develop in some lacustrine settings (e.g. Dean & Fouch, 1983; Platt & Wright, 1991), lakes are extremely sensitive to subtle climatic and other changes (Dean & Fouch, 1983; Talbot & Allen, 1996). Given the thickness and lateral continuity of the Watterson carbonates in the absence of extensive exposure surfaces, shallow water facies or evaporites, a connection to the open ocean is strongly suggested. We infer that the inner ramp, represented by the domal, stratiform and arkosic facies assemblages, was subjected to negligible tidal energy and, except during storms, experienced minimal wave and current action. In the distal (offshore) part of the inner ramp, stromatolitic bioherms were isolated within fields of microbial mats. Shoreward, we envisage a complex zone containing extensive fields of stratiform stromatolites with interspersed pools receiving siliciclastic mud and sand, transitional to coastal and fluvial environments (Ducker and Tavani formations). Offshore, the pelitic facies assemblage is inferred to have been deposited on the mid- to outer ramp, and to record normal background (below fair-weather wave base) suspension sedimentation that was interrupted by frequent storm events. Large gravity slides and associated megabreccias, described from examples of rimmed shelves (e.g. Spence & Tucker, 1997), are lacking, suggesting a uniform, gently inclined inner-ramp to mid-ramp slope, rather than a steep escarpment.

### *Inner ramp*

*Stromatolitic bioherms.* The domal stromatolite facies assemblage is considered to have formed in the deepest part of the inner ramp and to represent the build-up of isolated bioherms that were separated by fields of stratiform stromatolites. Although small domal stromatolites occur in a diverse range of environments (e.g. Hoffman, 1976b; Grotzinger, 1989; Braga *et al.*, 1995), a subtidal setting is favoured because the Watterson stromatolites lack features commonly described from Palaeoproterozoic peritidal deposits such as desiccation cracks, teepee structures, pisolites, fenestrae, beach rosette conglomerates and tufas with microdigitate stromatolites (e.g. Ricketts & Donaldson, 1979; Grotzinger, 1989; Sami & James, 1993). Prevailing low wave energy conditions are inferred because the Watterson stromatolites are symmetric, close-spaced, laterally linked, nodular to bulbous forms that lack interdome debris. These are in contrast to the large, branching, highly asymmetric channelized forms found at sites of relatively continuous wave and/or current attack (Hoffman, 1967, 1976a; Horodyski, 1976; Pelechaty & Grotzinger, 1989; Ricketts & Donaldson, 1989). Transitions in which dm- to m-scale domal stromatolites grade laterally or vertically to stratiform stromatolites with cm-scale domes may, in part, reflect decreasing water depth (see Fairchild & Herrington, 1989; Grotzinger, 1989).

The origin of the layering in the Watterson stromatolites is uncertain. Although they lack features that could indicate precipitation of relatively coarse bedded crystals on the sea-floor (herringbone textures; botryoidal fans; laminar fibrous crusts; e.g. Hofmann & Jackson, 1987; Kerans & Donaldson, 1989; Kah & Knoll, 1996; Sami & James, 1996; Sumner & Grotzinger, 1996a; Grotzinger & Knoll, 1999) the distinction between detrital micrite and micritic cements is problematic (e.g. Reid *et al.*, 1990). For the Watterson stromatolites, we lack the degree of preservation to differentiate reliably between growth of laminae by trapping-and-binding of lutitic ooze precipitated as whitings (e.g. Robbins *et al.*, 1997), and *in situ* growth by successive encrustation of sea-floor micritic cements (see Hoffman, 1973; Grotzinger, 1990; Grotzinger & Knoll, 1999). Regardless, the lutitic laminae in the Watterson stromatolites lack abundant sand-sized grains that occur in immediately adjacent siliciclastic-bearing intraclastic and oolitic beds, such as have been described from mixed siliciclastic–carbonate stromatolites elsewhere (Horodyski, 1976; Fairchild, 1991; Braga *et al.*, 1995). This implies a separation of process, with low-energy lutitic fairweather stromatolite growth punctuated by storm events (see below). Local siliciclastic silt to very fine sand grains dispersed in some stromatolites probably reflect aeolian deflation of the alluvial plain, and transport by offshore-directed winds.

*Storm deposits.* Siliciclastic sand-bearing dolorudite, oolitic–intraclastic dolarenite and dololutite beds within the bioherms are considered to be storm deposits. Erosional surfaces that cut into stromatolites and are capped by siliciclastic sand-rich intraclast breccias are inferred to record storm events; succeeding stromatolite beds (with dolomicrite and dolomicrospar laminae) probably represent recolonization during calm water periods. The abundance of siliciclastic sand grains implies that flows with an offshore component transported sediment derived from nearshore facies to the bioherms. The mechanisms whereby sand is transported from coastal zones, and the processes leading to ultimate deposition in offshore areas, remain controversial. As emphasized by Myrow & Southard (1996), a spectrum of processes is possible (e.g. sediment gravity flows, shore-parallel geostrophic currents, storm waves, storm-surge ebb currents, rip currents and combinations thereof). For the siliciclastic sand-bearing dolorudites, dolarenites and dololutites, the combination of upward-fining sequences (with erosional bases, graded beds, parallel stratification, ripple cross-stratification) and wave-rippled caps suggests deposition by seaward-returning bottom flows that waned to predominantly oscillatory flows, as has been inferred for similar beds regarded as tempestites (e.g. Aigner, 1985; Fairchild & Herrington, 1989; Ross & Donaldson, 1989; Tirsgaard, 1996). Lacking evidence of extensive oolitic sand shoals, we suggest that relatively weak, intermittent fairweather currents produced small oolitic patches that were reworked to provide ooids to the storm beds.

*Stromatolitic plains; siliciclastic pools.* The stratiform stromatolite assemblage is considered to have been deposited in the interior of the ramp, forming a broad transition zone that linked offshore, predominantly carbonate-accumulating environments to coastal and continental sites of predominately siliciclastic sedimentation. Thick intervals of stratiform stromatolites are interpreted to represent extensive plains of microbial mats. Similar to stromatolites in the distal segment of the inner ramp, these are regarded as subtidal fairweather deposits. The siliciclastic-free rudites, arenites and lutites that are locally interbedded with the stratiform stromatolites are interpreted to be tempestites, but in contrast to the siliciclastic-bearing beds associated with the bioherms, these are thought to be washover deposits (see, for example, Aigner, 1985) created when onshore storm waves and currents eroded and transported biohermal detritus shoreward (see Gagan *et al.*, 1990).

Arkose- and pelite-rich intervals within the stratiform stromatolite assemblage are considered to record a complex network of short-lived pools within the microbial plain. Sharp-based siliciclastic upward-fining sequences and rhythmites within these intervals are considered mass flow deposits formed by subaqueous plumes that were fed by unchannelized and poorly channelized subaerial floods. Associated carbonate intraclasts and oncoids were probably derived by local reworking of microbial mat surfaces. The extensive development of these rocks may reflect transverse (immediately adjacent to fluvial sources) and longitudinal (related to longshore oceanic currents) siliciclastic sediment dispersal, such as has been described from some modern siliciclastic–carbonate transitions (Roberts, 1987; Friedman, 1988; Acker & Stearn, 1990; Bush, 1991), and also frequent sheet floods on the vascular plant-free Tavani fluvial plain (see below). Beds of stratiform stromatolites within these intervals are thought to represent sites or times of relatively clear-water deposition. Given the abundance of siliciclastic debris in associated beds, the paucity of non-carbonate contaminants in these stromatolites suggests *in situ* carbonate precipitation, an argument presented by numerous workers (e.g. Donaldson, 1963, 1976; Hoffman, 1973, 1974; Ginsburg, 1991). Migration of sediment-laden plumes and/or point sources is considered to have resulted in numerous episodes during which siliciclastic debris completely buried microbial mats, followed by mat recolonization and carbonate plain build-up.

*Nearshore channelized flats.* The arkosic facies assemblage, locally preserved on the western side of the basin adjacent to Pork arch, is interpreted to represent the most proximal part of the inner ramp where microbial mats were frequently inundated by siliciclastic debris introduced from nearby rivers. Thick beds of parallel- and cross-stratified arkose, and stratified conglomeratic beds containing silicified dolostone clasts, are interpreted as flood channel deposits. Variable silicification of adjacent clasts and truncation of silicified zones at clast boundaries suggest derivation from variably silicified sources rather than *in situ* alteration, and imply that silicification was penecontemporaneous. The particular abundance of silicified horizons in the arkosic facies assemblage may represent nearshore mixing of meteoric and sea water (see Hesse, 1989; Knauth, 1994).

*Mid-ramp to outer ramp*

We infer that the pelitic facies assemblage records the deepest environment preserved on the Watterson ramp. Conventionally, the mid-ramp is defined as the zone between fairweather and storm wave base, and the outer ramp is regarded as the zone below storm wave base (Burchette & Wright, 1992). For the Watterson Formation, because we lack sufficient resolution to distinguish between fairweather and storm wave base, the mid-ramp and outer ramp have been grouped into a single depozone. The predominant depositional process in this zone was background suspension sedimentation periodically interrupted by siliciclastic and carbonate mass flow pulses. The very fine sandstone–siltstone to mudstone upward-fining sequences and rhythmites are regarded as low concentration turbidity currents (see Piper & Stow, 1991). The relative abundance of siliciclastic material may indicate a component of longitudinal flow. Storm-generated flows that moved across the inner ramp with an offshore-directed component are considered to have been responsible for introducing carbonate sediment to sites accumulating mainly siliciclastic turbidites. Carbonate clasts within some of the siliciclastic graded siltstones and filling gutter casts probably represent reworked inner ramp deposits. The gutter casts are thought to have formed as a consequence of particularly intense storm-generated return flows (see Whitaker, 1973; Fairchild & Herrington, 1989; Myrow, 1992). Local stratiform stromatolites suggest relatively deep microbial mat colonization such as inferred elsewhere (e.g. Fairchild & Herrington, 1989; Simonson *et al.*, 1993; Altermann & Siegfried, 1997). Presumably, water depths were photic (<*c*.100 m), and mat growth may have resulted from short-term decreases in sea-level or siliciclastic supply. However, the possibility that these beds formed due to non-photosynthetic microbial mats (see Simonson *et al.*, 1993) cannot be ruled out. Rare ironstones may have resulted from upwelling of anoxic basinal waters on to the outer ramp and mixing under oxygenated conditions.

## DUCKER FORMATION

The Ducker Formation (Eade & Chandler, 1975) is a unit of predominantly arkose, siltstone and mudstone that forms local lenses intervening between the Watterson and Tavani formations in the Watterson Lake, Griffin Lake, Ducker Lake and Bray Lake areas

(Fig. 5). It consists of: discrete dm-scale interbeds of arkose with parallel stratification and local trough cross-stratification; mm- to dm-scale, sharp-based upward-fining sequences of siltstone (parallel stratified ± ripple cross-stratified) to mudstone; and, near the base of the section, local dm-scale beds of stratiform stromatolites.

The Ducker Formation lenses are inferred to represent the remnants of relatively fine-grained shallow-water deltas fed by an unchannelized to poorly channelized braided fluvial system (see Tavani Formation below) that formed gently inclined 'shoal water' rather than 'Gilbert-type' profiles (see Postma, 1990). Interbedding of discrete sandstone beds and sharp-based upward-fining sequences probably reflects alternating episodes of subaerial sheetfloods and subaqueous mass flows.

## TAVANI FORMATION

The Tavani Formation (Heywood, 1973) consists predominantly of arkose (arkose facies) that contains local pelite-rich lenses (pelitic facies). A mixed siliciclastic–carbonate unit locally crops out above thick sections of arkose. Stratiform stromatolites at the base of this unit contain evaporite pseudomorphs (evaporite-bearing facies). Evaporite pseudomorphs disappear upsection and the Tavani Formation is capped by mixed arkoses and carbonate rocks (carbonate facies).

### Tavani Formation: facies description

#### *Arkose facies*

The predominant lithotype of the Tavani Formation is well cleaved to apparently structureless, texturally submature arkose. The paucity of primary sedimentary structures may be owing to high strain and/or recrystallization as well as to deposition as massive beds. The most commonly preserved sedimentary structure is parallel stratification, typically highlighted by heavy mineral concentrations or zones of carbonate cement. Trough cross-bedding, in isolated dm-scale sets, is locally exposed. Discontinuous cm-scale mudstone layers are locally abundant, commonly forming mudchip breccias and mudcurls.

Near the base of the Tavani Formation, conglomerates containing intraformational clasts of variably silicified dolomicrospar (± arkose) occur as m-scale layers interbedded with massive and parallel-stratified arkose, and are typically associated with m-scale stratiform stromatolites beds. Granule- and pebble-clasts predominate, but some clasts are as large as 30 cm. Clasts are angular to subangular, but some are rounded. The conglomerates are both clast and matrix supported, but in all cases the matrix consists of coarse arkose. Basal contacts are erosional, locally defining low-relief (<1 m) channels, and weak stratification arising from variations in clast abundance and size is locally developed.

#### *Pelitic facies*

Lenses in which dm- to m-scale pelitic intervals alternate with the m-scale sandstone beds are exposed in Windy River and North Henik Lake area (Figs 4 & 5). The pelitic intervals contain: (i) cm- to dm-scale, erosionally based, commonly graded, upward-fining sequences of sandstone to mudstone (Fig. 14A); (ii) cm-scale zones of microlaminated siltstone and mudstone rhythmites; and (iii) discrete cm- to dm-scale layers of mudstone. Intervening sandstones contain parallel-stratified heavy minerals, mudcurls and layers of mudchip breccia. Small, symmetric wave ripples and ladderback ripples are commonly displayed on the tops of sandstone bedding surfaces (Fig. 14B & C). These are draped by mudstones containing sand-filled desiccation cracks (Fig. 14B). Soft sediment deformation structures are common and include: sandstone dykes (Fig. 14D), ball and pillow structures, load casts and flame structures.

#### *Evaporite-bearing facies*

Evaporite-bearing siliciclastic and carbonate rocks are preserved in a structural depression north of Ducker Lake and in a downfaulted wedge at Mountain Lake (Figs 2 & 5). The evaporite-bearing unit ('T2') consists of thickly interbedded arkose, siltstone, stratiform and domal stromatolitic dolostone, and intraformational breccia and conglomerate. Pseudomorphs after gypsum, up to 3 cm long, filled with quartz, dolomite and muscovite, form lenticular prisms with euhedral terminations and swallowtail twins, as well as stubby prisms with pseudohexagonal cross-sections (Fig. 15; see Buick & Dunlop, 1990; Wilson & Versfeld, 1994 for illustrations of other Precambrian examples). The pseudomorphs are typically dispersed and randomly oriented, but are locally concentrated in cm-scale crystal mush layers in which long axes lie in bedding. Four-sided moulds after halite, some with a salt-hopper geometry, have been reported from this facies (Miller & Reading, 1993).

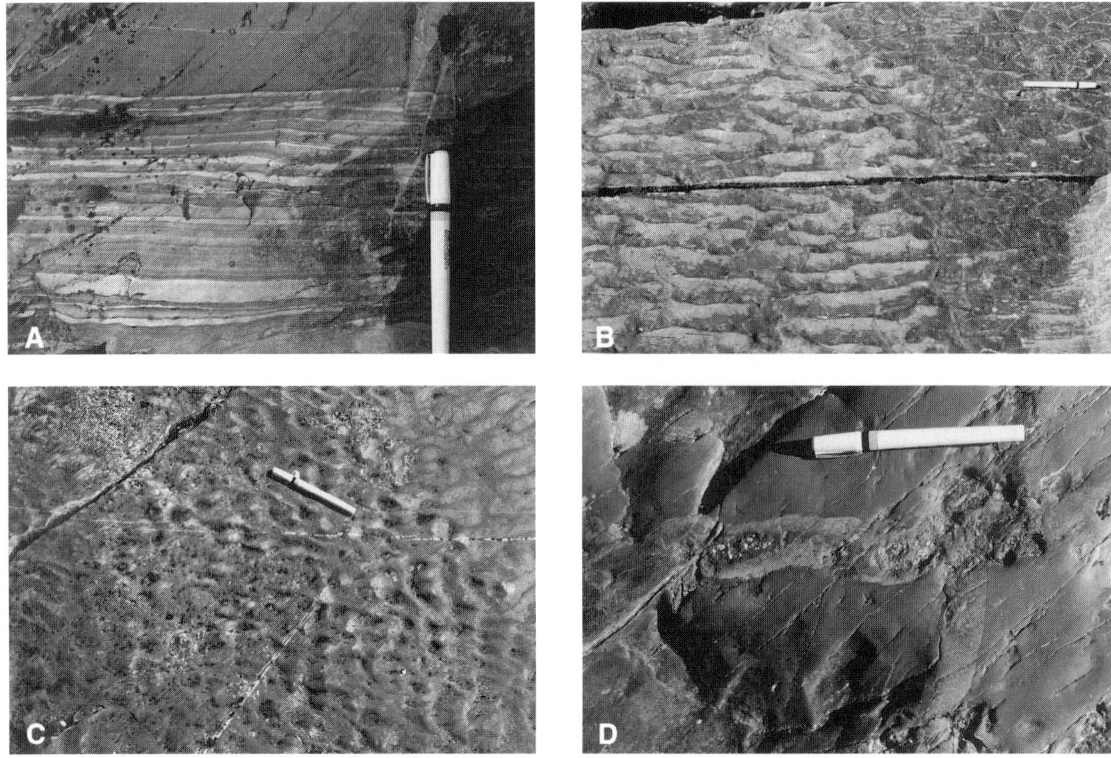

**Fig. 14.** Tavani Formation pelitic facies. (A) Graded sandstone to mudstone in sharp-based upward-fining sequences. (B) Bedding-parallel view of sandstone with small wave ripples draped by dark-toned mudstone with sand-filled desiccation cracks (right side of photo). (C) Bedding surface with small interference ripples. (D) Bedding surface of mudstone layer cut by polygonal sandstone dykes. Note coarse mud clasts in centre of dyke beneath pen, suggesting formation by injection rather than by passive infilling.

*Carbonate facies*

North of Ducker Lake, the evaporite component that defines the evaporite-bearing facies disappears upsection, and rocks in the uppermost subdivision of the Tavani Formation ('T3') are indistinguishable from the mixed siliciclastic–carbonate facies of the Watterson Formation. Additional exposures are close to the north-west margin of the basin, such as north of Bate Lake (Fig. 5), where Eade (1974) described a *c.*800 m-thick section of mixed carbonate–siliciclastic rocks above *c.*1200 m of Tavani Formation sandstones, and west of Bate Lake (Eade & Chandler, 1975).

## Tavani Formation: interpretation of depositional environments

The lower part of the Tavani Formation is inferred to have been deposited in a sandy alluvial to coastal wedge that led away from Pork arch and prograded over the Watterson ramp. The upper part of the formation is inferred to represent marine onlap and re-establishment of the ramp above a flooding surface.

Many studies of Precambrian alluvial depositional systems emphasize long-held views (Schumm, 1968; Cotter, 1978; Long, 1978) concerning differences in continental weathering, erosion, transport and deposition arising from the absence of vascular land plants (e.g. Eriksson *et al.*, 1998; Altermann, this volume, pp. 15–32). Without vascular plants to anchor loose sediment and stabilize slopes, any rainstorm would have produced flash floods. Such flashy behaviour is exhibited in the Tavani Formation by parallel-stratified texturally submature arkoses. These were probably deposited by high-energy sheetfloods that, unimpeded by land vegetation, spread rapidly across the fluvial plain, similar to examples described from sparsely vegetated modern environments (e.g. Williams, 1971;

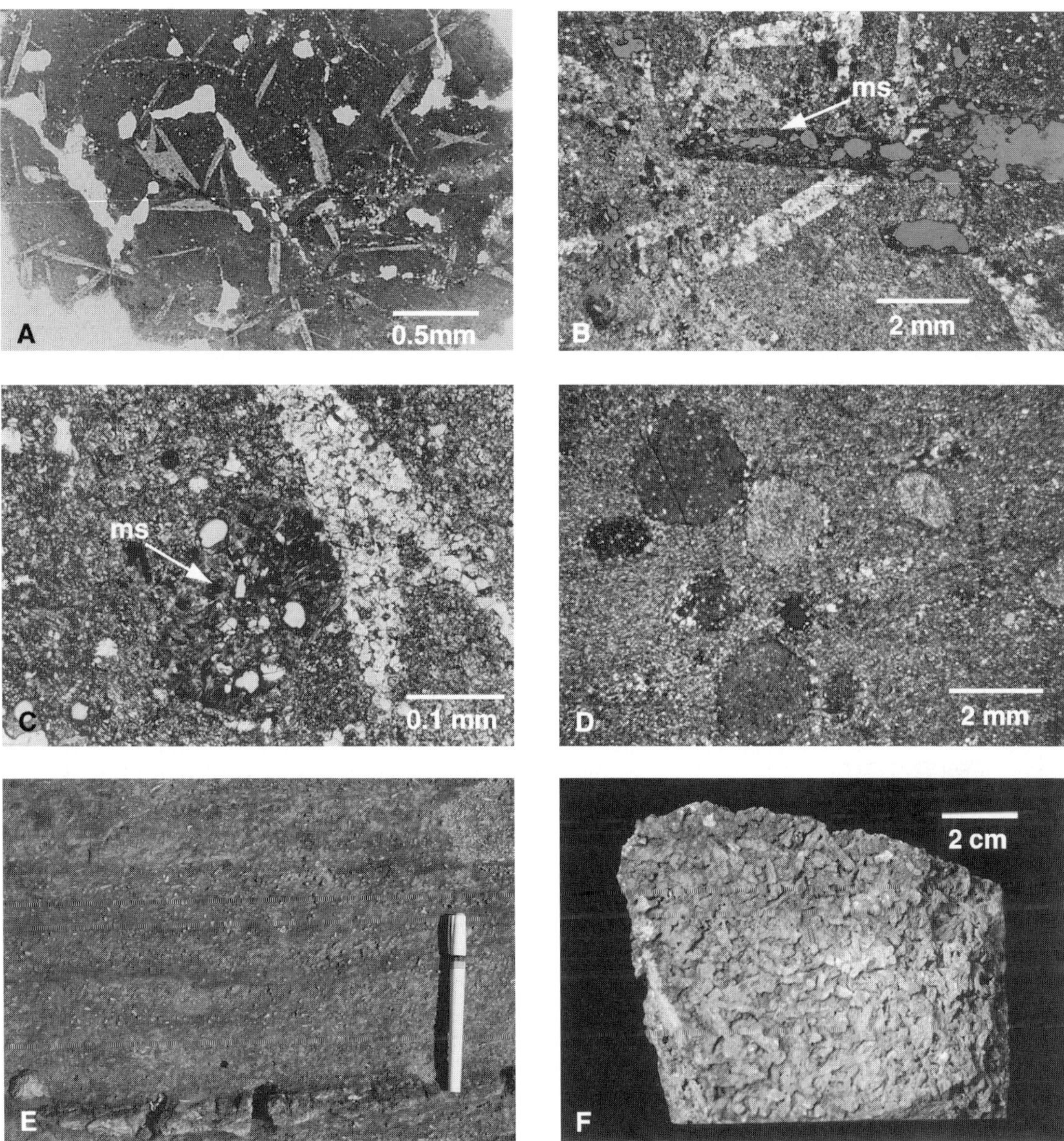


**Fig. 15.** Tavani Formation evaporite-bearing facies. (A) and (B) Haematite-stained dolomicrospar with dispersed pseudomorphs of microquartz and muscovite (ms) after lenticular gypsum. (C) Dolomicrospar with pseudomorph of microquartz after lenticular gypsum with swallowtail twin, and muscovite (ms) after stubby prismatic gypsum. (D) Dolomicrospar with pseuodomorphs displaying pseudohexagonal cross-section of dolomite and microquartz after gypsum. (E) Variably oriented, dispersed pseudomorphs of microquartz and dolomite after prismatic gypsum in laminated dolostone and siliciclastic siltstone. (F) Bedding surface showing crystal mush of microquartz and dolomite pseudomorphs after gypsum.

Tunbridge, 1983; Blair, 1999). Some of the structureless arkose beds may record highly concentrated, sand-charged debris floods (Miall, 1970), with suspension sedimentation yielding primary massive beds (see Martin & Turner, 1998). Mudstone layers that contain breccias and mudcurls are probable waning flood suspension deposits that underwent varying degrees of desiccation and reworking. The paucity of mudrocks probably reflects efficient by-passing of muds out of the vegetation-free landscape due to flash floods and aeolian deflation (see, for example, Long, 1978; Dalrymple *et al.*, 1985). The absence of conglomerates and well defined channels in the arkose facies (with the exception of intraformational conglomerates filling channels near the base of the section) is significant, and implies deposition on a low-gradient sand plain and minimal relief in nearby areas. The interbedded massive and parallel-stratified arkoses, stratiform stromatolite lenses and stratified, channelized conglomerates near the base of the section are interpreted as coastal deposits in which early cemented and silicified stromatolites growing in protected areas during times of relatively high water were reworked by channelized flows and flooded by fluvial sand sheets.

Fully enclosed by fluvial deposits, pelitic facies lenses are interpreted as having formed by alternating subaerial sheet flood and still-water sedimentation in ephemeral freshwater ponds that occupied local depressions on the Tavani fluvial plain. The m-scale sandstone beds are considered to record poorly channelized floods that entered the ponds and generated small mass flows. The pelitic beds are considered to be the products of mass-flow and suspension deposition within the ponds. Periodic emergence is indicated by the small-scale wave ripples, ladderback ripples and mudcracks.

The abrupt appearance of evaporite-bearing carbonate rocks above 1000–1200 m of predominantly arkose marks the sequence 3 and sequence 4 boundary, and is interpreted as a marine flooding surface. The occurrence of evaporite pseudomorphs, lacking in the remainder of the Tavani Formation and in the Watterson Formation, suggests a shift towards arid conditions during sequence 4. Like the interstratified arkoses, stratiform stromatolites and conglomerates near the base of the Tavani formation, the evaporite-bearing carbonates of sequence 4 are interpreted as coastal deposits. However, the randomly oriented lenticular gypsum psuedomorphs and the crystal mush layers, similar to fabrics observed in intertidal deposits of modern coastal sabkhas (e.g. Shearman, 1978), indicate relatively intense evaporation. These rocks are interpreted to have been deposited in coastal pans that experienced alternating episodes of fluvial flooding, marine submergence and hydrographic isolation. They are not considered indicators of Phanerozoic-like seawater compositions because mixing of meteoric and marine waters probably produced fluids that did not represent the chemistry of ambient sea water (see Grotzinger, 1990).

## UPPER HURWITZ GROUP AND POST-PRECAMBRIAN COMPARISON

Upper Hurwitz Group strata represent basin rejuvenation after a long period of apparent tectonic quiescence (*c.*2.1–1.9 Ga). Data presented above indicate that sequence 3 was deposited on a gently sloping mixed siliciclastic–carbonate ramp (Watterson Formation) that was overlain by prograding coastal and low-gradient fluvial deposits (Ducker Formation and lower Tavani Formation). Establishment of this continental–carbonate ramp depositional framework records uplift along a low-relief basement-cored arch, basin-centred subsidence and marine flooding. The ramp geometry, tectonic regime and environmental zonation of the upper Hurwitz Group are comparable to post-Precambrian mixed siliciclastic–carbonate systems, despite temporal changes in depositional processes arising from the evolution of land- and ocean-based life forms.

### Arch–ramp geometry and tectonic regime

The onset of carbonate sedimentation on many platforms has been related to a reduction of source-area relief and a diminished siliciclastic supply (e.g. Dott & Byers, 1981; Grotzinger & McCormick, 1988; Bond *et al.*, 1989). However, in Hurwitz Basin, carbonate sediments started to accumulate despite source area rejuvenation and influx of siliciclastic debris. Furthermore, the predominance of poorly channelized flash flood arkoses in the lower Tavani Formation, and the virtual absence of conglomerates, suggest deposition on a low-relief fluvial plain. Apparently, arching caused elevation of source areas, but with a relief that was inadequate to yield sediment volumes that were sufficient to overwhelm carbonate production or supply significant coarse-grained debris. Development of the upper Hurwitz mixed siliciclastic–carbonate ramp adjacent to a low-relief arch is similar to carbonate ramps, both Proterozoic and Phanerozoic, related to peripheral bulges in foreland basins (e.g. Grotzinger & McCormick, 1988; Dorobek, 1995), or to arches within intracratonic basins (Smith *et al.*, 1993; Choi *et al.*, 1999).

Intracratonic basins are persistent depocentres that initiate during supercontinent break-up and reactivate during supercontinent assembly. For example, the Michigan, Illinois, Williston and Hudson Bay intracratonic basins in the interior of North America formed during the Neoproterozoic–Early Cambrian break-up of Rodinia, and were reactivated during

Early Palaeozoic to Mesozoic orogenic events on the flanks of North America during the assembly of Pangaea (Sloss, 1988; Kominz & Bond, 1991; Klein, 1995; Burgess *et al.*, 1997; Howell & van der Pluijm, 1999). Whereas intracratonic sedimentation of sequences 1 and 2 in Hurwitz Basin is interpreted to have started during early break-up stages of the Neoarchaean supercontinent Kenorland, sequences 3 and 4 overlapped with assembly of at least the western part (present coordinates) of Laurentia, and hence tectonic controls on the evolution of Hurwitz Basin appear to have been similar to those that influenced the Neoproterozoic to Phanerozoic intracratonic basins in North America (Aspler *et al.*, 2001). Although new geochronologic and isotopic data indicate that basin rejuvenation occurred during post *c.*1.91 Ga shortening along the western margin of the Rae domain, current age constraints cannot distinguish between initiation of sequence 3 during late extension related to splitting of the southern Hearne domain margin, and shortening related to early convergence in western Trans-Hudson orogen (Fig. 1, Aspler *et al.*, 2001). None the less, the low relief continental arch–carbonate ramp geometry inferred for Hurwitz Basin is comparable to relatively low-relief flexures and carbonate ramps established in Phanerozoic intracontinental basins, and imposes significant limits on possible mechanisms (Aspler *et al.*, 2001).

### Depositional environments and siliciclastic–carbonate mixing

Continental to marine transitions in which coastal and offshore environments are linked by hybrid areas containing fluvially fed marine siliciclastic depozones and carbonate banks have been described from numerous modern settings (e.g. Roberts, 1987; Bush, 1991; Fay *et al.*, 1992; Gillespie *et al.*, 1998; Testa & Bosence, 1998; Woolfe & Larcombe, 1998). In ancient successions, migration of these localized carbonate and siliciclastic depocentres (resulting from short-term variations in sea level, sources and volumes of siliciclastic sediment, fluvial drainage patterns, carbonate production, ocean hydrodynamics and climate) translates into a complex interfingering of carbonate and siliciclastic rocks. This has been documented in many Phanerozoic basins (e.g. Mount, 1984; Blair, 1988; Lee & Kim, 1992; Osleger & Montañez, 1996; de Wet, 1998; Gallagher & Holdgate, 2000; Spalletti *et al.*, 2000), and is well expressed in proximal inner-ramp deposits of the Watterson Formation. Similarly, offshore and onshore 'punctuated mixing' (Mount, 1984) of siliciclastic and carbonate components by storm-generated currents has commonly been documented in modern (Aigner, 1985; Friedman, 1988) and Phanerozoic (e.g. Kelling & Mullin, 1975; Aigner, 1985; Lee & Kim, 1992; Hips, 1998) mixed systems, and is evident across the Watterson ramp.

An important difference relative to the post-Precambrian is that vascular plants would have been absent on the Tavani fluvial plain. Without plants to anchor loose sediment and stabilize slopes, transport by sheet floods would have been predominant. Thus, not only was the transfer of sediment to nearshore environments more efficient, but the delivery mechanism was probably more diffuse. As a result, large volumes of sediment appear to have entered the proximal inner ramp as unchannelized to poorly channelized floods generating subaqueous mass flows that led to the accumulation of extensive arkose- and pelite-rich units.

### Carbonate production and depositional processes

As emphasized in numerous papers (e.g. Grotzinger, 1989, 1990; Grotzinger & Knoll, 1995, 1999; Altermann, this volume, pp. 15–32), the most significant difference between Precambrian and younger carbonate platforms lies in the proliferation of Precambrian stromatolites and the absence of Precambrian carbonate-secreting skeletal organisms that could precipitate, be a source of and bind carbonate sediment. The general similarities between Precambrian and younger platforms suggests that direct precipitation of carbonate from sea water compensated for the lack of skeletal organisms. Precipitation processes were probably microbially mediated on all scales (e.g. Grotzinger, 1990; Grotzinger & Knoll, 1999; Wright & Altermann, 2000; Altermann, this volume, pp. 15–32) and may have been aided by higher saturation of marine waters with respect to calcium (e.g. Grotzinger, 1990; Fairchild, 1991; Grotzinger & Kasting, 1993; Grotzinger & Knoll, 1995, 1999), in part maintained by higher concentrations of $Fe^{2+}$ (particularly before *c.*2.2–1.9 Ga: Sumner & Grotzinger, 1996b). Carbonate units in the Watterson Formation contain extensive thick stromatolitic beds that display micritic internal laminae, but lack evidence of a coarse detrital component. Because it is unlikely that abrasion of pre-existing units could have produced only voluminous lutitic carbonate, precipitation from sea water was probably the predominant depositional process, although we are unable to distinguish growth by *in situ* cementation from growth due to trapping-and-binding of calcilutite precipitated in suspension as whitings. Relatively minor detrital dolarenites and dolorudites were ultimately derived from reworking of chemical precipitates.

## CONCLUSIONS

Sequences 3 and 4 in the upper Hurwitz Group record post *c.*1.91 Ga rejuvenation of Hurwitz Basin after a lacuna of at least 200 Myr. During rejuvenation, crustal flexure led to growth of an extensive, gently inclined, mixed siliciclastic–carbonate ramp that was paired to a low-gradient sandy alluvial plain adjacent to a low-relief basement-cored arch (Pork arch).

Sequence 3 strata (Watterson Formation, Ducker Formation and lower Tavani Formation) define an upward-shallowing profile that represents progradation of continental facies away from the Pork arch towards shallow- and deep-marine environments. Sequence 4 (upper Tavani Formation) reflects re-establishment of the ramp subsequent to marine flooding.

The Watterson Formation is inferred to have been deposited on a storm-dominated, microtidal, mixed siliciclastic–carbonate ramp. Outer- to mid-ramp environments (below wave base) are represented by siliciclastic siltstones, very fine-grained sandstones, mudstones, detrital carbonates and stratiform stromatolites. These were deposited by background suspension sedimentation that was periodically interrupted by siliciclastic and carbonate mass flows, in part derived from inner-ramp environments during storms. Distal inner-ramp stromatolitic bioherms and stratiform stromatolite fields formed by fairweather calcilutite sedimentation and/or *in situ* precipitation; siliciclastic sand-bearing dolorudite, oolitic-intraclastic dolarenite and dololutite interbeds are tempestites that record seaward-returning storm-generated bottom flows. Proximal inner-ramp environments included subtidal stromatolite fields interspersed with pools receiving siliciclastic sand and mud introduced by subaerial sheet floods.

Ducker Formation deltaic arkoses and pelites, and fluvial arkoses in the lower part of the Tavani Formation, created a sandy alluvial to coastal wedge that led away from Pork arch and prograded over the Watterson ramp. The predominance of parallel-stratified unchannelized arkoses in the Tavani Formation indicates deposition by sheet floods that, unimpeded by land vegetation, spread rapidly over a low-relief fluvial plain.

The ramp geometry and environmental zonation of the upper Hurwitz Group are comparable to post-Precambrian mixed siliciclastic–carbonate systems, despite contrasts in the relative importance of specific depositional processes arising from differences in land- and ocean-based life forms. Without plants to anchor loose sediment and stabilize slopes, sheet floods would have predominated on the fluvial plain, resulting in large volumes of sediment entering the proximal inner ramp as diffuse plumes that generated subaqueous mass flows and deposited extensive arkose and pelite beds. Without carbonate-secreting skeletal organisms that could precipitate, be a source of and bind carbonate sediment, chemical precipitation from sea water was the predominant process for carbonate deposition.

## ACKNOWLEDGEMENTS

Field work was funded by the Geology Office, Indian and Northern Affairs Canada (Yellowknife) and the Geological Survey of Canada. Preparation of the manuscript was supported by the Canada–Nunavut Geoscience Office. We thank Al Donaldson and Wlady Altermann for discussions and reviews of an initial draft, and appreciate the constructive comments made by journal reviewers Carol de Wet and Viviane Testa. This is a contribution to the Western Churchill NATMAP Project and the Canada–Nunavut Geoscience Office.

## REFERENCES

Acker, K.L. & Stearn, C.W. (1990) Carbonate–siliciclastic transition and reef growth on the northeast coast of Barbados, West Indies. *J. sediment. Petrol.*, **60**, 18–25.

Ahr, W.M. (1998) Carbonate ramps, 1973–1996: a historical review. In: *Carbonate Ramps* (Eds Wright, V.P. & Burchette, T.P.), Spec. Publs geol. Soc. London, No. 149, pp. 7–14. Geol. Soc. London, Bath.

Aigner, T. (1985) Storm depositional systems. *Lecture Notes in Earth Sciences*, **3**. Springer-Verlag, Berlin, 174 pp.

Altermann, W. & Nelson, D.R. (1998) Sedimentation rates, basin analysis and regional correlations of three Neoarchaean and Palaeoproterozoic sub-basins of the Kaapvaal craton as inferred from precise U–Pb zircon ages from volcaniclastic sediments. *Sediment. Geol.*, **120**, 225–256.

Altermann, W. & Siegfried, H.P. (1997) Sedimentology and facies development of an Archean shelf: carbonate platform transition in the Kaapvaal Craton, as deduced from a deep borehole at Kathu, South Africa. *J. Afr. Earth Sci.*, **24**, 391–410.

Aspler, L.B. & Bursey, T.L. (1990) Stratigraphy, sedimentation, dome and basin basement-cover infolding and implications for gold in the Hurwitz Group, Hawk

Hill–Griffin–Mountain Lakes area, District of Keewatin. *Paper 90-1C*, pp. 219–230. Geological Survey of Canada, Ottawa.

Aspler, L.B., Bursey, T.L. & LeCheminant, A.N. (1992) Geology of the Henik, Montgomery Lake, and Hurwitz groups in the Bray–Montgomery–Ameto lakes area, southern District of Keewatin, Northwest Territories. *Paper 92-1C*, pp. 157–170. Geological Survey of Canada, Ottawa.

Aspler, L.B., Bursey, T.L. & Miller, A.R. (1989) Sedimentology, structure and economic geology of the Poorfish–Windy thrust–fold belt, Ennadai Lake area, District of Keewatin, and the shelf to foreland transition in the foreland of Trans Hudson Orogen. *Paper 89-1C*, pp. 143–155. Geological Survey of Canada, Ottawa.

Aspler, L.B. & Chiarenzelli, J.R. (1996a) Stratigraphy, sedimentology and physical volcanology of the Ennadai–Rankin greenstone belt, Northwest Territories, Canada: Late Archean paleogeography of the Hearne Province and tectonic implications. *Precam. Res.*, **77**, 59–89.

Aspler, L.B. & Chiarenzelli, J.R. (1996b) Relationships between the Montgomery Lake and Hurwitz groups, southern District of Keewatin NWT and stratigraphic revision of the lower Hurwitz Group. *Can. J. Earth Sci.*, **33**, 1243–1255.

Aspler, L.B. & Chiarenzelli, J.R. (1997a) Initiation of *c.*2.45–2.1 Ga intracratonic basin sedimentation of the Hurwitz Group, Keewatin Hinterland, Northwest Territories, Canada. *Precam. Res.*, **81**, 265–297.

Aspler, L.B. & Chiarenzelli, J.R. (1997b) Archean and Proterozoic Geology of the North Henik Lake area, District of Keewatin, Northwest Territories. *Paper 1997-C*, pp. 145–155. Geological Survey of Canada, Ottawa.

Aspler, L.B. & Chiarenzelli, J.R. (1998) Two Neoarchean supercontinents? Evidence from the Paleoproterozoic. *Sediment. Geol.*, **120**, 75–104.

Aspler, L.B., Chiarenzelli, J.R. & Bursey, T.L. (1993a) Archean and Proterozoic geology of the Padlei belt, District of Keewatin, NWT. *Paper 93-1C*, pp. 147–158. Geological Survey of Canada, Ottawa.

Aspler, L.B., Chiarenzelli, J.R. & Bursey, T.L. (1994a) Ripple marks in quartz arenites of the Hurwitz Group, Northwest Territories, Canada: Evidence for sedimentation in a vast, Early Proterozoic, shallow, fresh-water lake. *J. sediment. Res.*, **64A**, 282–298.

Aspler, L.B., Chiarenzelli, J.R., Cousens, B.L., McNicoll, V.J. & Davis, W.J. (2001) Paleoproterozoic intracratonic basin processes, from breakup of Kenorland to assembly of Laurentia: Hurwitz Basin, Nunavut, Canada. *Sediment. Geol.*, **141–142**, 287–318.

Aspler, L.B., Chiarenzelli, J.R., Ozarko, D.L. & Powis, K.B. (1993b) Geological map of the Watterson Lake area, District of Keewatin, NWT. *Open File 2767*. Geological Survey of Canada, Ottawa.

Aspler, L.B., Chiarenzelli, J.R., Ozarko, D.L. & Powis, K.B. (1994b) Geology of Archean and Proterozoic supracrustal rocks in the Otter and Ducker lakes area, southern District of Keewatin, Northwest Territories. *Paper 1994-C*, pp. 165–174. Geological Survey of Canada, Ottawa.

Aspler, L.B., Höfer, C. & Harvey, B.J.A. (2000) Geology of the Henik, Montgomery and Hurwitz groups, Sealhole and Fitzpatrick lakes area, Nunavut. In: *Current Research, Paper 2000-C12*, 10 pp., CD-ROM. Geological Survey of Canada, Ottawa.

Bechstädt, T. & Schweizer, T. (1991) The carbonate–clastic cycles of the East-Alpine Raibl group: result of third-order sea-level fluctuations in the Carnian. *Sediment. Geol.*, **70**, 241–270.

Bell, R.T. (1968) Preliminary notes on the Proterozoic Hurwitz Group, Tavani (55K) and Kaminak Lake (55L) areas, District of Keewatin. *Paper 68-36*, 17 pp. Geological Survey of Canada, Ottawa.

Bell, R.T. (1970) The Hurwitz Group—a prototype for deposition on metastable cratons. In: *Symposium on Basins and Geosynclines of the Canadian Shield* (Ed. Baer, A.J.), *Paper 70-40*, pp. 159–169. Geological Survey of Canada, Ottawa.

Beukes, N.J. (1987) Facies relations, depositional environments and diagenesis in a major Early Proterozoic stromatolitic carbonate platform to basinal sequence, Campbellrand Subgroup, Transvaal Supergroup, southern Africa. *Sediment. Geol.*, **54**, 1–46.

Blair, T.C. (1988) Mixed siliciclastic-carbonate marine and continental syn-rift sedimentation, Upper Jurassic–lowermost Cretaceous Todos Santos and San Ricardo Formations, western Chiapas, Mexico. *J. sediment. Petrol.*, **58**, 623–636.

Blair, T.C. (1999) Sedimentary processes and facies of the waterlain Anvil Spring Canyon alluvial fan, Death Valley, California. *Sedimentology*, **46**, 913–940.

Bond, G.C., Kominz, M.A., Steckler, M.S. & Grotzinger, J.P. (1989) Role of thermal subsidence, flexure, and eustasy in the evolution of Early Paleozoic passive-margin carbonate platforms. In: *Controls on Carbonate Platform and Basin Development* (Eds Crevello, P.D., Wilson, J.L., Sarge, J.F. & Read, J.F.), Spec. Publs Soc. econ. Paleont. Miner., Tulsa, **44**, 39–61.

Braga, J.C., Martin, J.M. & Riding, R. (1995) Controls on microbial dome fabric development along a carbonate–siliciclastic shelf-basin transect, Miocene, SE Spain. *Palaios*, **10**, 347–361.

Buick, R. & Dunlop, J.S.R. (1990) Evaporitic sediments of Early Archean age from the Warrawoona Group, North Pole, Western Australia. *Sedimentology*, **37**, 247–277.

Burchette, T.P. & Wright, V.P. (1992) Carbonate ramp depositional systems. *Sediment. Geol.*, **79**, 3–57.

Burgess, P.M., Gurnis, M. & Moresi, L. (1997) Formation of sequences in the cratonic interior of North America by interaction between mantle, eustatic, and stratigraphic processes. *Geol. Soc. Am. Bull.*, **108**, 1515–1535.

Bush, D.M. (1991) Mixed carbonate/siliciclastic sedimentation: northern insular shelf of Puerto Rico. In: *Mixed Carbonate–Siliciclastic Sequences* (Eds Lomando, A.J. & Harris, P.M.). Society of Economic Paleontologists and Geologists, Tulsa, Core Workshop 15, pp. 447–484.

Choi, Y.S., Simo, J.A. & Saylor, B.Z. (1999) Sedimentologic and sequence stratigraphic interpretation of a mixed carbonate–siliciclastic ramp, midcontinent epeiric sea, Middle to Upper Ordovician Decorah and Galena Formations, Wisconsin. In: *Advances in Carbonate Sequence Stratigraphy: Applications to Reservoirs, Outcrops and Models* (Eds Harris, P.M., Saller, A.H. & Simo, J.A.), Spec. Publs Soc. econ. Paleont. Geol., Tulsa, **63**, 275–289.

Cotter, E. (1978) The evolution of fluvial style, with special reference to the central Appalachian Paleozoic. In: *Fluvial*

*Sedimentology* (Ed. Miall, A.D.), Mem. Can. Soc. petrol. Geol., Calgary, **5**, 361–383.

Dalrymple, R.W., Narbonne, G.M. & Smith, L. (1985) Eolian action and the distribution of Cambrian shales in North America. *Geology*, **13**, 607–610.

Davidson, A. (1970) Precambrian geology, Kaminak Lake map-area, District of Keewatin. *Paper 69-51*, 27 pp. Geological Survey of Canada, Ottawa.

Davis, W.J., Aspler, L.B., Rainbird, R.H. & Chiarenzelli, J.R. (2000) Detrital zircon geochronology of the Proterozoic Hurwitz and Kiyuk groups: a revised post-1.92 Ga age for deposition of the upper Hurwitz Group. *GeoCanada 2000 Conference CD*. Iron Leaf Communications, Calgary.

Dean, W.E. & Fouch, T.D. (1983) Lacustrine environment. In: *Carbonate Depositional Environments* (Eds Scholle, P.A., Bebout, D.G. & Moore, C.H.), Mem. Am. Ass. Petrol. Geol., Tulsa, **33**, 97–130.

de Wet, C.B. (1998) Deciphering the sedimentological expression of tectonics, eustasy, and climate: a basinwide study of the Corallian Formation, southern England. *J. sediment. Res.*, **68**, 653–667.

Dolan, J.F. (1989) Eustatic and tectonic controls on deposition of hybrid siliciclastic/carbonate basinal cycles: discussion with examples. *Am. Ass. Petrol. Geol. Bull.*, **73**, 1233–1246.

Donaldson, J.A. (1963) Stromatolites in the Denault Formation, Marion Lake, coast of Labrador, Newfoundland. *Geol. Surv. Can. Bull.*, **102**, 33 pp.

Donaldson, J.A. (1976) Paleoecology of cononphyton and associated stromatolites in the Precambrian Dismal Lakes and Rae groups, Canada. In: *Stromatolites* (Ed. Walter, M.R.), pp. 523–534. Elsevier, Amsterdam.

Dorobek, S.L. (1995) Synorogenic carbonate platforms in foreland basins: controls on stratigraphic evolution and platform/reef morphology. In: *Stratigraphic Evolution of Foreland Basins* (Eds Dorobek, S.L. & Ross, G.M.), Spec. Publs Soc. econ. Paleont. Miner., Tulsa, **52**, 127–147.

Dott, R.H. Jr & Byers, C.W. (1981) The orthoquartzite–carbonate suite revisited. *J. sediment. Petrol.*, **51**, 329–350.

Doyle, L.J. & Roberts, H.H. (Eds) (1988) *Carbonate–Clastic Transitions*. Developments in Sedimentology, 42. Elsevier, Amsterdam, 304 pp.

Eade, K.E. (1974) *Geology of Kognak River Area, District of Keewatin, Northwest Territories*. Mem. geol. Surv. Can., Ottawa, **377**, 66 pp.

Eade, K.E. & Chandler, F.W. (1975) Geology of Watterson Lake (west half) map-area, District of Keewatin. *Paper 74-64*, 10 pp. Geological Survey of Canada, Ottawa.

Eriksson, P.G., Condie, K.C., Tirsgaard, H. *et al.* (1998) Precambrian clastic sedimentation systems. *Sediment. Geol.*, **120**, 5–53.

Fairchild, I.J. (1991) Origins of carbonate in Neoproterozoic stromatolites and the identification of modern analogues. *Precam. Res.*, **53**, 281–299.

Fairchild, I.J. & Herrington, P.M. (1989) A tempestite–stromatolite–evaporite association (Late Vendian, East Greenland): a shoreface-lagoon model. *Precam. Res.*, **43**, 101–127.

Fay, M., Masalu, D.C.P. & Muzuka, A.N.N. (1992) Siliciclastic–carbonate transitions in surface sediments of a back-reef lagoon north of Dar es Salaam (Tanzania). *Sediment. Geol.*, **78**, 49–57.

Friedman, G.M. (1988) Case histories of coexisting reefs and terrigenous sediments: the Gulf of Elat (Red Sea), Java Sea, and Neogene Basin of the Negev, Israel. In: *Carbonate–Clastic Transitions* (Eds Doyle, L.J. Doyle & Roberts, H.H.). Developments in Sedimentology, 42. Elsevier, Amsterdam, pp. 77–98.

Gagan, M.K., Chivas, A.R. & Herczeg, A.L. (1990) Shelf-wide erosion, deposition, and suspended sediment transport during Cyclone Winifred, central Great Barrier Reef, Australia. *J. sediment. Petrol.*, **60**, 456–470.

Gallagher, S.J. & Holdgate, G. (2000) The paleogeographic and paleoenvironmental evolution of a Paleogene mixed carbonate–siliciclastic cool-water succession in the Otway Basin, southeast Australia. *Palaeogeogr. Palaeoclimatol. Palaeoecol.*, **156**, 19–50.

Gillespie, J.L., Nelson, C.S. & Nodder, S.D. (1998) Post-glacial sea-level control and sequence stratigraphy of carbonate–terrigenous sediments, Wanganui shelf, New Zealand. *Sediment. Geol.*, **122**, 245–266.

Ginsburg, R.N. (1991) Controversies about stromatolites: vices and virtues. In: *Controversies in Modern Geology* (Eds Müller, D.W., McKenzie, J.A. & Weissert, H.), pp. 25–36. Academic Press, London.

Grotzinger, J.P. (1986) Cyclicity and paleoenvironmental dynamics, Rocknest platform, northwest Canada. *Geol. Soc. Am. Bull.*, **97**, 1208–1231.

Grotzinger, J.P. (1989) Facies and evolution of Precambrian carbonate depositional systems: emergence of the modern platform archetype. In: *Controls on Carbonate Platform and Basin Development* (Eds Crevello, P.D., Wilson, J.L., Sarge, J.F. & Read, J.F.), Spec. Publ. Soc. econ. Petrol. Miner., Tulsa, **44**, 79–106.

Grotzinger, J.P. (1990) Geochemical model for Proterozoic stromatolite decline. *Am. J. Sci.*, **290A**, 80–103.

Grotzinger, J.P. & Kasting, J.F. (1993) New constraints on Precambrian ocean composition. *J. Geol.*, **101**, 235–243.

Grotzinger, J.P. & Knoll, A.H. (1995) Anomalous carbonate precipitates: is the Precambrian the key to the Permian? *Palaios*, **10**, 578–596.

Grotzinger, J.P. & Knoll, A.H. (1999) Stromatolites in Precambrian carbonates: evolutionary mileposts or environmental dipsticks. *Ann. Rev. Earth planet. Sci.*, **27**, 313–358.

Grotzinger, J.P. & McCormick, D.S. (1988) Flexure of the Early Proterozoic lithosphere and evolution of Kilohigok Basin (1.9 Ga), northwest Canadian Shield. In: *New Perspectives in Basin Analysis* (Eds Kleinspehn, K. & Paola, C.), pp. 405–431. Springer-Verlag, New York.

Hanmer, S., Aspler, L., Sandeman, H. *et al.* (2000) Henik–Kaminak–Tavani supracrustal belt, late Archean oceanic crust and island arc remnants; upper crust protected from Proterozoic reworking. *GeoCanada 2000 Conference CD*. Iron Leaf Communications, Calgary.

Heaman, L.M. (1994) 2.45 Ga global mafic magmatism: Earth's oldest superplume? In: *Abstracts of the Eighth International Conference on Geochronicity, Cosmochronicity and Isotope Geology* (Eds Lauphere, M.A., Dalrymple, G.B. & Turrin, B.D.), US geol. Surv. Circ., **1107**, 132.

Heaman, L.M. & LeCheminant, A.N. (1993) Paragenesis and U–Pb systematics of baddeleyite ($ZrO_2$). *Chem. Geol.*, **110**, 95–126.

Hesse, R. (1989) Silica diagenesis: origin of inorganic and replacement cherts. *Earth Sci. Rev.*, **26**, 253–284.

Heywood, W.W. (1973) Geology of Tavani map-area, District of Keewatin. *Paper 72-47*, 14 pp. Geological Survey of Canada, Ottawa.

Hips, K. (1998) Lower Triassic storm-dominated ramp sequence in northern Hungary: an example of evolution from homoclinal through distally steepened ramp to Middle Triassic flat-topped platform. In: *Carbonate Ramps* (Eds Wright, V.P. & Burchette, T.P.), Spec. Publ. geol. Soc. London, No. 149, pp. 315–338. Geol. Soc. London, Bath.

Hoffman, P. (1967) Algal stromatolites: use in stratigraphic correlation and paleocurrent determination. *Science*, **157**, 1043–1045.

Hoffman, P. (1973) Recent and ancient algal stromatolites: seventy years of pedagogic cross-pollination. In: *Evolving Concepts in Sedimentology* (Ed. Ginsburg, R.N.), Johns Hopkins Univ. Stud. Geol., **21**, 178–191.

Hoffman, P. (1974) Shallow and deepwater stromatolites in Lower Proterozoic platform-to-basin facies change, Great Slave Lake, Canada. *Bull. Am. Ass. Petrol. Geol.*, **58**, 856–867.

Hoffman, P. (1976a) Stromatolite morphogenesis in Shark Bay, Western Australia. In: *Stromatolites* (Ed. Walter, M.R.), pp. 261–271. Elsevier, Amsterdam.

Hoffman, P. (1976b) Environmental diversity of middle Precambrian stromatolites. In: *Stromatolites* (Ed. Walter, M.R.), pp. 599–611. Elsevier, Amsterdam.

Hofmann, H.J. & Davidson, A. (1998) Paleoproterozoic stromatolites, Hurwitz Group, Quartzite Lake area, Northwest Territories, Canada. *Can. J. Earth Sci.*, **35**, 280–289.

Hofmann, H.J. & Jackson, G.D. (1987) Proterozoic ministromatolites with radial-fibrous fabric. *Sedimentology*, **34**, 963–971.

Holmes, A.E. & Christie-Blick, N. (1993) Origin of sedimentary cycles in mixed carbonate-siliciclastic systems: an example from Canning Basin, Western Australia. In: *Carbonate Sequence Stratigraphy: Recent Developments and Applications* (Eds Loucks, R.G. & Sarg, J.F.), Mem. Am. Ass. Petrol. Geol., **57**, 181–212.

Horodyski, R.J. (1976) Stromatolites from the Middle Proterozoic Altyn Limestone, Belt Supergroup, Glacier National Park, Montana. In: *Stromatolites* (Ed. Walter, M.R.), pp. 585–597. Elsevier, Amsterdam.

Howell, P.D. & van der Pluijm, B.A. (1999) Structural sequences and styles of subsidence in the Michigan basin. *Geol. Soc. Am. Bull.*, **111**, 974–991.

Johnson, H.D. & Baldwin, C.T. (1996) Shallow siliciclastic seas. In: *Sedimentary Environments*, 3rd edn (Ed. Reading, H.G.), pp. 232–280. Blackwell Science, Oxford.

Kah, L.C. & Knoll, A.H. (1996) Microbenthic distribution of Proterozoic tidal flats: Environmental and taphonomic considerations. *Geology*, **24**, 79–82.

Kelling, G. & Mullin, P.R. (1975) Graded limestones and limestone–quartzite couplets: possible storm-deposits from the Moroccan Carboniferous. *Sediment. Geol.*, **13**, 161–190.

Kerans, C. & Donaldson, J.A. (1989) Deepwater conical stromatolite reef, Sulky Formation (Dismal Lakes Group), Middle Proterozoic, NWT. In: *Canadian Reef Research Symposium* (Eds Geldsetzer, H.H.J., James, N.P. & Tebbutt, G.E.), Mem. Can. Soc. petrol. Geol., **13**, 81–88.

Klein, G.D. (1995) Intracratonic basins. In: *Tectonics of Sedimentary Basins* (Eds Busby, C.J. & Ingersoll, R.V.), pp. 459–478. Blackwell Science, Oxford.

Knauth, L.P. (1994) Petrogenesis of chert. In: *Silica Physical Behavior, Geochemistry and Materials Applications* (Eds Heaney, P.J., Prewitt, C.T. & Gibbs, G.V.). *Rev. Mineral.*, **29**, 233–258.

Kominz, M.A. & Bond, G.C. (1991) Unusually large subsidence and sea-level events during middle Paleozoic time: new evidence supporting mantle convection models for supercontinent assembly. *Geology*, **19**, 56–60.

le Roux, J.P. (1997) Cycle hierarchy of a Neoproterozoic carbonate–siliciclastic shelf: Matjies River Formation of the Kango Group, South Africa. *S. Afr. J. Geol.*, **100**, 1–10.

Lee, Y.I. & Kim, J.C. (1992) Storm-influenced siliciclastic and carbonate ramp deposits, the lower Ordovician Dumugol Formation, South Korea. *Sedimentology*, **39**, 951–969.

Leighton, M.W. (1996) Interior cratonic basins: a record of regional tectonic influences. In: *Basins and Basement of Eastern North America* (Eds van der Pluijm, B.A. & Catacosinos, P.A.), Spec. Publ. geol. Soc. Am., Boulder, **308**, 77–93.

Long, D.G.F. (1978) Proterozoic stream deposits: some problems of recognition of ancient sandy fluvial systems. In: *Fluvial Sedimentology* (Ed. Miall A.D.), Mem. Can. Soc. petrol. Geol., Calgary, **5**, 313–341.

McCormick, D.S. & Grotzinger, J.P. (1993) Distinction of marine from alluvial facies in the Paleoproterozoic (1.9 Ga) Burnside Formation, Kilohigok Basin, NWT, Canada. *J. sediment. Petrol.*, **63**, 398–419.

Martin, C.A.L. & Turner, B.R. (1998) Origins of massive-type sandstones in braided river systems. *Earth Sci. Rev.*, **44**, 15–38.

Miall, A.D. (1970) Devonian alluvial fans, Prince of Wales Island, Arctic Canada. *J. sediment. Petrol.*, **40**, 556–571.

Miller, A.R. & Reading, K.L. (1993) Iron-formation, evaporite, and possible metallogenic implications for the Lower Proterozoic Hurwitz Group, District of Keewatin, Northwest Territories. *Paper 93-1C*, pp. 179–186. Geological Survey of Canada, Ottawa.

Mount, J.F. (1984) Mixing of siliciclastic and carbonate sediments in shallow shelf environments. *Geology*, **12**, 432–435.

Myrow, P.M. (1992) Pot and gutter casts from the Chapel Island Formation, southeast Newfoundland. *J. sediment. Petrol.*, **62**, 992–1007.

Myrow, P.M. & Southard, J.B. (1996) Tempestite deposition. *J. sediment. Res.*, **66**, 875–887.

Osleger, D.A. & Montañez, I.P. (1996) Cross-platform architecture of a sequence boundary in mixed siliciclastic–carbonate lithofacies, Middle Cambrian, southern Great Basin, USA. *Sedimentology*, **43**, 197–217.

Patterson, J.G. & Heaman, L.M. (1991) New geochronologic limits on the depositional age of the Hurwitz Group, Trans-Hudson hinterland, Canada. *Geology*, **19**, 1137–1140.

Pelechaty, S.M. & Grotzinger, J.P. (1989) Stromatolite bioherms of a 1.9 Ga foreland basin carbonate ramp, Beechey Formation, Kilohigok Basin, Northwest Territories. In: *Canadian Reef Research Symposium* (Eds Geldsetzer, H.H.J., James, N.P. & Tebbutt, G.E.), Mem. Can. Soc. petrol. Geol., **13**, 93–104.

Piper, D.J.W. & Stow, D.A.V. (1991) Fine-grained turbidites. In: *Cycles and Events in Stratigraphy* (Eds Einsele, G. & Seilacher, A.) pp. 360–376. Springer-Verlag, Berlin.

Platt, N.H. & Wright, V.P. (1991) Lacustrine carbonates: facies models, facies distributions and hydrocarbon aspects. In: *Lacustrine Facies Analysis* (Eds Anadón, P., Cabrera, Ll. & Kelts, K.), Spec. Publs int. Ass. Sediment., No. 13, pp. 57–74. Blackwell Scientific Publications, Oxford.

Postma, G. (1990) Depositional architecture and facies of river and fan deltas: a synthesis. In: *Coarse-grained Deltas* (Eds Colella, A. & Prior, D.B.), Spec. Publs int. Ass. Sediment., No. 10, pp. 13–27. Blackwell Scientific Publications, Oxford.

Read, J.F. (1985) Carbonate platform facies models. *Bull. Am. Ass. Petrol. Geol.*, **69**, 1–21.

Reid, R.P., Macintyre, I.G. & James, N.P. (1990) Internal precipitation of microcrystalline carbonate: a fundamental problem for sedimentologists. *Sediment. Geol.*, **68**, 163–170.

Ricketts, B.D. & Donaldson, J.A. (1979) Stone rosettes as indicators of ancient shorelines: examples from the Precambrian Belcher Group, Northwest Territories. *Can. J. Earth Sci.*, **16**, 1887–1891.

Ricketts, B.D. & Donaldson, J.A. (1989) Stromatolite reef development on a mud-dominated platform in the Middle Precambrian Belcher Group of Hudson Bay. In: *Canadian Reef Research Symposium* (Eds Geldsetzer, H.H.J., James, N.P. & Tebbutt, G.E.), Mem. Can. Soc. petrol. Geol., Calgary, **13**, 113–119.

Robbins, L.L., Tao, Y. & Evans, C.A. (1997) Temporal and spatial distribution of whitings on Great Bahama Bank and a new lime mud budget. *Geology*, **25**, 947–950.

Roberts, H.H. (1987) Modern carbonate–siliciclastic transitions: humid and arid tropical examples. *Sediment. Geol.*, **50**, 25–65.

Ross, G.M. & Donaldson, J.A. (1989) Reef development and facies geometry on a high-energy Early Proterozoic carbonate shelf (Hornby Bay Group, Northwest Territories, Canada). In: *Canadian Reef Research Symposium* (Eds Geldsetzer, H.H.J., James, N.P. & Tebbutt, G.E.), Mem. Can. Soc. petrol. Geol., Calgary, **13**, 120–128.

Sami, T.T. & James, N.P. (1993) Evolution of an early Proterozoic foreland basin carbonate platform, lower Pethei Group, Great Slave Lake, north-west Canada. *Sedimentology*, **40**, 403–430.

Sami, T.T. & James, N.P. (1994) Peritidal carbonate platform growth and cyclicity in an early Proterozoic foreland basin, upper Pethei Group, Northwest Canada. *J. sediment. Res.*, **B64**, 111–131.

Sami, T.T. & James, N.P. (1996) Synsedimentary cements as Paleoproterozoic building blocks, Pethei Group, Northwestern Canada. *J. sediment. Res.*, **66**, 209–222.

Schumm, S.A. (1968) Speculations concerning paleohydrologic controls of terrestrial sedimentation. *Geol. Soc. Am. Bull.*, **79**, 1573–1588.

Shearman, D.J. (1978) Evaporites of coastal sabkhas. In: *Marine Evaporites* (Eds Dean, W.E. & Schreiber, B.C.). Society of Economic Paleontologists and Geologists, Tulsa, Short Course 4, pp. 6–42.

Simonson, B.M., Schubel, K.A. & Hassler, S.W. (1993) Carbonate sedimentology of the early Precambrian Hamersley Group of Western Australia. *Precam. Res.*, **60**, 287–335.

Sloss, L.L. (1988) Tectonic evolution of the craton in Phanerozoic time. In: *Sedimentary Cover—North American Craton* (Ed. Sloss, L.L), The Geology of North America, D-2. Geological Society of America, Boulder, pp. 25–51.

Smith, G.L., Byers, C.W. & Dott, R.H. Jr (1993) Sequence stratigraphy of the lower Ordovician Prairie Du Chien Group on the Wisconsin Arch and in the Michigan Basin. *Bull. Am. Ass. Petrol. Geol.*, **77**, 49–67.

Spalletti, L.A., Franzese, J.R., Matheos, S.D. & Schwarz, E. (2000) Sequence stratigraphy of a tidally dominated carbonate–siliciclastic ramp; the Tithonian–Early Berriasian of the southern Nequén Basin, Argentina. *J. geol. Soc. London*, **157**, 433–446.

Spence, G.H. & Tucker, M.E. (1997) Genesis of carbonate megabreccias and their significance in carbonate sequence stratigraphic models: a review. *Sediment. Geol.*, **112**, 163–193.

Sumner, D.Y. & Grotzinger, J.P. (1996a) Herringbone calcite: petrography and environmental significance. *J. sediment. Res.*, **66**, 419–429.

Sumner, D.Y. & Grotzinger, J.P. (1996b) Were kinetics of Archean calcium carbonate precipitation related to oxygen concentration? *Geology*, **24**, 119–122.

Talbot, M.R. & Allen, P.A. (1996) Lakes. In: *Sedimentary Environments*, 3rd edn (Ed. Reading, H.G.), pp. 83–124. Blackwell Science, Oxford.

Tella, S., Annesley, I.R., Borradaile, G.J. & Henderson, J.R. (1986) Precambrian geology of parts of Tavani, Marble Island, and Chesterfield Inlet map areas, District of Keewatin: a progress report. *Paper 86–13*, 20 pp. Geological Survey of Canada, Ottawa.

Testa, V. & Bosence, D.W.J. (1998) Carbonate–siliciclastic sedimentation on a high-energy, ocean-facing, tropical ramp, NE Brazil. In: *Carbonate Ramps* (Eds Wright, V.P. & Burchette, T.P.), Spec. Publ. geol. Soc. London, No. 149, pp. 55–71. Geol. Soc. London, Bath.

Tirsgaard, H. (1996) Cyclic sedimentation of carbonate and siliciclastic deposits on a late Precambrian ramp: the Elisabeth Bjerg Formation (Eleonore Bay Supergroup), East Greenland. *J. sediment. Res.*, **66**, 699–712.

Tunbridge, I.P. (1983) Alluvial fan sedimentation of the Horseshoe Park flood, Colorado, USA, July 15th, 1982. *Sediment. Geol.*, **36**, 15–23.

Whitaker, J.H.McD. (1973) 'Gutter casts', a new name for scour-and-fill structures: with examples from the Llandoverian of Ringerike and Malmöya, southern Norway. *Norsk geol. tidsskr.*, **53**, 403–417.

Williams, G.E. (1971) Flood deposits of the sand-bed ephemeral streams of Central Australia. *Sedimentology*, **17**, 1–40.

Williams, H., Hoffman, P.F., Lewry, J.F., Monger, J.W.H. & Rivers, T. (1991) Anatomy of North America: thematic portrayals of the continent. *Tectonophysics*, **187**, 117–134.

Wilson, A.H. & Versfeld, J.A. (1994) The Early archean Nondweni greenstone belt, southern Kaapvaal Craton, South Africa, Part I. Stratigraphy, sedimentology, mineralization and depositional environment. *Precam. Res.*, **67**, 243–276.

Wilson, J.L. (1967) Cyclic and reciprocal sedimentation in Virgilian strata of southern New Mexico. *Bull. geol. Soc. Am.*, **78**, 805–818.

Woolfe, K.J. & Larcombe, P. (1998) Terrigenous sediment accumulation as a regional control on the distribution of reef carbonates. In: *Reefs and Carbonate Platforms in the*

*Pacific and Indian Oceans* (Eds Camoin, G.F. & Davies, P.J.), Spec. Publs int. Ass. Sediment., No. 25, pp. 295–310. Blackwell Science, Oxford.

Wright, D.T. & Altermann, W. (2000). Microfacies development in Late Archean stromatolites and oolites of the Ghaap Group of South Africa. In: *Carbonate Platform Systems: Components and Interactions* (Eds Insalaco, E., Skelton, P.W. & Palmer, T.J.), Spec. Publs geol. Soc. London, No. 178, pp. 51–70. Geol. Soc. London, Bath.

*Spec. Publs int. Ass. Sediment.* (2002) **33**, 323–338

# Aspects of Late Palaeoproterozoic fluvial style: the Uairén Formation, Roraima Supergroup, Venezuela

D. G. F. LONG

*Department of Earth Sciences, Laurentian University, Sudbury, Ontario, P3E 2C6, Canada*

## ABSTRACT

Strata of the Uairén Formation, at the base of the Late Palaeoproterozoic Roraima Supergroup in the Gran Sabana, near Santa Elena de Uairén, Bolívar, south-eastern Venezuela, can be interpreted in terms of deposition in a variety of vegetation-free fluvial environments in a foreland basin setting. Sandstones and small to medium pebble conglomerates at the base of the formation reflect erratic deposition from high-energy (proximal) gravelly braided streams in valley-fill and valley-marginal alluvial fan settings. Arid conditions are indicated by the presence of minor diamictites of debris flow origin. Overlying predominantly sandy strata were deposited by sheet floods in a low-relief basin characterized by flashy ephemeral sandy braided fluvial systems in a terminal-fluvial setting. Temporary development of playa lakes may be indicated by thin mud drapes and fine to very fine sandstones with symmetrical ripples. Low angle truncation of laminae may reflect erosion and deposition by sheet floods. Aeolian reworking may be significant in this setting and is indicated locally by ripple cross-stratified silt-free sandstone sets between more poorly sorted sandstones. Medium to very large pebble conglomerate at the base of the upper member of the formation rests on a regionally significant erosion surface, reflecting uplift and/or a dramatic change in climatic regime. Overlying sandstones reflect deposition by sandy and gravelly sandy perennial braided fluvial systems with high width to depth ratios. Strata of the Uairén Formation are succeeded by poorly exposed lacustrine mudstones of the Cuquenán Formation, which are overlain by fluvial and aeolian deposits of the Uaimapué Formation. In Proterozoic and Phanerozoic sequences, thick sets of flat laminated sandstones are commonly interpreted in terms of beach-face deposition. This study shows that this can be incorrect and that such sequences are a common feature of pre-vegetation fluvial settings, especially in arid to semi-arid terminal fluvial systems.

## INTRODUCTION

The Roraima Supergroup is a thick sequence of gently dipping to flat lying, late Palaeoproterozoic sandstones, siltstones and conglomerates, which include minor volcanic ash beds and are locally intruded by basic dikes and sills. Strata were deposited in an extensive east–west basin, at least 900 km long and 500 km wide, located on the northern part of the Guiana Shield (Suszczyniski, 1981a,b; Gibbs & Barron, 1993) and including parts of Venezuela, Guyana, Surinam and Brazil (Fig. 1). The main area of outcrop of the Roraima Supergroup is in the Gran Sabana Region of Bolivar State, Venezuela (Fig. 1), where the sequence forms a series of high, steep sided table-lands (mesas), known as 'tepuy' (tepuis). The sequence is of economic interest in that it appears to host low-concentration placer deposits which provide a local source of gold and diamonds in overlying Recent alluvial deposits (Lopez *et al.*, 1942; Hea, 1975; Gibbs & Barron, 1993; Sidder & Mendoza, 1995; Dohrenwend *et al.*, 1995).

This study focuses on aspects of the sedimentology of the lowermost part of the Roraima Supergroup in an area immediately north and north-west of the town of Santa Elena de Uairén, located at 4° 36′ N; 61° 07′ W, 10 km north of the border with Brazil, and 60 km west of the border with Guyana (Fig. 2). The lower member of the Uairén Formation contains significant thicknesses of flat-laminated sandstones, which are here interpreted as products of deposition by sheet floods. The significance of this facies and its relevance to the understanding of Precambrian and early

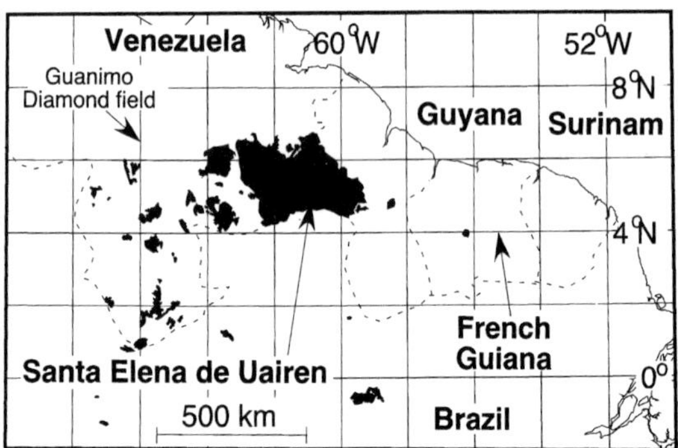


**Fig. 1.** Distribution of late Precambrian strata of the Roraima Supergroup (in black) in northern South America (after Gibbs & Barron, 1993).

Palaeozoic fluvial processes before the onset of rooted land vegetation is considered after discussion of the Uairén Formation.

## Previous work

Strata now assigned to the Roraima Supergroup were first described by Anderson & Dunn (1895) from Guyana. They were originally named the 'Roraima strata' by Dalton (1912), based on exposures of the sequence near Mount Roraima, on the shared borders of Venezuela, Guyana and Brazil. The name was subsequently changed to Roraima Formation by Aguerrevere *et al.* (1939) and to the Roraima Group by Reid (1974a,b, 1977), and raised to supergroup status by Reis *et al.* (1990).

While additional work has been done on the lithology, distribution (Lopez *et al.*, 1942; Gansser, 1954); Van de Putte, 1972, 1974; Reid, 1974a,b, 1977; Reid & Bisque, 1975; Ghosh, 1977, 1978; Yánez, 1972, 1985; Santos, 1985) and geomorphological expression (Tate, 1930; Martin-Kaye, 1952; Briceño & Schubert, 1990; Dohrenwend *et al.*, 1995) of the Roraima Supergroup, only a limited amount of attention has been paid to the sedimentary framework of the unit in Venezuela (Read, 1974a; Ghosh, 1985; Simon *et al.*, 1985). Examination of equivalent strata in adjacent parts of Brazil led Reis *et al.* (1985, 1990) to suggest deposition of parts of the sequence in fluvial, deltaic, marine, lacustrine and aeolian settings.

The geology of the Santa Elena district is shown on maps by Direccion de Geologia (1976), as well as on regional maps by Almeidam, (1978), Bellizzia, (1984) Sidder (1990), Gibbs & Barron (1993) and Sidder & Mendoza (1995).

## Geological framework

The geological framework of the area investigated is shown in Fig. 2, based on mapping by Direccion de Geologia (1976). In the Gran Sabana, strata of the Roraima Supergroup unconformably overlie Early Proterozoic acid to intermediate volcanic rocks and subvolcanic intrusives belonging to the 1.92–1.69 Ga Uatuma Supergroup (Gibbs & Barron, 1993). These developed above an Archaean to Early Proterozoic basement which was consolidated during the Trans-Amazonian Episode between 2.15 and 1.96 Ga, and were later deformed in an as yet unnamed orogeny between 1.86 and 1.73 Ga (Sidder & Mendoza, 1995).

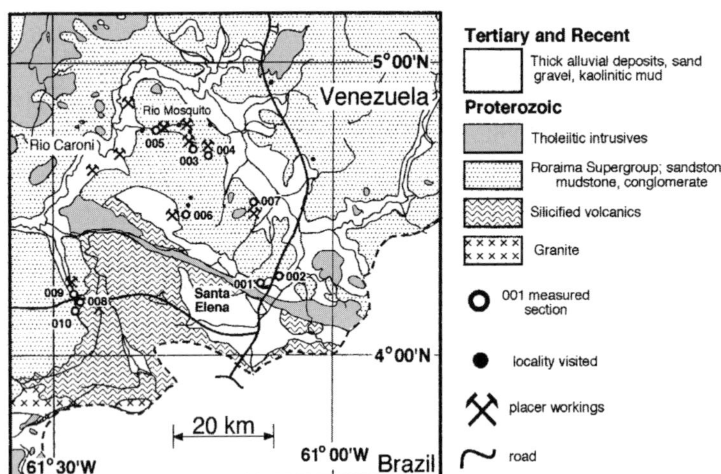


**Fig. 2.** Geology of the area around Santa Elena de Uairén (Direccion de Geologia, 1976).

In the vicinity of Santa Elena de Uairén, pre-Trans-Amazonian rocks include meta-volcanic, meta-igneous and associated meta-sediments of the Pastora Group and Los Caribes Formation, and granitic rocks of the 2.23–2.05 Ga Supamo Complex which are deeply weathered and support a lush jungle cover (Lopez *et al.*, 1942; Benaim, 1972; Ghosh, 1985). Strata of the overlying Uatuma Supergroup are included in the Cuchivero Group (Priem *et al.*, 1973; Hea, 1975; Gibbs & Barron, 1993; Sidder & Mendoza, 1995). These consist of a thick sequence of weakly metamorphosed felsic to intermediate subaerial volcanic rocks and associated granitic intrusions. Comprehensive descriptions of these rocks can be found in summary papers by Gibbs & Barron (1993) and Sidder & Mendoza (1995).

Kimberlite pipes have not as yet been discovered in the vicinity of Santa Elena de Uairén. Diamonds in the Roraima Supergroup and its reworked cover may have been derived directly or indirectly from a suite of 1.73 Ga kimberlites, similar to those discovered by Nixon *et al.* (1989, 1992, 1994) in the Guaniamo Region 450 km west-north-west. If so, this represents a maximum age for the base of the supergroup. As the Guaniamo kimberlites are not seen in direct contact with strata of the Roraima Group a maximum age for deposition at 1.8 Ga is also possible.

Strata of the Roraima Supergroup rest with a marked unconformity on the Early Proterozoic basement complex. In the study area, Reid (1974a) was able to document local palaeo-relief of approximately 50 m, and noted that volcanic rocks immediately below the contact are strongly altered, suggesting a period of intense weathering prior to deposition. In the vicinity of Santa Elena de Uairén the Roraima Supergroup consists predominantly of a gently folded sequence of fine- to very coarse-grained sandstone, with minor pebble and cobble conglomerate and mudstone, which are cut by mafic dikes and sills of the Avanavero Suite. These are dated at 1.61–1.66 Ga (Snelling, 1963; Snelling & McConnell, 1965; Gibbs & Barron, 1993) and are thought to be related to a period of continental tholeiitic magmatism which produced minor vitric ash-flow tuffs in the middle part of the Roraima sequence (Uaimapué Formation).

The stratigraphic nomenclature for the Roraima Supergroup is complicated by its political and physical geography (see Gibbs & Barron, 1993, for details). In the Santa Elena area, Reid (1974a) was able to divide the supergroup into four formations (Fig. 3). These include a basal sequence of sandstones with minor conglomerate (Uairén Formation: 850 m), overlain by

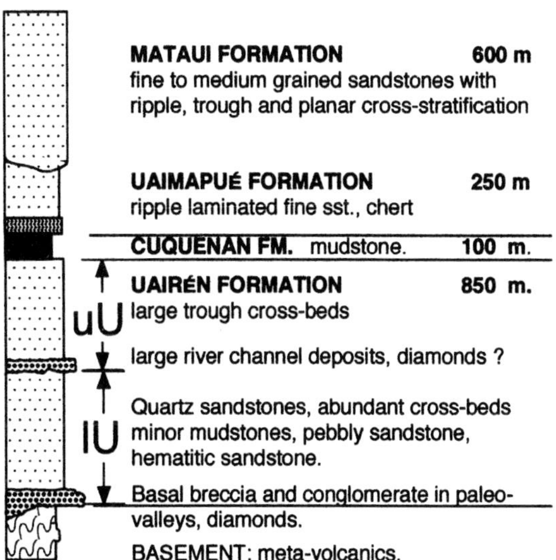


**Fig. 3.** Stratigraphy of the Roraima Supergroup, near Santa Elena de Uairén, after Reid (1974a). uU = upper member, and lU = upper member of Uairén Formation in this paper.

50–100 m of recessive weathering mudstones of the Cuquenán Formation (also known as the Kukenán Formation), which may include minor tuffaceous horizons, equivalent to those dated at 1.65 Ga by Priem *et al.* (1973) from Surinam. The Cuquenán Formation is capped by a second sequence of sandstone, with minor conglomerate (Uaimapué Formation: 250 m) and an upper sequence of quartzose sandstones (Mataui Formation: 600 m). The latter forms the scarps and peaks of the higher tepuys, such as Mount Roraima (elevation 2810 m), which give the Gran Sabana its distinctive appearance. Ferruginous rocks at the contact between the Uairén and Cuquenán Formation were interpreted as a Proterozoic palaeo-laterite by Reid (1974a). Subsequent observations by Dohrenwend *et al.* (1995) have shown that these are remnants of a regionally extensive middle Tertiary ferricrete, and are not of Proterozoic age.

In the study area (Fig. 2) most of the Roraima strata examined belong to the Uairén Formation (equivalent to the Arai Formation of Reis *et al.*, 1990). Strata of the Cuquenán Formation are very poorly exposed. They may underlie the valley of the Río Ququenán, from 61° 30′ W to 60° 00′ W. Strata of the Uaimapué Formation form the highlands to the north, north-west and north-east of the Río Ququenán, most of which lie within the 'Park Nacional Canaima' and were not examined during this study.

Strata of the Roraima Supergroup have been subjected to low-grade (subgreenschist) metamorphism associated with intrusion of basaltic dikes and sills of the Avanavero Suite. Evidence for this includes numerous quartz veins (some with optically pure quartz) in softer weathering parts of the sequence. The presence of parallel fractures in many of the quartz pebbles within the conglomerates of the Uairén Formation suggests that post-depositional shearing may also have taken place, prior to delithification of the sequence by extensive tropical and subtropical weathering, and not frost shattering as suggested by Hea (1975).

Early Phanerozoic strata have not been recognized in Gran Sabana. It is possible that the area was uplifted to its present level following opening of the South Atlantic in late Aptian times (Jones, 1987). In the last 100 Ma the Roraima Supergroup was subject to deep weathering, mostly in tropical to subtropical settings. This is probably responsible for the conversion of feldspars within much of the sequence to kaolinite, leaving much of the sequence poorly lithified. The Roraima Supergroup is mantled by a thin cover of Late Tertiary to Recent alluvial and residual deposits (Dohrenwend *et al.*, 1995).

## NEW OBSERVATIONS: UAIRÉN FORMATION

The stratigraphic distribution of strata within the Uairén Formation is shown in Fig. 4, based on ten sections measured during this study. In the area around Santa Elena de Uairén, the formation is dominated by medium- to very coarse-grained sandstones, containing variable amounts of quartz (monocrystalline

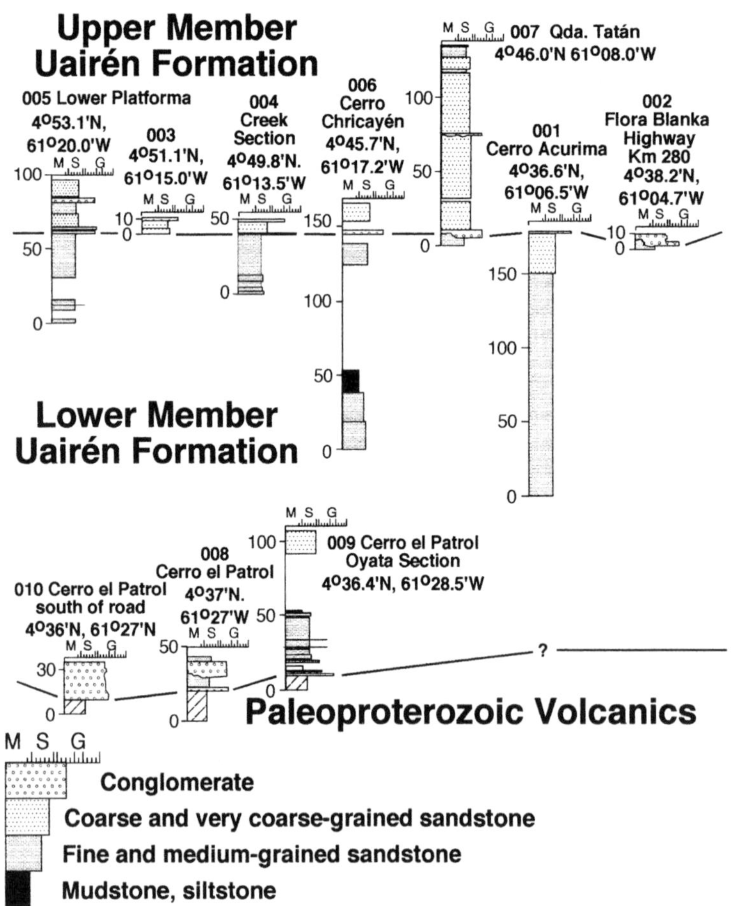


**Fig. 4.** Stratigraphy of the Uairén Formation, in the vicinity of Santa Elena de Uairén, based on sections measured during this study. Median grain size is indicated by width of the column: M, mudstone; S, sandstone; G, conglomerate. Small divisions, one phi divisions. Scale in metres. For location of sections see Fig. 2.

strained and unstrained, and polycrystalline), feldspar (largely altered to kaolinite) and volcanic rock fragments. Conglomerates are present locally at the base of the formation and at a second level, 200–300 m above the base of the formation (Figs 3 & 4). Mudstones are rare, and are limited to thin interbeds between sandstone units.

The Uairén Formation is here divided into upper and lower members, based on the presence of conglomeratic marker beds. The lithology of the formation is discussed in approximate stratigraphic order, prior to discussion of its sedimentology.

## Lower member of the Uairén Formation

The lower member of the Uairén Formation consists of up to 300 m of sandstone and mudstone with minor discontinuous conglomerates near its base (Fig. 4). The maximum thickness estimate is based on topographic reconstruction, as no continuously exposed section through the member is available.

### *Basal conglomerates*

Conglomeratic strata were examined at the base of the Uairén Formation in the vicinity of Cerro el Patrol (Fig. 4, sections 008–010). The conglomerates include massive to weakly stratified, well sorted, large to very large pebble conglomerates, with a clast supported framework of well rounded pebbles of quartz and acid volcanic rock, with lesser amounts of iron formation and feldspar (Fig. 5A). The matrix of these conglomerates consists of a coarse-grained sandstone (subvolcanilithic arenite). The clast supported conglomerates occur in composite sets from 3 to 25 m thick, the thicker, and in parts coarser, units being associated with palaeovalleys developed in the underlying Early Proterozoic basement (see Reid, 1974a). In section 008, which may have developed close to the side of a palaeovalley, the conglomerates are interbedded with, and are locally erosive into, massive to plane laminated medium- and fine-grained sandstone and minor mudstone.

In the lower 10 m of section 009 conglomeratic strata are found interbedded with laminated mudstones. Conglomeratic strata in this section are more poorly sorted than in sections 008 and 010. At the base of section 009 massive, poorly sorted cobble-grade conglomerates consist predominantly of subrounded acid volcanic clasts, set in a matrix of sandy mudstone. Maximum clast size is up to 62 cm intermediate diameter. Conglomerates higher in this sequence are in part clast supported (including one pebbly mudstone), with intraclasts of laminated mudstone, and some subrounded cobbles projecting above the tops of the bed.

Minor, massive, lenticular pebbly and cobbly medium-grained sandstones are present in the next 20 m of section 009 as thin interbeds between trough cross-stratified (Fig. 5C) and planar bedded sandstones (Fig. 5E).

### *Sandstones, mudstones*

The conglomeratic sequence at the base of the lower member of the Uairén Formation is overlain by 200–250 m of fine- to (pebbly) very coarse-grained sandstone, which ranges in composition from feldspathic arenite to subvolcanilithic arenite to quartz arenite. These include planar and trough cross-stratified varieties, some of which, at the 30–50 m level of section 009, contain abundant heavy mineral laminae (Fig. 5B & D).

Graded, massive to plane laminated medium- and fine-grained sandstones may form the bulk of the lower member. These are well developed at the base of section 006 and in section 001, where they occur in sets 40–80 cm thick, interbedded with thin sets of laminated mudstone (Fig. 6A, B & D). Large-scale bedforms may be indicated by low angle truncation of laminae (Fig. 6Aa). The base of these sets is typically sharp, and may be erosive as the sandstones often contain angular intraclasts of locally derived mudstone (Fig. 6B). Thicker (1–20 m) sequences of laminated mudstone are present locally in sections 009 and 006, where they may be associated with thin graded laminated units of fine- to very fine-sand grade, and oscillation ripple laminated fine sandstone (Fig. 6C & D).

Sandstones within the upper few tens of metres of the lower member are typically arkosic, with abundant planar and trough cross-stratification (Fig. 7A). Trough cross-stratification in silty, fine-grained sandstones in section 002 is locally overturned and shows signs of liquefaction (Fig. 6Ea & b). Large-scale trough cross-stratification is present locally in sections 001 and 006.

## Upper member of Uairén Formation

The upper member of the Uairén Formation consists of 150 m or more of sandstone, pebbly sandstone and conglomerate. The base of the member is marked by a regionally extensive conglomerate with a locally erosive base. The top of the member was not seen. It may be gradational with recessive weathering mudstones of the overlying Cuquenán Formation.

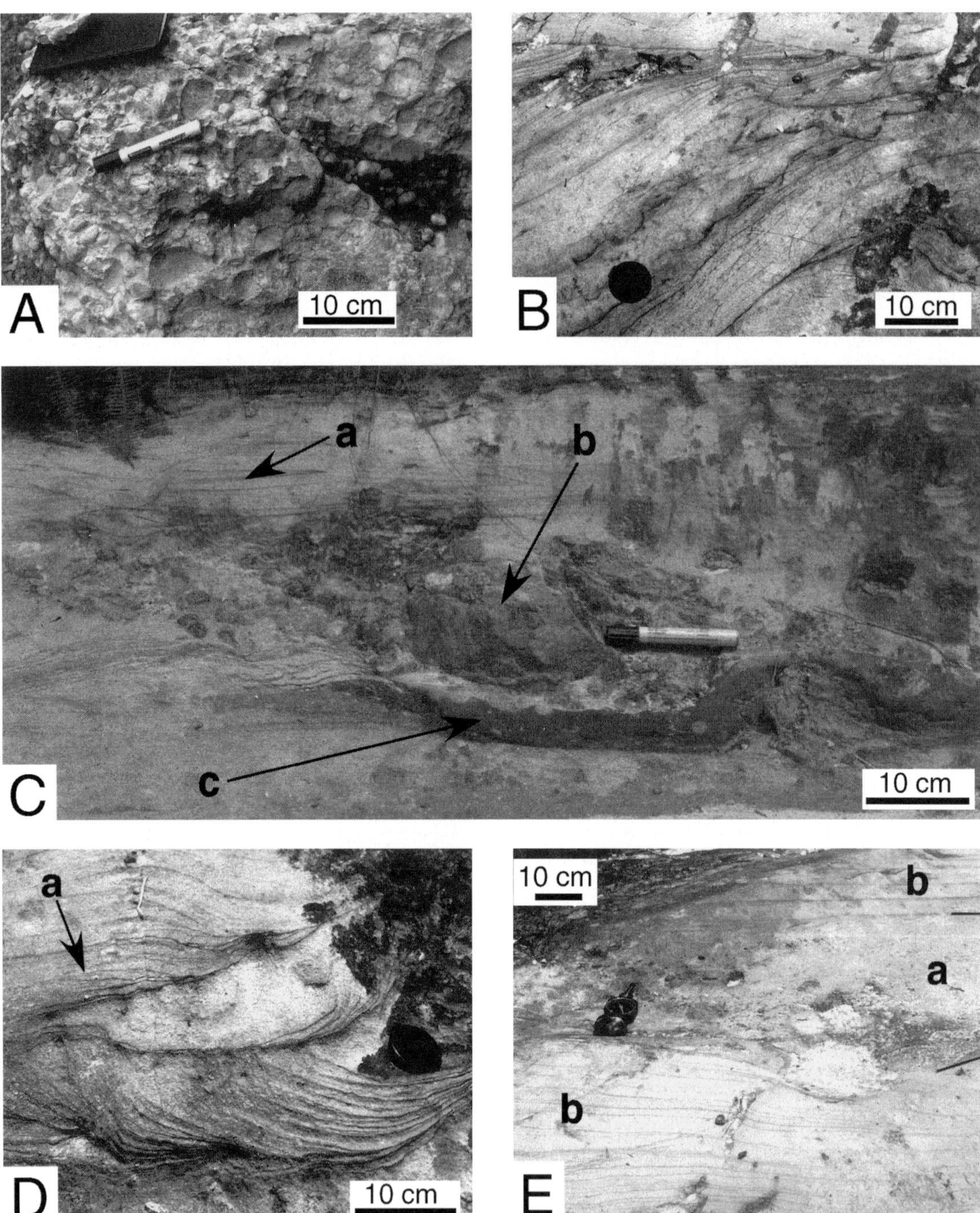


**Fig. 5.** Strata from the lower member of the Uairén Formation. (A) Moderately well sorted medium- to large-pebble conglomerate at base of lower member, section 008. (B) Planar cross-stratified sandstones with abundant heavy minerals on foresets, 30 m level, section 009; note deformation of foresets. (C) Matrix supported pebbly sandstone (b) and slightly pebbly mudstone (c) at the margin of a shallow channel in section 009, interpreted as a debris-flow levee deposit; note concentration of heavy minerals in associated sheet-wash deposits (a). (D) Trough cross-stratified sandstones with abundant heavy minerals on foresets, 29 m level, section 009; note small-scale truncation (a) indicating migration of ripples during formation of the sets from sinuous crested dunes. (E) Highly lenticular unit of pebbly medium-grained sandstone (a), with erosive base, 28 m level, section 009; thin sheet-wash deposits (b) are separated by thin bands of heavy minerals.

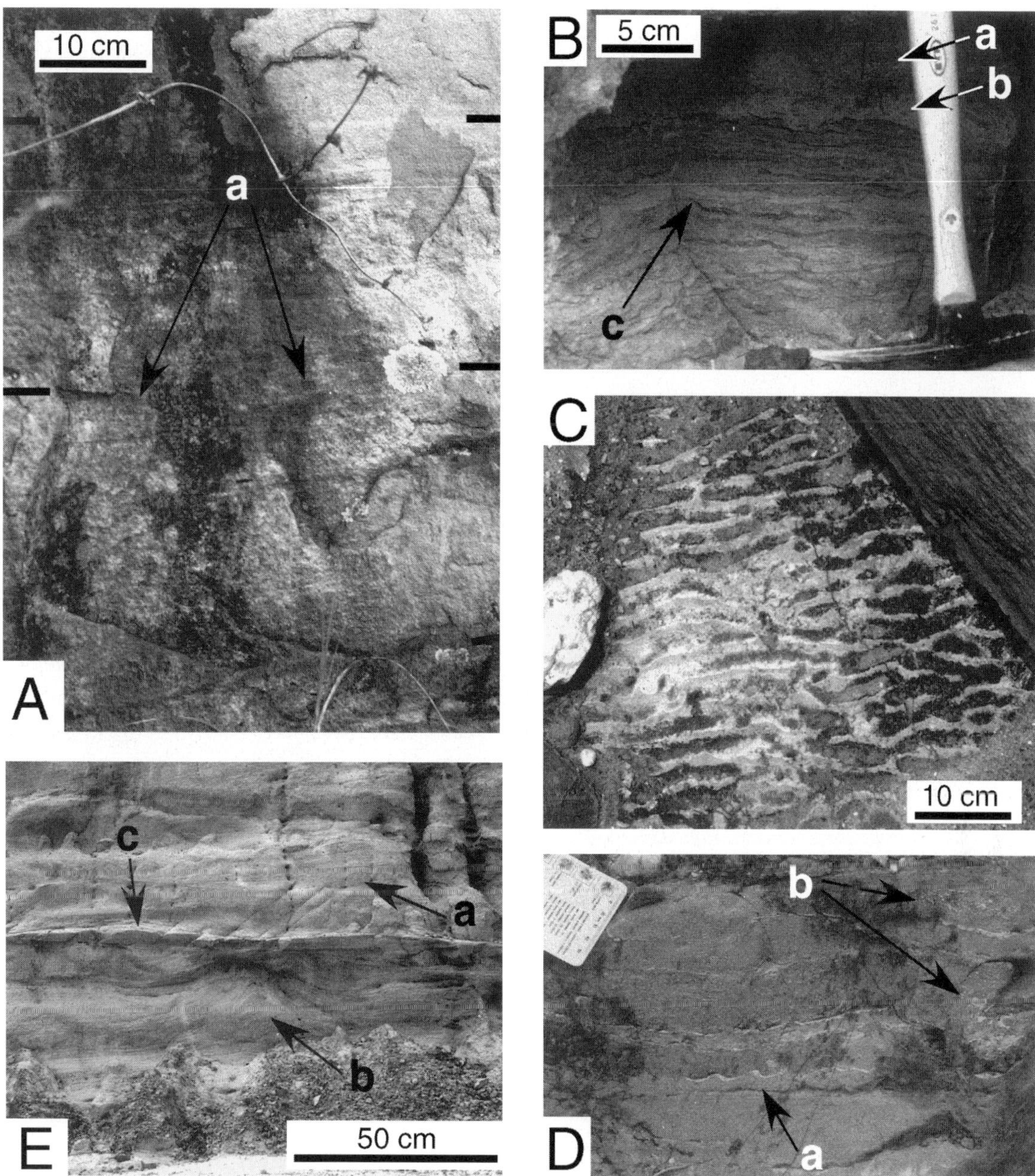


**Fig. 6.** Strata from the lower member of the Uairén Formation. (A) Graded–laminated medium sandstone, 10 m level, section 006; note low-angle truncation of laminae (a), which may be a characteristic feature of these unusual stream-flood deposits. (B) Distal sheet-wash deposits in section 001; note local cross-lamination (a), mudstone intraclasts (b) and irregular lamination (c). (C) Ripple marks on bed surface of fine-grained sandstone, 44 m level, section 006. (D) Thin sets of graded–laminated fine- and very fine-grained sandstone, with rippled tops. Sets are separated by laminated mudstones and locally include intraclasts derived from the underlying mudstone, 45 m level, section 006. (E) Pink trough cross-stratified silty fine-grained sandstones with evidence of overturned foresets (a) and liquefaction structures (b), section 002. White bands (c) are silt-free ripple laminated sandstone of possible aeolian origin.

**Fig. 7.** Strata from the upper member of the Uairén Formation. (A) Trough cross-stratified medium-grained sandstones, 5 m level section 004. (B) Trough cross-stratified large pebble conglomerate, 40 m level, section 004. Clasts are predominantly quartz, with minor volcanic pebbles, feldspathic pebbles and 3% iron formation. (C) Large-scale composite trough cross-stratification in coarse sandstones, section 004.

*Conglomerate*

The base of the upper member of the Uairén Formation is characterized by a regionally continuous sequence of massive planar and trough cross-stratified medium to very large pebble conglomerate, which is characterized by abundant, well rounded clasts of quartz (Fig. 7B). Volcanic clasts are less abundant than in the basal conglomerates of the lower member. Pebbles of iron formation and quartz arenite are present locally.

While the basal contact of the conglomerate appears conformable in some sections (004, 005, 006, 001), it is erosive in sections 002 and 007. Near Flora Blanka (section 002) the basal unit of the upper member consists of massive, stratified and planar cross-stratified medium and large pebble conglomerate and conglomeratic sandstone which has a clearly erosive base, with a relief of at least 4 m (Fig. 8). Evidence of lateral erosion at the base of this conglomerate includes steep local relief at its base and large blocks of fine-grained sandstone. Similar, near vertical channel margins were also found at the base of the upper member at Qda Tatán (section 007). Minor lenticular medium and small pebble conglomerates are present locally at higher levels in the member (section 005, 83 m level; section 007, 75 m level) but lack marked lateral continuity.

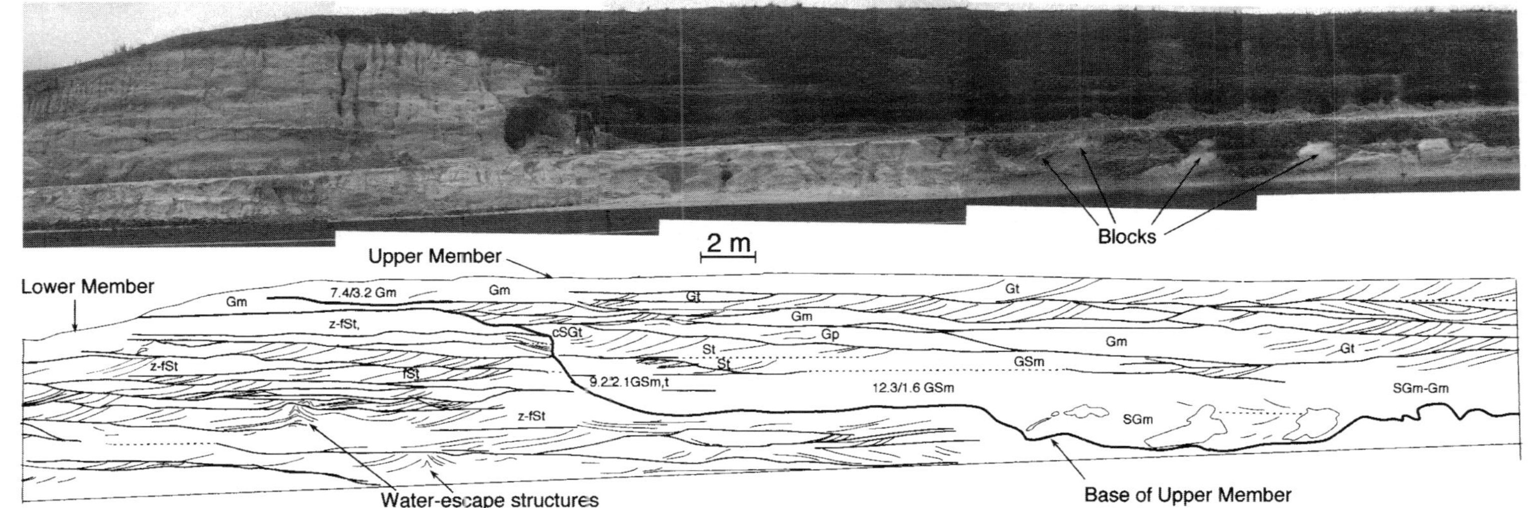


**Fig. 8.** Large-scale erosional features at contact between lower (light) and upper member (dark) of the Uarién Formation at section 002. Large blocks (right) were derived by lateral erosion of underlying strata. Lower sketch shows preliminary details of the sedimentary architecture (facies codes modified from Miall (1996): Gm, massive conglomerate; Gt, trough cross-stratified conglomerate; SG, sandy conglomerate; GS, conglomeratic sandstone; St, trough cross-stratified sandstone).

*Sandstones, pebbly sandstones*

The bulk of the upper member of the Uairén Formation consists of pebbly and non-pebbly planar and trough cross-stratified coarse- to very coarse-grained sandstones, which are well exposed in the hills above Qda Tatán (007) and in the Sabana between the Mosquito and Cuquenán rivers. Large-scale composite trough cross-beds are present locally (Fig. 7C), as are plane laminated pebbly sandstone beds and minor small pebble conglomerates.

## SEDIMENTOLOGY OF THE UAIRÉN FORMATION

Sandstones and conglomerates of the Uairén Formation are best interpreted in terms of deposition in braided rivers or amalgamated braidplains in arid to semi-humid climatic settings. Thicker mudstones may represent deposits of temporary lakes. Although Precambrian river deposits are thought to have had more flashy discharge characteristics and less stable banks than modern rivers (Long, 1978a,b, 2000; Winston, 1978; Fuller, 1985; Eriksson *et al.*, 1998), it is possible to compare their depositional profiles with modern rivers in order to gain insight into their architecture (Rust, 1978; Rust & Koster, 1984; Miall, 1978, 1996; Røe & Hermansen, 1993).

### Lower member of Uairén Formation

Well sorted clast-supported conglomerates at the base of the lower member are interpreted as the products of deposition in proximal, shallow gravel-bed braided rivers analogous to the modern Scott and Athabasca rivers (Miall 1978, 1996). Most of the massive to weakly imbricated conglomerate in this sequence was deposited on longitudinal in-channel gravel bars, with minor trough and planar cross-stratified conglomerates developing in chutes and at the downstream end of the bars. Minor sandstones represent bar-edge sand wedges and in-channel sand dunes deposited during falling stage flow conditions. The lenticular cross-section of these conglomerates, based on closely spaced sections (Fig. 4), indicates a valley-fill, rather than fan or braidplain origin. This is supported by local evidence of erosion of stream banks in overbank and tributary stream deposits in section 008.

The poorly sorted, in part framework-supported, conglomerates in the lower 2 m of section 009 represent debris flow deposits of local origin, which emanated from tributary streams or fans along the walls of the major valley systems. A debris flow origin is supported by the presence of large clasts which project above the base of individual beds, the poorly organized fabric of the beds and their erosive bases (Fisher, 1971; Bull, 1972). Debris flow deposits are common in arid region alluvial fan systems, where they occur most commonly in proximal- to mid-fan settings. The highly lenticular pebbly sandstones which occur interbedded with medium-grained sandstones higher in section 009 (Fig. 5C & E) may represent debris flow levees, left at the margins of a shallow, high-gradient, incised channel. The presence of a thin unit of slightly pebbly mudstone beneath one such unit (Fig. 5Cc) suggests that debris flow events may not all have been of similar magnitude. The laminated siltstones and fine sandstones which are interbedded with debris flow deposits in section 009 may represent sheet-flood deposits, developed in a mid-fan setting where streams emerged from entrenched channels. This is usually a locus for deposition of heavy minerals (Fig. 5Ca & Eb).

The trough cross-stratified and planar cross-stratified medium-grained sandstones which occur in the middle part of section 009 are the products of deposition from sinuous and straight crested dunes, in a shallow sandy braided system. The high concentration of heavy minerals in these sandstones (Fig. 5B & D) may reflect deposition in a mid-fan setting, where steam channel deposits are no longer confined to narrow incised channels but can migrate freely across the fan surface or alternatively could represent deposition in distal parts of an arid to semi-arid terminal fluvial system where rivers emerge from distinct channel systems as floodouts (Tooth, 1999). The general upward-fining profile seen in section 009 suggests a progressive reduction in topographic relief as basin filling (and subsidence) continued, and fluvial style shifted from valley fill to braidplain deposition. Thicker sequences of laminated mudrock in the middle of section 009 could have developed in small lakes.

The massive, graded to stratified medium- to fine-grained sandstones which form the bulk of the lower member (Fig. 10) represent deposition by superimposed ephemeral sheet-flood cycles on a low-relief sandy braidplain. Thin sets of flat laminated sandstones associated with planar and trough cross-stratified sets are common in modern and ancient fluvial deposits, where they are typically interpreted as the rarely preserved products of deposition under upper flow regime conditions (Røe & Hermansen, 1993; Eriksson *et al.*, 1993; Martins-Neto, 1994; Hjellbakk, 1997; Tirsgaard

& Øxnevad, 1998). The presence of extensive stacked sets of apparently flat laminated sandstones, in the absence of interbedded cross-stratified sands, is not typical of modern river deposits and may represent a distinctive pre-vegetative fluvial style produced by sheet floods or stream floods. Strata of this type have been described by Bhattacharyya & Morad (1993) from the Proterozoic Dhandraul Sandstone of central India. A key feature for their recognition is the presence of low angle discontinuities (Fig. 6Aa) which indicate the presence of large-scale low-relief bedforms. The presence of rare parting lineation in the Indian example (not yet seen in the Roraima examples) means that this facies could easily be confused with beachface deposits. The absence of water-escape structures in the associated massive sets, combined with the lack of high-relief scour on basal contacts, indicates that these were not produced by hyperconcentrated flow mechanisms, as has been suggested for massive-type sheet sandstones in Phanerozoic braided stream deposits by Martin & Turner (1998) and Proterozoic stream deposits by Røe & Hermansen (1993).

Thinner bedded sets may be analogous to deposits of Bijou Creek type rivers (Miall, 1978). The type example of this highly ephemeral river type is characterized by flashy discharge and, owing to source area characteristics, a lack of coarse grained material. Plane lamination may develop under upper flow regime (sheet-wash) conditions, with the cap of mudstone (often removed by aeolian activity) reflecting waning flow conditions (Fig. 6D). Local transition to lower-flow regime conditions is reflected by local scouring at the base of ripple cross-laminated sets with intraclasts (Fig. 6Ba, b & Db). Thin sets, with ripple laminated tops (Fig. 6C & Da), may represent distal variants of this type of flash-flood deposit, with oscillation ripple marks developed on the top of sets developed in shallow non-permanent lakes.

Planar and trough cross-stratified sandstones near the top of the lower member (Figs 6E & 7A) reflect migration of straight and sinuous crested bedforms in shallow semi-permanent to perennial braided systems (Miall, 1996), similar in style to the lower reaches of the Platte River (Miall, 1978; Long, 1978b, 2000; Crowley, 1983; Røe & Hermansen, 1993). Larger-scale composite bedforms in these sequences may reflect deposition on side bars or lateral migration of channels within the river bed. Overturned cross-beds and liquefaction features are common in finer-grained fluvial sandstones (Long, 1978a; McManus & Bajabaa, 1988; Røe & Hermansen, 1993). Thin sets of relatively silt free (white) ripple laminated sandstone (Fig. 6Ec), which separate (pink) beds with these features, may reflect periods of aeolian reworking, perhaps indicating a transition to more humid, semi-ephemeral conditions (see Tirsgaard & Øxnevad, 1998). Adhesion ripples, warts and inversely graded sand layers, characteristic of aeolian facies in the lower part of the Uaimapué Formation, were not seen in the lower member of the Uairén Formation.

Palaeocurrent observations for fluvial strata in the lower member (based on mean, tilt-corrected azimuths of maximum inclination of cross-stratification) are shown in Fig. 9. While southerly directed cross-stratification in the vicinity of Cerro el Patrol may reflect minor topographic variation near the base of the member, the bulk of the coarser-grained cross-stratified units near to the top of the member indicate development of an extensive south-westerly tilted braidplain. The local easterly flow at Chiricayén is influenced by inclusion of cross-stratification data from the middle of the member, and may suggest development of trunk streams along the axis of temporary lake beds.

### Upper member of Uairén Formation

Massive and planar cross-stratified conglomerates and associated pebbly sandstones at the base of the upper member are interpreted as products of deposition in shallow to deep gravel-bed braided to wandering rivers (Miall, 1996), for the most part with similar characteristics to the Scott River model of Miall (1978). In places, such as the Flora Blank section (002), where pebbly sandstones are more common the rivers may have been intermediate in character between the Scott and Donjek models of Miall (1978). These rivers were clearly capable of significant lateral erosion, as indicated by steep channel walls and large blocks of fine-grained sandstone which appear to have been cut from the channel walls (Fig. 8). Pebbly sandstones may have accumulated as in-channel bars and as side-bars, between coarser grained longitudinal bars. The thick sequence of pebbly planar and trough cross-stratified sandstones which cap the upper member are interpreted as a proximal variant of a Platte type river or braidplain (with characteristics intermediate between the Platte and Donjek models of Miall, 1978).

The presence of conglomerates at the base of the upper member indicates a marked change in depositional style within the Roraima basin. Both the erosive base and the laterally extensive character of the basal conglomerates suggest development on a broad high-energy, non-cohesive braidplain, similar to the Cadomin

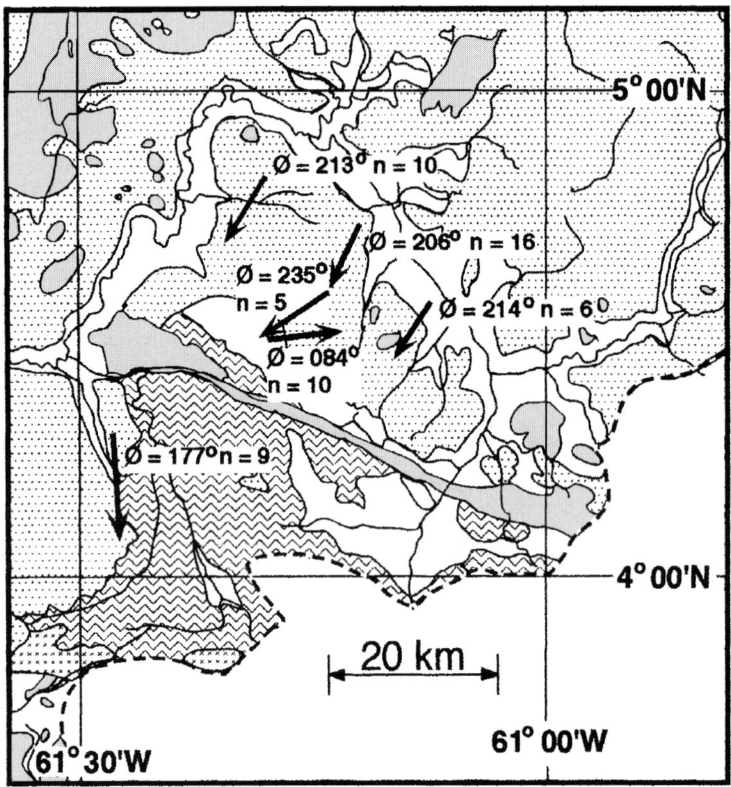


**Fig. 9.** Vector mean palaeoflow directions in the lower member of the Uarién Formation (*n*, number of cross-beds measured at each site).

Formation in Alberta. The latter conglomerate developed on a low-gradient pediment produced by uplift of the adjacent Rocky Mountains (McLean, 1977). A similar increase in relief of the source area is required to provide sufficient gradient to transport coarse debris into the Roraima basin, unless there was a marked increase in rainfall owing to a pronounced shift in climate. The latter is possible, given the absence of extensive sheet-flood deposits in the upper member.

Palaeocurrent observations in cross-stratified sandstones in the upper member suggest a palaeogeography similar to that in the lower member, with the braidplain dipping gently to the south-south-west (Fig. 10). The local easterly deviation in palaeoflow at the waterfall section at Qda Tatán may reflect local development of longitudinal (trunk) river systems, but insufficient observations were made to confirm this hypothesis.

## ECONOMIC ASPECTS

The strong spatial association of Recent placer deposits with conglomeratic strata in the Uairén Formation, combined with a lack of signs of marked abrasion on gold grains in the placer deposits, suggests that these conglomerates are the main source for gold and diamonds in the younger deposits. The ultimate source of detrital gold may be volcanic strata, both of local origin (east of Cerro el Patrol) and in volcanic highlands to the north-east. The diamonds may have been derived from the Guiana Shield to the north and east of the Roraima basin, although their ultimate source is enigmatic, and may reflect reworking of 1.73 Ga kimberlite pipes or unusual doleritic intrusives (Rod, 1962; Reid, 1974a,b,c, 1977; Schönberger & de Roever, 1974; Hastings, 1974; Nixon *et al.*, 1989, 1992, 1994; Meyer & McCallum, 1993; Channer & Cooper, 1997).

## IMPLICATIONS FOR MODELS OF PRE-VEGETATIVE FLUVIAL SYSTEMS

This study demonstrates that for sandy strata dominated by large-scale planar or trough cross-stratification, existing fluvial models, based largely on

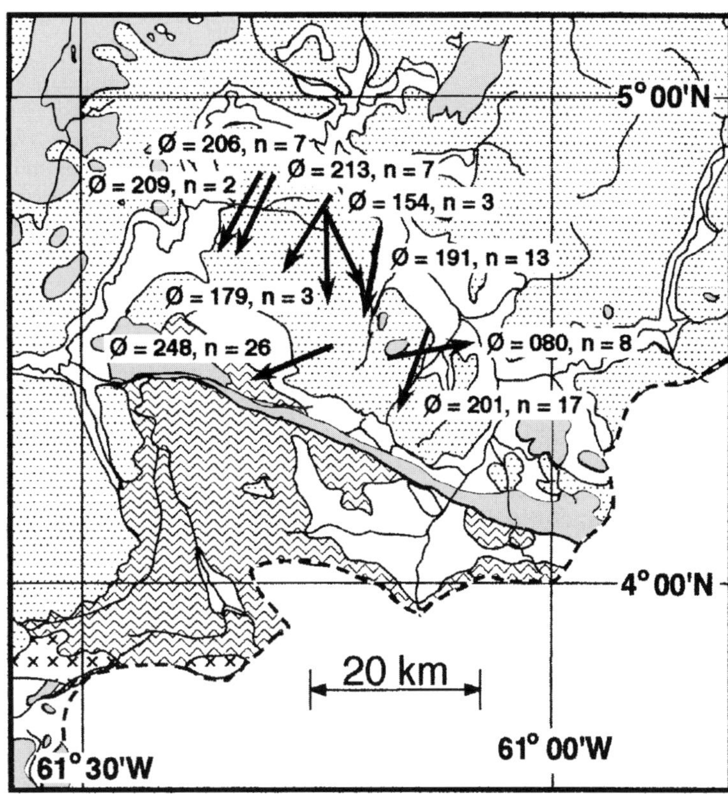


**Fig. 10.** Vector mean palaeoflow directions in the upper member of the Uarién Formation (*n*, number of cross-beds measured at each site).

the study of entrenched rivers in humid–temperate climatic regimes, can be useful in understanding the products of fluvial systems developed before the advent of vascular plants. The dramatic effect of vegetation on bank stability has been demonstrated by Smith (1976), who found that non-vegetated sediments can be 20 000 times less stable than comparable sediments containing 18% by volume of plant roots. Although microbial life-forms may have been present in Proterozoic terrestrial systems (Horodyski & Knauth 1994; Martini, 1994; Ohmoto, 1996), they are unlikely to have contributed significantly to bank stability. The immediate response to increased flow (flood events) in perennial and semi-perennial rivers is likely to have been a rapid increase in the width of channels, rather than significant deepening (Schumm, 1968; Long, 1978a; Fuller, 1985; Martins-Neto, 1994). This tends to produce bodies with marked planar geometry, reflecting development of broad braidplains with width to depth ratios probably far in excess of 1000:1 (Fuller, 1985; Martins-Neto, 1994). The presence of large blocks of reworked sandstone at the base of the incised channel at section 002 (Fig. 8) may indicate an extended period of non-deposition and soil development prior to the deposition of the upper member of the Uairén Formation, as blocks of this size are unlikely to have remained intact without some form of early cement or clay binding to provide cohesion.

Proterozoic strata dominated by flat (Sh) and low-angle inclined (Sl) stratification, as seen in parts of the lower member of the Uairén Formation, are less easily modelled using existing standard facies models (Miall, 1996) and may be easily confused with beach-face deposits. Flat and low-angle inclined stratification have been recognized in numerous Proterozoic fluvial deposits, both as a dominant facies (Bhattacharyya & Morad, 1993; Eriksson *et al.*, 1993; Hjellbakk, 1993, 1997; Sønderholm & Tirsgaard, 1998; Tirsgaard & Øxnevad, 1998) and as a minor facies associated with conglomeratic strata (Martins-Neto, 1994). Thick successions of these facies are typically related to deposition under upper flow regime conditions associated with flash floods in high-energy ephemeral streams in arid to semi-arid conditions (Williams, 1971; Frostick & Reid, 1977; Stear, 1985; Abdullatif, 1989; Tooth, 1999). The characteristics of modern flash floods, including sheet wash (broad flows, a few centimetres thick), sheet floods (several kilometres wide and 4–60 cm deep) and stream floods (less than 1.5 m deep and up to 4.8 km wide), are poorly understood due to a lack of direct measurements (Hogg, 1982; Graf, 1988). In modern dry-land river settings the presence of sparse to dense vegetation tends to break up sheet flows to produce turbulent flows in rills and small channels (Graf, 1988). This would not have been the case in Proterozoic settings, even in semi-humid to humid environments, and hence upper flow regime deposits from sheet floods might be expected to be more common. In terminal fluvial systems rapid water infiltration leads to increased sediment concentrations in the flows, deposition of massive beds or beds with poorly defined lamination (Hjellbakk, 1993).

Many modern ephemeral rivers terminate in desert environments (Bourke & Pickup, 1999), and hence the absence of identifiably aeolian deposits in many Proterozoic fluvial deposits is enigmatic (Ross, 1983; Eriksson *et al.*, 1998). Tirsgaard & Øxnevad (1998) have suggested that this may be a function of preservation owing to the position of the groundwater table, which would have been too low in arid and semi-arid conditions to allow preservation of aeolian deposits. They suggest that optimal conditions for preservation occur in semi-ephemeral systems, in more humid settings, where a seasonally high water table is combined with preservation space created by subsidence. Using this model, the presence of possible aeolian ripples between sets of trough cross-stratified fine-grained sandstones at the top of the Uairén Formation in section 002 (Fig. 6Ec) may herald a change from predominantly arid conditions to more humid conditions during deposition of the upper member.

Further understanding of the dynamics of Proterozoic fluvial systems might be achieved by detailed architectural element analysis (see Røe, 1987; Rainbird, 1992; Bhattacharyya & Morad, 1993; Hjellbakk, 1997; Eriksson *et al.*, 1998, tirsgaard & Øxnevad, 1998; Sønderholm & Tirsgaard, 1998; Chakraborty, 1999; Long, 2000), especially in combination with palaeohydraulic modelling (Els, 1990; Chakraborty, 1999; Van der Neut & Eriksson, 1999). This is not feasible in the Roraima basin, at least in the vicinity of Santa Elena de Uairén, due to the absence of accessible, laterally extensive exposures transverse to palaeoflow.

## ACKNOWLEDGEMENTS

I thank Luca Riccio for introducing me to the geology of the Roraima Supergroup, and Carson Mines of Vancouver, BC, Canada, for field support and permission to publish this paper. I thank Martin Gibling and Wullf Mueller for constructive reviews.

## REFERENCES

Abdullatif, O.M. (1989) Channel-fill and sheet-flood facies sequences in the ephemeral terminal river Gash, Kassala, Sudan. *Sediment. Geol.*, **63**, 171–184.

Aguerrevere, S.E., Lopez, V.H., Delgado, C. & Freeman, C.A. (1939) Exploración de la Gran Sabana. *Rev. Fomento*, **3**, 501–729.

Almeidan, F.F. (1978) Tectonic map of South America, scale 1:5 000 000, two sheets. Commission for the Geologic Map of the World.

Anderson, C.W. & Dunn, W.A. (1895) *The Conglomerate Prospecting Expedition*. Government of British Guiana Admiralty Report, Georgetown.

Bellizzia, N.P. de (1984) Mapa geologico y structural de Venezuela, escalea 1:2 500 000. Republica de Venezuela Ministrio de Energia y Minas, Caracas.

Benaim, N. (1972) Geologia de la region de Botanamo Edo, Bolivar. *Mem. IV Cong. geol. Venezolano*, **3**, 1292–1314.

Bhattachryya, A. & Morad, S. (1993) Proterozoic braided ephemeral fluvial deposits; an example from the Dhandraul Sandstone Formation of the Kaimur Group, Son Valley, central India. *Sediment. Geol.*, **84**, 101–114.

Bourke, M.C. & Pickup, G. (1999) Fluvial form and variability in arid central Australia. In: *Varieties of Fluvial Form* (Eds Miller, A.J. & Gupta, A.), pp. 249–271. John Wiley and Sons, Chichester.

Briceño, H.O. & Schubert, C. (1990) Geomorphology of the Gran Sabana, Guiana Shield, southeastern Venezuela. *Geomorphology*, **3**, 125–141.

Bull, W.B. (1972) Recognition of alluvial-fan deposits in the stratigraphic record. In: *Recognition of Ancient Sedimentary Environments* (Eds Rigby, J.K. & Hamblin, W.K.), Spec. Publ. Soc. econ. Paleont. Miner., Tulsa, **16**, 63–83.

Chakraborty, T. (1999) Reconstruction of fluvial bars from the Proterozoic Manchaeral Quartzite, Pranhita–Godavari Valley, India. In: *Fluvial Sedimentology IV* (Eds Smith, N.D. & Rogers, J.), Spec. Publs int. Ass. Sediment., No. 28, pp. 451–466. Blackwell Science, Oxford.

Channer, D. & Cooper, R.E.C. (1997) The Guanimo diamond region, Bolivar state, Venezuela: a new kimberlite province. *Mem. VIII Cong. geol. Venezolano*, **1**, 143–146.

Crowley, K.D. (1983) Large-scale bed configurations (macroforms), Platte River Basin, Colorado and Nebraska: primary structures and formulative processes. *Geol. Soc. Am. Bull.*, **94**, 117–133.

Dalton, L.V. (1912) On the geology of Venezuela. *Geol. Mag.* **5**(9).

Direccion de Geologia (1976) Mapa NB-20-11, Santa Elena, Edition 1-DG. Direccion de Geologia, Caracas.

Dohrenwend, J.C., Yánez, G.P. & Lowry, G. (1995) Cenozoic landscape evolution of the southern part of the

Gran Sabana, Southeastern Venezuela. Implications for the occurrence of gold and diamond placers. *US geol. Soc. Bull.*, **2124K**, 17 pp.

Els, B.G. (1990) Determination of some palaeohydraulic parameters for a fluvial Witwatersrand succession. *S. Afr. J. Geol.*, **93**, 531–537.

Eriksson, P.G., Condie, K.C., Tirsgaard, H. *et al.* (1998) Precambrian clastic sedimentation systems. *Sediment. Geol.*, **120**, 5–53.

Eriksson, P.G., Schreiber, U.M., van der Neut, M., Labuschagne, H., van der Schyff, W. & Potgieter, G. (1993) Alternative marine and fluvial models for the non-fossiliferous quartzitic sandstones of the Early Proterozoic Daspoot Formation, Transvaal Sequence of South Africa. *J. Afr. Earth Sci.*, **16**, 355–366.

Fisher, R.V. (1971) Features of coarse-grained, high-concentration fluids and their deposits. *J. sediment. Petrol.*, **41**, 916–927.

Frostick, L.E. & Reid, I. (1977) The origin of horizontal laminae in ephemeral stream channel-fill. *Sedimentology*, **24**, 1–9.

Fuller, A.O. (1985) A contribution to the conceptual modelling of pre-Devonian fluvial systems. *Trans. geol. Soc. S. Afr.*, **88**, 189–194.

Gansser, A. (1954) Observations of the Guiana Shield (S. America). *Eclogae geol. Helv.*, **47**, 77–112.

Ghosh, S.K. (1977) Geologia del Groupo Roraima en Territorio Federal Amazonas. *Mem. V Cong. geol. Venezalano*, **1**, 167–193.

Ghosh, S.K. (1978) The Roraima problem. A case of Precambrian cratonic sedimentation in S. America. *Xth int. Sediment. Cong., Jerusalem*, **1**, 246–247.

Ghosh, S.K. (1985) Geology of the Roraima Group and its implications. *Mem. I Symp. Amazonico*. Spec. Publ. Venez. Dir. Geol., Bol. Geol., **10**, 768.

Gibbs, A.K. & Barron, C.N. (1993) *The Geology of the Guiana Shield*. Oxford University Press, Oxford, 246 pp.

Graf, W.L. (1988) *Fluvial Processes in Dryland Rivers*. Springer-Verlag, Berlin, 346 pp.

Hastings, D.A. (1974) Proposed origin for Guianian diamonds: comment. *Geology*, **2**, 475–476.

Hea, J.P. (1975) Exploration needed for a new mining industry in Venezuela. *CIM Bull.*, March, 133–141.

Hjellbakk, A. (1993) A 'flash-flood' dominated braid delta in the upper Proterozoic Næ ringselva Member, Varanger Peninsula, northern Norway. *Sediment. Geol.*, **114**, 131–161.

Hogg, S.E. (1982) Sheetfloods, sheetwash, sheetflow, or–? *Earth Sci. Rev.*, **18**, 56–76.

Horodyski, R.J. & Knauth, L.P. (1994) Life on land in the Precambrian. *Science*, **263**, 496–498.

Jones, E.J.W. (1987) Fracture zones in the equatorial Atlantic and the breakup of western Pangea. *Geology*, **15**, 533–536.

Keats, W. (1973) *The Roraima Formation Project, 1971 to 1973, Final Report*. Guyana Geol. Surv. Dept Rep. WK 1/73, Georgetown.

Keats, W. (1974) The Roraima Formation in Guyana: a revided stratigraphy and proposed environment of deposition. *Mem. II Cong. Latinoamer. Geol., Caracas, 1973*. Spec. Publ. Venez. Min. Minas Hidroc., Dir. Geol., Bol. Geol., **7**(2), 901–940.

Long, D.G.F. (1978a) Proterozoic stream deposits: some problems of recognition and interpretation of ancient sandy fluvial systems. In: *Fluvial Sedimentology* (Ed. Miall, A.D.), Mem. Can. Soc. petrol. Geol., Calgary, **5**, 314–341.

Long, D.G.F. (1978b) Depositional environments of a thick Proterozoic sandstone: the (Huronian) Mississagi Formation of Ontario, Canada. *Can. J. Earth Sci.*, **15**, 190–206.

Long, D.G.F. (2000) Architecture of Proterozoic sand-dominated fluvial systems. *GeoCanada 2000, Abstracts and Program*, CD-ROM. Can. Soc. petrol. Geol., Calgary.

Lopez, V.M., Mencher, E. & Brineman, J.H. Jr (1942) Geology of south-eastern Venezuela. *Geol. Soc. Am. Bull.*, **53**, 849–872.

McLean, J.R. (1977) The Cadomin Formation: stratigraphy, sedimentology and tectonic implications. *Bull. Can. Soc. petrol. Geol.*, **25**, 792–827.

McManus, J. & Bajabaa, S. (1988) The importance of air escape processes in the formation of dish-and-pillar and tepee structures within modern and Precambrian fluvial deposits. *Sediment. Geol.*, **120**, 337–343.

Martin, C.A.L. & Turner, B.R. (1998) Origins of massive-type sandstones in braided river systems. *Earth Sci. Rev.*, **44**, 15–38.

Martini, J.E.J. (1994) A late Archean–Paleoproterozoic (2.6 Ga) paleosol on ultramafics in the eastern Transvaal, South Africa. *Precam. Res.*, **67**, 159–180.

Martin-Kaye, P. (1952) The Roraima Formation in British Guiana. *Geol. Surv. Br. Guiana Bull.*, **22**.

Martins-Neto, M.A. (1994) Braidplain sedimentation in a Proterozoic rift basin: the São João da Chapada Formation, southeastern Brazil. *Sediment. Geol.*, **89**, 219–239.

Meyer, H.O.A. & McCallum, M.E. (1993) Diamonds and their sources in the Venezualan portion of the Guiana Shield. *Econ. Geol.*, **88**, 989–998.

Miall, A.D. (1978) Lithofacies types and vertical profile models in braided river deposits: a summary. In: *Fluvial Sedimentology* (Ed. Miall, A.D.), Mem. Can. Soc. petrol. Geol., Calgary, **5**, 597–604.

Miall, A.D. (1996) *The Geology of Fluvial Deposits: Sedimentary Facies, Basin Analysis and Petroleum Geology*. Springer-Verlag, Berlin, 582 pp.

Nixon, P.H., Davies, G.R., Condliffe, E., Baker, R. & Brown, R.B. (1989) Discovery of ancient source rocks of Venezuela diamonds. *Extended Abstracts, Workshop on Diamonds, Geophysical Laboratories, Washington, DC*, pp. 73–75.

Nixon, P.H., Davies, G.R., Rex, D. & Gray, A. (1992) Venezuela kimberlites. *J. Volcanol. geotherm. Res.*, **50**, 101–115.

Nixon, P.H., Griffin, W.L., Davies, G.R. & Condliffe, E. (1994) Ct-garnet indicators in Venezuela kimberlites and their bearing on the evolution of the Guyana craton. *Proc. 5th Int. Kimberlite Conf., Araxa, Brazil*. Spec. Publ. CPRM, Brazilia, **1A**, 378–387.

Ohmoto, H. (1996) Evidence in pre-2.2 Ga paleosols for the early evolution of atmospheric oxygen and terrestrial biota. *Geology*, **24**, 1135–1138.

Priem, H.N.A., Boelrick, M.A.I.M., Herbeda, E.H., Verdurmen, E.A.Th. & Verschure, R.H. (1973) Age of the Precambrian Roraima Formation in Northeastern South America: evidence from isotopic dating of Roraima pyroclastic volcanic rocks in Surinam. *Geol. Soc. Am. Bull.*, **84**, 1677–1684.

Rainbird, R.H. (1992) Anatomy of a large-scale braid-plain quartzarenite from the Neoproterozoic Shaler Group,

Victoria Island, Northwest Territories, Canada. *Can. J. Earth Sci.*, **29**, 2537–2550.

Reid, A.R. (1974a) Stratigraphy of the type area of the Roraima Group. *Mem. 9th Inter-Guyana Geol. Conf.*, pp. 343–353. Venez. Min. Minas Hidrocarburos, Caracas.

Reid, A.R. (1974b) Proposed origin for Guianian diamonds. *Geology*, **2**, 67–68.

Reid, A.R. (1974c) Proposed origin for Guianian diamonds: reply. *Geology*, **2**, 476.

Reid, A.R. (1977) *An attempt to localize kimberlite source areas for Venezuelan diamonds from stratigraphy and analysis of diamond mineral inclusions*. PhD thesis, Colorado School of Mines, Golden, CO.

Reid, A.R. & Bisque, R.E. (1975) Stratigraphy of the diamond bearing Roraima Group, Estado Bolivar, Venezuela. *Q. J. Colo. Sch. Mines*, **70**(1), 13–16.

Reis, N.J., Pinheiro, S. da S. & Carvalho, J.E. (1985) *Subdivisão litostratigrafica da Formacão Suapi-Grupo Roraima, Territorio Federal de Roraima*. Anais do Symposio de Geologia da Amazonia, Belem.

Reis, N.J., Pinheiro, S. da S., Costi, H.T. & Costi, J.B.S. (1990) A cobertura sedimentar Proterozoica Media do Supergrupo Roraima no noreste do Estado de Roraima, Brasil: atribucoes aos seus sistemas deposicionais e esquema evolutivo da sua borda meridional. *Cong. Brasil. Geol.*, **36**. RN Brasil, Belem.

Rod, E. (1962) From where did the sand of the Roraima Formation come? *Bol. Inf. Asoc. Venez. geol. Miner. Petrol.*, **5**, 303–308.

Røe, S.-L. (1987) Cross-strata and bedforms of probable transitional dune to upper-stage plane-bed origin from a Late Precambrian fluvial sandstone, northern Norway. *Sedimentology*, **34**, 89–101.

Røe, S.-L. & Hermansen, M. (1993) *Processes and Products of Large, Late Precambrian Sandy Rivers in Northern Norway*. Spec. Publs int. Ass. Sediment, No. 17. Blackwell Scientific Publications, Oxford.

Ross, G.M. (1983) Proterozoic aeolian quartz arenites from the Hornby Bay Group, Northwest territories, canada: implications for Precambrian aeolian processes. *Precam. Res.*, **20**, 149–160.

Rust, B.R. (1978) Depositional models for braided alluvium. In: *Fluvial Sedimentology* (Ed. Miall, A.D.), Mem. Can. Soc. petrol. Geol., Calgary, **5**, 605–625.

Rust, B.R. & Koster, E.H. (1984) Coarse alluvial deposits. In: *Facies Models*, 2nd edn (Ed. Walker, R.G.). Geosci. Can. Reprint Ser. 1, 53–69.

Santos, J.O.S. (1985) A subdivisão estratigrafica de Grupo Roraima. *Anais do II simpósio do geologia da Amazonia*, Belem.

Schönberger, H. & de Roever, E.W.F. (1974) Possible origin of diamonds in the Guiana Shield: comment. *Geology*, **2**, 474–475.

Schumm, S.A. (1968) Speculations concerning paleohydrologic controls of terrestrial sedimentation. *Geol. Soc. Am. Bull.*, **79**, 1573–1588.

Sidder, G.B. (1990) Geologic province map of Venezuelan Guiana Shield. US geol. Surv. Open File Rep., OF 90-0073, 14 pp. + map.

Sidder, G.B. & Mendoza, V.S. (1995) Geology of the Venezuelan Guyana Shield and its relation to the geology of the entire Guyana Shield. *US geol. Surv. Bull.*, **2124B**, 41 pp.

Simon, C., Castrillo, J. & Munoz, N.G. (1985) Sedimentologia en zonas de Santa Elena de Uairén y Monte Roraima, Estado Bolivar, Venezuela. *Mem. Congr. geol. Venez.*, **6**, 1135–1163.

Smith, D.G. (1976) Effect of vegetation on lateral migration of anastomosed channels of a glacial meltwater river. *Geol. Soc. Am. Bull.*, **87**, 857–860.

Snelling, N.J. (1963) Age of the Roraima Formation, British Guiana. *Nature*, **198**, 1079–1080.

Snelling, N.J. & McConnell, R.B. (1965) Age determinations from British Guiana. *Proc. geol. Soc. London*, **1619**, 8–11.

Sønderholm, M. & Tirsgaard, H. (1998) Proterozoic fluvial styles: response to base-level changes and climate (Rivieradal sandstone, eastern North Greenland). *Sediment. Geol.*, **120**, 257–274.

Stear, W.O. (1985) Comparison of the bedform distribution and dynamics of modern and ancient sandy ephemeral flood deposits in the southwestern Karoo Region, South Africa. *Sediment. Geol.*, **45**, 209–230.

Suszczynski, E. (1981a) Structural framework. In: *Precambrian of the Southern Hemisphere* (Ed. Hunter, D.R.). Developments in Precambrian Geology, 2. Elsevier, Amsterdam, pp. 835–844.

Suszczynski, E. (1981b) The cratonic environment. In: *Precambrian of the Southern Hemisphere* (Ed. Hunter, D.R.). Developments in Precambrian Geology, 2. Elsevier, Amsterdam, pp. 851–861.

Tate, G.M.M. (1930) Notes on the Mount Roraima region. *Geogr. Rev.*, **20**, 31–52.

Tirsgaard, H. & Øxnevad, I.E.I. (1998) Preservation of pre-vegetational mixed fluvio-aeolian deposits in a humid climatic setting: an example from the Middle Proterozoic Rriksfjord Formation, Southwest Greenland. *Sediment. Geol.*, **120**, 295–317.

Tooth, S. (1999) Floodouts in central Australia. In: *Varieties of Fluvial Form* (Eds Miller, A.J. & Gupta, A.), pp. 219–247. John Wiley & Sons, Chichester.

van der Neut, M. & Eriksson, P.G. (1999) Palaeohydrological parameters of a Proterozoic braided fluvial system (Wilgerivier Formation, Waterberg Group, South Africa) compared with a Phanerozoic example. In: *Fluvial Sedimentology IV* (Eds Smith, N.D. & Rogers, J.), Spec. Publs int. Ass. Sediment., No. 28, pp. 381–392. Blackwell Science, Oxford.

van de Putte, H.M. (1972) Contribution to the stratigraphy and structure of the Roraima Formation, State of Bolivar, Venezuela. *Proc. 9th Inter-Guyana geol. Conf., Puerto Ordaz, Venezuela*, pp. 372–394.

Williams, G.E. (1971) Flood deposits of the sand-bed ephemeral streams of central Australia. *Sedimentology*, **17**, 1–40.

Winston, D. (1978) Fluvial systems of the Precambrian Belt Supergroup, Montana and Idaho, USA. In: *Fluvial Sedimentology* (Ed. Miall, A.D.), Mem. Can. Soc. petrol. Geol., Calgary, **5**, 343–359.

Yánez, G. (1972) Geologia del area de Santa Elena de Uaíren (Venezuela). *Mem. 9th Inter-Guyana geol. Congr., Ciudad Guayana*, pp. 660–667.

Yánez, P.G. (1985) El Grupo Roraima en el sureste de Venezuela. *Geominas*, **13**, 5–22.

*Spec. Publs int. Ass. Sediment.* (2002) **33**, 339–350

# Volcanogenic and sedimentary rocks within the Svecofennian Domain, Ylivieska, western Finland—an example of Palaeoproterozoic intra-arc basin fill

K. STRAND

*Thule Institute, University of Oulu, PO Box 7300, FIN-90014 Oulu, Finland*

## ABSTRACT

Palaeoproterozoic reworked volcaniclastic deposits, basaltic andesitic to rhyolitic pyroclastics and lava flows in the central part of the Fennoscandian Shield, western Finland, constitute part of the volcanic zones of the Svecofennian Domain, which formed in a convergent tectonic setting and accreted against the Archaean cratonic margin. In the Ylivieska area, western Finland, much of the detritus was deposited as sandy- and fine-grained turbidites in a shallowing sea, succeeded by the deposition of rhyolitic pyroclastics and lavas. The lavas contain spherical gas cavities, indicating relatively rapid cooling of a viscous flow near its vent area. Volcanic activity was terminated by the deposition of fragmental, shallow water to subaerial basaltic to andesitic tuffs of hydrovolcanic origin. Locally, erosional scours and sand wave-like structures are well preserved. The build-up of a volcanic complex was succeeded by deposition of predominantly fan-delta to fluvial coarse clastics. The lithic clasts in clast-supported conglomerates include acidic to intermediate volcanics, porphyrites and sand- to fine-grained turbidite clasts, and indicate erosion of part of the underlying volcanic complex. The sedimentation pattern in Ylivieska can be considered a Proterozoic example of a calc-alkaline volcanic complex of primary volcaniclastics and lavas, deposited between turbiditic sequences and fan-delta to fluvial successions that probably originated as part of an intra-arc basin fill behind an accreting oceanic island arc. The depositional model presented here demonstrates volcanism and volcaniclastic sedimentation coupled with relative sea-level fall due to active arc build-up and tectonic uplift of some areas. The rhyolite flows with high-K affinity indicate that the activity was related to the early formation of a juvenile continent. The depositional events are probably related to the late phase of early Svecofennian accretionary tectonics that occurred between *c*.1.90 and 1.87 Ga. The build-up of a micro-continental arc with its contemporaneous intrusive activity formed the base of a major crustal addition to the Palaeoproterozoic Svecofennian Domain.

## INTRODUCTION

This study describes volcaniclastic deposits and lava flows in the Palaeoproterozoic Svecofennian Domain in the Ylivieska area, western Finland, as examples of Precambrian successions related to plate convergence and initial crustal growth (Fig. 1). This accretion was associated with co-magmatic granodioritic, tonalitic and granitic intrusives (Korsman *et al.*, 1997). The general aim of this study is to determine whether the deposits were produced by splitting of Palaeoproterozoic subduction-related volcanic arc blocks (see Gaál & Gorbatschev, 1987; Gaál, 1990; Skiöld *et al.*, 1993). A subduction zone is inferred to have dipped towards the north (BABEL working group, 1990; Öhlander *et al.*, 1993). Rocks of this study were presumably generated during the late phase of the Svecofennian orogeny between 1.90 and 1.87 Ga. In the new division of Korsman *et al.* (1997), these rocks belong to the accretionary arc complex of central and western Finland.

Although the rocks are metamorphosed up to amphibolite facies, the undeformed deposits are suitable for a sedimentological analysis. Only well preserved undisturbed regional blocks were chosen for facies analysis. The bedding features are in many places well preserved, although recrystallization has resulted in porphyroblasts of aluminous silicates or their relics in the finest sediments.

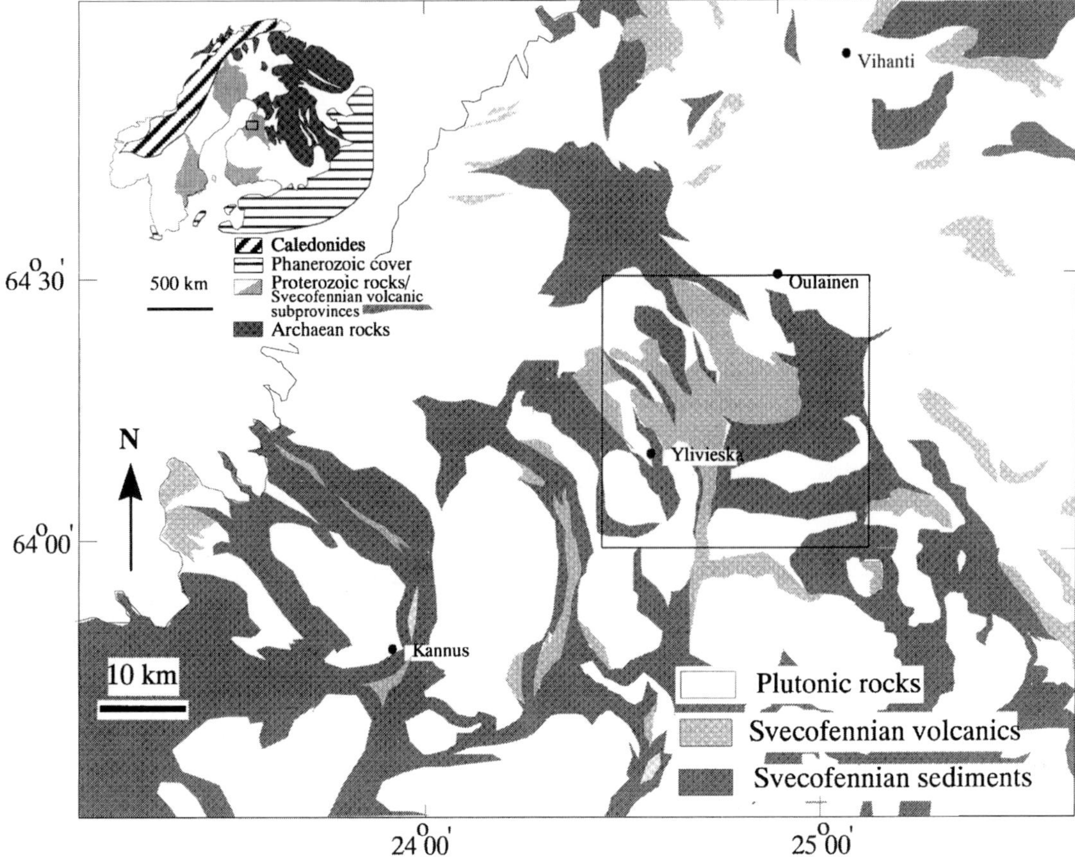


**Fig. 1.** General geological features of the Fennoscandian Shield and western Finland showing the location of the map in Fig. 2.

## GEOLOGICAL AND REGIONAL SETTING

The Palaeoproterozoic Svecofennian Domain within the Fennoscandian Shield (Fig. 1) shows the features of subduction-related processes and contains a series of sutures resulting from accretion of terranes, i.e volcanic arcs against the late Archaean continent between *c*.1.90 and 1.86 Ga (Gaál & Gorbatschev, 1987). It is presumed that the section studied in the Ylivieska area was preceded by the formation of a primitive island arc as identified and characterized by bimodal volcanism in the Pyhäsalmi area, 80 km to the south-east (Kousa, 1990), where most of the silicic members are classified as low-K rhyolites with an age of 1930 ± 15 Ma (Vaasjoki & Sakko, 1988). This is comparable with the U–Pb zircon age of 1921 ± 2 Ma from rhyolite at Riitavuori in the Pyhäsalmi area (Lahtinen, 1994; Kousa, Marttila & Vaasjoki, in preparation). The calc-alkaline felsic to mafic volcanics at Ylivieska, however, show typically high-K affinity (Kousa, 1990; Strand, 1993) and lithological similarity to felsic volcanics in the Pihtipudas area, 100 km to the south-east, which have an age of 1883 ± 20 Ma (Aho, 1979). In the Vihanti area, 50 km north of the study area, the younger phase of volcanic activity was terminated by the emplacement of 1878 ± 17 Ma hypabyssal plagioclase porphyries (Vaasjoki & Sakko, 1988). Thus, the timing of Svecofennian sedimentation in the Ylivieska area is constrained by 1.90–1.87 Ga volcanics (see Vaasjoki & Sakko, 1988; Claesson *et al.*, 1993). In Sweden, some shallow submarine and subaerial volcanics were extruded at 1880–1890 Ma (Welin, 1987), but in the Tampere region, southern Finland, a minimum age for turbidite sedimentation is provided by the associated 1904 Ma felsic volcanics

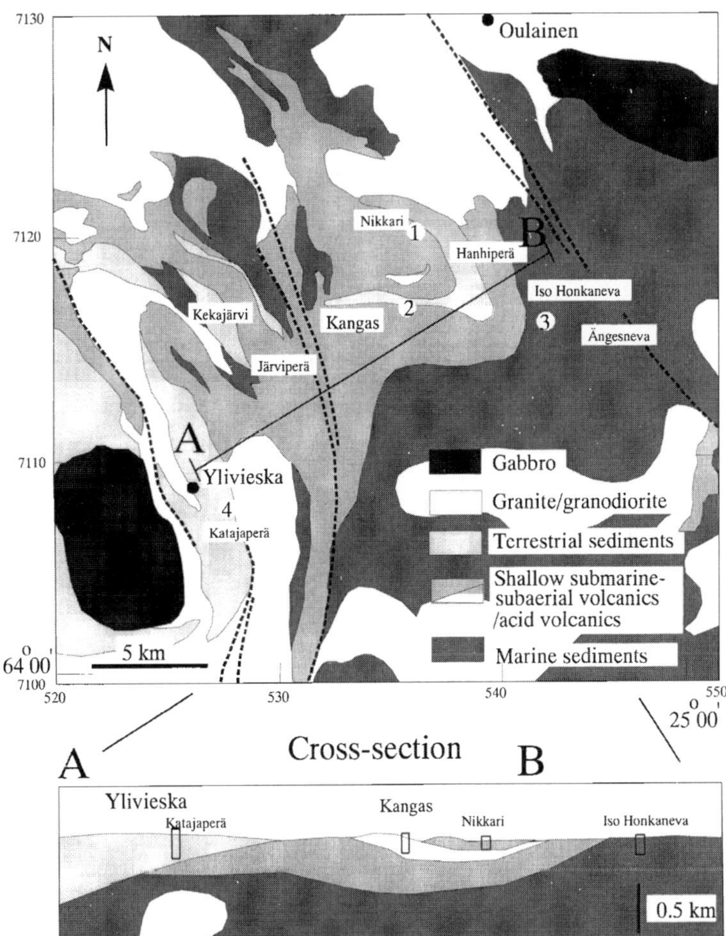


**Fig. 2.** General geological map of the Ylivieska area showing the locations of detailed study areas 1–4. Major shear zones are indicated on the map. The lower inset illustrates the general lithostratigraphy of the Ylivieska area along the cross-section A–B, and the location of studied sections.

(Kähkönen *et al.*, 1989), and the youngest zircon crystal in Svecofennian sediments yielded a mean $^{207}Pb/^{206}Pb$ age of $1884 \pm 6$ Ma (Claesson *et al.*, 1993).

In the northern part of the Ylivieska area, the rocks form a large, relatively tight synclinal belt that plunges to the north-west; the oldest rock units are located in the southern and south-eastern part of the study area (Fig. 2). The earliest stratigraphic framework and description of the rocks in this area was presented by Salli (1964). In his study, the primary volcanogenic rocks were located in the middle and upper portions of an upward-coarsening stratigraphic sequence. In general, plagioclase porphyrites occur as extensive extrusives, sills and dikes; in some areas they make construction of continuous sections difficult (Fig. 2).

## LITHOFACIES OF THE YLIVIESKA SECTION

The various lithofacies of the Ylivieska area can be grouped broadly into three major rock units, which are described here in stratigraphic order from base to top. In the following descriptions, facies codes are modified after Miall (1977, 1985). In the volcanic deposits, the terms tuff and ash are grain-size terms according to the classification of Fisher (1961, 1966).

### Graded to massive sandstones and laminated siltstones

#### *Description*

The graded to massive sandstones and laminated siltstones are best described from location 3 in the Iso

Honkaneva area (Fig. 2, Map sheet 2433 02A, $x$ = 7114.76, $y$ = 541.19). The Exploration Department at the Geological Survey of Finland has several drill holes (DH322 and DH346) about 12 km west from the town of Ylivieska through sedimentary rock types considered similar to the deposits in the Iso Honkaneva area (Västi, 1989). The drilled sequences dip 30° to the northwest. Given the dip, the total thickness of these deposits is up to 3 km. The rocks underlying the succession are unknown, as is the basal contact. No primary volcanic layers or lavas were observed within the sandstones and siltstones. The nature of the contact between the sediments and overlying pyroclastic deposits remains enigmatic.

The deposits typically consist of small-scale upward-fining sequences, 10–50 cm thick, that are graded to massive, sandstone-laminated siltstone couplets with partial Bouma sequences followed by poorly bedded to ungraded mudstones (Fig. 3A). The beds sometimes lack the basal or middle parts of a complete Bouma sequence ($T_{a,b,c,d,e}$), typically forming sequences such as $T_a$, $T_{a,b}$, $T_{a,c,d,e}$ and $T_{c,d,e}$. Locally the sandstones contain load casts and flame structures at their bases (Fig. 3B). Contacts between beds are erosive to non-erosive.

The framework grains in sandstones are angular to subangular quartz, feldspar and some rock fragments, and overall they are wacke-like, being relatively rich in fine-grained matrix. The rock fragments consist predominantly of angular quartz and feldspar grains. Locally, metamorphism has produced relics of andalusite, cordierite, staurolite and garnet porhyroblasts in fine-grained rocks. Notwithstanding this, the relic bedding structures in some areas are well preserved (Fig. 3).

*Interpretation*

The graded to massive sandstone and laminated siltstone represent normal 'background' sedimentation for a sedimentary basin. The deposits are interpreted as mass-flow beds and distal turbidites. The association typically shows graded beds with partial Bouma sequences interpreted to represent mid-fan to more distal depositional lobe portions of a submarine fan; alternatively, these turbidites could represent parts of slope aprons (see Pickering *et al.*, 1986). Graded beds formed during low-density turbidity flows and the finer-grained units between turbidite beds represent suspension sedimentation (Lowe, 1982). Water depth of more than 1000 m is suggested based on the overall thickness of this association, which is up to 3 km of relatively homogeneous distal turbidites. The thickness of similar homogeneous distal turbidites also indicates remarkable subsidence of the depositional basin coincident with deposition. The sources of the monocrystalline quartz and feldspar framework grains are plutonic or volcanic. Polycrystalline quartz grains are probably metamorphic grains, or recrystallized cherts. The rock fragments composed of large quartz and feldspar grains are presumably plutonic.

## Rhyolitic and basaltic andesitic volcanics

*Description*

The contact between the turbidites and the overlying volcanogenic deposits is not exposed. The contact is inferred to be located where the first occurrence of volcanogenic material and structures is evident. The total thickness of the volcaniclastics is estimated to be 500–1000 m. The pyroclastic rocks are divided into two volcanic sequences. The contact between these two sequences is sharp.

The first volcanic sequence contains rhyolitic flows and overlying acidic fragmentary material described as ash deposits. The rhyolitic flows at location 2 in Kangas (Fig. 2, Map sheet 2431 11D, $x$ = 7116.17, $y$ = 535.86) are 20–40 m thick, massive and characterized by cavities and large vesicles, 3–5 cm in diameter (Fig. 4). Vesicle content is up to 20%, although vesicle density varies from place to place. Internal fractures are common. The contacts between flows are not exposed. Petrographically, the flows contain fine-grained microcrystalline quartz, sericite, some biotite and minor potash feldspar according to X-ray diffraction results, although the feldspar is not well identified in thin section. Vesicles are filled with calcite and quartz. North-east from Kangas, silicic volcaniclastic material referred to as tuff, rather than flows, is dominant. No vesicles were observed. The total thickness of the tuff is several tens of metres and overall stratification is weak, represented by faint parallel lamination, which is disturbed by later subparallel fracturing. Laterally, these rocks can be followed for 2–3 km. About 55 km to the south-east of Kangas, in the Kuusanjärvi area, acid volcanics are predominantly horizontally bedded (Kousa, 1990). The beds are 5–50 cm thick with sharp contacts, and both inverse and normal grading is present. Petrographically these crystalline tuffs contain 1 cm sized euhedral quartz and feldspar phenocrysts in a fine-grained microcrystalline quartz matrix. The faint pseudomorphs after glass shards and pumice can be identified as of the same composition as the matrix. The rhyolitic

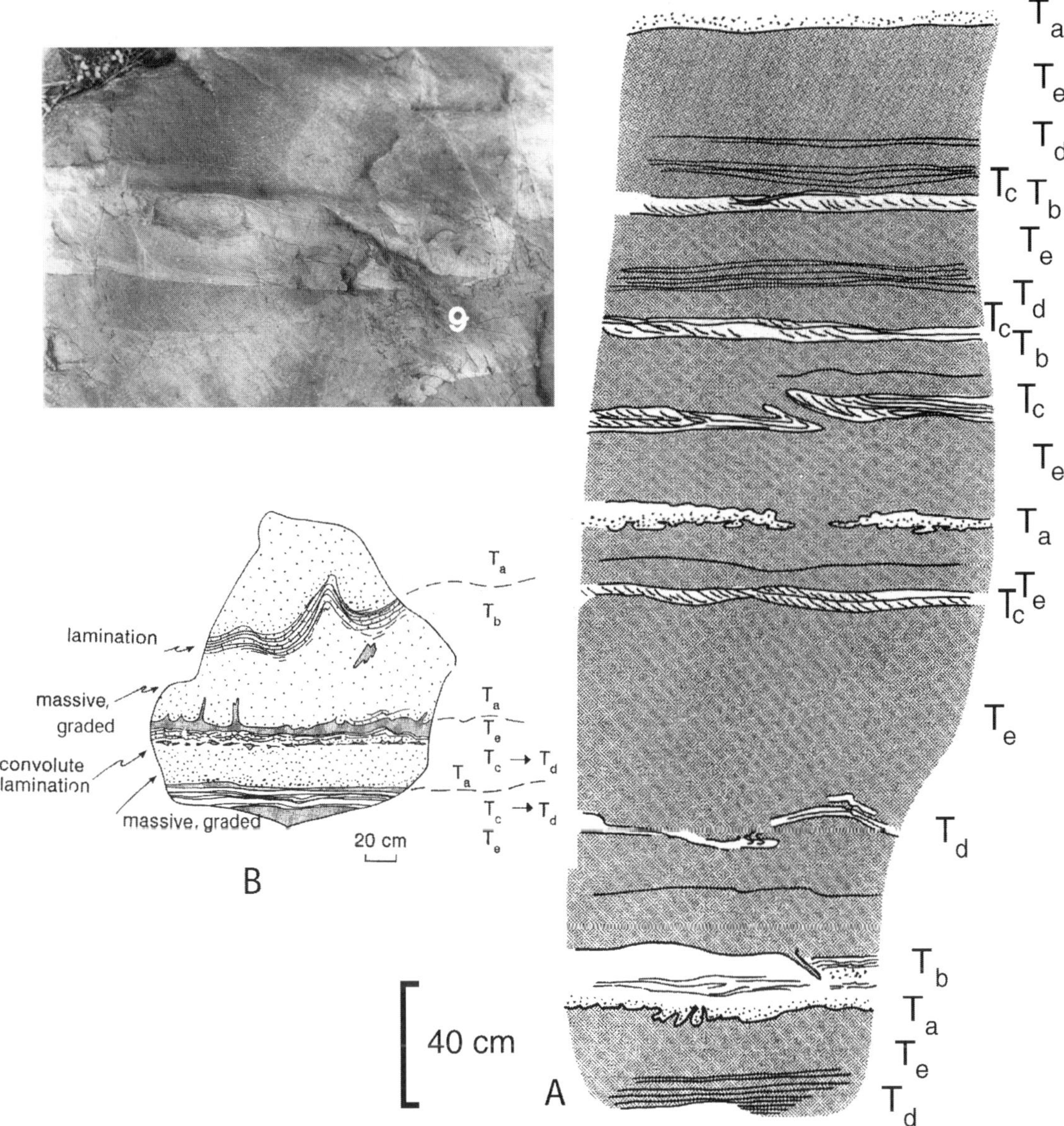


**Fig. 3.** Typical exposures of turbidites with partial Bouma sequences. (A) Beds locally lack the basal or middle parts of a complete Bouma sequence ($T_{a,b,c,d,e}$). (B) Some sandstones have load casts and flame structures. Inset photograph illustrates the bedding features preserved in the fine-grained deposits. The number '9' in the photograph is 1.2 cm high.

flows and tuffs show chemical composition of $SiO_2$ = 63.17–76.91 wt%; $K_2O$ = 3.89–10.58 wt%, typical for primary volcanic material (Cas & Wright, 1987).

The overlying volcanic sequence is dominated by bedded fragmentary material with a chemical composition of basaltic andesite: $SiO_2$ = 52.75–55.51 wt%; $Al_2O_3$ = 13.29–15.68 wt%; MgO = 5.26–7.44 wt%; $TiO_2$ = 0.78–0.88 wt%. These volcanic rocks are best represented at location 1 in Nikkari, (Fig. 2, Map sheet 2431 12C, $x$ = 7120.42, $y$ = 536.58), where basaltic to andesitic fragmentary deposits are bedded, 10–30 m thick, and are laterally traceable for roughly 500 m. These rocks contain large-scale planar cross-beds. The lower contacts show erosional scours

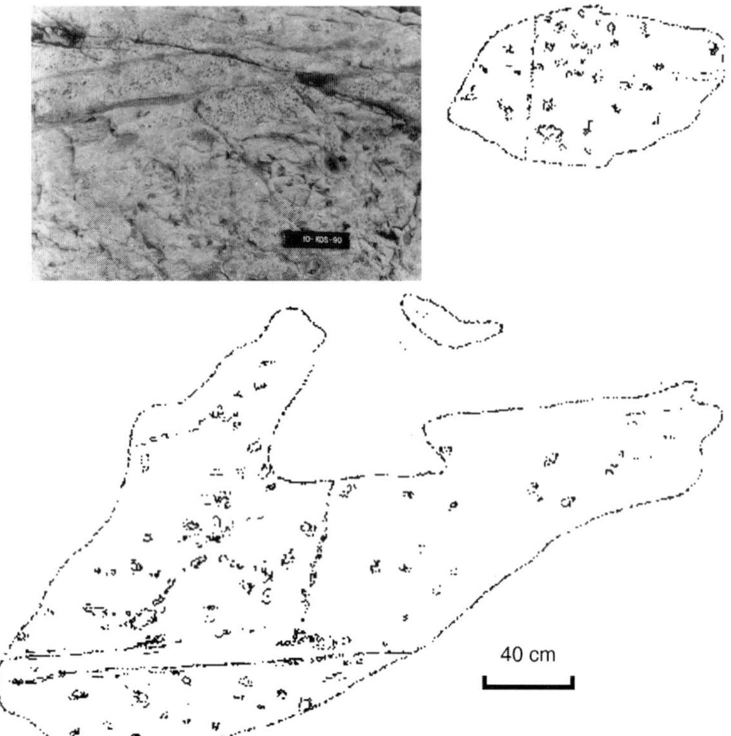


**Fig. 4.** Typical outcrops of rhyolitic lava. Inset photograph shows some relics of spherical gas cavities. Label in the photograph is 16.5 cm long.

(Fig. 5A, Lm, ACh). The set thickness of cross-beds varies from 0.5 to 1.5 m. A few hundred metres towards the north-west, coarser-grained counterparts (Bmm, Lm) are visible (Fig. 5B). The estimated thickness of these facies varies from 4 to 6 m. Fragmentary material consists only of blocky hornblende pseudomorphs, 0.2–1.5 cm in diameter.

*Interpretation*

The rhyolitic deposits are interpreted as flows because they contain longitudinal to semi-spherical gas cavities and large vesicles. These indicate relatively rapid cooling of a viscous flow near its vent area (see Bommischen & Kauffman, 1987). The weak stratification and parallel bedding of acid fragmentary material is consistent with ash fall deposition (Cas & Wright, 1987). These deposits have the same chemical composition as the associated rhyolitic lavas. The relatively rapid transition observed between the underlying deposits and the overlying pyroclastics of mostly shallow marine or subaerial rhyolites may indicate that significant regional uplift occurred before initial eruptions took place.

Volcanic activity was terminated by the deposition of fragmental, shallow water to subaerial tuffs (see Wohletz, 1983; Cas & Wright, 1987). The large-scale planar cross-bedded sets are indicative of sand waves. Parallel lamination commonly indicates weakening flows. The scour structures and large-scale sand waves support a pyroclastic surge origin for the tuffs at Nikkari (e.g. Chough & Sohn, 1990). Some primary blocky-shaped pyroclast particles are identified as hornblende pseudomorphs. Angular, blocky shapes and monomineralic characteristics of these fragments are further indications of their pyroclastic origin rather than reworked volcaniclastics.

## Massive conglomerates and cross-bedded sandstones

*Description*

At location 4 in Katajaperä, Ylivieska area (Fig. 2, Map sheet 2431 07D, $x = 7105.42$, $y = 528.74$), the rocks consist of relatively coarse-grained sandstones and conglomerates (Fig. 6A; i.e. the Ylivieska conglomerate of Salli, 1964) forming successions up to

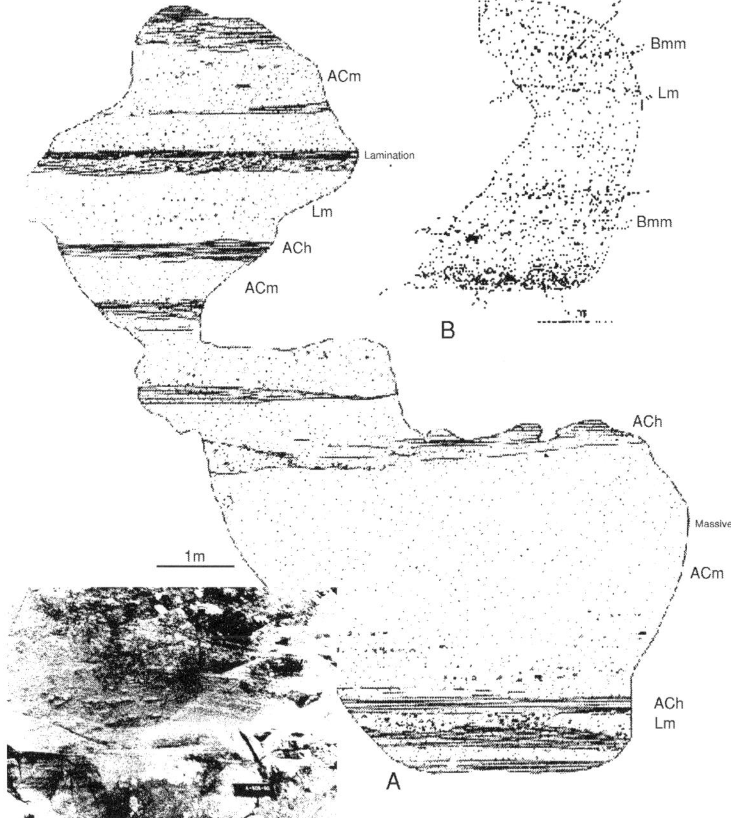


**Fig. 5.** Typical exposures of basaltic andesite tuff of hydrovolcanic origin. (A) Primary sedimentary structures such as erosional scours and sand wave-like structures are preserved. (B) Proximal areas are located a few hundred metres towards the north-west, where the coarser-grained counterparts are seen. Photograph shows erosional scour and scattered acid tuff fragments. Bmm, matrix-supported, massive pyroclastic breccia; Lm, clast-supported massive or crudely bedded lapilli; AC, coarse ash; h, horizontal bedded or laminated; m, massive. Label is 16.5 cm long.

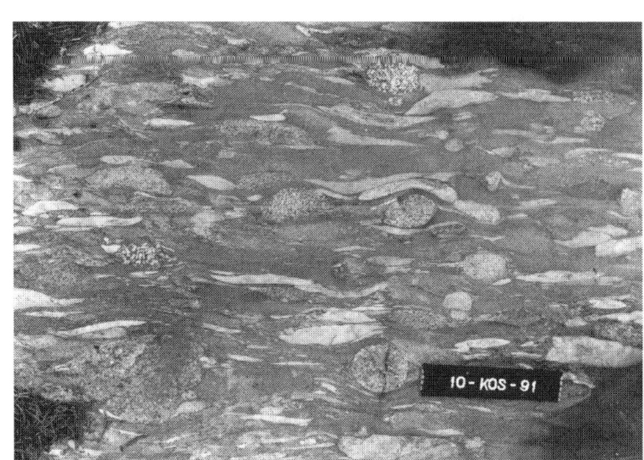

**Fig. 6.** Typical polymictic conglomerate of the fan-delta to fluvial association at Ylivieska. (A) The lithic clasts in clast-supported conglomerates include acidic to intermediate volcanics, porphyrites and sand- to fine-grained turbidite clasts. This outcrop demonstrates the dominant regional foliation ($S_{dom} = 250°/90°$). Label is 16.5 cm long. (B) Horizontal stratification and trough cross-bedding are the most common sedimentary structures in sandstone interbeds. Label is 16.5 cm long.

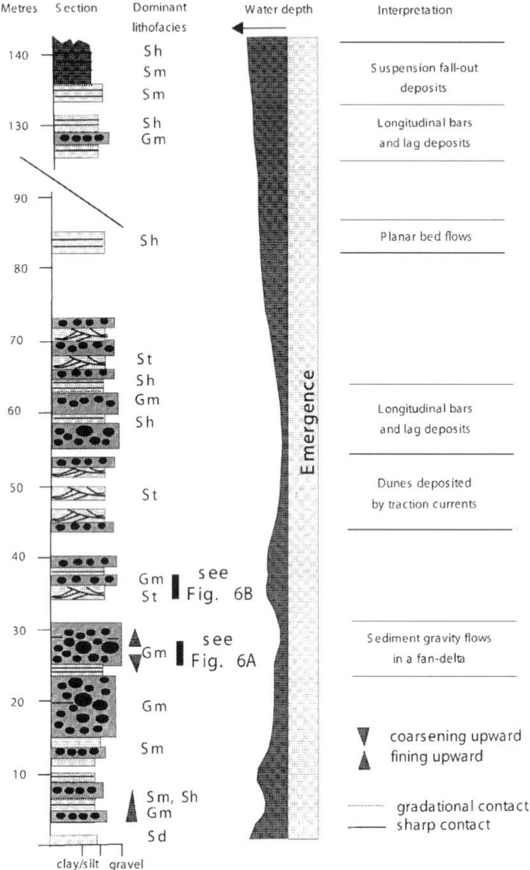


**Fig. 7.** Type section of the fan-delta to fluvial association at Katajaperä, Ylivieska. Gm, massive conglomerate; S, sandstone; t, trough cross-bedded; h, horizontally bedded; m, massive; g, normally graded; d, deformed.

several hundreds of metres thick with well defined facies characteristics. The Katajaperä section (Fig. 7) shows a coarsening upward character for the first 60 m; the upper 80 m is an upward-fining succession. Cross-stratified, poorly sorted sandstones occur as lensoid units associated with pebble conglomerates. Horizontal stratification (Sh) and trough cross-bedding (St) are the most common sedimentary structures in sandstone interbeds (Fig. 6B). The thickness of these interbeds is from 5 to 10 cm.

Dominant framework grains are angular to subangular quartz, oligoclase and scattered rock fragments. The sedimentary material is relatively immature. The lithic clasts in the clast-supported conglomerates include acidic (40 vol%) to intermediate volcanics (10 vol%), diabase (13 vol%), porphyrite (2 vol%) and sand- to fine-grained turbidite clasts (35 vol%) determined by counting 100 pebbles along a line on an outcrop surface. The clasts vary from rounded to well rounded (Fig. 6A), with the larger ones being the most rounded. These conglomerates do not contain any plutonic pebbles. The conglomerates are cut in some areas by plagioclase porphyry. Fine-grained sediments in the uppermost part of the section show graded bedding. The fine-grained sandstones and siltstones in the uppermost units are also quartz- and sericite-rich.

*Interpretation*

Coarse-clastics in the Ylivieska area indicate reworking of underlying units. The lithic clasts in clast-supported conglomerates include acidic to intermediate volcanics, porphyrites and sand- to fine-grained turbidite clasts and correlate well with the underlying rock types located in the north-east. As the conglomerates do not contain any plutonic pebbles, deeper crustal units were not exposed during erosion. Deposition is interpreted to have taken place in a fan-delta or fluvial environment (Nemec & Steel, 1988; Postma, 1990), where large areas of underlying volcanic rocks were uplifted, mostly to a subaerial position. It is suggested that the section at Katajaperä was formed after tectonic uplift and has characteristics of both subaerial and subaqueous conditions, favouring a fan-delta interpretation. The quartz-rich and aluminous character of the upper part of the section may be further evidence of subaerial chemical weathering conditions in the source area (see Nesbitt & Young, 1982). The abundance of coarse-grained lithofacies in the sections indicates proximity to sediment supply, but the rounded to well rounded clasts indicate significant rounding during transport, probably over tens of kilometres. Facies St and Sh reflect deposition by traction currents. Gms, Sm and Sg indicate transport by sediment gravity flows (see Postma, 1990).

## INTRA-ARC VOLCANISM AND SEDIMENTATION IN THE SVECOFENNIAN DOMAIN

In the stratigraphy of the Ylivieska area, the primary volcanogenic rocks form the middle and upper portions of the upward-coarsening stratigraphic succession overlying the major part of the succession, which comprises mostly sandy- and fine-grained turbidites up to 3 km thick.

Mature volcanic arc (c. 1.90 Ga)

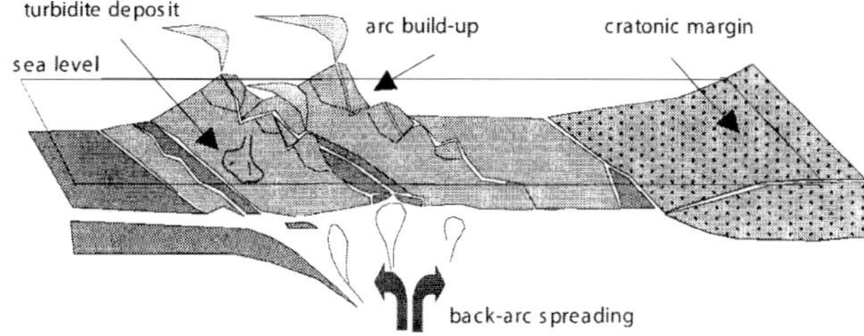


Marginal-basin stage during deposition of the studied section (c. 1.87 Ga)

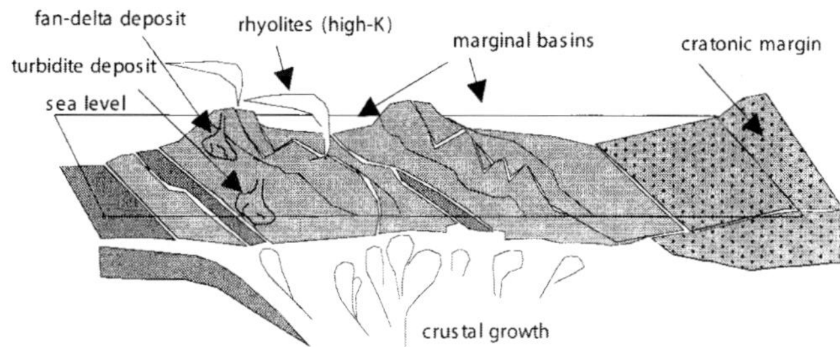


**Fig. 8.** Schematic construction of tectonic setting of the Ylivieska section.

The first eruptive products were primarily rhyolitic flows from small-volume rhyolitic eruptions. Later, calc-alkaline volcanism culminated in basaltic and andesitic eruptions of hydrovolcanic character. Hydrovolcanic eruptions occured in shallow-water to subaerial environments, resulting in the accumulation of predominantly surge deposits associated with some fine-grained, mostly laminated, beds. The fine-grained units between turbidites represent suspension sedimentation and overall relatively deep water.

The sediments, varying from marine (turbidites) to more or less continental (fan-delta or fluvial), are interpreted as an intra-arc basin fill associated with mature arc formation (Fig. 8). The effusive to explosive volcanism was probably associated with transitory extensional stages in a convergent plate setting, and the rhyolite flows, which show a high-K affinity, indicate that the activity was related to the development of a mature volcanic arc and to the formation of juvenile continental crust. The uppermost coarse-grained secondary volcaniclastics indicate that large areas of volcanic rocks were uplifted to a subaerial position. It is suggested that tectonic activity was the main cause and controlling factor for erosion and sedimentation by changing relative sea-level and topography in the source area. The sedimentation was overall volcaniclastic, with pyroclastics and lavas constituting a relatively small part of the total rock record, suggesting that mature Precambrian arcs may be mostly sedimentary, containing abundant reworked and resedimented volcanics. The depositional model presented here demonstrates volcanism and volcaniclastic sedimentation coupled with relative sea-level fall owing to active arc build-up and tectonic uplift of some areas.

The inferred tectonic setting largely resembles the situation in North Wales during the Ordovician, where Precambrian crust formed a part of a small microcontinent (Kokelaar *et al.*, 1984; Fritz & Howells, 1991; Orton, 1991). A very similar environment is present in the Taupo Volcanic Zone, New Zealand (Stern, 1985), where the volcanic zone is a back-arc or intra-arc rift coupled with active andesitic to rhyolitic volcanism and with volcaniclastic sedimentation occurring in shallow-water to subaerial environments. An Archaean example comes from the Hauy Formation, Canada, where a slope to shallow-marine succession is associated with K-rich volcanism in an integral part of an evolved arc setting (Mueller, 1991).

## COMPARISON WITH OTHER PARTS OF THE SVECOFENNIAN DOMAIN

The depositional events of the study area are considered to be related to the late phase of early Svecofennian accretionary plate tectonics occurring during the time interval *c*.1.90–1.87 Ga. The intra-arc basin series of the Ylivieska area are calc-alkaline felsic to mafic with typically high-K affinity. Kousa (1990) has named the coeval felsic volcanics as the Kuusaanjärvi Formation, which is lithologically similar to volcanics in the Pihtipudas area, 100 km south-east of Ylivieska (see Aho, 1979; Kousa 1990). Similar *c*.1875 Ma mature Svecofennian volcanic arc rocks and associated syn- and late-orogenic igneous rocks are described in northern Sweden by Skiöld *et al.* (1993).

It is also suggested that approximately 1880 Ma magmatism and a new phase of active convergence with intra-arc extension and rifting was made possible, probably because of the development of a new subduction system to the south (see Edelman & Janus-Järkkälä, 1983; Ehlers *et al.*, 1986). Marginal basins were formed in north-central Sweden and south-central Finland and detritus was derived from both surrounding continental areas and topographic highs, i.e. volcanic arc-blocks (see Claesson *et al.*, 1993). Tectonic activity was the main control on erosion and sedimentation by changing relative sea-level and topography in the source area. Finally, a continental marginal arc, i.e. Andean arc, was formed, which represents the typical orogenic product in the Svecofennian Domain (Gaál & Gorbatschev, 1987; Gaál, 1990). Within this arc, several internal terrane boundaries are visualized as conductive and reflective structures that extend at least to the middle of the thickened crust (Korja *et al.*, 1993). Crust-forming processes first generated a volcanic-sediment block some 15 km thick over the oceanic lithosphere. The block was then modified and the juvenile crust became substantially thicker (estimated 30–35 km). Final thicknesses due to collisions were greater than 50 km, as estimated by Korja *et al.* (1993). According to Kärki *et al.* (1993), the collision of the arc block with the craton margin first created NW–SE trending folds and thrusts, which were later modified by N–S trending shear zones between 1.93 and 1.80 Ga. A close connection of the arc-block complex with the cratonic margin is supported by the fact that clastic materials in the marginal basins have detrital zircons derived from both Archaean cratonic sources and arc-block sources (Claesson *et al.*, 1993). The youngest detrital ages constrain the minimum age of Svecofennian sedimentation at 1.80 Ga, but the zircons with ages between 2.12 and 1.88 Ga may also indicate pre-existing microcontinents providing detritus at 1.9 Ga (Cleasson *et al.*, 1993).

The Svecofennian massive sulphide deposits may be modelled by characterizing the types of volcanism, depositional depths and environments. Relevant hydrothermal vent areas and possible polymetallic sulphide deposits in the rock column may be located by using palaeovolcanological reconstructions and sedimentary facies analysis, together with more conventional exploration methods. The existing ore districts, e.g. the Pyhäsalmi ore district in Finland and the Skellefte ore district in Sweden, probably represent relatively deep-sea settings alongside and partly overgrown by later arc-blocks, which are the most suitable conditions for ore-forming processes. Submarine rhyolitic calderas are thought to be important sites for the formation of volcanic-hosted massive sulphide deposits (see Cas, 1992; Large, 1992), such as the Kuroko-type. Most volcanogenic massive sulphide deposits form in a submarine caldera setting. Miocene massive sulphide deposits in Japan are associated with rhyolitic lava domes and volcaniclastic rocks believed to have been emplaced at a minimum water depth of 1000 m (Ohmoto & Takahashi, 1983). In general, the water depth for these caldera systems is controversial. Recent evidence suggests substantial depths are possible, although the depth for the Kuroko caldera itself is relatively shallow (1000 m).

## CONCLUSION

The sedimentary and pyroclastic rocks of the Ylivieska area were deposited in subaqueous, transitional and subaerial environments. The primary volcanics were derived from subaerial eruptive columns and deposited as lavas, ash flows and minor ash falls. The rocks studied suggest initial sedimentation at shallow water depths, followed by turbidite sedimentation beneath the acid tuffs, to nearshore marine to subaerial environments associated with hydrovolcanic ash deposits. Volcanics were derived from subaerial sources or from subaerial eruptive columns. The presence of pumice implies shallow submarine or subaerial eruption columns. The hydrovolcanic processes imply shallow-water volcanism. The sedimentary rocks above and adjacent to the major volcanics were deposited in subaerial to transitional marine environments as fluvial deposits and fan deltas. The sediments were associated with mature arc formation and infilling of an intra-arc basin. It is suggested that the fill of an extensional marginal basin was one of the major components in the formation of juvenile crust of the Svecofennian Domain.

This study shows that the influences on sedimentation in volcanic terranes are complex, and include volcanism, normal surficial processes and tectonics. The relationship between volcanism and tectonic setting is not always fully understood, but evaluation of the interplay between these influences should be based on an appraisal of the total geological framework using a sedimentological facies analysis approach as a guiding principle (see Bergh & Torske, 1988; Mueller, 1991; Orton, 1991; Aspler & Chiarenzelli, 1996). In addition, careful documentation of the volcaniclastic depositional environment (e.g. bounding facies and water depths) has applications in volcanic-hosted mineral deposit exploration.

## ACKNOWLEDGEMENTS

The Research Council for Natural Sciences and Engineering of the Academy of Finland is thanked for financial support to study this area. Jukka Kousa and Kaj Västi from the Geological Survey of Finland are thanked for their advice and helpful comments related to the study area and regional stratigraphy and for a clarifying field excursion. This study is also a contribution to the IGCP Project 371 COPENA (Structure and Correlation of Precambrian in Northeast Europe and the North Atlantic Realm). The co-editor Patricia Corcoran and IAS Special Publication reviewers are thanked for constructive comments and improvements, especially R.W. Ojakangas for refining the paper.

## REFERENCES

Aho, L. (1979) Petrogenetic and geochronological studies of metavolcanic rocks and associated granitoids in the Pihtipudas area, Central Finland. *Geol. Surv. Finl. Bull.*, **300**, 23 pp.

Aspler, L.B. & Chiarenzelli, J.R. (1996) Stratigraphy, sedimentology and physical volcanology of the Henik Group, central Ennadai–Rankin greenstone belt, Northwest Territories, Canada: late Archean paleogeography of the Herne Province and tectonic implications. *Precam. Res.*, **77**, 59–89.

BABEL Working Group (1990) Evidence for Early Proterozoic plate tectonics from seismic reflection profiles in the Baltic Shield. *Nature*, **348**, 34–38.

Bommischen, B. & Kauffman, D.F. (1987) Physical features of rhyolite lava flows in the Snake River Plain. In: *The Emplacement of Silicic Domes and Lava Flows* (Ed. Fink, J.H.), Spec. Publ. geol. Soc. Am., Boulder, **212**, 119–145.

Bergh, S.G. & Torske, T. (1988) Palaeovolcanology and tectonic setting of a Proterozoic metatholeiitic sequence near the Baltic Shield margin, northern Norway. *Precam. Res.*, **39**, 227–246.

Cas, R.A.F. (1992) Submarine volcanism: eruption styles, products, and relevance to understanding the host-rock successions to volcanic-hosted massive sulfide deposits. *Econ. Geol.*, **87**, 511–541.

Cas, R.A.F. & Wright, J.V. (1987) *Volcanic Successions, Modern and Ancient.* Allen & Unwin, London.

Chough, S.K. & Sohn, Y.K. (1990) Depositional mechanics and sequences of base surges, Songaksan tuff ring, Cheju Island, Korea. *Sedimentology*, **37**, 1115–1135.

Claesson, S., Huhma, H., Kinny, P.D. & Williams, I.S. (1993) Svecofennian detrital zircon ages—implications for the Precambrian evolution of the Baltic Shield. *Precam. Res.*, **64**, 109–130.

Edelman, N. & Jaanus-Järkkälä, M. (1983) A plate tectonic interpretation of the Precambrian of the archipelago of southwestern Finland. *Geol. Surv. Finl. Bull.*, **325**, 4–33.

Ehlers, C., Lindroos, A. & Jaanus-Järkkälä, M. (1986) Stratigraphy and geochemistry in the Proterozoic mafic volcanic rocks of the Nagu–Korpo area, SW Finland. *Precam. Res.*, 32, 297–315.

Fisher, R.V. (1961) Proposed classification of volcanoclastic sediments and rocks. *Geol. Soc Am. Bull.*, **72**, 1409–1914.

Fisher, R.V. (1966) Rocks composed of volcanic fragments and their classification. *Earth Sci. Rev.*, **1**, 287–298.

Fritz, W.J. & Howells, M.F. (1991) A shallow marine volcaniclastic facies model: an example from sedimentary rocks bounding the subaqueously welded Ordovician Garth Tuff, North Wales, UK. *Sediment. Geol.*, **74**, 217–240.

Gaál, G. (1990) Tectonic styles of Early Proterozoic ore deposition in the Fennoscandian Shield. *Precam. Res.*, **46**, 83–114.

Gaál, G. & Gorbatschev, R. (1987) An outline of the Precambrian evolution of the Baltic Shield. *Precam. Res.*, **35**, 15–52.

Kähkönen, Y., Huhma, H. & Aro, K. (1989) U–Pb zircon ages and Rb–Sr whole-rock isotope studies of Early Proterozoic volcanic and plutonic rocks near Tampere, southern Finland. *Precam. Res.*, **45**, 27–43.

Kärki, A., Laajoki, K. & Luukas, J. (1993) Major Palaeoproterozoic shear zones of the central Fennoscandian Shield. *Precam. Res.*, **64**, 207–223.

Kokelaar, B.P., Howells, M.F., Bevins, R.E., Roach, R.A. & Dunkley, P.N. (1984) The Ordovician maginal basin of Wales. In: *Marginal Basin Geology: Volcanic and Associated Sedimentary and Tectonic Processes in Modern and Ancient Marginal Basins.* (Eds Kokelaar, B.P. & Howells, M.F.), Spec. Publ. geol. Soc. London, No. 16, pp. 245–270. Geol. Soc. London, Bath.

Korja, A., Korja, T., Luosto, U. & Heikkinen, P. (1993) Seismic and geoelectric evidence for collisional and extensional events in the Fennoscandian Shield—implications for Precambrian crustal evolution. *Tectonophysics*, **219**, 129–152.

Korsman, K., Koistinen, T., Kohonen, J. *et al.* (Eds) (1997) *Suomen kallioperäkartta—Berggrundskartra över Finland (Bedrock Map of Finland 1 : 1 000 000)*. Geological Survey of Finland, Espoo.

Kousa, J. (1990) Palaeoproterozoic metavolcanic rocks in the borderzone of Savo and Pohjanmaa, central Finland. *IGCP 217 Proterozoic Geochemistry, National Working Group of Finland Symposium 'Proterozoic Geochemistry, Helsinki '90', Abstacts*, pp. 29–30.

LAHTINEN, R. (1994) Crustal evolution of the Svecofennian and Karelian domains during 2.1–1.79 Ga, with special emphasis on the geochemistry and origin of 1.93–1.91 Ga gneissic tonalities and associated supracrustal rocks in the Rautalampi area, central Finland. *Geol. Surv. Finl. Bull.*, **378**, 128 pp.

LARGE, R.R. (1992) Australian volcanic-hosted massive sulfide deposits: features, styles and genetic models. *Econ. Geol.*, **87**, 471–510.

LOWE, D.R. (1982) Sediment gravity flows: II. Depositional models with special reference to the deposits of high-density turbidity currents. *J. sediment. Petrol.*, **52**, 279–297.

MIALL, A.D. (1977) A review of the braided river depositional environment. *Earth Sci. Rev.*, **13**, 1–62.

MIALL, A.D. (1985) Architectural-element analysis: a new method of facies analysis applied to fluvial deposits. *Earth Sci. Rev.*, **22**, 261–308.

MUELLER, W. (1991) Volcanism and related slope to shallow-marine volcaniclastic sedimentation: an Archean example near Chibougamau, Quebec, Canada. *Precam. Res.*, **49**, 1–22.

NEMEC, W. & STEEL, R.J. (Eds) (1988) *Fan Deltas, Sedimentology and Tectonic Settings*. Blackie and Son, London, 444 pp.

NESBITT, H.W. & YOUNG, G.M. (1982) Early Proterozoic climates and plate motions inferred from major element chemistry of lutites. *Nature*, **299**, 715–717.

ÖHLANDER, B., SKIÖLD, T., ELMING, S.-Å., BABEL WORKING GROUP, CLAESSON, S. & NISCA, D.H. (1993) Delineation and character of the Archaean–Proterozoic boundary in the northern Sweden. *Precam. Res.*, **64**, 67–84.

OHMOTO, H. & TAKAHASHI, T. (1983) Geological setting of Kuroko deposits, Japan. III: Submarine calderas and Kuroko genesis. In: *The Kuroko and Related Volcanogenic Massive Sulphide Deposits* (Eds Ohmoto, H. & Skinner, B.J.), Econ. Geol. Monogr., **5**, 39–54.

ORTON, G.J. (1991) Emergence of subaqueous depositional environments in advance of a major ignimbrite eruption, Capel Curig Volcanic Formation, Ordovician, North Wales. An example of regional volcanotectonic uplift? *Sediment. Geol.*, **74**, 251–286.

POSTMA, G. (1990) Depositional architecture and facies of river and fan deltas: a synthesis. In: *Coarse-grained Deltas* (Eds Colella, A. & Prior, D.B.) Spec. Publs int. Ass. Sediment., No. 10, pp. 13–27. Blackwell Scientific Publications, Oxford.

PICKERING, K.T., STOW, D.A.V., WATSON, M.P. & HISCOTT, R.N. (1986) Deep-water facies, processes and models: a review and classification scheme for modern and ancient sediments. *Earth Sci. Rev.*, **23**, 75–174.

SALLI, I. (1964) The structure and stratigraphy of the Ylivieska–Himanka schist area, Finland. *Bull. Comm. Géol. Finl.*, **211**, 67 pp.

SKIÖLD, T., ÖHLANDER, B., MARKKULA, H., WIDENFALK, L. & CLAESSON, L.-Å. (1993) Chronology of Proterozoic orogenic processes at the Archaean continental margin in northern Sweden. *Precam. Res.*, **64**, 225–238.

STERN, T.A. (1985) A back-arc basin formed within continental lithoshere: the central volcanic region of New Zealand. *Tectoniphysics*, **112**, 385–409.

STRAND, K.O. (1993) Volcaniclastic sedimentation in a convergence setting related to volcanic arc, a Palaeoproterozoic example from Oulainen, western Finland. *IGCP 275 Deep Geology of the Baltic Shield, Annual Meeting and Symposium 'The Svecofennian Domain', Abstracts*, p. 55.

VAASJOKI, M. & SAKKO, M. (1988) The evolution of the Raahe–Ladoga zone in Finland: isotopic constraints. *Geol. Surv. Finl. Bull.*, **343**, 7–32.

VÄSTI, K.J. (1989) Geology of the Rauhala stratiform massive Zn–Cu–Pb sulphide deposit. In: *The Early Proterozoic Zn–Cu–Pb Sulphide Deposit of Rauhala in Ylivieska, Western Finland* (Ed. Kojonen, K.), Spec. Publ. geol. Surv. Finland, **11**, 5–18.

WELIN, E. (1987) The depositional evolution of the Svecofennian supracrustal sequences in Finland and Sweden. *Precam. Res.*, **35**, 95–113.

WOHLETZ, K.H. (1983) Mechanism of hydrovolcanic pyroclast formation: grain-size, scanning electron microscopy, and experimental studies. *J. Volcanol. geothermal Res.*, **17**, 31–63.

*Spec. Publs int. Ass. Sediment.* (2002) **33**, 351–367

# Palaeoproterozoic epeiric sea palaeoenvironments: the Silverton Formation (Pretoria Group, Transvaal Supergroup), South Africa

P. G. ERIKSSON*, W. ALTERMANN†, L. EBERHARDT†, S. AREND-HEIDBRINCK† *and* A. J. BUMBY*

**Department of Earth Sciences, University of Pretoria, Pretoria 0002, South Africa; and*
*†Institut für Allgemeine und Angewandte Geologie, Ludwig-Maximilians-Universität, Luisenstrasse 37, D-80333 München, Germany*

## ABSTRACT

A predominant argillaceous lithofacies association in the Early Proterozoic Silverton Formation (Pretoria Group, Transvaal Supergroup, South Africa) comprises massive mudstones, laminated claystones, laminated siltstones, graded mudstones and graded siltstones. These fine-grained rocks are interpreted as having been laid down as substorm wave-base pelagic deposits within an epeiric embayment. Fine fluvial detritus was able to by-pass a high-energy sand belt and coastline setting (represented by the Magaliesberg Formation, above the Silverton rocks) owing to fluvial flood stages and a high concentration of suspended sediment. High-amplitude storm waves resulted in subordinate graded coarse siltstone and fine sandstone beds being laid down within the transitional zone and offshore mud belt of the shallow epeiric palaeoenvironment. The model proposed for the Silverton Formation is comparable to the classic Shaw–Irwin epeiric sea model, with the exception that a strongly tidal coastline was present. With the initial onset of the Silverton transgression, fluvial sandstones of the underlying Daspoort Formation were reworked in the east of the basin. Approximately coeval lower Silverton braid-deltaic sandstones in the west of the basin were rapidly drowned, to be followed initially by turbidity current sediments, and subsequently by the predominant mudrocks, deposited as transgressive and subsequent highstand systems tracts. Regressive, tidally reworked braid-delta sandstones of the Magaliesberg Formation accompanied retreat of the Silverton epeiric embayment off the Kaapvaal craton, as freeboard of the latter became thermally enhanced.

## INTRODUCTION

Epeiric seas were probably common depositional palaeoenvironments during the Palaeoproterozoic (Eriksson *et al.*, 1998). This inference is based on two likely characteristics of early Palaeoproterozoic natural systems: periodically elevated sea-levels globally and aggressive weathering regimes. High rates of mass wasting almost certainly characterized the early Precambrian, owing to elevated temperatures and concomitant humid palaeoclimates, high atmospheric $CO_2$ content and a lack of vegetation (Corcoran *et al.*, 1998). Topography would thus have been reduced and even peneplained owing to aggresive weathering and rapid erosion; this would have enabled widespread inundation of cratonic regions when marine transgressions occurred, forming large epeiric seas.

Continental freeboard, the maximum elevation of a continent above mean sea-level (Wise, 1972), eustacy and continental crustal growth rates can be considered as interdependent variables (Eriksson, 1999). Increased mid-ocean ridge activity associated with enhanced continental crustal growth would have reduced ocean volumes and raised sea-level globally (Arndt, 1999). Although there is still much debate as to whether accelerated crustal growth occurred prior to *c*.3.6 Ga (Armstrong, 1981; Arndt, 1999), or close to the Archaean–Proterozoic boundary (Eriksson, 1995), a strong school of thought supports the latter (Windley, 1995). Eriksson *et al.* (1999) discuss evidence in support of globally elevated sea-levels affecting African, Indian and Australian (present-day frame of reference)

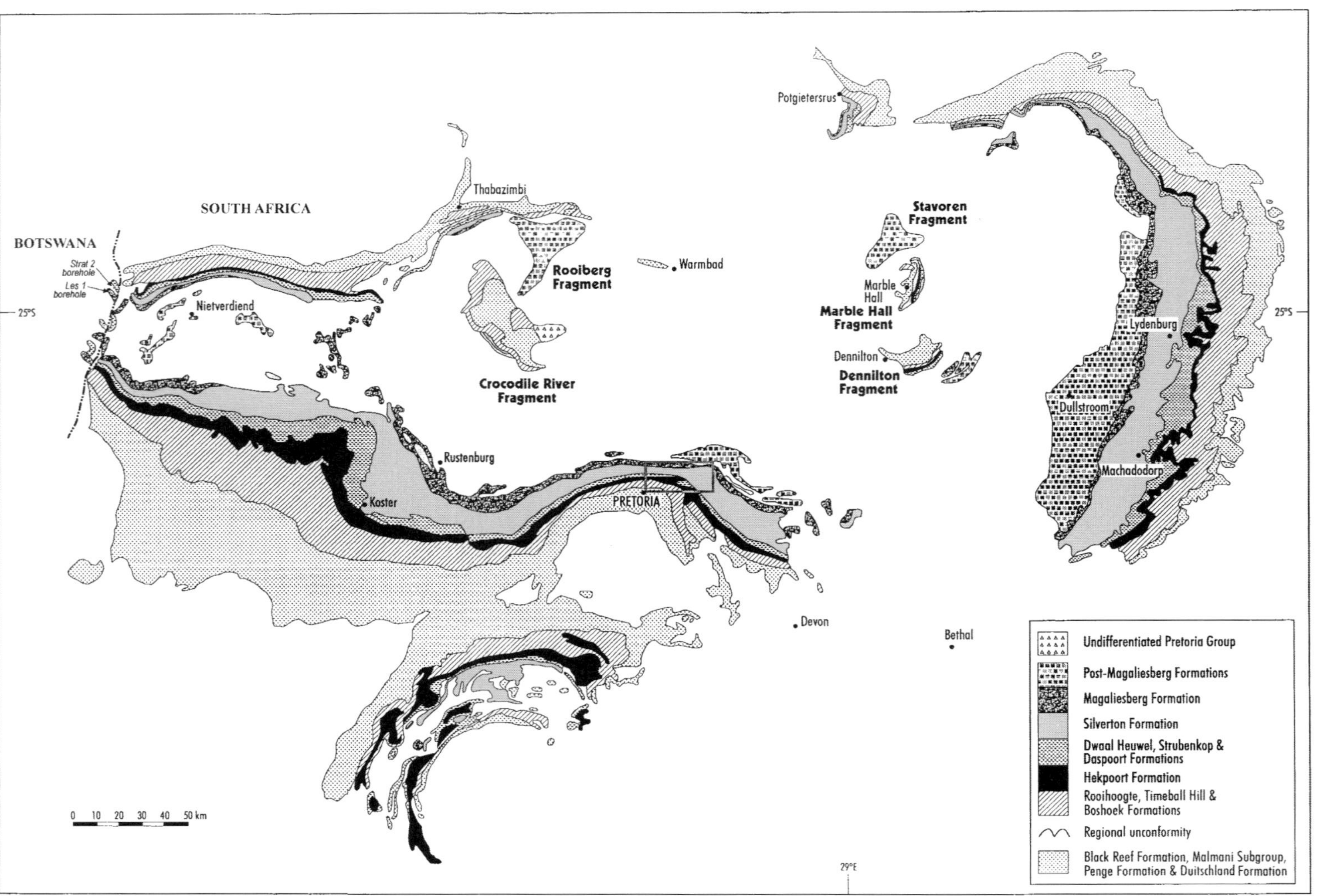


**Fig. 1.** Geological map of the Pretoria Group, Transvaal Supergroup. Note the Timeball Hill and Silverton formations, both inferred to have been deposited in epeiric marine palaeoenvironments. Note also Daspoort and Magaliesberg formations, which bound the Silverton Formation stratigraphically. Boreholes Les1 and Strat2 in eastern Botswana are summarized in Fig. 7.

cratons in the Late Archaean, during which major epeiric carbonate banded iron-formation (BIF) platforms formed. The same authors postulate a second global transgression at *c*.<2.2 to >2.15 Ga, owing to post-glacial climatic amelioration following the first global refrigeration event at *c*.2.4–2.2 Ga (Hambrey & Harland, 1985; Eyles & Young, 1994).

Within South Africa, the Late Archaean to early Palaeoproterozoic is best represented by deposits of the *c*.2.7–2.1 Ga Transvaal Supergroup (for a review, see Eriksson & Reczko, 1995). Within the upper subdivision of this succession, the Pretoria Group, there are two inferred transgressive epeiric marine deposits, the Timeball Hill and Silverton formations (Catuneanu & Eriksson, 1999) (Fig. 1). The postulated Timeball Hill epeiric sea has been modelled by Eriksson & Reczko (1998), who related a combination of pelagic, low-density turbidite, contourite and tidally reworked braid-delta facies to a relatively deep water basin, which resembled a scale model of a large modern ocean basin. This paper describes the geology of the inferred Silverton Formation epeiric sea and an attempt is made to define some of the more important characteristics of Palaeoproterozoic epeiric sea palaeoenvironments.

## EPEIRIC SEAS: GENERAL CONCEPTS

Epeiric seaways inundate large proportions of cratonic terranes, and closely resemble open ocean shelves (Brenner, 1980), with analogous sedimentation processes and products (Brenner & Davies, 1974; Spearing, 1976; Brenner, 1978, 1979; Klein & Ryder, 1978; Bouma *et al.*, 1982). Smaller equivalents, epeiric embayments, lack true shelf-breaks and ocean-type currents (Eriksson *et al.*, 1998). A lack of adequate modern examples (Friedman *et al.*, 1992) makes interpretation of Precambrian epeiric sea deposits problematic. Another problem with modern comparisons is one of scale, especially when contrasted with examples such as the Late Cretaceous seaway in North America, which at one time extended from the Gulf of Mexico to the Arctic Ocean (Swift & Rice, 1984). Estimating scale for Precambrian epeiric transgressions is difficult owing to incomplete preservation of basins. For the Archaean Witwatersrand epeiric basin, preserved outcrops give a minimum estimate of 600 by 400 km (Eriksson *et al.*, 1998), whereas that for the Palaeoproterozoic Timeball Hill Formation is 500 by 300 km (Eriksson & Reczko, 1998).

The Shaw (1964) and Irwin (1965) conceptual epeiric sea model, still widely accepted, is based on the Palaeozoic successions of North America, and encompasses three belts, X, Y and Z, which are summarized in Fig. 2. Certain qualifications need to be applied to this relatively simple conceptual model (Schopf, 1980), including water depths, which were probably shallower and equivalent to modern continental shelves (e.g. Eriksson & Reczko, 1998), rather than the maximum 200 m proposed in the model. Phanerozoic seaway successions include both widespread carbonates, indicating water depths up to about 30 m, and

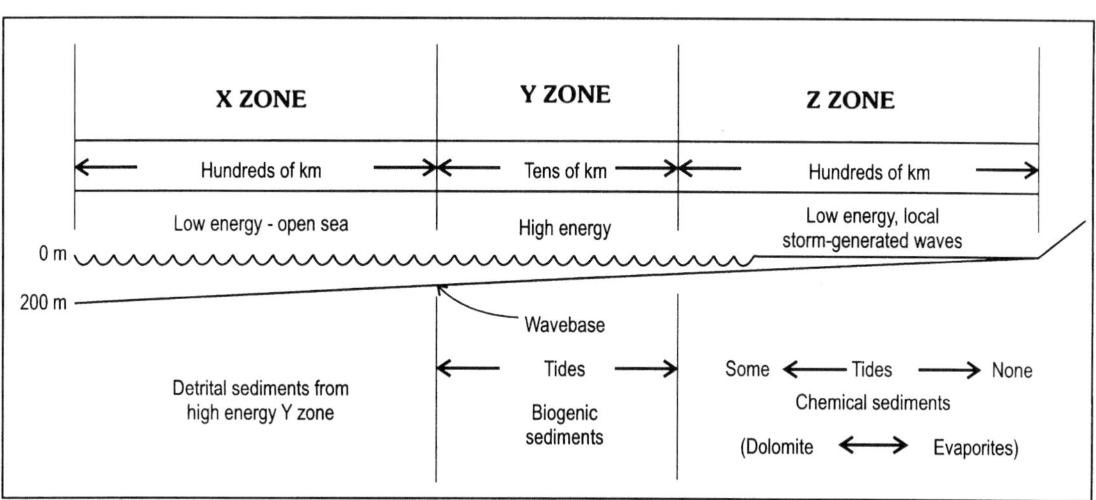


**Fig. 2.** Classic Shaw (1964) and Irwin (1965) epeiric sea model. Note wide and low-energy X (most seaward) and Z (most landward) zones, with narrow and high-energy Y zone in between.

phosphate nodules in black shales, suggesting the possibility of depths greater than 100 m (Friedman *et al.*, 1992, p. 391). As modern fairweather wave base is *c*.10 m and storm wave base seldom exceeds 200 m (Swift & Thorne, 1991; Johnson & Baldwin, 1996), most epeiric seas were probably subject to a measure of wave action. Thus, although long-period swells would have been dissipated at epeiric sea margins (distal X zone in Fig. 2) (Friedman *et al.*, 1992), storm waves may have been a significant influence on epeiric sea sedimentation further landwards. Tidal action was probably also more significant, as evidenced by extensive preserved peritidal flat deposits in Phanerozoic epeiric successions (e.g. Swett *et al.*, 1971; Pratt & James, 1986). Tidal height may even have increased as open ocean tides approached the shallow platform of an epeiric sea (Pratt & James, 1986).

## GENERAL GEOLOGY OF THE TRANSVAAL SUPERGROUP

The 2.7–2.1 Ga Transvaal Supergroup includes lowermost 'protobasinal' units, preserved in discrete, rift-related basins, and consisting of subaerial to basinal continental sedimentary facies and acid to basic volcanic rocks (Eriksson & Reczko, 1995). These were related to the widespread rifting and *c*.2.7 Ga volcanism of the Ventersdorp Supergroup on the Kaapvaal craton (Eriksson & Reczko, 1995). Subsequent thermal subsidence formed an intracratonic sag basin, in which the thin sheet sandstones of the Black Reef Formation were laid down, by braided fluvial channels and succeeding shallow marine and braid-delta systems (Catuneanu & Eriksson, 1999). Continued slow thermal subsidence led to the accumulation of almost 2 km of carbonate rocks and up to 700 m of succeeding BIF of the Chuniespoort Group (Eriksson & Altermann, 1998), under general highstand conditions (Catuneanu & Eriksson, 1999). This carbonate-BIF epeiric sea is estimated to have covered 600 000 km$^2$ (Beukes, 1987) and to have been 40–80 m deep (Eriksson & Altermann, 1998), and was probably a seaway rather than an embayment.

A sequence stratigraphic interpretation of the Pretoria Group (Fig. 3) suggests that thermal (?) uplift and erosion of the Chuniespoort chemical sedimentary rocks was followed by two cycles of syn-rift and post-rift sedimentation, separated by crustal doming. The lower Pretoria Group cycle comprises alluvial and lacustrine syn-rift deposits of the Rooihoogte Formation and post-rift clastic epeiric sea sediments of the Timeball Hill Formation (Fig. 3). The Timeball Hill Formation is described in detail by Eriksson & Reczko (1998), who propose that thermohaline contour currents reworked pelagic and turbidite deposits within this epeiric basin; thin stromatolitic carbonate beds interlayered with these facies suggest photic water depths up to about 100 m (Eriksson & Reczko, 1998). A relatively deep-water epeiric sea model is thus proposed for the Timeball Hill Formation (Fig. 4). Eriksson *et al.* (1999) note that the Timeball Hill epeiric transgression occurred within the period of the first global glaciation, at *c*.2.4–2.2 Ga, and that it thus probably reflects local tectonic subsidence rather than a major cratonic inundation event; an embayment model thus seems pertinent.

The upper Pretoria Group cycle followed crustal doming, probably thermal in origin, subsequent to which syn-rift alluvial to lacustrine deposits (Boshoek, Dwaalheuwel and Strubenkop formations) and the subaerial Hekpoort lavas were laid down (Fig. 3). Daspoort sedimentation began with alluvial gravel and fluvial sand lobes of syn-rift affinity, which became drowned in the east of the basin by a transgressive marine palaeoenvironment (Eriksson *et al.*, 1993). The thick Silverton shales which followed reflect the transgressive systems tract of an epeiric sea, which forms the subject of this paper; the regressive shoreline sands of the Magaliesberg Formation were laid down during the highstand systems tract which characterized upper Silverton deposition (Fig. 3). Post-Magaliesberg units formed under remnant marine coastline and continental sedimentation systems as base-level fall occurred owing to thermal uplift related to the *c*.2050 Ma Bushveld Complex intrusion (Catuneanu & Eriksson, 1999).

## GEOLOGY OF THE SILVERTON FORMATION

### Lithology, thickness and stratigraphy

This formation is characterized by various mudstones, with locally significant volcanic rocks, and minor carbonates, cherts and sandstones. Both contacts of the formation are sharp, but gradational in the sense that sandstone lenses are present in the predominantly argillaceous Silverton lithologies close to the contacts (Button, 1973; Schreiber, 1990; van der Neut, 1990). Total thickness of the Silverton Formation varies from *c*.2000 m in the east of the preserved basin to several hundred metres in the west; erosional removal

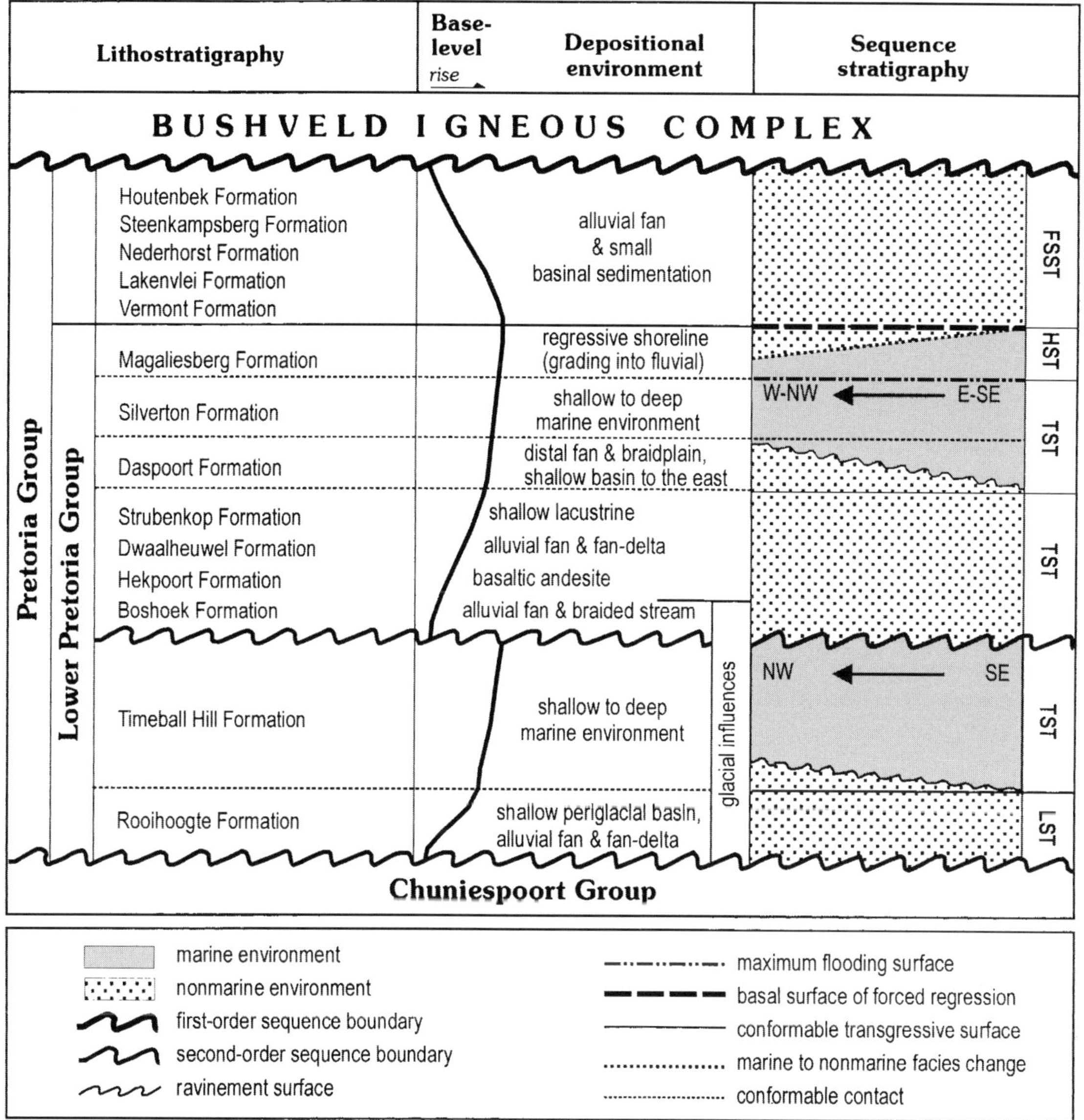


**Fig. 3.** Sequence stratigraphic interpretation for the Pretoria Group. Note two cycles of synrift and post-rift sedimentation (Rooihoogte–Timeball Hill and Boshoek–Houtenbek cycles) separated by a second-order sequence boundary reflecting crustal doming. Note also the two major marine transgressions, during Timeball Hill and Silverton epeiric sedimentation. Modified after Catuneanu & Eriksson (1999).

characterizes much of the southern and central preserved occurrences (Fig. 5).

Button (1973) discriminated between lower Boven Shale, medial Machadodorp Volcanic and upper Lydenburg Shale members in the eastern part of the basin. Their differentiation becomes problematic further westwards, due to poor outcrop quality and lack of boreholes (Fig. 5), but geochemical characteristics (Reczko, 1994) enable separation of the two shale members in this region (Fig. 6). Locally, in the far west, thin dolosiltites eroded from Chuniespoort lithologies cap the Silverton mudstones (Eriksson &

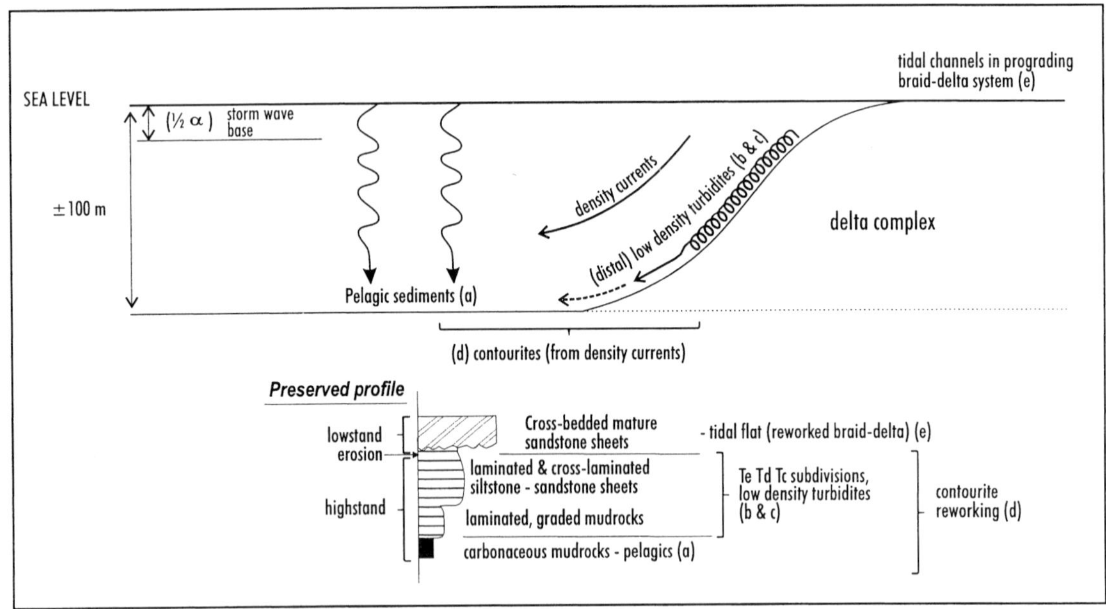


**Fig. 4.** Relatively deep water epeiric model postulated for the Timeball Hill Formation (modified after Eriksson & Reczko, 1998).

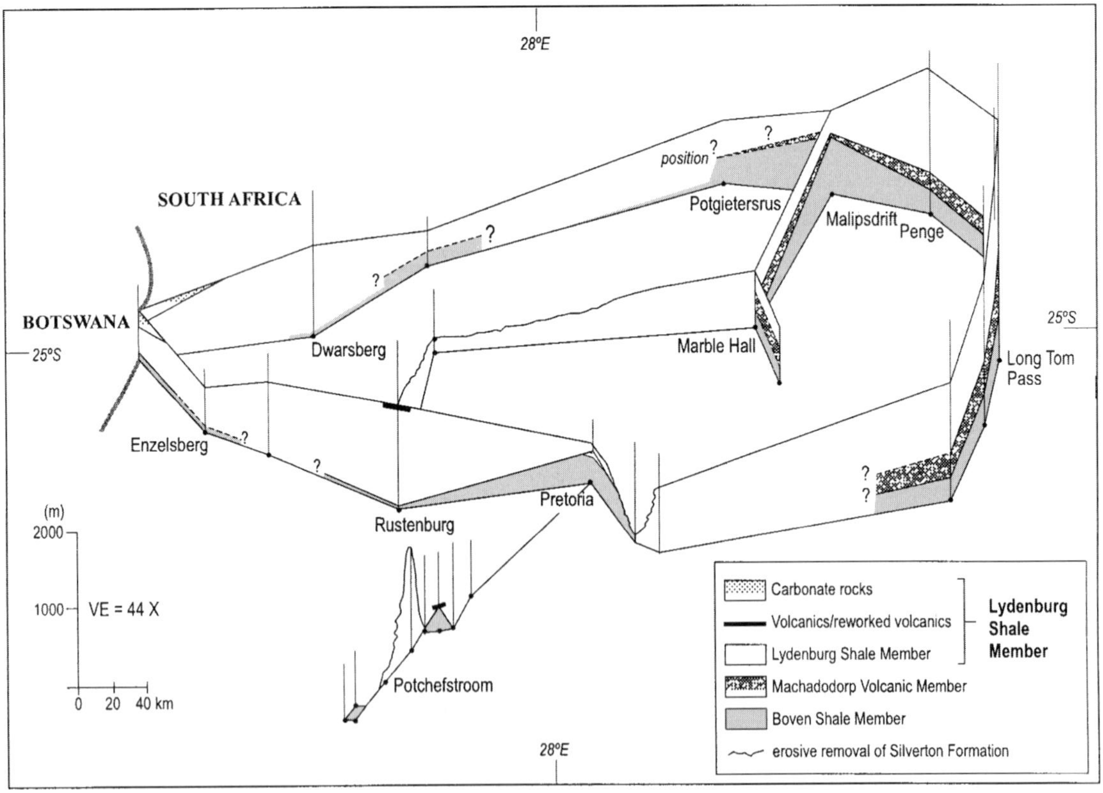

**Fig. 5.** Fence diagram illustrating preserved thicknesses of the Silverton Formation across the Pretoria Group basin. Note also threefold subdivision into Boven Shale, Machadodorp Volcanic and Lydenburg Shale members, and how this stratigraphy becomes impossible to apply to the western and southern parts of the preserved basin.

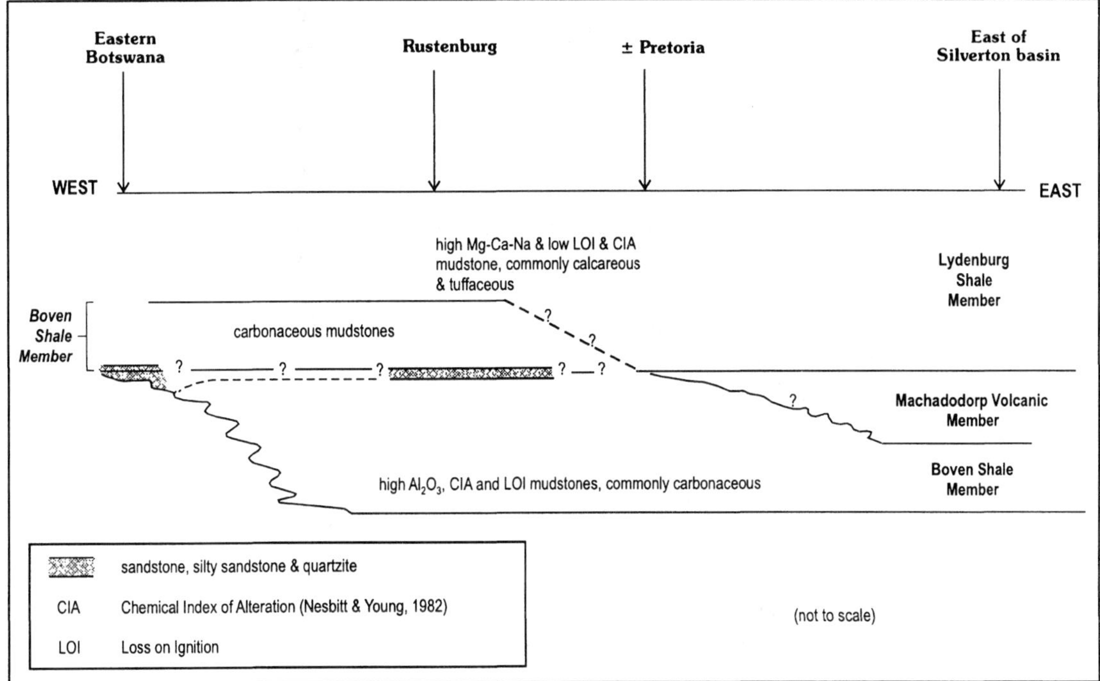


**Fig. 6.** Schematic (not to scale) east–west profile through the Silverton basin, illustrating a possible geochemically based stratigraphic subdivision applicable throughout the preserved basin (data from Reczko, 1994). Note basal arenaceous lithologies in the far west, recorded in boreholes Les1 and Strat2 (see also Fig. 7) of the basin.

Reczko, 1995). The Machadodorp volcanic rocks, up to 500 m in the east, comprise basal pyroclastic lithologies overlain by low potash tholeiitic basalts (Button, 1973; Reczko, 1994). Water-lain tuffs, locally pillowed basalts and minor cherts and carbonate rocks support subaqueous eruption (Button, 1973).

## Lithofacies

Lithofacies in the Silverton Formation can best be examined in the far west of the preserved basin. Here, despite the formation being thinner than in the east, two borehole (see Fig. 1 for location) cores are available for study at the Geological Survey of Botswana. Borehole Les1 starts in the lower Magaliesberg Formation and encompasses the upper part of the Silverton succession, whereas borehole Strat2 represents the lower part of the Silverton Formation. Detailed logging of the two cores (Eberhardt, 1997) suggests that the the basal part of Les1 and the upper part of Strat2 are stratigraphically approximately contiguous (Fig. 7).

### Arenaceous lithofacies association

#### *Description*

The lower 65 m of the Silverton Formation in borehole Strat2 encompasses an association of sandy and argillaceous lithofacies displaying rapid vertical variations (Fig. 8). The lower 36 m of this lithofacies association is characterized by inverse grading; within this interval, mudstone beds, 2–5 cm thick and each comprising laminated graphitic claystones passing up into cross-laminated, wavy laminated and normally graded quartz siltstones occur. These mudstone beds thicken (7–20 cm thick) and coarsen upwards to inversely graded fine sandstones (Fig. 8). The inversely graded beds have sharp to gradational lower contacts.

The lower, inversely graded interval is succeeded gradationally by 11.5 m of medium-grained cross-bedded, haematitic quartz sandstone beds, 30–40 cm thick, and containing internal normally graded quartz siltstone–claystone beds of 10–20 cm thickness. Upwards within the thickness of 11.5 m, the normally

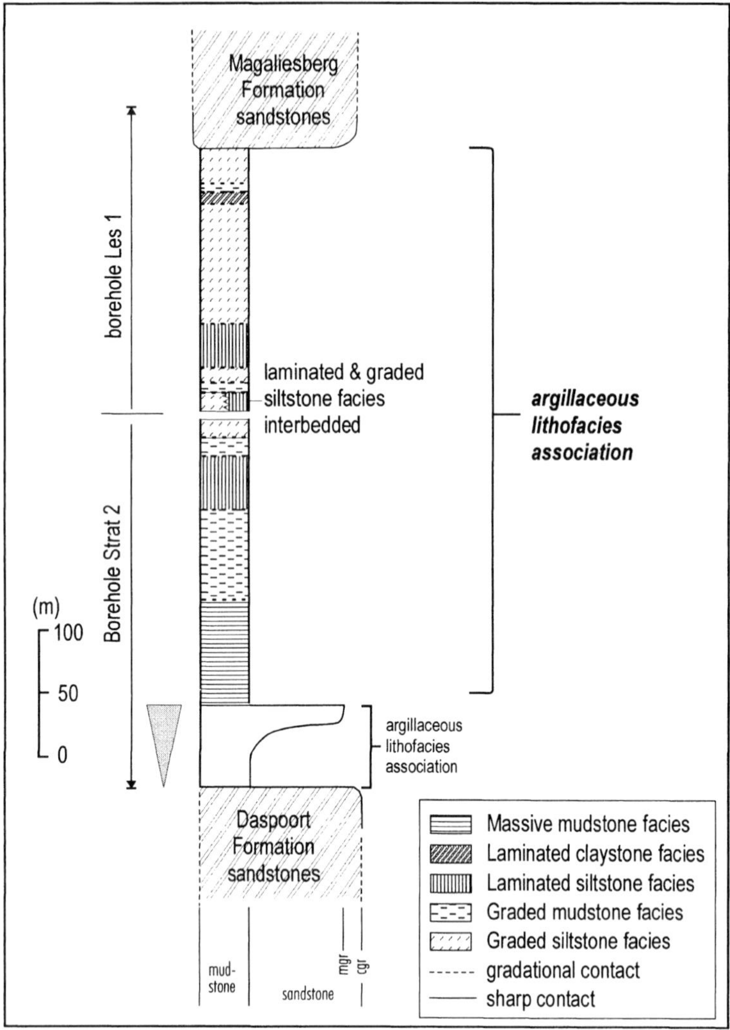


**Fig. 7.** Facies succession recorded in the two boreholes in eastern Botswana, considered to be stratigraphically approximately contiguous (Eberhardt, 1997). Note basal, relatively thin arenaceous lithofacies association, and overlying, predominant argillaceous lithofacies association. Modified after Eberhardt (1997).

graded internal beds thin, whereas the enclosing sandstones become fine-grained, more ferruginous and laminated; in the same direction, calcareous cement and micaceous matrix grains also increase within the sandstones, and isolated medium to coarse sand grains occur as well (Fig. 8).

Following 2 m of massive fine-grained sandstone, a 1.5 m thick succession of fine-grained normally graded oolitic ironstone beds, 8–35 cm thick and with basal pebbles of quartz and chert occurs (detailed section shown in Fig. 8). These basal pebbles resemble traction carpets and one of the graded ironstone beds has basal cobbles of ironstone, claystone and pyrite (Fig. 8). These pyrite cobbles themselves comprise goethitic/haematitic iron oolites embedded in pyrite; the rims of the oolites are not altered and the pyrite was thus most likely syngenetic (Reczko *et al.*, 1995). The oolites in the ironstones are up to 3 cm in diameter and polished sections reveal a polycyclic development of the oolitic grains. Normally graded medium- to coarse-grained sandstone–mudstone beds characterize the next 1.2 m above the 1.5 m thick oolitic ironstone interval (Fig. 8). The upper 13 m of the sandy facies association is made up of normally graded, mature, medium-coarse sandstone beds which alternate with massive calcareous sandstone beds (Fig. 8). The entire vertical succession of the sandy lithofacies association thus coarsens upwards and contacts

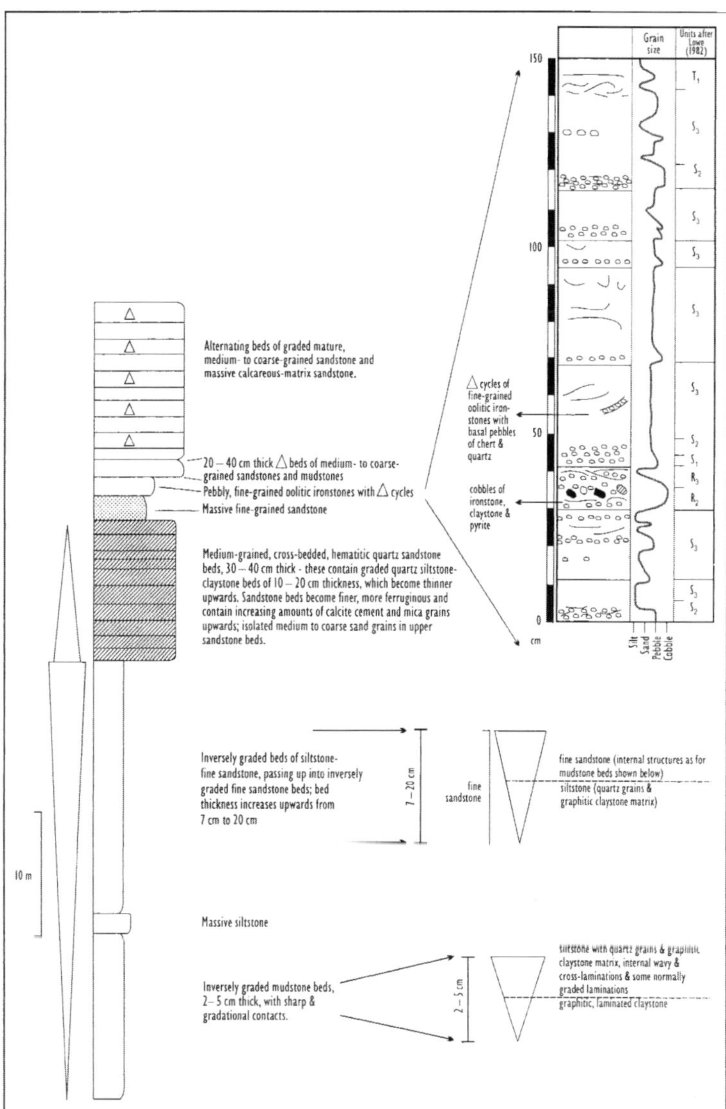


**Fig. 8.** Detailed sedimentary profile of the lower 65 m of borehole Strat2, illustrating the characteristics of the arenaceous lithofacies association of the Silverton Formation. Note common inversely graded beds in the lower 36 m interval, and, in contrast, the predominant normal grading above this. Detailed profile at top right shows 1.5 m interval of oolitic ironstone beds (inferred turbidite intervals, T, S and R, after Lowe, 1982).

between the facies tend to be gradational rather than sharp (Figs 7 and 8).

*Interpretation*

The lower predominant facies, inversely graded mudstones passing up into inversely graded fine sandstones, is problematic to interpret. Inverse grading is an uncommon sedimentary structure, and is often related to grain flows (e.g. Collinson & Thompson, 1982), where high dispersive pressure between the larger clasts may be responsible for smaller particles being displaced upwards, such as in density-modified grain flows (Lowe, 1982). However, in the Silverton facies discussed here, a debris flow origin is excluded owing to the laminated nature of the claystones and the cross-lamination and wavy lamination in the siltstones/sandstones together making up the inversely graded beds (Fig. 8); quite clearly, suspension sedimentation and low-energy traction current processes must have been operative.

Common inverse grading is typical of aeolian deposition (e.g. Hunter, 1977), and examples of this structure may also be found at the base of some coarse-grained turbidite beds (e.g. Stow *et al.*, 1996). It is well known that deltaic deposits often form upward-coarsening successions, from fine prodeltaic mudrocks, through silty and fine-sandy distal mouthbars, up into coarser sandstones of the proximal mouth bar and distributary channels, and that such successions may be repeated due to progradation of the deltaic system seawards (e.g. Reading & Collinson, 1996). Owing to the characteristic braided nature of Palaeoproterozoic fluvial systems (e.g. Eriksson *et al.*, 1998), a large number of channels of various sizes would have flowed into seas such as the Silverton epeiric basin, as is apparent for the coeval Magaliesberg braid-delta systems (Eriksson *et al.*, 1995). It is thus possible that the lower 36 m of the arenaceous lithofacies association identified in the Silverton borehole represents stacking of thin upward-coarsening cycles of distal braid-deltaic mudrocks and fine sandstones, formed as the Silverton epeiric sea transgressed across the Kaapvaal craton and slowly drowned the Daspoort braided rivers (Eriksson *et al.*, 1993) systems. The increasing sediment calibre and greater inversely graded bed thickness observed with increasing stratigraphic height in this lower part of the Silverton Formation presumably reflect a relative increase in energy of numerous small prograding braid-delta complexes debouching into the shallow advancing epeiric sea.

The medium- to coarse-grained sandstones that make up most of the upper portion of the lower 65 m of the Silverton Formation (Fig. 7) are typically either graded themselves or contain thinner graded siltstone–claystone beds (Fig. 8). These sandstones are compatible with the deposits of high-density, coarse-grained turbidity currents; successive pulses of surging turbidity flows could have produced the discrete graded beds observed in the lower Silverton Formation (Lowe, 1982). The cross-bedded sandstones in the arenaceous lithofacies association (Fig. 8) probably reflect lower flow regime sandy ripple and megaripple bedform migration, equivalent to the B interval of Bouma's (1962) medium-grained turbidites. Massive sands, as found interbedded with the above Silverton graded arenitic beds, are commonly found associated with turbidite deposits (Stow *et al.*, 1996).

Similar surging, high-density turbidity current deposition is inferred for the 1.5 m thick interval of oolitic ironstone beds (detailed profile in Fig. 8; note also T, S and R turbidite intervals after Lowe (1982)). This interpretation is supported by the inferred traction carpet layers, cross-lamination and graded bedding in this interval (Reczko *et al.*, 1995). Syngenetic freezing of goethitic/haematitic ooids in rapidly precipitating pyrite to form the pyrite cobbles observed at the base of one of the ironstone beds (Fig. 8) was most probably related to localized exhalative processes, occurring at fault intersections along the western margin of the Silverton basin (Reczko, 1994). The goethitic/haematitic oolites probably formed along the transgressive margin of the inferred epicontinental Silverton sea and thus could have marked transgressive surfaces in the preserved rock record. However, sulphide cobbles and ferruginous oolites are assumed to have been redeposited in deeper water by high-density turbidity currents as the Silverton epeiric sea advanced westwards and rapidly drowned (Catuneanu & Eriksson, 1999; Fig. 3) the braid-deltaic inversely graded deposits discussed above. Highly mature reworked arenites with bipolar palaeocurrent trends occur in the upper Daspoort Formation in the east of the Pretoria Group basin (Eriksson *et al.*, 1993); they support an interpretation that preceding Daspoort fluvial sandstones were drowned as the Silverton epeiric sea transgressed on to the Kaapvaal craton, with an active coastline advancing rapidly westwards to include the present-day western preserved margin of the basin where boreholes Les1 and Strat 2 were drilled.

## Argillaceous lithofacies association

### *Description*

These facies make up the rest of the Silverton succession recorded in the two boreholes and comprise massive mudstones, laminated claystones, laminated siltstones, graded mudstones and graded siltstones (Fig. 7). Thick (86 m), apparently massive mudstones (Fig. 9) immediately follow the arenaceous facies association discussed above; the contact zone in the core is brecciated and thus of uncertain affinity. These mudstones comprise carbonaceous and pyritic quartz siltstone and claystone, with minor interbeds, 5–10 cm thick, of laminated quartz siltstone. The pyrite occurs primarily as intergranular cement. The laminated claystone facies has laminae of 1–5 mm thickness and forms a 9 m thick unit near the top of the Silverton succession (Fig. 7). The laminated siltstone facies is characteristic of the medial part of the succession where two intervals of *c.*45 m thickness occur (Fig. 7); these siltstones and minor fine-grained sandstone laminae comprise clasts of quartz as well as calcareous grains set within a graphitic matrix.

The graded siltstone facies occurs in the upper half of the Silverton succession where intervals of 20–91 m thickness are found (Fig. 7). These beds of siltstone

**Fig. 9.** Massive mudstone facies from an outcrop in the east of the Silverton basin, analogous to that described from the boreholes to the west. Note spheroidal weathering, a common feature of field occurrences of this facies.

**Fig. 10.** Graded siltstone facies, illustrated from an outcrop in the Pretoria region, and analogous to that described from the western boreholes. Note predominant thin graded beds, a few centimetres in thickness (pen for scale), with tops of individual beds marked by more voluminous dark coloured graphitic matrix material. Also note thicker bed of fine-grained sandstone, about 20 cm thick, in the middle of the photograph.

and subordinate fine-grained sandstone are mostly a few centimetres to 10 cm thick, uncommonly up to 20 cm (Fig. 10), and comprise mainly calcareous clasts and lesser quartz grains within a micaceous to graphitic matrix, as well as minor calcite and haematite cement. Many of the beds are sharply based, some with obvious scoured bases, and internal structures in the beds include soft sediment deformation and compaction features, cross-lamination and lenticular lamination.

The graded mudstone facies, spread throughout the Silverton succession in the boreholes, with intervals from 8 to 71 m in thickness (Fig. 7), comprises calcareous and quartz siltstones (matrix graphitic with pyritic cement) which pass up into graphitic mudstones. These graded beds and laminae vary from a few millimetres to 20 cm in thickness (Fig. 11), contain similar internal structures to the graded siltstone facies and are separated locally by 1–5 cm thick graphitic mudstones.

*Interpretation*

In view of the Silverton Formation being bounded by underlying largely braided fluvial sandstones of the Daspoort Formation (Eriksson *et al.*, 1993) and by overlying Magaliesberg sandstones ascribed to regressive tidally reworked braid-delta complexes (Eriksson *et al.*, 1995), large quanitities of fine fluvial detritus would have been supplied to the Silverton basin. Enhanced weathering regimes, typical of the early Precambrian due to atmospheric composition (Corcoran *et al.*, 1998), and rapid erosion in the absence of land vegetation, would have promoted large amounts of fine fluvial detritus, transported in enormous braided systems (Eriksson *et al.*, 1998). Swift *et al.* (1991) recognize supply-dominated shelves, analogous to the Silverton epeiric sea, and accommodation-dominated shelf settings. For supply-dominated shelves, terrestrial sediment supply via rivers and deltas exceeds the combination of accommodation space creation and sediment dispersal (Johnson & Baldwin, 1996); a high concentration of suspended load is needed for muddy sediment to be laid down on the shelf, with >150 mg $l^{-1}$ probably being required (Friedman *et al.*,

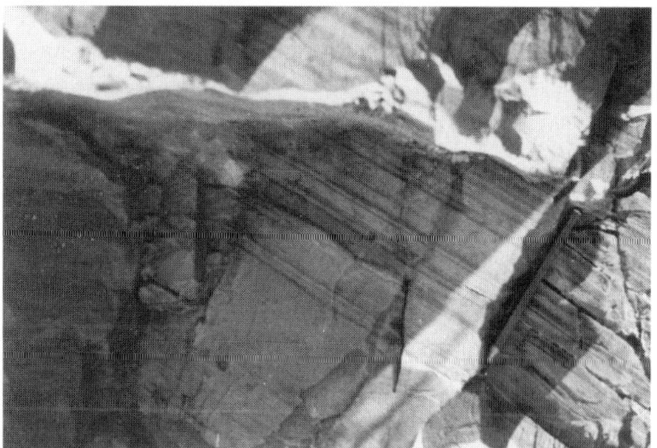

**Fig. 11.** Field outcrop (Pretoria region) of graded mudstone facies, similar to that described from western boreholes. Individual graded beds and laminae separated by thin, dark coloured graphitic mudstones; note variable thickness of graded beds, between a few millimetres and several centimetres (pencil for scale).

1992, p. 391). However, in the Palaeoproterozoic, with an inferred absence of filter-feeding invertebrates to remove significant quantities of mud from the water column, lower concentrations of suspended load may have sufficed for mud to escape into deeper water (B.M. Simonson, personal communication, 2000). On modern supply-dominated shelf settings, fine fluviodeltaic detritus will by-pass the high-energy coastal sand belt and be deposited on the medial to outer shelf regions, as, for example, found in the modern shelf off the Amazon River (e.g. Kuehl *et al.*, 1986); the offshore transport is dominated by river flood events, aided by rip currents, ebb-tidal currents and downwelling storm currents (Johnson & Baldwin, 1996). Thick successions (tens to hundreds of metres) of essentially homogeneous fine sediments will accumulate on supply-dominated shelves as a result, similar to the facies succession observed in the two Silverton boreholes (Fig. 7).

A transition zone on the shelf environment, between coastal sands and offshore muds, at an average depth of 8–10 m, is characterized by clayey silt and silty clay (Reineck & Singh, 1975), analogous to the predominant facies in the Silverton argillaceous association. Thin interbedded silts or fine sands in these predominant muds, characterized by graded bedding or appearing as graded rhythmites, and ascribed to storm suspension processes (e.g. Hayes, 1967; Reineck & Singh, 1975), strongly resemble the graded mudstone and graded siltstone facies of the Silverton Formation. Outer shelf deposits, both Precambrian and Phanerozoic, are typically dominated by massive or laminated mudstones, comparable to the massive mudstone, laminated claystone and laminated siltstone facies of the Silverton succession (Fig. 7), with subordinate graded siltstone to fine sandstone storm beds (e.g. Schieber, 1989; Chakraborty & Bose, 1992; Tirsgaard & Sønderholm, 1997; Eriksson *et al.*, 1998). Whether the massive mudstone facies of the Silverton Formation is really primarily massive or only apparently so, due to secondary processes such as diagenesis (bioturbation being excluded due to the age of these rocks), is uncertain; outcrops of this facies tend to be spheroidally weathered (Fig. 9), which, although not proving a primary massive nature, is compatible with it.

Applying conventional wisdom, the common presence of graphitic matrix material and pyrite cement in the Silverton mudrocks also supports below wave base suspension sedimentation within a generally anoxic and low-energy palaeoenvironment. However, modern shallow water examples of graphitic–pyritic mudrocks are known (e.g. Devonian Chattanooga Shale, North America; J. Schieber, personal communication, 2000), and in the Precambrian, with significantly less oxygen in the atmosphere, carbonaceous mudrocks may have been likely to accumulate in shallow water. The graded storm beds in muddy shelf settings may also exhibit cross-lamination and are often sharply based or even erosional (Tirsgaard & Sønderholm, 1997); similar features occur within the argillaceous facies association of the Silverton Formation, and may be ascribed to unidirectional bottom currents related to storm events (e.g. Walker & Plint, 1992). The apparent absence of wave-formed structures in the Silverton mudrock facies supports deposition having taken place beneath storm wave base.

The argillaceous facies association of the Silverton Formation, as identified in the two boreholes Les1 and Strat2, thus supports outer shelf mud-type low-energy deposition of abundant fluvial fines with intermittent graded storm beds and laminae also being preserved. This below wave base facies association suggests that rapid drowning of the Silverton epeiric coastline (arenaceous facies association, above) occurred and that the argillaceous facies association reflects the transition from a transgressive systems tract to a highstand condition (Fig. 3). The tidally reworked braid-delta sandstones of the Magaliesberg Formation represent the coeval fluvial feeder systems supplying abundant fine sediment to the adjacent Silverton epeiric basin. Coarse Magaliesberg fluvial sands were trapped in the high-energy coastal belt by strong tidal currents and storm wave activity.

## Basinal trends

Although there are no further borehole cores available for study from elsewhere in the preserved Silverton Formation, limited outcrop data, described by Button (1973), Schreiber (1990) and van der Neut (1990), exist for the central and eastern parts of the basin. Although argillaceous lithofacies, analogous to those described here, account for most outcrops examined, there are subtle differences across the basin. Within the south-central portion, around Pretoria (Fig. 1), van der Neut (1990) notes that channel-fills of laminated to cross-laminated mudrocks, ripple marks and interbedded sandstone lenses and thin beds occur within massive, laminated and graded mudrock facies. These combined features suggest the action of directed, eroding bottom currents, and deposition depths closer to storm wave base than the facies from the Botswana boreholes examined here (Fig. 7). The bottom currents may reflect downwelling storm processes, ebb-tidal

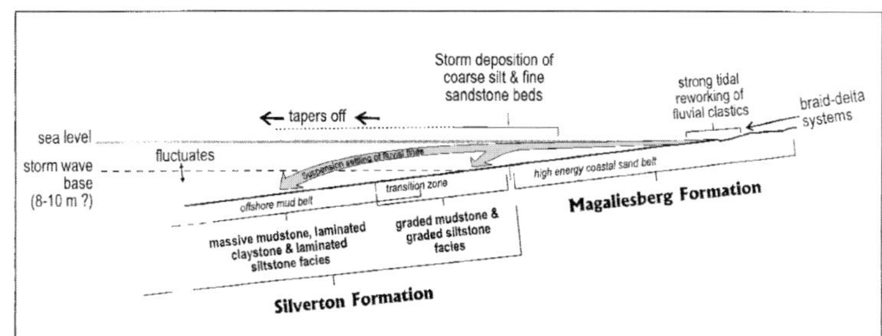


**Fig. 12.** Sedimentary model postulated for the Silverton Formation epeiric sea. See text for discussion.

currents, rip currents or, possibly, thermohaline geostrophic current action (e.g. Eriksson & Reczko, 1998).

In contrast, outcrops from the east of the preserved basin exhibit carbonate rocks and cherts associated with predominant graded, laminated and massive mudrock facies, and also more common high-graphite mudrocks (Button, 1973; Schreiber, 1990). These attributes, together with the greatly enhanced thickness of the Silverton Formation in the east (Figs 5 & 6), point to an outer shelf-like depositional realm further removed from fluvial feeder systems than those in the west of the basin. As Daspoort sedimentation, which preceded Silverton deposition, is characterized by eastern shallow marine sandstones, coeval with western fluvial deposits (Eriksson *et al.*, 1993), a westerly fluvial source for the Silverton is also assumed. Magaliesberg regressive coastal sandstones, interpreted as a lateral facies equivalent of deeper water Silverton sediments, show a similar western provenance (Eriksson *et al.*, 1995). With less clastic input and, most likely, slower sedimentation rates further from the fluvial feeder systems situated in the west, chemical sedimentation (carbonates and chert) became more prominent towards the east. Greater subsidence presumably accommodated more pelagic material and subordinate graded storm beds. The calcareous siltstones and fine sandstones, which are common in the facies identified in the western boreholes, are fully clastic rock types derived from Chuniespoort Group dolomitic source areas, and should not be confused with the chemically/biochemically precipitated carbonate muds and marls in the east. The latter also exhibit vague stromatolitic structures (Button, 1973), thereby suggesting relatively shallow depths, above the photic limit, albeit predominantly below storm wave base.

## Depositional model

A shallow epeiric sea model is proposed for the Silverton Formation (Fig. 12). Approximately coeval sheet sandstones of the Magaliesberg Formation were laid down by widespread braid-delta systems which debouched into the Silverton basin from an extensive western hinterland. Strong tidal reworking occurred in the braid-delta channels (Eriksson *et al.*, 1995), thus forming a higher energy sand belt above storm wave base at the margins of the Silverton basin. Aggressive Precambrian weathering, allied to a lack of terrestrial vegetation, provided high concentrations of fine fluvial sediment, which by-passed the high-energy coastal sand belt through a combination of river flood events, rip, ebb-tidal and downwelling storm (wave) currents, and laid down a thick succession of argillaceous sediments, essentially below wave base (Fig. 12). Graded mudstone and graded siltstone facies of the Silverton Formation were deposited from suspension in a transition zone between the coastal sand belt and an offshore mud belt. The latter accumulated massive mudstone, laminated claystone and laminated siltstone facies as outer shelf-like deposits in the Silverton basin (Fig. 12). Subordinate, thin-graded storm beds/laminae of fine sandstone and coarse siltstone occur within the argillaceous Silverton facies, attesting to intermittent storm wave deposition, probably reflect a fluctuating depth to storm wave base in the transition zone. Storm wave base probably fluctuated owing to the varying combination of wave lengths and wave periods (Friedman *et al.*, 1992).

The model proposed (Fig. 12) is based largely on the borehole cores examined from the far western portion of the preserved Silverton basin. Outcrops in the south-central part of this depository suggest deposition closer to storm wave base, whereas those from the east of the basin point to deeper water outer shelf-like conditions, further removed from the fluvial feeder systems. Greater subsidence in the east led to a much thicker accumulation of pelagic beds there than in the west (Fig. 5).

Silverton deposition was preceded by largely fluvial sedimentation of the Daspoort Formation. More mature Daspoort sandstones with bimodal and polymodal palaeocurrent trends in the east of the preserved basin support transgressive marine reworking of fluvial detritus as Silverton epeiric transgression began. In the west of the basin, in the boreholes examined, initial Silverton sandy deposits (Fig. 8) point to coastline deposition, largely by numerous small braid-deltaic systems which formed inversely graded beds. Subsequent rapid drowning of this prograding braid-delta coastline led to the formation of coarse, sandy, high-density turbidite deposits, including re-sedimented ferruginous oolites and pyritic cobbles. The oolites probably developed along a trangressive surface and were redeposited in deeper water by turbidity currents as the epeiric basin deepened. Subsequent deposition of predominant mudrocks, below wave base, formed the later transgressive and high-stand facies tracts of the Silverton Formation (Fig. 3). Coeval tidally reworked braid-delta Magaliesberg sandstones reflect a regressive shoreline which supplied fluvial fines for offshore suspension sedimentation, and which gradually advanced eastwards as the Silverton epeiric sea retreated off the Kaapvaal craton.

## DISCUSSION

The small scale of the preserved Silverton basin, approximately 500 by 300 km (Fig. 1), compares poorly with the vast epeiric seas of the Phanerozoic, such as the Late Cretaceous seaway of North America, which measured approximately 5000 by 1400 km at maximum transgression (Reading, 1978). The Silverton sea was thus probably an epeiric embayment rather than a seaway.

Water depths in the Silverton embayment are likely to have been shallower than those inferred for the Cretaceous seaway (Fig. 12). Although it is difficult to estimate accurate water depths in such settings, common black, pyritic mudrocks in the west of the Silverton basin (boreholes examined here) are compatible with sub-wave base deposition and point to anoxic deeper waters; however, as pointed out above, there are Phanerozoic examples of pyritic black shales which formed in depths significantly less than 100 m, and even shallower depths would probably have been possible in the Precambrian with less oxygen in the atmosphere. Locally stromatolitic carbonates and cherts in the east of the basin (Button, 1973) suggest photic water depths between approximately 100 and 30 m (Eriksson & Reczko, 1998). Unlike the Timeball Hill Formation, where evidence for common turbidite deposits and contour currents supports depths close to 100 m (Eriksson & Reczko, 1998) (Fig. 4), most of the Silverton Formation (possibly excluding the lower inferred turbidite deposits) was probably laid down nearer the upper depth limit envisaged above.

At such shallow depths, normal waves would have become dampened within about 100 km from the open ocean (Eriksson & Reczko, 1998) and the Silverton epeiric embayment would have been subject to less frequent storm wave action. Storm wave base fluctuates according to combined changes in wave length and wave period (Friedman *et al.*, 1992), with waves having the greatest lengths and amplitudes deepening effective storm wave base (Reading & Collinson, 1996). The middle shelf or transition zone (Swift & Thorne, 1991) and outer shelf zones are affected by storm waves at depths of approximately >10 m and >20 m, respectively (Johnson & Baldwin, 1996). The relative scarcity of thin-graded coarse siltstone and fine sandstone beds in the predominant Silverton mudrock facies suggests either that large storm waves were uncommon in the epeiric palaeoenvironment, or that water depths were in excess of the 10–20 m proposed above. Much of the Silverton basin was thus probably analogous to a middle to outer shelf setting, and characterized by predominant pelagic settling of fluvial fine material, with limited storm wave deposition of coarse silt and fine sandstone beds (Fig. 12). As such relatively shallow depths (<30 m?) appear to have applied over the central and eastern parts of the preserved basin too, the Silverton epeiric basin floor must have been characterized by a very shallow gradient.

Comparing the inferred Silverton model with that of Shaw (1964) and Irwin (1965) (Fig. 2), a very extensive X zone probably characterized the Silverton, with the Z zone not being preserved in the present basin. The high-energy Y zone in this classical model compares favourably with the tidal braid-delta setting inferred for the Magaliesberg Formation sandstones, thought to be the coeval high-energy sand belt and coastline of the Silverton Formation (Fig. 12). Strong tidal currents reworked the Magaliesberg braid-delta sandstones (Eriksson *et al.*, 1995). The preserved stacking pattern, of regressive Magaliesberg shoreline sandstones overlying offshore, transgressive to highstand-type epeiric mudrocks of the Silverton Formation (Fig. 3), resulted from retreat of the epeiric embayment off the Kaapvaal craton. This retreat may well reflect thermal elevation of the central Kaapvaal craton towards the end of Transvaal Supergroup sedimentation, as the Bushveld magma plume (intrusion at *c*.2.05 Ga; Eriksson & Reczko, 1995) became active (Catuneanu & Eriksson, 1999).

The model proposed for the Silverton epeiric embayment (Fig. 12) thus strongly resembles a modern clastic, supply-dominated shelf (Swift *et al.*, 1991), but with one significant difference, namely a scarcity of fair weather wave activity. However, with the broad shelf palaeoenvironment envisaged for the Silverton deposition, tidal range would have been enhanced and preservation of fair weather wave deposits would be unlikely (e.g. Hayes, 1979). In the absence of significant fair weather wave activity, strong tidal currents characterized the high-energy sand belt and coastline, and would have reworked the storm wave beach deposits that formed intermittently. In a modern shelf setting, tides would generally predominate in the coastal zone where barrier islands absorbed fair weather wave energy, leaving tidal lagoonal settings behind them. A general lack of preserved barrier island deposits is noted in many Precambrian shallow marine palaeoenvironments (Eriksson *et al.*, 1998). As tidally influenced epeiric seas appear to have been much more common in the Precambrian (Eriksson *et al.*, 1998; Eriksson & Reczko, 1998), probably reflecting the influence of continental crustal growth (and therefore freeboard) on eustacy (e.g. Eriksson, 1999), this provides a possible explanation for the perceived paucity of Precambrian barrier island deposits.

The Silverton epeiric embayment model (Fig. 12) may thus be added to the classic Shaw–Irwin epeiric sea model, as reflecting a common Precambrian permutation thereof. The Silverton model was probably more common in this regard than that proposed for the Timeball Hill Formation (Fig. 4) of the Transvaal Supergroup. The latter epeiric sea has a preserved size compatible with an embayment, but facies indicate water depth up to *c*.100 m and hence support an alternative seaway interpretation. However, in view of the Timeball Hill epeiric sea being coeval with the first global glaciation known (Eyles & Young, 1994), eustatic sea levels would logically have been reduced and large-scale flooding of the Kaapvaal craton unlikely. This suggests that that main control on the Timeball Hill transgression (Fig. 3) was structural rather than eustatic. The Timeball Hill epeiric palaeoenvironment (Fig. 4), with evidence for deeper water, and a combination of turbidite, pelagic, contourite and deltaic sedimentation (Eriksson & Reczko, 1998) within a shallow marine setting, is thus probably not a widely applicable epeiric sea model. The Silverton model (Fig. 12) discussed here appears to have a much wider relevance for the early Precambrian, when the interplay between continental freeboard, continental crustal growth rates and eustacy is thought to have been important (Eriksson, 1999).

## CONCLUSIONS

The Silverton Formation comprises a predominant argillaceous facies association of massive mudstones, laminated claystones, laminated siltstones, graded mudstones and graded siltstones, and a subordinate arenaceous lithofacies association. The latter is ascribed to braid-deltaic and turbidity current deposition under transgressive conditions. The argillaceous facies are interpreted as sub-storm wave base pelagic deposits, laid down in an epeiric embayment under transgressive and highstand conditions. The muddy detritus laid down in the transitional and offshore mud belt zones of this shallow epicontinental sea was derived from fine fluvial sediment which was able to by-pass a high-energy coastal sand belt, represented by the Magaliesberg Formation which overlies the Silverton rocks. Storm wave action locally produced graded coarse siltstone and fine sandstone beds.

The model suggested for the Silverton Formation epeiric embayment (Fig. 12) compares reasonably with the Shaw–Irwin classic model applied to Phanerozoic epeiric seaways, but with some significant differences. In the Silverton model, the inshore low-energy Z zone (Fig. 2) was either absent or not preserved, and a strongly tidal coastline (Y zone) appears to have been present much of the time. The Silverton epeiric model may also be favourably compared to a modern clastic, supply-dominated shelf, but with the difference that fair weather wave action was limited. As a result, strong tidal action reworked fair weather wave deposits and the development of wave-built barrier islands would have been uncommon. The depositional environment proposed for the Silverton Formation was probably quite widespread in the early Precambrian, when extensive inundation of continental margins, thereby forming epeiric embayments, occurred owing to enhanced continental crustal growth rates.

## ACKNOWLEDGEMENTS

The authors are grateful to RTZ (Africa Division) for generous financial and logistical assistance during field work in the Silverton basin; we are particularly grateful to Scott Jenkins and Rob Taylor in this regard. We also thank the staff of the Geological

Survey of Botswana, who facilitated examination of the cores from boreholes Les1 and Strat2; Roger Key and Read Mapeo were particularly helpful in this. Boris Reczko is thanked for his advice, especially on the geochemical stratigraphy of the Silverton Formation. P.G.E. is very grateful to the National Research Foundation of South Africa, and the University of Pretoria for research funding. Referees Bruce Simonson and Jürgen Schieber are acknowledged for their insight and thanked for their very constructive advice. Book editors Patricia Corcoran and Wlady Altermann were towers of strength throughout.

## REFERENCES

ARMSTRONG, R.L. (1981) Radiogenic isotopes: the case for crustal recycling on a near-steady-state no-continental-growth Earth. *Phil. Trans. R. Soc. London*, **A301**, 443–472.

ARNDT, N. (1999) Why was submarine volcanism on submerged continental platforms so common in the Precambrian? *Precam. Res.*, **97**, 155–164.

BEUKES, N.J. (1987) Facies relations, depositional environments and diagenesis in a major Early Proterozoic stromatolitic carbonate platform to basinal sequence, Campbellrand Subgroup, Transvaal Supergroup, Southern Africa. *Sediment. Geol.*, **54**, 1–46.

BOUMA, A.H. (1962) *Sedimentology of Some Flysch Deposits: a Graphic Approach to Facies Interpretation*. Elsevier, Amsterdam, 168 pp.

BOUMA, A.H., BRENNER, R.L. & KNEBEL, H.J. (1982) Continental shelf and epicontinental seaways. In: *Sandstone Depositional Environments* (Eds Scholle, P.A. & Spearing, D.), pp. 281–327. American Association of Petroleum Geologists, Tulsa.

BRENNER, R.L. (1978) Sussex sandstone of Wyoming: an example of Cretaceous offshore sedimentation. *Bull. Am. Ass. Petrol. Geol.*, **62**, 181–200.

BRENNER, R.L. (1979) A sedimentologic analysis of the Sussex Sandstone, Powder River Basin, Wyoming. *Earth Sci. Bull.*, **12**, 36–47.

BRENNER, R.L. (1980) Construction of process–response models for ancient epicontinental seaway depositional systems using partial analogs. *Bull. Am. Ass. Petrol. Geol.*, **64**, 1223–1244.

BRENNER, R.L. & DAVIES, D.K. (1974) Oxfordian sedimentation in Western Interior United States. *Bull. Am. Ass. Petrol. Geol.*, **58**, 444–467.

BUTTON, A. (1973) *A regional study of the stratigraphy and development of the Transvaal Basin in the eastern and northeastern Transvaal.* PhD thesis, University of the Witwatersrand, Johannesburg.

CATUNEANU, O. & ERIKSSON, P.G. (1999) The sequence stratigraphic concept and the Precambrian rock record: an example from the 2.3–2.1 Ga Pretoria Group, Kaapvaal craton. *Precam. Res.*, **97**, 215–251.

CHAKRABORTY, C. & BOSE, P.K. (1992) Rhythmic shelf storm beds: Proterozoic Kaimur Formation, India. *Sediment. Geol.*, **77**, 259–268.

COLLINSON, J.D. & THOMPSON, D.B. (1982) *Sedimentary Structures*. George Allen & Unwin, London, 194 pp.

CORCORAN, P.L., MUELLER, W.U. & CHOWN, E.H. (1998) Climatic and tectonic influences on fan deltas and wave- to tide-controlled shoreface deposits: evidence from the Archaean Keskarrah Formation, Slave Province, Canada. *Sediment. Geol.*, **120**, 125–152.

EBERHARDT, L. (1997) *Die frühproterozoischen Gesteine der Transvaal Supergruppe im nordwestlichen Teil des Transvaal Beckens in Südafrika und Botswana.* MSc thesis, Ludwig-Maximilians University, Munich.

ERIKSSON, K.A. (1995) Crustal growth, surface processes, and atmospheric evolution on the early Earth. In: *Early Precambrian Processes* (Eds Coward, M.P. & Ries, A.C.), Spec. Publs geol. Soc. London, No. 95, pp. 11–25. Geol. Soc. London, Bath.

ERIKSSON, P.G. (1999) Sea level changes and the continental freeboard concept: general principles and application to the Precambrian. *Precam. Res.*, **97**, 143–154.

ERIKSSON, P.G. & ALTERMANN, W. (1998) An overview of the geology of the Transvaal Supergroup dolomites (South Africa). *Environ. Geol.*, **36**(1/2), 179–188.

ERIKSSON, P.G. & RECZKO, B.F.F. (1995) The sedimentary and tectonic setting of the Transvaal Supergroup floor rocks to the Bushveld Complex. *J. Afr. Earth Sci.*, **21**, 487–504.

ERIKSSON, P.G. & RECZKO, B.F.F. (1998) Contourites associated with pelagic mudrocks and distal delta-fed turbidites in the Lower Proterozoic Timeball Hill Formation epeiric basin (Transvaal Supergroup), South Africa. *Sediment. Geol.*, **120**(1–4), 319–335.

ERIKSSON, P.G., SCHREIBER, U.M., VAN DER NEUT, M., LABUSCHAGNE, H., VAN DER SCHYFF, W. & POTGIETER, G. (1993) Alternative marine and fluvial models for the non-fossiliferous quartzitic sandstones of the Early Proterozoic Daspoort Formation, Transvaal Sequence of southern Africa. *J. Afr. Earth Sci.*, **16**, 355–366.

ERIKSSON, P.G., RECZKO, B.F.F., BOSHOFF, A.J., SCHREIBER, U.M., VAN DER NEUT, M. & SNYMAN, C.P. (1995) Architectural elements from Lower Proterozoic braid-delta and high energy tidal flat deposits in the Magaliesberg Formation, Transvaal Supergroup, South Africa. *Sediment. Geol.*, **97**, 99–117.

ERIKSSON, P.G., CONDIE, K.C., TIRSGAARD, H. *et al.* (1998) Precambrian clastic sedimentation systems. *Sediment. Geol.*, **120**(1–4), 5–53.

ERIKSSON, P.G., MAZUMDER, R., SARKAR, S., BOSE, P.K., ALTERMANN, W. & VAN DER MERWE, R. (1999) The 2.7–2.0 Ga volcano-sedimentary record of Africa, India and Australia: evidence for global and local changes in sea level and continental freeboard. *Precam. Res.*, **97**, 269–302.

EYLES, N. & YOUNG, G.M. (1994) Geodynamic controls on glaciation in Earth history. In: *Earth's Glacial Record* (Eds Deynoux, M., Miller, J.M.G., Domack, E.W., Eyles, N., Fairchild, I.J. & Young, G.M.), pp. 1–28. Cambridge University Press, Cambridge.

FRIEDMAN, G.M., SANDERS, J.E. & KOPASKA-MERKEL, D.C. (1992) *Principles of Sedimentary Deposits*. Macmillan, New York, 717 pp.

HAMBREY, M.J. & HARLAND, W.B. (1985) The late Proterozoic glacial era. *Palaeogeogr. Palaeoclimatol. Palaeoecol.*, **51**, 255–272.

HAYES, M.O. (1967) *Hurricanes as Geological Agents: Case Studies of Hurricanes Carla, 1961, and Cindy, 1963*. Report of Investigation, Bureau of Economic Geology, Austin, TX, **61**, 54 pp.

Hayes, M.O. (1979) Barrier island morphology as a function of tidal and wave regime. In: *Barrier Islands—from the Gulf of St Lawrence to the Gulf of Mexico* (Ed. Leatherman, S.P.), pp. 1–27. Academic Press, New York.

Hunter, R.E. (1977) Basic types of stratification in small eolian dunes. *Sedimentology*, **24**, 361–387.

Irwin, M.L. (1965) General theory of epeiric clear water sedimentation. *Bull. Am. Ass. Petrol. Geol.*, **49**, 445–459.

Johnson, H.D. & Baldwin, C.T. (1996) Shallow clastic seas. In: *Sedimentary Environments: Processes, Facies and Stratigraphy*, 3rd edn (Ed. Reading, H.G.), pp. 232–280. Blackwell Science, Oxford.

Klein, G. de V. & Ryder, T.A. (1978) Tidal circulation patterns in Precambrian, Paleozoic, and Cretaceous epeiric and mioclinal shelf areas. *Bull. Geol. Soc. Am.*, **89**, 1050–1058.

Kuehl, S.A., DeMaster, D.J. & Nittrouer, C.A. (1986) Nature of sediment accumulation on the Amazon continental shelf. *Continent. Shelf Res.*, **6**, 209–225.

Lowe, D.R. (1982) Sediment gravity flows: II. Depositional models with special reference to the deposits of high-density turbidity currents. *J. sediment. Petrol.*, **52**, 279–297.

Pratt, B.R. & James, N.P. (1986) The St George Group (Lower Ordovician) of western Newfoundland: tidal flat island model for carbonate sedimentation in shallow epeiric seas. *Sedimentology*, **33**, 313–343.

Reading, H.G. (Ed.) (1978) *Sedimentary Environments and Facies*. Blackwell, Oxford, 557 pp.

Reading, H.G. & Collinson, J.D. (1996) Clastic coasts. In: *Sedimentary Environments: Processes, Facies and Stratigraphy*, 3rd edn (Ed. Reading, H.G.), pp. 154–231. Blackwell Science, Oxford.

Reczko, B.F.F. (1994) *The geochemistry of the sedimentary rocks of the Pretoria Group, Transvaal Sequence*. PhD thesis, University of Pretoria, Pretoria.

Reczkó, B.F.F., Eriksson, P.G. & Snyman, C.P. (1995) Some evidence for the base-metal potential of the Pretoria Group: stratigraphic targets, tectonic setting and REE patterns. *Miner. Deposita*, **30**, 162–167.

Reineck, H.-E. & Singh, I.B. (1975) *Depositional Sedimentary Environments*. Springer-Verlag, Berlin, 439 pp.

Schieber, J. (1989) Facies and origin of shales from the Mid-Proterozoic Newland Formation, Belt basin, Montana, USA. *Sedimentology*, **36**, 203–219.

Schopf, T.J.M. (1980) *Paleoceanography*. Harvard University Press, Cambridge, MA, 341 pp.

Schreiber, U.M. (1990) *A palaeoenvironmental study of the Pretoria Group in the eastern Transvaal*. PhD thesis, University of Pretoria, Pretoria.

Shaw, A.B. (1964) *Time in Stratigraphy*. McGraw-Hill, New York, 365 pp.

Spearing, D.R. (1976) Upper Cretaceous Shannon Sandstone: an offshore shallow-marine sand body. *Ann. Guidebook Wyoming geol. Ass.*, **28**, 65–72.

Stow, D.A.V., Reading, H.G. & Collinson, J.D. (1996) Deep seas. In: *Sedimentary Environments: Processes, Facies and Stratigraphy*, 3rd edn (Ed. Reading, H.G.), pp. 395–453. Blackwell Science, Oxford.

Swett, K., Klein, G. de V. & Smit, D.E. (1971) A Cambrian tidal sand body—the Eriboll Sandstone of Northwest Scotland: an ancient–recent analog. *J. Geol.*, **79**, 400–415.

Swift, D.J.P., Philips, S. & Thorne, J.A. (1991). Sedimentation on continental margins, IV: lithofacies and depositional systems. In: *Shelf Sand and Sandstone Bodies: Geometry, Facies and Sequence Stratigraphy* (Eds Swift, D.J.P., Oertel, G.F., Tilman, R.W. & Thorne, J.A.), Spec. Publs int. Ass. Sediment., No. 14, pp. 89–152. Blackwell Scientific Publications, Oxford.

Swift, D.J.P. & Rice, D.D. (1984) Sand bodies on muddy shelves: a model for sedimentation in the western interior Cretaceous seaway, North America. In: *Siliciclastic Shelf Sediments* (Eds Tillman, R.W. & Siemers, C.T.), Spec. Publ. Soc. econ. Miner. Paleont., Tulsa, **34**, 43–62.

Swift, D.J.P. & Thorne, J.A. (1991) Sedimentation on continental margins, I: a general model for shelf sedimentation. In: *Shelf Sand and Sandstone Bodies: Geometry, Facies and Sequence Stratigraphy* (Eds Swift, D.J.P., Oertel, G.F., Tilman, R.W. & Thorne, J.A.), Spec. Publs int. Ass. Sediment., No. 14, pp. 3–31. Blackwell Scientific Publications, Oxford.

Tirsgaard, H. & Sønderholm, M. (1997) The sedimentary evolution and lithostratigraphy of the Late Precambrian Lyell Land Group (Eleonore Bay Supergroup), East and Northeast Greenland. *Bull geol. Surv. Denmark Greenland*, **178**, 60 pp.

van der Neut, M. (1990) *Afsettingstoestande van die Pretoria Groep gesteentes in die Pretoria–Bronkhorstspruit–Delmas gebied*. MSc thesis, University of Pretoria, Pretoria.

Walker, R.G. & Plint, A.G. (1992) Wave- and storm-dominated shallow marine systems. In: *Facies Models: Response to Sea-level Change* (Eds Walker, R.G. & James, N.P.), pp. 219–238. Geological Association of Canada, Waterloo, Ontario.

Windley, B.F. (1995) *The Evolving Continents*. Wiley, Chichester, 526 pp.

Wise, D.U. (1972) Freeboard of continents through time. *Mem. geol. Soc. Am.*, **132**, 87–100.

*Spec. Publs int. Ass. Sediment.* (2002) **33**, 369–381

# Facies sequence and cryptic imprint of sag tectonics in the late Proterozoic Sirbu Shale, Central India

S. SARKAR*, S. CHAKRABORTY*, S. BANERJEE† *and* P. K. BOSE*

**Department of Geological Sciences, Jadavpur University, Calcutta-700 032, India; and*
*†Department of Earth Sciences, IIT Bombay, Powai, Mumbai-400 076, India*

## ABSTRACT

This paper examines the upper section (<185 m) of the late Proterozoic Sirbu Shale, central India, laid down in a storm-dominated shelf and lagoonal palaeoenvironment. Five of the six facies are laterally extensive, cyclical, interbedded shale-siltstone/fine sandstone. They generally have gradational mutual transitions and are amenable to palaeogeographical interpretation. The sixth facies is a coarse sandstone of low textural/mineralogical maturity, confined to a localized occurrence and encased by one of the deeper shelf facies. The coarser interbeds bear storm signatures in all the facies. Current-formed features below storm beds record dominantly shore-parallel flow, although a shore-normal component is also evident. In the westward-opening intracratonic sag basin where deposition took place, the shelf succession built up as an overall prograding highstand systems tract that is divisible into a number of metre-scale parasequences. Slump features on top of each of the parasequences correlate the intervening marine flooding events with NE–SW extensional events and resultant landward subsidence. Fischer plots of parasequences reveal another, cryptic low-frequency depositional cyclicity. Troughs in the Fischer curves, denoting the longer cycle, roughly coincide with selective occurrence of NE–SW trending slide planes, implying superimposition of a NW–SE extension on the more frequent NE–SW extension. Enhanced subsidence is thus suggested at longer intervals owing to simultaneous orthogonal extension. The NE–SW extension reflects occasional readjustment of underlying rift blocks, whereas the NW–SE extension was possibly related to a plate-margin process. The resultant shelf basin probably assumed a NW–SE elongated, oval geometry with a dominant NE or landward-sloping flank. Plotting of shelf succession thickness downwards from a datum plane at a large number of locations simulates a three-dimensional basin configuration, as predicted from the inferred basin dynamics.

## INTRODUCTION

Epeiric seas were common during Precambrian and early Palaeozoic times (Shaw, 1964; Dott & Batten, 1976; Eriksson *et al.*, 1998). Slow subsidence and the low-gradient depositional slope of these basins might have introduced a preferred sequential trend. Epeiric seas do not have unequivocal modern analogues (Shaw, 1964; Eriksson *et al.*, 1998) and therefore one can test the related issues only in ancient successions. Eustatic sea-level changes could cause large-scale inundations or exposures, but relatively insignificant palaeobathymetric changes in these basins. As a result, subtidal successions may not show significant facies changes in response. A tangible illustration of this effect is provided by the late Proterozoic intracratonic Sirbu Shale in the Son Valley, Central India (Fig. 1). This paper undertakes a process-related facies analysis (Driese *et al.*, 1991) and further analyses the facies succession over a 3500 km$^2$ area against a background of syn-sedimentation deformation history.

Sedimentologists generally treat shale formations summarily and thereby leave a major source of geological information unexplored. Process-related facies analysis in terms of physical characteristics of shales is seldom carried out (e.g. Schieber, 1986). Consequently, palaeogeographical interpretations of these formations are based on association, and hence are tenuous and uncertain. The present reappraisal of the Sirbu Shale generates a better understanding of

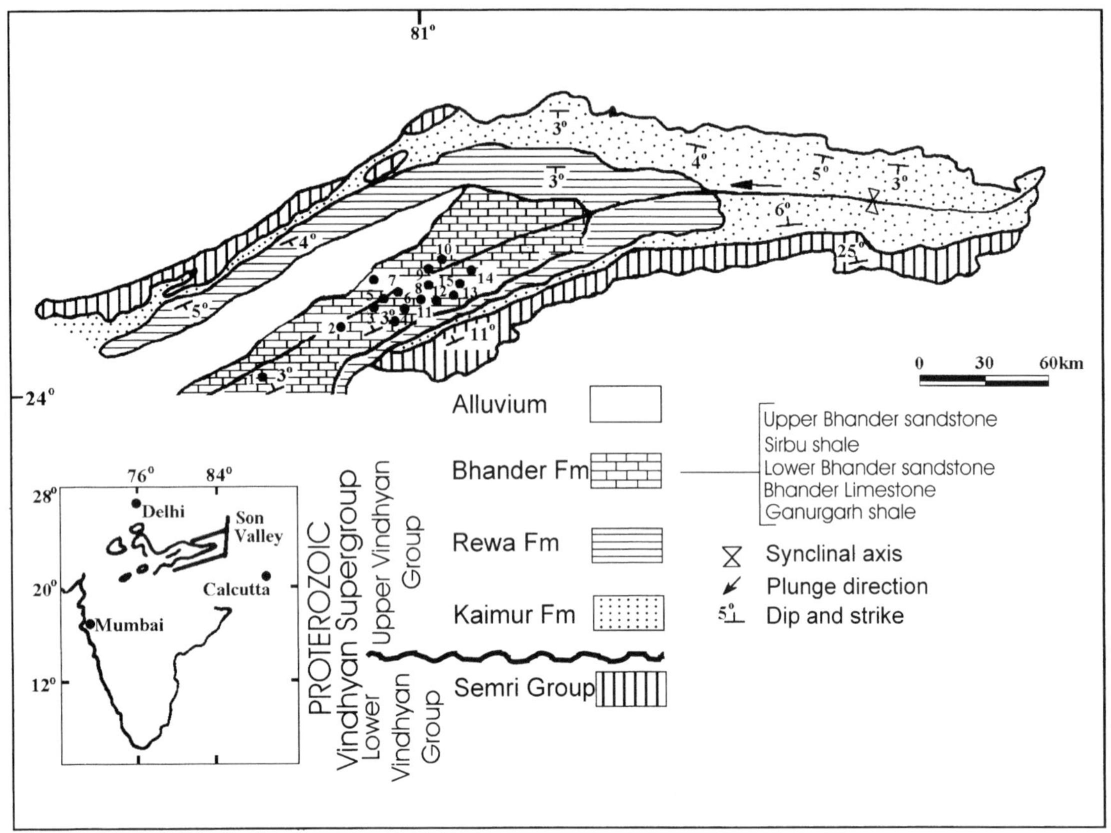


**Fig. 1.** Geological map of the Vindhyan Supergroup showing locations of sections of the Sirbu Shale (encircled numbers) utilized for detailed analyses. Map of India within inset. Vindhyan stratigraphy on the right.

palaeohydraulics and a greater confidence in palaeogeographical interpretation. Facies succession and penecontemporaneous deformation structures also reveal depositional cyclicities related to basin dynamics.

## GEOLOGICAL SETTING

The Sirbu Shale is almost at the top of the two-tiered Meso- to Neoproterozoic Vindhyan Supergroup, Central India (Venkatchala *et al.*, 1996), and is bounded by the Lower Bhander Sandstone of coastal playa origin below (Chanda & Bhattacharyya, 1982; Bose & Chaudhuri, 1990) and the marginal marine to terrestrial Upper Bhander Sandstone above (Bose *et al.*, 1999; Fig. 1). Both the bounding members are red in colour, while the Sirbu Shale is grey.

The Vindhyan Supergroup formed in an epeiric basin that opened westward (Chanda & Bhattacharyya, 1982, and references therein). The basin developed in the Southern Hemisphere (Williams & Schmidt, 1996). In this paper we present all vectors in their present latitudinal context. The lower Vindhyans developed during active rifting accompanied by a dextral shear (Bose *et al.*, 1997). The upper Vindhyan sedimentation, in contrast, took place in an intracratonic sag basin (Chanda & Bhattacharyya, 1982; Chakraborty *et al.*, 1998) wherein deformation and volcaniclastic deposition were severely restricted (Chakraborty, 1996). The upper Vindhyan succession is free from extraneous conglomerates, and fine-grained quartz arenites, shales and limestones are its main constituents.

The mud-rich Sirbu Shale is prone to weathering and forms steep-sided conical hills, unless protected by the Upper Bhander Sandstone. Workable sections are few and far between, and bedding plane sections are very rare. It is entirely terrigenous, except for a few oolitic and stromatolitic carbonate patches at its base. Beds are subhorizontal and only broadly warped locally. The upward transitions from the Lower Bhander Sandstone to the lagoonal part of the Sirbu Shale and upward into the shelf part of the Sirbu Shale evoke transgressions, whereas the gradational upward passage of the latter into the Upper Bhander Sandstone required progradation.

## FACIES

The basal grey shale of the Sirbu Shale incorporating the carbonate patches is *c*.7.2 m thick, bears features of emergence and forms a lagoonal facies assemblage (Singh, 1974). It is immediately overlain by a dark shale containing only a subordinate amount of siltstone (facies A, described below). The contact is demarcated, in places, by a two to three clast thick, clast-supported mud pebble conglomerate. The clasts are, on average, 3 cm long and their interstitial spaces are filled dominantly by calcite cement. This paper focuses on the facies assemblage lying above this sharp contact.

The younger assemblage, up to 185 m thick, is devoid of emergent features and is divisible into the following six facies:

**1** Facies A, shale. Dark green shale attaining thickness up to 22 cm characteristically contains submillimetre thick planar fine silt interlaminae. The shale laminae often incorporate millimetres thick, darker and wrinkled shaly wisps of length about 14 cm. The facies also contains lenticular siltstone beds of thickness 9 mm to 3 cm at vertical spacing about 10 cm (Fig. 2A). These siltstone beds have lateral extent up to about 6 cm, and are internally massive or planar laminated and locally finely cross-laminated. Their bases are sharp and planar, while their tops may be

**Fig. 2.** Facies comprising the Sirbu Shale. (A) Facies A dark shale with thin detached lenses of siltstone (pen marked by a circle for scale). (B) Facies B dark shale with siltstone lenses larger in dimension than those in facies A. (C) Facies C shale with laterally persistent siltstone interbeds. Note broadly undulating top of the siltstone beds (pocket knife marked by a circle for scale). (D) Wavy and hummocky laminae within fine sandstone interbeds of facies D. (*continued on p. 372*)

E

F

G

**Fig. 2.** (*cont'd*) (E) Siltstone interbeds containing lenticles of rippled fine grained sandstone of facies E (part of a thick sandstone interbed is at the top). (F) Coarse-grained sandstone with sheet-like geometry comprising facies F encased by facies C above and below. (G) Three-dimensional undulating basal surface of facies F exposed in orthogonal sections.

slightly undulated because of fossilized ripple forms of amplitude about 2 mm. Overall grading, although weak, is locally discernible within the beds. Soles of the beds bear tool marks, including groove casts and prod marks, oriented in an E–W direction (Fig. 3).

**2** Facies B, shale with minor interbedding of siltstone. This facies attains a thickness of up to 45 cm, and differs from facies A only in details of the siltstone beds. The siltstone lenses range in thickness from 5 to 8.5 cm, and in lateral extent from 8 to 12 cm (Fig. 2B). The beds may be massive or planar laminated and wave rippled at the top. Grading is conspicuous within them and some ripples resemble the fading ripples of Pedersen (1985), having drapes that become increasingly muddy upwards. Casts of gutters as well as grooves and prods are present at the soles of the beds. Gutters (width *c*.4 cm, depth *c*.3 cm) and ripples (width *c*.2 cm, height *c*.1.5 cm) are larger than their counterparts in facies A. The tool marks are roughly parallel to the gutters, maintaining a consistent N–S

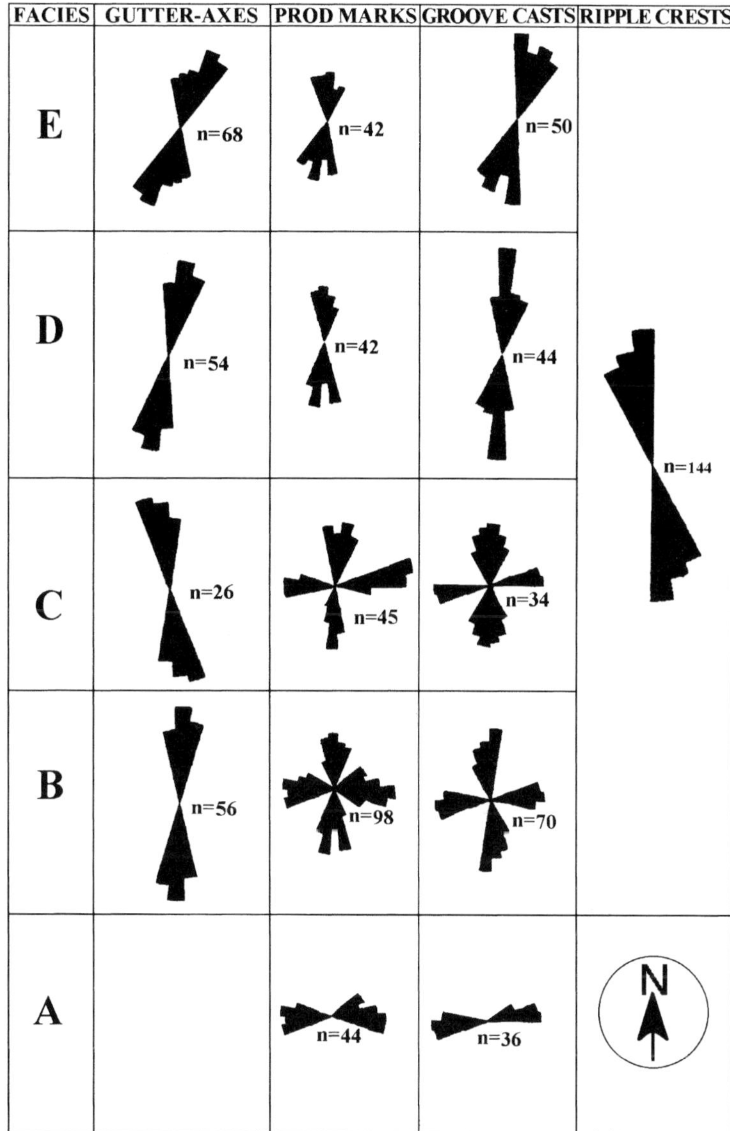


**Fig. 3.** Orientation of sole features with respect to wave ripple crest orientation in the principal constituent facies of the Sirbu Shale. Note the dominance of sole features parallel to the wave ripple crests, although orthogonal relationships also exist.

trend; prod marks are, however, bipolar (Fig. 3). The facies grades downwards into facies A and upwards into facies C.

**3** Facies C, shale–siltstone. Alternate shale and siltstone beds of comparable thicknesses comprise this facies, which attains a thickness of up to 56 cm. The shales have internal characters the same as their counterparts in facies A and B. The siltstones, however, form persistent beds of about 9 cm thickness and have very sharp bases, but with tops grading into mud. They, like their counterparts in facies B, are massive and planar laminated in their lower parts and ripple laminated above (Fig. 2C). Wave ripples on top of them are larger than those in facies B (width *c*.4 cm, height *c*.2.5 cm). Fading ripples are common. Gutters on the soles of beds are also larger (width *c*.10.5 cm; depth *c*.6.5 cm) and generally asymmetric, and adjacent gutters may have axial planes that dip in opposite directions. The majority of prod marks are parallel to the gutters and have the same N–S trend; but others are normal to gutters (Fig. 3). This facies, while underlying facies F, exhibits minute reclined folds of amplitude 1.2 cm between undisturbed layers. There are broad and shallow channels up to 2.5 m wide and up to 0.5 m deep. Parallel laminated shale that locally spills over the banks fills many of these channels and none of the channels has disparately coarser fill, as expected in lagoons. Sandstone beds amalgamate, but rarely.

**4** Facies D, sandstone–shale interbedded. This facies attains a thickness of up to 72 cm and is characterized by interbedded very fine-grained sandstone and silty shale. The silty shale beds have an average thickness of 14 cm and comprise minute (mm amplitude) ripples encased in thin stringers of mud. The sandstone beds are sheet-like and range in thickness from 10.5 to 15 cm. Gutters on the base of beds have widths and depths of 14 and 12 cm respectively. Wave ripples (width 8 cm, height *c*.4.5 cm) occur on top of the beds. Grooves and prod marks generally parallel to the gutters are also present. Mud clasts are concentrated at the base of the beds. Notwithstanding mud clast concentration, the basal surfaces of some sandstone beds conform to the configuration of bedforms on top of the underlying amalgamated sandstone beds. Internally the beds show wavy laminae (Fig. 2D) and moderate- to high-angle cross-stratification. The lower parts of the beds may be planar laminated or massive. The wavy laminae have amplitude and wave length, on average, of 14 and 20 cm respectively. Tool marks under the sandstone beds are both parallel and normal to the gutter axes (Fig. 3).

**5** Facies E, sandstone–sandy siltstone interbedded. The major components of this facies, which has a thickness of up to 90 cm, are laterally extensive sandstone beds that generally have flat bases with comparatively larger gutters up to 19 cm wide and 14.5 cm deep. Above the gutters, sediments may be massive or planar laminated and ripple laminated towards the top. Preserved ripples (width *c*.10 cm, height *c*.6 cm) on bed tops locally show profound asymmetry in profile. On the bed surfaces, there are numerous syneresis cracks that cross-cut each other. Sandstone beds are laterally persistent, range in thickness up to 22 cm and often amalgamate. The most marked difference from facies D, however, arises from the fact that the silty interbeds (Fig. 2E) incorporate thin wave rippled lenticles of very fine sandstone. These interbeds, however, do not persist laterally more than 3–4 m. At the sole of sandstone beds tool marks are parallel to the gutters (Fig. 3).

**6** Facies F, sandstone. This facies is distinctive, being composed of coarse, poorly sorted sandstones. Only one, 3.1 m thick, unit occurs near Sankargarh (Fig. 1). It is encased by facies C and both of its contacts are sharp (Fig. 2F). A large fraction of the constituent grains of this facies is of medium to coarse sand grade. While sandstones and siltstones in all other facies are mud-free and quartzose, in this facies they contain about 12% mud. Sand and mud clasts constitute 18% by volume and the majority of them concentrate at the base of the beds. Bed thickness is laterally variable and ranges up to 12 cm. Thin mud lenticles separate the sandstone beds in places. Every bed shows pronounced grading. The base of the facies unit is very sharp, but undulated due to the presence of alternate scours and mounds, resembling hummocks and swales (Fig. 2G). The swales are up to 25 cm deep and up to 1.6 m wide. The 1–2 cm thick basal portion of the sandstone bed immediately overlying the hummocky and swaley muddy substratum is characterized by parallel laminae conforming to the basal topography. Laminae dip centrifugally on the hummocks and centripetally within the swales. Upwards the parallel laminated set gives way to sets of wavy laminations, including hummocky cross-stratifications. The hummocky cross-stratifications have amplitude and wavelength of about 18 and 90 cm respectively. High-angle cross-stratifications are also locally present. Bed tops show multiple sets of sinuous-crested, asymmetric mega-ripples of amplitude and wavelength 2.5 and 14 cm, respectively, indicating palaeoflow NNW. On some bed surfaces, there are wave ripples (width *c*.15 cm, height *c*.4.5 cm). This facies forms an anomalous element within facies C and is laterally traceable up to half a kilometre in a river section.

Wave ripples are not measurable in facies A, but in all other facies their crests trend consistently about N–S.

## INTERPRETATION

All upward transitions from facies A to B, B to C, C to D and D to E are gradational, suggesting deposition in adjacent palaeoenvironments. Only facies F, confined to a single stratigraphical level and geographical locality, appears to be atypical. The general facies descriptors are consistent with those of numerous storm-dominated shelf successions described in the literature (e.g. Dott & Bourgeois, 1982; Bose *et al.*, 1988). The mud-filled channels represent by-passing storm channels on shelves (Goldring & Aigner, 1982). Broad palaeogeographical reconstruction and the trend of the wave ripple crests suggest a N–S palaeoshoreline alignment, while the land lay on the east. Palaeocurrent data (Fig. 3) indicate that the storm-induced flows were dominantly shore-parallel geostrophic, although evidence of shore-normal particle movement is not lacking.

Among all the facies, A is inferred to be the most distal offshore facies, but bipolar sole marks indicate its location above storm wave base. Gutters are absent below siltstone interbeds in this facies and imply relative weakening of the flows through distal travel. The wispy darker laminae within shales of this facies, and of facies B and C were probably deposited in microbial mats (Schieber, 1986, 1998). Where the bases of sandstone beds conform to bedforms on an underlying bed, substrate binding by a microbial mat is also indicated.

Grading in the coarser clastic beds of all the facies indicates deposition from steadily waning flows. Reversals in prod orientations, the presence of wave ripples and hummocky cross-stratification reveal an oscillatory component to the flow. The axial plane divergence of adjacent gutters in facies C indicates operation of paired spiral eddies in the flow (Swift *et al.*, 1973). The orientation of the sole features indicates that the seaward or westward downwelling flow was deflected to the left and was dominantly shore-parallel. This is consistent with the Coriolis force in the original southern hemisphere location of the depositional basin (Williams & Schmidt, 1996). Nevertheless, the flow also had shore-normal components, particularly in the relatively offshore region (Fig. 3).

In transitions from facies A to E through B, C and D, the coarser components gain in lateral continuity, grain size, thickness and bedform dimensions. The A to E transition reflects increasing proximity to the shore. The presence of wave ripples within the relatively finer interbeds in facies D and E suggests deposition within fair weather wave base, whereas facies A, B and C were deposited largely between fair weather and storm wave base.

In the unusual facies F, the undulatory base, internal hummocky cross-stratification and wave ripples suggest deposition from storm waves of distinctly greater intensity. Earthquakes could have induced the waves. The occurrence of all the sandstone beds of this facies juxtaposed one above another in a single stratigraphic interval suggests rapid recurrence of the same unusually strong storm surges as may happen in the event of an earthquake and aftershocks. Reclined folds in facies C underlying it further strengthen the contention. The sharp erosional base creating alternate domal hummocks and bowl-like swales indicate predepositional wave erosion. The basal low-angle draping sand laminae are essentially hummocky and swaley cross-stratifications of Walker & Plint (1992). The deposition evidently took place above storm wave base. Despite the presence of hummocky cross-stratification, most of the ripples indicate the dominance of a unidirectional current during the waning stage of storms. However, local wave ripples possibly relate to residual oscillatory motion of waning storms because of the fact that facies F is both underlain and overlain by C formed below fair weather wave base. Nevertheless, the seismite interpretation of the facies remains rather tentative. Because of this uncertainty, the following analysis of general implications does not include the section containing this localized aberrant facies.

## FACIES PACKAGE

The shelf facies succession is overall upward-coarsening. Facies A, B and C dominate the base, while facies D and E dominate towards the top of the succession. The succession is, however, divisible into a number of vertically stacked upward-shoaling facies packages/cycles ranging in thickness from 1.5 to 11.5 m (Fig. 4). A complete cycle, though never found, would show gradual upward transitions from A to E. The majority of these cycles are top-truncated in the lower level and bottom-truncated in the upper level, but each of them shows part of the A to E facies succession.

The basal 7.2 m of the Sirbu Shale are a likely result of limited transgression on the coastal playa deposits of the Lower Bhander Sandstone. The sharp contact between the lagoonal and the shelfal deposits of the Sirbu Shale, in contrast, accounts for a substantial transgression. The conglomerate with early pore cement on its top suggests an omission surface on which the thin openwork conglomerate is a transgressive lag. The succession above the lag, however, records an overall progradation. The lag thus denotes maximum flooding and the overlying succession is the product of a highstand systems tract (Posamantier *et al.*, 1988; van Wagoner *et al.*, 1988), although the upward-shoaling cycles or parasequences within the shelf succession indicate intermittent relatively minor flooding.

## DEFORMATION STRUCTURES

Two kinds of penecontemporaneous deformation structures are conspicuous and common within the

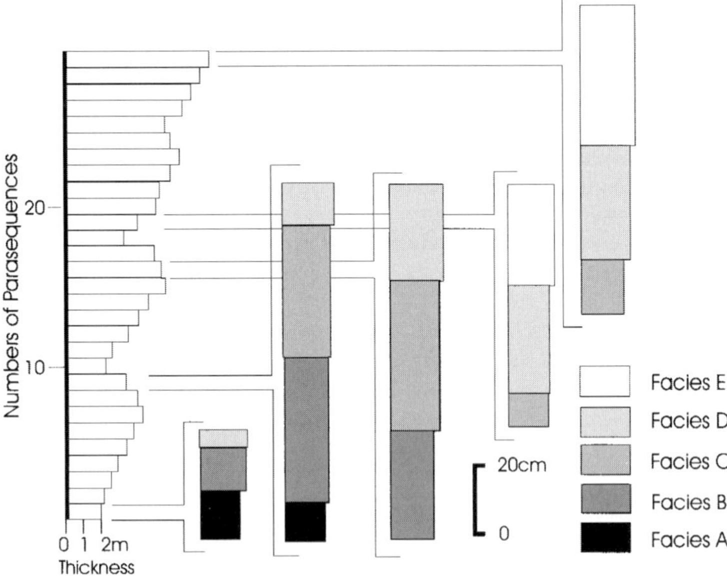


**Fig. 4.** Thickness variation in successive numbers of parasequence in measured succession at Shankargarh (7 in Fig. 1). Facies succession varies within individual parasequences at different levels. Note the dominance of facies A and B near the base and D and E in the upper part of the succession.

shelf succession. First are slump folds of various magnitudes (12–25 cm; Fig. 5A), along with local slump scars often bearing striations on their polished surfaces (Fig. 5B). Second are inclined slide planes that are broadly straight and slightly sinuous in plan and bear dip-parallel striations (Fig. 5C). They dip about 20° and range in vertical height to about 17 cm. The beds and laminae under them thin downslope. Both the slump scars and slide planes are associated downslope with intraclast-rich wedges confined between laminated sediments.

The axial planes of the slump folds, irrespective of their locations, dip SW, albeit with wide dispersion, and the slump scars, in consistency, dip NE (Fig. 6). In contrast, the slide planes strike, again irrespective of their locations, consistently NE–SW, but dip divergently (Fig. 6). The slump features conspicuously concentrate at the top of successive parasequences. Slide planes occur on top of a few, but not all, parasequences. The vertical height of the disturbed zones at the top of the parasequences is usually about a metre.

A

B

**Fig. 5.** Penecontemporaneous deformation structures within the Sirbu Shale. (A) Top-truncated slump folds. (B) Inclined and polished slump scar bounded above and below by horizontal strata. *(continued)*

**Fig. 5.** (*cont'd*) (C) Striated slide plane. Note clast-bearing deposit adjacent to the plane. Also note the rippled sandstone cover on the slide-affected strata. Match for scale.

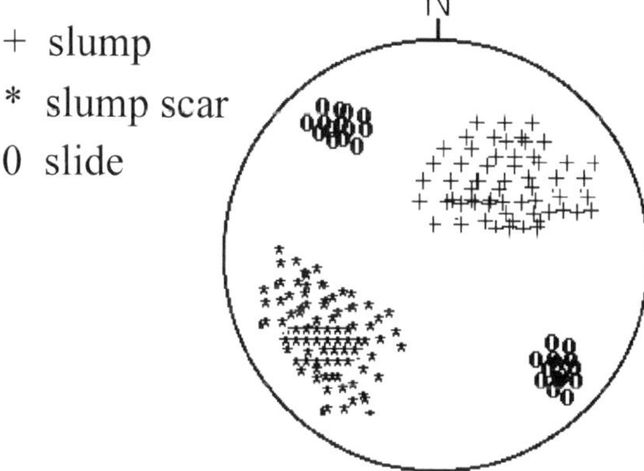


**Fig. 6.** Poles of the axial planes of slump folds, slump scars and slide planes.

The preferential occurrence of penecontemporaneous deformation structures on top of all parasequences suggests that basin subsidence caused intermittent flooding. The dip directions of slump scars and axial planes of slump folds both suggest a NE basin-floor tilt during the subsidence events. NE–SW extension is suggested and this direction matches one of the directions of extensional components recorded in the lower Vindhyans (Bose *et al.*, 1997). On the other hand, the divergent slide planes indicate NW–SE extension of the basin that is not consistent with any of the two extensional components developed during rifting of the Vindhyan basin (Bose *et al.*, 1997). The common occurrence of slide planes of essentially similar nature and orientation in the formations bounding the Sirbu Shale as well, however, indicates the long persistence of this extensional stress.

## BASIN DYNAMICS

The vertical succession at Dureha reveals 29 parasequences and fewer parasequences are seen at Ucharhera, Sankargarh, Pala and Hardua (Fig. 4; for locations see Fig. 1). In all these sections, parasequences tend to thicken upwards (Fig. 4). Concentration of deformation structures at the top of every parasequence suggests that basin subsidence caused the termination of parasequences. Slump features ubiquitously present along all these levels indicate north-easterly or landward slope generation owing to this subsidence. Since none of the parasequences bears evidence of emergence, the parasequence thickness variation is not correlatable with accommodation space generation (Wilgus *et al.*, 1988). Overall progradation implies either progressive decrease in magnitude of subsidence in successive events of extension or shortening of time intervals between successive events.

A Fischer plot (Fischer, 1964; Goldhammer *et al.*, 1987) of the parasequences, assuming their uniform frequency, reveals another, more subtle, depositional cyclicity of lower frequency. Several parasequences constitute each of these cycles measured from one trough to the next. Since storms are products of meteorological perturbations, the two orders of depositional cycles may be mutually correlatable with climatic cycles (see Banerjee, 1998). However, wide variability in the number of parasequences constituting individual longer cycles makes such a correlation rather untenable. The parasequences are probably tectonic cycles, as discussed above. Slide planes tend to concentrate at the levels of deflections, particularly the troughs, of the Fischer curves (Fig.7). It therefore appears that, like the shorter cyclicity of parasequences, the longer cyclicity also owes its origin to basinal extension, but in a NW–SE direction. On occasions of coincidence of the two extensional forces, the magnitude of subsidence was presumably accentuated.

Orthogonal extensions on such infrequent occasions presumably led to flattening deformation, as is common for intracratonic sag basins (Ayudin & Nur,

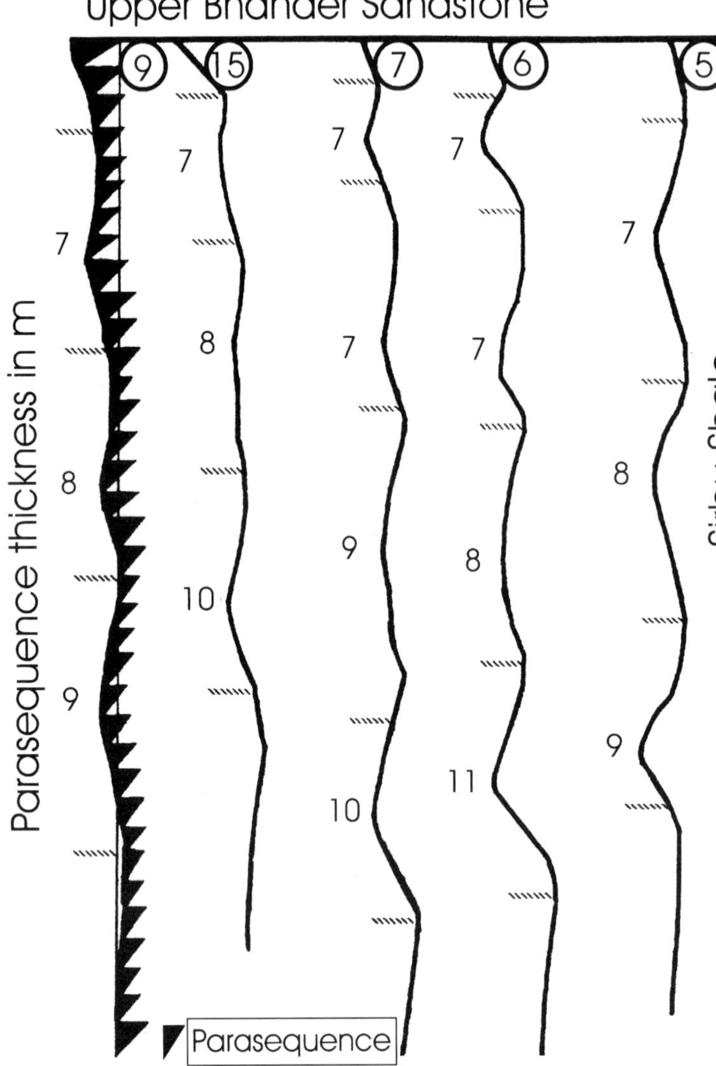


**Fig. 7.** Fischer plot of parasequence thickness along sections measured down from the contact of the Sirbu Shale with the Upper Bhander Sandstone. Locations of the sections are indicated by encircled numbers with reference to Fig. 1. Other numbers alongside the curves indicate number of parasequences comprising successively larger cycles measured from trough to trough of the Fischer curves. Horizons with abundant slides are indicated by cross-hatches. Note their rough coincidence with the troughs of the Fischer curves. Upward, a slight leftward shift of the curves is indicative of overall increase in parasequence thickness, i.e. progradation.

1982). Yet, because of the more frequent occurrence of NE–SW extension, the depositional basin is likely to have assumed a NW–SE elongation, unless the magnitude of relatively infrequent NW–SE extension could obviate its effect. Bipolarity of sole marks, testifying to wave action, and evidence of well developed shore-parallel geostrophic flow suggest that connection of the depositional basin to the open sea was well maintained, despite the formation of a high on the seaward side because of inland differential subsidence.

In all the studied sections, the number of parasequences constituting individual longer cycles decreases upwards (Fig. 7). Either the frequency of the longer cyclicity diminished, or the frequency of the shorter cyclicity increased progressively. In both cases, the difference in frequency between the two cyclicities would decrease progressively and the effect of NW–SE extension on basin configuration is likely to have been more pronounced through time. However, the longer cycles have a maximum to minimum thickness ratio close to 1 (varying between 1.07 and 1.22) and thus possibly they, rather than the parasequences, maintained a near-uniform frequency.

A reasonable reconstruction of the shelf basin configuration at the end of Sirbu Shale deposition has been made by plotting the vertical distance of the

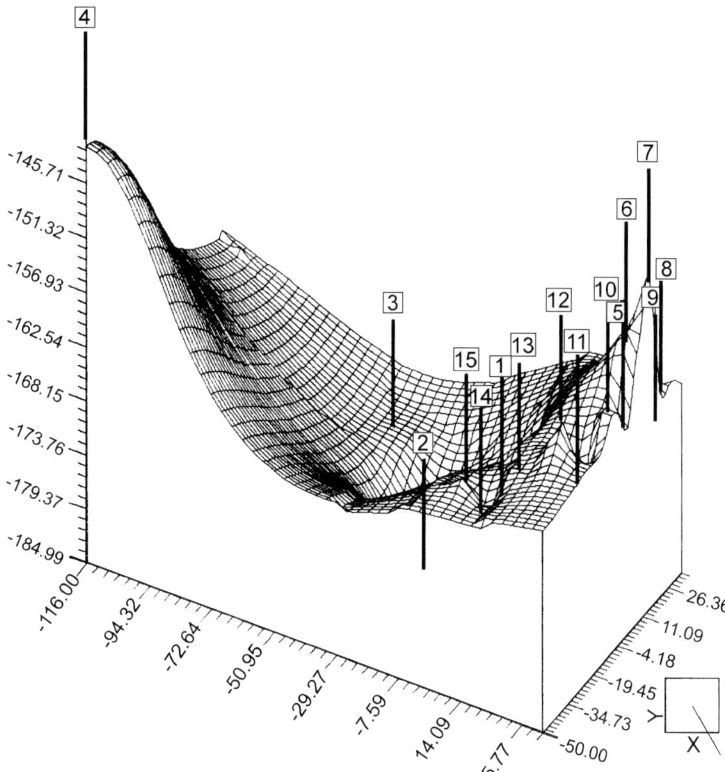


**Fig. 8.** Reconstructed configuration of the shelf basin. The vertical scale indicates vertical inverse distance (in metres) of the maximum flooding surface down from the datum plane (see text) in different locations numbered with reference to Fig. 1. The horizontal scale indicates distance of the measured sections with respect to Maihar (8). Direction of view given at right bottom.

maximum flooding surface down from the contact between the Sirbu Shale and the Upper Bhander Sandstone at a number of localities situated not on a straight line (Fig. 8). Maximum flooding surfaces are generally planar and of very gentle slope (Thorne & Swift, 1991), unless the sedimentation rate is very high (Siggerud & Steel, 1999). Facies characteristics in the Sirbu Shale rule out high depositional rate during fair weather periods. The maximum flooding surface near the base of the Sirbu Shale is thus presumed to be primarily planar. Its non-planar configuration in Fig. 8 reveals the configuration that the shelf basin assumed at the close of the Sirbu Shale deposition.

The reconstructed basin is, indeed, elongated in a NW–SE direction, but asymmetric because of the dominant slope north-eastwards, i.e. landwards. Another NE–SW elongated, shallow depression presumably owing to NW–SE extension is superimposed on this general configuration. The NE–SW extension that dominantly controlled the Sirbu basin configuration also played the major role during early rifting in the Vindhyan basin, as documented by Bose *et al.* (1997). However, the signature of NW–SE extension is non-existent in the lower Vindhyan succession. It is therefore reasonable to assume that the Sirbu shelf basin configuration was principally controlled by subsidence owing to occasional readjustment of underlying rift blocks. On the other hand, the NW–SE extension had an impact only after the rifting ceased. The slide planes that also occur in the formations bounding the Sirbu Shale are clearly products of a relatively weaker, but steady, regional stress regime maintained during most of the upper Vindhyan time. The stress was probably tied to a plate-margin process (Allen & Allen, 1993). Sediment compaction was certainly a minor additional factor in basin subsidence (Eriksson *et al.*, 2001).

## CONCLUSIONS

The late Proterozoic Sirbu Shale, central India, evolved as a highstand systems tract on a storm-dominated siliciclastic shelf, following flooding on top of the initial lagoonal deposit. Differential subsidence owing to a high-frequency NE–SW extension, due to readjustment of underlying rift blocks, punctuated the overall progradational trend of the shelf succession. The rate of subsidence accelerated at times as a result of comparatively infrequent NW–SE extension, possibly caused by a plate-margin process. A progressive decline in the frequency of the NE–SW extension and consecutive fall in the rate of subsidence caused the progradation. The depositional basin assumed an oval shape elongated in a NW–SE direction. Tectonism, albeit operating at a low level, was a major control on the succession architecture.

## ACKNOWLEDGEMENTS

The authors are grateful to Jeff Chiarenzelli and Pat Eriksson for reviewing the manuscript. P.K.B. thankfully acknowledges the financial support from DST. S.S., S.C. and P.K.B. acknowledge the infrastructural facilities provided by the Department of Geological Sciences, Jadavpur University. S.B. acknowledges the infrastructural facilities provided by IITB.

## REFERENCES

Allen, P.A. & Allen, J.R. (1993) *Basin Analysis: Principles and Applications*. Blackwell Scientific Publications, Oxford, 451 pp.

Ayudin, A. & Nur, R. (1982) Evolution of pull apart basins and their scale independence. *Tectonics*, **1**, 91–105.

Banerjee, S. (1998) *Facets of the Mesoproterozoic Semri sedimentation in Son valley, India*. PhD thesis, Jadavpur University, Calcutta.

Bose, P.K., Banerjee, S. & Sarkar, S. (1997) Slope controlled seismic deformation and tectonic framework of deposition: Koldaha Shale, India. *Tectonophysics*, **269**, 151–169.

Bose, P.K., Chakrabarty, S. & Sarkar, S. (1999) Recognition of ancient aeolian longitudinal dunes: a case study from Upper Bhander Sandstone, Son Valley, India. *J. sediment. Res.*, **69**, 72–81.

Bose, P.K. & Chaudhuri, A.K. (1990) Tide versus storm in epieric coastal deposition: two Proterozoic sequences, India. *Geol. J.*, **25**, 81–101.

Bose, P.K., Chaudhuri, A. & Seth, A. (1988) Facies, flow and bedform patterns across a storm-dominated inner continental shelf: Proterozoic Kaimur Formation, Rajasthan, India. *Sediment. Geol.*, **59**, 275–293.

Chakraborty, P.P. (1996) *Facies and sequence development in some late Proterozoic Formations, India with some clues for basin evolution*. PhD thesis, Jadavpur University, Calcutta.

Chakraborty, P.P., Sarkar, S. & Bose, P.K. (1998) A view point on intracratonic chenier evolution: clue from a reappraisal of the Proterozoic Ganurgarh Shale, central India. In: *The Indian Precambrian* (Ed. Paliwal, B.S.), pp. 61–72. Scientific Publishers (India), Jodhpur.

Chanda, S.K. & Bhattacharyya, A. (1982) Vindhyan sedimentation and paleogeography: post Auden developments. In: *Geology of Vindhyanchal* (Eds Valdiya, K.S., Bhatia, S.B. and Gaur, V.K.), pp. 88–101. Hindustan Publishing Corporation, Delhi.

Dott, R.H. & Batten, R.L. (1976) *Evolution of the Earth*, 2nd edn. McGraw-Hill, New York, 649 pp.

Dott, R.H. Jr & Bourgeois, J. (1982) Hummocky stratification: significance of its variable bedding sequences. *Geol. Soc. Am. Bull.*, **93**, 663–680.

Driese, S.G., Fischer, M.W., Easthouse, K.A., Marks, G.T., Gogola, A.R. & Schoner, A.E. (1991) Modes for genesis of shoreface and shelf sandstone sequences, southern Appalachians: paleoenvironmental reconstruction of an early Silurian shelf system. In: *Shelf Sand and Sandstone Bodies: Geometry, Facies and Sequence Stratigraphy* (Eds Swift, D.J.P., Oertel, G.F., Tillman, R.W. & Thorne, J.A.), Spec. Publs int. Ass. Sediment., No. 14, pp. 309–338. Blackwell Scientific Publications, Oxford.

Eriksson, P.G., Condie, K.C., Trisgaard, S. *et al.* (1998) Precambrian clastic sedimentation systems. *Sediment. Geol.*, **120**, 5–53.

Eriksson, P.G., Martin-Neto, M., Nelson, D.R. *et al.* (2001) An introduction to Precambrian basins: their characteristics and genesis. *Sediment. Geol.*, **141**(2), 1–35.

Fischer, A.G. (1964) The Lofer cyclothems of the Alpine Triassic. *Bull. Kansas geol. Surv.*, **169**, 107.

Friedman, G.M., Sanders, J.E. & Kopaska-Merkel, D.C. (1992) *Principles of Sedimentary Deposits*. Macmillan, New York, 792 pp.

Goldhammer, R.K., Dunn, P.A. & Hardie, T. (1987) High frequency glacio-eustatic sea-level oscillations with Milankovich characteristics recorded in Middle Triassic platform carbonates in northern Italy. *Am. J. Sci.*, **287**, 853–892.

Goldring, R. & Aigner, T. (1982) Scour and fill: the significance of event separation. In: *Cyclic and Event Stratification* (Eds Einsele, G. & Seilacher, A.), pp. 354–362. Springer-Verlag, Berlin.

Hallam, A. (1981) *Facies Interpretation and Stratigraphic Record*. W.H. Freeman, Oxford, 231 pp.

Hendersen, G.K. (1985) Thin, fine-grained storm layers in a muddy shelf sequence: an example from the Lower Jurassic in the Stenlille1. *J. geol. Soc. London*, **142**, 357–374.

Posamentier, H.W., Jervy, M.T. & Vail, P.R. (1988) Eustatic controls on clastic deposition I. Conceptual framework. In: *Sea-level Changes: an Integrated Approach* (Eds Wilgus, C.K., Hastings, B.S., Kendall, C.G.StC., Posamentier, H.W., Ross, C.A. & van Wagoner, J.C.), Spec. Publ. Soc. econ. Paleont. Miner., Tulsa, **42**, 109–124.

Reading, H.G. (1978) *Sedimentary Environments and Facies*. Blackwell Scientific Publications, Oxford, 569 pp.

Schieber, J. (1986) The possible role of benthic microbial mats during the formation of carbonaceous shales in shallow Proterozoic basins. *Sedimentology*, **33**, 521–536.

Schieber, J. (1998) Possible indicators of microbial mat deposits in shales and sandstones: examples from the Mid-Proterozoic belt Supergroup, Montana, USA. *Sediment. Geol.*, **120**, 105–124.

Shaw, A.B. (1964) *Time in Stratigraphy*. McGraw-Hill, New York, 365 pp.

Siggerud, E.I.H. & Steel, R.J. (1999) Architecture and trace-fossil characteristics of a 10 000–20 000 year, fluvial to marine sequence, Ebro Basin, Spain. *J. sediment. Res.*, **69**, 365–383.

Singh, I.B. (1974) Depositional environment of the Upper Vindhyan sediments in the Satna–Maihar area, Madhya Pradesh and its bearing on the evolution of the Vindhyan sedimentation basin. *J. palaeont. Soc. Ind.*, **19**, 48–70.

Swift, D.J.P., Duane, D.B. & McKinney, T.F. (1973) Ridge and swale topography of the Middle Atlantic Bight, North America: secular response to the Holocene hydraulic regime. *Mar. Geol.*, **15**, 227–247.

Thorne, J.A. & Swift, D.J.P. (1991) Sedimentation on continental margins VI: regime model for depositional sequences, their component system tracts and bounding surfaces. In: *Shelf Sand and Sandstone Bodies: Geometry, Facies and Sequence Stratigraphy* (Eds Swift, D.J.P., Oertel, G.F., Tillman, R.W. & Thorne, J.A.), Spec. Publs int. Ass. Sediment., No. 14, pp. 189–225. Blackwell Scientific Publications, Oxford.

van Wagoner, J.C., Posamentier, H.W., Mitchum, R.M. *et al.* (1988) An overview of the fundamentals of sequence stratigraphy and key definitions. In: *Sea-level Changes: an Integrated Approach* (Eds Wilgus, C.K., Hastings, B.S., Kendall, C.G.StC., Posamentier, H.W., Ross, C.A. & van Wagoner, J.C.), Spec. Publ. Soc. econ. Paleont. Miner., Tulsa, **42**, 39–45.

Venkatachala, B.S., Sharma, M. & Shukla, M. (1996) Age and life in Vindhyans—facts and conjectures. In: *Recent Advances in Vindhyan Geology* (Ed. Bhattacharyya, A.), Mem. geol. Soc. India, **36**, 137–165.

Walker, R.G. & Plint, A.G. (1992) Wave and storm dominated shallow marine system. In: *Facies Models, Response to Sealevel Change* (Eds Walker, R.G. & James, N.P.), pp. 9–238. Geol. Ass. Can.

Williams, G.E. & Schmidt, P.W. (1996) Origin and paleomagnetism of the Mesoproterozoic Gangau tilloid (basal Vindhyan Supergroup), Central India. *Precam. Res.*, **79**, 307–325.

Wilgus, C.K., Hastings, B.S., Kendall, C.G.StC., Posamentier, H.W., Ross, C.A. & van Wagoner, J.C. (Eds) (1988) *Sealevel Changes: an Integrated Approach.* Spec. Publ. Soc. econ. Paleont. Miner., Tulsa, **42**, 407 pp.

*Spec. Publs int. Ass. Sediment.* (2002) **33**, 383–403

# Sedimentation and tectonic setting of Early Neoproterozoic glacial deposits in south-eastern Brazil

M. A. MARTINS-NETO *and* C. M. HERCOS*

*Departamento de Geologia, Escola de Minas, Universidade Federal de Ouro Preto Caixa Postal 173, 35 400–000, Ouro Preto/MG, Brazil*

## ABSTRACT

Glaciogenic deposits of the Neoproterozoic Macaúbas Megasequence, Jequitaí area, south-eastern Brazil, comprise two facies associations. The diamictite–sandstone–pelite facies association includes ice-proximal till, alluvial-fan and lacustrine deposits, and the outwash-plain facies association consists of proglacial fluvial sandstones. Stratigraphic, sedimentologic and structural data indicate that syn-depositional structural highs separated unique depocentres. The structural highs probably formed due to Early Neoproterozoic extensional tectonics during the break-up of Rodinia and may have been aided by plume-related doming. Tectonically induced topographic highlands may have played an important role in generating the Macaúbas glaciation.

## INTRODUCTION

The causes and controls of glaciation remain controversial, particularly for Precambrian examples (for recent reviews see Eyles & Young, 1994; Young, 1995a; Eriksson *et al.*, 1998). Fundamental questions centre on the roles of palaeoaltitude, atmospheric $CO_2$ concentration, tectonically induced uplift and whether or not the entire Earth was in a 'snowball' state (e.g. Young & Gostin, 1991; Eyles, 1993; Williams, 1994, 1996; Kaufman & Knoll, 1995; Young, 1995b; Kaufman, 1997; Park, 1997; Evans *et al.*, 1997; Eriksson *et al.*, 1998; Hoffman *et al.*, 1998; Martins-Neto, 1998a; Hoffman, 1999). Early Neoproterozoic (*c*.900 Ma) glaciogenic deposits occur in many parts of Brazil (Fig. 1; Trompette, 1994). The primary goal of this paper is to describe glaciogenic deposits in the Macaúbas Megasequence of the Jequitaí area, south-eastern Brazil (Fig. 2). Stratigraphic, sedimentologic and structural data indicate ice-proximal and proglacial outwash deposition in sub-basins partitioned by rifting processes, and suggest that rift-related uplift contributed to glaciation.

* Present address: PETROBRAS, E&P Campos Basin district, Macaé/RJ, Brazil.

## REGIONAL SETTING

The Jequitaí area is at the transition between the São Francisco craton and the external zone of the Araçuaí fold belt in south-eastern Brazil (Fig. 2). Three unconformity-bounded successions are recognized (Fig. 3; Martins-Neto & Pedrosa-Soares, 2000; Martins-Neto *et al.*, 2001): the Espinhaço Megasequence (Palaeoproterozoic to Mesoproterozoic); the Macaúbas Megasequence (Neoproterozoic); and the Bambuí Megasequence (Neoproterozoic).

The Espinhaço Megasequence (*c*.1720–1500 Ma) was deposited in an intracratonic rift-sag basin that evolved in four stages (pre-rift, rift, transitional and flexural), each of which is bounded by a regional unconformity (Martins-Neto, 1998b, 2000).

The Macaúbas Megasequence represents rift to drift sedimentation related to the break-up of Rodinia (Pedrosa-Soares *et al.*, 1992, 1998; Martins-Neto *et al.*, 2001). Glaciogenic deposits have long been recognized (e.g. Branner, 1919; Moraes & Guimarães, 1930; Isotta *et al.*, 1969; Karfunkel & Hoppe, 1988) and grade to distal glacially influenced marine sediment gravity flow deposits (Pedrosa-Soares *et al.*, 1992; Martins-Neto *et al.*, 1997, 2001; Uhlein *et al.*, 1999). Intermediate to felsic volcanic rocks interpreted

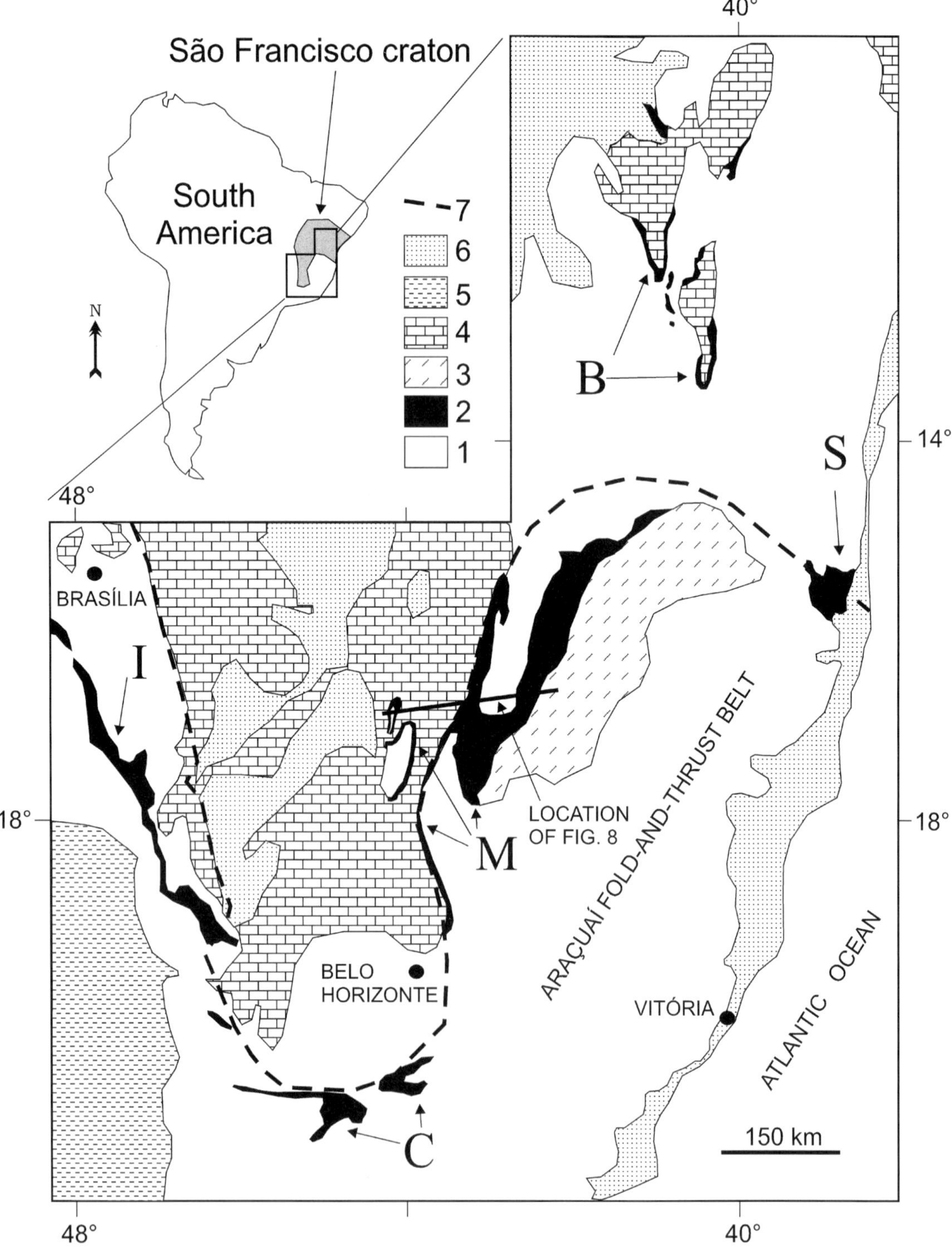


**Fig. 1.** Simplified geological map (modified from D'Agrella-Filho *et al.*, 1990) showing location of Early Neoproterozoic glaciogenic units in the São Francisco craton, Brasília and Araçuaí fold belts. (1) >1.1 Ga basement rocks; (2) glaciogenic units (M, Macaúbas deposits; B, Bebedouro Group; S, Salobro Formation; C, Carandaí Formation; I, Ibiá and Carmo do Rio Claro formations); (3) passive-margin deposits of the Macaúbas Megasequence; (4) Mid to Late Neoproterozoic Bambuí Group and Salitre Formation; (5) Phanerozoic Paraná basin; (6) Mesozoic and Cenozoic covers; (7) cratonic limit after Alkmim *et al.* (1993).

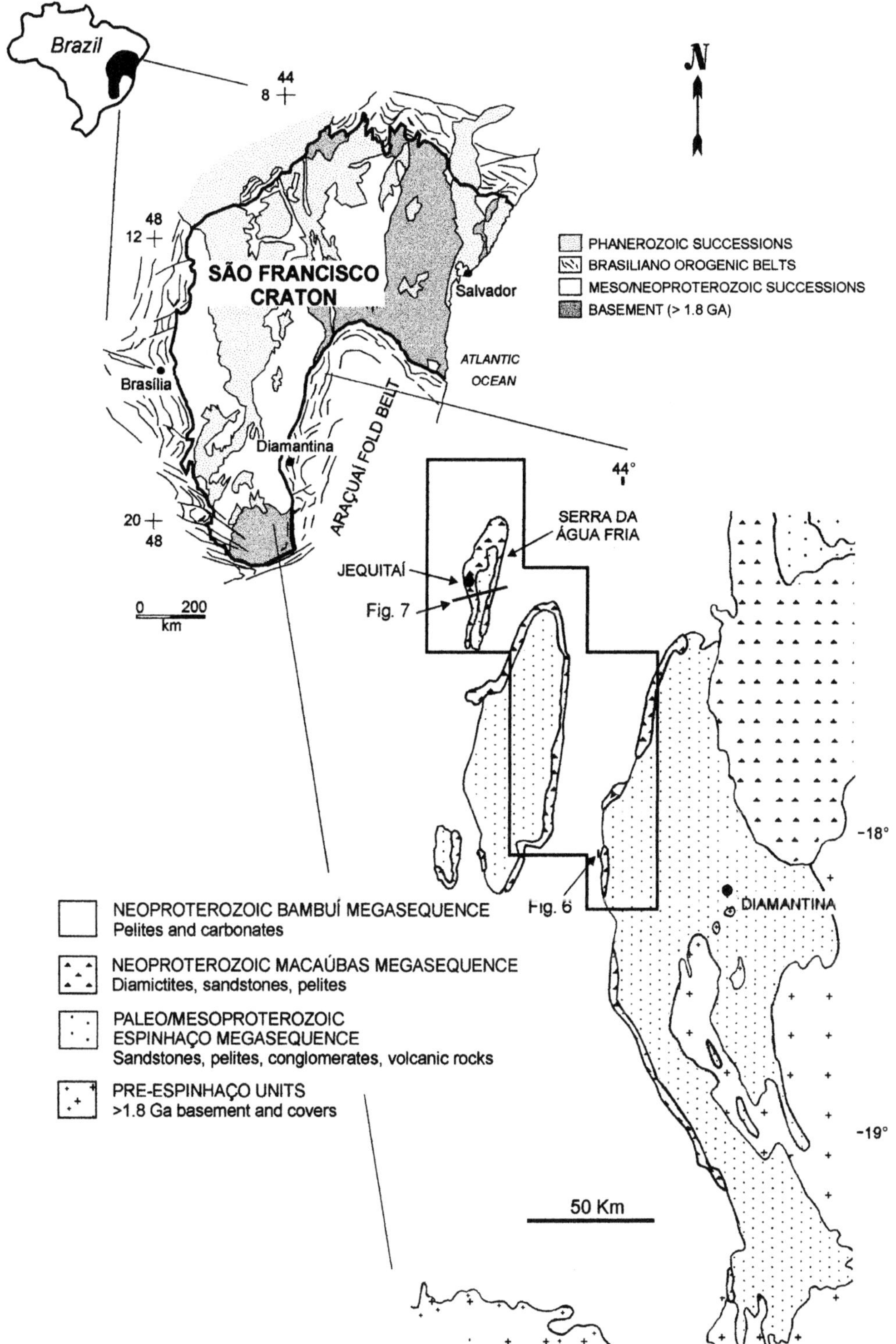


**Fig. 2.** Location of the studied area (inset) in the transitional domain between the São Francisco craton and the Araçuaí fold belt. Regional map of the São Francisco craton simplified from Alkmim & Marshak (1998). Geology of the detailed area simplified from Pedrosa-Soares *et al.* (1994).

| **Stratigraphy** | | **Age (Ma)** | **Basin type** | **Tectonic Regime** | **Depositional environment** | **Ref.** |
|---|---|---|---|---|---|---|
| **Megasequence** | **Tectonosequence** | | | | | |
| **Bambuí** | Três Marias | ~ 600 | Foreland | Flexural by tectonic load | Continental Shallow marine | 10,12 |
| | Paraopeba | < 790 | | | Shallow marine | 8,12,13 |
| *Paraconformity to erosional unconformity (locally angular)* 14,15,16,17,19,20 | | | | | | |
| **Macaúbas** | Superior | > 650 | Rift-Passive margin | Flexural/Thermal | Marine (glacio-eustatic sea-level rise) | 9,17, 19 |
| | Inferior | ~ 900 | | Extensional | Glacio-continental Glacio-marine | 9,11,18, 19, 23 |
| *Erosional unconformity (locally angular)* 17,18,19,21,22 | | | | | | |
| **Espinhaço** | Flexural Stage Conselheiro Mata | 1500 ? | Rift-Sag | Flexural/Thermal | Shallow marine Eolian | 6,7,22 |
| | Transitional Stage Galho do Miguel | | | | | 2,5,6 |
| | Rift Stage Sopa-Brumadinho São João da Chapada Natureza | ~ 1715 | | Extensional | Alluvial fan Fan delta Lacustrine Fluvial Eolian | 1,2,3, 4,6 |
| | Pre-rift Stage Olaria TS | ~ 1720 | | | | 1, 6 |

1)Reis(1999); 2,3,4,5,6)Martins-Neto(1993,1994,1996a, b,1998b); 7,8)Dupont(1995,1996); 9)Hercos & Martins-Neto(1997); 10)Chiavegatto(1992); 11)Uhlein (1991); 12)Dardenne (1978); 13)Schöll (1972); 14)Braun et al. (1993); 15)Teixeira et al. (1993); 16)Martins et al.(1993); 17,18) Martins-Neto et al.(1997, 1999); 19)Hercos(2000); 20)Romeiro Silva(1997); 21)Souza F°(1995); 22)Espinoza(1996); 23)Buchwaldt et al.(1999).

**Fig. 3.** Stratigraphic framework and tectono-depositional features of the Palaeo/Meso and Neoproterozoic metasedimentary covers of the São Francisco craton and Araçuaí fold belt (modified after Martins-Neto, 1998b).

as products of initial stretching have U–Pb zircon ages of *c.*1000 Ma (Brito Neves *et al.*, 1993) and badelleyite from mafic rocks yielded an age of 906 ± 2 Ma (Machado *et al.*, 1989). Macaúbas diamictites contain 1000–900 Ma detrital zircons (Buchwaldt *et al.*, 1999). The passive-margin wedge includes shelf, slope and deep-sea deposits (Lima *et al.*, in preparation; Martins-Neto *et al.*, 2001). Rocks inferred to represent related oceanic crust (northern termination of the Adamastor Ocean) have an Sm/Nd age of 816 ± 72 Ma (Pedrosa-Soares *et al.*, 1992, 1998).

The closing of the Brazilide Ocean (which began at *c.*800 Ma west of the São Francisco craton: Pimentel *et al.*, 1995) generated the Brasília fold belt and a foreland basin in which the Bambuí Megasequence (composed mainly of pelites, limestones and marls, and subordinate sandstones, arkoses and conglomerates) was deposited (Dardenne, 1978; Castro, 1997; Martins-Neto *et al.*, 1997). The subsequent closing of the northern branch of the Adamastor Ocean began at about 650 Ma east of the São Francisco craton (Araçuaí fold belt: Pedrosa-Soares *et al.*, 1992), deforming the eastern domains of the Bambuí Megasequence (Martins-Neto *et al.*, 1997).

## SUB-MACAÚBAS MEGASEQUENCE UNCONFORMITY

A prominent unconformity marks the base of the Macaúbas Megasequence throughout its outcrop area (Pflug & Renger, 1973; Martins-Neto *et al.*, 1997). Glacially abraded pavements (Fig. 4A & B) are common above the Espinhaço Megasequence basement, where structures varying from fine lineations to large grooves can be observed (Isotta *et al.*, 1969; Karfunkel and Hoppe, 1988; Espinoza, 1996; Hercos, 2000). Crescentic fractures and grooves showing steps (Fig. 4C) indicate the ice flow direction. Breccias can be found in some places mantling the grooved and striated surfaces (Fig. 5). The Macaúbas Megasequence fills palaeovalleys incised into successive stratigraphic levels of the Espinhaço Megasequence (Fig. 6).

On the western side of the Serra da Água Fria, an angular relationship between the Macaúbas and Espinhaço megasequences is preserved (Fig. 7). In contrast, on the eastern side, Macaúbas and Espinhaço strata are subparallel. Interpretation of reflection seismic sections (Hercos, 2000) and field data indicate that block tilting during rifting controlled the character of the basal unconformity, such that angular relationships are preserved on the faulted western margin and non-angular relationships on the flexed eastern margin.

**Fig. 4.** Subglacial erosion structures at sub-Macaúbas Megasequence unconformity; substrate consists of shallow-marine sandstones (Espinhaço Megasequence). (A) Grooved and (B) striated pavements. (C) Wide groove showing inside steps (arrowed), which give the palaeodirection of glacier movement (285°). A and C are from the western border of the Serra da Água Fria and B is from its eastern border (see Figs 2 & 7).

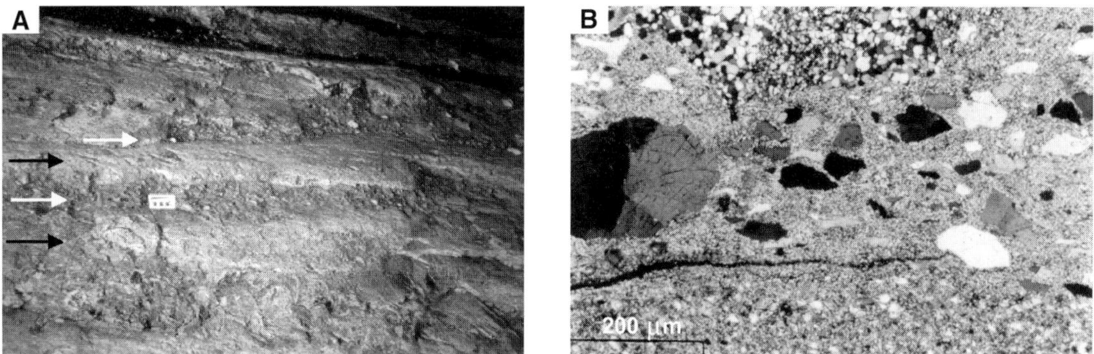


**Fig. 5.** (A) Unconformity surface above Espinhaço Megasequence aeolian deposits: glaciogenic breccia fills and mantles furrows (elongated left to right, white arrows) above aeolian-dune foresets (black arrows) in basement. (B) Thin section of poorly sorted breccia with angular fragments floating in matrix of quartz, sericite, biotite and opaque minerals. Note large aeolian clast with bimodal grain-size banding in upper part of photo. Crossed polarizers. Outcrop located at the pavement shown in Fig. 12.

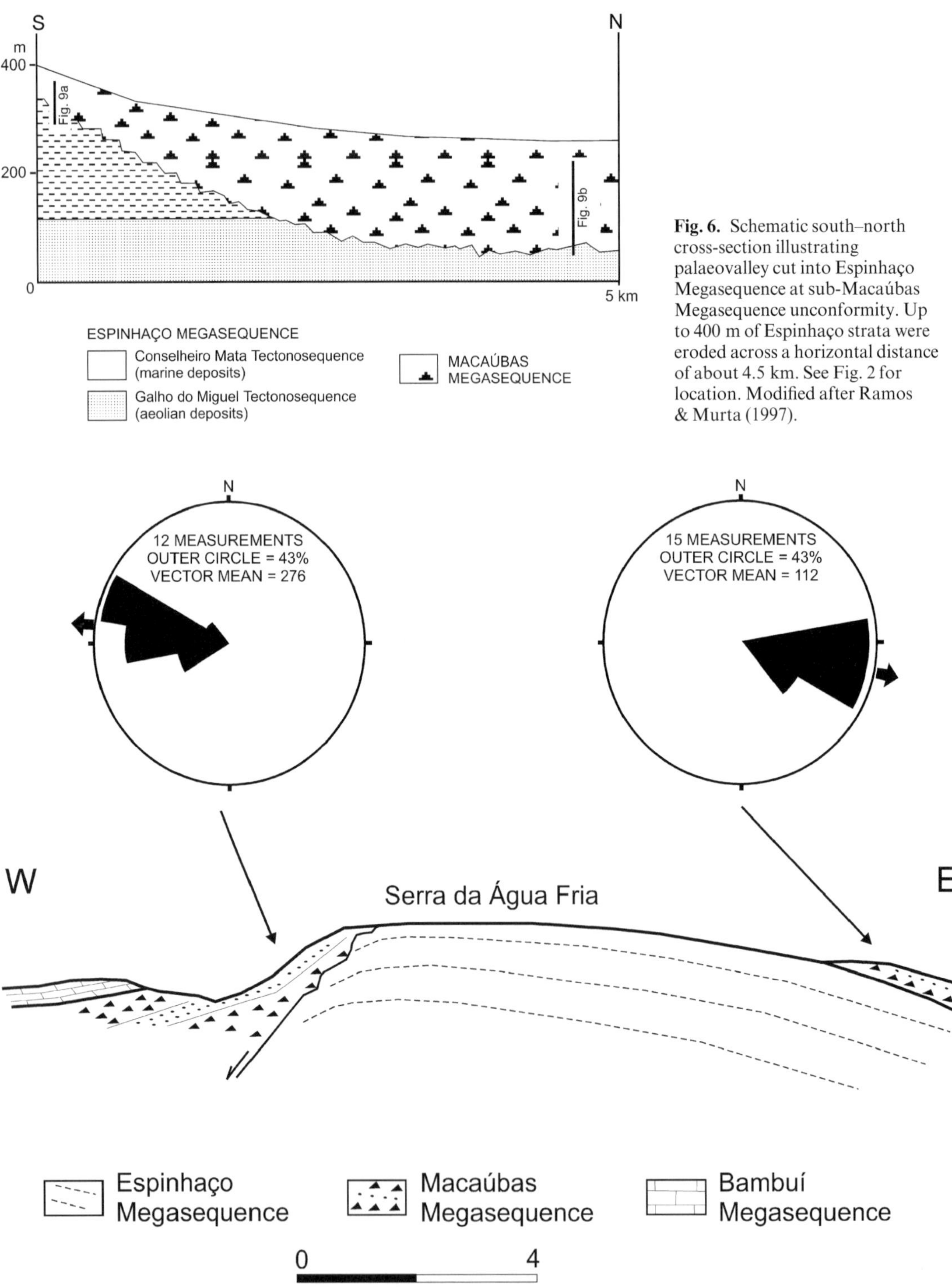


**Fig. 6.** Schematic south–north cross-section illustrating palaeovalley cut into Espinhaço Megasequence at sub-Macaúbas Megasequence unconformity. Up to 400 m of Espinhaço strata were eroded across a horizontal distance of about 4.5 km. See Fig. 2 for location. Modified after Ramos & Murta (1997).

**Fig. 7.** Simplified cross-section through the Serra da Água Fria. On the western side, an angular discordance separates Macaúbas beds from Espinhaço basement and palaeocurrents are to the west, in contrast to the eastern side where units are subparallel and palaeocurrents are to the east. See Fig. 2 for location.

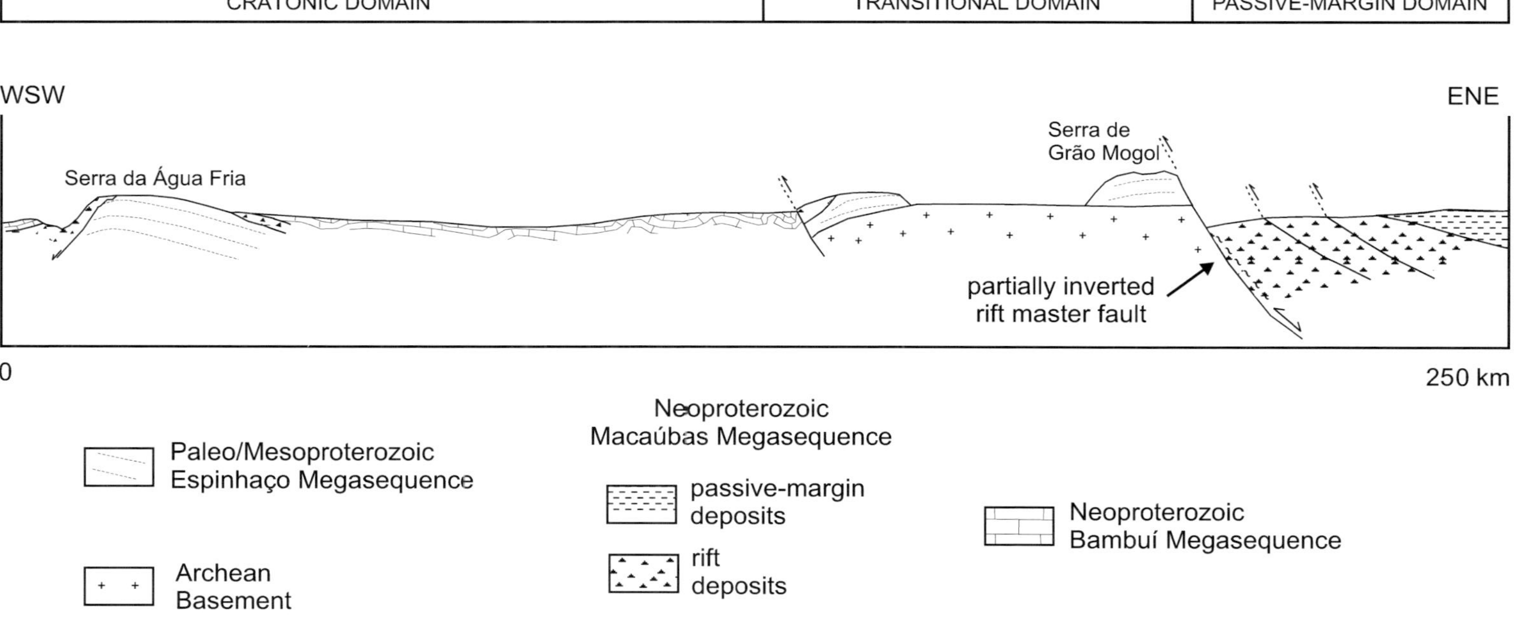


**Fig. 8.** Simplified regional cross-section along the São Francisco craton border and Araçuaí fold belt illustrating the cratonic/transitional and passive-margin domains of the Macaúbas-Salinas basin. The partially inverted rift master fault defines the boundary between the cratonic/transitional domain and the passive-margin domain. The Serra de Grão Mogol is considered to represent a preserved rift shoulder. Shortening structures in the passive-margin domain are greatly simplified. See Fig. 1 for location.

## DEPOSITIONAL DOMAINS OF THE MACAÚBAS–SALINAS BASIN AND THE CONTEXT OF GLACIOGENIC RIFT DEPOSITS

The Macaúbas-Salinas basin defines three depositional domains (Fig. 8). In the east, the passive-margin domain contains up to 12 km (Uhlein *et al.*, 1999) of rift and passive-margin wedge deposits. The rift deposits consist of marine diamictites, greywackes and pelites deposited by sediment gravity flows (Uhlein *et al.*, 1999; Martins-Neto *et al.*, 2001). The passive-margin wedge comprises greywackes, sandstones and pelites, that accumulated in a shelf–slope–deep-sea depositional system (Lima *et al.*, in preparation; Martins-Neto *et al.*, 2001). In addition, it contains a volcanic-exhalative sedimentary unit that is associated with ultramafic rocks and is interpreted to be the remnant of the northernmost branch of the Adamastor ocean (Pedrosa-Soares *et al.*, 1998).

In the west, the cratonic and transitional domains consist of relatively thin (to several hundred metres) glaciogenic rift deposits. This paper focuses on ice-proximal alluvial-fan, fluvial and lacustrine facies. In response to glacio-eustatic sea-level fluctuations, these interfingered with shallow-marine deposits, characterized by sandstones, siltstones and pelites with parallel-lamination, wave ripple marks and hummocky cross-stratification (Karfunkel & Hoppe, 1988; Espinoza, 1996; Ramos & Murta, 1997; Martins-Neto *et al.*, 2001).

## GLACIOGENIC DEPOSITS OF THE CRATONIC AND TRANSITIONAL DOMAINS, MACAÚBAS MEGASEQUENCE

Two facies associations can be recognized in the cratonic and transitional domains of the Macaúbas Megasequence: the diamictite–sandstone–pelite facies association (Table 1) and the outwash-plain facies association (Table 2).

### The diamictite–sandstone–pelite facies association

#### *Facies description and interpretation*

*Diamictites (facies Dm).* Massive diamictites crop out throughout the study area, forming dm- to m-scale sheet-like to lenticular bodies (Figs 9, 10 & 11). Although most diamictites are interbedded with sandstones, amalgamated intervals up to 100 m thick are common, particularly near the base of the unit, in the deeper parts of the palaeovalleys (Figs 6 & 9A & B). Some bodies increase in thickness down-palaeoslope. Contacts between diamictite bodies are sharp, varying from relatively flat to strongly irregular.

Clasts are angular to subrounded and range in size from granules to large boulders, including metre to decametre bedrock rafts (Fig. 11C & D). Clasts float in a texturally and compositionally immature sandstone matrix that typically contains significant proportions of clay minerals and sericite. The clast/matrix ratio is variable, with the clast percentage ranging between 10 and 50% (Fig. 11A & B). Scratched, faceted and polished clasts have been reported (e.g. Fig. 6 in Karfunkel & Hoppe, 1988). Clast types reflect diverse sources, comprising predominantly sandstones and carbonates (mostly dolostones), subordinate granites, gneisses, vein quartz and mafic and felsic volcanic rocks. Some bodies contain exclusively sandstone clasts, whereas others contain exclusively or predominantly granitic and gneissic clasts. According to Karfunkel & Hoppe (1988), crystalline-basement clasts tend to be predominant in diamictites at higher stratigraphic levels, reflecting deep levels of erosion. However, some diamictites containing exclusively granitic and gneissic clasts also occur near the middle of the measured sections, between polymictic diamictites (e.g. Fig. 10 at about 130 m). Hence, although clast compositions record a general trend of cover to crystalline basement stripping, glacial debris was derived from unique source rocks at different times. Carbonate clasts are more common near the base of the section (Karfunkel & Hoppe, 1988; Ramos & Murta, 1997), suggesting initial erosion of a carbonate platform.

The question as to whether ancient diamictites such as those in the Macaúbas Megasequence are true 'tillites' deposited directly by glacial ice, rather than mass flow deposits of resedimented debris, remains controversial (e.g. Eriksson *et al.*, 1998, and references therein). In the present example, abundant well sorted sandstones intercalated with the diamictites indicate concurrent fluvial outwash and diamictite sedimentation, and some of the diamictites that thicken down palaeoslope may be debris flow deposits. However, in view of the diverse clast suite, polished, faceted and striated clasts, and occurrence of diamictites directly above striated pavements (Fig. 12), minimal reworking is indicated and at least some of the diamictites are probably true continental tillites in the sense of Dreimanis & Schlüchter (1985).

**Table 1.** Diamictite–sandstone–pelite facies association (codes modified after Miall, 1978).

| Facies | Key lithologies | Sedimentary features | Depositional processes |
|---|---|---|---|
| Dm, Diamictites | Diamictites containing granules to large boulders (and some metric to decametric bedrock rafts), composed of sandstones, carbonates, pelites, granites, gneisses, vein quartz and volcanic rocks | Massive, sheet-like to lenticular bodies, high variation in clast/matrix ratio | Direct deposition by glaciers and resedimented mass flows |
| SSp, Parallel-stratified beds | Cm to dm thick interbedding between sandstones and siltstones, locally with thin intervening mudstone layers; with scattered sandstone, carbonates and vein quartz pebbles and cobbles | Parallel-stratification, undulating bedding, soft-sediment deformation structures | Sheet floods |
| Sm, Massive sandstones | Medium- to coarse-grained to pebbly, poorly sorted sandstones, with scattered sandstone, carbonates and vein quartz pebbles and cobbles | Structureless, homogeneous, sheet-like bodies, local soft-sediment deformation structures | Heavily sediment-laden (high concentration) mass flows |
| Sh, Horizontally stratified sandstones | Fine- to coarse-grained, poorly sorted sandstones, local scattered sandstone, carbonates and vein quartz pebbles and cobbles | Parallel-stratification, low-angle truncations, local soft-sediment deformation structures | Sheet floods |
| St, Trough cross-stratified sandstones | Fine- to coarse-grained, poorly sorted sandstones, sometimes with scattered sandstone, carbonates and vein quartz pebbles and cobbles | Trough cross-stratification, soft-sediment deformation structures | 3-D dune migration |
| Sg, Graded sandstones | Fine- to medium-grained sandstones | Sheet-like bodies with normal grading, parallel-lamination and ripple cross-lamination | Turbidity currents |
| Rh, Silt/mud rhythmites | White siltstones and claystones, with local isolated outsized clasts | Mm rhythmic alternation | Varves with dropstones |
| P, Pelites | Mudstones | Structureless or parallel-laminated | Vertical accretion |

**Table 2.** Outwash-plain facies association (codes modified after Miall, 1978).

| Code | Characteristics | Interpretation |
|---|---|---|
| Sh | Parallel-laminated/stratified sandstones | Sheetfloods under upper-flow regime |
| St/Sp | Cross-bedded sandstones | Subaqueous dune migration under lower-flow regime |
| Sr | Cross-laminated sandstones | Current-ripple migration |
| P | Pelites | Vertical-accretion deposits |

*Parallel-stratified beds (facies SSp).* Locally intervening between the basal unconformity and diamictite beds is a thin (to 5 m) package of sandstone, siltstone and mudstone (e.g. Figs. 9B & 12). The sandstones form cm- to dm-thick parallel-stratified layers containing scattered sandstone, carbonate and vein quartz pebbles and cobbles. In-phase, undulating bedding is locally developed (Fig. 13A) and may reflect deposition by supercritical flows. Soft sediment deformation structures are common, including convolute-bedding (Fig. 13B), dewatering structures and load structures. The sandstone beds were probably

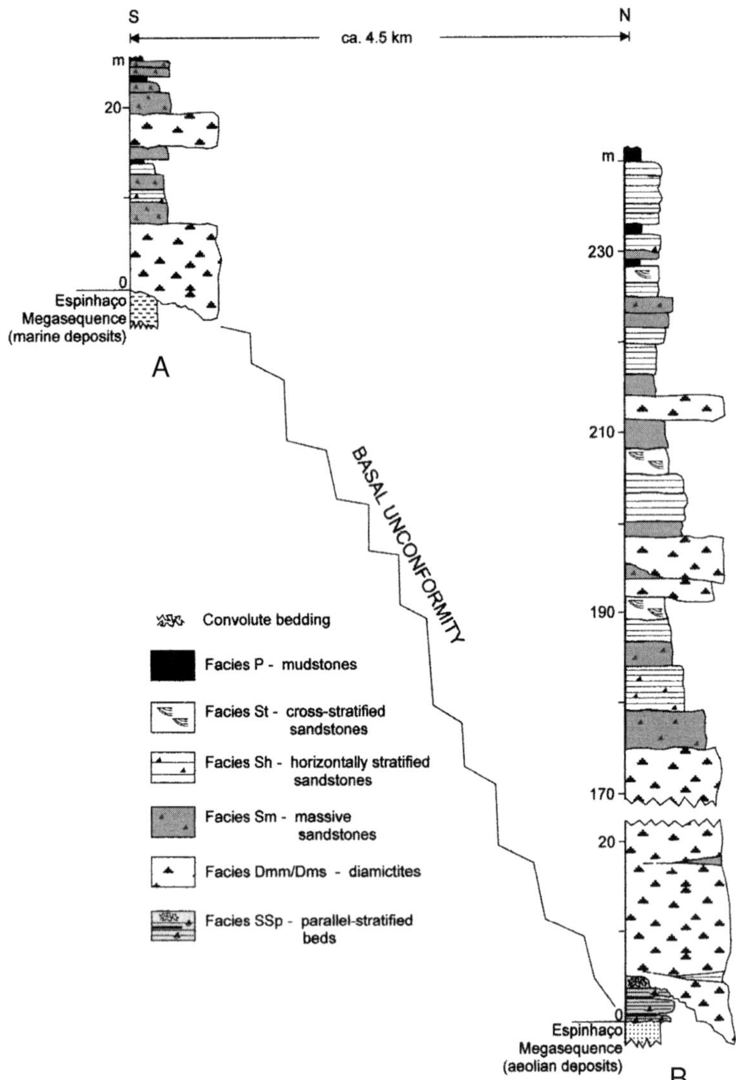


**Fig. 9.** Measured sections of the diamictite–sandstone–pelite facies association. See Figs 2 & 6 for location. Modified after Ramos & Murta (1997).

deposited by high-energy sheet floods generated by melting ice. Their position between the basal unconformity and diamictite beds suggests proglacial or subglacial processes; abundant soft-sediment deformation structures may reflect loading by ice and/or till sheets.

*Massive sandstones (facies Sm).* Massive, sheet-like to lenticular bodies of medium- to coarse-grained, poorly sorted sandstones and pebbly sandstones are interbedded with diamictites (Figs 9A, B & 10). Convolute bedding is locally observed, suggesting rapid deposition from flows of declining velocity at high sedimentation rates (Lowe, 1975). Because the sandstones lack structures produced by stream-flow processes such as stratification or internal erosion surfaces, each bed probably represents a single depositional event, involving the rapid settling of a high-concentration mass flow. This is also indicated by the massive, poorly sorted texture of the beds and the occurrence of framework-supported pebbles and cobbles.

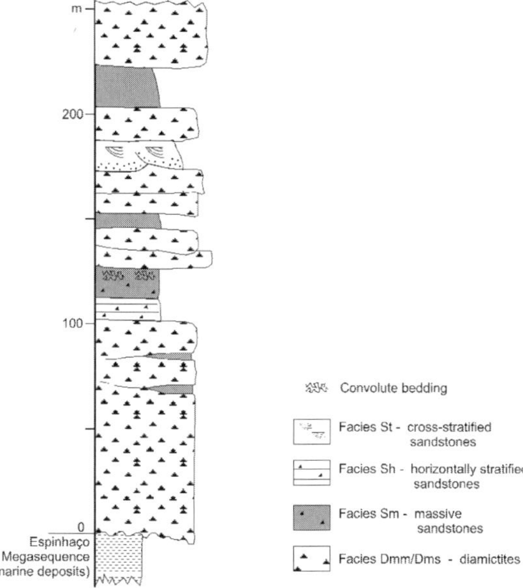


**Fig. 10.** Measured section of the diamictite–sandstone–pelite facies association near the western border of the Serra da Água Fria. See Figs 7 & 8 for location. Modified after Hercos & Martins-Neto (1997).

*Horizontally stratified sandstones (facies Sh).* Interbedded with massive sandstones (facies Sm) and cross-stratified sandstones (facies St) (Figs 9A, B, 10 & 13C), sheet-like beds of facies Sh are composed of fine- to coarse-grained to pebbly, poorly sorted sandstones. Internal erosion surfaces are common and granules and pebbles of sandstone, carbonate and vein quartz are concentrated along these surfaces. Slump and dewatering structures are locally developed. These horizontally stratified sandstones probably represent upper flow regime sheet floods.

*Trough cross-stratified sandstones (facies St).* Facies St comprises fine- to coarse-grained poorly sorted trough cross-stratified sandstones. Local pebbly beds are at the base of upward-fining sequences (e.g. Fig. 10; *c.*180 m). Facies St is interpreted as the product of subaqueous three-dimensional dune migration under lower flow regime conditions. Reliable palaeocurrent measurements can be taken from facies St. The obtained data allowed the interpretation that sediment dispersion was controlled by a complex palaeogeography composed of structural highs that supplied sediments and depocentres that captured them (e.g. rose diagrams on Fig. 7).

*Graded sandstones (facies Sg).* Facies Sg consists of sandstone to pelite Bouma-like upward-fining sequences. Basal, medium- to fine-grained sandstones (1–15 cm thick) are sheet-like and normally graded (Fig. 13D), and display sharp, erosional contacts with local granule concentrations. They pass upwards into sandstones with crude parallel-stratification, ripple cross-stratification and pelite (Fig. 13D). The upward-fining sequences are probably subaqueous mass flow deposits, reflecting low-concentration turbidity currents.

*Silt/mud rhythmites (facies Rh).* Rhythmites composed of mm-scale siltstone and claystone alternations form local m-scale intervals (Karfunkel & Hoppe, 1988;

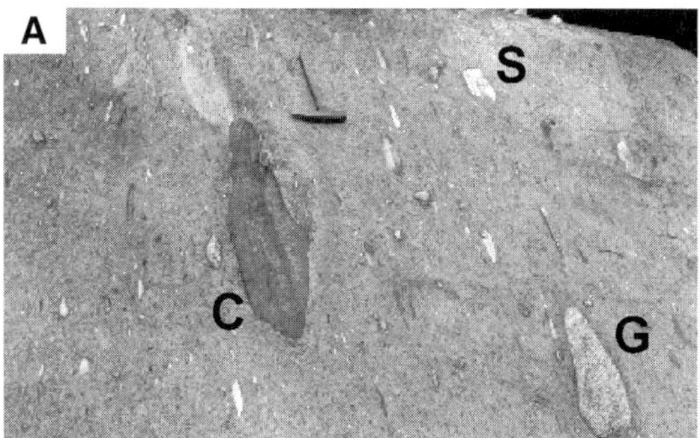


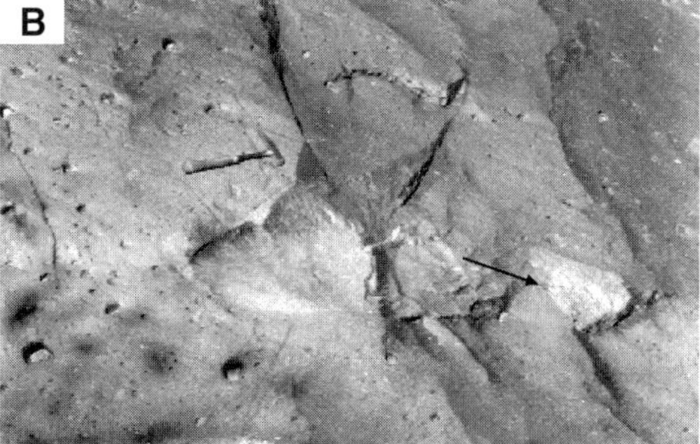


**Fig. 11.** (A) Diamictite of facies Dm with carbonate (C), sandstone (S) and granitoid (G) clasts (preferred clast orientation is along cleavage). (B) Diamictite of facies Dm with large granitoid boulder, which broke during deposition (two pieces are linked by the arrow). (*continued on p. 394*)

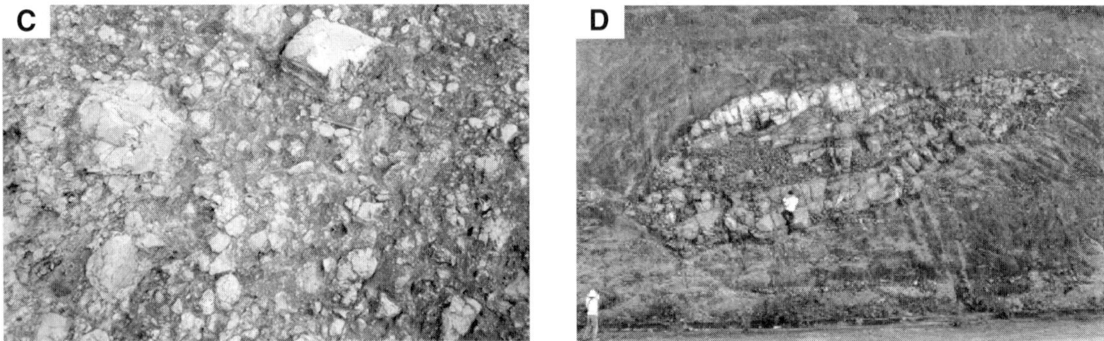


**Fig. 11.** (*cont'd*) (C) Clast-rich diamictite of facies Dm. (D) Large raft of aeolian Espinhaço sandstone, probably derived from the Serra da Água Fria to the south-south-west (photo is from a road cut situated *c.*500 m to the north-east of the outcrop located in Fig. 15).

**Fig. 12.** Wedge of parallel-stratified sandstone (facies SSp) above glacial pavement at the sub-Macaúbas unconformity, located at the base of the vertical section shown in Fig. 9B. Dark-toned area of facies SSp forms part of cliff face that is in shadow. Tillites (facies Dm) lie directly above pavement at the right of the photo. Note man for scale (circled).

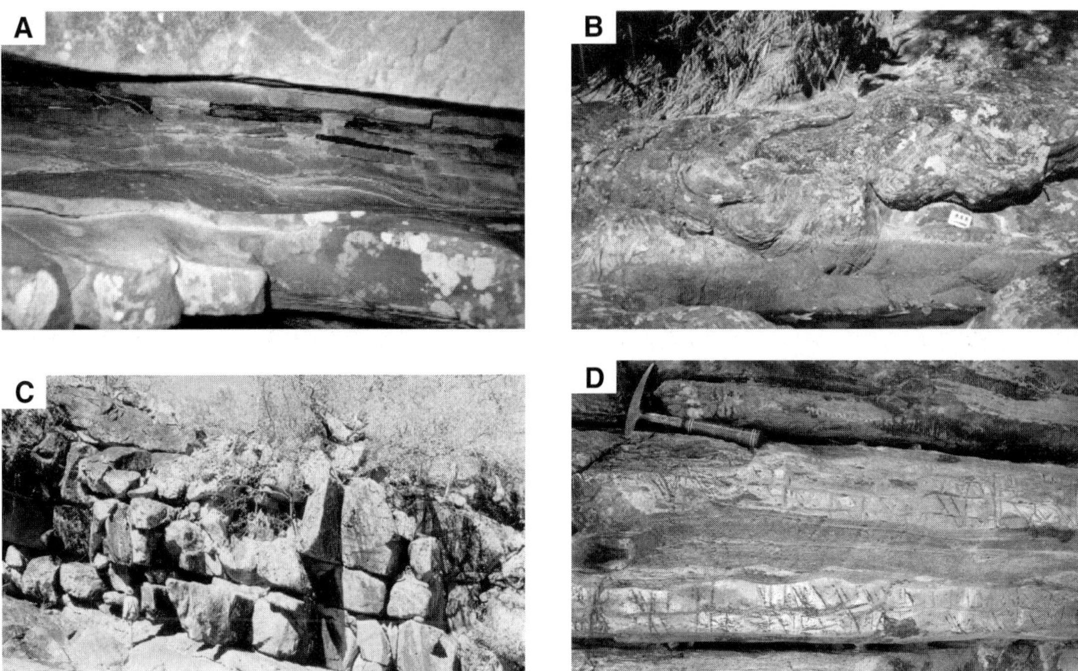

**Fig. 13.** (A) Undulating, in-phase beds of facies SSp. (B) Convolute bedding in sandstones of facies SSp. (C) Sandstone beds of the diamictite–sandstone–pelite facies association. Man is standing on top of a diamictite bed. (D) Turbidite beds of facies Sg.

Ramos & Murta, 1997). Local granule and pebble lonestones deform bedding (Fig. 8 in Karfunkel & Hoppe, 1988), and probably represent dropstones derived from melt-out of lacustrine icebergs.

*Pelites (facies P).* Structureless or parallel-laminated mudstones are interbedded with the sandstone facies, particularly in more distal parts of the facies association. Poor exposure precludes an accurate reconstruction of the geometry and areal distribution of this facies. However, the association with the sandstone facies suggests deposition by vertical accretion, probably on flood plains.

## The outwash-plain facies association

### *Facies description and interpretation*

Four facies are recognized in the outwash-plain facies association (Table 2): parallel-stratified sandstones, cross-stratified sandstones, ripple cross-laminated sandstones and pelites.

*Parallel-stratified sandstones (facies Sh).* Facies Sh forms sheet-like, locally amalgamated bodies up to 1 m thick. These have high lateral extent/thickness ratios. Lower surfaces are typically erosional; parallel-stratified sandstones locally grade upwards to cross-stratified sandstone facies. The sandstones are generally fine- to medium-grained and moderately sorted; but coarse-grained, poorly sorted bands can be observed locally. The sandstones are interpreted to be upper flow regime sheet flood deposits.

*Cross-stratified sandstones (facies St/Sp).* Facies St/Sp is composed of fine- to medium-grained, moderately sorted sandstones, organized in 10–30 cm thick cross-stratified sets. The 30–70 cm thick cosets are lenticular with high lateral extent/thickness ratios. Both trough (St) and planar–tabular (Sp) cross-stratification is developed. Overturned cross-beds can be locally observed (see Fig. 17b), suggesting rapid deposition and flows with highly oscillating power (e.g. Allen, 1984). Facies St is interpreted to represent the migration of subaqueous three-dimensional dunes under

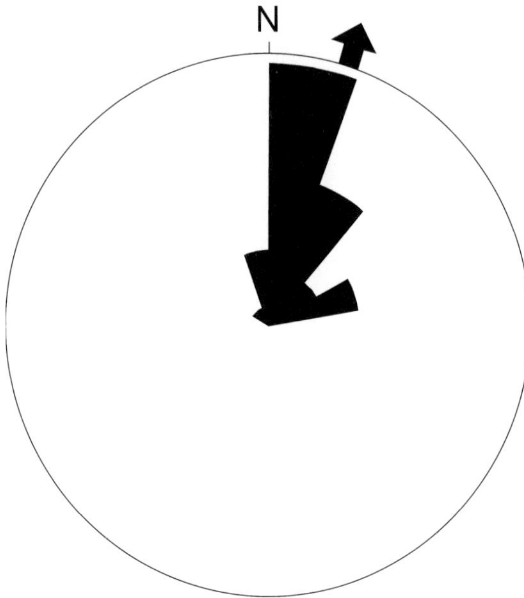


**Fig. 14.** Palaeocurrents from the outwash-plain facies association at Serra da Água Fria.

lower flow regime conditions, whereas facies Sp is the product of two-dimensional dune migration (see Ashley, 1990). The cross-beds indicate unimodal NNE palaeocurrents (Fig. 14).

*Ripple cross-laminated sandstones (facies Sr).* Locally developed cm-scale cross-laminated fine-grained sandstone intercalations reflect current ripple migration in the lower part of the lower flow regime.

*Pelites (facies P).* Thin (mm- to cm-scale) locally continuous argillaceous siltstones cap some sandstone cosets. These siltstones are probably vertical accretion deposits formed during the end stages of flood cycles. Some of the sandstones contain intraclasts derived from reworking of desiccated pelite layers.

### *Depositional architecture*

In order to characterize the depositional architecture of the outwash-plain facies association, a *c.*120 m long outcrop was studied in detail (Figs 15, 16, 17A & 18). Waning flood sequences vary in thickness from 0.3 to 1.2 m (0.5 m average), each of them representing a depositional/aggradational event, characterized by gradual loss of energy with time. A complete sequence begins with the products of upper flow regime (parallel-stratified sandstones of facies Sh), followed by sets

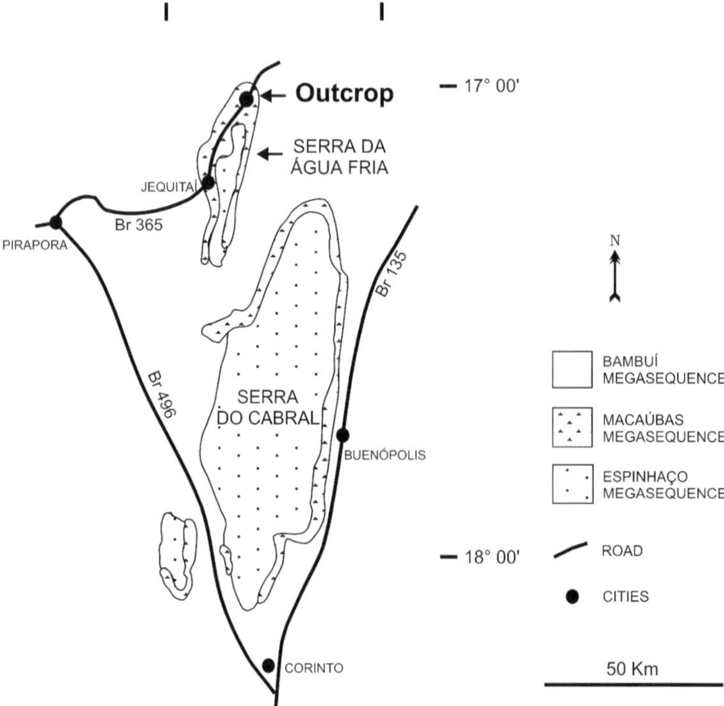


**Fig. 15.** Location of outcrop used for detailed study of outwash-plain facies association (see Figs 16–18). Geology modified after Pedrosa-Soares *et al.* (1994).

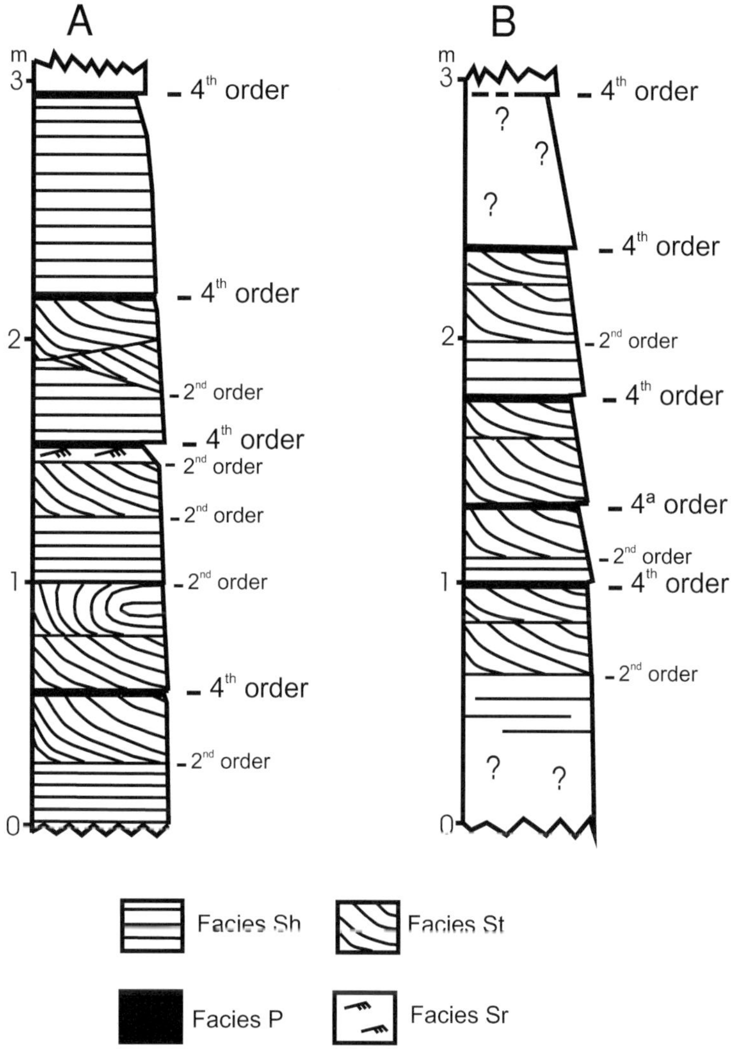

**Fig. 16.** Measured sections, outwash-plain facies association, Serra da Água Fria. Hierarchy of the bounding surfaces according to Miall (1990, p. 214). See Fig. 15 for location, Fig. 18 for context.

**Fig. 17.** (A) Waning-flood sequence with horizontally stratified sandstones at base (facies Sh) followed by cross-stratified sandstones (facies St/Sp) that are capped by a thin mudstone drape (facies P). (B) Overturned cross-bed in facies St/Sp.

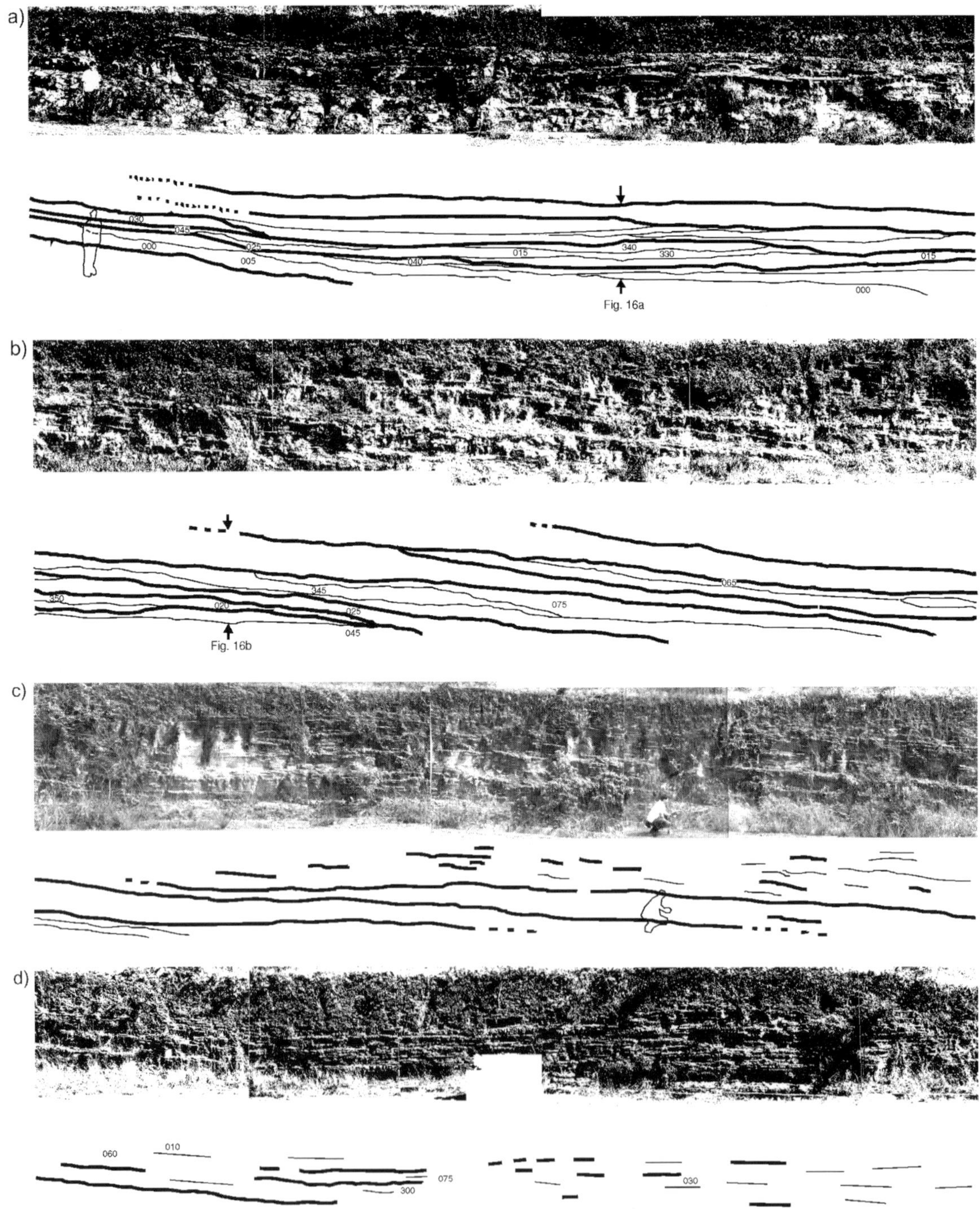


**Fig. 18.** Photomosaic of outwash-plain facies association, Serra da Água Fria. Segments a, b, c, d are continuous viewed towards 220–040°. Thick lines represent fourth-order surfaces; thin lines represent second-order surfaces. Note decrease in the definition of surfaces and depositional elements from a to d. Three-digit numbers indicate palaeocurrent azimuths. Vertical arrows show the base and top of the measured sections illustrated in Fig. 16.

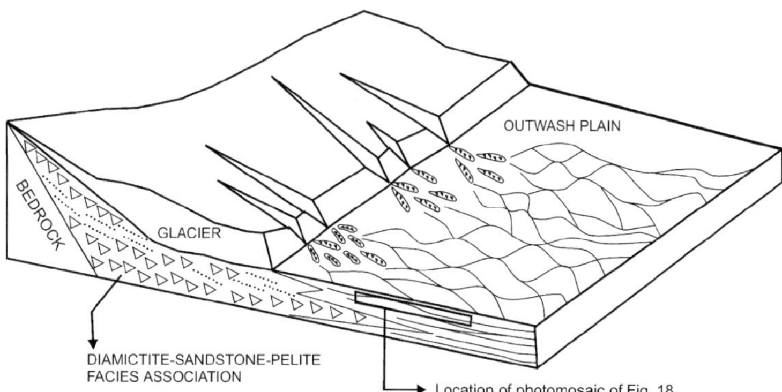


**Fig. 19.** Schematic palaeogeographical reconstruction of the outwash-plain facies association of the Macaúbas Megasequence, illustrating its relationship to the diamictite–sandstone–pelite facies association (not to scale).

of cross-stratified sandstones (facies St/Sp), ripple cross-laminated fine-grained sandstone (facies Sr) and mudrocks of vertical accretion origin (facies P). However, most of the waning-flood cycles are incomplete: some cycles do not have a basal facies Sh, indicating that the entire flood was lower flow regime; other cycles lack Sr at their top, suggesting rapid waning of the flow. The presence of pelites of facies P directly over sandstones of facies Sh shows deposition of mud over upper flow regime flat beds, suggesting particularly rapid deceleration.

Waning flood sequences define sheet-like bodies with high lateral continuity and low relief. These are separated by fourth-order bounding surfaces (terminology of Miall, 1990) that locally completely remove cosets of underlying sequences. Second-order bounding surfaces separate cosets comprising sets of facies Sh and St/Sp (Figs 16, 17 & 18). Some separate cosets of facies St/Sp form cosets of facies Sr. Bodies enclosed by second-order bounding surfaces are normally lenticular and show thickness compensation with the neighbouring cosets where they pinch out laterally. First-order bounding surfaces, not detectable in the photomosaic of Fig. 18, can be recognized in the field, where they separate cross-bedded sets of facies St/Sp. Significantly, bounding surfaces become diffuse in a downcurrent direction (Fig. 18), probably indicating proximal to distal expansion of meltwater floods which spread beyond the limits of wide, shallow braided channels.

### Depositional and palaeogeographical model

We envisage that the diamictite–sandstone–pelite facies association reflects ice-proximal glacial, mass flow, alluvial fan, fluvial and lacustrine processes and that the outwash-plain facies association represents proglacial braidplain deposition (Fig. 19) controlled by high-discharge meltwater floods. The diamictite–sandstone–pelite facies association was deposited above a high-relief basement palaeotopography in which depocentres were partitioned by structural highs. The outwash-plain facies association contains characteristics typical of braidplains in which ephemeral flood processes were predominant, such as: an abundance of upper flow regime parallel stratification; a simplicity of facies; unimodal, low-dispersion palaeocurrents; high width to depth ratio of depositional elements; a loss of bounding surfaces downcurrent; and an absence of well defined channels, macroforms and banks. Although differences between glacial and non-glacial braidplain deposits are far outweighed by their similarities (Smith, 1985), lateral interfingering between the diamictite–sandstone–pelite facies association and the outwash-plain facies association suggests a genetic link. This interfingering is probably a consequence of advance and retreat of the glacial front.

## DISCUSSION

As outlined above, ice-proximal and proglacial sedimentation in the cratonic and transitional domains of the Macaúbas Megasequence occurred in sub-basins that were separated by basement palaeohighs. Deep palaeovalleys were filled with tillites, mass-flow diamictites, fluvial sandstones and lacustrine rhythmites. These interfinger with proglacial outwash deposits that prograded axially. Block tilting, indicated by asymmetric angular relationships at the sub-Macaúbas unconformity across the Serra da Água Fria (Fig. 7), was probably responsible for basin partitioning, and reflects Early Neoproterozoic extensional tectonics during Rodinia break-up (Martins-Neto *et al.*, 1997).

Previous workers (e.g. Karfunkel & Hoppe, 1988; D'Agrella-Filho *et al.*, 1990; Uhlein *et al.*, 1995, 1999) have interpreted Macaúbas glaciogenic deposits to represent a uniformly eastwards moving continental ice-sheet. However, data presented herein indicate multiple depocentres and a more complex dispersion pattern. The sub-basins preserved in the cratonic and transitional domains represent subsidiary structures to the main rift depocentre of the Macaúbas-Salinas basin, and the Serra de Grão Mogol represents an Early Neoproterozoic rift shoulder (Fig. 8). We infer that uplift related to the elastic rebound of the rift shoulders contributed to the Macaúbas glaciation (Martins-Neto *et al.*, 1997). In addition, the rifting process which generated the Macaúbas–Salinas Basin may be related to a mantle plume (Martins-Neto, 1998a). Correa-Gomes & Oliveira (1997) have proposed an early Neoproterozoic mantle plume in the region where the Araçuaí and West Congo fold belts are presently located, based on reconstruction of flow directions in mafic dykes in east-south-east Brazil and the Congo craton, as well as on lithogeochemical analysis. The initial regional domal uplift due to the plume activity would have generated elevated rift structures and, consequently, favoured the generation of glaciers on the rift shoulders (Martins-Neto, 1998a).

## CONCLUSIONS

Early Neoproterozoic (*c*.900 Ma) glaciogenic deposits (Macaúbas Megasequence) are well exposed in the Jequitaí area, south-eastern Brazil. The Macaúbas glaciation corresponds in age to the Lower Congo glaciation in West Africa (Hambrey & Harland, 1985), whose products are preserved at the base of the West Congo Supergroup (Trompette, 1994).

A prominent unconformity, including incised palaeovalleys, marks the base of the Early Neoproterozoic Macaúbas Megasequence throughout its outcrop area. Three depositional domains can be defined in the Macaúbas–Salinas basin. In the east, the passive-margin domain contains up to 12 km (Uhlein *et al.*, 1999) of rift and passive-margin wedge deposits. In the west, the cratonic and transitional domains are composed of significantly thinner (up to several hundred metres thick) syn-rift glaciogenic deposits.

Two facies associations can be recognized in the deposits of the cratonic and transitional domains: the diamictite–sandstone–pelite facies association, which includes ice-proximal till, alluvial-fan and lacustrine deposits; and the outwash-plain facies association, which consists of proglacial fluvial sandstones. Stratigraphical, sedimentological and structural data indicate that syn-depositional structural highs separated unique depocentres. These structures probably formed due to Early Neoproterozoic extensional tectonics during the break-up of Rodinia, and may have been aided by plume-related doming.

Palaeomagnetic studies on *c*.1.0 Ga basic dykes in the eastern São Francisco craton (D'Agrella-Filho *et al.*, 1990) indicate that the glaciated area would have been located between 40° and 60° latitude at the start of Macaúbas rifting. The Macaúbas glaciation probably occurred on an elevated supercontinent (Rodinia), where larger areas could have been covered by a continental ice-sheet. In the Jequitaí area, tectonically induced elevations in topography during the rifting process may have played an important role in generating the glaciers and in controlling sediment dispersion and preservation.

As emphasized by Eriksson *et al.*, (1998, Section 10), in contrast to other depositional systems, processes of erosion, transportation and deposition in Precambrian glaciated systems may have been quite similar to modern settings. The data and results presented above, with the exception of the difficulty in recognizing true continental till deposits, confirm that statement when compared to those of Quaternary and modern settings (e.g. Menzies, 1995; Miller, 1996).

## ACKNOWLEDGEMENTS

Frst we would like to thank Wlady Altermann for the invitation to contribute to this volume. Larry Aspler is especially thanked for careful review and constructive suggestions on an earlier draft. We are also grateful to R. Pflug and H. Miller for their reviews of the final manuscript. The research which resulted in this article (which includes the MSc thesis of C.M.H.) has been funded by grants to M.A. Martins-Neto from FAPEMIG–Research Support Foundation of Minas Gerais State, Brazil, and from CNPq–Brazilian National Research Council, through a research fellowship.

## REFERENCES

Alkmim, F.F., Brito Neves, B.B. & Alves, J.A.A. (1993) Arcabouço tectônico do Cráton do São Francisco: uma revisão. In: *O Cráton do São Francisco* (Eds Dominguez, J.M.L. & Misi, A.), pp. 45–62. SBG/SGM/CNPq, Salvador.

ALKMIM, F.F. & MARSHAK, S. (1998) Transamazonian orogeny in the southern São Francisco craton region, Minas Gerais, Brazil: evidence for Paleoproterozoic collision and collapse in the Quadrilátero Ferrífero. *Precam. Res.*, **90**, 29–58.

ALLEN, J.R.L. (1984) *Sedimentary Structures: Their Character and Physical Basis*, Vol. 1, 2nd edn. Developments in Sedimentology, 30. Elsevier, Amsterdam, 593 pp.

ASHLEY, G.M. (1990) Classification of large-scale subaqueous bedforms: a new look at an old problem. *J. sediment. Petrol.*, **60**, 160–172.

BRANNER, J.C. (1919) Outlines of the geology of Brazil to accompany the Geological Map of Brazil. *Geol. Soc. Am. Bull.*, **30**, 189–338.

BRAUN, O.P.G., MARTINS, M. & OLIVEIRA, W.J. (1993) Continuidade das sequências rifeanas sob a Bacia do São Francisco constatada por levantamentos geofísicos em Minas Gerais. *II Simp. do Cráton do São Francisco, Anais, SBG, Salvador*, pp. 164–166.

BRITO NEVES, B.B., VAN SCHMUS, W.R., BABINSKI, M. & SABIN, T. (1993) O evento de magmatismo de 1.0 Ga nas faixas móveis ao norte do Cráton do São Francisco. *II Simp. do Cráton do São Francisco, Anais, SBG, Salvador*, pp. 243–245.

BUCHWALDT, R., TOULKERIDIS, T., BABINSKI, M., NOCE, C.M., MARTINS-NETO, M.A. & HERCOS, C.M. (1999) Age determination and age related provenance analysis of the Proterozoic glaciation event in central eastern Brazil. *II South American Symp. on Isotope Geology, Argentina*, pp. 387–390.

CASTRO, P.T.A. (1997) *Os conglomerados da borda SW do Craton do São Francisco junto à porção S da Faixa Brasília: sedimentologia e relações estratigráficas com as rochas do Grupo Bambuí*. PhD thesis, University of Brasília.

CHIAVEGATTO, J.R.S. (1992) *Análise estratigráfica das sequências tempestíticas da Formação Três Marias (Proterozóico Superior), na porção meridional da Bacia do São Francisco*. MSc thesis, Federal University of Ouro Preto, Brazil.

CORREA-GOMES, L.C. & OLIVEIRA, E.P. (1997) O enxame radial gigante de diques máficos da província Bahia-Congo: implicações reológicas e tectônicas da presença de uma pluma mantélica na intefácie América do Sul-África, 1.0 Ga atrás. *V SNET—Simp. Nac. Estudos Tectônicos, Brazil*, pp. 52–55.

D'AGRELLA-FILHO, M.S., PACCA, I.G., TEIXEIRA, W., ONSTOTT, T.C. & RENNE, P.R. (1990) Paleomagnetic evidence for the evolution of Meso- to Neoproterozoic glaciogenic rocks in central-eastern Brazil. *Palaeogeogr. Palaeoclimatol. Palaeoecol.*, **80**, 255–265.

DARDENNE, M.A. (1978) Síntese sobre a estratigrafia do Grupo Bambuí no Brasil central. *SBG 30 Cong. Bras. Geol., Recife, Anais*, **2**, 597–610.

DREIMANIS, A. & SCHLÜCHTER, C. (1985) Field criteria for the recognition of till or tillite. *Palaeogeogr. Palaeoclimatol. Palaeoecol.*, **51**, 7–14.

DUPONT, H. (1995) O Grupo Conselheiro Mata no seu quadro paleogeográfico e estratigráfico. *Bol. Soc. Bras. Geol. Núcleo Minas Gerais*, **13**, 9–10.

DUPONT, H. (1996) O Supergrupo São Francisco entre a Serra do Cabral e as serras do Espinhaço e de Minas: estudo estratigráfico e estrutural e relações de contato com o Supergrupo Espinhaço. *SBG 39 Cong. Bras. Geol, Salvador, Anais*, **5**, 489–493.

ERIKSSON, P.G., CONDIE, K.C., TIRSGAARD, H. *et al.*, (1998) Precambrian clastic sedimentation systems. In: *Precambrian Clastic Sedimentation Systems* (Eds Eriksson, P.G., Tirsgaard, H. & Mueller, W.U.). *Sediment. Geol.* spec. issue, **120**, 5–53.

ESPINOZA, J.A.A. (1996) *Sistemas deposicionais e relações estratigráficas da Tectonosseqüência Conselheiro Mata, na borda leste da Serra do Cabral, Minas Gerais, Brasil*. MSc thesis, Federal University of Ouro Preto, Brazil.

EVANS, D.A., BEUKEF, N.J. & KIRFCHVINK, J.L. (1997) Low-latitude glaciation in the Paleoproterozoic era. *Nature*, **386**, 262–266.

EYLES, N. (1993) Earth's glacial record and its tectonic setting. *Earth Sci. Rev.*, **35**, 1–248.

EYLES, N. & YOUNG, G.M. (1994) Geodynamic controls on glaciation in Earth history. In: *Earth's Glacial Record* (Eds Deynoux, M., Miller, J.M.G., Domack, E.W., Eyles, N., Fairchild, I.J. & Young, G.M.), pp. 1–28. Cambridge University Press, Cambridge.

HAMBREY, M.J. & HARLAND, W.B. (1985) The late Proterozoic glacial era. *Palaeogeogr. Palaeoclimatol. Palaeoecol.*, **51**, 255–272.

HERCOS, C.M. (2000) *Estratigrafia da Bacia Neoproterozóica do São Francisco, na região entre Pirapora e a Serra da Água Fria, MG, com base em dados de campo e sísmica de reflexão*. MSc thesis, Federal University of Ouro Preto, Brazil.

HERCOS, C.M. & MARTINS-NETO, M.A. (1997) Considerações sobre os supergrupos Espinhaço e São Francisco na borda oeste da Serra da Água Fria (MG). *Bol. Soc. Bras. Geol. Núcleo Minas Gerais*, **14**, 19–21.

HOFFMAN, P.F. (1999) The break-up of Rodinia, birth of Gondwana, true polar wander and the snowball Earth. *J. Afr. Earth Sci.*, **28**, 17–33.

HOFFMAN, P.F., KAUFMAN, A.J., HALVERSON, G.P. & SCHRAG, D.P. (1998) A Neoproterozoic snowball Earth. *Science*, **281**, 1342–1346.

ISOTTA, C.A.L., ROCHA-CAMPOS, A.C. & YOSHIDA, R. (1969) Striated pavement of the Upper Precambrian glaciation in Brazil. *Nature*, **222**, 467–468.

KARFUNKEL, J. & HOPPE, A. (1988) Late Proterozoic glaciation in central-eastern Brazil: synthesis and model. *Palaeogeogr. Palaeoclimatol. Palaeoecol.*, **65**, 1–21.

KAUFMAN, A.J. (1997) An Ice age in the tropics. *Nature*, **386**, 227–228.

KAUFMAN, A.J. & KNOLL, A.H. (1995) Neoproterozoic variations in the C-isotopic composition of seawater: stratigraphic and biogeochemical implications. *Precam. Res.*, **73**, 27–49.

LOWE, D.R. (1975) Water-escape structures in coarse-grained sediments. *Sedimentology*, **22**, 157–204.

MACHADO, N., SCHRANK, A., ABREU, F.R., KNAUER, L.G. & ALMEIDA ABREU, P.A. (1989) Resultados preliminares da geocronologia U/Pb na Serra do Espinhaço Meridional. *Bol. Soc. Bras. Geol. Núcleo Minas Gerais*, **10**, 171–174.

MARTINS, M., TEIXEIRA, L.B. & BRAUN, O.P.G. (1993) Considerações sobre a estratigrafia da Bacia do São Francisco com base em dados de subsuperfície. *II Simp. do Cráton do São Francisco, Anais, SBG, Salvador*, pp. 167–169.

MARTINS-NETO, M.A. (1993) *The sedimentary evolution of a Proterozoic rift basin: the basal Espinhaço Supergroup, southern Serra do Espinhaço, Minas Gerais, Brazil*. PhD Thesis, Albert-Ludwigs Universität, Freiburg, Germany.

MARTINS-NETO, M.A. (1994) Braidplain sedimentation in a Proterozoic rift basin: the São João da Chapada Formation, southeastern Brazil. *Sediment. Geol.*, **89**, 219–239.

MARTINS-NETO, M.A. (1996a) Lacustrine fan-deltaic sedimentation in a Proterozoic rift basin: the Sopa-Brumadinho Tectonosequence, southeastern Brazil. *Sediment. Geol.*, **106**, 65–96.

MARTINS-NETO, M.A. (1996b) Aspectos tectono-deposicionais da Tectonossequência Galho do Miguel, Bacia Espinhaço (MG). *SBG 39 Cong. Bras. Geol., Salvador, Anais*, **5**, 391–394.

MARTINS-NETO, M.A. (1998a) Mantle plume, rifting and the early Neoproterozoic glaciation in the São Francisco craton and Araçuaí fold belt, southeastern Brazil. *International Conference on Precambrian and Craton tectonics/ 14th International Conference on Basement Tectonics, Ouro Preto, Brazil, abstracts*, pp. 32–34.

MARTINS-NETO, M.A. (1998b) O Supergrupo Espinhaço em Minas Gerais: registro de uma bacia rifte-sag do Paleo/Mesoproterozóico. *Rev. Brasil. Geociênc.*, **28**(2), 151–168.

MARTINS-NETO, M.A. (2000) Tectonics and sedimentation in a Paleo/Mesoproterozoic rift-sag basin (Espinhaço Basin, southeastern Brazil). *Precam. Res.*, **103**, 147–173.

MARTINS-NETO, M.A., CASTRO, P.T.A. & HERCOS, C.M. (1997) O Supergrupo São Francisco (Neoproterozóico) no Cráton do São Francisco em Minas Gerais. *Bol. Soc. Bras. Geol. Núcleo Minas Gerais*, **14**, 22–24.

MARTINS-NETO, M.A., GOMES, N.S., HERCOS, C.M. & REIS, L.A. (1999) Fácies glaciocontinentais (*outwash plain*) na Megasseqüência Macaúbas, norte da Serra da Água Fria (MG). *Rev. Brasil. Geociênc.*, **29**(2), 281–292.

MARTINS-NETO, M.A. & PEDROSA-SOARES, A.C. (2000) Superposition of sedimentary basin cycles from late Paleoproterozoic to late Neoproterozoic in the São Francisco craton and Araçuaí fold belt, eastern Brazil. *31 Int. Geol. Congr., Rio de Janeiro, Brazil.*

MARTINS-NETO, M.A., PEDROSA-SOARES, A.C. & LIMA, S.A.A. (2001) Tectono-sedimentary evolution of sedimentary basins from Late Paleoproterozoic to Late Neoproterozoic in the São Francisco craton and Araçuaí fold belt, eastern Brazil. In: *The influence of magmatism, tectonics, sea level change and palaeoclimate on Precambrian basin evolution: change over time* (Eds Eriksson, P.G., Catuneanu, O., Aspler, L., Chiareznelli, J. & Martins-Neto, M.A.). *Sediment. Geol.* spec. issue, **141–142**, 343–370.

MENZIES, J. (Ed.) (1995) *Modern Glacial Environments*. Butterworth Heinemann, Oxford.

MIALL, A.D. (1978) Lithofacies types and vertical profile models in braided river deposits: a summary. In: *Fluvial Sedimentology* (Ed. Miall, A.D.), Mem. Can. Soc. petrol. Geol., Calgary, **5**, 597–604.

MIALL, A.D. (1990) *Principles of Sedimentary Basin Analysis*, 2nd edn. Springer-Verlag, Heidelberg, 668 pp.

MILLER, J.M.G. (1996) Glacial sediments. In: *Sedimentary Environments* (Ed. Reading, H.G.), pp. 454–484. Blackwell Science, Oxford.

MORAES, L.J. & GUIMARÃES, D. (1930) Geologia da região norte de Minas Gerais. *An. Acad. Brasil. Ciênc.*, **2**, 153–186.

PARK, J.K. (1997) Paleomagnetic evidence for low-latitude glaciation during deposition of the Neoproterozoic Rapitan Group, Mackenzie Mountains, NWT, Canada. *Earth planet, Sci. Lett.*, **34**, 129–139.

PEDROSA-SOARES, A.C., DARDENNE, M.A., HASUI, Y., CASTRO, F.D.C., CARVALHO, M.V.A. & REIS, A.C. (1994) Mapa Geológico do Estado de Minas Gerais, escala 1:1 000 000 e Nota Explicativa. COMIG, Belo Horizonte, 97 pp.

PEDROSA-SOARES, A.C., NOCE, C.M., VIDAL, P., MONTEIRO, R.L.B.P. & LEONARDOS, O.H. (1992) Toward a new tectonic model for the Late Proterozoic Araçuaí (SE Brazil)—West Congolian (SW Africa) belt. *J. S. Am. Earth Sci.*, **6**, 33–47.

PEDROSA-SOARES, A.C., VIDAL, P., LEONARDOS, O.H. & BRITO NEVES, B.B. (1998) Neoproterozoic oceanic remnants in eastern Brazil: further evidence and refutation of an exclusively ensialic evolution for the Araçuaí–West Congo orogen. *Geology*, **26**, 519–522.

PFLUG, R. & RENGER, F.E. (1973) Estratigrafia e evolução geológica da margem SE do Cráton Sanfranciscano. *SBG 27 Cong. Bras. Geol., Aracajú, Anais*, **2**, 5–19.

PIMENTEL, M.M., FUCK, R.A., DARDENNE, M.A., SILVA, L.J.H.D. & MENEZES, P.R. (1995) O magmatismo ácido peraluminoso associado ao Grupo Araxá na região entre Pires do Rio e Ipamerí, Goiás: Características geoquímicas e implicações geotectônicas. *Simp. Geol. Centro-Oeste, Anais, Goiânia*, pp. 68–71.

RAMOS, M.L.S. & MURTA, C.R. (1997) *Caracterização das relações de contato entre os supergrupos Espinhaço e São Francisco no front da Serra do Espinhaço meridional, região entre Santa Bárbara e Curimataí, MG*. BSc thesis, Federal University of Ouro Preto, Brazil.

REIS, L.A. (1999) *Mapeamento geológico em unidades tectono-estratigráficas (1:25 000) na Serra do Espinhaço meridional, região de Sopa/Guinda, Minas Gerais*. BSc thesis, Federal University of Ouro Preto, Brazil.

ROCHA-CAMPOS, A.C., YOUNG, G.M. & SANTOS, P.R. (1996) Re-examination of a striated pavement near Jequitaí, MG: implications for Proterozoic stratigraphy and glacial geology. *An. Acad. Brasil. Ciênc.*, **68**, 593.

ROMEIRO SILVA, P.C. (1997) A passagem do Mesoproterozóico para o Neoproterozóico no centro-leste do Brasil e o estilo estrutural envolvido. *Bol. Soc. Bras. Geol. Núcleo Minas Gerais*, **14**, 9.

SCHÖLL, W.U. (1972) Der südwestliche Randbereich der Espinhaço-Zone, Minas Gerais, Brasilien. *Geol. Rundsch.*, **61**, 201–216.

SMITH, N.D. (1985) Proglacial fluvial environments. In: *Glacial Sedimentary Environments* (ed. by G.M. Ashley, J. Shaw & N.D. Smith), *SEPM Short Course*, **16**, 85–134.

SOUZA F.R.C. (1995) *Arcabouço estrutural da porção externa da Faixa Araçuaí na Serra do Cabral (MG) e o contraste de estilos deformacionais entre os supergrupos Espinhaço e São Francisco*. MSc thesis, Federal University of Ouro Preto, Brazil.

TEIXEIRA, L.B., MARTINS, M. & BRAUN, O.P.G. (1993) Evolução geológica da Bacia do São Francisco com base em sísmica de reflexão e métodos potenciais. *II Simp. do Cráton do São Francisco, Anais, SBG, Salvador*, pp. 179–181.

TROMPETTE, R. (1994) *Geology of Western Gondwana (2000–500 Ma): Pan-African–Brasiliano Aggregation of South America and Africa*. A.A. Balkema, Rotterdam, 350 pp.

UHLEIN, A. (1991) *Transição cráton–faixa dobrada: exemplo do Cráton do São Francisco e da Faixa Araçuaí (Ciclo Brasiliano) no Estado de Minas Gerais. Aspectos Estrati-*

*gráficos e Estruturais.* PhD thesis, University of São Paulo, Brazil.

UHLEIN, A., TROMPETTE, R. & ALVARENGA, C.J.S. (1999) Neoproterozoic glacial and gravitational sedimentation on a continental rifted margin: the Jequitaí–Macaúbas sequence (Minas Gerais, Brazil). *J. S. Am. Earth Sci.*, **12**, 435–451.

UHLEIN, A., TROMPETTE, R. & EGYDIO-SILVA, M. (1995) Rifteamentos superpostos e tectônica de inversão na borda sudeste do Cráton do São Francisco. *Geonomos*, **3**, 99–107.

WILLIAMS, G.E. (1994) The enigmatic Late Proterozoic glacial climate: an Australian perspective. In: *Earth's Glacial Record* (Eds Deynoux, M., Miller, J.M.G., Domack, E.W., Eyles, N., Fairchild, I.J. & Young, G.M.), pp. 146–164. Cambridge University Press, Cambridge.

WILLIAMS, G.E. (1996) Soft-sediment deformation structures from the Marinoan glacial succession, Adelaide fold belt: implications for the paleolatitude of late Neoproterozoic glaciation. *Sediment. Geol.*, **106**, 165–175.

YOUNG, G.M. (1995a) Glacial environments of pre-Pleistocene age. In: *Past Glacial Environments, Sediments, Forms and Techniques* (Ed. Menzies, J.), pp. 239–252. Butterworth Heinemann, Oxford.

YOUNG, G.M. (1995b) Are Neoproterozoic glacial deposits preserved on the margins of Laurentia related to the fragmentation of two supercontinents? *Geology*, **23**, 153–156.

YOUNG, G.M. & GOSTIN, V.A. (1991) Late Proterozoic (Sturtian) succession of the North Flinders Basin, South Australia; an example of temperate glaciation in an active rift setting. *Spec. Pap. geol. Soc. Am.*, **261**, 207–222.

*Spec. Publs int. Ass. Sediment.* (2002) **33**, 405–436

# New evidence of glacial abrasion of the Late Proterozoic unconformity around Varangerfjorden, northern Norway

K. LAAJOKI

*Mineralogical–Geological Museum, University of Oslo, Sars'gate 1, 0562 Oslo, Norway; and*
**Department of Geology, University of Oulu, Box 3000, 90401 Oulu, Finland*

## ABSTRACT

The regional angular unconformity that separates the late Proterozoic (Riphean–Vendian) Vadsø and Tanafjorden groups from the overlying Varangerian (late Vendian) glacigenic rocks (the Smalfjord, Nyborg and Mortensnes formations) of the Vestertana Group can be mapped on both sides of south-western Varangerfjorden. New observations of late Proterozoic glacial striations and small grooves were made at Handelsneset, Skjåholmen and Vieranjarga, 5–20 km east of Bigganjargga. The striations and grooves, in addition to the fact that the classical Bigganjargga surface is not a plane, but part of a wider striated trough, support glacial abrasion of the Varangerfjorden unconformity and the Bigganjargga surface being an essential part of it. Indicator stone observations support transport from the present south-east to north-west. At palaeotopographically higher parts of Vieranjarga, the subglacially abraded unconformity was destroyed in part by syn-Smalfjord *in situ* brecciation attributed to periglacial frost shattering. A large part of the unconformity at East Vieranjarga and Skjåholmen was incised by intra-Smalfjord (interglacial) erosion, which went several metres down to the underlying Vadsø Group basement. Diagenetic processes modified the classical Bigganjargga surface. There are no features on the unconformity that permit differentiation between late Proterozoic glacial abrasion processes and those of Palaeozoic or later ice ages. Special care is needed to distinguish glacigenic soft-sediment striations and small grooves from those inscribed in hard rock. The former mark the presence of a hiatus or disconformity within a stratigraphic sequence and may be produced subglacially or by floating ice, whereas the latter indicate, in most cases, a marked angular unconformity, of which the Bigganjargga surface is a classical example, or a non-conformity

## INTRODUCTION

The lower boundary of the late Proterozoic Smalfjord Formation of the Vestertana Group, exposed at Varangerfjorden, northern Norway, is one of the most magnificent and important angular unconformities preserved within the geological record. Rosendahl (1945) considered its significance so great that he marked it as the base of the Palaeozoicum. This surface is referred to here as the Varangerfjorden unconformity (VFU) after the fjord at the eastern head where it is best exposed and is the main target of this study.

The VFU has received much attention at Bigganjargga (Oaibaččanjar'ga) (traditional locality and geological unit names are used in this paper, current locality names on topographic maps are given in parentheses), where Reusch (1891) first described glacial striations 'from a period much older than the ice age'. The glacial origin and regional significance of the VFU were later confirmed by Bjørlykke (1967), who showed that it represents a glacial palaeovalley or trough running approximately parallel to the present Varangerfjorden, and by Edwards (1972, 1975, 1984), who carried out extensive sedimentological studies on the diverse glacial rocks of the Vestertana Group.

There are, however, a group of authors who deny the glacial origin of the Bigganjargga tillite and the glacial abrasion of the underlying surface (most recently, Jensen & Wulff-Pedersen, 1996, 1997). This discrepancy between the glacial and non-glacial schools led to re-examination of the classical

* Permanent and mailing address.

Bigganjargga surface and other outcrops where the VFU is known to be exposed. The main emphasis of this study is placed on new field evidence concerning the late Proterozoic glacial abrasion of the VFU.

## GEOLOGICAL SETTING AND REGIONAL STRATIGRAPHY

Bedrock from the Varanger Peninsula to Laksefjorden in East Finnmark, northern Norway, is composed of late Proterozoic (Riphean–Vendian)–Cambro-Ordovician sedimentary rocks (Fig. 1) that comprise three main entities (Siedlecka & Roberts, 1996): (i) the autochthonous formations of the southern half of the Varanger Peninsula, which pinch out nonconformably upon the crystalline late Archaean–early Proterozoic basement of the Fennoscandian (Baltic) Shield in the south (Fig. 2); (ii) the allochthonous–parautochthonous formations of the Caledonian Gaissa Nappe Complex in the Laksefjordvidda–Tanafjorden area (Fig. 1); and (iii) the allochthonous formations of the northern half of the Varanger Peninsula, the Barents Sea Region, north-east of the Trollfjorden–Komagelva Fault Zone (Fig. 1). The submarine-fan–deltaic–shallow marine units of the last area (the Barents Sea Group) are not considered in this study.

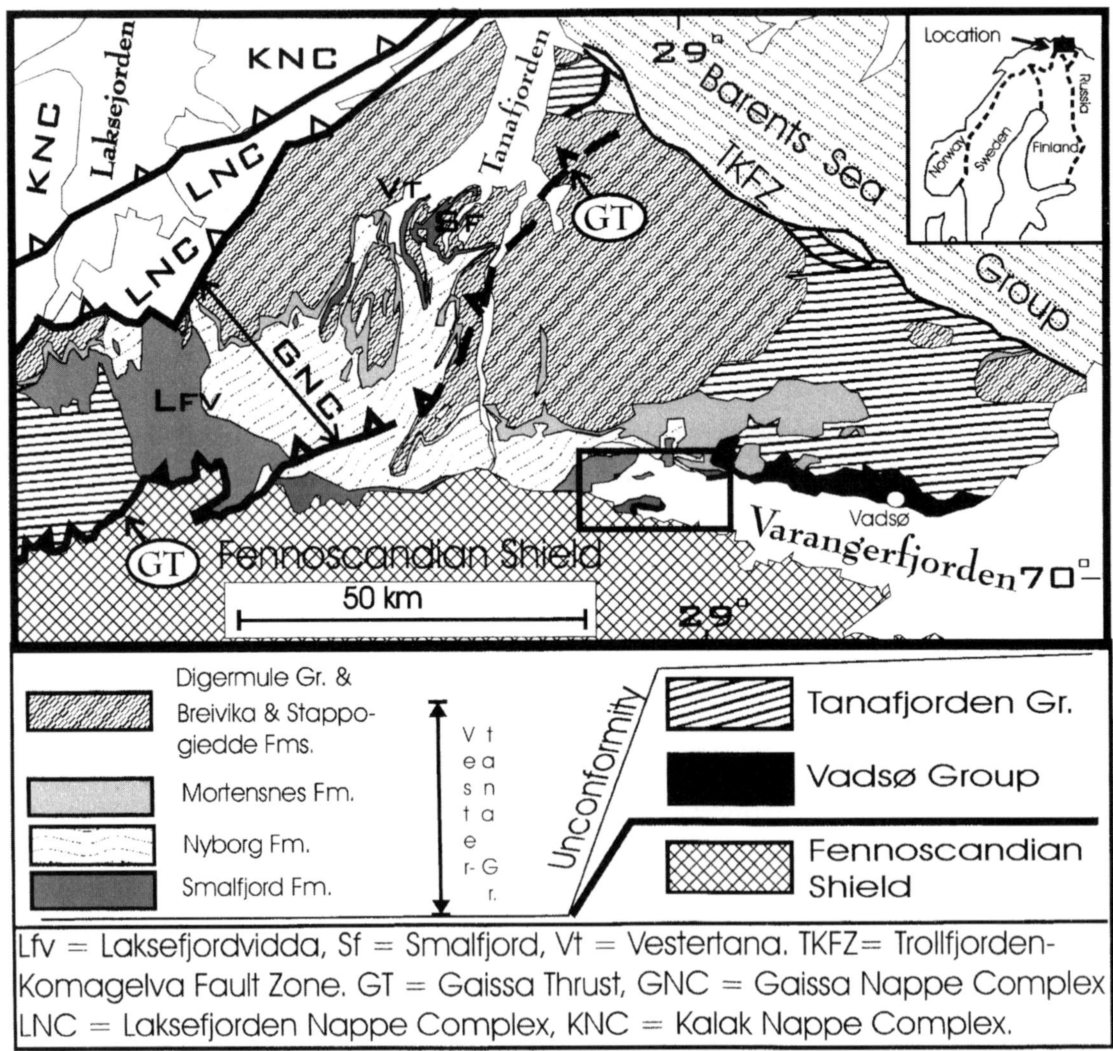


**Fig. 1.** Geological map of the Varangerfjorden–Laksefjordvidda area (simplified from Siedlecka & Roberts, 1996). The area of Fig. 2 is framed.

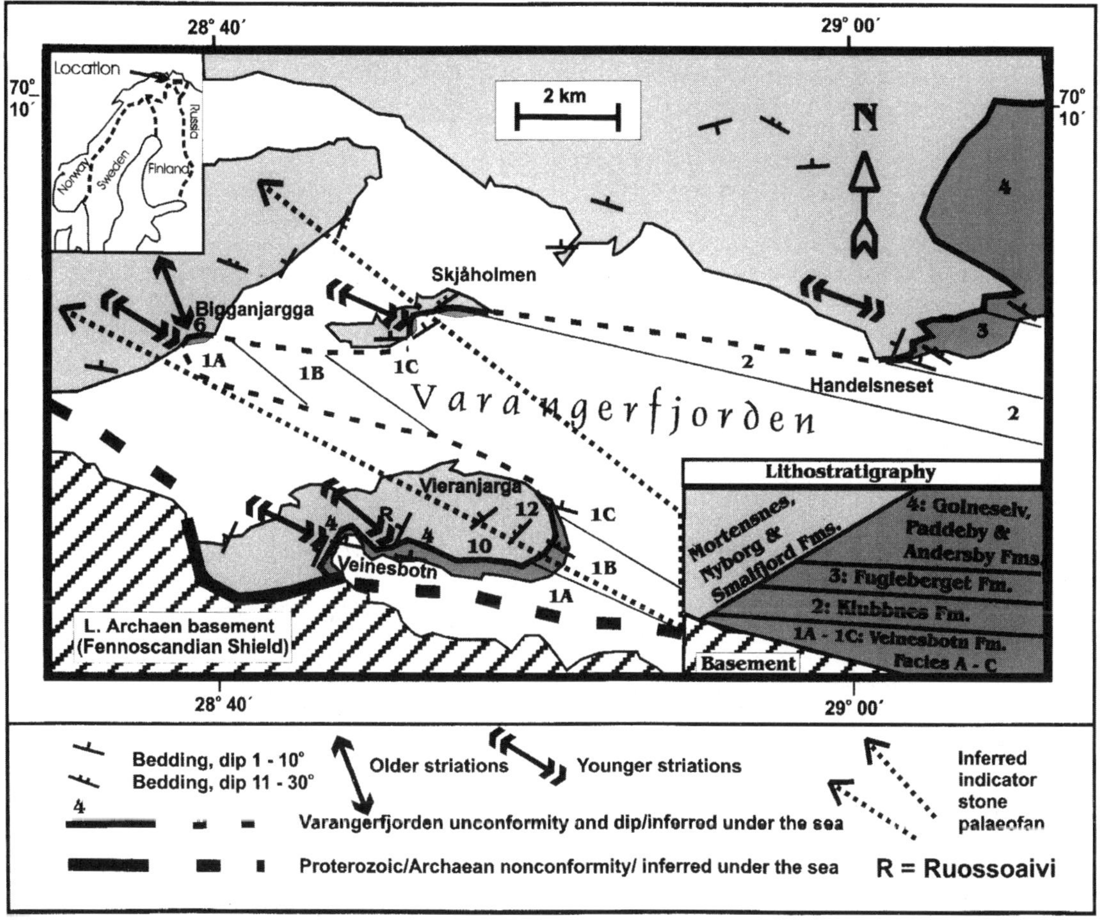


**Fig. 2.** Geological map of the Varangerfjorden area (modified after Siedlecka, 1990, 1991, with complementary observations) and simplified lithostratigraphic sketch. Lithostatigraphic nomenclature follows Siedlecka (1990, 1991). The dip measurements on the Varangerfjorden unconformity are from relatively flat (non-channnelized) parts of the surface. Striation observations are from Table 2. The inferred indicator stone palaeofan is based on this study.

The autochthonous formations of the southern Varanger Peninsula are divided into two series by the VFU. The formations under the unconformity dip shallowly (2–20°) to the north and are made up of the fluvial–shallow marine Vadsø Group and the overlying shallow marine Tanafjorden Group (Table 1). The Vadsø Group is found on both sides of Varangerfjorden and as a narrow occurrence east of Tanafjorden (Fig. 1). The Vestertana Group overlies the Vadsø and Tanafjorden groups unconformably and is famous for the tillites in the Smalfjord and Mortensnes formations, separated by the interglacial Nyborg Formation (Fig. 3). The sedimentary rocks are unmetamorphosed and only locally affected by the Caledonian deformation. Fossil studies give an upper Riphean age for the bulk of the Vadsø Group, whereas the Tanafjorden Group and the bulk of the Vestertana Group formed in Vendian time (Siedlecka & Roberts, 1992). Rb–Sr dating of illite fractions indicates that the age of burial diagenesis of the Tanafjorden Group is *c.*650 Ma, the Varangerian glaciation in the type area occurred between ≤630 and *c.*590 Ma and the burial diagenesis in the Vestertana Group occurred at around 560–570 Ma (Roberts *et al.*, 1997).

The boundary between the late Proterozoic formations of the southern Varanger Peninsula and those of the Laksefjordvidda–Tanafjorden area is defined by the Gaissa Thrust, the sole thrust of the north-east Caledonian Gaissa Nappe Complex (Fig. 1). The thrust is well defined under the south-west part of the complex, where it separates the rocks of the Tanafjorden Group and the Smalfjord Formation from the autochthonous

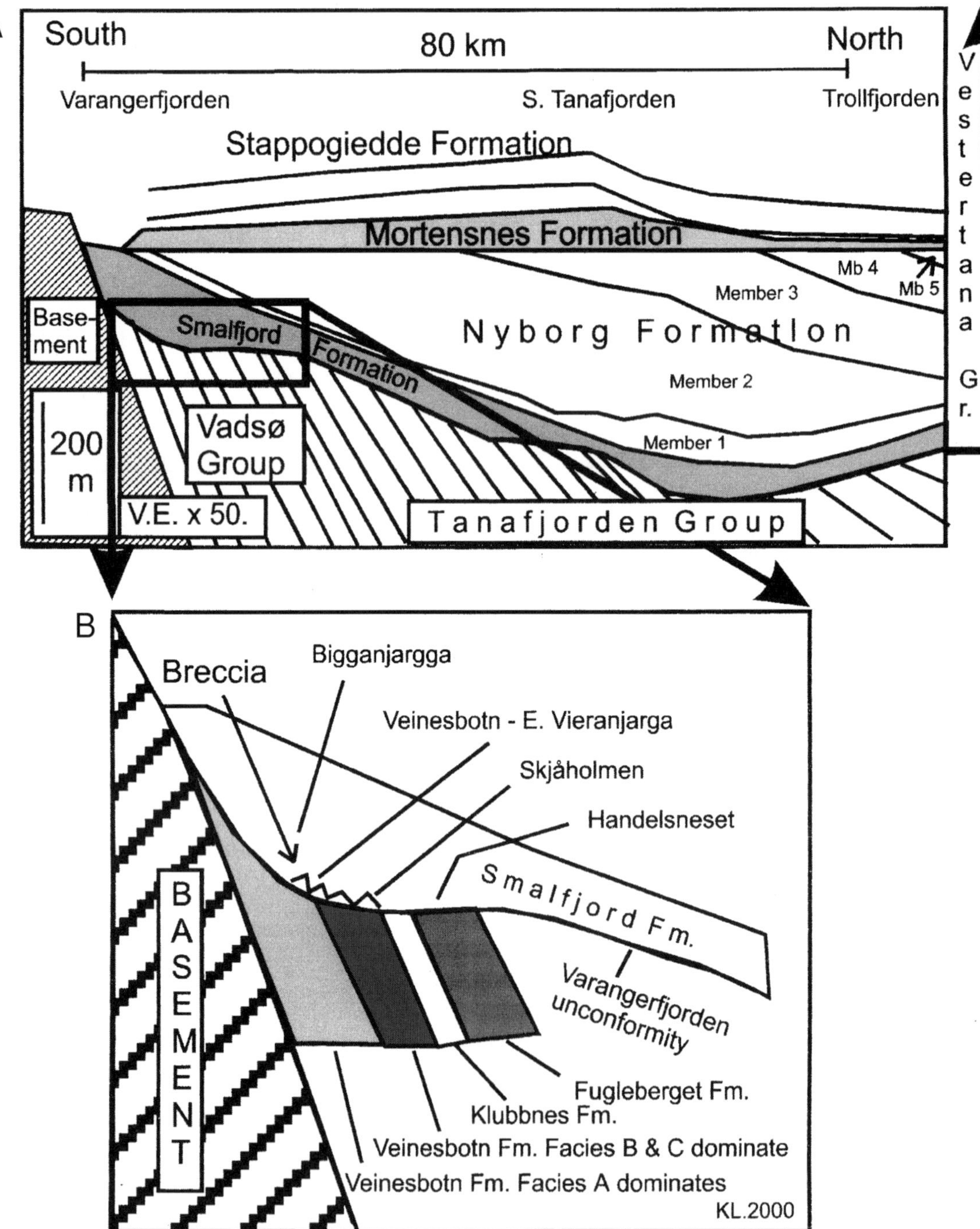


**Fig. 3.** (A) North–south cross-section across the outcrop belt of late Precambrian rocks in East Finnmark, using the base of the Mortensnes Formation as a datum (modified from Edwards & Føyn, 1981, Fig. 2). (B) Blow-up showing how the Varangerfjorden unconformity erodes different formations of the Vadsø Group at each study locality. Facies A–C of the Veinesbotn Formation refer to Hobday's (1974) classification. The barbed line indicates the brecciated part of the unconformity.

**Table 1.** Layer cake presentation of the lithostratigraphy of the Vadsø, Tanafjorden and Vestertana groups after Siedlecka & Roberts (1992). In reality, the Varangerfjorden unconformity cuts obliquely all the underlying formations down to the late Archaean basement (Figs 2 & 3). Formations essential to this study are set in bold (Veinesbotn, Klubbnes, Fugleberget and Smalfjord formations).

| Group (age)/formation | Thickness (m) |
|---|---|
| *Vestertana Group (**upper Vendian**)* | 1317–1655 |
| Breivika Fm | 600 |
| Stappogiedde Fm | 505–545 |
| Mortensnes Fm | 10–60 |
| Nyborg Fm | 200–400 |
| **Smalfjord** | 2–50 |
| ***Varangerfjorden unconformity*** | |
| *Tanafjorden Group (**Vendian**)* | 1448–1665 |
| Grasdale Fm | 280 |
| Hanglečærro Fm | 200 |
| Vagge Fm | 80 |
| Gamasfjellet Fm | 280–300 |
| Dakkovarre Fm | 273–350 |
| Stangeneset Fm | 205–255 |
| Grønneset Fm | 130–200 |
| *Vadsø Group (**Lower Vendian–upper Riphean**)* | |
| Ekkerøya Fm | 590–960 |
| Golneselva Fm | 15–190 |
| Paddeby Fm | 50–135 |
| Andersby Fm | 25–120 |
| **Fugleberget Fm** | 25–40 |
| **Klubbnes Fm** | 125 |
| **Veinesbotn Fm** | 50 |
| | 300 |
| ***Non-conformity*** | |
| ***Late Archaean** basement of the Fennoscandian Shield* | |
| Total thickness of Vestertana Group + Digermulen Group | 2827–3210 |
| Total thickness of Vadsø Group + Tanafjorden Group | 2038–2625 |

Dividal Group (too thin to be shown in Fig. 1) rimming the Fennoscandian Shield along the Caledonian front. This part of the Gaissa Nappe Complex contains a thick sequence of the Smalfjord Formation in the Krokvatnet palaeovalley (Føyn & Siedlecki, 1980). The Gaissa Thrust dies out towards the east, passing into a weakly folded zone, which separates the parautochthonous formations of the Vestertana and Digermule groups south-west of Tanafjorden from the autochthonous formations to the east (Føyn, 1985). Type Smalfjord Formation sections occur between the fjords of Smalfjord and Vestertana (Edwards, 1984).

## PREVIOUS STUDIES OF THE VARANGERFJORDEN UNCONFORMITY

The VFU is best exposed at the following localities from the west to the east: Bigganjargga (Oaibaččanjar'ga), the southern and eastern parts of the Vieranjarga (Vieranjar'ga, Kvalnes in older literature) peninsula, the south-east margin of the island of Skjåholmen and at Handelsneset (treated also under the name Mortensnes in older literature) (Fig. 2).

The Varangerfjorden unconformity exposed at Bigganjargga (Reusch, 1891) has been studied for over 100 years and numerous papers have been published discussing the origins of both the tillite and its pavement. Ideas concerning the origin and significance of the Bigganjargga surface can be divided into six major groups:

**1** Pre-1918 interpretations, i.e. from the time before Holtedahl (1918) recognized the regional unconformity. All authors (Reusch, 1891, 1898; Strahan, 1897; Schiøtz, 1898; Dal, 1900) agree that the surface is glacial, but only Schiøtz (1898) considers it Quaternary.

**2** Holtedahl (1918, 1919, 1960) considers the surface glacial, but does not correlate it with the regional unconformity. In his early paper, Føyn (1937) agrees with this concept.

**3** The surface is both glacial and part of the regional ('pre-tillitic') unconformity (Rosendahl, 1931, 1945; Bjørlykke, 1967; Edwards, 1972, 1974, 1984, 1997; Banks *et al.*, 1974; Edwards & Føyn, 1981; Føyn 1985; Siedlecka & Roberts, 1992; Laajoki, 1999a,b).

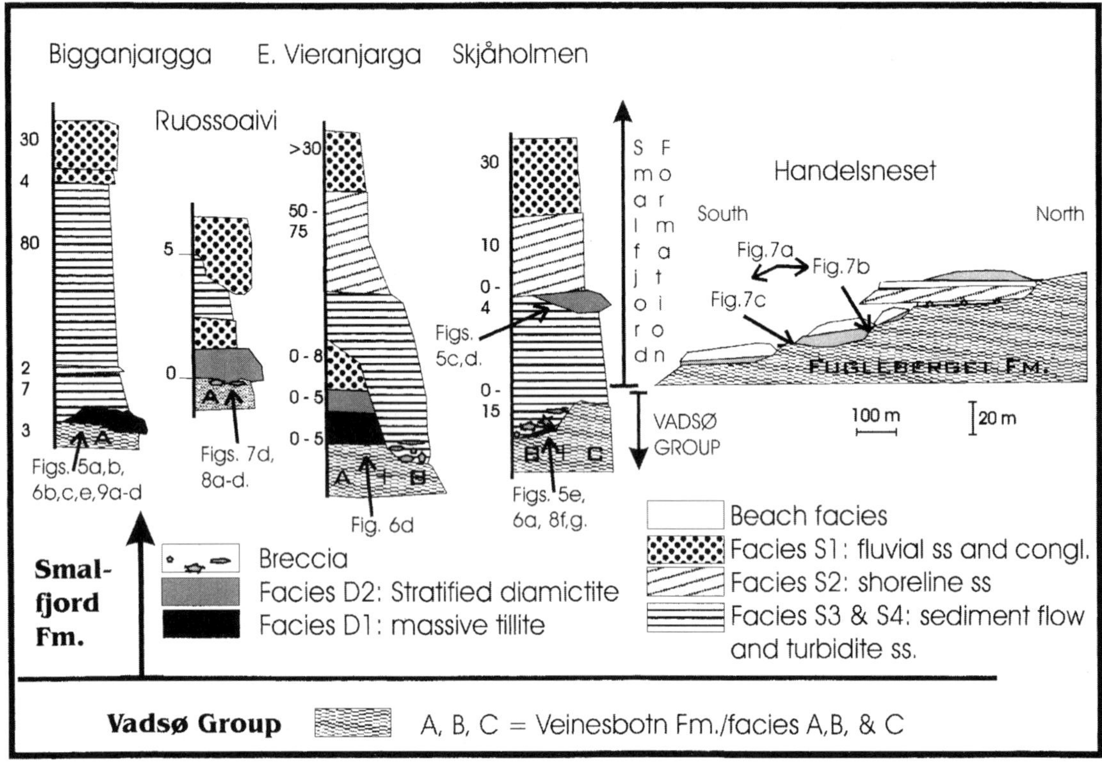


**Fig. 4.** Stratigraphic sections of the areas studied. Sections of Bigganjargga, Hadelsneset, Skjåholmen and East Vieranjarga and the cross-section of Handelsneset are modified from Edwards (1984). The Ruossoaivi section was measured by the author. Facies codes of the Veinesbotn and Smalfjord formations refer to Hobday (1974) and Edwards (1984), respectively. Unit thicknesses in metres. Locations of the photographs are indicated.

**4** The surface is glacial, but the authors are not sure whether the pavement was consolidated before deposition of the Bigganjargga tillite (Strahan, 1897; Dal, 1900, von Gaertner, 1944; Spjeldnæs, 1964).
**5** The surface is glacial, but was not necessarily formed by the overlying tillite (von Gaertner, 1944; Spjeldnæs, 1964; Edwards, 1972, 1975, 1984; Laajoki, 1999a; Rice & Hofmann, 2000).
**6** The non-glacial school denies glacial origin of both the Bigganjargga tillite and the striated surface and attributes their origins to sediment gravity flow (mud flow, slump, debris flow etc.; Crowell, 1964, 1999; Harland, 1964; Harland & Rudwick, 1964; Harland *et al.*, 1966; Schermerhorn, 1971, 1974; Jensen & Wulff-Pedersen, 1996, 1997; Arnaud & Eyles, 1999). This concept, which has caused some confusion, is discussed in more detail in a later section.

Bjørlykke (1967, Fig. 11) described a glacially striated unconformity at the western end of the bay of Veinesbotn (Vieravuodna), but unfortunately the exact location of this occurrence is not known. The well exposed unconformity with step-like erosion at Skjåholmen was first detected by Bjørlykke (1967, Fig. 4). Edwards (1984) described the local strong relief indicated by near vertical walls up to 4 m high, and concluded that the smooth truncation of the beds represents glacial erosion, and that the irregular relief of the unconformity may reflect bedding control of that erosion. Holtedahl (1918) first described the unconformity at Handelsneset. Rosendahl (1931) considered it glacial and correlated it with the Bigganjargga surface. Edwards (1984) studied the locality in greater detail and produced a cross-section showing the dramatic erosion of the underlying sandstone: the unconformity erodes about 140 m of the Vadsø Group at the distance of 800 m to the south (see the cross-section in Fig. 4). The extent of the erosional relief is about 40 m. Edwards (1984) concluded that the smooth but sharp truncation of the underlying beds is suggestive of glacial erosion.

## LITHOSTRATIGRAPHY ACROSS THE VARANGERFJORDEN UNCONFORMITY AND MORPHOLOGY OF THE GLACIAL SURFACES AROUND VARANGERFJORDEN

Figure 4 provides simplified stratigraphic columns of the sections studied. The underlying and overlying rocks belong to the Veinesbotn, Klubbnes and Fugleberget formations of the Vadsø Group and to the Smalfjord Formation of the Vestertana Group, respectively (Fig. 3, lower cross-section). The descriptions, interpretations and facies names and codes of Hobday (1974) and Edwards (1972, 1975, 1984) are used in this paper. It should be noted that at that time, the term facies was used in a broader sense than is commonly done today; their facies resemble more closely a facies association in modern terminology. Rice & Hofmann (2000, Fig. 2f) have recently described a striated surface at Saranjargåp'pi, on the southern side of Varangerfjorden. Because they are not quite sure of its Proterozoic age and the present author has not visited this locality, this surface is not yet considered part of the VFU.

### Facies under the Varangerfjorden unconformity

The oldest underlying sedimentary unit is the Veinesbotn Formation (300 m), at the base of the Vadsø Group (lithostratigraphic inset in Fig. 2). The type locality is along the southern shore of the Vieranjarga peninsula (Banks *et al.*, 1974) and is composed of three lithological units, from the oldest to the youngest (Hobday, 1974): facies A, pink, mostly trough cross-bedded and fluvial arkosic–quartzarenitic sandstone exposed at south Vieranjarga and at Bigganjargga; facies B, shallow shelf quartz arenitic sandstones and siltstones, exposed at east Vieranjarga; and facies C, a heterolithic unit of intertidal sandstones, siltstones and mudstones exposed at the north-east tip of Vieranjarga and at north-east Skjåholmen (Fig. 2). The Klubbnes Formation (*c.*50 m) consists of an upward-coarsening succession of micaceous siltstones and fine-grained sandstones exposed at the north-east tip of Skjåholmen and at Mortensnes. This formation is unexposed in the areas represented by the stratigraphic columns in Fig. 4. The part of the VFU exposed at Mortensnes is underlain mostly by the subarkosic–arkosic, large-scale, cross-bedded, flat-bedded or massive sandstones of the Fugleberget Formation (*c.*100 m thick), which Hobday (1974) interpreted as braided stream deposits (his facies Y).

### Facies of the Smalfjord Formation

The Smalfjord Formation around Varangerfjorden is composed mostly of (glaci)fluvial–shallow marine sandstones and conglomerates, but contains tillitic/diamictitic units in its lower parts. The only locality where a tillite deposit has been described as lying directly above the VFU is at Bigganjargga. This paper will report another case from Skjåholmen. The tillites probably serve as the lowermost Smalfjord deposits at Vieranjarga, but the unconformity is unexposed. At higher palaeovalley levels, Smalfjord turbiditic sandstones and fluvial–deltaic conglomerates and sandstones overlie the VFU. In their preliminary notice Arnoud & Eyles (1999) question the glacial origin of the Smalfjord tillites (and the Bigganjargga surface), but here the descriptions and nomenclature at hand in proper research papers are followed.

The manifold lithologies of the lowermost Smalfjord Formation are best revealed by the vertical section approximately 20 m in height along east Vieranjarga, described in detail by Edwards (1975, 1984). The Smalfjord Formation is composed of a massive tillite, up to 5 m thick (unit A: subglacial lodgment or melt-out till, facies D1), and a tillite, up to 6.5 m thick, with abundant lenses of stratified sandstone and conglomerate (unit B: supraglacial/proglacial drift, facies D2), erosively overlain by polymictic conglomerates and sandstones (unit C, up to 8 m thick: fluvial channel deposits, facies S1 and facies S2) and medium bedded sandstones (unit D: marine sandstones, facies S3 and facies S4). In the north-west corner of the section, a sequence approximately 3 m thick of intercalated sandstones and siltstones lies directly on the VFU (unit X: turbidites). There is also a small body of orthoconglomerate at the northern end of units A and B (C1: local winnowing product of unit A). Applying this terminology, the deposits and part of the VFU exposed in the study area are described in the following section. Figure 4 gives the relevant stratigraphic columns.

### Bigganjargga

The Veinesbotn facies A trough-cross bedded quartzite* is overlain unconformably by the Bigganjargga

* This rock is in most cases so hard that it is called quartzite in the sense defined in Bates & Jackson (1995). The overlying Smalfjord arenites are less indurated and poorer in quartz than the Veisnesbotn and are called sandstones.

**Table 2.** Localities and measurements (raw data, the readings give the average direction of inferred ice-flow) of late Proterozoic glacial striations (S), small grooves (G) and ridges (R) on the unconformity around the Varangerfjorden. For comparison sporadic data of Pleistocene ice-flow direction around Veinesbotn are included. Small grooves and striations are not separated in the Bigganjargga data. The measurements by the author are corrected for declination by +8°.

| Locality | Coordinates* | Older set | Younger set | Measurements | Comments | Reference |
|---|---|---|---|---|---|---|
| *Late Proterozoic* | | | | | | |
| Bigganjargga (classical surface, if not otherwise commented) | 0563090/7781497 | 325° (G & S) | | $n = 69$ | Modified slightly by recent erosion | Rice & Hofmann (2000) |
| | | | 283° (G & S) | $n = 17$ | | |
| | | | | | | This study, Figs 5B & 9A–C |
| | | 325.1° ± 1.9° (G & S) | | $n = 90$ | Some of the small grooves are internally striated, Fig. 9A | |
| | | | 285.9° (G & S) | $n = 17$ | | |
| | | | 304° (R) | $n = 1$ | In trough | |
| | | 322.5° (S) | | $n = 6$ | Under the winnowed Smalfjord conglomerate | |
| Handelsneset | 0577129/7781292 | | 291° (G) | $n = 3$ | Striated internally | This study, Fig. 7A–C |
| | | | 287° (S) | $n = 16$ | Thin, 1–2 mm, and parallel | |
| | | | 293° (R) | $n = 1$ | | |
| Skjåholmen | 0567375/7781762 | | 288° (S) | | Faint, several parallel | This study, Fig. 5E |
| Veinesbotn | Not known† | | 288° (S) | From Fig. 11 in Bjørlykke (1967) | Unclear photograph | Bjørlykke (1967, Fig. 11) |
| Ruossoaivi Vieranjarga | 0566982/7777608 | | 306° (S) | $n = 5$ | Excavated | This study, Fig. 7D |
| *Pleistocene striations* | | Older? | Dominant set | | | |
| Ruossoaivi on the VFU | 0568613/7777352 | | 90° (S) | $n = 11$ | Morphologically identical to Proterozoic striations | This study, Fig. 7E |
| Veinesbotn on Veinesbotn quartzite split surface | 0566850/7777600 | | 64° (S) | $n = 7$ | Palaeoflow to 64°/crescentic fractures | This study |
| | | 6° (S) | | $n = 1$ | | |
| Nesseby on Smalfjord sandstone | 0570352/7783157 | | 120° (S) | $n = 51$ | Palaeoflow to 120°/*roches moutonnées* | This study |

*GPS readings of UTM coordinates according to 1 : 50 000 topographic map sheets of Norges Geografiske Oppmåling.
†According to Bjørlykke's information this locality is close to the main road at the end of Veinesbotn (Veinesbugten) (see Bjørlykke, 1967, Fig. 2).

tillite/?diamictite and the turbiditic Smalfjord sandstone (facies S4) on the eastern and western parts of the classical outcrop, respectively. Facies S4 sandstone is overlain by fluvial conglomerates and sandstones of facies S1.

The Bigganjargga surface has traditionally been described as a plane or platform: 'a perfectly planar striated pavement' (Edwards, 1972). This holds good for the classical western part (Fig. 5A & B), although this is not exactly a plane, but has a dip varying slightly from 22°/18° (dip direction/dip) to 58°/12°, with an average of 32°/14° ($n = 12$, spacing *c*.5 m). This gives an average strike of 310° to the surface, a value that lies between the average ice-flow directions indicated by the older and younger sets of striations and small grooves (Table 2). A new observation is that the eastern part of the Bigganjargga surface, under algae, is not planar, but forms part of the western margin of a smoothly irregular erosional palaeosurface with a minimum depth of 0.8 m and a width of over 5 m. This structure is called a trough without any genetic connotation and the trough margin disappears into the sea in the east. The base of the trough cuts three trough cross-bedded sets across the Veinesbotn Formation (Fig. 5A). The trough axis plunges shallowly (probably only a few degrees) to *c*.330°, which is close to average ice-flow direction indicated by the older striation set. The smooth threshold within the trough is faintly striated, with indicated ice-flow directions varying from 342 to 283°. As a whole the Bigganjargga surface can be seen to rise 2 m to the classical platform at a distance of 30 m from its eastern margin and it intersects five cross-bedded sets of the Veinesbotn sandstone. Microscopically, the unconformity is sharp and does not show any evidence of deformation (Rice & Hofmann, 2000).

West of Bigganjargga, the unconformity is unexposed, but it can easily be located as a small topographic threshold, and first of all as a lithological boundary between a pink, trough cross-bedded Veinesbotn facies A sandstone and a brownish, medium-bedded turbiditic Smalfjord facies S4 sandstone. At a distance of 500 m it rises to 10 m above the present sea-level where it turns south and goes into the sea. This change of dip direction is attributed to open folding, but there may also be a hidden fault between Vieranjarga and Bigganjargga. Thus the Bigganjargga surface is part of the shallowly dipping western margin of a wide channel or palaeovalley, the axis of which probably plunges shallowly to *c*.310° and lies north-east of the Bigganjargga outcrop.

### Ruossoaivi (Ruosso'aivi), southern Vieranjarga

The VFU can be followed for a distance of about 6 km on the peninsula of Vieranjarga. It is well exposed around the head of Veinesbotn and in east Vieranjarga. The surface can, however, be mapped precisely on the basis of the lithological contrast between the pink Veinesbotn facies A quartzite (Hobday, 1974) and the turbiditic–fluvial sandstones and conglomerates of the Smalfjord Formation. The latter correspond to Bjørlykke's (1967) Kvalnes conglomerate. The small but distinctive regional angular unconformity between these formations can be seen even from a distance.

The VFU was uncovered under the conglomeratic Smalfjord sandstone on the southern flank of the small hill of Ruossoaivi in the middle part of Vieranjarga (Fig. 2). The Veinesbotn trough cross-bedded quartzite (facies A) is overlain unconformably by a *c*.2 m thick facies D2 diamictite. This contains granitoid, gneiss and quartzite boulders in addition to 50 cm long fragments of massive sandstone of Smalfjord facies S4 type and fragments of the Veinesbotn quartzite at its basal parts. Similar diamictite is interbedded with turbiditic sandstone in nearby outcrops. These observations indicate that the diamictite was deposited into water and was eroded by facies S4 sandstones. The latter fact suggests that the diamictite is younger than the massive tillites, which lie directly upon the VFU and are overlain by facies S4 sandstones at Bigganjargga and Skjåholmen. The diamictite is overlain by facies S1 and facies S4 sandstones and conglomerates. The unconformity is sharp and mostly planar. The excavated part dips 3–10° to 290–320°, whereas the underlying Veinesbotn quartzite dips 3–9° to 350–20° (trough cross-beds make it difficult to measure the bedding plane exactly). Beds in the overlying diamictite could not be measured. The unconformity has been eroded locally by intra-Smalfjord *in situ* brecciation.

### Eastern Vieranjarga

The underlying bedrock is mainly composed of shallow-marine siltstones and sandstones of Veinesbotn facies B. These are overlain unconformably by facies D1 massive tillite of the Smalfjord Formation, but the contact itself is unexposed. Smalfjord facies D2 diamictites and fluvial and turbiditic sandstones and conglomerates of Smalfjord facies S1 and facies S4 overlie the tillite. In the north, the tillites and diamictites are missing due to deep intra-Smalfjord

A

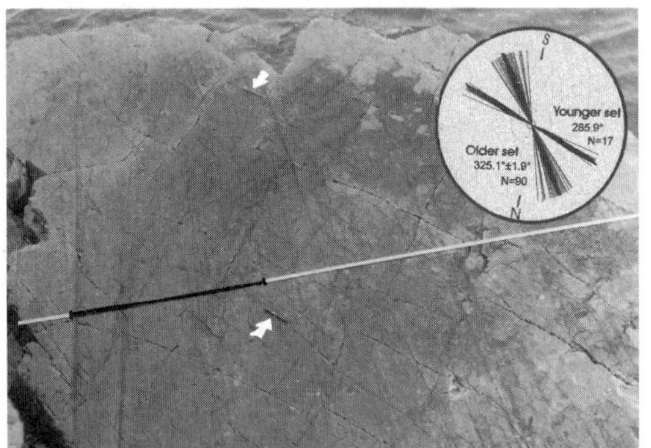


B

C

**Fig. 5.** Glacially striated and grooved hard-rock surfaces (A, B, E) and an intraformational erosional surface (C, D). (A) The Varangerfjorden unconformity at Bigganjargga seen from the east. The classical planed surface (see B) is visible on the far left. The channelized part (arrows and pens right of the tape) erodes three trough cross-bedded sets of the Veinesbotn facies A quartzite. Note the sharp nature of the major part of the unconformity (under and left of the tape). (B) The classical striated surface of the Veinesbotn Formation quartzite at Bigganjargga. Deepening and widening of many of the more prominent older striations indicate, but do not prove (see Iverson, 1991) that ice moved to *c*.325° (down on the photograph) when the older set was formed. Older striations cut each other without any clear systematics. The direction of fainter striations of the younger set is parallel to the pens (arrows). The dark portion of the tape is 1 m. The rose diagram is orientated approximately parallel to true north. Individual lines represent individual measurements; histograms were calculated at 5° intervals; set orientations are presented as mean azimuths. (C) An intra-Smalfjord erosional contact between evenly bedded (turbiditic) Smalfjord facies S4 sandstone and diamictite. The overlying sandstone represents facies S2. The largest clasts in the diamictite are *c*.15 cm long. Note the irregular, non-planar, in part stepped surface (see also Edwards, 1984, Fig. 11). (*continued*)

**Fig. 5.** (*cont'd*) (D) Close-up of the contact in C. Note the irregularly scoured appearance of the contact surface and the lack of streamlines and planing. (E) A photomosaic of the Varangerfjorden unconformity between alternating siltstone and sandstone beds of the Veinesbotn Formation (dipping 5–23° at the foot of the outcrop) and a thin relic of a massive Smalfjord tillite (under the upper head of the 140 cm long stick). The tillite is overlain by Smalfjord Formation sandstones and conglomerates dipping shallowly away from the viewer. Faint striations (Table 2) can be seen on the excavated part of the unconformity under the hammer on the right. The largest boulder (arrow on the left) in the tillite is 40 × 60 cm. Seen from the east.

erosion, which cuts a non-glacial channel down into the Veinesbotn Formation. The bases of the erosional channels are filled with sedimentary breccias and turbiditic sandstones of facies S4.

### Skjåholmen

The Varangerfjorden unconformity can be traced for over 2 km on the eastern part of Skjåholmen island, where it is mostly sharp, but irregular (Edwards, 1984, Fig. 9). The underlying Veinesbotn Formation consists of alternating siltstone and sandstone beds of shallow-marine facies B and C (Hobday, 1974). The outcrops are mostly two-dimensional and so the exact nature of the unconformity surface cannot be seen. The unconformity is mostly overlain by Smalfjord facies S4 sandstones and associated breccias, which are overlain by a local diamictite (Fig. 5C & D), which Edwards (1984) considered a tillite.

In the southern part of the section, a thin body (relic) of massive tillite lies directly on the unconformity (Fig. 5E). The tillite contains both granitoid and quartzite clasts in a micaceous, quartz–feldspar matrix (Fig. 6A), and corresponds to the upper part of Edwards's (1975) unit A massive facies D1 tillite at eastern Vieranjarga in terms of both its stratigraphic position and its lithology. South of, and topographically below, this locality, the glacial unconformity has been eroded by the intra-Smalfjord hiatus and the sequence begins with S4 turbiditic sandstones and associated breccias.

### Handelsneset

The underlying bedrock is made up of cross-bedded quartzites of the Fugleberget Formation. At the lower, southern part of the *c.*800 m long, high-relief section (Fig. 4), the Smalfjord Formation starts with coarse diamictite/conglomerate overlain by beach-type sandstones (Edwards, 1984). In the northern part of the section, the Smalfjord Formation contains breccia and facies S2 sandstones.

## GLACIAL VERSUS NON-GLACIAL ABRASION OF THE BIGGANJARGGA SURFACE

The first workers who suspected the glacial origin and regional significance of the Bigganjargga surface were Crowell (1964), Harland (1964), Harland & Rudwick (1964) and Harland *et al.* (1966). Their brief statements were based on short visits to Bigganjargga and should be considered as qualified proposals reflecting the historical period, when 'great efforts were made to discount alleged tillites as having other origins' (Harland & Herod, 1975, p. 190). After Bjørlykke's (1967) and Edwards's (1975, 1984) studies, Harland (1983), for instance, seems to have accepted the glacial origin of the Bigganjargga tillite. Crowell (1999, p. 59 and Fig. 60), however, seems to hold to his early concepts (Crowell, 1964).

The only authors who have more seriously questioned the glacial origin of the Bigganjargga surface are Jensen & Wulff-Pedersen (1996, 1997). Their main arguments include: (i) the underlying Veinesbotn quartzite and the overlying Smalfjord sandstone are so identical that they form a continuous succession within the Vadsø Group; (ii) in thin section, the two rocks are both mineralogically and texturally indistinguishable; (iii) the Veinesbotn quartzite was not consolidated before the deposition of the Bigganjargga tillite; (iv) the unconformity under the Bigganjargga tillite is a local feature and it is not possible to identify it outside the diamictite lens; (v) the striations and imprints on the surface are soft-sedimentary features caused by the Bigganjargga diamictite which the authors interpret as a debris flow deposit; (vi) there is no sign of extensive pressure solution features in the succession. In summary, the authors consider the Bigganjargga unconformity as an intraformational (intra-Smalfjord) soft-sediment slide- and scour-marked surface.

Edwards (1997) discussed Jensen's and Wulff-Pedersen's statements and remarked that the authors do not take into account the several stratigraphical and sedimentological studies that were published both before and after Crowell's 1964 paper and which correlate the Bigganjargga surface with the regional unconformity under the Vadsø Group. Rice & Hofmann (2000) are also against Jensen's and Wulff-Pedersen's interpretations and accept the glacial origin of the Bigganjargga. Some of their criteria are treated below. The following comments are based on the observations made by the present author.

The underlying Veinesbotn quartzite at Bigganjargga is different from the overlying Smalfjord sandstone in terms of sedimentary facies, lithology and petrography. The former is pink on its weathered surface and trough cross-bedded throughout. The latter starts with a brownish-grey, thin-bedded (10–20 cm), massive and in part pebbly sandstone with load casts, current ripples and flutes. In addition, the palaeocurrent directions differ. Most of these marked contrasts and differences were documented by Bjørlykke (1967). Hobday (1974) and Edwards (1975) interpreted the two units as fluvial and turbiditic, respectively. Another striking lithological difference is that solitary quartzite pebbles and angular siltstone fragments are common in the basal parts of the lowermost Smalfjord sandstone beds above the VFU west of the Bigganjargga tillite, whereas only few quartz, feldspar and granitoid granules were found in the Veinesbotn quartzite. Petrographical differences are also distinct; the Veinesbotn quartzite is a microcline-bearing quartz-rich arenite with quartz cement and is both texturally and mineralogically rather mature (Fig. 6B), whereas the lowermost Smalfjord sandstone is both texturally and mineralogically heterogeneous and immature, containing both arenite- and wacke-textured parts and being rich in lithic fragments (Fig. 6C).

Jensen & Wulff-Pedersen (1996, 1997) did not give any interpretation of the depositional environment of the two sandstone sequences at Bigganjargga, and did not take into account the fact that the underlying quartzite is almost identical to the type Veinesbotn quartzite at Veinesbotn, only 5 km from Bigganjargga to the south-east along bedding strike (Fig. 2), and that the Smalfjord Formation is not known to contain Veinesbotn-type quartzite. They do not refer to the most detailed study available on the Veinesbotn Formation (Hobday, 1974), where the quartzite at Bigganjargga is considered as one of the type localities of the lower part (facies A) of the Veinesbotn formation. At Veinesbotn, the quartzite is separated by a distinctive angular unconformity from the overlying Smalfjord turbiditic sandstones and conglomerates, resembling those at Bigganjargga. Similar relationships can be seen at Vieranjarga and Skjåholmen. These regional observations confirm that the underlying quartzite and the overlying sandstone at Bigganjargga belong, respectively, to the basal part of the Vadsø Group (the Veinesbotn Formation), and to the basal part of the Vestertana Group (the Smalfjord Formation), as is shown on recent 1 : 50 000 bedrock geological maps (Siedlecka, 1990, 1991).

Because the Veinesbotn quartzite at Bigganjargga represents the lowermost unit of the Vadsø Group–Tanafjorden Group package, *c.*2300 m thick (Table 1),

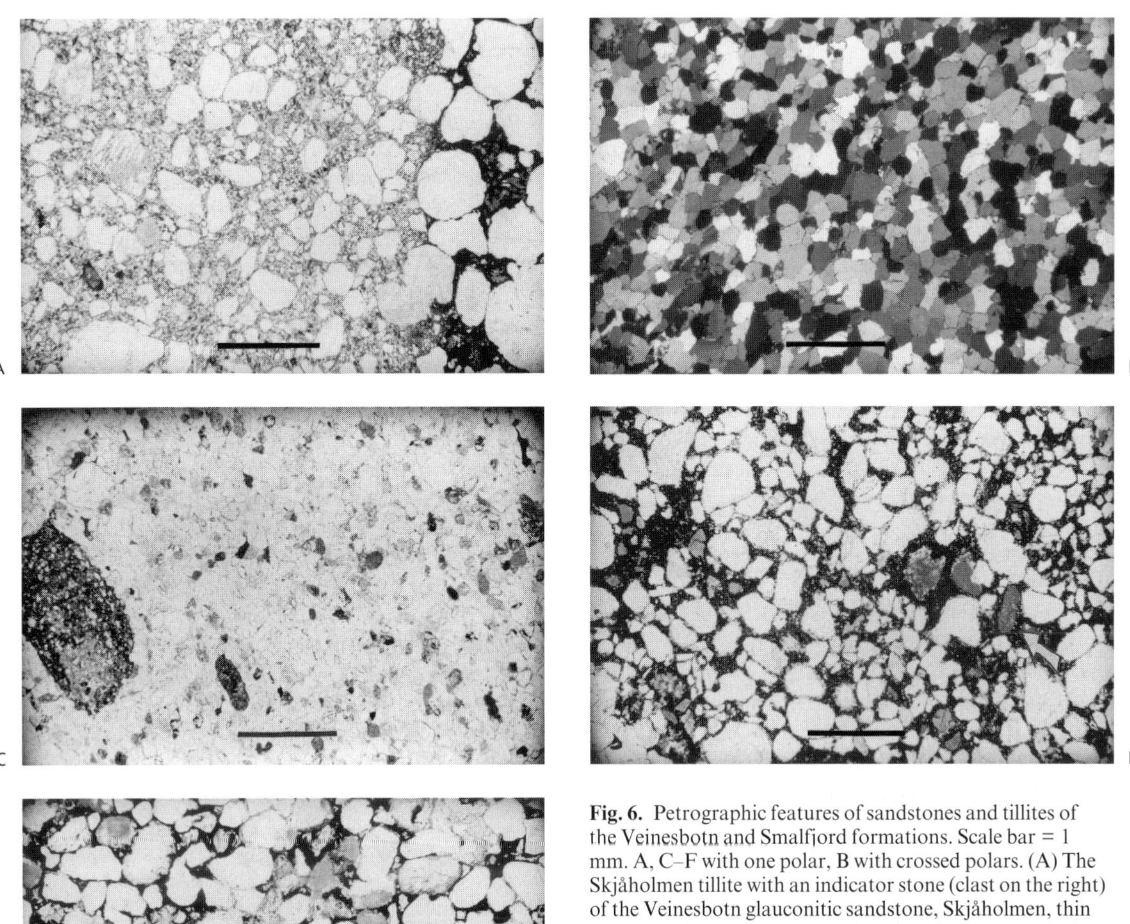

**Fig. 6.** Petrographic features of sandstones and tillites of the Veinesbotn and Smalfjord formations. Scale bar = 1 mm. A, C–F with one polar, B with crossed polars. (A) The Skjåholmen tillite with an indicator stone (clast on the right) of the Veinesbotn glauconitic sandstone, Skjåholmen, thin section (TS) 18 457. (B) Granoblastic Veinesbotn quartzite from 10 cm under the Varangerfjorden unconformity at Bigganjargga, TS 17 475. (C) Lithic Smalfjord sandstone from 50 cm above the Varangerfjorden unconformity at Bigganjargga, TS 17 486. (D) A distinctive Veinesbotn facies B clastic sandstone with glauconitic aggregates (arrow) and carbonate–iron oxide cement, east Vieranjarga, TS 17 496. (E) An indicator stone of the Veinesbotn glauconitic sandstone in the Bigganjargga tillite, Bigganjargga, TS 17 476.

it was probably buried deep enough to be lithified before it was exhumed. Roberts *et al.* (1997) give an age of *c.*650 Ma for this first diagenesis in the overlying Stangenes Formation of the Tanafjorden Group. The pre-Smalfjord lithification of the Veinesbotn Formation is proven by the presence of abundant sharp-edged fragments of the Veinesbotn Formation in the breccias above the unconformity at Vieranjarga and Skjåholmen and the indicator stones in the Bigganjargga and Skjåholmen tillites (Fig. 6A, D & E). The straight, unstepped unconformity on the siltstone-dominated part of the Veinesbotn Formation under the Skjåholmen tillite (Fig. 5E) indicates that even the softest parts of the Veinesbotn Formation were indurated enough before the deposition of the Smalfjord Formation to be planed obliquely to the bedding. The morphology of the Bigganjargga surface differs from a true intra-Smalfjord soft-sediment erosional hiatus exposed at Skjåholmen *c.*5 m above the VFU. Here a thin diamictite unit erodes the underlying turbiditic facies S4 sandstone beds. This contact is scoured and highly uneven (not planed), in part stepped, and is not streamlined (Fig. 5C & D; see also Edwards, 1984, Fig. 11).

These observations should be sufficient to annul the myth of a soft-sediment or semi-consolidated state of the Veinesbotn quartzite during deposition of the Bigganjargga tillite. They also contradict the statements that 'the unconformity is just another bed surface in the succession' (Jensen & Wulff-Pedersen, 1997, p. 875) and that the Bigganjargga surface is a bedding-plane surface (Crowell, 1999, Fig. 60). On the contrary, the present study agrees with Rosendahl, who concluded in 1931 that the underlying rock was quartzite when it was scoured by ice and that the Bigganjargga surface is part of the regional unconformity. Consequently, the Bigganjargga striations and small grooves were inscribed in hard rock and the depressions and imprints on the classical Bigganjargga surface must be attributed to post-depositional pressure solution and later processes, as is discussed briefly in a later section.

## NEW STRIATION AND SMALL GROOVE OBSERVATIONS

New striation and small groove (terminology after Laverdière *et al.*, 1985) measurements by the author together with those by Rice & Hofmann (2000) from Bigganjargga are provided in Table 2. Some of these new Late Proterozoic glacial observations are shown in Fig. 7A–D.

The relatively good matches (Fig. 2) of the inferred ice-flow directions, based on measurements (Table 2) of the younger striations and small grooves at Bigganjargga, the striations at Ruossoaivi and Skjåholmen, and the striations and small grooves at Handelsneset, support a glacial origin for these bedrock-inscribed structures and their being formed by the same ice flow. Unfortunately, no small-scale ice-flow direction indicators were detected with them. The crescentic gouge mentioned by Bjørlykke (1967, Fig. 6b) could be a secondary feature.

The older striations and small grooves are found only at Bigganjargga. This may be the only place they formed, or these older structures escaped the erosion of later ice flow(s). The Bigganjargga surface is the only larger part of the VFU exposed. This question and the question of whether or not these two sets of striations were produced by two separate ice flows or just a local shift of initially more northward ice flow to more westward ice flow cannot be answered because of the small number and sizes of the outcrops and two-dimensionality of the exposed parts of the VFU (see Veillette & Roy, 1995). The fact that the older striations and small grooves can be seen everywhere on the Bigganjargga surface and its western extension, and that the surface is flat and seems not to have been a protected site during the deposition of the overlying rocks, supports a local shift in ice flow. The lack of any truncated surfaces, of which Veillette (1983, 1986) gives good Pleistocene examples, also supports this interpretation. The relation of the eastern trough to the striations at the classical Bigganjargga surface is also problematic because only a small fraction of the trough surface can be studied. It can be concluded that the trough was scoured before the striations were formed, but whether it represents an erosional trough older than the glaciation that abraded the classical surface, and which process formed it, are questionable.

A mud or debris flow origin for the striations and grooves of the classical Bigganjargga surface is discounted because it is not known how these flows could create two subparallel sets of pervasive striations and small grooves in a hard rock. The glacial-like 'striated floor', interpreted by Harrington (1971) to have originated by debris-laden torrential water flows, was developed on a comparatively soft tuff boulder. In addition, the geological setting excludes the snowslide (Dyson, 1937) and volcanic sandblast (Hovey, 1909) models as possible causes of the Bigganjargga striations. As shown by McCarrol *et al.* (1989), a catastrophic subglacial drainage can produce 'striations' (their quote marks) with cross-cutting relationships even on a hard rock surface. These can be recognized by the occurrence of diagnostic short, wide, tapering 'striations', indicating transport of boulders in saltation. These structures are missing at the Bigganjargga surface and there is no evidence of any catastrophic events in the rock record around Varangerfjorden. Glacial striations formed on soft or semi-consolidated sediments are not discussed in this connection, because the sub-Smalfjord basement was made of hard rock, but they are reviewed below.

The striations at Bigganjargga are so similar to Pleistocene striations that they caused Schiøtz (1898) to propose that they were Quaternary in age. Pleistocene striations developed on the Veinesbotn Formation facies A quartzite were detected by the present author on several outcrops in the surroundings of Ruossoaivi. They occur on parts of the VFU that have been only slightly abraded by the Pleistocene glaciation. The only reason these striations are not accepted as Precambrian is that their directions match the Pleistocene striations (Table 2). Figure 7E shows that the Pleistocene striations are not as pronounced as the late Proterozoic striations at Bigganjargga (Fig. 5D),

**Fig. 7.** Glacial features of the Varangerfjorden unconformity at Handelsneset (A–C), Ruossoaivi (D) and Vieranjarga (E). (A) The high-relief Varangerfjorden unconformity (follows the base of the outcrop) between the Fugleberget Formation (dips 20–40° to the left) and the Smalfjord Formation (bulk of the outcrop wall, dips away from the viewer). The black and white arrows show close-ups in B and C, respectively. The hammer handle (close to the black arrow) is 70 cm long. Seen from the east. (B) Small grooves (arrows) on a part of the Varangerfjorden unconformity with steep relief. The small groove on the right is striated internally, whereas recent weathering has modified the surface of the small groove on the left. The scale lies on Fugleberget sandstone. Seen to 300°. (C) Smoothly grooved surface of the Fugleberget Formation under the Smalfjord conglomerate. The small grooves (arrows) are parallel to the marker pen. Seen to the west. (D) Striations on a newly uncovered part of the unconformity at Ruossoaivi, Vieranjarga. The surface is smoothly convex up. The overlying Smalfjord rock is tillitic. The pen head points to 298°. (E) Pleistocene striations on the Veinesbotn quartzite at the Varangerfjorden unconformity east of Ruossoaivi. Ice-flow towards the viewer (east).

which can be attributed to differences in ice dynamics and angles of abrasion (the Pleistocene ice flow was to the north-east), increased hardness of the Veinesbotn quartzite since the Varangian time due to its reburial under the Vestertana Group and differences in recent erosional histories of these two surfaces. No crescentic fractures were detected on the VFU surfaces abraded by the Pleistocene ice, but they are common on the 'fresh' post-VFU surfaces cleft into the Veinesbotn Quartzite at the head of Veinesbotn, topographically a few metres below the imagined extension of the VFU.

No s-forms/p-forms (Dahl, 1965; Kor *et al.*, 1991) can be seen on the Bigganjargga surface, which indicates that erosion by subglacial water was not significant when the striations and small grooves were formed. It might be that the base of the glacier was close to its pressure-melting point, but not significantly above it (see Bourman & Alley, 1999), or that there were no subglacial channels, but the meltwater discharged in a thin water layer between the glacier and its pavement.

## INDICATOR STONE OBSERVATIONS

It is generally accepted that sources of the clasts in the Bigganjargga tillite were located somewhere in the east or south-east (e.g. Bjørlykke, 1967; Edwards, 1975). Granitoid and gneiss clasts were probably derived from the metamorphosed Precambrian basement exposed in south Varangerfjorden. Source areas of most of the sedimentary rock clasts cannot be located due to the lack of outcrops. Clast types indicate that the majority were derived from the Vadsø and Tanafjorden groups. There is one distinct sandstone type easy to identify both in the field and under a microscope. This is a greenish or reddish quartz-rich sandstone where well rounded quartz clasts cemented by carbonate and iron oxide define a distinct clastic texture. Subrounded glauconitic aggregates form an additional distinctive feature (Fig. 6D). The only locality where this rock is known to crop out is at east Vieranjarga, but it occurs as clasts in both the Bigganjargga and Skjåholmen tillites (Fig. 6A & E). The indicator-stone clasts are well rounded, which indicates that the Veinesbotn quartzite was consolidated during Smalfjord glaciation. The indicator stone palaeofan in Fig. 2 is drawn on the basis of these three points and demonstrates that the flow that deposited these rocks came from the south-east.

## POSSIBLE PERIGLACIAL *IN SITU* BRECCIAS

### Description

At Ruossoaivi, only 1 m to the east of the striated part of the VFU in Fig. 7D, the unconformity surface has been destroyed by *in situ* brecciation. The brecciated part is *c.*1.5 m wide and 25 cm deep at its deepest point. The Veinesbotn quartzite under the breccia contains both vertical and horizontal fractures, a few mm thick, filled with rusty material (Fig. 8A). At the base of the breccia the fractures are wider, open upwards and are filled with rusty iron oxide pigmented carbonate material, small rock fragments and mineral grains of the Veinesbotn quartzite (Fig. 8B & C).

A

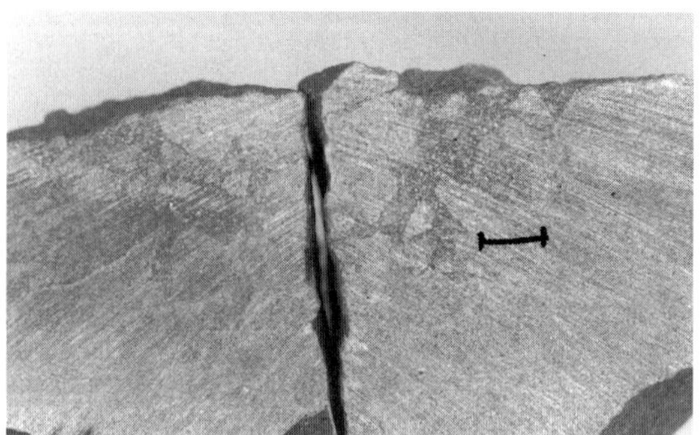
B

**Fig. 8.** Destruction of the Varangerfjorden unconformity by syn-Smalfjord brecciation and erosion of the Veinesbotn Formation at Ruossoaivi (A–D), Veinesbotn (E) and Skjåholmen (F, G). (A) *In situ* breccia in facies A quartzite. Smalfjord detritus fills thin subvertical and subhorizontal (white arrow) cracks. The scale stands on the unconformity, which is shown also by the black arrow on the right. (B) A vertical cut of the *in situ* breccia (top up). The white angular fragments and the bulk of the sample are Veinesbotn facies A quartzite. The Smalfjord detritus fill (darker) between the quartzite fragments is rich in iron oxide and carbonate. Thin section in C was made of the part above the bar (1 cm). (*continued*)

**Fig. 8.** (*cont'd*) (C) Vertical cut of *in situ* breccia at the Veinesbotn/Smalfjord contact. Relatively fresh Veinesbotn quartzite with minor secondary carbonate occupies the left part of the thin section (TS 18 777). The dark matrix surrounding the Veinesbotn mineral and rock fragments is made of carbonate and iron oxide. Scale bar = 1 mm, one polar. (D) The *in situ* breccia surface seen from above. The bar on the Veinesbotn quartzite fragment is 1 cm. (E) *In situ* breccia (behind and right of the scale) and slightly transported breccia fragments filling erosional trough between the Veinesbotn facies A quartzite (below the scale) and the Smalfjord Formation. Arrows on the right show the erosional contact. (F) breccia wedge (black arrow) and a slightly moved fragment (white arrow) of cross-bedded Veinesbotn quartzite at the Veinesbotn Formation/ Smalfjord Formation (main upper part) contact. Massive–pebbly turbiditic sandstone of the Smalfjord Formation fills the space between the bulk Veinesbotn quartzite and the wedge (thin black arrows). Seen from the east. Scale 8 cm. (G) A steep-walled erosional channel (on the left) cut into the Veinesbotn Formation with alternating quartzite and siltstone beds. The channel plunges to 302° and is filled with massive Smalfjord Formation facies S4 sandstone (above the head of the 1 m long stick). The black arrow points to the vertical contact between the two formations under which the siltstone unit of the Veinesbotn Formation has been split off.

Figure 8D shows *in situ* fragments of the Veinesbotn quartzite protruding from their pavement through the infiltrated Smalfjord matrix. The basal part of the breccia trough contains angular fragments of the Veinesbotn quartzite in diamictic Smalfjord matrix. At the eastern margin of the breccia, the Smalfjord matrix intrudes into the Veinesbotn quartzite to a distance of 50 cm along a fracture parallel to the bedding plane.

Similar *in situ* breccia was found at Veinesbotn, where the Veinesbotn quartzite fragments fill a *c*.20 cm deep erosional trough (Fig. 8E). The *in situ* contact is visible, but most of the quartzite fragments in the trough are redeposited and there are solitary granitoid clasts with them.

### Interpretation

The breccias are considered younger than the striations because it is not likely that they could have preserved the subglacial abrasion, which formed the striations practically at the same palaeogeographic level and surface, and because they do not show any evidence of subglacial deformation or orientation. The sharp edges of the fragments and the lack of kaolin or other clay minerals in the breccia fractures (Fig. 8C) indicate that the weathering that caused the brecciation was physical. Because the breccias are under the formation that contains glacigenic rocks and the near-by unbrecciated parts of the same surface contain glacial striations in both Ruossoaivi and Veinesbotn, it is inferred that periglacial physical weathering formed the breccia. This is supported by the occurrence of a probable periglacial sand wedge at the upper contact of the D2 facies tillite at Vieranjarga (Edwards, 1975). This locality is situated only 3 km east of Ruossoaivi and probably represents roughly the same palaeotopographic and palaeoerosional/permafrost level as the Ruossoaivi breccia.

Descriptions of ancient periglacial breccias are lacking. Williams & Tonkin (1985) and Williams (1986) described the late Precambrian Cattle Grid Breccia, South Australia, formed during the Marinoan Glaciation, *c*.680 Ma. The *in situ* part of this breccia consists of only a minor matrix of sandstone, in which preferred orientation of tabular clasts reflects the original bedding of the quartzite. Associated sand-wedge polygons, sand wedges and other periglacial structures support the *in situ* brecciation of the quartzite through frost shattering in an open system under seasonal freeze–thaw cycles. This breccia was developed on a disconformity and cannot as such serve as an exact analogue to the Ruossoaivi and Veinesbotn breccias, which were developed on a consolidated and tilted quartzite. Young & Gostin (1988) described a thin basal breccia on a dolostone at the base of the Sturtian rocks. They attribute this breccia to a period of subaerial periglacial weathering prior to the onset of the Sturtian glaciation.

Mustard & Donaldson (1987a,b) recorded unconformity surface breccias under the Coleman Member of the early Proterozoic Gowganda Formation. The breccias show an *in situ* relationship to the fractured Archaean basement and have been preserved in palaeotopographic lows. Freeze–thaw action may have separated blocks in the upper part of the breccia, allowing later infiltration of sediments of the Coleman Member. Those breccias preserved in steep-sided angular depressions are considered subglacial and related to plucking and lee-side quarrying. Davison & Hambrey (1996) considered some of the breccias under the Proterozoic Stoer Group to be subglacial, but Young (1999) questions the glacial origin of these deposits.

At the Gabberas Mountain, in the type Smalfjord Formation area, Føyn (1937) described a breccia between the underlying Vagge shale and the overlying Smalfjord tillite. It was formed by a glacier or floating icebergs. The location of the breccia above the pre-tillitic angular unconformity and its close association with the Smalfjord tillite indicate that periglacial weathering could be another possibility.

Pharaoh (1985, p. 351) described an irregular unconformity with *in situ* breccia under the Nyvoll Tillite Member in the Repparfjord–Komagfjord tectonic window, some 200 km west of Varangerfjorden. This breccia is developed upon folded early Proterozoic metasandstones of the Svecokarelian basement. The eroded stumps of metasandstone stick up into the tillite. In places, angular blocks of metasandstone have been moved only a few centimetres from their source rocks. Some of the blocks are faceted with rare striations, but a striated basement has not been found. Pharaoh (1985) does not give any interpretation for the breccia, but prefers to correlate the Nyvoll Tillite Member with the tillites of the Mortensnes Formation.

## OTHER BRECCIAS ABOVE THE VANGERFJORDEN UNCONFORMITY

Edwards (1984, Figs 8–10) has described several breccias composed of fragments of the underlying

Veinesbotn quartzite at Skjåholmen. These have been preserved against the palaeoscarps above the unconformity with irregular relief. Edwards (1984) proposed that the irregular unconformity represents a glacially scoured surface, which may have been locally modified by wave and current action, and that the steep slopes may represent palaeowave-cut cliffs. He regards the breccias as reworked glacial drift.

The author studied the southernmost part of the unconformity at Skjåholmen, which Edwards did not visit. Here, large Veinesbotn quartzite fragments up to 2 m long are located at the Veinesbotn/Smalfjord contact. Some of the fragments are still attached to the Veinesbotn pavement, but are intruded by Smalfjord facies S4 turbidite sandstone along the fractures parallel or subparallel to the bedding planes in the Veinesbotn quartzite (Fig. 8F). The intruding sandstone contains small, angular fragments of the Veinesbotn quartzite. Almost *in situ* breccia tails also occur. This locality is topographically a few metres below the tillite in Fig. 5E and represents the deepest part of the unconformity exposed at Skjåholmen. Between these two localities, a steep-walled erosional palaeoscarp or channel filled by massive Smalfjord sandstone was observed (Fig. 8G). These observations indicate that this part of the unconformity is not glacial, but was formed by later intra-Smalfjord erosion, and that the primary glacial surface is probably preserved only under the small tillite relic in Fig. 5E.

Breccias are also abundant in north Vieranjarga, north of Edwards's (1975) unit A and B tillites, where they underlie the thick fluvial sandstones and conglomerates of Edwards's unit C. The thickest breccia is over 2 m thick and consists mainly of Veinesbotn quartzite fragments. Breccias either lie directly on the unconformity flooring erosional palaeochannels or are interbedded with turbiditic sandstones in the basal parts of the palaeochannel fill. These breccias indicate that the glacial period that produced the lowermost tillites and diamictites was followed by a period of intra-Smalfjord erosion. This erosion was deep and went through the glacial sediments several metres down to the Veinesbotn Formation at East Vieranjarga and Skjåholmen, as is shown in Fig. 4. The close association of these breccias with turbiditic sandstones indicates (glaci)fluvial–submarine erosion and deposition, but more exact interpretations would demand a detailed sedimentological study of the whole Smalfjord Formation in the area.

## POST-VFU SOLUTION FEATURES AT THE BIGGANJARGGA SURFACE

At the very eastern part of the classical Bigganjargga surface, under the Bigganjargga tillite, relics of a thin (*c*.0.5 mm) quartz coating can be seen (Fig. 9A), similar to a slickenside. Because there is no evidence of tectonic deformation in the Bigganjargga rocks and it is unknown whether a wet-based ice sheet could form a slickenside, it is probable that diagenetic low temperature metamorphic fluids deposited the quartz coating. Its presence indicates that the unconformity was a well defined boundary surface along which these fluids could have flowed. Careful investigation of nearby parts of the exposed surface reveals that a semi-vitreous and microcrystalline film covers the bottoms of some of the striations with rugged margins, indicating post-striation modification (Fig. 9B). Rice & Hofmann (2000) interpret this film as primary glacial polish, but the rugged morphology of the margins of the striations contradicts this interpretation. To the west of the classical surface, the striations become gradually fainter and the unconformity is rugged, with abundant irregular pits of varying depths (Fig. 9C & D). Further to the west, the rugged surface leaves the unconformity and continues into the Veinesbotn Formation, indicating that there was probably a diagonal fracture within the Veinesbotn quartzite along which fluids were able to flow.

Rice & Hofmann (2000) interpret imprints across striations at Bigganjargga as relics of mud-flakes (their Fig. 2b & e) within the Veinesbotn Formation which were cut across and quarried out during the formation of the pavement. However, they do not give any direct evidence of the existence of mud-flakes in the Veinesbotn quartzite at Bigganjargga. Morphologically the pits in their Fig. 2b are similar to those in Fig. 9D of this study.

The presence of a quartz coating, quartz cement in the overlying Smalfjord sandstone (Fig. 6C) and the summed thickness (2827–3210 m, Table 1) of the Vestertana and Digermulen groups indicate that the VFU was buried well within the depth of quartz cementation. The depth where quartz cementation takes place varies from 1 km to more than 2.5–3 km in different sedimentary basins (Bjørlykke & Egeberg, 1993). Because no signs of reactions between silicate minerals have been observed in thin section, it is likely that the silica was released mostly by pressure solution (see Bjørlykke & Egeberg, 1993). Roberts *et al.* (1997) believe that burial diagenesis in the Vestertana Group succession occurred at around 560–570 Ma, i.e. in the later stages of the Vendian.

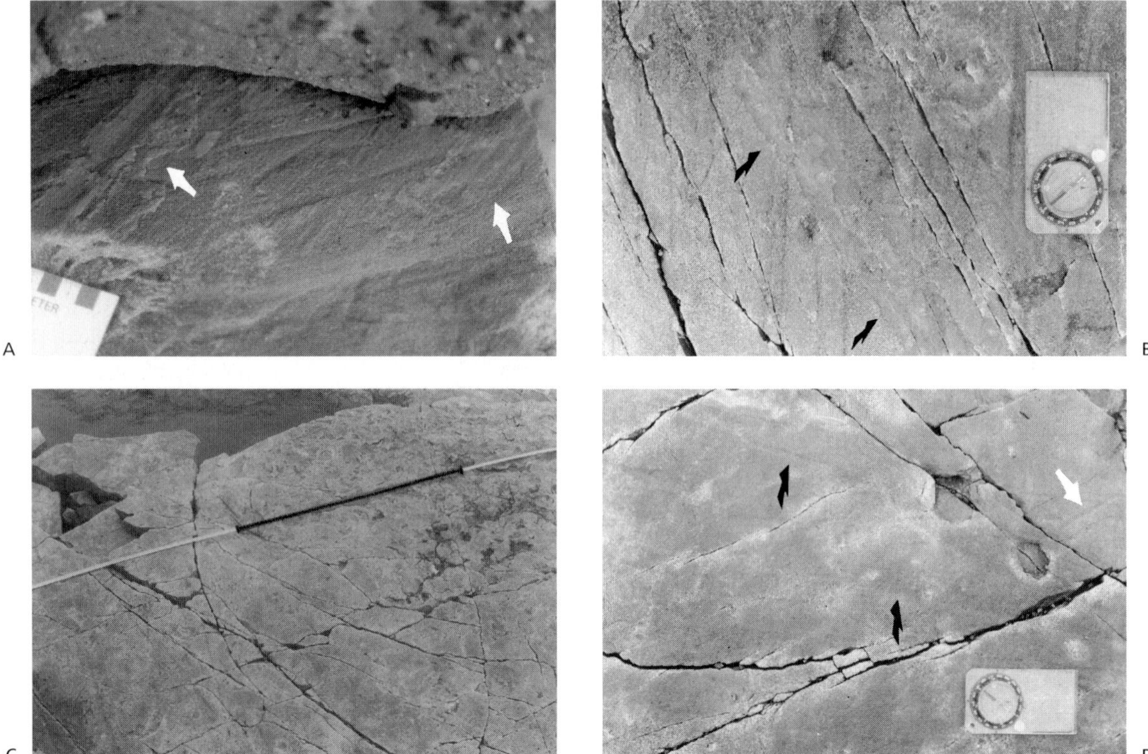


**Fig. 9.** Destruction of glacial features at the classical Bigganjargga surface. (A) Relics of quartz coating on thin younger striations (arrow on the right) and an internally striated small groove (arrow on the left) at the unconformity. Seen from the west. Scale in centimetres. (B) Younger striations destroyed in part by later solution. Notice the irregular margins of the striations (arrows). A very shallow and flat-bottomed solution pit occurs left of the compass (12 cm long and parallel to older striations). (C) Destruction of the western part of the classical striated unconformity caused by post-depositional solution. Outside the right upper corner of the photograph, the solution-pitted surface leaves the unconformity and continues within the Veinesbotn Formation. Dark portion of the tape is 1 m. (D) Close-up of almost completely destroyed older striations (black arrows) and younger striations (white arrow) at the transitional zone in C. The surface is mat and contains two solution pits with mat bases. The 12 cm long compass is parallel to the older striations.

## REGIONAL CORRELATION OF THE VARANGERFJORDEN UNCONFORMITY

The nature of the sub-Smalfjord unconformity west of Varangerfjorden is poorly known and no Varangerian striated pavements have been reported there. The unconformity cannot be directly observed in the Tana area, but on the basis of the way 'the younger sandstones' progressively overstep 'the older sandstone series', Føyn (1937) concluded that the unconformity dips shallowly (1–2°) and that the direction of the largest unconformity is 30° N–40° E.

Using additional previous work by Føyn (1967) and Edwards *et al.* (1973), Føyn & Siedlecki (1980) mapped the sub-Smalfjord Krokvatn palaeovalley on Laksefjordvidda, *c.*50 km west of Varangerfjorden. The NW–SE trending valley is about 15 km wide and over 30 km long, and cuts down several hundred metres into the underlying Tanafjorden Group. They considered the Bigganjargga tillite contemporaneous with the lower or middle glacial conglomerate of the Krokvatn palaeovalley. The Smalfjord succession in the Varanger palaeovalley is, however, dominantly non-tillitic and differs in this respect from both the Krokvatn palaeovalley and the type Smalfjord area in the Vestertana area, where thick Smalfjord tillite successions are located (Reading & Walker, 1966; Banks *et al.*, 1971; Føyn & Siedlecki, 1980; Edwards, 1984).

Føyn (1964) described an unconformity between a tillite of the Boroas Group and the underlying quartzite of the Bossekop Group in the Alta–

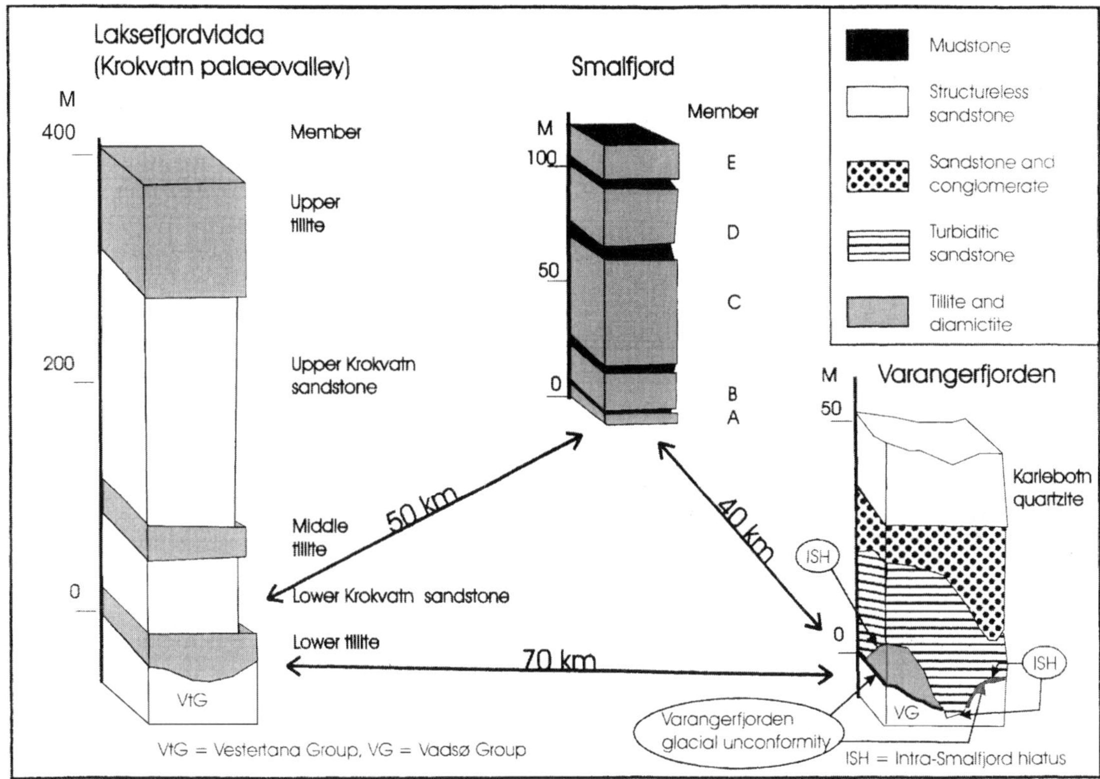


**Fig. 10.** Simplified stratigraphic columns of the Smalfjord Formation at its type area, Smalfjord (Edwards, 1984), Laksefjordvidda (Føyn & Siedlecki, 1980) and Varangerfjorden (lower part simplified from Fig. 4). Vertical scales in Smalfjord and Laksefjordvidda are reduced two and four times, respectively, in comparison with the Varangerfjorden column.

Kvænangen window (see Holtedahl, 1918, Fig. 4). The tillite is correlated with the Mortensnes Formation. Within the same window, a solitary tillite body was found, which lies unconformably on the Karelian basement (Roberts & Fareth, 1974). Holmsen (1955, 1956) described in detail how thin (0.5–4 m) erosional remnants of tillites lie directly on the Precambrian basement in the Carajavre area. The Precambrian peneplain is said to be very even, with a constant dip of nearly 7° to the north-west. How this surface continues from this locality via Finland and Sweden down to southern Norway is documented in the review article by Kumpulainen & Nystuen (1985).

Regional correlation of ancient glacially striated surfaces (or unconformities in general) and associated deposits is not a simple task. For instance, in the Karoo Basin, South Africa, striated surfaces can represent the onset of glaciation, maximum glaciation or deglaciation, and thus can imply an age difference of up to 20 Ma (Visser, 1983, p. 688). Second, the striated surface can be older than the rock lying on it, as was noted by Edwards (1975) and Spjeldnæs (1964) in their discussion of the Bigganjargga case. Third, even near-by surfaces may be of different age and origin. Furthermore, the preservation potential of glacial surfaces and deposits of different tectonic-depositional settings vary greatly (Bjørlykke, 1985; Nystuen, 1985).

The Smalfjord Formation crops out in three disconnected areas in east Finnmark: Varangerfjorden, Laksefjordvidda and Smalfjord (Fig. 1). The lithostratigraphic columns of these areas are quite different (Fig. 10): the Smalfjord Formation in the Varangerfjorden area contains only relics of tillitic rock, whereas three tillite units and two intervening interglacial units occur in Laksefjordvidda. The type Smalfjord section around the fjord of Smalfjord is represented by five mudstone-capped tillite units with negligible sandstone and conglomerate. It is impossible to correlate the Varangerfjorden column in any greater detail with the type Smalfjord column on the basis of the sedimentological data available. In the Laksefjordvidda area, both the upper and lower Krokvatn sandstones are structureless, light-coloured sandstones whose sedimentation history is not known (Føyn & Siedlecki, 1980). These seem to be similar to the structureless Karlebotn quartzite (Bjørlykke, 1967) that overlies the turbiditic and fluvial sandstones and conglomerates of the lower part of the Smalfjord Formation west of Bigganjargga. This quartzite might be correlated with either the lower or upper Krokvatn sandstone. If the first alternative holds, the glacial pavement around Varangerfjorden could counterpart the regional unconformity under the lower Krokvatn tillite. In the latter case, the Bigganjargga surface may have been abraded during the second Smalfjord glaciation. This uncertainty demonstrates that unconformities are not good candidates for large-scale correlation, as is discussed in Christie-Blick *et al.* (1995).

## REVIEW OF GLACIALLY STRIATED AND GROOVED PRE-PLEISTOCENE PAVEMENTS

The ages of known glacially abraded pre-Pleistocene surfaces with striations and small grooves range from early Proterozoic to late Palaeozoic (Table 3) and they vary from non-conformities developed upon metamorphic–plutonic ('crystalline') basement to intraformational ('intertillite') hiatus on soft sediment (several entries in Hambrey & Harland, 1981). Striated boulder pavements (Eyles, 1988, 1994) and furrowed pavements (von Brunn, 1977; Visser, 1990) are not reviewed.

### Terminological comments

Many of the basic terms in glaciology are used in different ways. For example, the use of glacial striation (often called simply striation) is restricted by most authors to striations inscribed by a glacier on a bedrock surface (e.g. Allen, 1984; Laverdière *et al.*, 1985; Bates & Jackson, 1995; Bennet & Glasser, 1996; Ehlers, 1996), whereas some authors also apply it to glacial striations formed on compact sediment (e.g. Flink, 1971, p. 90) or on 'firm but unindurated sand' (Crowell, 1999, p. 62). Woodworth-Lynas & Dowdeswell (1994, p. 241) apply it to ice-formed striations on 'bedding planes in unconsolidated unfrozen sediment'. Other loosely used terms are 'pavement' and 'glacial pavement'. The *Glossary of Geology* (Bates & Jackson, 1995) defines these respectively as 'a bare rock surface that suggests a paved road in smoothness, hardness, horizontality, surface extent, or close packing of its units' and 'a polished, striated, relatively smooth, planed-down rock surface produced by glacial abrasion'. The Ordovician glaciated surfaces in the Sahara (Beuf *et al.*, 1971; Fairbridge, 1974) are possibly the best ancient examples that fulfil these definitions, as does the classical Bigganjargga surface (Fig. 5A & B) and the glaciated bedrock surface on the Cambrian Smith Bay Shale (Bourman & Alley, 1999, Fig. 3e). The latter could also be classified as an undulating surface (Laverdière *et al.*, 1985). Planar parts of glacial surfaces under the Dwyka Formation have also been considered glacial pavements (e.g. Frakes & Crowell, 1975, Fig. 10; Tankard *et al.*, 1982, Fig. 11.4; von Brunn & Marshall, 1989).

On the basis of Slater's (1932) and Visser & Loock's (1988) descriptions, the Nooitgedacht glacial surface resembles a hummocky Pleistocene surface on Precambrian basement in Scandinavia and North America, rather than a typical glacial pavement. Slater (1932) avoided calling the surfaces pavements, but used the expression 'glacial surfaces', whereas Visser & Loock (1988) defined them as 'drumlinoid complexes'. Du Toit (1926, p. 210) used the expression 'undulating floor upon which the tillite rests'.

Identified soft-sediment glacial surfaces have been given various names, including 'glaciated pavement' (Schwellnus, 1941), 'intraformational striated surface' (Young, 1996, Fig. 7.2), 'soft-sediment pavement' (Visser, 1985, Fig. 6), 'soft-sediment glacial surface' (Visser, 1990, Fig. 2), 'intratillite pavement' (von Brunn, 1977; Gravenor & Monteiro, 1983; Visser & Hall, 1984, Fig. 8) and 'glacially striated, soft-sediment surface' (Rocha Campos *et al.*, 1999). Woodworth-Lynas & Dowdeswell (1994, p. 241) referred 'to ice-formed, subparallel ridges, grooves and striations on bedding planes in unconsolidated unfrozen sediments as *soft-sediment striated surfaces*'. This definition is good, but the term itself overemphasizes striation in cases where the surface is predominantly grooved. This has happened with the Oorlogskloof surface (Savage, 1972, Fig. 2), in which the striated grooves are simply termed striations by Woodworth-Lynas (1996, Fig. 4.8). It is difficult to determine how hard, consolidated, compacted, indurated or lithified a glaciated surface is because these terms are general and relative (see Bates & Jackson, 1995). They are used loosely and have different meanings to authors with different research backgrounds.

**Table 3.** Review of pre-Pleistocene (from early Proterozoic to late Palaeozoic) glacially striated and grooved pavements (for terminology see the text).

| Locality | Salient characteristics and interpretations | Reference |
|---|---|---|
| *Early Proterozoic* | Only one hard-rock pavement from Canada is known | |
| Ontario, Canada | Striated non-conformity under Gowganda Fm | Schenk, 1965 |
| Ontario, Canada | Intraformational striated (soft-sediment) surface on mudstone of Gowganda Fm | Young, 1996, Fig. 7.2 |
| *Late Proterozoic* | Examples from every continent, Antarctica excluded | Hambrey & Harland, 1981; Crowell 1999 |
| Kimberley Region, West Australia | Quartzite or highly indurated micaceous sandstone pavement up to 250 square feet exposed over distance of about mile. Pavement has been polished, grooved, striated and plucked | Dow, 1965; Perry & Roberts, 1968, Roberts *et al.*, 1972; Sweet *et al.*, 1974 |
| | 20 separate pavements have been recorded | Plumb, 1981 |
| North Flinders Basin, Adelaide Geosyncline, South Australia | A cast of striated pavement reported from one locality. Daily *et al.* (1973) reinterpreted surface as tectonic. Precisely parallel and regular ridges and microridges ('striations') support this interpretation | Mirams, 1964, cited in Young & Gostin, 1991 |
| Taoudeni Basin, West Africa | Striated pavements and roches moutonnées within the glacial Jbeliat Group and on its substratum (from Archaean to Upper Proterozoic) | Deynoux & Trompette, 1981; Biju-Duval & Gariel, 1969 |
| | Valuable tools in palaeogeographic reconstruction | Deynoux 1985 |
| Henan Province, China | Striated late Proterozoic pavements at several localities. Striated surfaces at various levels on disconformity (parallel unconformity) below the Luoquan Formation and on different rock types. Unusual relative age relationship between faint striations (older) and much larger and deeper grooves (younger) (Guan Baode *et al.* 1986, Fig. 5a) may be attributed to floating ice (see Woodworth-Lynas & Dowdeswell, 1984) | Mu Yongji, 1981; Wang Yuelen *et al.*, 1981; Guan Baode *et al.*, 1986 |
| Aksu area, Tarim Basin, China | Striated pavement on angular unconformity at the base of the Sinian System. A striation pattern (variability in thickness, depth and length and lack of precise parallelism of the striations) similar to Pleistocene striations on an even hard rock surface | Lu Songnian & Gao Zhenjia, 1994, Fig. 7.4 |
| Varangerfjorden, Norway | See the text | See the text* |
| Central East Greenland | Intraformational striated pavements within the Tillite Group | Moncrieff & Hambrey, 1988 |
| Mineral Fork, Utah, USA | Unconformity with a few examples of poorly preserved striated surfaces | Ojakangas & Matsch, 1980; Christie-Blick 1983 |
| | Glacial origin of some structures has been discussed | Christie-Blick, 1982; Ojakangas & Matsch, 1982 |
| Minas Gerais, Brazil | Famous glacial surface under the (?) late Proterozoic Jequitaí Formation | Isotta *et al.*, 1969 |
| | Re-examination of surface near Requital strongly suggests a 'soft sediment' origin for the glacial striations and grooves. Main lines of evidence: '(a) striae inside grooves laterally covered slumped plow ridges, (b) clasts inside furrows partially embedded in the quartzite, (c) sinuosity of furrows, (d) skip (crescentic-like) marks and ridges transverse to furrows, (e) occurrence of striated surfaces on two bedding planes that are separated by a 25 cm thick bed of quartzite, and (f) contiguous striated and ripple-marked areas on the same bedding plane; ripple marks a few centimetres below the striated surface are, however, not deformed.' These features, together with great areal | Rocha Campos *et al.*, 1996 |

*continued on p. 428*

**Table 3.** *(cont'd)*

| Locality | Salient characteristics and interpretations | Reference |
|---|---|---|
| | extension, generally planar nature of striated surface, and consistent orientation of striae and grooves (Isotta *et al.*, 1969, Fig. 2) favour formation in the fluctuating grounding zone of marine ice sheet or tongue just grazing the sediment surface in shallow-marine, tidal, glacier margin setting (Rocha Campos *et al.*, 1996, 1999) | |
| | Poorly developed intertillite striations and grooves are located approximately 4 km east of Jequitaí | Gravenor & Monteiro, 1983 |
| *Late Devonian–Early Carboniferous* | Two examples from Brazil | Crowell, 1999 |
| Parnaiba Basin, Brazil | Striated pavement on late Devonian Capeças sandstone in Brazil. Striations are parallel and there seems to be only one set. Surfaces could represent soft-sediment striated pavements. This is supported by Caputo's (1985) remark that the pavement, the top of the Capeças Formation, was probably frozen when striations were developed on it | Caputo, 1985 |
| *Ordovician–Silurian* | All occurrences are in Africa | Crowell, 1999 |
| Central Sahara, Africa | Huge and continuous Late Ordovician glacial surfaces under the Tamadjert Formation. Lower boundary is erosional, locally an angular unconformity. Striated glacial pavements and glacial palaeogeomorphological features (fluted surfaces, roches moutonnées, glacial valleys) common on unconformity and along intraformational erosional unconformities (disconformities) | Beuf *et al.*, 1971; Biju-Duval *et al.*, 1981. See also reviews by Bennacef *et al.*, 1971; Allen, 1975; Rapp, 1975 |
| Hodh area, Taoudeni Basin, West Africa | Several glaciated pavements, roches moutonnées with striations, furrows and some crescentic. Both areas played a role in reconstruction of the palaeogeography of the Late Ordovician glaciation in West Africa (Deynoux 1985, Fig. 3). Beuf *et al.* (1971) interpreted soft-sediment striated surfaces exposed in over 1200 km of outcrop from the Tamadjert Formation to have been generated by clastic material embedded in the sole of an active terrestrial glacier. Woodworth-Lynas & Dowdeswell (1994) and Woodworth-Lynas (1996) suggested that they may represent possible ice scour marks | Deynoux & Trompette, 1981 |
| Saudi Arabia | Upper Ordovician soft-sediment striated surfaces in glacial palaeovalley deposits of the Sarah Formation | Vaslet, 1990 |
| Ethiopia | A glacially striated unconformity under the Endaga Arbi Tillite. Likely late Ordovician in age | Dow *et al.*, 1971, Fig. 1b; Hambrey, 1981 |
| South Africa | A distinctly furrowed and striated pavement under the Pakhuis tillite skimming the Oskop sandstone interpreted as having been produced by moving but barely grounded ice floe | Visser, 1962; Rust, 1967 (theses cited in Rust, 1981) |
| | Also attributed to wind-driven ice floes floating in a shallow marine embayment or | Rust, 1981 |
| | to glacier advancing into a sea across sediment layers that were firm but not lithified | Crowell, 1999 |
| *Late Palaeozoic* | Products of the Gondwana glaciation the Karoo Basin as the type example<br>In addition to Africa, several occurrences are found in Australia and Brazil | Wanless & Cannon 1966; Frakes & Crowell, 1970; Crowell & Frakes, 1971a,b, 1972, 1975; Hambrey & Harland, 1981; Crowell, 1999 |
| Karoo and Kalahari basins, southern Africa (in general) | More than 100 localities with glacial striations reported in the Karoo region | Du Toit, 1921; Crowell & Frakes, 1972; Martin, 1981a,b; von Brunn & Stratten, 1981 |

*continued*

**Table 3.** (*cont'd*)

| Locality | Salient characteristics and interpretations | Reference |
|---|---|---|
| | Numerous and widely distributed striations, grooves, roches moutonnées and other ice-flow indicators have produced relatively reliable palaeogeographic reconstructions in the Karoo Basin and other Dwyka basins in southern Africa | Du Toit, 1921; Crowell & Frakes, 1972; Stratten, 1977; Visser, 1991, 1997a,b; von Brunn, 1994, 1996 |
| Occurrences in southern Africa | Hard-rock surfaces | |
| Nooitgedacht 66 | Classical 'pavements' of the national monument on Nooitgedacht 66 on early Proterozoic metalavas of the Ventersdorp Group are roches moutonnées or dome-like rock knobs, e.g. Visser & Loock's (1988) drumlinoid complexes | Du Toit, 1926; Slater, 1932 |
| | Some striations on bedrock ('hard rock') exceed 50 m in length. Other structures include chattermarks, nail-head striations and crescentic gouges. Nail-head striations on the Nooitgedacht 66 pavement developed on resistant quartzite of the Natal Group | von Brunn & Stratten, 1981 |
| Lynmouth pavement | The Lynmouth pavement forms a slightly undulatory smooth streambed covering an area of some 1500 m$^2$ | von Brunn & Marshall, 1989 |
| KwaZulu/Natal | Striations on volcanic rock and quartzite have constant pattern over some 60 × 100 km | von Brunn, 1996, Fig. 2 |
| Roode Mond 392, Namaqualand | Hard-rock grooves and striations on quartz–feldspar schist and garnetiferous gneiss bedrock | Visser, 1985 |
| Kaokofeld | Polished and striated valley walls and striated pavements of hard quartzite of the Otavi formation. Several striation sets | Martin & Schalk, 1957 |
| | Soft-sediment surfaces | |
| Riverton (intra-Dwyka) pavement | Grooves are described as 'picture-frame moulding both in form and smoothness'. Material, when moulded, was unfrozen | Slater, 1932, Plate 8a |
| Wolfschlucht 93 near Kimberley | Unique soft-sediment striated and grooved pavements on Kuibis quartzite under the Nama tillite. Solitary groove with regularly-spaced chatter marks believed to have been caused by a soft piece of shale pushed over unconsolidated pavement. Another solitary groove with accumulated pavement material on its sides; concluded that pavement was made of 'semi-plastic or rather unconsolidated rock' and 'the impressions are much more like grooves produced on a muddy or sandy surface by dragging the point of a stick over it'. Various directions and relative ages of grooves were attributed to a floating glacier that locally touched a previously glaciated pavement and that smaller ice raft would merely bump and touch the already smoothed surface. May have been the first time that ice-scour origin was suggested for ancient glacigenic grooves | Schwellnus, 1941 |
| | Restudies attribute the grooves to floating and occasionally grounding ice-floes or to sea-ice | Martin, 1965, p. 109; von Brunn, 1977 |
| Oorlogskloof pavement near Nieuwoundville | Several smooth glacial surfaces with grooves about 12 inches wide and distinctly V-shaped, and can be traced locally 50 yards or more. Gouging erratics are found *in situ* | Rust, 1963; Visser, 1990, Fig. 2 |
| | Contemporaneous slumping and pushing of groove margins is considered evidence for soft-sediment origin of the grooves | Savage, 1972 |
| | At Koperfontein, Die Mond, and Elandslei, a grooved arenaceous layer contains extrabasinally-derived clasts with sediment commonly bulldozed into bow waves on downflow side of clasts. Grooves at Oorlogskloof and Elandsvlei were gouged out of soft sediment by erratics carried along by the ice when it overrode a thin subaqueous ice-marginal apron formed primarily during periods of ice front retreat. | Visser, 1990 |

*continued on p. 430*

**Table 3.** (*cont'd*)

| Locality | Salient characteristics and interpretations | Reference |
|---|---|---|
| | Preservation of bedforms, with their delicate slump fans, is attributed to separation of the glacier sole from substrate during sudden rise in sea-level | |
| | Gouging out by a moving ice sheet formed the Riverton grooves | von Brunn, 1977 |
| | Grooves represent possible fossil ice-scour marks. One interesting feature in the photographs published by Rust (1963, Fig. 1) and Savage (1972, Fig. 2) is that the two adjacent largest grooves are not symmetrically V-shaped, but their left walls are shorter and steeper dipping than the right walls. Both walls are straight. This seems to favour ice-scour origin | Woodworth-Lynas & Dowdeswell, 1994; Woodworth-Lynas, 1996 |
| Hanaus 43 near Gibeon. | A solitary wide, flat-bottomed groove with spilled margins, swath and scraping on rippled bedding planes in Fish River sandstones interpreted as drag marks caused by thin ice floes, which were moved by wind in very shallow water | Hälbich, 1964, Plate 1 |
| Kenhardt District | Small soft-sediment pavements with shallow grooves and striations on bedding plane. Floating-ice origin is suggested for some of the grooves | Visser, 1985, Figs 6 & 7 |
| Troubridge Basin, South Australia | Famous glacial hard rock pavements at Inman Valley, Hallet Cove and Kangaroo Island | Harris, 1981 |
| | Permian glaciated bedrock ('hard rock') surfaces with striations, grooves, chatter marks, friction cracks, crescentic gouges, p-forms, sichelwannen, and miniature rock crag-and-tail and roches mouttonnées at Kangaroo Island. Similarity with Pleistocene glacially polished hard rock surfaces (e.g. Press, 1983). Glacial erosional forms and presence of thick lodgement till suggest that the local basal ice was at pressure-melting point during their formation | Bourman & Alley, 1999 |
| Parana Basin, Brazil | Twenty striated surfaces on pre-Gondwana (Devonian) Furnas Sandstone, and on periglacial deposits of the Itarare Formation (upper Carboniferous). Striations have been disposed in an almost perfect parallel pattern, without the interference of different age striations. These are not very conclusive in terms of sense of ice movement. Chattermarks, crescentic gouges, nailhead striations, and related structures are not clearly preserved. Striations are attributed to the Gondwana ice sheets. The remarkable surface, at least 2 × 3 m in size, on the Furnas Sandstone in Fig. 2 shows long striations most of which continue across the whole length of the outcrop and are precisely parallel. There seem to be unstriated domains, gradual change in amplitude, and lateral migration of some of the ridges across adjacent grooves. These are all distinctive features of soft-sediment striated surfaces produced by floating ice (Woodworth-Lynas & Dowdeswell, 1994, Rocha-Campos *et al.*, 1996, Woodworth-Lynas, 1996) | Bigarella *et al.*, 1967 |
| Bahia, NE Brazil | Floating ice origin is considered possible for the multiple soft-sediment striated surfaces in the late Palaeozoic Curitiba Formation | Rocha-Campos *et al.*, 1997 |
| Irai, India | A striated pavement on limestone. Frakes *et al.* (1975) mentioned that the surfaces resemble structural slickensides. Presence of cross-striations favours Smith's interpretation | Smith, 1963 |
| Ohio Range, Transantarctic Mountains, Antarctica | Buckeye Formation contains at least five striated pavements within the diamictite sequence | Aitchison *et al.*, 1988, Fig. 3 |
| | Considered soft-sediment striated surfaces | Woodworth-Lynas & Dowdeswell, 1994 |

* See also the complementary paper to the present study: K. Laajoki (2001) Additional observations on the late Proterozoic Varangerfjorden unconformity, Finnmark, northern Norway. *Bull. geol. Soc. Fin.*, **73**(1/2).

### Discussion

Most of the known late Proterozoic glacial pavements represent unconformities on hard rock surfaces similar to their Pleistocene counterparts, but soft-sedimentary striated pavements have been described recently in an increasing number of localities. The huge late Ordovician–early Silurian glacial pavements in North Africa represent an extremely well preserved record of soft-sediment surfaces of either subglacial or ice scour origin, and late Palaeozoic glacial sequences contain both abraded hard rock surfaces and intraformational soft-sediment striated surfaces. There are no significant morphological differences between pre-Pleistocene and Pleistocene glacially abraded surfaces, which could suggest that significant changes in glacial abrasion processes took place since late Proterozoic time. The differences observed may be attributed to preservation/exposure/identification of the surfaces and deposits, the differences in sizes and palaeogeographical/tectonic positions and dynamics of the ancient glacier systems.

The study of glacially striated surfaces has declined since the classical African and other studies referred to in Table 3. Re-examination of the Jequitaí pavement (Rocha Campos *et al.*, 1996), recent ice scour studies and reinterpretations (Woodworth-Lynas & Dowdeswell, 1994; Woodworth-Lynas, 1996) and detailed bedrock surface studies (Bourman & Alley, 1999) demonstrate that these surfaces could significantly increase our understanding of regional lithostratigraphy and sequence stratigraphy, ancient glacial history and ice-mass dynamics. For example, the excellently preserved striations and other glacial abrasion structures under the Moonlight Valley Tillite (see Plates XVI–XVIII in Dow, 1965; Fig. 11.4 in Williams, 1994) would deserve a more detailed documentation in an easily accessible publication (see Perry & Roberts, 1968; Bourman & Alley, 1999). Figure 4 in Martin & Schalk (1957) shows a complex striation system of at least three relative age groups on a quartzite, but this feature is not treated more closely in the text. The same holds for Fig. 1b in Dow *et al.* (1971), which depicts multidirectional striations on metavolcanics. Careful analysis of these and other striated outcrops in the area could enlighten the age problem of the Endaga Arbi Tillite (late Ordovician versus Permo-Carboniferous; Hambrey 1981). The glacial pavements under the Luoquan Formation (Mu Yongji, 1981; Wang Yuelun *et al.*, 1981; Guan Baode *et al.*, 1986) offer a good opportunity to study how different lithological units with varying degrees of hardness were glacially abraded.

Less spectacular pavements are also worthy of detailed study because even small surfaces can unveil valuable palaeoglacial information as shown by Schenk (1965, Figs 7 & 8), Visser & Hall (1984, Fig. 8), Miller (1989, Fig. 5c) and the present study (Fig. 7B–D). Furthermore, glacial striations could have been preserved only in specific palaeotopographic sites, such as flanks of incised valleys (von Brunn, 1994).

Glacial striations and grooves can serve as valuable tools in reconstruction of palaeoice-flow, but they must be supported by other studies (e.g. Menzies, 1995; Bennet & Glasser, 1996). This is well demonstrated by studies of Pleistocene abraded glacial surfaces, which often contain more than two sets of striations (e.g. Press, 1983; Veillette & Roy, 1995, Fig. 1; Mattsson, 1997). Their causal relationship with the overlying tillite is not always an unequivocal question (Catto, 1998). Furthermore, although there are data sets of several thousands or even tens of thousands of measurements available in the Pleistocene record (e.g. Glückert, 1974), establishment of palaeoice-sheet flow patterns (e.g. Klassen & Thompson, 1989; Kleman, 1990; Syverson, 1995) and reconstruction of glacier dynamics (Sharp *et al.*, 1989) may be difficult.

## CONCLUSIONS

New striation and small groove observations support Bjørlykke's (1967) glacial palaeovalley model for the Varangerfjorden unconformity and confirm the sub glacial origin of the Bigganjargga surface.

After deposition of the lowermost tillites, the Varangerfjorden unconformity was incised by deep intra-Smalfjord interglacial erosion.

No features indicating that late Proterozoic glacial abrasion processes were different from those of Palaeozoic or later ice ages can be seen on the Varangerfjorden unconformity.

## ACKNOWLEDGEMENTS

Permission from the Environmental Department of the Regional Commissioner of Finnmark to collect samples and clean up the algae-covered part of the Bigganjargga outcrop is gratefully acknowledged. Prof. Knut Bjørlykke kindly lent the author his Varangerfjorden thin sections. Fruitful discussions with him of the study topic are also acknowledged. About two weeks of fieldwork was carried out in the summers of 1997 and 1999, when the author was visiting NorFA Professor at the University of Oslo. The paper was completed at the Mineralogical–Geological Museum of the University of Oslo in winter 1999/2000, where the author was a Visiting Senior Scientist of the Research Council of Norway (Project 'The Significance of Exogenic Processes for the Development of the Precambrian Crust' led by Prof. Tom Andersen) and a Senior Scientist of the Academy of Finland (Project 'Ore Processes and Geotectonics'). The thin sections used were made at the University of Oulu. Drs Malcolm Hicks and Patricia Corcoran checked the English of the manuscript. The author is indebted to P. Corcoran, H. Tirsgaard and an anonymous reviewer for helpful reviews that improved the manuscript significantly and Profs A.C. Rocha-Campos and G.M. Young for helpful information on some references.

## REFERENCES

AITCHISON, J.C., BRADSHAW, M.A. & NEWMAN, J. (1988) Lithofacies and origin of the Buckeye Formation: Late Palaeozoic glacial and glaciomarine sediments, Ohio Range, Transantarctic Mountains, Antarctica. *Palaeogeogr. Palaeoclimatol. Palaeoecol.*, **64**, 93–104.

ALLEN, J.R.L. (1984) *Sedimentary Structures. Their Character and Physical Basis.* Unabridged one-volume edition. Developments in Sedimentology, 30. Elsevier, Amsterdam. 593 + 663 pp.

ALLEN, P. (1975) Ordovician glacials of the central Sahara. In: *Ice Ages: Ancient and Modern* (Eds Wright, A.E. & Moseley, F.), pp. 275–286. Seel House Press, Liverpool.

ARNAUD, E.V. & EYLES, C.H. (1999) Identifying the sedimentological signature of a Neoproterozoic glaciation: the Smalfjord Formation, northern Norway. *Geol. Soc. Am. Abstr.*, **31**, A487.

BANKS, N.L., EDWARDS, M.B., GEDDES, W.P., HOBDAY, D.K. & READING, H.G. (1971) Late Precambrian and Cambro-Ordovician sedimentation in east Finnmark. *Norges geol. unders.*, **269**, 197–236.

BANKS, N.L., HOBDAY, D.K., READING, H.G. & TAYLOR, P.N. (1974) Stratigraphy of the late Precambrian 'older sandstone series' of the Varangerfjord area, Finnmark. *Norges geol. unders.*, **303**, 1–18.

BATES, R.L. & JACKSON, J.A. (Eds) (1995) *Glossary of Geology*, 3rd edn, CD-ROM. American Geological Institute, Alexandria, VA.

BENNACEF, A., BEUF, S., BIJU-DUVAL, B., DE CHARPAL, O., GARIEL, O. & ROGNON, P. (1971) Example of cratonic sedimentation; Lower Paleozoic of Algerian Sahara. *Am. Ass. Petrol. Geol. Bull.*, **55**, 2225–2245.

BENNET, M.R. & GLASSER, N.F. (1996) *Glacial Geology: Ice Sheets and Landforms.* Wiley, Chichester, 364 pp.

BEUF, S., BIJU-DUVAL, B., DE CHARPAL, O., ROGNON, P., GARIEL, O. & BENNACEF, A. (1971) Les grès du Paléozoïque Inférieur au Sahara: sédimentation et discontinuités—évolution structurale d'un craton. Publ. Inst Franç. Petrole, Collection 'Science et Technique du Petrole', **18**, 464 pp.

BIGARELLA, J.J., SALAMUNI, R. & FUCK, R.A. (1967) Striated surfaces and related features, developed by the Gondwana ice sheets (State of Parana, Brazil). *Palaeogeogr. Palaeoclimatol. Palaeoecol.*, **3**, 265–276.

BIJU-DUVAL, B., DEYNOUX, M. & ROGNON, P. (1981) Late Ordovician tillites of the Central Sahara. In: *Earth's Pre-Pleistocene Glacial Record* (Eds Hambrey, M.J. & Harland, W.B.), pp. 99–107. Cambridge University Press, Cambridge.

BIJU-DUVAL, B. & GARIEL, O. (1969) Nouvelles observations sur les phénomènes glaciaires 'Éocambriens' de la bordure nord de la synéclise de Taoudeni, entre le Hank et le Tanezrouft, Sahara Occidental. *Palaeogeogr. Palaeoclimatol. Palaeoecol.*, **6**, 283–315.

BJØRLYKKE, K. (1967) The Eocambrian 'Reusch Moraine' at Bigganjargga and the geology around Varangerfjord, northern Norway. *Norges geol. unders.*, **251**,18–44.

BJØRLYKKE, K. (1985) Glaciations, preservation of their sedimentary record and sea level changes. A discussion based on the Late Precambrian and Lower Palaeozoic sequence in Norway. *Palaeogeogr. Palaeoclimatol. Palaeoecol.*, **51**, 197–207.

BJØRLYKKE, K. & EGEBERG, P.K. (1993) Quartz cementation in sedimentary basins. *Am. Ass. Petrol. Geol. Bull.*, **77**, 1538–1548.

BOURMAN, R.P. & ALLEY, N.F. (1999) Permian glaciated bedrock surfaces and associated sediments on Kangaroo Island, South Australia: implications for local Gondwana ice-mass dynamics. *Aust. J. Earth Sci.*, **46**, 523–531.

CAPUTO, M.V. (1985) Late Devonian glaciation in South America. *Palaeogeogr. Palaeoclimatol. Palaeoecol.*, **51**, 291–317.

CATTO, N.R. (1998) Comparative study of striations and basal till clast fabrics, Malpeque–Bedeque region, Prince Edward Island, Canada. *Boreas*, **27**, 259–274.

CHRISTIE-BLICK, N.H. (1982) Upper Precambrian (Eocambrian) Mineral Fork Tillite of Utah; a continental glacial and glaciomarine sequence; discussion. *Geol. Soc. Am. Bull.*, **93**, 184–186.

CHRISTIE-BLICK, N.H. (1983) Glacial-marine and subglacial sedimentation, upper Proterozoic Mineral Fork Formation, Utah. In: *Glacial-marine Sedimentation* (Ed. Molnia, B.F.), pp. 703–776. Plenum Press, New York.

CHRISTIE-BLICK, N.H., DYSON, I.A. & VON DER BORCH, C.C. (1995) Sequence stratigraphy and the interpretation of Neoproterozoic earth history. *Precam. Res.*, **73**, 3–26.

CROWELL, J.C. (1964) Climatic significance of sedimentary deposits containing dispersed megaclasts. In: *Problems in Palaeoclimatology* (Ed. Nair, A.E.M.), pp. 86–99. John Wiley & Sons, London.

CROWELL, J.C. (1999) *Pre-Mesozoic Ice Ages: Their Bearing on Understanding the Climate System.* Mem. geol. Soc. Am., Boulder, **192**, 106 pp.

CROWELL, J.C. & FRAKES, F.A. (1971a) Late Palaeozoic glaciation: Part IV, Australia. *Geol. Soc. Am. Bull.*, **82**, 2515–2540.

CROWELL, J.C. & FRAKES, F.A. (1971b) Late Palaeozoic glaciation of Australia. *J. geol. Soc. Aust.*, **17**, 115–155.

CROWELL, J.C. & FRAKES, F.A. (1972) Late Palaeozoic glaciation: Part V, Karoo Basin, South Africa. *Geol. Soc. Am. Bull.*, **83**, 2887–2912.

CROWELL, J.C. & FRAKES, F.A. (1975) Late Palaeozoic glaciation. In: *Gondwana Geology* (Ed. Campbell, K.S.W.), pp. 313–331. Australian National University Press, Canberra.

DAHL, R. (1965) Plastically sculptured detail forms on rock surfaces in northern Nordland, Norway. *Geogr. Ann.*, **47A**, 83–140.

DAILY, B., GOSTIN, V.A. & NELSON, C.A. (1973) Tectonic origin of for an assumed glacial pavement of Late Proterozoic age, South Australia. *J. geol. Soc. Aust.*, **20**, 75–78.

DAL, A. (1900) Geologiske iagttagelser omkring Varangerfjorden. *Norges geol. unders.*, **28**(5), 1–16.

DAVISON, S. & HAMBREY, M.J. (1996) Indications of glaciation at the base of the Proterozoic Stoer Group (Torridonian), NW Scotland. *J. geol. Soc. London*, **153**, 139–149.

DEYNOUX, M. (1985) Terrestrial or waterlain glacial diamictites? Three case studies from the Late Precambrian and Late Ordovician glacial drifts in West Africa. *Palaeogeogr. Palaeoclimatol. Palaeoecol.*, **51**, 97–141.

DEYNOUX, M. & TROMPETTE, R. (1981) Late Ordovician tillites of the Taoudeni Basin, West Africa. In: *Earth's Pre-Pleistocene Glacial Record* (Eds Hambrey, M.J. & Harland, W.B.), pp. 89–96. Cambridge University Press, Cambridge.

DOW, D.B. (1965) Evidence of a late pre-Cambrian glaciation in the Kimberley Region of Western Australia. *Geol. Mag.*, **102**, 407–414.

DOW, D.B., BEYTH, M. & HAILU, T. (1971) Palaeozoic glacial rocks recently discovered in northern Ethiopia. *Geol. Mag.*, **108**, 53–60.

DU TOIT, A.L. (1921) The Carboniferous glaciation of South Africa. *Trans. geol. Soc. S. Afr.*, **24**, 188–227.

DU TOIT, A.L. (1926) *The Geology of South Africa*, 1st edn. Oliver & Boyd, Edinburgh, 445 pp.

DYSON, J.D. (1937) Snowslide striations. *J. Geol.*, **45**, 549–552.

EDWARDS, M.B. (1972) *Glacial, interglacial, and postglacial sedimentation in Late Precambrian shelf environment, Finnmark, north Norway*. DPhil dissetation, Oxford University.

EDWARDS, M.B. (1975) Glacial retreat sedimentation in the Smalfjord Formation, Late Precambrian, North Norway. *Sedimentology*, **22**, 75–94.

EDWARDS, M.B. (1984) Sedimentology of the upper Proterozoic glacial record, Vestertana Group, Finnmark, North Norway. *Norges geol. unders. Bull.*, 394, 76 pp.

EDWARDS, M.B. (1997) Discussion of glacial or non-glacial origin for the Bigganjargga tillite, Finnmark, northern Norway. *Geol. Mag.*, **134**, 873–876.

EDWARDS, M.B., BAYLIS, P., GIBLING, M., GOFFE, W., POTTER, M. & SUTHREN, R.J. (1973) Stratigraphy of the 'older sandstone series' (Tanafjord Group) and Vestertana Group north of Stallogaissa, Laksefjord district, Finnmark. *Norges geol. unders.*, **294**, 25–41.

EDWARDS, M.B. & FØYN, S. (1981) Late Precambrian tillites in Finnmark, North Norway. In: *Earth's Pre-Pleistocene Glacial Record* (Eds Hambrey, M.J. & Harland, W.B.), pp. 606–610. Cambridge University Press, Cambridge.

EHLERS, J. (1996) *Quaternary and Glacial Geology*. John Wiley & Sons, Chichester. 578 pp.

EYLES, C.H. (1988) A model for striated boulder pavement formation on glaciated, shallow-marine shelves; an example from the Yakataga Formation, Alaska. *J. sediment. Petrol.*, **58**, 62–71.

EYLES, C.H. (1994) Intertidal boulder pavements in the northeastern Gulf of Alaska and their geological significance. *Sediment. Geol.*, **88**, 161–173.

FAIRBRIDGE, R.W. (1974) Glacial grooves and periglacial features in the Saharan Ordovician. In: *Glacial Geomorphology* (Ed. Coates, D.R.), pp. 315–327. George Allen & Unwin, Boston.

FLINK, R.F. (1971) *Glacial and Quaternary Geology*. John Wiley & Sons, New York, 892 pp.

FØYN, S. (1937) The Eo-Cambrian series of the Tana district, northern Norway. *Norsk geol. Tidsskr.*, **17**, 65–164.

FØYN, S. (1964) Den tilliteførende formasjongruppe i Alta —en jevnføring med Øst-Finnmark og indre Finnmark. Summary: The tillite-bearing formations of the Alta districk—a correlation with eastern Finnmark and the interior of Finnmark. *Norges geol. unders.*, **228**, 139–150.

FØYN, S. (1967) Dividal-gruppen ('Hyolithus-sonen') i Finnmark og dens forhold til eokambrisk–kambriske formasjoner. (Summary: The Dividal Group ('the Hyolithus Zone') in Finnmark and its relations to the Eocanbrian–Cambrian formations). *Norges geol. unders.*, **249**, 1–84.

FØYN, S. (1985) The late Precambrian in northern Scandinavia. In: *The Caledonide Orgen-Scandinavia and Related Areas* (Eds Gee, D.G. & Sturt, B.A.), pp. 233–246. John Wiley & Sons, London.

FØYN, S. & SIEDLECKI, S. (1980) Glacial stadials and interstadials of the Late Precambrian Smalfjord Tillite on Laksefjordvidda, Finnmark, North Norway. *Norges geol. unders.*, **358**, 31–45.

FRAKES, L.A. & CROWELL, J.C. (1970) Late Proterozoic glaciation: Part II. Africa exclusive the Karroo Basin. *Geol. Soc. Am. Bull.*, **86**, 2261–2286.

FRAKES, L.A., KEMP, E.M. & CROWELL, J.C. (1975) Late Proterozoic glaciation: Part VI. Asia. *Geol. Soc. Am. Bull.*, **86**, 454–464.

GLÜCKERT, G. (1974) Map of glacial striation of the Scandinavian ice sheet during the last (Weichsel) glaciation in northern Europe. *Bull. geol. Soc. Finl.*, **46**, 1–8.

GRAVENOR, C.P. & MONTEIRO, R.L.B.P. (1983) Ice-thrust features and a possible intertillite pavement in the Proterozoic Macaubas Group, Jequitai area, Minas Gerais, Brazil. *J. Geol.*, **91**, 113–116.

GUAN BAODE, WU RUITANG, HAMBREY, M.J. & GENG WUCHEN (1986) Glacial sediments and erosional pavements near the Cambrian–Precambrian boundary in western Henan Province, China. *J. geol. Soc. London*, **143**, 311–323.

HÄLBICH, I.W. (1964) Observations on primary features in the Fish River series and the Dwyka series in South West Africa. *Trans. geol. Soc. S. Afr.*, **67**, 95–108.

HAMBREY, M.J. (1981) Palaeozoic tillites in northern Ethiopia: In: *Earth's Pre-Pleistocene Glacial Record* (Eds Hambrey, M.J. & Harland, W.B.), pp. 38–40. Cambridge University Press, Cambridge.

HAMBREY, M.J. & HARLAND, W.B. (Eds) (1981) *Earth's Pre-Pleistocene Glacial Record* Cambridge University Press, Cambridge, 1004 pp.

HARLAND, W.B. (1964) Critical evidence for a great infra-Cambrian glaciation. *Geol. Rundsch.*, **54**, 45–61.

HARLAND, W.B. (1983) The Proterozoic glacial record. *Mem. geol. Soc. Am.*, **161**, 279–288.

Harland, W.B. & Herod, K.N. (1975) Glaciation through time. In: *Ice Ages: Ancient and Modern* (Eds Wright, A.E. & Moseley, F.), pp. 189–216. Seel House Press, Liverpool.

Harland, W.B., Herod, K.N. & Krinsley, D. (1966) The definition and identification of tills and tillites, *Earth Sci. Rev.*, **2**, 225–256.

Harland, W.B. & Rudwick, M.J.S. (1964) The great Infra-Cambrian ice age. *Sci. Am.*, **211**, 28–36.

Harrington, H.J. (1971) Glacial-like 'striated floor' originated by debris-laden torrential water flows. *Am. Ass. Petrol. Geol. Bull.*, **55**, 1344–1347.

Harris, W.K. (1981) Permian diamictites of South Australia. In: *Earth's Pre-Pleistocene Glacial Record* (Eds Hambrey, M.J. & Harland, W.B.), pp. 469–473. Cambridge University Press, Cambridge.

Hobday, D.K. (1974) Interaction between fluvial and marine processes in the lower part of the late Precambrian Vadsö Group, Finnmark. *Norges geol. unders.*, **303**, 39–56.

Holmsen, P. (1955) Hyolithus-sonens basale lage I Vest-Finnmark (Summary: The basal layers of the 'Hyolithus-zone' in western Finnmark). *Norges geol. unders.*, **195**, 65–72.

Holmsen, P. (1956) De eokambriske lag under hyolithussonen mellem Carajavrre og Caskias, Vestfinnmark (Summary: The Eocambrian beds below the Hyolithus-zone between Carajavrre and Caskias, Western Finnmark). *Norges geol. unders.*, **200**, 47–50.

Holtedahl, D. (1918) Bidrag til Finmarkens geologi. *Norges geol. unders.*, **84**, 311 pp.

Holtedahl, O. (1919) On the Paleozoic formations of Finmarken in northern Norway. *Am. J. Sci.*, **47**, 85–107.

Holtedahl, O. (1960) Stratigraphy of the Sparagmite Group, including 'the sandstone division of Finnmark'. In: *Geology of Norway* (Ed. Holtedahl, O.), pp. 111–127. *Norges geol. unders.*, **208**.

Hovey, E.O. (1909) Striations and U-shaped valleys produced by other than glacial action. *Bull. geol. Soc. Am.*, **20**, 409–416.

Isotta, C.A.L., Rocha-Campos, A.C. & Yoshida, R. (1969) Striated pavement of the Upper pre-Cambrian glaciation in Brazil. *Nature*, **222**, 466–468.

Iverson, N.R. (1991) Morphology of glacial striae: implications for abrasion of glacier beds and fault surfaces. *Geol. Soc. Am. Bull.*, **103**, 1308–1616.

Jensen, P.A. & Wulff-Pedersen, E. (1996). Glacial or non-glacial origin for the Bigganjargga tillite, northern Norway. *Geol. Mag.*, **133**, 137–145.

Jensen, P.A. & Wulff-Pedersen, E. (1997) Discussion of glacial or non-glacial origin for the Bigganjargga tillite, Finnmark, northern Norway. *Geol. Mag.*, **134**, 873–876.

Klassen, R.A. & Thompson, F.J. (1989) Ice flow history and glacial dispersal patterns, Labrador. *Paper 89–20*, pp. 21–29. Geological Survey of Canada, Ottawa.

Kleman, J. (1990) On the use of glacial striae for reconstruction of palaeo-ice sheet flow patterns. *Geogr. Ann.*, **72A**, 217–236

Kor, P.S.G., Shaw, J. & Sharpe, D.R. (1991) Erosion of bedrock by subglacial meltwater, Georgia Bay, Ontario: a regional view. *Can. J. Earth Sci.*, **28**, 623–642.

Kumpulainen, R. & Nystuen, J.P. (1985) Late Proterozoic basin evolution and sedimentation in the westernmost part of Baltoscandia. In: *The Caledonide Orgen-Scandinavia and Related Areas* (Eds Gee, D.G. & Sturt, B.A.), pp. 213–232. John Wiley & Sons, London.

Laajoki, K. (1999a) Late Proterozoic glacial unconformity at Bigganjargga, Finnmark, Norway. EUG 10, 28 March to 1 April 1999, Strasbourg, France. *J. Conf. Abstr.*, **4**(1), 732.

Laajoki, K. (1999b) Late Proterozoic glacial unconformity around Varangerfjorden, northern Norway. *Abstract Volume, 19th Regional European Meeting of Sedimentology, Copenhagen, 22–27 August, 1999*, p. 149.

Laverdière, C., Guimont, P. & Dionne, J.-C. (1985) Les formes et les marques de l'érosion glaciaire du plancher rocheux: signification, terminologie, illustration (Forms and marks of glacial erosion on bedrock: signification, terminology, illustration). *Palaeogeogr. Palaeoclimatol. Palaeoecol.*, **51**, 365–387.

Lu Songnian & Gao Zhenjia (1994) Neoproterozoic tillite and tilloid in the Aksu area, Tarim Basin, Xinjiang Uygur Autonomous Region, northwest China. In: *Earth's Glacial Record* (Eds Deynoux, M., Miller, J.M.G., Domack, E.W., Eyles, N., Fairchild, I.J. & Young, G.M.), pp. 95–100. Cambridge University Press, Cambridge.

McCarrol, D., Matthews, J.A. & Shakesby, R.A. (1989) 'Striations' produced by catastrophic subglacial drainage of a glacier-dammed lake, Mjølkedalsbreen, southern Norway. *J. Glaciol.*, **35**, 193–196.

Martin, H. (1965) *The Precambrian Geology of South West Africa and Namaqualand.* Precambrian Research Unit, University of Cape Town, 159 pp.

Martin, H. (1981a) The Late Palaeozoic Dwyka Group of the South Kalahari Basin in Namibia and Botswana and the subglacial valleys of the Kaokoveld in Namibia. In: *Earth's Pre-Pleistocene Glacial Record* (Eds Hambrey, M.J. & Harland, W.B.), pp. 61–66. Cambridge University Press, Cambridge.

Martin, H. (1981b) The Late Palaeozoic Dwyka Group of the Karasburg Basin, Namibia. In: *Earth's Pre-Pleistocene Glacial Record* (Eds Hambrey, M.J. & Harland, W.B.), pp. 67–70. Cambridge University Press, Cambridge.

Martin, H. & Schalk, K. (1957) Gletscherschliffe an der Wand eines U-Tales im noerdlichen Kaokofeld, Suedwestafrika. *Geol. Rundsch.*, **2**, 271–575.

Mattsson, Å. (1997) Glacial striae, glacigenous sediments and Weichselian ice movements in southernmost Sweden. *Sediment. Geol.*, **111**, 285–311.

Menzies, J. (Ed.) (1995) *Modern Glacial Environments: Processes, Dynamics and Sediments*. Butterworth Heinemann, Oxford, 621 pp.

Miller, J.M.G. (1989) Glacial advance and retreat sequences in a Permo-Carboniferous section, central Transantarctic Mountains. *Sedimentology*, **36**, 419–430.

Moncrieff, A.C.M. & Hambrey, M.J. (1988) Late Precambrian glacially-related grooved and striated surfaces in the Tillite Group of central East Greenland. *Palaeogeogr. Palaeoclimatol. Palaeoecol.*, **65**, 183–200.

Mustard, P.S. & Donaldson, J.A. (1987a) Substrate quarrying and subglacial till deposition by an early Proterozoic ice sheet: evidence from the Gowganda Formation at Cobalt, Ontario, Canada. *Precam. Res.*, **34**, 347–368.

Mustard, P.S. & Donaldson, J.A. (1987b) Early Proterozoic ice-proximal glaciomarine deposition: the lower Gowganda Formation at Cobalt, Ontario, Canada. *Geol. Soc. Am. Bull.*, **98**, 373–387.

MU YONGJI (1981) Luoquan Tillite of the Sinian System in China. In: *Earth's Pre-Pleistocene Glacial Record* (Eds Hambrey, M.J. & Harland, W.B.), pp. 402–413. Cambridge University Press, Cambridge.

NYSTUEN, J.P. (1985) Facies and preservation of glaciogenic sequences from the Varanger ice age in Scandinavia and other parts of the North Atlantic Region. *Palaeogeogr. Palaeoclimatol. Palaeoecol.*, **51**, 209–229.

OJAKANGAS, R.W. & MATSCH, C.L. (1980) Upper Precambrian (Eocambrian) Mineral Fork tillite of Utah: a continental glacial and glaciomarine sequence. *Geol. Soc. Am. Bull.*, **91**, 495–501.

OJAKANGAS, R.W. & MATSCH, C.L. (1982) Upper Precambrian (Eocambrian) Mineral Fork Tillite of Utah; a continental glacial and glaciomarine sequence: reply. *Geol. Soc. Am. Bull.*, **93**, 186–187.

PERRY, W.J. & ROBERTS, H.G. (1968) Late Precambrian glaciated pavements in the Kimberley region, Western Australia. *J. geol. Soc. Australia*, **15**, 51–56.

PHARAOH, T. (1985) The stratigraphy and sedimentology of autochthonous metasediments in the Repparfjord–Komagfjord tectonic window, west Finnmar. In: *The Caledonide Orgen-Scandinavia and Related Areas* (Eds Gee, D.G. & Sturt, B.A.), pp. 347–357. John Wiley & Sons, London.

PLUMB, K.A. (1981) Late Proterozoic (Adelaidean) tillites of the Kimberley–Victoria River region, Western Australia and Northern Territory. In *Earth's Pre-Pleistocene Glacial Record* (Eds Hambrey, M.J. & Harland, W.B.), pp. 504–514. Cambridge University Press, Cambridge.

PRESS, V.K. (1983) Canada's heritage of glacial features. *Geol. Surv. Can. Miscell. Rep.*, **28**, 119 pp.

RAPP, A. (1975) Some views on the Ordovician palaeoglaciation in Saharan Africa. *Geol. Fören. Stockh. Förh.*, **97**, 142–150.

READING, H.G. & WALKER, R.G. (1966) Sedimentation of Eocambrian tillites and associated sediments in Finnmark, northern Norway. *Palaeogeogr. Palaeoclimatol. Palaeoecol.*, **2**, 177–212.

REUSCH, H. (1891) Skuringsmerker og morenegrus eftervist I Finnmarken fra en periode meget elder end 'istiden' (Summary: Glacial stria and boulder-clay in Norwegian Lapponie from a period much older than the last ice age). *Norges geol. unders.*, **1**, 78–85, 97–100.

REUSCH, H. (1898) Professor Schiøtz's bemerkninger om de preglaciale skuringsmerker I Finnmarken. *Nyt mag. naturv.*, **36**, 11–12.

RICE, A.H.N. & HOFMANN, C.-C. (2000) Evidence for glacial origin of Neoproterozoic III striations at Oaibaččannjar'ga, Finnmark, northern Norway. *Geol. Mag.*, **137**, 355–366.

ROBERTS, D. & Fareth, E. (1974) Correlation of autochthonous stratigraphical sequences in the Alta–Reppafjord region, west Finnmark. *Norsk. geol. Tidsskr.*, **54**, 123–129.

ROBERTS, H.G., GEMUTS, I. & HALLIGAN, R. (1972) Adelaidean and Cambrian stratigraphy of the Mount Ramsay 1 : 250 000 sheet area, Kimberley Region, Western Australia. *Dept nat. Develop., BMR, geol. geophys. Rep.*, **150**, 72 pp.

ROBERTS, D., GOROKHOV, I.M., SIEDLECKA, A. *et al.* (1997) Rb–Sr dating of illite fractions from Neoproterozoic shales on Varanger Peninsula, North Norway. *Norges geol. unders., Bull.*, **433**, 24–25.

ROCHA-CAMPOS, A.C., DA CRUZ, F.E., SOLAK SATO, P.E. & SAITO, M.M. (1997) Late Palaeozoic glacial and floating ice striated surfaces in northeastern Brazil. *An. Acad. Brasil. Cienc.*, **69**, 271.

ROCHA-CAMPOS, A.C., DOS SANTOS, P.R. & CANUTO, J.R. (1999) Multiple glacially striated soft sediment surface in the Late Palaeozoic and Proterozoic of Brazil. *An. Acad. Brasil. Cienc.*, **71**, 841.

ROCHA-CAMPOS, A.C., YOUNG, G.M. & DOS SANTOS, P.R. (1996) Re-examination of a striated pavement near Jequitai; implications for Proterozoic stratigraphy and glacial geology. *An. Acad. Brasil. Cienc.*, **68**, 593.

ROSENDAHL, H. (1931) Bidrag til Varangernesets geologi. *Norsk. geol. tidsskr.*, **12**, 487–506.

ROSENDAHL, H. (1945) Praekambrium–Eokambrium i Finnmark. *Norsk. geol. tidsskr.*, **25**, 327–349.

RUST, I.C. (1963) Note on a glacial pavement near Nieuwoudtville. *S. Afr. J. Sci.*, **59**, 12.

RUST, I.C. (1981) Early Palaeozoic Pakhuis Tillite, South Africa. In: *Earth's Pre-Pleistocene Glacial Record* (Eds Hambrey, M.J. & Harland, W.B.), pp. 113–117. Cambridge University Press, Cambridge.

SAVAGE, N.M. (1972) Soft-sediment glacial grooving of Dwyka age in South Africa. *J. sediment. Petrol.*, **42**, 307–308.

SCHENK, P.E. (1965) Depositional environment of the Gowganda Formation (Precambrian) at the south end of Lake timagami, Ontario. *J. sediment. Petrol.*, **36**, 309–318.

SCHERMERHORN, L.J.G. (1971) Upper Ordovician glaciation in northwest Africa? Discussion. *Geol. Soc. Am. Bull.*, **82**, 265–268.

SCHERMERHORN, L.J.G. (1974) Late Precambrian mixtites: glacial and/or nonglacial. *Am. J. Sci.*, **274**, 673–824.

SCHIØTZ, O.E. (1898) Om Dr. Reusch's präglaciale skuringsmerker. *Nyt mag. naturv.*, **36**, 1–10.

SCHWELLNUS, C.M. (1941) The Nama Tillite in the Klein Kharas Mountains, SWA. *Trans. geol. Soc. S. Afr.*, **44**, 19–33.

SHARP, M., DOWDESWELL, J.A. & GEMMELL, J.C. (1989) Reconstructing past glacier dynamics and erosion from glacial geomorphic evidence: Snowdon, North Wales. *J. Quat. Sci.*, **4**, 115–130.

SIEDLECKA, A. (1990) Varangerbotn berggrunnskart 2335 3, 1 : 50 000, föreløpig utgave. *Norges geol. unders.*

SIEDLECKA, A. (1991) Nesseby berggrunnskart 2335 2, 1 : 50 000, föreløpig utgave. *Norges geol. unders.*

SIEDLECKA, A. & ROBERTS, D. (1992) The bedrock geology of Varanger Peninsula, Finnmark, North Norway; an excursion guide. *Norges geol. unders.*, spec. paper, **5**, 45 pp.

SIEDLECKA, A. & ROBERTS, D. (1996) Finnmarks Fylke. Berggrundsgeologi M 1 : 500 000. *Norges geol. unders.*

SLATER, G. (1932) The glaciated surfaces of Nooitgedacht, near Kimberlay, and the upper Dwyka boulder shakes of eastern part of Griqualand West (Cape Province), 1929. *Trans. R. Soc. S. Afr.*, **20**, 301–325.

SMITH, A.J. (1963) Evidence for a Talchir (lower Gondwana) glaciation; striated pavement and boulder bed at Irai, central India. *J. sediment. Petrol.*, **33**, 739–750.

SPJELDNÆS, N. (1964) The Eocambrian glaciation in Norway. *Geol. Rundsch.*, **54**, 24–45.

STRAHAN, A. (1897) On glacial phenomena of Palaeozoic age in the Varanger Fjord. *Q.J. geol. Soc.*, **53**, 137–146.

STRATTEN, T. (1977) Conflicting directions of Dwyka ice flow in the western Cape Province and southern South West Africa. *Trans. geol. Soc. S. Afr.*, **80**, 79–86.

SWEET, I.P., PONTIFEX, I.R. & MORGAN, C.M. (1974) The geology of the Auvergne 1 : 250 000 Sheet Area, Northern Territory. Geol. geophys. Rep., Bur. Min. Res. Aust., **161**, 98 pp.

SYVERSON, K.M. (1995) The ability of ice-flow indicators to record complex, historic deglaciation events, Burroughs Glacier, Alaska. *Boreas*, **24**, 232–244.

TANKARD, A.J., JACKSON, M.P.A., ERIKSSON, K.A., HOBDAY, D.K., HUNTER, D.R. & MINTER, W.E.L. (1982) *Crustal Evolution of South Africa; 3.8 Billion Years of Earth History*. Springer-Verlag, New York, 523 pp.

VASLET, D. (1990) Upper Ordovician glacial deposits in Saudi Arabia. *Episodes*, **13**, 147–161.

VEILLETTE, J.J. (1983) Les polis glaciaires au Témiscaminque: une chrologie relative. *Comm. géol. Can. Etude*, **83-1A**, 187–196.

VEILLETTE, J.J. (1986) Former southwesterly ice flows in the Abitibi–Timiskaming region: implication for the configuration of the late Wisconsinan ice sheet. *Can. J. Earth Sci.*, **23**, 1724–1741.

VEILLETTE, J.J. & ROY, M. (1995) The spectacular cross-striated outcrops of James Bay, Quebec. *Current Research 1995-C*, pp. 243–248. Geological Survey of Canada, Ottawa.

VISSER, J.N.J. (1983) Glacial-marine sedimentation in the late Palaeozoic Karoo Basin, Southern Africa. In: *Glacial-marine Sedimentation* (Ed. Molnia, B.F.), pp. 667–701. Plenum Press, New York.

VISSER, J.N.J. (1985) The Dwyka Formation along the north-western margin of the Karoo Basin in the Cape Province, South Africa. *Trans. geol. Soc. S. Afr.*, **88**, 37–48.

VISSER, J.N.J. (1990) Glacial bedforms at the base of the Permo-Carboniferous Dwyka Formation along the western margin of the Karoo Basin, South Africa. *Sedimentology*, **37**, 231–245.

VISSER, J.N.J. (1991) The paleoclimatic setting of the late Paleozoic marine ice sheet in the Karoo Basin of southern Africa. Spec. Paper geol. Soc. Am., **261**, 181–189.

VISSER, J.N.J. (1997a) Controls of Early Permian shelf deglaciation in the Karoo Basin of South Africa. *Palaeogeogr. Palaeoclimatol. Palaeoecol.*, **125**, 129–139.

VISSER, J.N.J. (1997b) A review of the Permo-Carboniferous glaciation in Africa. In: *Late Glacial and Postglacial Environmental Changes; Quaternary, Carboniferous–Permian, and Proterozoic* (Ed. Martini, I.P.), pp. 169–191. Oxford University Press, Oxford.

VISSER, J.N.J. & HALL, K.J. (1984) A model for the deposition of the Permo-Carboniferous Kruitfontein boulder pavement and associated beds, Elandsvlei, South Africa. *Trans. geol. Soc. S. Afr.*, **87**, 161–168.

VISSER, J.N.J. & LOOCK, J.C. (1988) Sedimentary facies of the Dwyka Formation associated with the Nooitgedacht glacial pavements, Barkly West District. *S. Afr. J. Geol.*, **91**, 38–48.

VON BRUNN, V. (1977) A furrowed intratillite pavement in the Dwyka Group of northern Natal. *Trans. geol. Soc. S. Afr.*, **80**, 125–130.

VON BRUNN, V. (1994) Glaciogenic deposits of the Permo-Carboniferous Dwyka Group in the eastern region of the Karoo Basin, South Africa. In: *Earth's Glacial Record* (Eds Deynoux, M., Miller, J.M.G., Domack, E.W., Eyles, N., Fairchild, I.J. & Young, G.M.), pp. 60–69. Cambridge University Press, Cambridge.

VON BRUNN, V. (1996) The Dwyka Group in the northern part of Kwazulu/Natal, South Africa: sedimentation during late Palaeozoic glaciation. *Palaeogeogr. Palaeoclimatol. Palaeoecol.*, **125**, 141–163.

VON BRUNN, V. & MARSHALL, C.G.A. (1989) Glaciated surfaces and the base of the Dwyka Formation near Pietermaritzburg, Natal. *S. Afr. J. Geol.*, **92**, 420–426.

VON BRUNN, V. & STRATTEN, T. (1981) Late Palaeozoic tillites of the Karoo basin of South Africa. In: *Earth's Pre-Pleistocene Glacial Record* (Eds Hambrey, M.J. & Harland, W.B.), pp. 71–79. Cambridge University Press, Cambridge.

VON GAERTNER, H.R. (1944) Bemerkungen über den Tillit von Bigganjarga am Varangerfjord. *Geol. Rundsch.*, **34**, 226–231.

WANG YUELUN, LU SONGNIAN, GAO ZHENJIA, LIN WEIXING & MA GUOGAN (1981) Sinian tillites in China. In: *Earth's Pre-Pleistocene Glacial Record* (Eds Hambrey, M.J. & Harland, W.B.), pp. 386–401. Cambridge University Press, Cambridge.

WANLESS, H.R. & CANNON, J.R. (1966) Late Palaeozoic glaciation. *Earth Sci. Rev.*, **1**, 247–286.

WILLIAMS, G.E. (1986) Precambrian permafrost horizons as indicators of palaeoclimate. *Precam. Res.*, **32**, 233–242.

WILLIAMS, G.E. (1994) The enigmatic late Proterozoic glacial climate: an Australian perspective. In: *Earth's Glacial Record* (Eds Deynoux, M., Miller, J.M.G., Domack, E.W., Eyles, N., Fairchild, I.J. & Young, G.M.), pp. 146–164. Cambridge University Press, Cambridge.

WILLIAMS, G.E. & TONKIN, D.G. (1985) Periglacial structures and palaeoclimate significance of a late Precambrian block field in the Cattle Grid copper mine, Mount Gunson, South Australia. *Aust. J. Earth Sci.*, **32**, 287–300.

WOODWORTH-LYNAS, C.M.T. (1996) Ice scours as an indicator of glaciolacustrine environment. In: *Past Glacial Environments: Sediments, Forms and Techiques* (Ed. Menzies, J.), pp. 161–178. Butterworth Heinemann, Oxford.

WOODWORTH-LYNAS, C.M.T. & DOWDESWELL, J.A. (1994) Soft-sediment striated surfaces and massive diamicton facies produced by floating ice. In: *Earth's Glacial Record* (Eds Deynoux, M., Miller, J.M.G., Domack, E.W., Eyles, N., Fairchild, I.J. & Young, G.M.), pp. 241–259. Cambridge University Press, Cambridge.

YOUNG, G.M. (1996) Glacial environments of pre-Pleistocene age. In: *Past Glacial Environments: Sediments, Forms and Techniques* (Ed. Menzies, J.), pp. 239–252. Butterworth Heinemann, Oxford.

YOUNG, G.M. (1999) Some aspects of the geochemistry, provenance and palaeoclimatology of the Torridonian of SW Scotland. *J. geol. Soc. London*, **156**, 1097–1111.

YOUNG, G.M. & GOSTIN, V.A. (1988) Stratigraphy and sedimentology of Sturtian glacigenic deposits in the western part of the North Flinders Basin, South Australia. *Precam. Res.*, **39**, 151–170.

YOUNG, G.M. & GOSTIN, V.A. (1991) Late Proterozoic (Sturtian) succession of the North Flinders Basin, South Australia: an example of temperate glaciation in an active rift setting. *Spec. Paper geol. Soc. Am.*, **261**, 207–222.

# Index

Page numbers in *italics* refer to figures; those in **bold** refer to tables.